physica status solidi

VOLUME 28 · NUMBER 2 · 1968

AKADEMIE-VERLAG · BERLIN

Classification Scheme

1. Structure of Solids
   1.1 Alloys. Metallurgy
   1.2 Solid-State Phase Transformations
   1.3 Surfaces
   1.4 Films

2. Non-Crystalline State

3. Crystallography
   3.1 Crystal Growth
   3.2 Interatomic Forces

4. Microstructure of Solids

5. Perfectly Periodic Structures

6. Lattice Mechanics. Phonons
   6.1 Mössbauer Investigations

7. Acoustic Properties of Solids

8. Thermal Properties of Solids

9. Diffusion in Solids

10. Defect Properties of Solids (Irradiation Defects see 11)
    10.1 Defect Properties of Metals
    10.2 Photochemical Reactions. Colour Centres

11. Irradiation Effects in Solids

12. Mechanical Properties of Solids (Plastic Deformations see 10)
    12.1 Mechanical Properties of Metals (Plastic Deformations see 10.1)

13. Electron States in Solids
    13.1 Band Structure. Fermi Surfaces
    13.2 Excitons
    13.3 Surface States
    13.4 Impurity and Defect States

14. Electrical Properties of Solids. Transport Phenomena
    14.1 Metals. Conductors
    14.2 Superconductivity. Superconducting Materials and Devices
    14.3 Semiconductors
        14.3.1 Semiconducting Films
        14.3.2 Semiconducting Devices. Junctions (Contact Problems see 14.4.1)
    14.4 Dielectrics
        14.4.1 High Field Phenomena, Space Charge Effects, Inhomogeneities, Injected Carriers (Electroluminescence see 20.3; Junctions see 14.3.2)
        14.4.2 Ferroelectric Materials and Phenomena

15. Thermoelectric and Thermomagnetic Properties of Solids

16. Photoconductivity. Photovoltaic Effects

17. Emission of Electrons and Ions from Solids

18. Magnetic Properties of Solids
    18.1 Paramagnetic Properties
    18.2 Ferromagnetic Properties
    18.3 Ferrimagnetic Properties. Ferrites
    18.4 Antiferromagnetic Properties

(Continued on cover three)

physica status solidi

Volume 28 · Number 2 · Pages 429 to 850, K97 to K182, and A5 to A10

August 1, 1968

AKADEMIE-VERLAG · BERLIN

Subscriptions and orders for single copies should be addressed to

AKADEMIE-VERLAG GmbH, 108 Berlin, Leipziger Straße 3 – 4 or to Buchhandlung KUNST UND WISSEN, Erich Bieber, 7 Stuttgart 1, Wilhelmstr. 4 – 6 or to Deutsche Buch-Export und -Import GmbH, 701 Leipzig, Postschließfach 160

Editorial Note:

"physica status solidi" undertakes that an original paper *accepted for publication* before the 8th of any month will be published within 50 days of this date unless the author requests a postponement. In special cases there may be some delay between *receipt* and *acceptance* of a paper due to the review and, if necessary, revision of the paper.

Schriftleiter und verantwortlich für den Inhalt: Professor Dr. Dr. h. c. P. Görlich, 102 Berlin, Neue Schönhauser Str. 20 bzw. 69 Jena, Humboldtstr. 26. Redaktionskollegium: Dr. S. Oberländer, Dr. E. Gutsche, Dr. W. Borchardt. Anschrift der Schriftleitung: 102 Berlin, Neue Schönhauser Str. 20, Fernruf: 42 67 88. Verlag: Akademie-Verlag GmbH 108 Berlin, Leipziger Str. 3–4. Fernruf: 22 04 41, Telex-Nr. 112 020. Postscheckkonto: Berlin 35 021. – Die Zeitschrift „physica status solidi" erscheint jeweils am 1. des Monats. Bezugspreis eines Bandes M 90,– (Sonderpreis für die DDR M 60,–). Bestellnummer dieses Bandes 1068/28. Jeder Band enthält zwei Hefte. Gesamtherstellung: VEB Druckerei „Thomas Müntzer" Bad Langensalza. – Veröffentlicht unter der Lizenznummer 1310 des Presseamtes beim Vorsitzenden des Ministerrates der Deutschen Demokratischen Republik.

Contents

Short Notes

Pre-printed Titles

Systematic List

The Author Index of Volume 28 Begins on Page 851

(It will be delivered together with Volume 29, Number 1.)

Review Article

phys. stat. sol. 28, 437 (1968)

Subject classification: 13 and 20; 13.3; 20.1; 21; 21.2; 21.6

Sektion Physik der Universität München

Optical Plasma Resonances in Solids

By

W. Steinmann[1])

Contents

1. Introduction

The conduction electrons of a metal, together with the ion cores, form a plasma. If the charge equilibrium of this system is disturbed, the electron gas performs density oscillations, the so-called plasma oscillations. The simplest model of a metal is the free electron gas. The ion cores are assumed to be smeared out, resulting in a homogeneous positive background, and the electrons are, in the first approximation, considered as completely undamped. In this system, the frequency of the plasma oscillation is given by

$$\omega_{\mathrm{p}} = \left(\frac{4 \pi n e^2}{m}\right)^{1/2}, \tag{1.1}$$

where n is the density of the electron gas and e and m are the electronic charge and mass.

The quanta $\hbar \omega_{\mathrm{p}}$ of the plasma oscillations are called plasmons. Insertion of typical figures for metallic electron densities into (1.1) yields plasmon energies between 3 and 30 eV. The theory of plasmons in solids has been developed extensively during the past two decades and can now be found in the textbooks and monographs [1].

---

[1]) Present address: European Space Research Laboratory, Noordwijkerhout, the Netherlands.

The free electron gas is too crude a model to describe the electronic properties of any material completely and accurately. It is useful only in certain frequency ranges and around the plasma frequency for only certain materials such as Al and the alkali metals. Generally one has to take into account the interband transitions which are in the same energy range as the plasmons, i.e., one has to give up the assumption of the homogeneous background. This shifts the actual plasma frequency from the figure given by (1.1). A general expression for the plasma frequency is obtained from Maxwell's equations. It is easy to show that a necessary condition for a longitudinal electric wave to be sustained in a medium is that the complex dielectric constant must vanish [2]:

$$\varepsilon(\omega_p) = 0 \,. \tag{1.2}$$

Thus, if $\varepsilon(\omega)$ or the optical constants of a material are known, the plasma frequency can be determined from (1.2). Since in general ε is complex ($\varepsilon = \varepsilon_1 + i\varepsilon_2$), ω_p will also be complex. Therefore (1.2) yields not only the energy but also the lifetime of the plasmon.

To excite a plasmon it is necessary to disturb the charge equilibrium in the solid. This can be done by charged particles, and in fact the first and main experimental evidence for the plasmons has been obtained from measurements of energy losses of energetic electrons, which had been reflected from, or passed through, solid targets. Many of the characteristic energy losses could be directly related to the excitation of plasmons in the target. These investigations have been recently reviewed in a number of articles [3, 4, 5].

The energy loss spectrum can also be expressed by the dielectric constant [2, 6, 7]. One finds for the probability P for a charged particle to lose energy $\hbar\,\omega$ by interacting with a medium characterized by a dielectric constant $\varepsilon(\omega) = \varepsilon_1(\omega) + i\,\varepsilon_2(\omega)$ that

$$P(\hbar\,\omega) \propto \operatorname{Im}\frac{1}{\varepsilon(\omega)} = \frac{\varepsilon_2}{\varepsilon_1^2 + \varepsilon_2^2} \,. \tag{1.3}$$

For a free electron gas (1.3) represents a resonance curve at the plasma frequency, the half-width of which corresponds to the imaginary part of the frequency determined by (1.2). This can be seen by substituting into (1.3) the dielectric constant of a free electron gas

$$\varepsilon_1 = 1 - \frac{\omega_p^2}{\omega^2 + \gamma_d^2}\,, \qquad \varepsilon_2 = \frac{\gamma_d}{\omega}\,\frac{\omega_p^2}{\omega^2 + \gamma_d^2}\,; \qquad \gamma_d \ll \omega_p\,, \tag{1.4}$$

where ω_p is given by (1.1) and γ_d is a damping constant related to the imaginary part of the solution of (1.2) by

$$\operatorname{Im}\omega_p = \frac{\gamma_d}{2}\,.$$

The half-width of (1.3) is $\gamma_d/\omega_p = \varepsilon_2$.

The plasma waves in the bulk of an optically isotropic and homogeneous material, which we have considered so far, are purely longitudinal. Therefore they do not interact with a transversal electromagnetic wave. This means that these plasmons cannot decay into photons nor can they be excited by light. However, there exists at the surface of a medium a vibrational mode which is distinct from the bulk mode. In a thin film the two surface waves are coupled. This leads to a number of vibrational modes, some of which can radiate. As a result, metallic films irradiated with energetic electrons emit light at the plasma frequency. The reverse process is the excitation of the radiating plasmon by light, and can be observed in the optical transmission and reflection spectrum and also in the spectral dependence of the photoelectric emission. Finally, it has been recently observed that the plasma radiation can also be excited by light. This article reviews the work which has been done in studying these phenomena.

There are three metals, Al, K, and Ag, which have mainly been used in the investigations of optical plasma resonances. Al and K show the optical behaviour of a free electron gas in the neighbourhood of their plasma frequency. In this energy range their dielectric constant is well represented by the Drude formula (1.4). However, ω_p and γ_d have to be determined phenomenologically from (1.2). ω_p is slightly shifted from the figure obtained from (1.1) due to non-negligible ion core polarizability. The optical properties of Ag are quite different [8]. The plasmon energy is drastically shifted from the free electron value 9 eV to 3.78 eV by a strong interband transition which starts at 4 eV, just above the actual plasma frequency. Nevertheless the plasma frequency is a solution of (1.2) and has a very small imaginary part. In fact, the relative half-width of the loss function (1.3) is smaller than any other plasma resonance we know so far. This can be explained as follows: The dielectric constant of Ag, although it differs largely from the Drude formula (1.2), shows certain similarities. Its real part passes almost linearly from negative values below ω_p to positive values above ω_p. The imaginary part varies slowly compared with the real part. Thus in the neighbourhood of ω_p the dielectric constant is given by [9]

$$\varepsilon_1 = \varepsilon_1'(\omega_p)\,(\omega - \omega_p); \qquad \varepsilon_1' = \frac{d\varepsilon_1}{d\omega}, \qquad \varepsilon_2(\omega) = \text{const}. \tag{1.5}$$

Substitution of (1.5) into (1.3) yields a Lorentzian curve with a half-width

$$\frac{\Delta\omega}{\omega_p} = \frac{2}{\varepsilon_1'}\frac{\varepsilon_2}{\omega_p}.$$

Note that for the free electron gas $\varepsilon_1'(\omega_p) = 2/\omega_p$ which leads to $\Delta\omega/\omega_p = \varepsilon_2$. For Ag, however, ε_1' is much larger . Thus the extremely sharp plasma resonance is not due to ε_2 being extraordinarily small but to the very steep slope of $\varepsilon_1(\omega)$ at the plasma frequency.

Among the free electron like materials, Al has the smallest relative half-width of the plasma resonance. Furthermore, Al has been most frequently and extensively studied in characteristic energy loss experiments. However, optical experiments with Al at ω_p are difficult because the corresponding wavelength λ_p (828 Å) lies in the extreme ultraviolet beyond the LiF cut-off. Thus no windows and transparent substrates are available, and there is also a lack of suitable light sources and polarizers. Moreover a thin oxide layer, while it does not influence the electron energy loss, may well affect the optical plasma resonance. Finally, in the experiments on optical plasma resonances, the film thickness has to be small compared with λ_p. For Al this means films about 100 Å thick, which are rather hard to prepare and to work with.

From an optical point of view, K is a much more convenient free electron like material, since its plasma wavelength (3280 Å) is in the middle of the quartz ultraviolet. However, K is chemically very reactive, requiring ultra-high vacuum in the target chamber, and furthermore tends to form island films with optical anomalies if condensed on substrates not kept at liquid nitrogen temperature [10].

Ag has neither of the disadvantages of Al and K. It is chemically stable and the plasma wavelength is equal to that of K. In addition, the plasma resonance is the most pronounced, and this facilitates the experimental detection. Hence most of the optical plasma resonance phenomena have first been studied experimentally with Ag films. However, Ag is not a free electron like material. The plasma resonance has been called a hybrid resonance, as distinct from the free electron resonance in Al and K, because it is largely influenced by an interband transition, as mentioned above. Therefore the experiments have been repeated with Al and/or K.

Table 1

Plasmon energy $\hbar\,\omega_p$, relative half-width of the resonance line, $\Delta\omega/\omega_p$, and plasma wavelength in vacuum, $\lambda_p = 2\,\pi\,c/\omega$ of Al, K, and Ag

	$\hbar\,\omega_p$ (eV)	$\frac{\Delta\omega}{\omega_p}$	λ_p (Å)	Reference
Al	14.97	4×10^{-2}	828	[11]
K	3.78	7×10^{-2}	3280	[12]
Ag	3.78	2×10^{-2}	3280	[13]

Table 1 summarizes the plasma frequency, wavelength, and relative half-width of Al, K, and Ag. The measurement of the optical plasma resonance yields a direct determination of the plasma frequency while in the electron energy loss experiments the plasmon energy appears as a small difference between two large quantities. Therefore the optical method is principally more accurate than the energy loss method. As regards the half-width no accurate results for the sharp losses could be obtained from energy loss experiments until very recently, when predispersers which reduced the thermal half-width of the primary electrons were introduced. The optical method requires only relatively modest resolution to measure the half-width of even the sharpest resonance peaks. The figures quoted in Table 1 for Al and Ag are taken from the most recent and refined energy loss experiments, while for K results of optical measurements are given, since no equally reliable figures from energy losses exist yet.

2. Plasma Modes in Thin Metal Films

At the interface between two media there exists an additional plasma mode distinct from the bulk plasma oscillation treated in the preceeding section [14]. In this surface oscillation the E-field is periodic in the surface plane and exponentially decreasing normal to the surface. The frequency of this mode is given by [15]

$$\varepsilon_a(\omega_s) = -\varepsilon_b(\omega_s)\,, \tag{2.1}$$

where ε_a and ε_b denote the dielectric constants of medium "a" and "b" respectively. In the case of a solid bounded by vacuum (2.1) reduces to

$$\varepsilon(\omega_s) = -1\,. \tag{2.2}$$

For a free electron gas, by substituting (1.4) into (2.2) and neglecting γ_d,

$$\omega_s = \frac{\omega_p}{\sqrt{2}}\,. \tag{2.3}$$

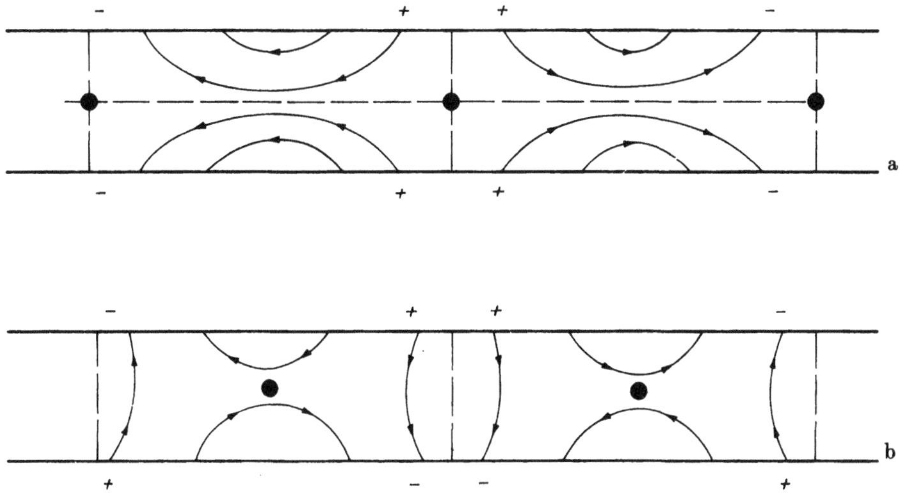

Fig. 1. a) Symmetric and b) antisymmetric mode of surface plasma oscillation in a thin foil. The arrows represent electric field lines, the + and − signs surface charges (after Stern and Ferrell [15])

Note that, in contrast to the bulk frequency ω_p, the surface frequency ω_s is influenced by surface layers, as follows from (2.1).

The surface plasmons have been verified experimentally in energy loss spectra [3]. The surface losses differ from the volume losses in their angular dependence, in that their intensity is independent of the film thickness, and in the above mentioned sensitivity with respect to surface layers. The surface loss of Al, which is 10.3 eV for an unoxidized surface, has been found to be shifted to 7 eV during oxidation [16, 17]. In a thin film the two surface modes interact provided the film thickness is not larger than the penetration depth of the E-field normal to the surface. This coupling leads to a splitting of the resonance frequency, i.e. to two different surface modes: a symmetric one in which the two surface waves are in phase, and an antisymmetric one in which the waves are displaced by π (see Fig. 1). The frequencies of these modes depend on the film thickness t, the wave number k parallel to the surface, and the dielectric constant ε. A dispersion relation is obtained by solving Maxwell's equations, including retardation, for the proper boundary equations. This has first been done by Ritchie and Eldridge [18] (see also Otto [19]), which found the following relations:

$$\varepsilon = -\sqrt{\frac{k^2 - \varepsilon \frac{\omega^2}{c^2}}{k^2 - \frac{\omega^2}{c^2}}} \coth\left(\frac{t}{2}\sqrt{k^2 - \varepsilon \frac{\omega^2}{c^2}}\right) \tag{2.4}$$

for the symmetric mode and

$$\varepsilon = -\sqrt{\frac{k^2 - \varepsilon \frac{\omega^2}{c^2}}{k^2 - \frac{\omega^2}{c^2}}} \tanh\left(\frac{t}{2}\sqrt{k^2 - \varepsilon \frac{\omega^2}{c^2}}\right) \tag{2.5}$$

for the antisymmetric mode.

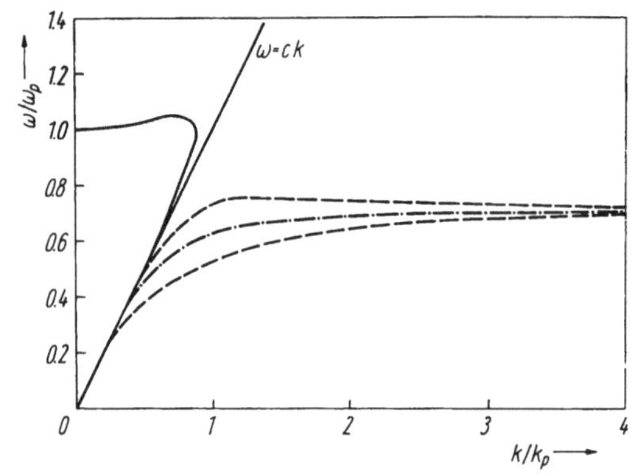

Fig. 2. Dispersion of surface plasmons in an undamped free electron gas, according to Kliewer and Fuchs [20]. $k_p = 2\pi/\lambda_p = \omega_p/c$. Dashed curves: non-radiative coupled surface modes in a thin foil (solution of (2.4) (lower curve) and (2.5) (upper curve) for $t = c/\omega_p = \lambda_p/2\pi$). Dash-dotted curve: decoupled surface mode in a thick film (solution of (2.4) and (2.5) for $t \to \infty$). Solid curve above $\omega = c\,k$ line: radiative surface plasmon in a thin foil (solution of (2.7) for $t = \lambda_p/2\pi$). For this curve, ω and k denote the real part of these quantities

These relations have recently been studied in detail by Kliewer and Fuchs [20].

For a free electron gas without damping ($\varepsilon = 1 - \omega_p^2/\omega^2$), the solution of (2.4) and (2.5) is shown by the dashed curves in Fig. 2 for a fairly thin foil ($t = c/\omega_p = \lambda_p/2\pi$). For thick foils, the oscillations become decoupled (dash-dotted curve in Fig. 2). Peaks corresponding to the two frequencies of the coupled surface mode have been found in the energy loss spectrum using thin Al foils [21, 22, 85, 88]. The E-fields of these modes decrease exponentially outside the foil, and hence the oscillations do not radiate. This can also be seen from Fig. 2. For a radiative mode, the phase velocity ω/k must be greater than c (the velocity of light). Thus the straight line $\omega = c\,k$ in Fig. 2 divides the radiative and non-radiative regions. Modes below this line are non-radiative and only those above this line can radiate.

For radiative waves, either $\sqrt{k^2 - \frac{\omega^2}{c^2}}$ or $\sqrt{k^2 - \varepsilon\,\frac{\omega^2}{c^2}}$ must be imaginary. While the non-radiative modes in an undamped free electron gas ($\varepsilon_2 = 0$) are stationary, the radiative modes are always damped, since energy is radiated away from the foil. Therefore the frequency of these "virtual modes" [20] is complex, the imaginary part corresponding to the radiative decay rate of the plasmon.

Kliewer and Fuchs [20] have shown that there exists a large number of virtual modes in a metallic film. For thin foils ($t \lessapprox \lambda_p/2\pi$) most of the modes have frequencies far remote from ω_p and are strongly damped. There is only one mode for which $\mathrm{Re}\,\omega \approx \omega_p$ and $\mathrm{Im}\,\omega \ll \mathrm{Re}\,\omega$, at least if the angle of emission $\theta \lessapprox 45°$. This mode, which is shown in Fig. 2, as the solid curve above the $\omega = c\,k$ line, causes the optical plasma resonance phenomena dealt with in the following sections.

The angle of emission θ is determined by

$$\frac{k\,c}{\omega} = \sin\theta \tag{2.6}$$

which ensures that the condition $\omega > c\,k$ is always fulfilled. Note that, since ω is complex, k must also be complex in order that θ be real.

The dispersion relation for the radiative plasma mode is [20]

$$\varepsilon = -\sqrt{\frac{\varepsilon \frac{\omega^2}{c^2} - k^2}{k^2 - \frac{\omega^2}{c^2}}} \left(\tan \frac{t}{2} \sqrt{\varepsilon \frac{\omega^2}{c^2} - k^2}\right) \tag{2.7}$$

which corresponds to (2.5). Hence this mode is of the antisymmetric type. Furthermore the electric polarization has no component perpendicular to the plane expanded by the foil normal and the $\boldsymbol{k}$-vector which by definition, is parallel to the surface. Therefore the radiation must be polarized in the plane of emission (p-polarized).

Finally, we note that for thin foils ($t \ll \lambda_p/\pi$) Kliewer and Fuchs have shown that

$$\operatorname{Im} \omega \approx \frac{\omega_p^2 t}{4 c} \sin \theta \tan \theta . \tag{2.8}$$

Thus the radiation damping increases linearly with t and, for small θ, is proportional to θ^2, vanishing at $\theta = 0$.

3. Plasma Radiation

In 1958 Ferrell [23] predicted that, in thin foils, plasmons excited by energetic electrons should be able to decay into photons, resulting in an emission of light of the bulk plasma frequency ω_p. Although his derivation was incomplete, since it did not include retardation, his results were essentially correct, anticipating many of the findings later obtained in more rigorous theories. His paper provided a very elucidating microscopic picture and acted as a stimulus not only for the investigation of the plasma radiation but for the work on optical plasma resonances in general.

Ferrell recognized that the radiating mode must be of the antisymmetric type, from which it follows that the radiation is p-polarized. For $k = 0$, the electronic motion is strictly normal to the foil. The charge flows back and forth betweeen the surfaces and the foil represents a dipole. It is thus reasonable to expect this oscillation to radiate.

Ferrell's results may be summarized in the following relation, which gives the intensity I of the radiation as a function of ω, the angle of emission θ, the foil thickness t, and the velocity of the incident electrons v:

$$I = C\, Y(x)\, F(\theta)\, G(\omega) , \tag{3.1}$$

$$Y(x) = \frac{\sin^2 x}{x} ; \qquad x = \frac{\omega_p t}{2 v} = \frac{\pi t}{\lambda_p \beta} , \qquad \beta = \frac{v}{c} , \tag{3.1 a}$$

$$\left.\begin{aligned} F(\theta) &= \frac{\gamma_r}{\gamma_t} \cos \theta ; \qquad & \gamma_r &= \frac{\omega_p^2 t}{2 c} \sin \theta \tan \theta , \\ \gamma_t &= \gamma_r + \gamma_d , & \gamma_d &= \omega_p\, \varepsilon_2(\omega_p) , \end{aligned}\right\} \tag{3.1 b}$$

$$G(\omega) = \frac{\gamma_t^2}{\gamma_t^2 + 4 (\omega - \omega_p)^2} . \tag{3.1 c}$$

C is a constant with respect to t, ω, and θ.

$Y(x)$ is related to the excitation of a radiative plasmon by an incident electron. It predicts an oscillatory dependence of the intensity on the foil thickness, with a period depending also on the velocity of the electrons. The optimal thickness, at which Y has the first and highest maximum, is $t_{\text{opt}} = 0.4\,\beta\,\lambda_{\text{p}}$. Y becomes zero at $t_0 = 0, \beta\,\lambda_{\text{p}}, 2\,\beta\,\lambda_{\text{p}}$, etc. This may be understood as follows: For $t = t_{\text{opt}}$ the flying time of an electron through the foil is slightly shorter than $T/2$, where T is the period of the plasma oscillation. This means that the electron is moving almost in phase with the plasma wave during the whole passage. On the other hand, for $t = \beta\,\lambda_{\text{p}}$, the flying time of the electron is equal to the period of the oscillation. Thus it is moving parallel to the plasma wave during half of passage and anti-parallel during the other half. The excitation taking place during the first half is cancelled by the damping in the second half. This oscillatory thickness dependence is one of the most striking features of plasma radiation and facilitates its distinction from other sources of radiation, e.g. bremsstrahlung.

$F(\theta)\,G(\omega)$ describes the decay of the plasmon into a photon. The radiative decay rate γ_{r} is related to the imaginary part of the resonance frequency in a loss-free plasma ($\gamma_{\text{d}} = 0$) by $\text{Im}\,\omega = \gamma_{\text{r}}/2$. Note that Ferrell's result (3.1b) is identical with the expression (2.7) as obtained by Kliewer and Fuchs [20] for thin films. Plasmons can decay not only by radiation but also by intrinsic damping, characterized by the decay rate γ_{d}, which we have already introduced in Section 1. The two decay mechanisms compete, and the one having the larger decay rate prevails. This affects the angular distribution $F(\theta)$. The cosine distribution characteristic for a plane light source is modified by the factor $\gamma_{\text{r}}/\gamma_{\text{t}}$. Because γ_{r} decreases with decreasing angle and vanishes at $\theta = 0$, the intrinsic decay predominates at low angles. At large angles $\gamma_{\text{r}} \gg \gamma_{\text{d}}$ and the radiative decay prevails. Consequently, $F(\theta) = 0$ at $\theta = 0$ increases with increasing θ, passes a maximum, and approaches the cosine curve for $\theta \to \pi/2$.

The spectral distribution $G(\omega)$ represents a Lorentzian curve centered at ω_{p}, with a half-width equal to γ_{t}, the total damping rate. Thus the half-width depends on the angle. In order to determine from the plasma radiation the quantity characteristic of the material, γ_{d}, the measured half-width has to be extrapolated to $\theta = 0$.

The dispersion, i.e. the dependence of ω on θ, cannot be derived within the framework of Ferrell's theory, since it is largely determined by retardation. Furthermore, (3.1) has been derived for a free electron gas and in the limit of very thin films ($t \ll \lambda_{\text{p}}/2\,\pi$). A general expression for the radiation, which is not subject to these restrictions, can be obtained by a phenomenological method. One considers an electron passing the interface between two media of different dielectric constants and calculates the arising electromagnetic field by applying Maxwell's equation. This problem has been solved by Ginzburg and Frank in 1945 [24]. They called this phenomenon "transition radiation". Their theory has been verified by extensive experimental work of Boersch and co-workers, who measured the radiation from thick targets of various materials bombarded with energetic electrons [25 to 29]. The work on transition radiation up to 1964 is reviewed in an article by Frank [30].

For a comparison with plasma radiation, the theory of transition radiation has to be extended to include the case of a thin film. This has been done by Pafomov [31], Stern [32], and Ritchie and Eldridge [18]. For the radiation from a foil, bounded by vacuum on both sides, they obtained the following expression:

$$\left.\begin{aligned}
I(\omega,\ \theta,\ t,\ v)\ \mathrm{d}\omega\ \mathrm{d}\Omega &= \frac{e^2\,\beta^2}{\pi^2\,c}\sin^2\theta\cos^2\theta\left|\frac{A}{B}\right|^2\mathrm{d}\omega\ \mathrm{d}\Omega\ ;\\
A = (\varepsilon - 1)\Big[&(\beta\,s + 1)\,(\beta\,s - 1 + \beta^2)\,(s + u)\,\mathrm{e}^{-\frac{i\,\omega\,t\,s}{c}} + \\
&+ (\beta\,s - 1)\,(\beta\,s + 1 - \beta^2)\,(s - u)\,\mathrm{e}^{+\frac{i\,\omega\,t\,s}{c}} - \\
&- 2\,s\,(\beta\,|\cos\theta| + 1)\,(\beta\,|\cos\theta| - 1 + \beta^2\,\varepsilon)\,\mathrm{e}^{-\frac{i\,\omega\,t}{\beta\,c}}\Big],\\
B = (\beta^2\cos^2\theta - 1)\,(\beta^2\,s^2 - 1)&\left[(s + u)^2\,\mathrm{e}^{-\frac{i\,\omega\,t\,s}{c}} - (s - u)^2\,\mathrm{e}^{+\frac{i\,\omega\,t\,s}{c}}\right];\\
\beta = \frac{v}{c}\,,\qquad s &= \sqrt{\varepsilon - \sin^2\theta}\,,\qquad u = \varepsilon\,|\cos\theta|\,.
\end{aligned}\right\}\quad (3.2)$$

Substitution of the free electron gas expression (1.4) for ε into (3.2) yields results which are essentially identical with those obtained by Ferrell [33] (see Fig. 3). The angular distribution of the intensity shows only qualitative agreement. The dispersion is the same as the one shown in Fig. 2 for the radiative mode. The thickness dependence of the intensity, not shown in Fig. 3, agrees with (3.1a) except for minor deviations. Hence the plasma radiation is a plasma resonance in the transition radiation of thin foils. Transition radiation and plasma radia-

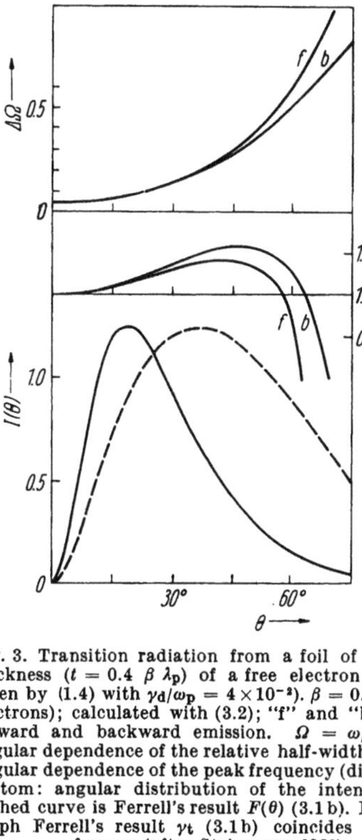

Fig. 3. Transition radiation from a foil of optimum thickness ($t = 0.4\ \beta\ \lambda_p$) of a free electron gas (ε as given by (1.4) with $\gamma_d/\omega_p = 4\times10^{-2}$). $\beta = 0.3$ (25 keV electrons); calculated with (3.2); "f" and "b" denote forward and backward emission. $\Omega = \omega/\omega_p$. Top: angular dependence of the relative half-width; middle: angular dependence of the peak frequency (dispersion); bottom: angular distribution of the intensity. The dashed curve is Ferrell's result $F(\theta)$ (3.1b). In the top graph Ferrell's result γ_t (3.1b) coincides with the f-curve (after Steinmann [33])

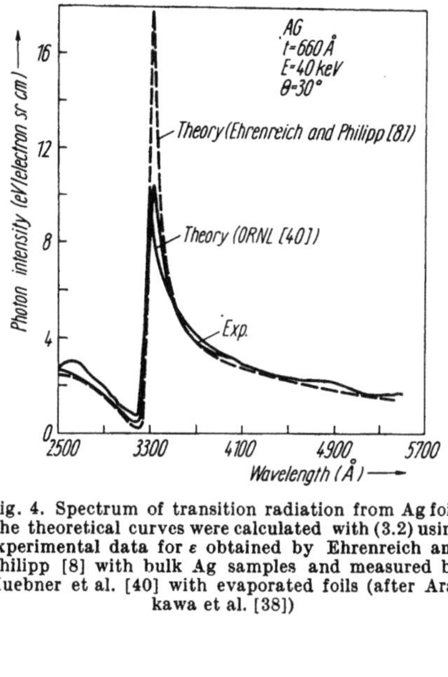

Fig. 4. Spectrum of transition radiation from Ag foil. The theoretical curves were calculated with (3.2) using experimental data for ε obtained by Ehrenreich and Philipp [8] with bulk Ag samples and measured by Huebner et al. [40] with evaporated foils (after Arakawa et al. [38])

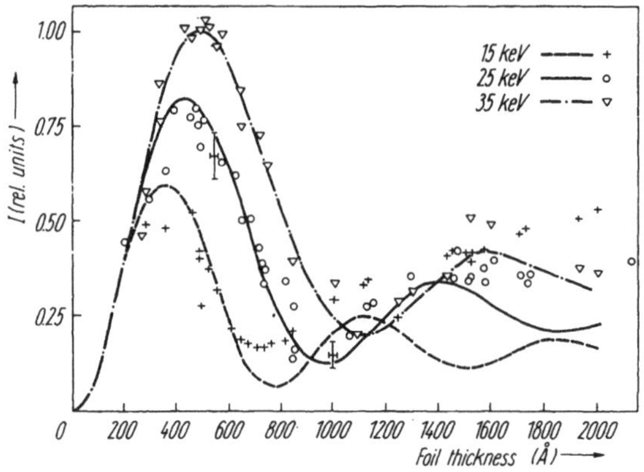

Fig. 5. Thickness dependence of the peak intensity of the transition radiation from Ag for three electron energies. p-polarized component corrected for bremsstrahlung. The curves were calculated with (3.2) using ε-data of evaporated Ag foils [41] (after Bürker and Steinmann [39])

tion are two ways of describing the same phenomenon, transition radiation being more comprehensive but less explicit.

The first experiments on plasma radiation were made with Ag films. They showed a peak at the plasma frequency and the expected dependence of the intensity on foil thickness [34] and angle [35]. The results shown in Figs. 4 and 5 were obtained later in more refined experiments [36 to 39]. The spectrum Fig. 4 does not show the Lorentzian line expected from (3.1c) because Ag is not a free electron like material. The measured peak intensity, both absolute and relative to the intensity of the long-wavelength continuum, does not agree with the intensity calculated with (3.2) if, for ε, data obtained with bulk polished samples [8] are used. Perfect agreement is found if ε-values measured with evaporated thin films [40, 41] are substituted [38, 39]. They differ mainly in that $\varepsilon_2(\omega_p)$ is about twice as large for thin films. This shows how sensitive the plasma radiation is to changes in ε and that, in principal, it can be used to determine the optical constants at the plasma frequency.

The thickness dependence, Fig. 5, shows the expected oscillatory behaviour characteristic for the plasma radiation. The period depends on the energy of the incident electrons in quantitative agreement with (3.1a). This may be seen most clearly from the variation of the minimum, for which (3.1a) yields a thickness of 780 Å for 15 keV, 1000 Å for 25 keV, and 1160 Å for 35 keV.

The radiation is almost perfectly polarized in the plane of emission for high electron energies and thin films [37, 39]. With decreasing energy and increasing thickness the degree of polarization decreases. This can be explained by bremsstrahlung, which increases under these conditions and has components both parallel and perpendicular to the plane of emission.

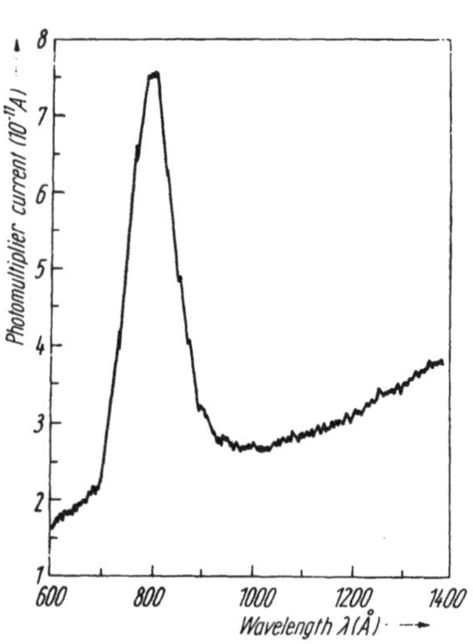

Fig. 6. Spectrum of the transition radiation from a 320 Å Al foil bombarded with 80 keV electrons. Backward emission, $\theta = 20°$ (after Herickhoff et al. [44])

Fig. 7. Thickness dependence of the peak intensity of the transition radiation from Al (after Herickhoff et al. [44])

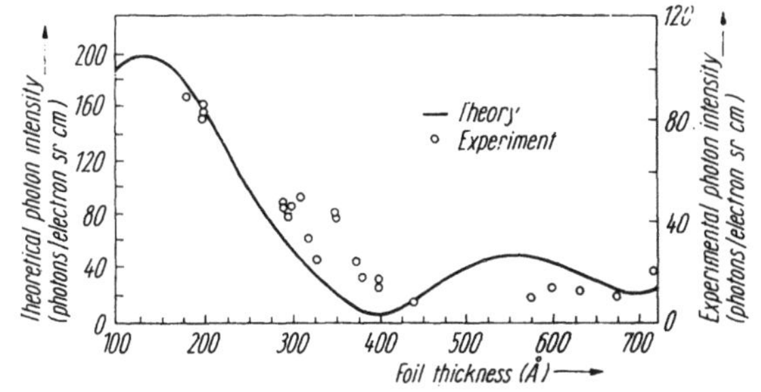

More recently transition radiation in the vacuum ultraviolet has been measured from a number of metals, and more or less pronounced plasma resonances have been found in all spectra [42]. The most important experiment on plasma radiation was made with Al films [43, 44, 45]. For the 80 keV electrons used in this experiment the optimal thickness is only 150 Å. As it is rather difficult to work with unsupported films of this thickness, most of the measurements have been made with films 320 Å thick. Although this is close to the thickness of the minimum (Fig. 7), the spectrum exhibits the resonance line at the plasma wavelength (Fig. 6). For a quantitative comparison with theory, the oxide layers have to be taken into account. Al_2O_3 is strongly absorbing at the plasma wavelength of Al. Thus a marked influence of the oxide layers on the radiation can be expected. To calculate this effect, the expression (3.2) has been extended for the case of a three-layered foil [46]. The calculation showed that essentially the oxide layers act merely as filters, reducing the intensity but leaving the position and the width of the line unchanged. When the oxide layers were taken into account, the difference in the two intensity scales in Fig. 7 disappeared

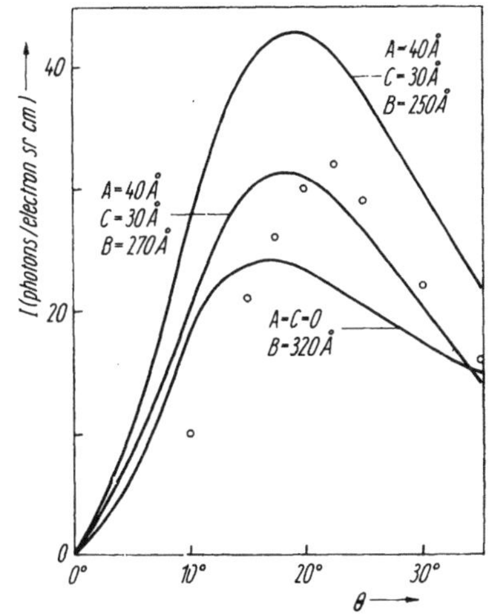

Fig. 8. Angular distribution of the peak intensity of the transition radiation from Al. The points are experimental data of Herickhoff et al. [44], the curves were calculated with the formula for three-layered foils. A, C thicknes of oxide layers; B thickness of Al (after Steinmann and Wille [46])

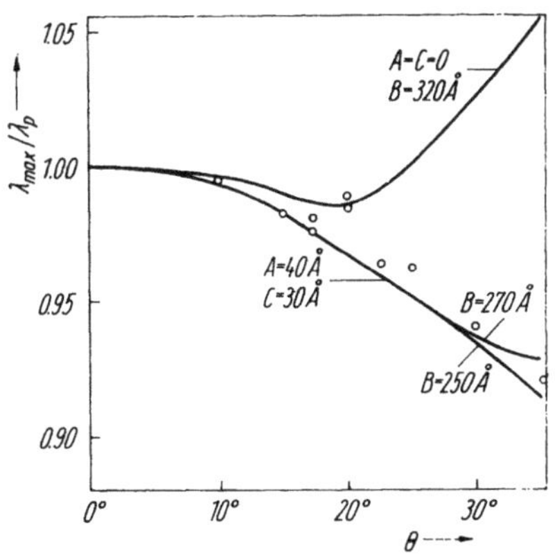

Fig. 9. Angular dependence of the peak wavelength of the transition radiation from Al. Notation as in Fig. 8 (after Steinmann and Wille [46])

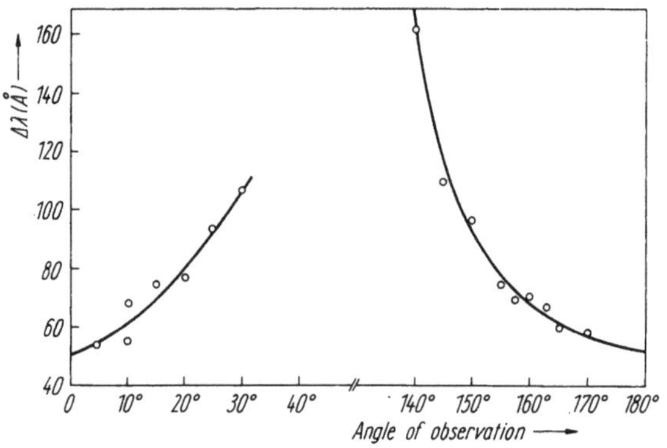

Fig. 10. Angular dependence of the half-width of the plasma peak in the transition radiation from Al (after Arakawa et al. [45])

and the measured angular dependence of the intensity (Fig. 8) and wavelength (Fig. 9) agreed quantitatively with theory. The half-width (Fig. 10) shows the expected angular dependence. Extrapolation to $\theta = 0$ should yield γ_d. In Fig. 10 $\Delta\lambda = 50$ Å which corresponds to $\gamma_d/\omega_p = 0.06$, somewhat higher than the value listed in Table 1. However, energy loss experiments [47] have shown that the half-width of the plasma resonance in Al depends rather sensitively on the crystallite size, which in turn depends on the foil preparation. This may explain the deviation. Recently Bürker [48] has measured the plasma radiation of Al foils and found half-widths which correspond to $\gamma_d/\omega_p = 0.04$. In conclusion, the experimental results on the radiation emitted by electron bombarded Ag and Al foils can be quantitatively described by the theory of transition radiation and have verified all the predictions derived from the model of the radiative plasmon decay.

4. Optical Plasma Resonance Absorption

The plasmon which decays into a photon can also be excited by a photon. The proper conditions for this optical plasmon excitation follow from the characteristics of the inverse process, the plasma radiation. Fairly thin films have to be irradiated by p-polarized light at non-normal incidence. A complete reversion of the plasma radiation experiment would be the measurement of an energy gain of an electron passing through a foil in which a large plasmon density is optically produced. This experiment has been proposed [54] but not yet made, due to the lack of a sufficiently intense light source and to the obvious problem connected with heat production in the foil.

It is not necessary, however, to reverse the process completely in order to prove the optical excitation of plasmons. Once excited by light the plasmon will decay through one of the two channels considered in the preceeding section. If the plasmon decays by intrinsic damping, the photon is absorbed. This leads to a minimum in the transmittivity at ω_p [55, 56]. If the plasmon decays into a photon again, this can only be emitted in the direction of the reflected or transmitted beam, according to equation (2.6) which is a consequence of momentum conservation in the plane of the film. Hence plasmon re-radiation leads to a peak in the reflectivity at ω_p [57].

The intrinsic damping is caused by interband transitions. If the plasmon energy is larger than the work function, the electron excited by the decaying plasmon may be emitted. Thus photon absorption via plasmon excitation and subsequent intrinsic decay may result in an enhanced photoemission at ω_p.

4.1 Plasma resonance in the transmittivity and reflectivity

General expressions for the transmittivity $T(\omega)$ and the reflectivitiy $R(\omega)$ of thin films have been derived in the framework of classical optics [58] and are well known and widely used. For a film irradiated by p-polarized light they take the following form in the notation of equation (3.2):

$$R_p = \left| \frac{(s^2 - u^2)\left(1 - e^{\frac{2 i \omega t s}{c}}\right)}{(s+u)^2 - (s-u)^2 e^{\frac{2 i \omega t s}{c}}} \right|^2 , \tag{4.1 a}$$

$$T_p = \left| \frac{4 s u e^{\frac{i \omega t s}{c}}}{(s+u)^2 - (s-u)^2 e^{\frac{2 i \omega t s}{c}}} \right|^2 . \tag{4.1 b}$$

For a very thin film ($t \ll \lambda_p/2\pi$) of a free electron gas (ε given by (1.4)) in the neighbourhood of ω_p (where $|\varepsilon| \ll \sin\theta$ holds if θ is not too small) expansion of the exponentials in (4.1) gives

$$R_p = \frac{\gamma_r^2}{\gamma_t^2 + 4(\omega - \omega_p)^2} , \tag{4.2 a}$$

$$T_p = \frac{\gamma_d^2 + 4(\omega - \omega_p)^2}{\gamma_t^2 + 4(\omega - \omega_p)^2} , \tag{4.2 b}$$

where the notation of equation (3.1) has been used.

The resonance peak in R_p is obvious from the denominator in (4.2a), the minimum in T_p is more easily recognized by considering

$$1 - T_p = \frac{\gamma_d^2 - \gamma_t^2}{\gamma_t^2 + 4(\omega - \omega_p)^2} \tag{4.2 c}$$

which exhibits the same resonance as R_p.

In the formula (3.1) for the plasma radiation, $F(\theta)\, G(\omega)$ represents the probability for radiative decay of a plasmon. This must be equal to the inverse process, i.e. the optical excitation of a plasmon. Comparison of (3.1) and (4.2) yields

$$\frac{2 F G}{\cos\theta} = A_p + 2 R_p , \tag{4.3}$$

where $A_p = 1 - T_p - R_p$ is the absorptivity. The right-hand side of (4.3) represents the optical excitation probability in terms of the subsequent decay processes. The photon is either absorbed (A_p) or re-emitted ($2 R_p$). The factor 2, in front of R_p and $F G$, arises because the plasmon can radiate into the forward and backward directions. The $\cos\theta$ on the left-hand side of (4.3) accounts for the fact that $F G$ refers to the unit area of the foil, while A and R refer to the unit area of the cross-section of the light beam. Thus (4.3) demonstrates that plasma radiation and plasma resonance absorption are in fact exactly inverse processes.

Experimental evidence for the optical plasma resonance was first found in the transmission spectrum of thin Ag films [59, 60, 89] (Fig. 11). Note that the dip is only exhibited with p-polarized light. The peak in the spectrum of R_p does not show up because it is situated on the steep slope of R_p at ω_p. The same effect has been measured in thin K foils [61]. This experiment has recently been repeated [84, 12] (see Fig. 12) in order to obtain precise data for the plasma frequency and half-width of K. This was done by a fit of calculated and measured data of the T_p/T_s spectrum. The calculations were based on (4.1b) and the corresponding expression for T_s, using (1.4) and varying ω_p and γ_d until the best fit was achieved. The result is shown in Fig. 13. Fig. 14 demonstrates how sensitively T_p/T_s depends on the half-width. The data for ω_p and γ_d/ω_p are listed in Table 1, and are believed to be more precise than the energy loss data so far published [17].

Similarly, the plasma resonance in the transmission spectra of thin Mg films has been measured [86]. The minimum was found at 10.1 eV ($\lambda = 1225$ Å) which is in fair agreement with the electron energy loss measured at 10.5 eV [102].

Experiments with Al films were hampered by the lack of a suitable light source and polarizer. Ejiri and Sasaki [62, 63] have used the partial polarization effectuated by the grating in a Seya-Namioka monochromator, where light is reflected at an angle of 35°. They measured the dip in the transmission through films of Al and various other materials.

An ideal light source for these experiments is synchrotron radiation, which originates from orbiting, highly relativistic electrons, e.g. as in circular accelerators [64]. The spectrum is a broad, pure continuum and the radiation is almost perfectly polarized. The synchrotron radiation of the 6 GeV electron synchrotron in Hamburg (DESY) has recently been used to measure plasma resonances in Al [65, 66, 67]. The transmission spectra (Fig. 15) show the dip in the p-component at the plasma wavelength and a reasonable agreement between measured

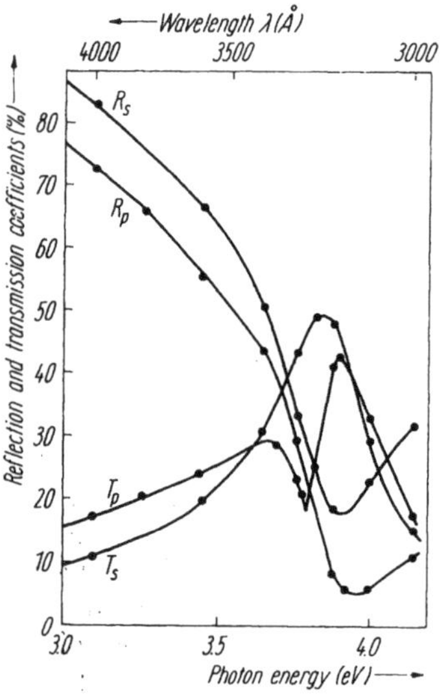

Fig. 11. Transmission and reflection spectra of a 424 Å Ag film irradiated by p- and s-polarized light at $\theta = 45°$ (after Yamaguchi [59])

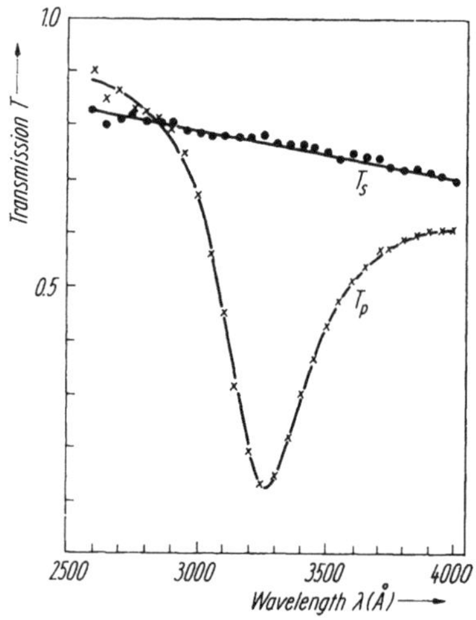

Fig. 12. Transmission spectra of a 300 Å K film on sapphire for p- and s-polarized light incident at $\theta = 30°$ (after Brambring [12])

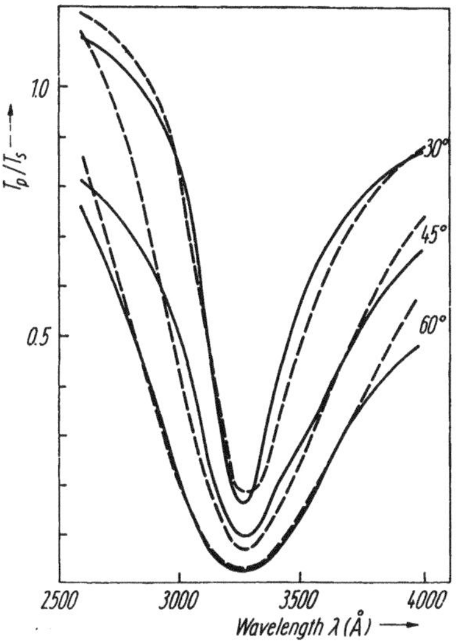

Fig. 13. Experimental (solid) and calculated (dashed) curves for the spectra of T_p/T_s of a 300 Å K film on sapphire for three angles of incidence (after Brambring [12])

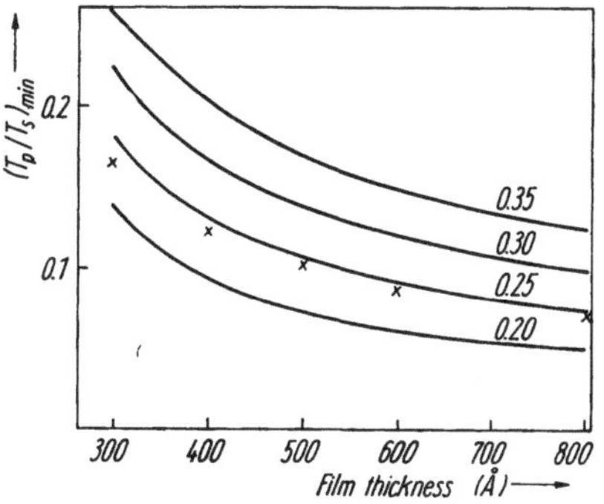

Fig. 14. Thickness dependence of the spectral minimum of T_p/T_s of K films; $\theta = 30°$. The crosses are experimental data, the curves are calculated using ε as given by (1.4) with various damping constants γ_d. The numbers refer to the half-width $\hbar\gamma_d$ in eV (after Brambring [12])

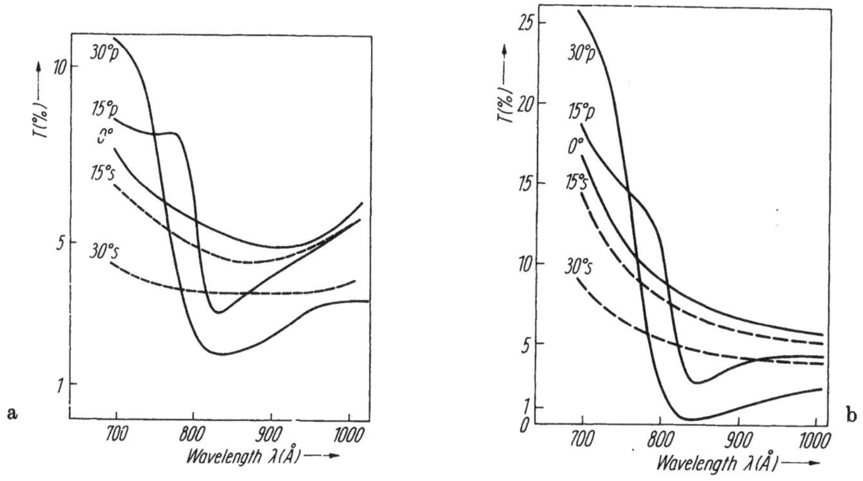

Fig. 15. Transmission spectra of unsupported Al films for p- and s-polarized light incident at 15 and 30°. a) Experimental data for a 280 Å foil; b) calculated curves for a 250 Å Al film covered on each side with 40 Å Al_2O_3 (after Skibowski [66])

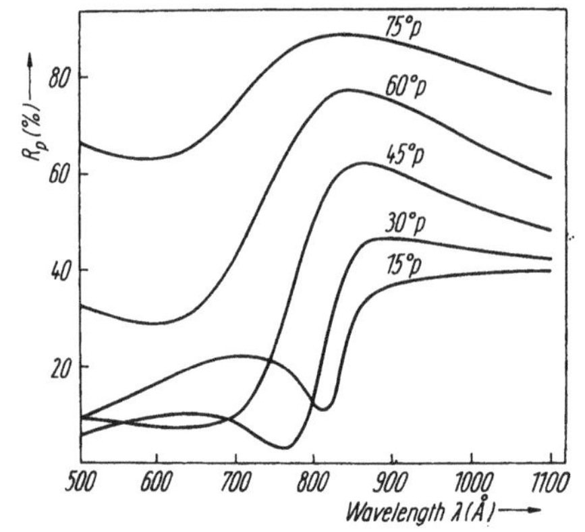

Fig. 16. Calculated reflection spectra for p-polarized light incident on an unoxidized 100 Å Al film on glass (after Feuerbacher [91])

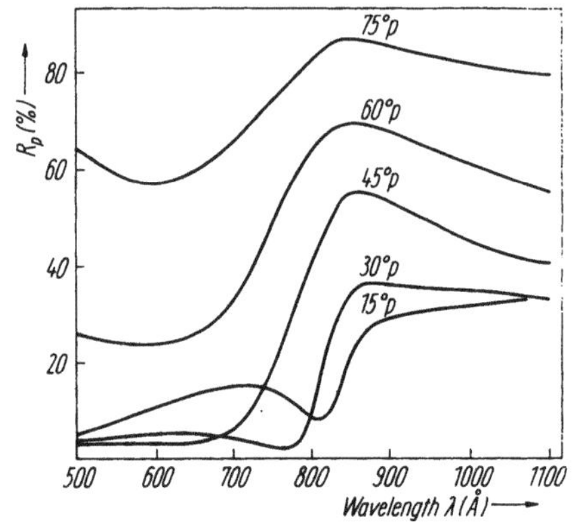

Fig. 17. Measured p-reflectance of an Al film 100 Å thick on glass for various angles of incidence (after Feuerbacher [91])

and calculated curves. However, the resonance is less pronounced than in K. This is due to oxide layers, unavoidable on unsupported films which have to be used in this spectral range.

The resonance in reflection is completely masked by an oxide layer [67]. However, oxidation can be avoided in reflection experiments, since the film can be evaporated and subsequently measured in situ. Ultra-high vacuum is necessary to allow the measurements to be made before an oxide layer is formed.

The expected resonance in the reflectance of a thin Al film on glass (Fig. 16) is not very pronounced. This is due to the optical absorption of the support, glass. Non-absorbing supports are not available in this spectral region. The experimental results (Fig. 17) [90, 91] agree fairly well with the calculated curves. A quantitative evaluation of these data could yield the plasma frequency ω_p and the damping rate γ_d of Al, but the thickness of the Al film and the optical constants of the support must be known rather accurately.

The dip just below the plasma wavelength in the curves of Figs. 16 and 17 is the well known Brewster case, where R_p vanishes in non-absorbing media. The Brewster angle θ_B is given by $\tan\theta_B = n$, where n is the refractive index. Al, as all free electron like materials, is weakly absorbing below λ_p. Hence $\varepsilon = n^2 - k^2 \approx n^2$ and from (1.4), neglecting γ_d, $n = (\lambda_p^2 - \lambda^2)^{1/2}/\lambda_p$. Thus, just below λ_p, $n \ll 1$ and increases with decreasing λ. Consequently θ_B is much smaller than in normal transparent media, where $n > 1$, and θ_B increases with decreasing wave-length, or, as is more easily seen in Fig. 17, the Brewster wavelength decreases with increasing angle.

4.2 Plasma resonance in photoemission

The plasma resonance absorption should lead to a corresponding resonance peak in photoemission, if $\hbar\,\omega_p > W$, where W is the work function. However, this resonance is not necessarily proportional to the absorptivity of the foil, i.e. the total amount of energy absorbed, since the escape depth of the photoelectrons is of the order of 10 Å [68, 69], small compared with the film thickness normally used in these experiments. To derive a phenomenological formula similar to (4.1) one calculates the energy density E^2, where E is the electric field strength just below the surface. The number of photoelectrons emitted is then [70]

$$N \propto P(\omega)\, E^2(\omega)\,,$$

where $P(\omega)$ is the transition probability of an electron to a state above vacuum level. If $P(\omega)$ varies slowly with ω, the spectral distribution of the photoyield will be approximately proportional to the spectrum of $E^2(\omega)$. For a thin foil irradiated by p-polarized light the following expression [66, 67] is derived:

$$\frac{E^2}{E_0^2} = |1 - r_p|^2 \cos^2\theta + |1 + r_p^2|^2 \frac{\sin^2\theta}{|\varepsilon|^2}\,, \tag{4.4}$$

where E_0 is the amplitude of the incident wave and $|r_p|^2 = R_p$ (see equation 4.1a)). The first term on the right-hand side represents the portion due to the E-component parallel to the surface, the second term the portion due to the perpendicular component. The latter should give rise to the plasma resonance. In fact, the factor $|\varepsilon|^{-2}$ has a peak at the plasma frequency, since $\varepsilon(\omega_p) = 0$. Under the conditions which have led from (4.1) to (4.2), (4.4) reduces to

$$\frac{E^2}{E_0^2} = \frac{4\,(\omega - \omega_p)^2 + (\gamma_d + 2\,\gamma_r)^2}{4\,(\omega - \omega_p)^2 + \gamma_t^2} \cos^2\theta + \frac{\omega_p^2}{4\,(\omega - \omega_p)^2 + \gamma_t^2} \sin^2\theta\,. \tag{4.5}$$

The first term on the right-hand side is small compared with the second term and can thus be neglected. The second term has the well known denominator representing the plasma resonance.

Experimentally this effect was observed more than 30 years ago, when Ives and Briggs [70, 71] measured pronounced spectral maxima in the photoyield of thin films of K, Rb, and Cs irradiated with p-polarized light at 60°. In addition they showed that the photoyield was proportional to the absorption spectrum, which they calculated, using the optical constants they had measured in a separate experiment with a polarimetric reflection analysis. Mayer and Thomas [72] measured the photoemission and the absorption of a thin K film on quartz simultaneously, directly proving the proportionality (see Fig. 18). In these experiments the films were only a few Å thick, comparable with or even smaller than the escape depth of the photoelectron. Here the absorption is proportional to the energy density just below the surface.

Recently, the photoelectric plasma resonance has been measured with thin Al layers, using the synchrotron radiation as a polarized light source. Even with oxide coat-

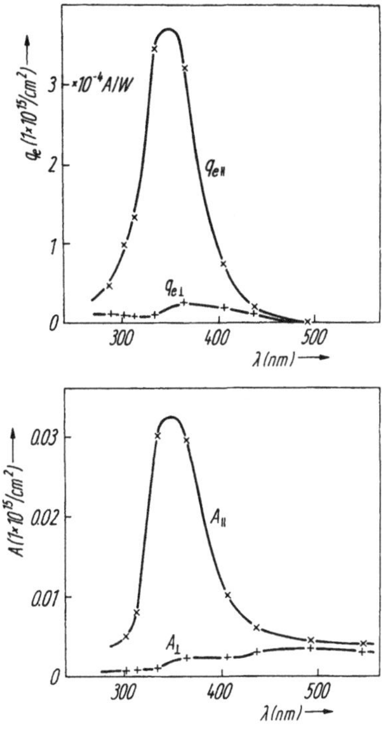

Fig. 18. Spectral dependence of the photoelectric yield q_e and the optical absorption u A of a 7.2 Å K layer on quartz irradiated by s- (⊥) and p- (||) polarized light at 45° (after Mayer and Thomas [72])

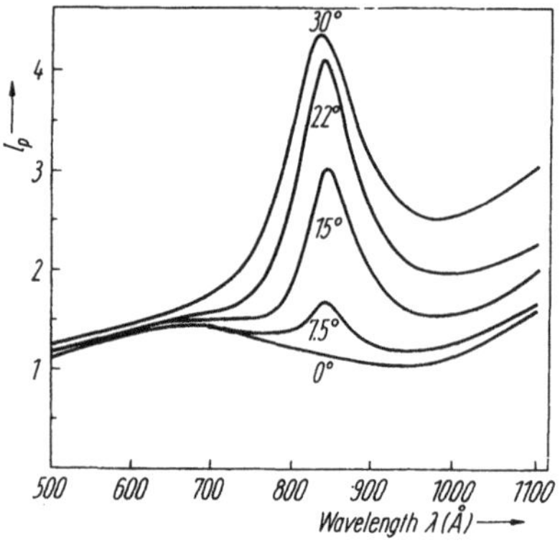

Fig. 19. Photoemission (photocurrent in arbitrary units normalized with respect to incident intensity) of a 50 Å Al film irradiated with p-polarized light at various angles of incidence (after Feuerbacher [91])

ed films a noticeable peak at ω_p was detected [66, 67, 73]. Subsequently, in situ measurements showed that the peak height decreases rapidly whilst the surface is oxidized [74]. When the experiment was repeated in utra-high vacuum, a pronounced resonance was found [91, 92] (Fig. 19). If the sample was irradiated with s-polarized light, spectra similar to the 0°-curve in Fig. 19 were measured at all angles of incidence.

In order to compare the experimental data with the calculated results, the spectra measured with p-polarized light were divided by the ones obtained with s-polarization (Fig. 20a). The calculations (Fig. 20b) were made using (4.4) and the corresponding equation for the case of s-polarization. A finite escape depth of 50 Å was assumed.

The measured and calculated data agree in the position of the maximum and in the dependence of the peak height on the angle of incidence. The discrepancies in the absolute height and width of the peak may be due to uncertainties in film thickness, escape depth of the photoelectrons, and optical constants of the support and of Al. If the other parameters are known well enough, the plasma frequency ω_p and damping rate γ_d can be determined by this experi-

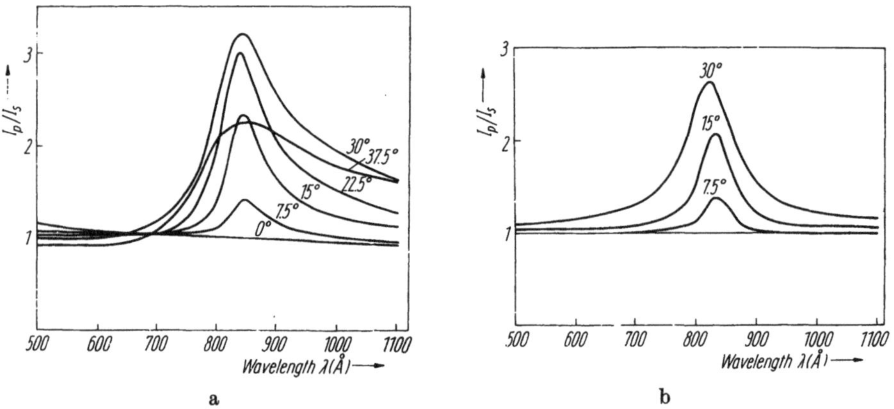

Fig. 20. a) Measured and b) calculated spectral dependence of the normalized photocurrent emitted from a 50 Å Al film on glass at various angles of incidence (after Feuerbacher [91])

ment. As compared with the effect in reflectance, the photoemission has the advantage that the resonance is more pronounced, but the disadvantage that an additional parameter, the escape depth, enters into the problem.

A precise optical determination, particularly of the damping rate γ_d of Al, would be highly desirable, since there seems to be a systematic discrepancy between the damping rates found in optical experiments and in electron energy loss spectra. Optical measurements [45, 103, 104] yield damping rates which are about 50% higher than those obtained from the line width of the electron energy loss [11, 47]. This may be explained by the effect detected by Festenberg [47] already mentioned at the end of Section 3. He found that with Al films with very small crystallites a line broadening occurred when the reciprocal $\boldsymbol{k}$-vector of the plasmon became larger than the grain size. Optical plasmons always have much smaller $\boldsymbol{k}$-vectors than those resolved in energy loss experiments. Therefore, even in films with normal grain size this additional damping may be effective for optical plasmons. The optical plasma resonances described above may facilitate a more precise determination of the damping rate.

In Ag the photoelectric plasma resonance is not immediately measurable, since $\hbar\,\omega_p$ is smaller than the work function. This problem could be overcome by coating the Ag surface with a monolayer of Cs. This would sufficiently lower the work function and yet no interference with the plasma resonance of Cs would be expected, since the plasma frequency of Cs is considerably lower. However, no relevant exerimental work has yet been reported.

5. Optical Plasma Resonances in which Momentum is not Conserved

In the resonance phenomena considered so far it was always assumed that the $\boldsymbol{k}$-vector of the plasmon parallel to the surface is conserved. This is quite obvious for the resonances in transmission and reflection, where momentum conservation is equivalent to the law of geometrical optics. Consequently, non-conservation of momentum leads to scattered light and this is indeed the first phenomenon to be treated in this section.

Secondly, the distinction between radiative and non-radiative plasmons is based on the $\boldsymbol{k}$-vectors, and it is apparent from Fig. 2 that by elastic scattering processes a non-radiative plasmon may become radiative and vice versa. This opens the possibility of interaction between photons and non-radiative surface plasmons. The resulting resonances are described in the second part of this section.

5.1 Plasma radiation excited by light (PREL)

Recently, Brambring and Raether [75] have measured plasma radiation from Ag films, which was excited by light rather than by electrons. They irradiated the sample with a spectral continuum of p-polarized light and measured the spectral distribution of the scattered light, thus outside the reflected or transmitted beam (Fig. 21). They found a sharp, intense peak at the plasma wavelength (Fig. 22).

Fluorescence could be ruled out in subsequent experiments [76 to 78] where the incident light employed was monochromatic and no frequency shift in the emitted light was measured. Furthermore, the radiation was also observed outside the plane of incidence (Fig. 21), and was always found to be polarized in the plane of emission. The angular distribution (Fig. 23) follows closely the curve for plasma radiation as given by equation (3.1b). Subsequently, the

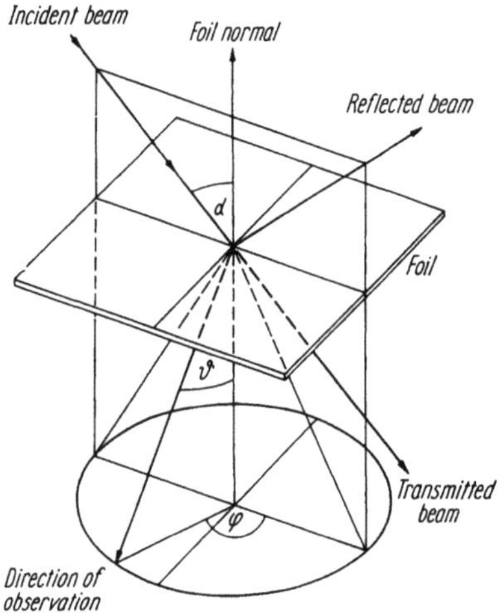

Fig. 21. Schematic diagram of the experiments on PREL (after Schreiber and Raether [77])

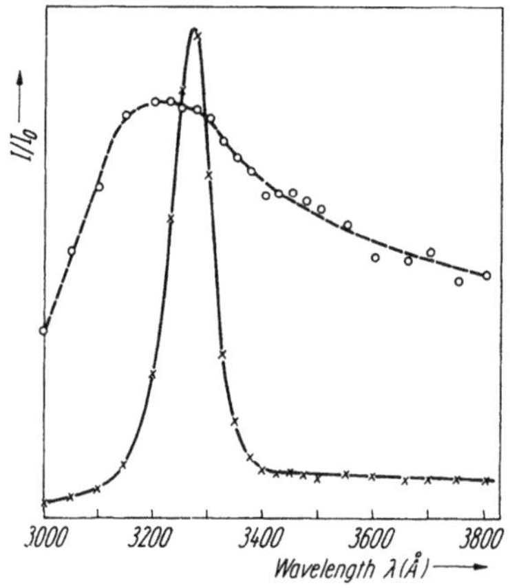

Fig. 22. Spectrum of PREL from a 300 Å Ag foil, irradiated with p-polarized light at $\alpha = 30°$, $\theta = 40°$, $\varphi = 90°$ (plane of incidenceperpendicular to plane of observation). Scattered light polarized parallel (solid curve) and perpendicular (dashed curve) to the plane of observation (after Steinmann et al. [76])

same effect was measured with thin K films [79, 87] (Fig. 24), proving that it is a pure plasma resonance and not due to the peculiar optical properties of Ag.

Phenomenologically, light scattering is attributed to inhomogeneities and surface irregularities. Stern [80] has given a classical theory of the PREL effect on the basis of surface roughness. He derived a formula containing terms which are closely related to the expression for the transition radiation (3.2) and, in addition, parameters related to the statistics of the surface roughness. Kretschmann and Raether [82] have re-derived Stern's formula in terms of the Fresnel coefficients of reflection and transmission and have demonstrated the relationship between plasma radiation excited by electrons and by light [97].

Numerical evaluation of the PREL formula and comparison with experi-

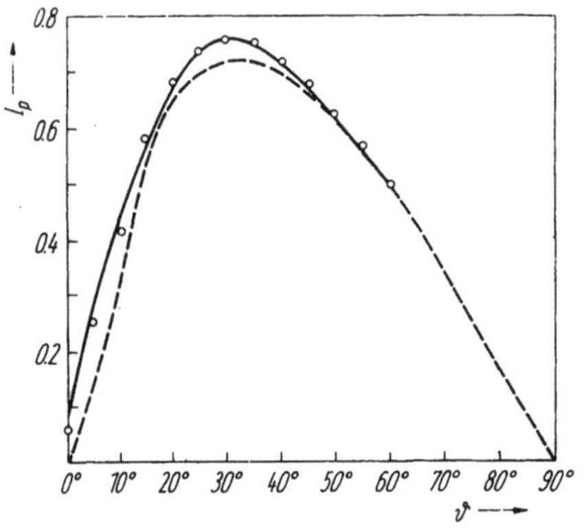

Fig. 23. Angular dependence of PREL from a 400 Å Ag foil; $\alpha = 60°$. Solid curve: experimental data; dashed curve: calculated with $F(\theta)$ (3.1b) using $\hbar\omega_p = 3.78$ eV, $\hbar\gamma_d = 0.1$ eV (after Brambring and Raether [78])

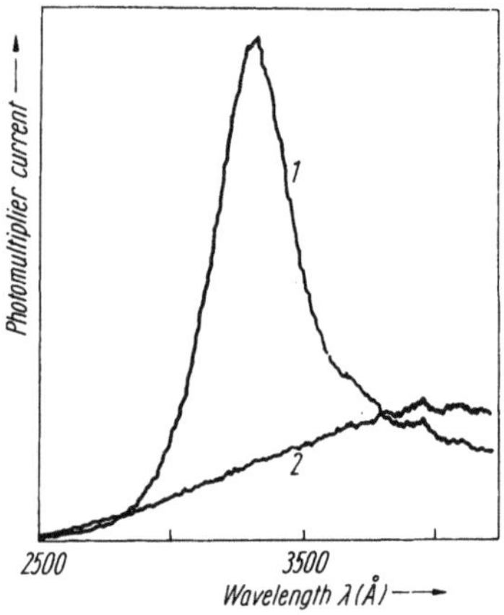

Fig. 24. Spectrum of PREL from a thin K film on quartz, irradiated with p-polarized light (curve 1). Curve 2: incident radiation, s-polarized (after Bösenberg and Raether [79])

Fig. 25. PREL intensity as a function of azimuth angle φ for an Ag film 600 Å thick ($\alpha = \theta = 30°$). The curves were calculated for three different correlation lengths σ, the experimental points at $\varphi = 90°$ were used for normalization (after Kretschmann and Raether [82])

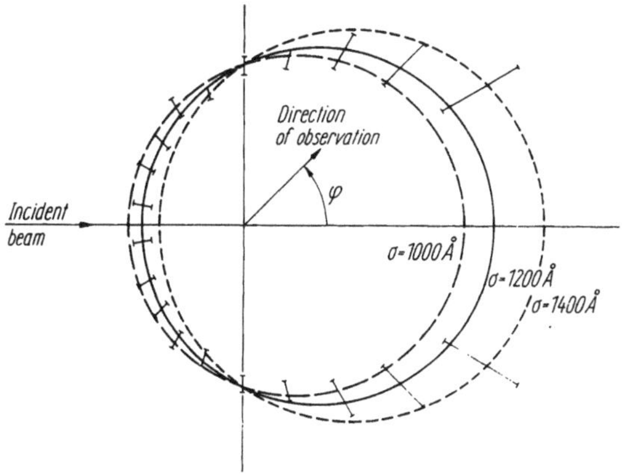

mental data [82, 83] showed good agreement in the spectrum, the angular distribution, and the dependence on azimuth angle φ (Fig. 25), confirming the correctness of the theory. Furthermore, quantitative information on the surface roughness can be obtained from these experiments. The results shown in Fig. 25 indicate that the correlation length σ, corresponding to the mean distance of the surface irregularities, is about 1200 Å.

Direct experimental evidence of the influence of inhomogeneities on the PREL effect was obtained by Bösenberg [87] with K films. He found no emission from films deposited at very low temperatures, having small electrical resistivity. When the deposition temperature was raised, the PREL effect was observed, but the resistivity was much higher, indicating structural imperfections. The influence of the surface roughness was demonstrated explicitly by Schreiber [83] with Ag samples. He studied the surface roughness with electron micrographs and found the PREL intensity to be closely correlated with the roughness.

In a microscopic theory of the PREL effect, an elastic scattering of the plasmon has to be assumed which changes the momentum parallel to the surface [76]. Wilems and Ritchie [81] have developed such a theory. The elastic scattering is ascribed to the interaction between a plasmon and a static density fluctuation which, in addition to surface irregularities, may represent any other inhomogeneity including phonons.

To illustrate the equivalence of the phenomenological and the microscopic theory, a surface with periodic irregularities, e.g. a grating, may be considered. It is conceivable that the $\boldsymbol{k}$-vector of a plasmon can be changed by adding or subtracting the reciprocal grating vector. A corresponding experiment [53] is described below.

5.2 Optical excitation and radiative decay of non-radiative surface plasmons

As pointed out in Section 2, plasma modes with phase velocity smaller than that of light are non-radiative, i.e. they cannot be excited by or emit plane electromagnetic waves. This does not mean that they cannot be excited by light. Otto [99] has shown that electromagnetic waves having the same phase velocity as the non-radiative surface plasmons can be realized. They occur at the vacuum side of a glass–vacuum interface in the case of total reflection. As is well known, their amplitude decreases exponentially in the direction normal to the interface much as in the case of a surface plasmon. If a metal surface is brought sufficiently close to the glass–vacuum interface, the non-radiative surface plasma mode is excited, provided the optical wave has the right frequency and wave-number. As a result, the reflection is drastically reduced. In this way, Otto could verify very precisely the low-frequency part of the dispersion curve shown in Fig. 2. Note that the momentum of the plasmon is conserved in this experiment.

If the $\boldsymbol{k}$-vector can be changed by elastic scattering, a direct interaction between photons and non-radiative plasmons becomes possible. The best established effect of this sort is the radiative decay of a non-radiative plasmon excited by electrons. Boersch and co-workers [27 to 29, 93] have measured radiation from bulk Ag samples bombarded with electrons at grazing incidence. They found an intense and narrow peak at 3500 Å. This corresponds to the energy of the surface plasmon as determined by equation (2.2). The position of the peak was shifted when the surface was covered with a dielectric layer, in agreement with (2.1). Stern [49] has explained the effect by surface roughness. Jones et al. [50] confirmed these results, but Cram and Arakawa [51], when repeating the experiment with thin Ag foils, found evidence that the radiation was bremsstrahlung [52]. The interpretation of the results on Ag is difficult, since volume and surface plasma frequency, as given by (1.2) and (2.2) respectively, lie very close together. In Al, the energies of the bulk plasmon (15 eV) and the surface plasmon (10 eV) are much better separated, and therefore some of the ambiguities existing for Ag do not occur. No evidence for the radiative decay of the surface plasmon was found, when oxide-coated Al films were irradiated with electrons at grazing incidence [98]. However, when the experiment was repeated with unoxidized samples, a pronounced peak at 10 eV was measured [100]. Ritchie [101] has given a microscopic theory of this effect by employing the same formalism which was used to describe the PREL effect [81].

The theoretical explanations [49, 100] of the radiative decay of the surface plasmon have received strong support by an experiment of Teng and Stern [53]. They irradiated an optical reflection grating which was coated with Al or Ag, with normally incident electrons. In this case the surface roughness was given by the grating rulings and hence the structure was known. They observed radiation in the visible range of the spectrum and measured two maxima in the angular distribution of the intensity. The position of these maxima, their dependence on the wavelength, and the direction of observation relative to the grating rulings could be quantitatively accounted for by assuming that the radiation was caused by the non-radiative surface mode, the wave vector of which had been changed by adding or subtracting a reciprocal grating vector.

Experimental evidence for the inverse process, the excitation of non-radiative surface plasmons by photons (plane waves), has recently been found in the reflectance spectra of Ag, which exhibited a dip at 3500 Å [93 to 95]. The minimum became deeper as the surface roughness was increased and was shifted when the surface was overcoated with quartz [94]. Finally, the optical excitation and subsequent radiative decay of non-radiative surface plasmons has been observed [96]. A number of pronounced peaks were found in the spectral yield of reflection gratings for p-polarized light. The variation of the position of these peaks with angle of incidence could be quantitatively explained with the dispersion curve (Fig. 2). This effect involves a twofold scattering, which is less probable than the single scattering necessary to explain the previously mentioned phenomena. This is compensated by the fact that the radiation is concentrated in the directions of diffraction.

6. Summary

Table 2 summarizes the phenomena described in this article. A number of different effects, all of which have been confirmed experimentally, can be classified and described within the framework of optical plasma resonances. This shows the usefulness of this concept. On the other hand, phenomenological formulae exist for most of the effects, which describe the expected results accurately and are generally applicable. The only parameter specific to the material in these expressions is the complex dielectric constant. Hence experiments on optical plasma resonances, particularly those which conserve momentum, are just methods of measuring optical constants. The advantage of these methods over normal techniques is that the spectra show pronounced structure at the

Table 2

Optical plasma resonances

Plasmon type	Excited by	k-conservation	Decay into	Effect	References
radiative	electron	yes	photon	plasma radiation	Ag: [34 to 39] Al: [43 to 45]
radiative	photon	yes	photon or electron	optical plasma resonance in transmission	Ag: [40, 59, 60, 89] K: [12, 61, 72] Al: [62, 63, 66, 67]
radiative	photon	yes	photon	optical plasma resonance in reflection	Al: [90, 91]
radiative	photon	—	electron	optical plasma resonance in photoemission	K: [70 to 72] Al: [66, 67, 73, 91, 92]
radiative	photon	no	photon	plasma radiation excited by light	Ag: [75 to 78, 83] K: [79, 87]
non-radiative	photon	yes	electron	frustrated total reflection	Ag: [99]
non-radiative	electron	no	photon	radiative decay of tangential surface plasmon	Ag: [27 to 29, 50, 51, 53, 93] Al: [53, 100]
non-radiative	photon	no	electron	dip in reflectivity	Ag: [93 to 95]
non-radiative	photon	no	photon	polarization of light by reflection grating	Al: [96]

plasma frequency. Thus the experiments yield a direct, sensitive determination of the plasmon energy and lifetime in the limit $k \to 0$. Some of the resonances in which momentum is not conserved may provide a useful tool for studying surface roughness.

References

[1] See, e.g., D. PINES, Elementary Excitation in Solids, W. A. Benjamin, Inc., New York 1963.
[2] H. FRÖHLICH and H. PELTZER, Proc. Phys. Soc. (London) **A68**, 525 (1955).
[3] H. RAETHER, Springer Tracts mod. Phys. **38**, 84 (1965).
[4] F. PRADAL, C. GOUT, and D. FABRE, J. Phys. Radium **26**, 372 (1965).
[5] O. KLEMPERER and J. P. G. SHEPHERD, Adv. Phys. **12**, 355 (1963).
[6] H. FRÖHLICH, Max-Planck-Festschrift 1958, VEB Deutscher Verlag der Wissenschaften, Berlin 1958 (p. 277).
[7] C. B. WILSON, Proc. Phys. Soc. (London) **76**, 481 (1960).
[8] H. EHRENREICH and H. R. PHILIPP, Phys. Rev. **128**, 1622 (1962).
[9] R. A. FERRELL and E. A. STERN, Amer. J. Phys. **30**, 810 (1962).
[10] H. DEICHSEL, Z. Phys. **174**, 136 (1963).
[11] J. GEIGER and K. WITTMAACK, Z. Phys. **195**, 44 (1966).
[12] J. BRAMBRING, Z. Phys. **200**, 186 (1967).
[13] J. DANIELS, Z. Phys. **203**, 235 (1967).
[14] R. H. RITCHIE, Phys. Rev. **106**, 874 (1957).
[15] E. A. STERN and R. A. FERRELL, Phys. Rev. **120**, 130 (1960).
[16] C. J. POWELL and J. B. SWAN, Phys. Rev. **118**, 640 (1960).
[17] C. KUNZ, Z. Phys. **196**, 311 (1966).
[18] R. H. RITCHIE and H. B. ELDRIDGE, Phys. Rev. **126**, 1935 (1962).
[19] A. OTTO, Z. Phys. **185**, 232 (1965).
[20] K. L. KLIEWER and R. FUCHS, Phys. Rev. **153**, 498 (1967).
[21] H. BOERSCH, J. GEIGER, A. IMBUSCH, and N. NIEDRIG, Phys. Letters (Netherlands) **22**, 146 (1966).
[22] H. RAETHER, Surface Sci. **8**, 233 (1967).
[23] R. A. FERRELL, Phys. Rev. **111**, 1214 (1958).
[24] V. GINZBURG and I. FRANK, J. Phys. (SU) **9**, 363 (1945).
V. GINZBURG and I. FRANK, Zh. eksper. teor. Fiz. **16**, 15 (1946).
[25] H. BOERSCH, C. RADELOFF, and G. SAUERBREY, Z. Phys. **165**, 464 (1961).
[26] G. SAUERBREY, Proc. 5th Internat. Congr. Electron Microscopy, Vol. 1, Academic Press, New York 1961 (p. AA-13).
[27] P. v. BLANCKENHAGEN, H. BOERSCH, D. FRITZSCHE, G. SAUERBREY, and H. G. SEIFERT, Phys. Letters (Netherlands) **11**, 298 (1964).
[28] H. BOERSCH, P. DOBBERSTEIN, D. FRITZSCHE, and G. SAUERBREY, Z. Phys. **187**, 97 (1965).
[29] H. BOERSCH and G. SAUERBREY, Optical Properties and Electronic Structure of Metals and Alloys, Ed. F. ABELÈS, North-Holland Publ. Comp., Amsterdam 1966 (p. 386).
[30] I. M. FRANK, Uspekhi fiz. Nauk **87**, 189 (1965); Soviet Phys. – Uspekhi **8**, 729 (1966).
[31] V. E. PAFOMOV, Zh. eksper. teor. Fiz. **33**, 1074 (1957); Soviet Phys. – J. exper. theor. Phys. **6**, 829 (1958).
[32] E. A. STERN, Phys. Rev. Letters **8**, 7 (1962).
[33] W. STEINMANN, Optical Properties and Electronic Structure of Metals and Alloys, Ed. F. ABELÈS, North-Holland Publ. Comp., Amsterdam 1966 (p. 399).
[34] W. STEINMANN, Phys. Rev. Letters **5**, 470 (1960); Z. Phys. **163**, 92 (1961).
[35] W. R. BROWN, P. WESSEL, and E. P. TROUNSON, Phys. Rev. Letters **5**, 472 (1960).
[36] A. L. FRANK, E. T. ARAKAWA, and R. D. BIRKHOFF, Phys. Rev. **126**, 1947 (1962).
[37] E. T. ARAKAWA, N. O. DAVIS, L. C. EMERSON, and R. D. BIRKHOFF, J. Phys. Radium **25**, 129 (1964).

[38] E. T. Arakawa, N. O. Davis, and R. D. Birkhoff, Phys. Rev. **135**, A224 (1964).
[39] U. Bürker and W. Steinmann, phys. stat. sol. **12**, 853 (1965).
[40] R. H. Huebner, E. T. Arakawa, R. A. McRae, and R. N. Hamm, J. Opt. Soc. Amer. **54**, 1434 (1964).
[41] J. Hofmann and W. Steinmann, Proc. Coll. Thin Films, Budapest 1965 (p. 185).
[42] R. J. Herickhoff, E. T. Arakawa, and R. D. Birkhoff, Phys. Rev. **137**, A1433 (1965).
[43] E. T. Arakawa, R. J. Herickhoff, and R. D. Birkhoff, Phys. Rev. Letters **12**, 319 (1964).
[44] R. J. Herickhoff, W. F. Hanson, E. T. Arakawa, and R. D. Birkhoff, Phys. Rev. **139**, A1455 (1965).
[45] E. T. Arakawa, R. N. Hamm, W. F. Hanson, and T. M. Jelinek, Optical Properties and Electronic Structure of Metals and Alloys, Ed. F. Abelès, North-Holland Publ. Comp., Amsterdam 1966 (p. 374).
[46] W. Steinmann and H. Wille, phys. stat. sol. **15**, 507 (1966).
[47] C. v. Festenberg, Phys. Letters (Netherlands) **23**, 293 (1966); Z. Phys. **207**, 47 (1967).
[48] U. Bürker, to be published.
[49] E. A. Stern, Optical Properties and Electronic Structure of Metals and Alloys, Ed. F. Abelès, North-Holland Publ. Comp., Amsterdam 1966 (p. 396).
[50] G. E. Jones, L. S. Cram, and E. T. Arakawa, Phys. Rev. **147**, 515 (1966).
[51] L. S. Cram and E. T. Arakawa, Phys. Rev. **153**, 455 (1967).
[52] J. C. Ashley, Phys. Rev. **155**, 208 (1967).
[53] Ye-Yung Teng and E. A. Stern, Phys. Rev. Letters **19**, 511 (1967).
[54] R. H. Ritchie, Phys. Letters (Netherlands) **24A**, 348 (1967).
[55] N. Matsudaira, J. Phys. Soc. Japan **18**, 380 (1963).
[56] M. Hattori, K. Yamada, and H. Suzuki, J. Phys. Soc. Japan **18**, 203 (1963).
[57] R. A. Ferrell and E. A. Stern, Amer. J. Phys. **30**, 810 (1962).
[58] See, e. g., H. Wolter, Encyclopedia of Physics, **24**, Springer-Verlag, Berlin/Göttingen/Heidelberg 1956 (p. 461).
[59] S. Yamaguchi, J. Phys. Soc. Japan **18**, 266 (1963).
[60] A. J. McAlister and E. A. Stern, Phys. Rev. **132**, 1599 (1963).
[61] H. Deichsel, Z. Phys. **174**, 136 (1963).
[62] A. Ejiri and T. Sasaki, J. Phys. Soc. Japan **20**, 876 (1965).
[63] T. Sasaki and A. Ejiri, Optical Properties and Electronic Structure of Metals and Alloys, Ed. F. Abelès, North-Holland Publ. Comp., Amsterdam 1966 (p. 417).
[64] R. Haensel and C. Kunz, Z. angew. Phys. **23**, 276 (1967).
[65] M. Skibowski and W. Steinmann, J. Opt. Soc. Amer. **57**, 112 (1967).
[66] M. Skibowski, Thesis, University of Munich, 1967.
[67] M. Skibowski, B. Feuerbacher, W. Steinmann, and R. P. Godwin, Z. Phys. **211**, 329 (1968); **211**, 342 (1968).
[68] H. Thomas, Basic Problems in Thin Film Physics, Ed. R. Niedermayer and H. Mayer, Vandenhoeck & Ruprecht, Göttingen 1966 (p. 307).
[69] F. J. Piepenbring, Optical Properties and Electronic Structure of Metals and Alloys, Ed. F. Abelès, North-Holland Publ. Comp., Amsterdam 1966 (p. 316).
[70] H. E. Ives and H. B. Briggs, J. Opt. Soc. Amer. **26**, 247 (1936).
[71] H. E. Ives and H. B. Briggs, J. Opt. Soc. Amer. **28**, 330 (1938).
[72] H. Mayer and H. Thomas, Z. Phys. **147**, 419 (1957).
[73] W. Steinmann and M. Skibowski, Phys. Rev. Letters **16**, 989 (1966).
[74] B. Feuerbacher, M. Skibowski, W. Steinmann, and R. P. Godwin, J. Opt. Soc. Amer. **58**. 137 (1968).
[75] J. Brambring and H. Raether, Phys. Rev. Letters **15**, 882 (1965).
[76] W. Steinmann, J. Hofmann, and K. Stettmaier, Phys. Letters (Netherlands) **23**, 234 (1966).
[77] P. Schreiber and H. Raether, Z. Naturf. **21a**, 2116 (1966).
[78] J. Brambring and H. Raether, Z. Phys. **199**, 118 (1967).

[79] J. Bösenberg and H. Raether, Phys. Rev. Letters **18**, 397 (1967).
[80] E. A. Stern, Phys. Rev. Letters **19**, 1321 (1967).
[81] R. E. Wilems and R. H. Ritchie, Phys. Rev. Letters **19**, 1325 (1967).
[82] E. Kretschmann and H. Raether, Z. Naturf. **22a**, 1623 (1967).
[83] P. Schreiber, Z. Phys. **211**, 257 (1968).
[84] J. Brambring and H. Raether, Z. Naturf. **21a**, 1527 (1966).
[85] J. B. Swan, A. Otto, and H. Fellenzer, phys. stat. sol. **23**, 171 (1967).
[86] H. O. Tittel, Phys. Letters (Netherlands) **26A**, 145 (1968).
[87] J. Bösenberg, Phys. Letters (Netherlands) **26A**, 74 (1967).
[88] T. Kloos, Z. Phys. **208**, 77 (1968).
[89] D. Schulz and M. Zurheide, Z. Phys. **211**, 165 (1968).
[90] B. Feuerbacher, R. P. Godwin, and M. Skibowski, Phys. Letters (Netherlands) **26A**, 595 (1968).
[91] B. Feuerbacher, Thesis, University of Munich, 1968.
[92] B. Feuerbacher, M. Skibowski, and R. P. Godwin, II. Internat. Conf. Vacuum – Ultraviolet Radiation Physics — Interaction with Solids, Gatlinburg 1968 (to be published in Bull. Amer. Phys. Soc.).
[93] P. Dobberstein, A. Hampe, and G. Sauerbrey, II. Internat. Conf. Vacuum Ultraviolet Radiation Physics — Interaction with Solids, Gatlinburg 1968 (to be published in Bull. Amer. Phys. Soc.).
[94] J. L. Stanford, H. E. Bennet, J. M. Bennett, E. J. Ashley, and E. T. Arakawa, II. Internat. Conf. Vacuum Ultraviolet Radiation Physics — Interaction with Solids, Gatlinburg 1968 (to be published in Bull. Amer. Phys. Soc.).
[95] S. E. Schnatterly, II. Internat. Conf. Vacuum Ultraviolet Radiation Physics — Interaction with Solids, Gatlinburg 1968 (to be published in Bull. Amer. Phys. Soc.).
[96] E. T. Arakawa, R. N. Hamm, R. H. Ritchie, and J. J. Cowan, II. Internat. Conf. Vacuum Ultraviolet Radiation Physics — Interaction with Solids, Gatlinburg 1968 (to be published in Bull. Amer. Phys. Soc.).
[97] E. Kretschmann and H. Raether, Z. Naturf. **23a**, 615 (1968).
[98] J. C. Ashley, L. S. Cram, and E. T. Arakawa, Phys. Rev. **160**, 313 (1967).
[99] A. Otto, phys. stat. sol. **26**, K99 (1968).
[100] U. Bürker and W. Steinmann, to be published.
[101] R. H. Ritchie, to be published.
[102] P. Schmüser, Z. Phys. **180**, 105 (1964).
[103] H. Ehrenreich, H. R. Philipp, and B. Segall, Phys. Rev. **132**, 1918 (1963).
[104] R. C. Vehse, E. T. Arakawa, and J. L. Stanford, J. Opt. Soc. Amer. **57**, 551 (1967).

(Received June 18, 1968)

Original Papers

phys. stat. sol. **28**, 463 (1968)

Subject classification: 10

Laboratoire de Physique des Solides[1]), *Faculté des Sciences, Orsay*

L'activation thermique des déviations sous faibles contraintes dans les structures h.c. et c.c.

Par

B. ESCAIG

Duesbery et Hirsch ont suggéré récemment que dans les structures c.c. et h.c., la déviation d'une dislocation hors de son plan de dissociation, doit se produire à faible contrainte par un mécanisme de double décrochement. On montre ici que ce mécanisme doit se produire en dessous d'une contrainte $\sigma_c = R/b^2$, où R est l'énergie de recombinaison et b le vecteur de Burgers de la dislocation; de plus, il n'introduit aucune discontinuité de contrainte, d'énergie, ou de volume d'activation. On calcule l'énergie d'activation maximum de la déviation correspondant à une composante thermique de la contrainte nulle, pour laquelle on retrouve l'expression donnée par Duesbery et Hirsch. En tenant compte de l'interaction élastique des décrochements au moyen d'une tension de ligne qui varie logarithmiquement avec leur distance de séparation, on obtient un volume d'activation qui varie en σ^{-1} à faible contrainte σ.

It is shown that the double kink mechanism proposed by Duesbery and Hirsch for the cross slip in b.c.c. and h.c.p. structures should be operative for stresses smaller than $\sigma_c = R/b^2$, where R is the recombination energy, and b the Burgers vector of the dislocation. The variation of the stress, of the energy or of the activation volume should show no discontinuity at the transition. The cross slip energy at zero stress is calculated, in agreement with Duesbery and Hirsch's results. Taking into account the elastic kink interaction through a line tension logarithmically dependent on the kink separation, one finds an activation volume which goes as the inverse of the stress at low stresses.

1. Introduction

Une dislocation est plus stable dans le plan où une certaine dissociation est possible; elle a donc tendance à s'y trouver. Si elle doit le quitter pour glisser dans un plan différent, elle doit auparavant passer par un état d'énergie supérieure, ou col. L'activation thermique d'un tel processus de déviation conditionne dans ce cas la mobilité de la dislocation, dans un réseau cristallin autrement parfait. Friedel [1] a ainsi expliqué le glissement sur les plans prismatiques des métaux hexagonaux compacts, où la dissociation des dislocations de glissement a lieu dans le plan de base qui, dans ces expériences, est un plan où les contraintes appliquées sont nulles. Plus récemment, le même mécanisme de déviation a été suggéré [2 à 4] pour rendre compte de la limite élastique des métaux cubiques centrés à basse température; dans cette structure la dissociation des vis a lieu dans plusieurs plans à la fois, et bloque ainsi les dislocations de glissement en position sessile. La discussion qui suit s'applique aussi bien à l'un ou à l'autre de ces phénomènes.

[1]) Laboratoire associé au C.N.R.S..

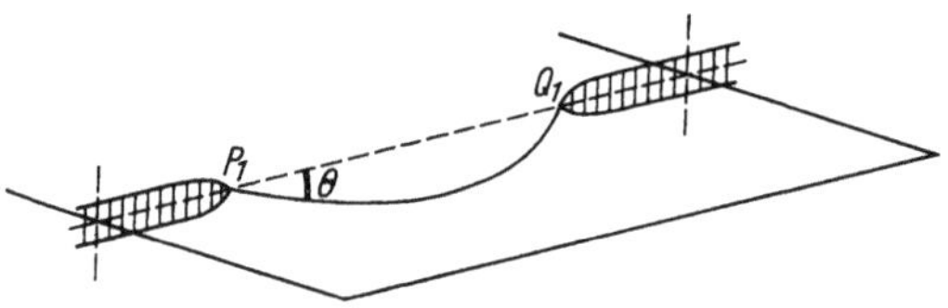

Fig. 1. Col de type A

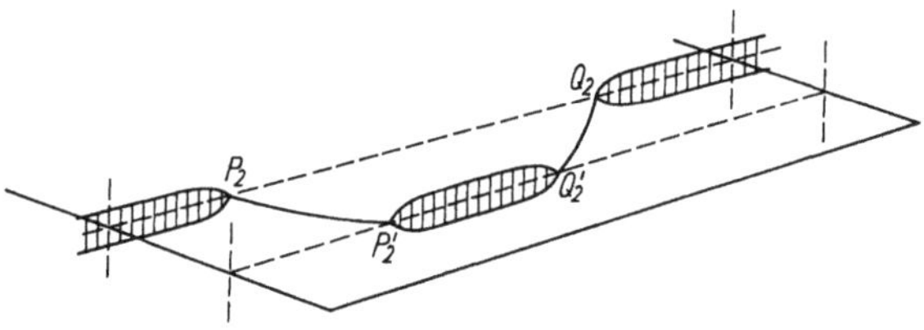

Fig. 2. Col de type B

Dans ce type d'analyse, il est essentiel de déterminer la configuration de col qui sépare les états sessiles et glissiles de la dislocation. Dans les analyses qui précédent [1 à 4], le col est obtenu en recombinant la dislocation sur une longueur critique, $P_1 Q_1$ (position de col A, Fig. 1). Cette longueur doit être assez grande pour que l'arc $P_1 Q_1$ puisse se développer sous l'action des contraintes appliquées sur le plan de glissement, entrainant la recombinaison totale de la dislocation, et son déplacement sur une certaine distance.

Duesbery et Hirsch [5] ont remarqué que ce col A peut n'être pas toujours atteint. Aux faibles contraintes, l'arc critique nécessaire $P_1 Q_1$ est de plus en plus grand, et pour l'atteindre l'arc recombiné PQ doit franchir plusieurs plans atomiques premiers voisins, et parallèles au plan de dissociation initial. PQ peut alors se dissocier de nouveau dans un de ces plans, hors de son plan de glissement, et redevenir ainsi sessile. Dans ce cas, la position de col A n'est jamais atteinte. Une position de col (d'énergie plus basse) est obtenue pour une séparation critique $P_2 Q_2$ des crans PP′ et QQ′ (position de col B, Fig. 2).

On se propose ici d'obtenir l'énergie d'activation du col B, et de préciser la transition du col A au col B. Deux facteurs déterminent l'évolution de l'énergie du col B:

(i) l'interaction élastique des jonctions PP′ et QQ′,

(ii) l'interaction des constrictions P′ et Q′.

Duesbery et Hirsch [5] ont donné un traitement simplifié du modèle en prenant des jonctions rectilignes d'inclinaison constante, interagissant en $(PQ)^{-1}$, et des constrictions bien formées et indépendantes. Ce traitement vaut à la limite des très faibles contraintes, où l'interaction élastique des jonctions prédomine, mais non à la transition du col A au col B, où nous montrons que c'est l'interaction des constrictions naissantes qui domine.

Nous avons pris un modèle de double décrochement dans lequel:

(i) les jonctions PP′ et QQ′ sont des segments de dislocation parfaite, courbés par la contrainte, et dont l'inclinaison est calculée en fonction de leur séparation; ils interagissent par l'intermédiaire d'une tension de ligne en log (l/b) où $l = PQ$, et où b est la période du réseau. Un tel facteur logarithmique est en effet habituel dans l'énergie d'une petite déviation à la ligne droite faite sur une dislocation rectiligne (cf. par exemple Hirth et al. [6] pour une petite déformation triangulaire);

(ii) l'interaction des constrictions P′ et Q′ est obtenue en calculant l'énergie de la boucle mal dissociée P′ Q′; ce problème a été résolu récemment par l'auteur [7] en utilisant l'approximation classique de Stroh pour décrire l'interaction élastique des partielles.

On montre ainsi que la contrainte critique de transition du col A au col B est celle qui donne, dans le col A, une flèche égale à b, soit

$$\sigma_c b^2 = R$$

où R est l'énergie de recombinaison par unité de longueur de la dislocation. De plus la configuration critique étant à la fois un col de type A et un col de type B, il n'existe à la transition aucune discontinuité de contrainte, d'énergie, ni de volume d'activation.

Le calcul de l'énergie du col B conduit en général à une résolution numérique seulement, sauf au cas limite voisin de la transition, $\sigma b^2 \lessapprox R$, ou voisin du domaine athermique de la contrainte, tel que $\sigma b^2 \ll R$. Des expressions asymptotiques sont obtenues dans chacun de ces cas. A la limite $\sigma = 0$, on retrouve une énergie égale à celle calculée par Duesbery et Hirsch. Mais la forme différente utilisée pour l'interaction élastique entre les décrochements par ces auteurs les conduit à un volume d'activation en $\sigma^{-1/2}$, alors que nous obtenons à cette limite un volume en σ^{-1}.

Nous présentons au Chapitre 2 le calcul de l'énergie du col B, et nous rassemblons nos résultats au Chapitre 3, en les appliquant essentiellement à l'interprétation des mesures de limite élastique des métaux cubiques centrés. En Annexe nous donnons les coefficients élastiques qui ont rapport plus spécifiquement à la structure hexagonale compacte, ou cubique centrée.

2. Énergie d'activation

Dans ce calcul nous supposons que les jonctions PP′ et QQ′ joignent deux plans atomiques adjacents. L'énergie nécessaire pour créer sur la dislocation dissociée initiale un double décrochement tel que PP′ Q′Q (Fig. 2), s'écrit

$$W = 2\, W_j(l', l') - \sigma b^2 l' + W_p(l') + 2\, W_c. \tag{1}$$

- b vecteur de Burgers de la parfaite; $l = PQ$, $l' = P'Q'$;
- W_j énergie nécessiare pour créer une jonction PP′ ou QQ′, compte tenu du travail des contraintes, σ, correspondant;
- W_c énergie d'une constriction bien formée comme en P et Q (donnée Annexe A.1);
- W_p énergie nécessaire pour former la partie mal dissociée P′Q′ à partir de l'état dissocié initial; on a [7]

$$W_p = \beta\, d\, (2\, U + E\, L'), \qquad l' = d\left(\frac{\beta}{\alpha}\right) L' . \tag{2}$$

α, β coefficients élastiques dépendant de la structure, calculés Annexe A.1; d largeur de dissociation de la dislocation rectiligne; U, E, L' sont des fonctions de $H_0 = h_0/d$, h_0 étant la largeur au centre de la boucle P′Q′ ($0 < H_0 \leqq 1$):

$$U(H_0) = H_0 \int_{t_0}^{1} [-\log t + H_0 (t-1)]^{1/2}\, dt\,, \quad t_0 = X/H_0\,, \quad X = b/d\,, \qquad (2a)$$

$$L'(H_0) = H_0 \int_{t_0}^{1} [-\log t + H_0 (t-1)]^{-1/2}\, dt\,, \qquad (2b)$$

$$E(H_0) = -\log H_0 + H_0 - 1\,. \qquad (2c)$$

Le paramètre $t_0 = b/h_0$ est introduit pour tenir compte de la "largeur" de la dislocation parfaite, qu'on prend égale à b. Lorsque l' croît de 0 à l'infini, H_0 croît de X à 1, et W_p croît de 0 à une limite finie, $2\,W_c$, représentant la self-énergie de deux constrictions complètes.

L'équation (1) définit l'énergie W comme fonction de deux variables indépendantes, l et l'. Le col B s'obtient en annulant ses deux dérivées partielles. Au Chapitre 2.1 nous étudions la condition $\partial W/\partial l' = 0$, c'est-à-dire que nous cherchons la forme d'équilibre et l'énergie W_j des jonctions PP′ et QQ′ pour une séparation l donnée. Au Chapitre 2.2 nous obtenons la condition de col en écrivant $\partial W/\partial l = 0$, c'est-à-dire en choisissant parmi tous les décrochements en équilibre précédents, celui dont l'énergie est maximum. Au Chapitre 2.3 nous donnons l'expression asymptotique de l'énergie du col, à la transition A → B, et aux faibles contraintes.

2.1 *Énergie des décrochements en équilibre*

Déterminons d'abord l'équilibre d'un double décrochement PP′Q′Q, de longueur donnée l = PQ. On montre Annexe A.2 qu'on peut associer au segment mal dissocié P′Q′ une tension de ligne:

$$\tau^* = \tau - R + \alpha\, E(H_0)\,.$$

τ est la tension de ligne du double décrochement PP′Q′Q supposé recombiné sur toute sa longueur (c'est-à-dire où la portion P′Q′ ne serait pas dissociée).

R est l'énergie de recombinaison de la dislocation dissociée initiale, par unité de longueur.

α est le coefficient élastique défini par

$$R = \alpha\, E(X) \quad \text{avec} \quad X = \frac{b}{d}\,, \quad E(X) = X - 1 - \log X \qquad (3)$$

et dépend de la structure (Annexe A.1).

La condition d'équilibre s'exprime par l'équilibre des tensions de ligne au point P′ ou Q′ (Fig. 3):

$$\frac{\partial W}{\partial l'} = \tau\,(1 - \cos\theta) - [R - \alpha\, E(H_0)]\,.$$

L'angle θ est petit pour $2b \ll l - l'$, ce qui est toujours le cas; faisant $1 - \cos\theta \approx \approx \theta^2/2$ et exprimant θ en fonction de la contrainte appliquée σ et de $(l - l')$, on obtient pour $\partial W/\partial l' = 0$:

$$l - l' = (2\,\tau)^{1/2}\lambda; \quad \lambda = \left(\frac{2}{\sigma b}\right)\left([R - \alpha\, E(H_0) + \sigma b^2]^{1/2} - [R - \alpha\, E(H_0)]^{1/2}\right). \qquad (4)$$

On calcule ensuite l'énergie $2\,W_j$ des jonctions PP′ et QQ′ à l'équilibre:

$$2\,W_j = (l - l')\, w; \quad w = R + (b^2/\lambda^2) - (\sigma b^2/2) - (1/12)\,(\sigma b\, \lambda/2)^2. \qquad (5)$$

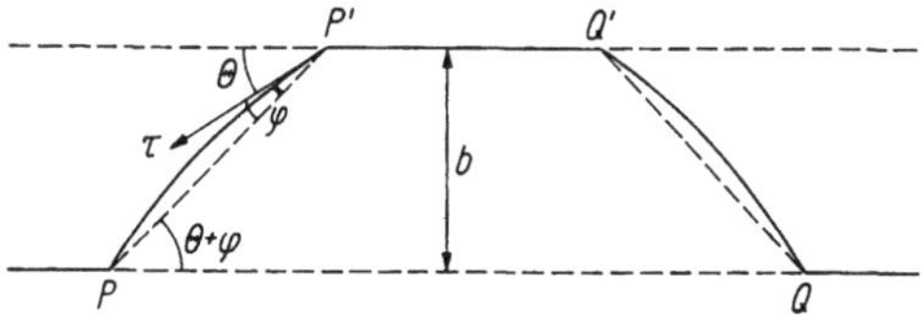

Fig. 3. Géométrie du col B

Les trois premiers termes de w tiennent compte de la recombinaison, de l'allongement, et du travail des contraintes nécessaires dans un modèle de jonctions rectilignes; le dernier terme représente le gain d'énergie obtenu en permettant aux jonctions de se courber sous la contrainte.

Les équations (1), (2), (4) et (5) résolvent le problème de l'énergie totale W du décrochement à l'équilibre. La séparation H_0 des partielles au centre de P'Q' détermine l', $E(H_0)$ — équation (2) —, $(l - l')$ — équation (4) —, donc aussi l, et enfin W_j et W. Dans ce qui suit nous considérons W comme fonction de l. Pour obtenir la position de col, il reste à annuler $\partial W/\partial l$.

Enfin, remarquons que la tension de ligne τ des jonctions n'intervient dans l'énergie totale que par sa racine carrée (équations (4) et (5)); cette énergie est donc peu sensible au détail du terme logarithmique à utiliser pour τ, que nous écrivons

$$\tau = (\mu b^2/4\pi) \log (l/b) \approx (\mu b^2/4\pi) \log\left((l'/b) + \lambda \sqrt{\mu}\right) \tag{6}$$

où μ est le coefficient de cisaillement.

2.2 *Configuration de col. Énergie d'activation*

Nous dérivons à présent W par rapport à l, l' restant constant; on obtient selon (1) et (5) la condition de col:

$$\frac{\partial W}{\partial l} = \frac{\partial (2\, W_j)}{\partial\, (l - l')} = w + \lambda \frac{\mathrm{d}w}{\mathrm{d}\lambda} - \lambda^2 \frac{\mathrm{d}w}{\mathrm{d}\lambda} \frac{\mathrm{d}(2\,\tau)^{1/2}}{\mathrm{d}l} = 0 \,. \tag{7}$$

Les deux premiers termes de la dernière égalité seraient seuls obtenus dans une approximation de tension de ligne constante, c'est-à-dire sans interaction élastique entre les jonctions. Ils s'écrivent, compte tenu de (4) et (5):

$$w + \lambda \frac{\mathrm{d}w}{\mathrm{d}\lambda} = \alpha\, E(H_0) - \sigma\, b^2 \,.$$

Ce résultat s'interpréte facilement. Si la tension de ligne τ est constante, la condition de col est simplement (cf. équation (1)):

$$\frac{\mathrm{d}}{\mathrm{d}l'} [-\sigma\, b^2\, l' + W_p(l')] = 0$$

car une fois atteint cette position, on diminue l'énergie totale en éloignant par translation les jonctions PP', QQ', ce qui laisse invariant $2\, W_j$, mais gagne de l'énergie en poursuivant la dissociation incomplète de P'Q', et en accroissant le travail des contraintes. Par ailleurs (cf. Annexe A.2):

$$\mathrm{d}\, W_p = \alpha\, E(H_0)\, \mathrm{d}l' \tag{8}$$

d'où la condition de col:

$$\alpha\, E(H_0) = \sigma\, b^2 \,. \tag{9}$$

Mais nous montrons Chapitre 2.3.1 que négliger l'interaction élastique des jonctions devant celle des constrictions P' et Q' ne peut se justifier qu'à leur naissance, au moment de la transition A $\rightarrow$ B. Sitôt que P'Q' atteint la valeur de quelques largeurs de dissociation, les constrictions sont pratiquement complètes et indépendantes; seule reste l'interaction des jonctions. Il est essentiel alors de tenir compte dans (7) du dernier terme en $(\mathrm{d}\tau/\mathrm{d}l)$.

L'équation (7) détermine la valeur de H_0 au col qui correspond à une contrainte donnée. On en déduit l'énergie par les équations (1), (2), (4) et (5). Mais la forme des fonctions $w(H_0)$, $\lambda(H_0)$ et surtout $l(H_0)$ est trop complexe pour tenter autre chose qu'une résolution numérique. Toutefois une solution analytique peut être trouvée dans deux types d'approximations:

(i) pour P'Q' $\lesseqqgtr d$, d étant la largeur de dissociation; comme on voit au Chapitre 2.3.1, on peut négliger l'interaction élastique des jonctions et prendre (9) comme condition de col;

(ii) pour P'Q' $\gg d$, au contraire, on voit au Chapitre 2.3.2 qu'on peut négliger l'interaction des constrictions, c'est-à-dire écrire $\alpha\, E(H_0) \approx 0$, ce qui correspond à des constrictions bien formées et indépendantes (8), et à des jonctions pratiquement rectilignes, et de longueur constante.

C'est ce sujet que nous abordons maintenant.

2.3 Formes asymptotiques de l'énergie d'activation

Le paramètre λ (équation (4)) est essentiel pour le calcul de l'énergie des jonctions W_j. On peut l'écrire

$$\lambda = (2/\sigma\, b)\,(R - \alpha\, E)^{1/2}\left(\left(1 + \frac{\sigma\, b^2}{R - \alpha\, E}\right)^{1/2} - 1\right).$$

De là deux développements sont possibles selon la valeur de $\sigma\, b^2/(R - \alpha\, E)$ par rapport à 1.

(i) $\sigma\, b^2 \gg R - \alpha\, E(H_0)$; dans ce cas P'Q' $\ll d$; en effet pour $R - \alpha\, E(H_0) = 0$, $H_0 = X$ (équation (3)) et $l' = $ P'Q' $= 0$ (équation (2)). L'équation (9), alors valable — voir Chapitre 2.3.1 — indique que dans ce domaine $\sigma\, b^2 \approx \alpha\, E(H_0) \lesssim R$

(ii) $\sigma\, b^2 \ll R - \alpha\, E(H_0)$; dans ce cas la contrainte est très faible; en effet $\alpha\, E \ll R - \sigma\, b^2$, donc a fortiori $\alpha\, E \ll R$, et on a $\sigma\, b^2 \ll R$.

2.3.1 Transition du col A au col B; $\sigma\, b^2 \lesssim R$

Considérons le cas où $\sigma\, b^2 \gg R - \alpha\, E$. Pour écrire la relation de col (7) d'une part nous développons λ par rapport à $t = [(R - \alpha\, E)/\sigma\, b^2]^{1/2}$:

$$\lambda \approx 2\, \sigma^{-1/2}\, [1 - t + (t^2/2)]\,,$$

d'autre part nous développons la tension de ligne au voisinage de $l_0 = 2(2\,\tau/\sigma)^{1/2}$, valeur de l lorsque t et l' sont nuls (cf. équation (4)):

$$\lambda\, \mathrm{d}(2\,\tau)^{1/2}/\mathrm{d}l \approx k\, [1 - t + (t^2/2)]\,,$$

$$k = (2 \log\, (l_0/b))^{-1} = [\log\, (8\, \mu\, b^2/R)]^{-1}$$

d'où la relation de col:

$$0 = \alpha\, E(H_0) - \sigma\, b^2\, [1 - (2\, k/3)] + k\, \sigma\, b^2\, t^2\,.$$

Dans les cas pratiques k est toujours petit: il varie de 0,12 pour les cubiques centrés peu dissociés ($d \approx 2\, b$) à 0,2 pour les hexagonaux compacts très dissociés ($d \approx 50\, b$). On voit que l'interaction élastique des jonctions indroduit dans la relation de col un terme de l'ordre de $\sigma\, b^2\, (2\, k/3) \approx 0{,}1\, \sigma\, b^2$, tout à fait négligeable.

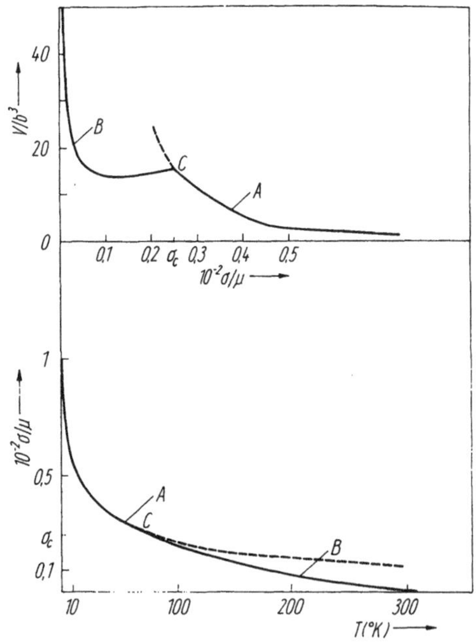

Fig. 4. Limite élastique et volume d'activation calculé pour une énergie de faute $\gamma = \mu\, b/192$ sur (110), et une largeur $2\,b$ de dislocation parfaite. Jusqu'au point C, pour $\sigma < \sigma_c$, col A; pour $\sigma < \sigma_c$, col. B. En pointillé, les valeurs obtenues avec le seul col A. Enfin, on a utilisé la valeur $\mu\, b^3 \approx 7$ eV, qui correspond aux cas du fer, du vanadium et du niobium

Le col correspondant à $l' = 0$ s'obtient pour une contrainte qu'on déduit de la relation précédente où $t = 0$, soit $\sigma\, b^2 = R$, en accord avec (9). Or une telle configuration est en fait une dislocation parfaite de P à Q, courbée sous la contrainte en un arc de cercle dont la flèche est b, et la corde PQ $= l_0 = 2\,(2\,\tau/\sigma)^{1/2}$. On la retrouve exactement parmi les configurations de col du type A, et pour la même valeur de l contrainte, R/b^2. En d'autres termes cette configuration, pour $\sigma\, b^2 = R$, est à la fois un col A et un col B. A ce moment, la contrainte, l'énergie, et le volume d'activation sont identiques dans les processus A ou B, comme on pourrait le constater avec les équations (1), (4) et (5). En dessous de $\sigma\, b^2 = R$, le volume d'activation du col B diminue[2]), tandis que celui du col A continue d'augmenter; de ce fait, le col B devient énergétiquement préférable au col A. Ainsi, la transition du type de col A au type de col B se produit pour

$$\sigma_c\, b^2 = R \tag{10}$$

et ne s'accompagne d'aucune discontinuité dans la variation de la limite élastique, de l'énergie, ou du volume d'activation.[3]) Ceci est illustré Fig. 4, où sont représentées les variations de la limite élastique $\sigma(T)$ et du volume d'activation $V(\sigma)$ calculées pour une énergie de faute d'empilement $\gamma = \mu\, b/192$ sur les plans (110), et une largeur $2\,b$ de dislocation parfaite. Jusqu'au point C, pour $\sigma < \sigma_c$ le modèle A a été employé; pour $\sigma < \sigma_c$ le modèle B a été employé (pour les équations correspondantes, voir [9] en Annexe).

2.3.2 *Contraintes faibles;* $\sigma\, b^2 \ll R$

Supposons maintenant $\sigma\, b^2 \ll R - \alpha\, E < R$; on développe alors λ par rap-

[2]) A mesure que P' et Q' se séparent, la tension de ligne de P'Q' diminue, donc aussi $(l - l')$ (Fig. 3) et le volume d'activation si $P'Q' \ll d$; en toute rigueur le volume devrait donc marquer un point anguleux à la transition, tandis que la courbe énergie–contrainte présenterait une inflexion. Mais l'augmentation du volume d'activation due à celle de P'Q' vient rapidement masquer ce phénomène, et le volume devient à nouveau faiblement croissant (Fig. 4).

[3]) Si on tient compte de l'interaction élastique des jonctions, pour $k \neq 0$, le col $l' = 0$ a lieu pour $\sigma_1\, b^2\,(1 - 0{,}7\,k) = R$. Mais la contrainte pour laquelle se produit le col de type A, ayant une flèche égale à b, dans la *même* approximation de tension de ligne variable (équation (6)) est aussi égale à σ_1. Cela ne change donc en rien la discussion, si ce n'est que la contrainte critique est en fait σ_1, telle que $\sigma_1\, b^2\,(1 - 0{,}7\,k) = R$.

port à $\sigma b^2/(R - \alpha E)$, et on obtient en négligeant αE devant R:

$$\lambda \approx b R^{-1/2} \left(1 - (\sigma b^2/4 R)\right) , \tag{11a}$$

$$l - l' \approx b (2 \tau/R)^{1/2} \left(1 - (\sigma b^2/4 R)\right) , \tag{11b}$$

$$2 W_j \approx 2 b (2 \tau R)^{1/2} \left(1 - (\sigma b^2/4 R)\right) . \tag{11c}$$

La relation de col donne la variation de $l = PQ$ avec σ; en développant (7) au premier ordre en $\sigma b^2/4 R$:

$$\sigma b^2 = \alpha E(H_0) + b (\mu b^2 R/2 \pi)^{1/2}/l \sqrt{\log (l/b)} .$$

Pour une contrainte assez faible, $\alpha E(H_0)$ devient très inférieur au second terme[4]) ($\sigma \lessapprox 5 \times 10^{-4} \mu$, limite peu sensible à la largeur de dissociation, ou à la structure). D'où:

$$\frac{l}{b} = \left(\frac{R}{2 \tau}\right)^{1/2} \frac{\mu}{2 \pi \sigma} \approx \left(\frac{R}{\mu b^2}\right)^{1/2} \frac{\mu}{2 \pi \sigma} . \tag{12}$$

Cette relation est intermédiaire entre celle donnée par Duesbery et Hirsch ($\sigma \sim l^{-2}$) et celle obtenue pour un col A ($\sigma \sim l^{-1}$).

L'énergie d'activation dépend alors de σ à travers la tension de ligne τ. En appliquant (1) au cas où $\alpha E(H_0) \approx 0$, les équations (11) et (12), conduisent à l'énergie de la déviation:

$$W = 4 W_c + 2 \mu b^3 \left(\frac{R}{\mu b^2}\right)^{1/2} \left(\frac{1}{2 \pi} \log \left(\frac{(\mu R)^{1/2}}{2 \pi \sigma b}\right)\right)^{1/2} \left(1 + \frac{\sigma b^2}{4 R}\right) . \tag{13a}$$

Enfin, pour des raisons physiques évidentes, la longueur l, équation (12), et avec elle le terme logarithmique de l'énergie (13a), ne deviennent pas infinis lorsque $\sigma = 0$; l est alors de l'ordre de la taille du réseau de dislocation, et la tension de ligne τ devient de l'ordre de $\mu b^2/2$, ce qui correspond à $\log (l/b) \approx 2 \pi$. On obtient ainsi pour $\sigma = 0$ l'énergie maximum du décrochement:

$$W_0 = \mu b^3 \left(2 \left(\frac{R}{\mu b^2}\right)^{1/2} + \sqrt{2} \frac{\beta}{\mu b^2} \frac{(1 - X)^2}{X}\right) ; \quad X = \frac{b}{d} . \tag{13b}$$

3. Conclusions

Nous pouvons maintenant résumer le mécanisme d'activation des déviations dans les structures h.c. ou c.c.. Nous avons montré que ce mécanisme diffère selon que la contrainte appliquée est supérieure ou inférieure à la contrainte critique $\sigma_c = R/b^2$, où R est l'énergie de recombinaison de la dislocation.

a) $\sigma > R/b^2$. L'activation thermique de la déviation nécessite le passage d'une position de col A (Fig. 1), où la dislocation doit être recombinée sur une longueur critique. Ce processus conduit [1 à 5] à une énergie qui décroît essentiellement en σ^{-1}, donc à un volume en σ^{-2}.

b) $\sigma < R/b^2$. La position de col du type A précédente n'est plus atteinte, car avant d'y parvenir la dislocation forme un double décrochement, dû à la redissociation sessile d'une partie du segment recombiné qui essaye de se développer. La position critique devient alors une configuration de double décrochement, ou col B (Fig. 2) comme l'ont remarqué les premiers Duesbery et Hirsch.

Enfin, nous avons montré que pour la contrainte $\sigma = R/b^2$, la configuration

[4]) Il suffit pour cela d'exprimer l en fonction de H_0, $l = l' + (l - l')$, avec $l - l'$ donné équation (11) et $l' \approx -d \log (2 E(H_0))$, expression valable [7] pour $E \ll 1$.

critique est commune à la fois aux deux types de cols; la transition d'un type à l'autre n'entraîne donc aucune discontinuité de contrainte, d'énergie, ou de volume d'activation.

L'activation du col B modifie les caractéristiques du glissement, telles qu'elles pourraient être prévues [2] si le col A était toujours atteint. De façon générale, la contrainte nécessaire pour activer la déviation, c'est-à-dire physiquement la limite élastique prévue, est inférieure à celle qu'on obtiendrait avec le type A de col, et s'annule plus rapidement avec la température. Ceci vient de ce que l'énergie nécessaire pour franchir le col B est plus basse que pour le col A. Avant de discuter plus en détail ces modifications, il est nécessaire de rappeler que la variation thermique de la contrainte se déduit directement de l'énergie de déviation $W(\sigma)$, lorsqu'on peut supposer que seul ce processus est activé de façon prépondérante. Dans ce cas, la vitesse de déformation plastique $\dot{\varepsilon}$ varie comme $\exp(-W/kT)$, d'où on déduit en inversant l'exponentielle:

$$T = \frac{W}{k\,C}, \qquad C = \log\left[\frac{\nu}{\dot{\varepsilon}}\,\varrho\, A \left(\frac{b}{l_c}\right)^2\right]. \tag{14}$$

C est essentiellement lié à A, aire balayée après chaque libération thermique, et à l_c, longueur critique de nucléation du glissement. Les autres symboles ont un sens classique: k est la constante de Boltzmann, ν une fréquence atomique, ϱ la densité de dislocation.

Nous discutons maintenant les conséquences de l'activation de col B, au voisinage de la transition, et à la limite des faibles contraintes.

(i) Voisinage de la transition, $\sigma \lessapprox R/b^2$

Au point $\sigma = R/b^2$, la courbe $\sigma(T)$ qui décrit le col B se raccorde tangentiellement à celle qui décrit le col A. En effet, pour cette valeur de la contrainte la configuration critique est commune aux deux cols. Ceux-ci ont donc même volume d'activation, ce qui entraîne une pente $\mathrm{d}\sigma/\mathrm{d}T$ identique, puisque l'aire de glissement A, et avec lui le facteur C équation (14), a la même valeur dans les deux types de cols. Cette pente commune peut donc aussi bien se déduire des équations du col A, que de celles du col B. On obtient ainsi, en négligeant la variation avec la contrainte de la largeur de dissociation, et celle de la tension de ligne τ:

$$\left.\begin{aligned}
&\left(\frac{\mathrm{d}W}{\mathrm{d}\sigma}\right)_{\sigma=\sigma_c} = \frac{4\sqrt{2}}{3}\,\frac{\tau}{b}^{1/2} R^{3/2}\,\frac{\mathrm{d}}{\mathrm{d}\sigma}\left(\frac{1}{\sigma}\right)_{\sigma=\sigma_c},\\
&\left(\frac{\mathrm{d}(\sigma/\mu)}{\mathrm{d}T}\right)_{\sigma=\sigma_c} = -\frac{k\,C}{\mu b^3}\left(\frac{\sigma_c}{\mu}\right)^{1/2}\left(\frac{9\pi}{4\log\left(4\,\frac{\mu}{\sigma_c}\right)}\right)^{1/2}, \qquad \sigma=\sigma_c=\frac{R}{b^2}\\
&\text{avec}\\
&A \approx b\,\varrho^{-1/2} \quad \text{et} \quad C \approx \log\left(\frac{\nu}{\dot{\varepsilon}}\,b\sqrt{\varrho}\,\frac{\sigma_c}{\mu}\right).
\end{aligned}\right\} \tag{15}$$

Dans le cas de métaux cubiques centrés, avec $\nu/\dot{\varepsilon} \approx 10^{18}$, $\varrho \approx 10^8\ \mathrm{cm}^{-2}$ et $\sigma_c/\mu \approx 2\times10^{-3}$, $C \approx 25$.

(ii) Très faibles contraintes, $\sigma \ll R/b^2$

Les équations (13a) et (14) permettent le calcul de $\sigma(T)$, avec une constante C peu dépendante de la contrainte, de l'ordre de 25. D'où, avec $X = b/d$, d étant la largeur de dissociation:

$$T \approx \frac{\mu b^3}{k C}\left\{2{,}2 \frac{\beta}{\mu b^2} \frac{(1-X)^2}{X}+2\left(\frac{\sigma_c}{\mu}\right)^{1/2}\left[\frac{1}{2\pi} \log \left(\frac{(\mu \sigma_c)^{1/2}}{2 \pi \sigma}\right)\right]^{1/2}\right\}, \quad \sigma_c b^2 = R . \qquad (16)$$

C'est essentiellement une variation de T en $\log \sigma$. D'ailleurs si nous calculons le volume d'activation $V = -\mathrm{d}W/\mathrm{d}\sigma$, à partir de (13a), nous obtenons un volume en σ^{-1}:

$$V = [l - 0{,}5\,(l - l')]\, b^2 \approx (b^3/2\pi)\,(\mu\, b^2/R)^{1/2}\,[(R/\sigma\, b^2\,) - \pi] \qquad (17)$$

(à condition d'estimer $[\log(l/b)]^{1/2} \approx (2\pi)^{1/2}$, c'est-à-dire $\tau \approx \mu\, b^2/2$).
Ceci provient d'une interaction élastique entre les décrochements qui varie comme le logarithme de leur séparation; à cette limite nous avons montré en effet qu'elle était due à un terme de tension de ligne en $(\log l/b)$.

On retrouve donc ici des variations analogues à celles formulées par Seeger [10] pour vaincre le frottement de Peierls-Nabarro, au moyen d'un double décrochement semblable. Un volume d'activation en σ^{-1} semble plus vraisemblable [8] qu'un volume en $\sigma^{-1/2}$ auquel Duesbery et Hirsch arrivent [5] à partir d'une interaction entre décrochements inversement proportionelle à leur séparation. Une telle approximation n'est d'ailleurs sans doute pas très bonne dans le cas présent, où les jonctions PP′, QQ′ (Fig. 2) sont très allongées sur la direction moyenne de la dislocation.

(iii) Enfin, lorsque la composante thermique de la contrainte est nulle, l'énergie d'activation atteint une valeur maximum finie W_0, et non une valeur infinie comme dans le cas du col A. W_0 est donnée par (13), et est identique à celle proposée par Duesbery et Hirsch [5].

Dans le cas des métaux cubiques centrés de transition, on observe en général [8] $W_0 \approx 0{,}1\,\mu\, b^3$; ceci est en accord avec (13b) pour une largeur de dissociation $d \approx 2\,b$.

Pour conclure, nous voyons que la valeur de la contrainte $\sigma_c = R/b^2$ est un paramètre important pour décrire la variation thermique de la limite élastique dans les structures c.c. ou h.c. (glissement prismatique). Par exemple, dans le cas du tungstène, un mécanisme de type A suffit à décrire correctement les mesures de limite élastique faites entre 20 et 465 °K par Beardmore et Hull, et par Rose et al. (cf. [2b], Fig. 4), avec une énergie de faute d'empilement de l'ordre de $\mu\, b/100$. En effet, dans ce cas, on calcule une contrainte critique égale à $\sigma_c = R/b^2 \approx 1{,}9 \times 10^{-3}\,\mu$. Cette valeur tombe dans le domaine où la contrainte devient presque athermique, et par conséquent, joue un rôle négligeable pour la plus grosse partie du phénomène.

Il n'en est pas toujours ainsi, et σ_c peut correspondre à des contraintes qui sont encore largement dans le domaine de forte variation thermique. Nous renvoyons pour une comparaison plus précise avec l'expérience, aux mesures de Conte et Groh [9] sur des poils de fer, publiées dans ce même numéro. On peut constater que dans le cas du fer σ_c est atteint vers 100 °K; dans le niobium [5], il l'est vers 150 °K. Enfin dans le cas de largeur de dissociation plus grande, disons trois ou quatre b, l'équation (10) montre que σ_c serait de l'ordre de $10^{-2}\,\mu$, et correspondrait à des températures encore plus basses.

Annexe

A.1 Calcul des coefficients α et β

Le coefficient α est celui qui est lié à l'énergie de recombinaison de la dislocation, R selon

$$R = \alpha\,(X - 1 - \log X)\,, \quad X = b/d$$

où d est la largeur de dissociation.

Le coefficient β est lié à l'énergie d'une constriction de la dislocation dissociée selon

$$W_c = \left(\frac{1}{2\sqrt{2}}\right)\beta\, b\,\frac{(1-X)^2}{X}\,.$$

Le facteur $(1 - X)^2$ a été introduit [2b] en développant dans l'équation (2a) l'intégrant de U, pour $H_0 = 1$ au voisinage de $t = 1$; il n'est important que pour les faibles dissociations.

Les valeurs de α et β qui suivent sont calculées dans le cas d'une dislocation vis, les seules qui peuvent dévier. Le coefficient β est calculé à l'aide de l'approximation classique de Stroh [2b], et en négligeant la dépendance de l'énergie de ligne τ avec le caractère de la dislocation. On obtient le tableau suivant:

Structure h.c.	dissociation dans le plan de base en dislocations de Shockley	$\alpha = \frac{\mu b^2}{16\pi}$	$\beta = \alpha\,[1{,}5 \log(1{,}7/x)]^{1/2}$
Structure c.c. [2b]	dissociation sessile dans les plans (112)	$\alpha = \frac{\mu b^2}{18\pi}$	$\beta = \alpha\,[\log(3/x)]^{1/2}$
Structure c.c. [2b]	dissociation sessile dans les plans (110)	$\alpha = \frac{\mu b^2}{34}$	$\beta = \alpha\,[0{,}53 \log(5/x)]^{1/2}$

A.2 Tension de ligne du segment mal dissocie P'Q' (Fig. 2)

Considérons un segment de dislocation mal dissocié, tel que P'Q' (Fig. 2) c'est-à-dire pincé aux deux extrémités et de longueur P'Q' $= l'$. Nous montrons qu'on peut lui associer une ,,tension de ligne'':

$$\tau^* = \tau - R + \alpha\, E(H_0)\,, \quad E(H_0) = -\log H_0 + H_0 - 1\,.$$

τ est la tension de ligne de la parfaite, R l'énergie de recombinaison; $\alpha\, E(H_0)$ représente la variation d'énergie d'interaction des deux partielles, rectilignes et parallèles, lorsqu'on fait passer leur séparation de la largeur d'équilibre initiale, d, pour un ruban de faute rectiligne, à la séparation h_0 obtenue ici au centre du segment P'Q'; h_0 est représenté ici par la variable sans dimension $H_0 = h_0/d$. Ainsi $\alpha\, E(H_0)$ est compris entre zéro et R, et la tension τ^* varie avec H_0, depuis $\tau - R$, tension de la dislocation bien dissociée obtenue pour $h_0 = d$ (et l', infini), à τ, tension de la parfaite, obtenue pour $h_0 = b$ (et l', nul).

En effet, l'énergie du segment P'Q' s'écrit

$$W(\mathrm{P'Q'}) = l'\,(\tau - R) + W_p(l')\,.$$

$W_p(l')$ est l'énergie nécessaire pour former P'Q' à partir de l'état dissocié initial. W_p est donné par l'équation (2). Si l' varie de $\mathrm{d}l'$:

$$\mathrm{d}W_p = \beta\, d\, (2\,\mathrm{d}U + L'\,\mathrm{d}E + E\,\mathrm{d}L')$$

dans les notations de l'équation (2). Or, on a la relation [7] $2\,\mathrm{d}U + L'\,\mathrm{d}E = 0$ entre les fonctions U, L', et E de H_0. Il vient alors:

$$\mathrm{d}W_p = \beta\, d\, E\,\mathrm{d}L' = \alpha\, E(H_0)\,\mathrm{d}l'$$

et

$$\mathrm{d}W(\mathrm{P'Q'}) = \tau^*\,\mathrm{d}l' = [\tau - R + \alpha\, E(H_0)]\,\mathrm{d}l'\ .$$

Bibliographie

[1] J. FRIEDEL, Internal Stresses and Fatigue in Metals, Elsevier P.C., Amsterdam 1959 (p. 220).
[2] B. ESCAIG, a) J. Phys. **27**, 205 (1966). b) J. Phys. **28**, 171 (1967).
[3] V. VITEK, phys. stat. sol. **18**, 687 (1966).
[4] R. A. FOXALL, M. S. DUESBERY et P. B. HIRSCH, Canad. J. Phys. **2**, 607 (1966).
[5] M. S. DUESBERY et P. B. HIRSCH, Battelle Coll. on Dislocation Dynamics, Seattle 1967, McGraw-Hill Publ. Comp. (sous presse).
[6] J. P. HIRTH, T. JØSSANG et J. LOTHE, J. appl. Phys. **37**, 110 (1966).
[7] B. ESCAIG, Battelle Coll. on Dislocation Dynamics, Seattle 1967, McGraw-Hill Publ. Comp. (sous presse); J. Phys. **29**, 225 (1968).
[8] H. CONRAD, The Relation between the Structural and Mechanical Properties of Metals, N.P.L. Teddington, H.M. Stat. Office, 1963 (p. 476).
[9] R. CONTE, P. GROH et B. ESCAIG, phys. stat. sol. **28**, 475 (1968).
[10] A. SEEGER, Phil. Mag. **1**, 651 (1956).

(Received April 11, 1968)

phys. stat. sol. 28, 475 (1968)

Subject classification: 10.1; 21.1.1

Centre de Recherches sur les Très Basses Températures, C.N.R.S., Grenoble (a), et Laboratoire de Physique des Solides[1]*), Faculté des Sciences, Orsay (b)*

Étude de l'écoulement plastique de monocristaux filamentaires de fer entre 300 et 4 °K

Par

R. CONTE (a), P. GROH (a) et B. ESCAIG (b)

On étudie du point de vue expérimental l'influence de la température et de la vitesse de traction sur la plasticité de monocristaux filamentaires de fer, dans un domaine de température allant de 4 à 300 °K. Les résultats des mesures sont comparés aux prédictions du modèle de Dorn et Rajnak (forces de Peierls) et à celles du modèle d'Escaig (dissociation–recombinaison). Ce dernier modèle semble mieux rendre compte de l'expérience, indiquant que les "barrières de potentiel de Peierls" sont probablement dues, dans la structure cubique centrée, à un effet de dissociation des dislocations.

Strain rate and temperature change experiments are made on iron whiskers from 4 to 300 °K. The elastic limit, the activation energy, and the activation volume are determined. The experimental results are compared with the Dorn-Rajnak model (Peierls force) and with the Escaig model (dissociation–recombination). The experimental data support the last model and suggest that the Peierls friction in b.c.c. structure may be related to a dissociation of dislocations.

1. Introduction

Le mécanisme de déformation plastique des métaux de structure cubique centrée, dans son tout premier stade (limite élastique) n'a pas encore été nettement éclairci. Deux types d'obstacles au glissement des dislocations ont été avancés: les impuretés, ou le réseau lui-même.

Selon le taux d'impuretés présentes dans la matrice, les dislocations peuvent interagir avec celles-ci sous forme d'atomes interstitiels (voir par exemple Stein et al. [1]) ou sous forme de fins précipités dans le cas des métaux moins purs (voir Mordike et Haasen [2]). W. Frank [3] a récemment calculé la variation thermique de la limite élastique correspondant à ces deux possibilités. Toutefois pour les métaux de grande pureté (quelques 10 ppm d'impuretés), et en dessous de 300 °K environ, plusieurs raisons permettent de penser que le rôle des impuretés, s'il existe toujours, n'est pas prépondérant. D'une part, Conrad [4, 5] (voir aussi [18] et [34]) a remarqué que la variation thermique de la limite élastique semble ne pas dépendre sensiblement du degré de pureté de l'échantillon. D'autre part le durcissement observé à 4 °K semble trop important (au moins un facteur 3) pour être dû aux concentrations très faibles d'impuretés, réparties au hasard, en solution solide. Si, par contre, celles-ci se répartissent de préférence sur les dislocations, alors la limite élastique et la cission d'écoulement plastique doivent avoir des énergies d'activation différentes. Conrad et Schoeck [6] ont vérifié que les deux phénomènes présentent au contraire la même énergie, montrant ainsi une origine commune.

[1]) Laboratoire associé au C.N.R.S..

Pour un métal de grande pureté, la déformation plastique semble plutot être contrôlée par le réseau cristallin lui-même. Un tel frottement a été le plus souvent envisagé selon la description habituelle en forces de Peierls [7, 8, 9], surtout par Conrad [4, 10]; Dorn et Rajnak [11], Guyot et Dorn [18] ont développé plus récemment ce modèle. D'un autre côté, l'étude des très faibles déformations, inférieures au millième, a été à l'origine d'une nouvelle formulation de ce frottement de réseau. Kossowsky et Brown [12] ont montré que les dislocations commencent à se déplacer au-dessus d'une limite "micro-élastique" τ_E, faible et peu sensible à la température, donc très inférieure à la limite élastique macroscopique conventionelle τ_M. D'autre part de nombreuses observations [13, 14, 15] en microscopie électronique montrent que de longues dislocations vis se développent après une faible déformation à basse température. Ces faits s'expliquent facilement si on remarque que les dislocations vis dans cette structure se dissocient de façon sessile [33]. Ainsi τ_E semble lié au mouvement initial des portions coins de dislocations préexistantes [12, 15], tandis que τ_M est lié à celui des parties vis développées au cours de la microdéformation, et qui doivent être recombinées en partie avant de redevenir mobiles. Vitek [16] et Escaig [17] ont ainsi calculé la limite élastique correspondante, τ_M.

Nous voulons souligner ici qu'à notre avis il s'agit d'une nouvelle façon d'exprimer le même effect de réseau. La présence d'une dissociation sessile rend seulement plus profondes et plus aigues les "vallées" $\langle 111 \rangle$ de Peierls. D'ailleurs dans le mécanisme de dissociation–recombinaison, on doit tenir compte de la redissociation ultérieure de l'arc de dislocation [35], une fois libéré après une première recombinaison, chaque fois qu'il traverse une rangée atomique $\langle 111 \rangle$. La périodicité des positions stables de la dislocation est donc analogue à celle de la description habituelle en forces de Peierls.

Nous avons donc entrepris de confronter à l'expérience le modèle de Dorn et Rajnak, et celui d'Escaig. Ce dernier modèle a été modifié pour tenir compte d'une part d'une largeur de dislocation parfaite, d'autre part de la remarque précédente [17b]. Du point de vue expérimental nous avons étudié l'influence de la température et de la vitesse de traction sur la plasticité de monocristaux filamentaires de fer. Peu de résultats expérimentaux sont connus en effet dans le domaine 4 à 300 °K, et pour des monocristaux de bonne pureté. Une étude de relaxation de la force appliqueé est en cours [19] ainsi qu'un complément à des mesures de fluage [20], dans le but de préciser les résultats obtenus dans le présent travail.

L'analyse des résultats obtenus ici montre que le modèle de dissociation–recombinaison donne une bonne représentation des mesures dans tout le domaine, de 4 à 300 °K. Il prévoit des largeurs de dissociation d'environ 3 b (b étant la période du réseau), la dislocation étant considérée comme recombinée lorsque sa largeur est de l'ordre de 2 b. Ceci correspond à une énergie de faute d'empilement d'environ 90 erg/cm^2 sur les plans (110), ou 170 erg/cm^2 sur (112).

Il est évident que le choix entre ce modèle et celui de Dorn et Rajnak est difficile à faire sur la base des seules mesures, car d'après ce que nous avons dit plus haut, les deux modèles ne diffèrent que par des détails secondaires à un certain point de vue, disons la forme des barrières de Peierls, qui est très différente dans les deux cas. Les prédictions du modèle de Dorn ne peuvent donc être très différentes de ce qu'on observe. Effectivement ce dernier modèle représente les mesures mais jusqu'à 160 °K environ. Au delà un désaccord systematique a lieu. Le modèle de dissociation semble donc mieux correspondre à la réalité, indiquant que les "barrières de Peierls" sont probablement dues, dans la structure cubique centrée, à un effet de dissociation.

Enfin, nous analysons dans la suite la cission appliquée τ comme somme [4, 5] d'une composante thermique τ^* ne dépendant que de la température, et d'une composante athermique τ_μ caractéristique de l'état de l'échantillon, soit

$$\tau = \tau^* + \tau_\mu .$$

2. Technique expérimentale

Les échantillons utilisés dans nos expériences étaient de gros „whiskers" de fer. Ils ont été préparés suivant la méthode de Brenner [21]: réduction de chlorure ferreux par l'hydrogène vers 650 °C. Les whiskers obtenus, de „diamètres" compris entre quelques microns et quelques dizaines de microns, étaient sélectionnés à l'aide d'un microscope métallographique. Nous n'avons utilisé que ceux qui avaient des faces exemptes de défauts (ramifications, marches, points d'oxydation, . . .), une section carrée (ou rectangulaire voisine d'un carré), et dont les côtés étaient compris entre 60 et 120 μm avec une variation relative inférieure à 2% sur 1 cm de longueur. Le choix d'un diamètre supérieur à 50 μm était destiné à éliminer un effet de taille observé précédemment [22, 23] sur ce type d'échantillons.

Des mesures de frottement interne [24] entre 15 et 120 °C n'ont pas permis de mettre en évidence la présence de carbone ou d'azote en inclusion. Des mesures de résistivité à champ magnétique nul, avec un courant de 1 mA donnent un rapport

$$\frac{\varrho(273\,°\mathrm{K})}{\varrho(4{,}2\,°\mathrm{K})} \approx 10^3$$

qui correspondrait à un taux d'impuretés voisin de 10 ppm [36, 37, 38]. L'axe de croissance et de traction est $\langle 100 \rangle$. N'ayant pu mettre en évidence après traction, les plans de glissement, nous avons adopté un facteur de Schmid de 0,47 [25] qui correspond au plan (112) de contrainte de maximale avec une direction de glissement $\langle 111 \rangle$.

Les essais de traction ont été effectués sur une machine construite au laboratoire et décrite par ailleurs [26]. La longueur utile de l'échantillon, de l'ordre de 5 mm entre les deux soudures, était mesurée au microcomparateur optique. Trois vitesses de traction ont été utilisées, correspondant à des vitesses d'allongement relatif de $1{,}5 \times 10^{-5}$, 5×10^{-5} et $1{,}5 \times 10^{-4}\ \mathrm{s}^{-1}$. Les températures autres que celles déterminées par les points fixes, mesurées contre l'échantillon, étaient régulées à mieux que $\pm$ 1°.

3. Résultats expérimentaux

3.1 Courbes de consolidation et variation thermique de la limite élastique

Quelques courbes de consolidation $\tau(\gamma)$ pour des vitesses de cisaillement relatif $\dot{\gamma} \approx 10^{-4}\ \mathrm{s}^{-1}$ sont présentées Fig. 1. Aucun effet, caractéristique du maclage ou de la propagation de bandes de Lüders [25], n'est observé. La rupture, parfaitement ductile jusqu'à 100 °K environ, semble présenter une

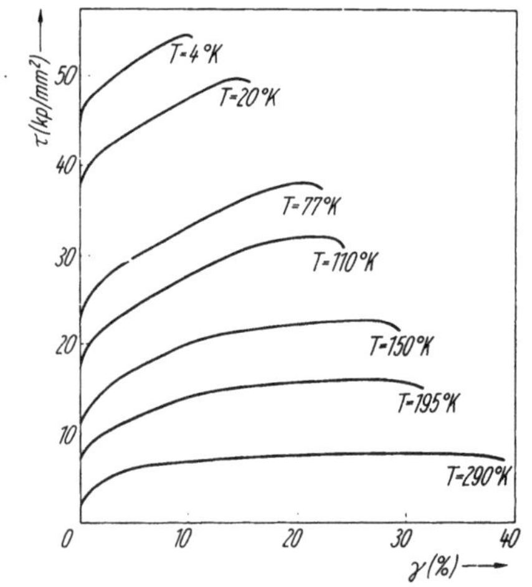

Fig. 1. Courbes de consolidation à vitesse de traction constante $\dot{\varepsilon} = 5 \times 10^{-5}\ s^{-1}$

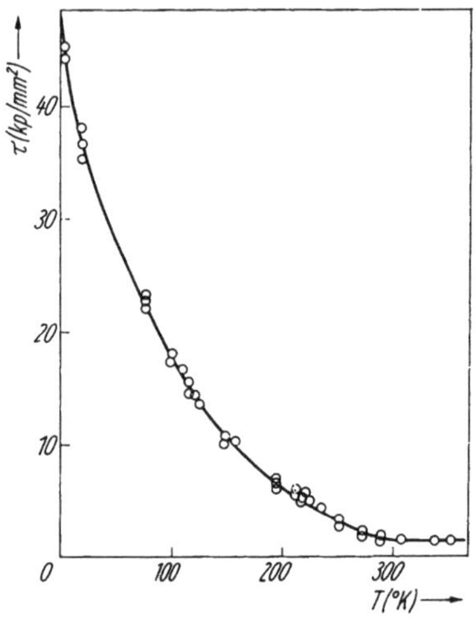

Fig. 2. Variation thermique de la limite élastique mesurée pour $\dot{\varepsilon} = 5 \times 10^{-5}\ s^{-1}$

amorce de fragilité pour des températures plus basses: la striction amorcée est limitée par une cassure nette. Cependant, pour des vitesses de traction 100 fois plus élevées, on observe des bandes de Lüders pour des températures voisines de l'ambiante, alors qu' à basse température, les courbes présentent souvent des décrochements caractéristiques du maclage. Ce dernier phénomène semble favorisé par la présence de divers reliefs à la surface de l'échantillon.

De l'ensemble des courbes de consolidation faites à diverses températures, nous avons déduit la variation thermique de la limite élastique pour $\dot{\gamma} = 10^{-4}\ s^{-1}$, rapportée sur la Fig. 2. La limite élastique ne varie pratiquement plus au-dessus de 300 °K environ, ce qui nous permet de déduire $\tau_\mu = 2$ kp/mm². Au-dessous de 300 °K, apparaît la forte variation thermique $\tau^*(T)$ de la limite élastique, caractéristique des métaux de structure cubique centrée, jusqu'à τ^* (0 °K) voisin de 46 kp/mm².

3.2 Changements de la vitesse de traction

Au cours de la déformation plastique, nous avons fait varier la vitesse de traction: des courbes de consolidation du type de celle présentée Fig. 3, pour 290 °K, ont été décrites à diverses températures. Les variations de la cission appliquée $\Delta\tau$ obtenues pour une variation de vitesse de $\dot{\varepsilon}_1 = 1{,}5 \times 10^{-5}\ s^{-1}$ à $\dot{\varepsilon}_2 = 1{,}5 \times 10^{-4}\ s^{-1}$ sont reportées en fonction de la cission appliquée (Fig. 4) et en fonction de la température (Fig. 5).

La Fig. 4, notamment, montre que pour des températures supérieures à 290 °K ou inférieures à 77 °K $\Delta\tau$ ne varie pas avec τ et ne dépend que de la température, alors que dans l'intervalle 290 à 77 °K, on observe une variation de $\Delta\tau$ avec τ voisine d'une loi de Cotrell-Stokes généralisée [27], $\Delta\tau/(\tau - \tau_0) =$ = const. Un maximum très étalé de $\Delta\tau$ vers 200 °K apparaît sur la Fig. 5 quel que soit le taux d'écrouissage, fait déjà signalé par Mordike et Haasen [2].

Fig. 3. Courbe de consolidation avec changement de la vitesse de traction $\dot{\varepsilon}$. $T = 290$ °K. De 0 à A $\dot{\varepsilon}_0 = 4{,}8 \times 10^{-5}\ s^{-1}$, après A alternativement $\dot{\varepsilon}_1 = 1{,}45 \times 10^{-5}\ s^{-1}$ ou $\dot{\varepsilon}_2 = 1{,}45 \times 10^{-4}\ s^{-1}$

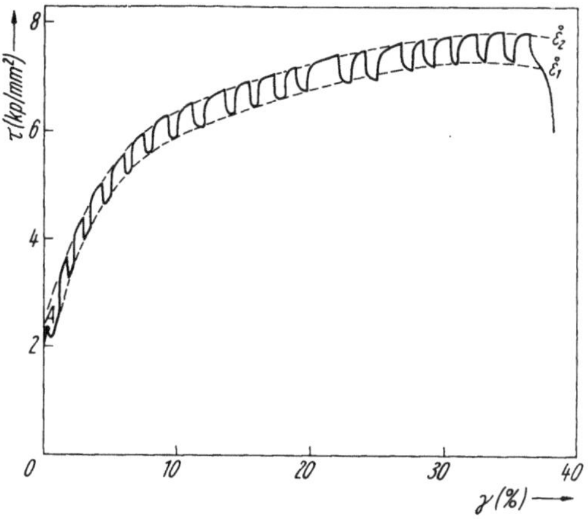

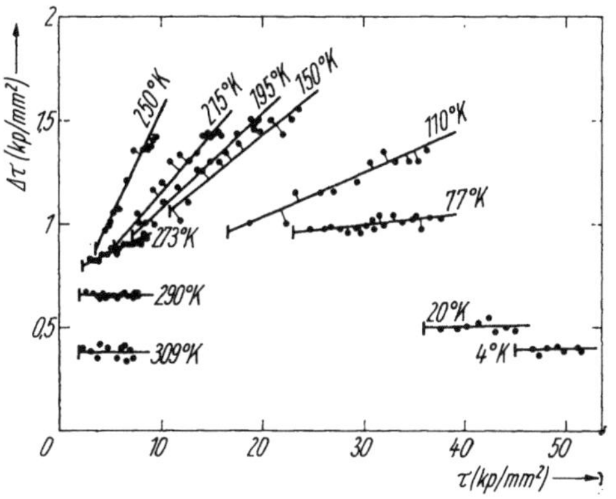

Fig. 4. Variation Δτ de contrainte produite par une variation de vitesse d'un facteur 10. | Limite élastique ● points expérimentaux

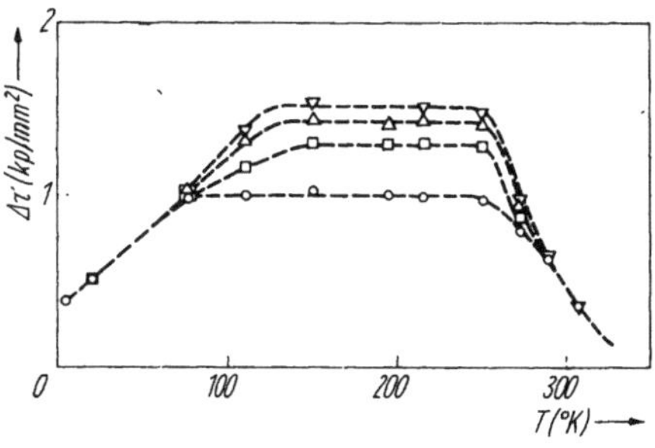

Fig. 5. Effet de la température sur la variation de contrainte produite par une variation de vitesse d'un facteur 10. ○ $\gamma = 1\%$, □ $\gamma = 10\%$, △ $\gamma = 20\%$, △ $\gamma = 30\%$

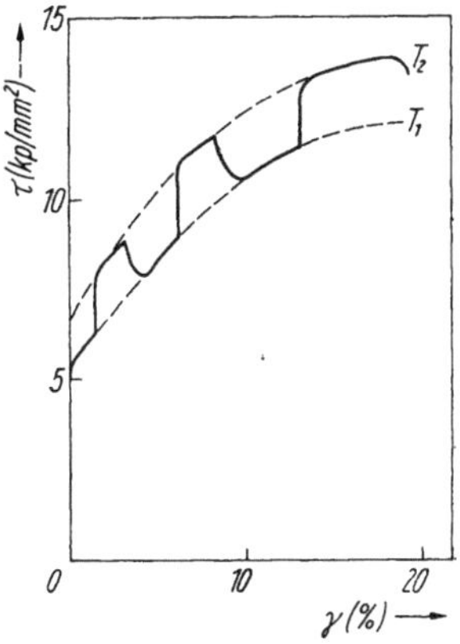

Fig. 6. Courbe de consolidation avec changement de la température. $T_1 = 215$ °K, $T_2 = 195$ °K, $\dot{\varepsilon} = 5 \times 10^{-5}\ s^{-1}$

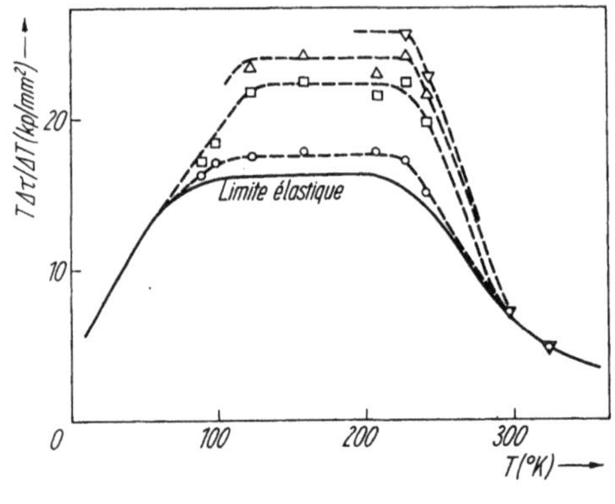

Fig. 7. Variation de $T\,\frac{\Delta T}{\Delta\tau}$ à vitesse constante avec la température. $\dot{\varepsilon} = 5\times10^{-5}\ s^{-1}$, $\Delta T = 20°$. ○ $\gamma = 1\%$, □ $\gamma = 10\%$, △ $\gamma = 20\%$, ▽ $\gamma = 30\%$

3.3 Changements de la température en cours de traction

Au cours d'essais de traction à vitesse constante ($\dot{\varepsilon} \approx 5\times10^{-5}\,s^{-1}$), nous avons soumis l'échantillon à des variations de température ΔT de 20°. La Fig. 6 représente une courbe de consolidation ainsi tracée, la température variant de $T_1 = 215$ °K à $T_2 = 195$ °K. La variation $\Delta\tau$ correspondante augmente avec le taux d'écrouissage pour des températures comprises entre 77 et 290 °K, alors que vers 300 °K et au-dessus aucune variation de $\Delta\tau$ n'est observée. Ces résultats sont résumés sur la Fig. 7 représentant les variations de $T\,\frac{\Delta\tau}{\Delta T}$ avec la température; la courbe $T\,\frac{\partial\tau}{\partial T}$ correspondant à la limite élastique et déduite de la Fig. 2, y est également portée. On remarque l'analogie de ces courbes avec les courbes $\Delta\tau(T)$ précédentes, et en particulier le même maximum étalé vers 200 °K.

3.4 Paramètres de l'activation thermique

Il est maintenant admis que, si la déformation plastique est contrôlée par un seul mécanisme activé thermiquement, celui-ci est caractérisé par la relation générale [28]

$$\dot{\varepsilon} = \dot{\varepsilon}_0 \exp\left(-\frac{\Delta G}{kT}\right). \tag{1}$$

$\dot{\varepsilon}$ est la vitesse relative de cisaillement, $\dot{\varepsilon}_0$ un facteur de fréquence, ΔG l'enthalpie libre de Gibbs du processus, fonction de la contrainte τ^* définie précédemment. De l'expression (1) on déduit la chaleur d'activation ΔH, souvent désignée par U, ,,enérgie d'activation thermique'':

$$U = \Delta H = kT^2\left(\frac{\partial \log\frac{\dot{\varepsilon}}{\dot{\varepsilon}_0}}{\partial T}\right)_{\tau^*}. \tag{2}$$

U peut alors se déduire de nos expériences par

$$U = kT^2\left(\frac{\log\frac{\dot{\varepsilon}_1}{\dot{\varepsilon}_2}}{\Delta\tau}\right)_T\left(\frac{\Delta\tau}{\Delta T}\right)_{\dot{\varepsilon}}. \tag{3}$$

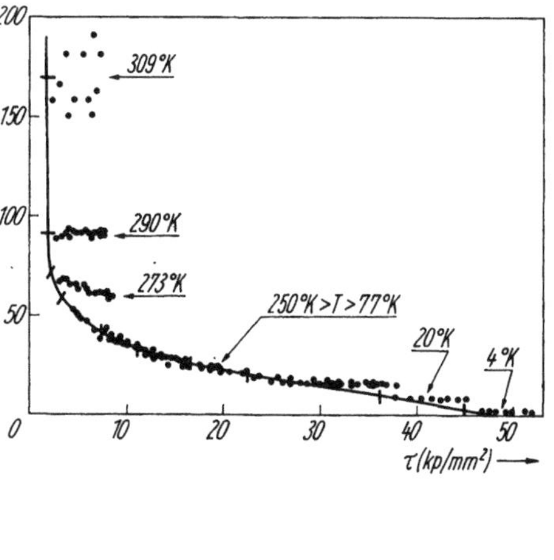

Fig. 8. Variation du volume d'activation avec la contrainte appliquée. ● Points expérimentaux; + valeur de V/b^3 extrapolées à la limite élastique

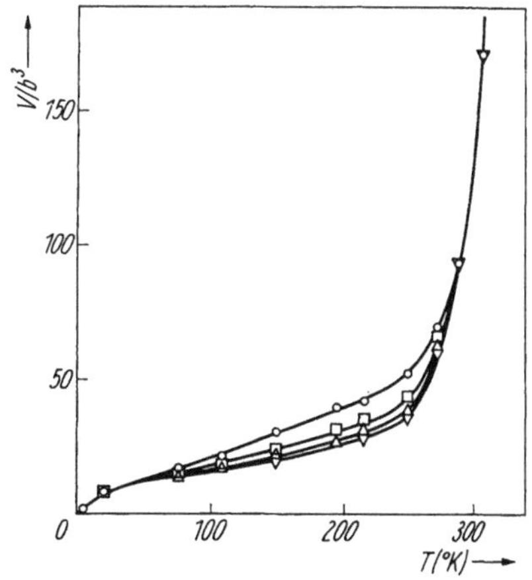

Fig. 9. Variation du volume d'activation avec la température. ○ $\gamma = 1\%$, □ $\gamma = 10\%$, △ $\gamma = 20\%$, ▽ $\gamma = 30\%$

On définit également un autre paramètre, le volume d'activation:

$$V = -\left(\frac{\partial\, \Delta G}{\partial \tau^*}\right)_T = kT\left(\frac{\log \frac{\dot\varepsilon_2}{\dot\varepsilon_1}}{\Delta \tau}\right)_T. \tag{4}$$

En pratique, dans la mesure où μ, module de cisaillement, ne varie pas, ou très peu, avec la température, on peut confondre [28] ΔH et ΔG, donc $V \approx \approx -\mathrm{d}U/\mathrm{d}\tau^*$. Les paramètres U et V sont enfin liés par la relation

$$U = -\,V\,T\left(\frac{\Delta\tau}{\Delta T}\right)_{\dot\varepsilon}. \tag{5}$$

Les volumes d'activation V, déduits de nos expériences par la relation (4) sont donnés Fig. 8 en fonction de la cission appliquée et Fig. 9 en fonction de la température. Les énergies d'activation, déduites des expériences à l'aide de (3), sont présentées Fig. 10 en fonction de la température. On remarque que U semble ne plus dépendre de l'écrouissage, les variations de V et de $T\left(\frac{\Delta\tau}{\Delta T}\right)_{\dot\varepsilon}$ se compensant mutuellement.

La valeur U_0 de 0,72 eV est analogue à celle que nous avoins déduite d'expériences de fluage et de revenu vers 300 °K sur des échantillons semblables [20, 29]. Les volumes d'activation déduits de ces mêmes mesures sont aussi

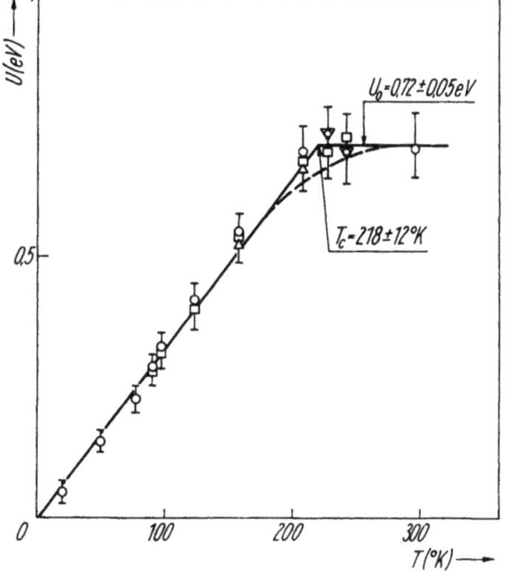

Fig. 10. Variation de l'énergie d'activation avec la température. ○ $\gamma = 1\%$, □ $\gamma = 10\%$, △ $\gamma = 20\%$, ▽ $\gamma = 30\%$

du même ordre que ceux que nous venons de présenter. Des mesures récentes de relaxation de la force appliquée [19] ont permis de vérifier que les volumes d'activation, déduits de celles-ci, sont sensiblement identique set varient de la même manière avec l'écrouissage que ceux déterminés ici.

4. Discussion des résultats

L'ensemble des résultats, en particulier la Fig. 10, montre qu'un seul mécanisme activé thermiquement agit, au moins jusqu'à 200 °K. Entre 200 et 300 °K par contre, les mesures deviennent difficiles à interpréter. Sans doute plusieurs mécanismes sont actifs. U n'est plus proportionnel à T, mais croît beaucoup plus lentement; le volume d'activation devient nettement sensible à l'écrouissage; vers la même température de 200 °K, nous avons précédemment observé une transformation de la loi de fluage [20] qui, logarithmique à plus basse température, devient linéaire. Stein [34] a également observé dans le même domaine de température, que la limite élastique (comme la cission d'écoulement plastique) d'échantillons de fer monocristallins n'est pas contrôlée par un processus unique d'activation thermique. Il est vraisemblable que les impuretés, les arbres de la forêt de dislocation, contribuent au-dessus de 200 °K au durcissement de façon concurrentielle avec le méchanisme unique qui semble agir à plus basse température.

Nous comparons ci-dessous les résultats expérimentaux avec le modèle de Dorn et Rajnak d'abord, puis avec le modèle d'Escaig.

4.1 Modèle de Dorn et Rajnak

Sur la base d'un mécanisme de Peierls, Dorn et Rajnak [11] ont calculé l'énergie d'un décrochement isolé, U_k, l'énergie de formation d'une paire de décrochements, U_n, et l'équation de leur mouvement. La composante thermique τ^* de la limite élastique est donnée par [11]

$$\frac{\tau^*}{\tau_p^*} = f_1\left(\frac{U_n}{2\,U_k}\right) \approx f_1\left(\frac{T}{T_c}\right). \tag{6}$$

τ_p^* est la force de Peierls, T la température, T_c celle pour laquelle $\tau^* = 0$. f_1 est une fonction donnée par Dorn et Rajnak, qui dépend faiblement de la forme de la barrière de Peierls considérée [18], cette forme étant caractérisée par un facteur de perturbation α, que nous avons pris égal à zéro. La contrainte à 0 °K, τ_p^*, et la température T_c sont ici deux paramètres choisis de façon à représenter au mieux les résultats expérimentaux.

La Fig. 11 indique les points expérimentaux, et la courbe théorique correspondant à $\tau_p^* = 46$ kp/mm², et $T_c = 210$ °K ($U_k =$ $= 0{,}36$ eV). L'accord est bon jusqu'à environ 160 °K, mais cesse au delà. On observe

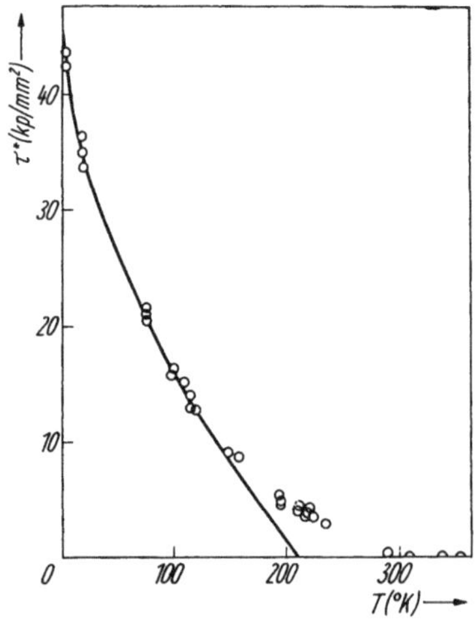

Fig. 11. Points expérimentaux et courbe théorique de Dorn et Rajnak

un accord de même nature pour le volume d'activation qu'on peut également déduire du modèle [11]. Ce genre de désaccord est fréquent [18]: il a été observé sur différents alliages cubiques centrés par Wynblatt et al. [30], Rawling et al. [31] pour des températures équivalentes.

Le modèle permet de calculer, selon les équations de Guyot et Dorn [18]: l'énergie de ligne Γ_0 des décrochements mis en jeu, l'ordre de grandeur de la séparation critique ω entre deux décrochements au moment de leur formation, et des valeurs extrêmes de la densité de dislocation, ϱ. Les valeurs déduites des nos mesures conduisent à

$$\Gamma_0 \approx \frac{2{,}9\,\mu\, b^2}{2}\,, \qquad \omega \approx 35\, b\,, \qquad 10^{16}\ \mathrm{cm}^{-2} < \varrho < 10^{21}\ \mathrm{cm}^{-2}\,.$$

L'énergie de ligne déduite, Γ_0, est du bon ordre de grandeur, quoiqu'un peu forte si on considère que le décrochement se fait sur 35 b de long, et 1 b de large. La densité de dislocation est, elle, nettement trop forte. Mais surtout que se passe-t-il au delà de 160 °K? Guyot et Dorn [18] ont avancé qu'à partir de telles températures le mécanisme de Peierls serait „remplacé" par le mécanisme décrit par Escaig. Nous pensons, au contraire, que ces deux mécanismes sont deux façons différentes de décrire *le même* effet de réseau. Si on veut cependant considérer simultanément les deux mécanismes, il est clair qu'ils doivent être *additifs*, la force de Peierls sur les partielles devant s'ajouter à la force de recombinaison. Si le mécanisme d'Escaig existe à haute température, pourquoi disparaîtrait-il à basse température, où on sait qu'il correspond à des contraintes supérieures à τ_p^*?

4.2 Modèle d'Escaig

Dans ce modèle on tient compte de la dissociation sessile des dislocations vis au cours de la microdéformation ($\varepsilon < 10^{-3}$); cette dissociation est sessile parce qu'elle a lieu dans plusieurs plans de glissement sécants à la fois, de type (110) ou (112). Pour redevenir mobiles, ces dislocations doivent être recombinées sur un ou plusieurs des plans de dissociation. Ceci a lieu à la limite élastique usuelle. Les équations dérivées de ce modèle ont été publiées précédemment [17a]; nous renvoyons donc à cette référence pour les détails.

Ces équations [17a] ont été modifiées à deux points de vue. D'une part on a tenu compte de la remarque proposée par Duesbery et Hirsch: une boucle glissile une fois libérée, peut à nouveau se redissocier et devenir sessile lorsqu'elle franchit une rangée atomique (111). Nous avons montré [17b] que ceci se produit dès la première rangée rencontrée, pour une contrainte égale ou inférieure à $\tau_c = R/b^2$ ou R est l'énergie de recombinaison mise en jeu par unité de ligne, et nous avons développé les équations correspondantes [17b].

D'autre part nous avons laissé comme paramètre à déterminer la largeur D_0 de la dislocation parfaite, c'est-à-dire à partir de laquelle on estime les partielles recombinées, soit

$$D_0 = \delta\, b_i\,, \tag{7}$$

b_i étant le vecteur de Burgers de la dislocation imparfaite qui se recombine (soit $\frac{1}{8}\langle 110\rangle$ dans les plans (110), soit $\frac{1}{6}\langle 111\rangle$ dans (112)).

Les équations qui donnent la limite élastique sont développées en annexe. Dans ce modèle deux paramètres doivent être choisis pour tenter de représenter les mesures: la largeur δ de la dislocation parfaite (équation (7)), et la contrainte

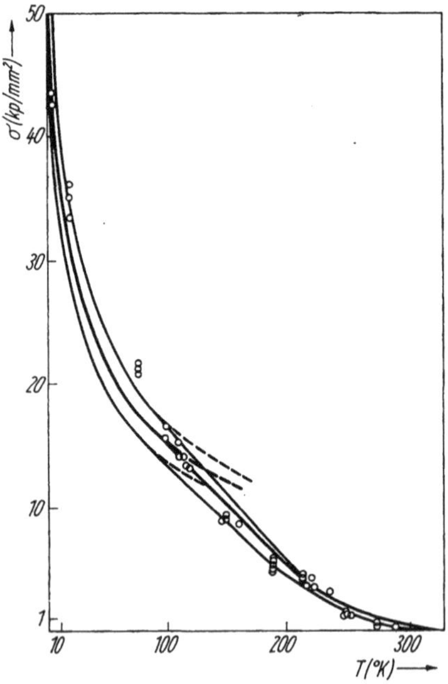

Fig. 12. Points expérimentaux et courbe théorique d'Escaig. En pointillé, on a représenté ce qu'on obtiendrait si on négligeait, en dessous de la contrainte R/b^2, la redissociation sessile ultérieure de l'arc libéré. La courbe centrale (traits forts) est la courbe choisie pour représenter les résultats. Les courbes voisines (traits fins) montrent ce qu'on obtient en prenant des valeurs un peu différentes des paramètres (voir Tableau 1)

τ_0 à 0 °K. Choisir τ_0 revient d'ailleurs à choisir une énergie de faute d'empilement γ puisqu'on a

$$\left.\begin{aligned} \frac{\tau_0}{\mu} &= \frac{0{,}32}{\delta} - 4{,}12\,\frac{\gamma}{\mu\, b}\ \text{sur}\,(110), \\ &\text{ou} \\ \frac{\tau_0}{\mu} &= \frac{3}{8\,\pi\,\delta} - 1{,}5\,\frac{\gamma}{\mu\, b}\ \text{sur}\,(112)\,. \end{aligned}\right\} \quad (8)$$

On choisit τ_0 de façon à représenter les résultats observés à 4 °K. D'un autre côté, l'énergie d'activation maximum obtenue, U_0 dépend essentiellement du produit $\delta\ \tau_0$ (cf. Annexe, équation (A9)). On est ainsi conduit à choisir $\delta\tau_0$ pour représenter U_0. Ceci permet ensuite, connaissant δ et τ_0, de comparer la courbe théorique $\tau(T)$ avec les points expérimentaux. Ceci est montré Fig. 12, ou on a pris $\mu\, b^3 = 7$ eV et $\mu = 8000$ kp/mm². La courbe centrale est tracée pour $\tau_0 = 72$ kp/mm² et $\delta = 11{,}5$, pour un glissement (110), ou aussi bien pour $\tau_0 = 70$ kp/mm² et $\delta = 5{,}5$ pour un glissement (112); on a représenté aussi les courbes obtenues pour des valeurs un peu différentes des paramètres afin de montrer la sensibilité de leur détermination. On a rassemblé (Tableau 1) les valeurs des paramètres correspondantes à ces courbes ainsi que les énergies de faute mises en jeu. En particulier, on a calculé ici la contrainte critique $\tau_c = R/b^2 \approx 2\times10^{-3}$ µm, correspondant à $T \approx 100$ °K, en dessous de laquelle on doit tenir compte de redissociations sessiles ultérieures de l'arc libéré.

L'accord obtenu est convenable. Les valeurs de δ correspondent à des largeurs de parfaites de l'ordre de $2\,b$, valeur raisonnable dans une telle approximation élastique du coeur des dislocations. La courbure des courbes théoriques est sans doute trop grande à basse température, entraînant une valeur de τ_0 trop élevée

Tableau 1

	110			112		
σ_0 (kp/mm²)	64	72	84	64	70	82
$\mu\, b/\gamma$	234	219	112	118	116	101
γ (erg/cm²)	85	90	104	170	172	200
larg. parf. /b	2,55	2,35	2	1,9	1,8	1,6
larg. imparf./b	3,7	3,5	3	3	3	2,7
σ_c (kp/mm²)		15,8			16	
W_0 (eV)		0,73			0,73	

d'une dizaine de kp/mm². Là aussi, nous pensons qu'à très basse température, où la contribution des pincements nécessaires devient importante [17 a], l'approximation élastique linéaire utilisée ne saurait donner un excellent accord. Rappelons que pour des dissociations aussi faibles que celles que nous prévoyons, il s'agit de la structure du coeur des dislocations, où visiblement l'approximation élastique est assez grossière.

D'un autre côté, entre 200 et 300 °K, les volumes d'activation calculés sont environ la moitié de ceux qu'on mesure, en liaison avec le fait que le modèle prévoit une énergie d'activation proportionnelle à T jusqu'à 300 °K, ce qui n'est pas observé. Dans ce domaine, la situation est délicate à interpréter quantitativement, puisque plusieurs mécanismes semblent entrer en concurrence.

Dans ces conditions on peut raisonnablement penser que le modèle de dissociation–recombinaison fournit une bonne explication de l'essentiel de la forte dépendance de la limite élastique avec la température observée dans le fer, même si tous les détails ne sont pas réductibles à une explication unique. Ceci revient à dire que les „barrières de Peierls" dans la structure cubique centrée auraient pour origine un phénomène de dissociation.

Annexe

Calcul de la limite élastique dans le modèle de dissociation-recombinaison

Les équations données ont été déjà établies, dans leur plus grande part, dans les références [17a] et [17b], auxquelles nous renvoyons pour les détails. La seule modification apportée ici est d'y expliciter le paramètre δ qui, voir équation (7), représente la largeur donnée à la dislocation parfaite, et qu'on laisse comme paramètre à déterminer.

On rappelle que les équations diffèrent selon que la cission τ est supérieure ou inférieure à R/b^2, R étant l'énergie de recombinaison. Dans le premier cas on se reportera à (17a), dans le second à (17b). Enfin, dans tout ce qui suit, c'est évidemment τ^* qu'il faut comprendre à la place de τ.

1. $\tau \geqq R/b^2$

b est la période du réseau; $R = \alpha\,(x - 1 - \log x)$, où α et x dépendent du plan de glissement considéré, et sont donnés cidessous.

$$T = \frac{U}{k\,C},$$

k, constante de Boltzmann, $C = \log\,(\dot{\varepsilon}_0/\dot{\varepsilon})$, est un nombre qui varie peu avec la contrainte τ (τ_0 est la contrainte à 0 °K):

$$C = \log\left(K\,\frac{\nu}{\dot{\varepsilon}}\,\varrho\,A\right) + 2\log\left(\frac{\tau}{\delta\,(\tau_0 - \tau)}\right). \tag{A1}$$

K facteur numérique égal à 2 si le plan de glissement est (110), ou 0,4 si c'est (112); ν fréquence atomique; A aire balayée par la dislocation après chaque libération thermique; ϱ densité de dislocations. Ici nous avons pris $A = b\,\varrho^{-1/2}$; avec $\nu = 10^{13}\,\mathrm{s}^{-1}$, $\dot{\varepsilon} = 5\times10^{-5}\,\mathrm{s}^{-1}$, $\varrho \approx 10^8\,\mathrm{cm}^{-2}$, le premier terme de C (indépendant de τ) vaut 32,4 pour (110) ou 30,8 pour (112).

a) Pour un glissement sur (110):

$$\left.\begin{aligned} &\alpha = \frac{\mu\, b^2}{34}\,, \quad x = 1 - 3{,}12\,\delta\,\frac{\tau_0 - \tau}{\mu}\,,\\ &U = 0{,}2\,\mu\, b^3\,\delta^3 \left(\frac{\tau_0 - \tau}{\mu}\right)^2 \left[\frac{0{,}147}{x}\left(\log\frac{5}{x}\right)^{1/2} + \frac{g(x)\,(\tau_0 - \tau)}{\tau}\right];\\ &g(x) = (x - 1 - \log x)^{3/2}\,(1 - x)^{-3}\,. \end{aligned}\right\} \quad (A2)$$

Le volume d'activation est

$$\left.\begin{aligned} \frac{V}{b^3} = 0{,}064\,\delta &\left[0{,}147\,\delta\,\frac{1 - x^2}{x^2}\left(\log\frac{5}{x}\right)^{1/2}\left(1 + \frac{1 - x}{2\,(1 + x)\log\frac{5}{x}}\right) + \right.\\ &\left. + 0{,}48\,\frac{\mu}{\tau}\,u^{3/2}\left(\frac{1 - x}{x\,u} + \frac{0{,}21}{\delta}\,\frac{\mu}{\tau}\right)\right];\\ u = (x - 1 - \log x)\,. & \end{aligned}\right\} \quad (A3)$$

b) Pour un glissement sur (112):

$$\left.\begin{aligned} &\alpha = \frac{\mu\, b^2}{18\,\pi}\,, \quad x = 1 - 8{,}38\,\delta\,\frac{\tau_0 - \tau}{\mu}\,,\\ &U = 1{,}87\,\mu\, b^3\,\delta^3 \left(\frac{\tau_0 - \tau}{\mu}\right)^2 \left[\frac{0{,}161}{x}\left(\log\frac{3}{x}\right)^{1/2} + \frac{g(x)\,(\tau_0 - \tau)}{\tau}\right], \end{aligned}\right\} \quad (A4)$$

et le volume d'activation:

$$\begin{aligned} \frac{V}{b^3} = 0{,}17\,\delta &\left[0{,}215\,\delta\,\frac{1 - x^2}{x^2}\left(\log\frac{3}{x}\right)^{1/2}\left(1 + \frac{1 - x}{2\,(1 + x)\log\frac{3}{x}}\right) + \right.\\ &\left. + \frac{3\,\mu}{4\,\pi\,\tau}\,u^{3/2}\left(\frac{1 - x}{x\,u} + \frac{\mu}{\tau}\,\frac{1}{4\,\pi\,\delta}\right)\right]. \end{aligned} \quad (A5)$$

$g(x)$ et $u(x)$ sont identiques au cas (110).

2. $\tau \leqq R/b^2$

Dans ce cas les équations restent celles données dans (17 b) avec les valeurs de x et de α données ci-dessus; on a

$$T = \frac{W}{k\,C}\,, \quad \text{avec ici} \quad C = \log\left(\frac{\nu}{\dot{\varepsilon}}\,\varrho\,A\,\frac{R}{\mu\,b^2}\right).$$

C ne dépend pratiquement pas de la contrainte. Une évaluation analogue à la précédente fournit $C = 25$ avec $R \approx 10^{-3}\,\mu\,b^2$, ce qui est justifié par la suite.

L'énergie d'activation W s'obtient en fonction de τ, dans les notations définies dans (17 b), selon

$$W = (l - l')\,w - \tau\,b^2\,l' + 2\,\beta\,\delta\,b_{\mathrm{i}}\,\frac{U}{x} + \alpha\,l'\,E + \frac{1}{\sqrt{2}}\,\beta\,\delta\,b_{\mathrm{i}}\,\frac{(1 - x)^2}{x}\,, \quad (A6)$$

où l, l', w, U, E sont des fonctions d'une variable H_0 comprise entre x et 1 et liée à τ selon

$$\alpha\,E - \tau\,b^2 + \frac{\mu\,b^4}{2\,\pi\,l\,(l - l')}\left(1 + \frac{(\tau\,\lambda^2)^2}{48}\right) = 0\,. \quad (A7)$$

Pour calculer $W(\tau)$ on se donne une valeur de H_0' comprise entre x et 1. Puis on calcule la contrainte correspondante en résolvant par itération l'équation (7); pour cela on part de la solution approchée $\tau\, b^2 = \alpha\, E(H_0)$. Une fois qu'on a τ, on peut alors calculer avec (6) l'énergie W correspondante. Ce calcul a été fait à Orsay, sur un ordinateur UNIVAC 1108.

Le volume d'activation se calcule par la relation

$$V = b^2\left(l - (l - l')\left(0{,}5 - \frac{\tau\,\lambda^2}{24}\right)\right). \tag{A8}$$

Enfin, l'énergie d'activation maximum devient

$$U_0 = \mu\, b^3\left[2\left(\frac{\alpha}{\mu\, b^2}\right)^{1/2}(x_0 - 1 - \log x_0)^{1/2} + \sqrt{2}\,\frac{\beta}{\mu\, b^3}\,\delta\, b_i\,\frac{(1 - x_0)^2}{x_0}\right]. \tag{A9}$$

$x_0 = 1 - 3{,}12\,\delta\,\tau_0/\mu$ pour (110), ou $x_0 = 1 - 8{,}38\,\delta\,\tau_0/\mu$ pour (112).

En pratique le second terme de la somme entre parenthèse, qui représente un terme de constrictions est très inférieur au premier terme. Ainsi U_0 dépend essentiellement du produit $\delta\,\tau_0$.

Bibliographie

[1] D. E. Stein, J. R. Low et A. V. Seybolt, Acta metall. **11**, 1253 (1963).
[2] B. L. Mordike et P. Haasen, Phil. Mag. **7**, 469 (1962).
[3] W. Frank, phys. stat. sol. **19**, 239 (1967).
[4] H. Conrad, Acta metall. **15**, 147 (1967).
[5] H. Conrad, J. Metals **16**, 582 (1964).
[6] H. Conrad et G. Shoeck, Acta metall. **8**, 791 (1960).
[7] J. Heslop et N. J. Petch, Phil. Mag. **1**, 866 (1956).
[8] Z. S. Basinski et J. W. Christian, Austral. J. Phys. **13**, 299 (1960).
[9] F. R. N. Nabarro, Z. S. Basinski et D. B. Holt, Adv. Phys. **13**, 193 (1964).
[10] H. Conrad et W. Hayes, Trans. ASM **56**, 944 (1963).
[11] J. E. Dorn et S. Rajnak, Trans. AIME **230**, 1052 (1964).
[12] R. Kossowsky et N. Brown, Acta metall. **14**, 131 (1966).
[13] A. S. Key, Direct Observations of Imperfections in Crystals, Interscience Publishers, 1962 (p. 213).
[14] J. R. Low et A. M. Turkalo, Acta metall. **10**, 215 (1962).
[15] A. Lawley et H. L. Gaigher, Phil. Mag. **10**, 15 (1964).
[16] V. Vitek, phys stat. sol. **18**, 687 (1966).
[17] B. Escaig, a) J. Phys. **28**, 171 (1967); b) phys. stat. sol. **28**, 463 (1968).
[18] P. Guyot et J. E. Dorn, Canad. J. Phys. **45**, 983 (1967).
[19] R. Conte et P. Groh, à paraître.
[20] E. Brouzet et R. Conte, C. R. Acad. Sci. (France) **260**, 4746 (1965).
[21] S. S. Brenner, Acta metall. **4**, 62 (1956).
[22] E. Brouzet et R. Conte, C. R. Acad. Sci. (France) **260**, 1089 (1965).
[23] S. S. Brenner, J. appl. Phys. **27**, 1484 (1956).
[24] R. Conte, B. Dreyfus et L. Weil, Acta metall. **10**, 1125 (1962).
[25] B. Jaoul, Étude de la plasticité et application aux métaux, Dunod, Paris 1965 (p. 388).
[26] R. Conte et P. Groh, Cryogenics, April 1967 (p. 81).
[27] A. H. Cottrell et R. J. Stokes, Proc. Roy. Soc. **A233**, 17 (1955).
[28] G. Schoeck, phys. stat. sol. **8**, 499 (1965).
[29] E. Brouzet et R. Conte, C. R. Acad. Sci. (France) **261**, 429 (1965).
[30] P. Wynblatt, A. Rosen et J. E. Dorn, Trans. AIME **233**, 651 (1965).
[31] R. D. Rawlings et C. W. A. Newey, Acta metall. **15**, 440 (1967).

[32] S. S. LAU, S. RANDJI, A. K. MUKERJEE, G. THOMAS et J. E. DORN, Acta metall. **15**, 237 (1967).
[33] P. B. HIRSCH, Fifth Internat. Congr. Cristallography, Cambridge 1960.
F. KROUPA, phys. stat. sol. **3**, K391 (1963).
[34] D. E. STEIN, Acta metall. **14**, 99 (1966).
[35] M. S. DUESBERY et P. B. HIRSCH, Battelle Coll. Dislocation Dynamics, Seattle 1967, McGraw-Hill Publ. Comp. (sous presse).
[36] W. GLAESER, H. IMGRUND et H. WEVER, phys. stat. sol. **24**, K5 (1967).
[37] G. R. TAYLOR, A. ISIN et R. V. COLEMAN, Phys. Rev. **165**, 621 (1968).
[38] S. ARAJS, B. F. OLIVER et G. R. DUNMYRE, J. appl. Phys. **36**, 2210 (1965).

(Received January 24, 1968)

phys. stat. sol. 28, 489 (1968)

Subject classification: 20.3; 2; 14.3; 22.8

A. F. Ioffe Physico-Technical Institute, Academy of Sciences of the USSR, Leningrad

Investigations of Some Properties of Vitreous and Crystalline $CdGeP_2$

By

N. A. Goryunova, S. M. Ryvkin, G. P. Shpenikov,
I. I. Tichina, and V. G. Fedotov

Recombination radiation is observed in vitreous and crystalline $CdGeP_2$. Some electric properties of vitreous and crystalline specimens are also studied. n- and p-type crystals with concentrations $n = 6 \times 10^{10}$ to 7×10^{15} cm^{-3} and $p = 7 \times 10^{14}$ to 4×10^{15} cm^{-3} at 300 °K and p-type vitreous $CdGeP_2$ are used. The effect of the recrystallisation of vitreous $CdGeP_2$ as well as the effect of impurities on the recombination radiation of crystalline $CdGeP_2$ is investigated.

In glasartigem und kristallinem $CdGeP_2$ wird Rekombinationsstrahlung beobachtet; einige elektrische Eigenschaften von glasartigen und kristallinen Proben werden untersucht, wobei n- und p-leitende Kristalle mit Konzentrationen $n = 6 \times 10^{10}$ bis 7×10^{15} cm^{-3} und $p = 7 \times 10^{14}$ bis 4×10^{15} cm^{-3} bei 300 °K und p-leitendes glasartiges $CdGeP_2$ verwendet werden. Der Einfluß der Rekristallisation von glasartigem $CdGeP_2$ und der Einfluß von Verunreinigungen auf die Rekombinationsstrahlung von kristallinem $CdGeP_2$ werden untersucht.

It is known that for the substances with diamond-like structure vitrification is not characteristic. The compounds of A^3B^5, A^2B^6, A^1B^7 type as well as the elements of the fourth group have not been obtained in the vitreous phase. But some semiconducting compounds of $A^2B^4C^5_2$ type with diamond-like structure have been obtained as vitreous [1, 2].

In this work recombination radiation was observed in vitreous and crystalline $CdGeP_2$. The absolute value of the radiation quantum efficiency was estimated. The surfaces of cleavages and crystals were investigated electron-microscopically. Some electric properties of vitreous and crystalline specimens were also studied.

$CdGeP_2$ single crystals were obtained by the method of directed cooling of the stoichiometric melt in a constant temperature gradient applied along the ingot.

The doping of the single crystals was realized during the growth.

Vitreous samples of $CdGeP_2$ were prepared from a melt with a constant degree of superheating [2]. Some of these samples were annealed at $T = 400$ °C for 3 hours and at $T = 500$ °C for some days.

The recombination radiation was excited by a beam of high velocity electrons with an energy of 25 keV. The repetition frequency amounted to 250 Hz, the impulse length to 0.8 s. The density of the exciting current varied up to 2 A/cm^2.

The radiation was analysed using the spectrometer IKS-12 with the prism F-1 and the photomultiplier FEU-62.

The specific conductivity and the Hall constant of vitreous and crystalline $CdGeP_2$ were measured with the dc electrometric amplifier U 1-2 and the megohmmeter MOM-3.

The electron-microscopic analysis was carried out in the electron microscope BS-242A "Tesla" using the replica method. The samples were prepared by the two-phase method (collodion–aluminium) with the precipitation of chromium for better contrast. The results of electric measurements are listed in Table 1.

Table 1

$CdGeP_2$-crystalline (cr) or vitreous (vitr) samples and types of conductivity	Specific resistivity at 300 °K (Ωcm)	Carrier concentration (cm^{-3})	Mobility of majority carriers (cm^2/Vs)	Energy gap (eV)
vitr, p-type	10^9-10^{10}			1.1
Sb-doped, cr, n-type	50	7×10^{15}	30	
Zn-doped, cr, n-type	10^{10}			1.8
In-doped, cr, n-type	10^3	$4\times10^{12}-10^{15}$	80–130	
Ga-doped, cr, p-type	10^7			
Bi-doped, cr-, p-type	10^2-10^3	$7\times10^{14}-4\times10^{15}$	25	
cr with non-controlled impurities	10^5-10^7	$6\times10^{10}-8\times10^{12}$	400	

More than 10 vitreous samples were investigated. All the samples showed hole conductivity with values of 10^{-9} to 10^{-10} Ω^{-1} cm^{-1}. The high specific conductivity the low mobility of the majority carriers in vitreous $CdGeP_2$ made it difficult to measure the carrier concentration. The temperature dependence of the electric conductivity of vitreous $CdGeP_2$ in the temperature region 300 to 800 °K obeys the well known exponential law being characteristic of semiconductors. From the temperature dependence of the electric conductivity the energy gap of vitreous $CdGeP_2$ was estimated and found to be 1.1 eV [2].

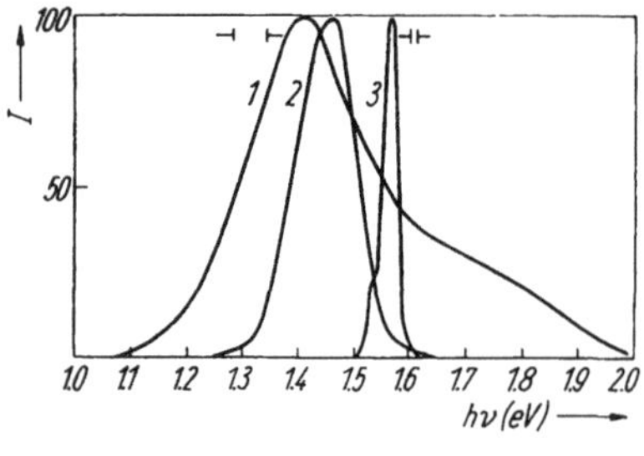

Fig. 1. Recombination radiation spectra of $CdGeP_2$. (1) non-annealed glass; (2) glass annealed at 400 °C for 3 hours; (3) crystalline samples, obtained by long-duration annealing of glass at 500 °C

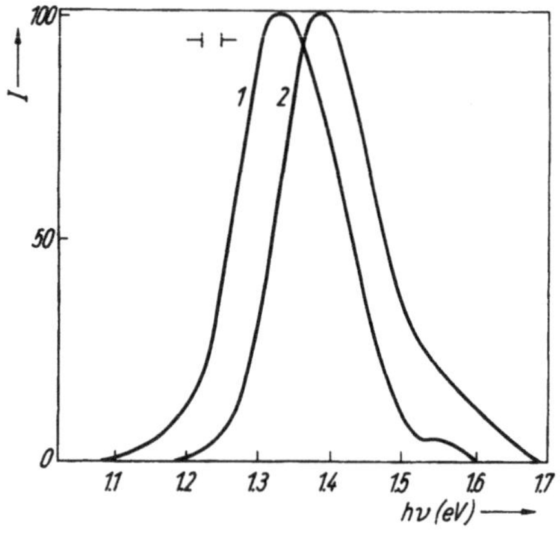

Fig. 2. Recombination radiation of crystalline $CdGeP_2$. (1) Ga-doped, (2) In-doped

The energy gap of vitreous $CdGeP_2$ was found to be smaller than for crystalline $CdGeP_2$ [3]. This agrees with certain theoretical concepts [4]. In all crystals a comparatively small carrier concentration was observed. This may be connected with deep donor levels or with partial compensation of shallow donors of the samples. A change in the conductivity sign in crystalline samples is observed if doped with gallium and bismuth.

The spectral distribution of the recombination radiation in $CdGeP_2$ at $T = 80$ °C is shown in Figs. 1 and 2.

The emission spectrum of vitreous $CdGeP_2$ (curve 1 of Fig. 1) has a maximum at the energy $h\nu \approx 1.4$ eV. The spectrum spreads over a wide energy range and its half-width greatly surpasses kT. The radiation intensity is very small. The emission spectrum of vitreous $CdGeP_2$ annealed at $T = 400$ °C for 3 hours (curve 2 of Fig. 1) has a maximum at $h\nu = 1.46$ eV. The half-width of the spectrum is equal to 0.13 eV, that is nearly two times less than in non-annealed samples. The radiation intensity is 20 times higher. This difference in the emission spectra of vitreous samples may be related to some structural changes of the glass which depend on temperature and duration of annealing and display themselves in the electron microscopic analysis (Fig. 3a, b). These changes are

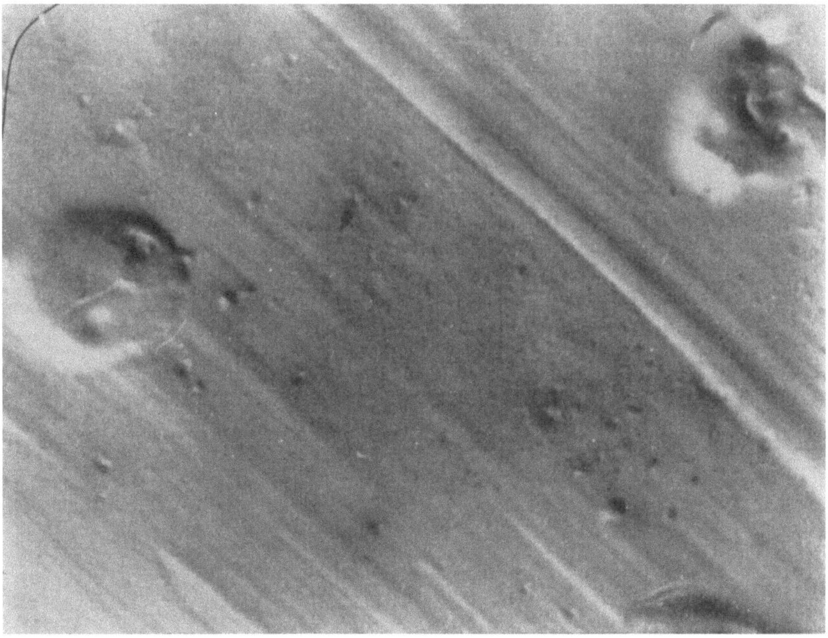

Fig. 3a

Fig. 3. Electron microscopic photos of sample surfaces (5000 ×).

a) non-annealed glass; b) glass annealed at 400 °C for 3 hours; c) crystalline samples, obtained by long-duration annealing of glass at 500 °C

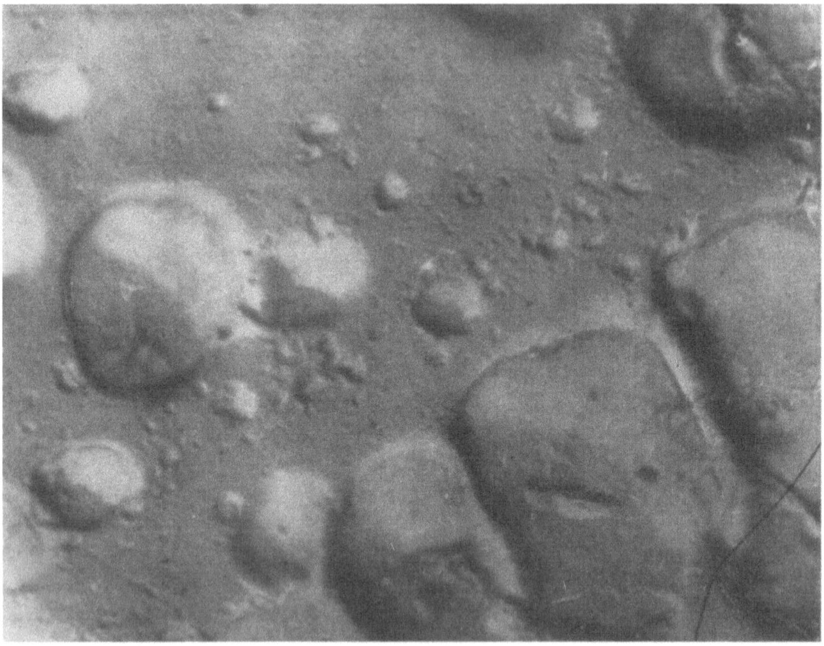

Fig. 3b

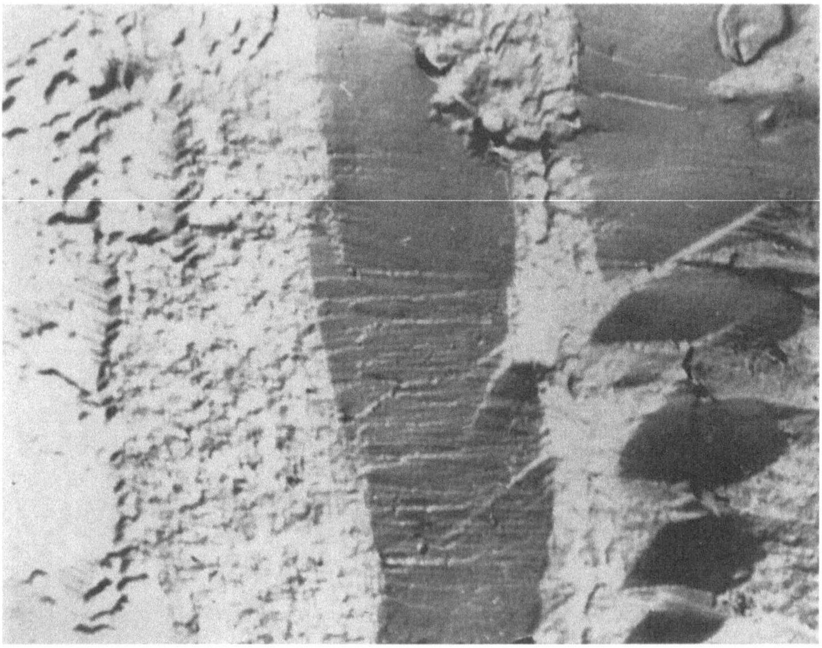

Fig. 3c

accompanied by a multiplication of crystalline regions after annealing. The X-ray structure analysis showed that the samples (curve 1 of Fig. 1) remained vitreous.

We may assume that these "crystalline" regions are responsible for the recombination radiation in vitreous $CdGeP_2$, which are always present in glass and whose number increases with annealing.

Long-duration annealing of a vitreous sample at $T = 500$ °C causes the transition to the crystalline phase which was confirmed by X-ray structure and electronographic analysis (Fig. 3c). The X-ray structure analysis showed in this sample a chalcopyrite structure with parameters corresponding to the parameters of crystalline $CdGeP_2$. The recombination radiation spectrum of this sample is shown in Fig. 1 (curve 3). The radiation spectrum has its maximum at an energy of $h\nu = 1.57$ eV. The extrapolation of the most steep part of the long-wavelength edge gives $h\nu = 1.54$ eV. On the long-wavelength edge a band of extrinsic radiation is superimposed. On the graph this band has a characteristic back with a energy maximum of about 1.54 eV. A small half-width of the principal line, $\Delta\lambda \approx 5\,kT$, being characteristic of the interband recombination radiation, attracts one's attention.

The emission spectra of crystalline $CdGeP_2$ obtained by directed crystallization differ very much from the spectra of vitreous $CdGeP_2$ and of recrystallized vitreous samples. In these samples only extrinsic radiation is observed, with an energy of the maximum not higher than E_g of the crystalline $CdGeP_2$. The recombination radiation of the crystals obtained by the directed crystallization method is shown in Fig. 2, curves 1 and 2. The maximum of the radiation in the crystals p-$CdGeP_2$ doped with gallium is located at $h\nu = 1.32$ eV. In the In-doped n-$CdGeP_2$ crystals $h\nu = 1.38$ eV. The crystals with non-controlled impurities and the crystals doped with Sb, Bi, Zn, and Ge have an emission maximum at 1.3 to 1.4 eV. In this way admixtures of Sb, Ga, Bi, Sn, Zn, Ge and the non-controlled impurities create in $CdGeP_2$ extrinsic levels with an activation energy within the limits of 0.4 to 0.5 eV from the nearest band. In the spectral curve 1 of Fig. 2 and in the majority of curves discussed above one may see simultaneously as on the n-type samples and the p-type samples a short-wavelength band with an energy of 1.55 eV. But the intensity of this band is by one order weaker than that of the long-wavelength band. The most intense radiation was observed in In-doped n-$CdGeP_2$ crystals; it was by more than one order more intense than in n-type $CdGeP_2$ with non-controlled admixtures.

The quantum efficiency of radiation related to the radiation output of the crystal, evaluated by the same method as in [5], is equal to 0.05% for the best single crystals. For the crystalline $CdGeP_2$ obtained by long-duration annealing of vitreous samples the quantum efficiency is equal to about 0.05%, too. For the vitreous $CdGeP_2$ in Fig. 1 (curve 1) the quantum efficiency is 10^{-4}%.

References

[1] A. A. Vaipolin, N. A. Goryunova, E. O. Osmanov, and Yu. V. Rud, Dokl. Akad. Nauk SSSR **160**, 633 (1965).

[2] I. I. Tichina, V. G. Fedotova, and I. I. Ivanova, Theses of Reports, Symp. Chemical Bonds in Semiconductors, Minsk 1967.

[3] I. I. Tichina, N. A. Goryunova, and R. Yu. Khansevarov, Fiz. Tekh. Poluprov. **1**, 111 (1967).

[4] A. I. Gubanov, Quantum-Electronic Theory of Amorphous Conductors, Izd. Akad. Nauk SSSR, Moscow 1963.
[5] F. M. Berkovskii, D. Z. Garbusov, N. A. Goryunova, G. V. Loshakova, S. M. Ryvkin, and G. P. Shpenikov, Fiz. Tekh. Poluprov. (in the press).

(Received April 26, 1968)

phys. stat. sol. 28, 495 (1968)

Subject classification: 19; 13; 18

Laboratoire de Magnétisme et de Physique du Solide, C.N.R.S., Bellevue

Quantum-Mechanical Treatment of the Transient Processes with Three-Magnon Non-Linear Transitions under Perpendicular Pumping

By

H. LE GALL

The transient growth of spin-wave populations with three-magnon non-linear processes under perpendicular pumping is studied. Calculations are made on the basis of second quantization of the magnetic system and the pumping microwave field, assuming that the magnons are in thermal distribution before the electromagnetic pulse is applied. General expressions describing transient evolution of the uniform mode and the parametric states are established in the case where the coincidence condition is satisfied (occurrence of the three-magnon transitions at resonance). The stationary state of the parametric magnons is also determined. To test experimentally the validity of the theory an expression for the transient power absorbed by the crystal at the beginning of the parametric excitation of the quasiparticles is derived. Basing on experimental data, the limits of the applicability of the theory are discussed.

Nous avons étudié la croissance transitoire des populations d'onde de spin en présence de processus non-linéaires à trois magnons en pompage perpendiculaire. Les calculs ont été établis sur la base de la seconde quantification du système magnétique et du champ microonde de pompage en supposant que les magnons sont à l'équilibre thermique avant que l'impulsion électromagnétique ne soit appliquée. Les expressions générales décrivant l'évolution transitoire du mode uniforme et des états paramétriques ont été établies dans le cas où la condition de coïncidence est satisfaite (apparition des transitions à trois magnons à la résonance). L'état stationnaire des magnons paramétrique a aussi été déterminé. Afin de vérifier expérimentalement la validité de la théorie nous avons calculé l'expression de la puissance transitoire absorbée par le cristal au début de l'excitation paramétrique des quasi-particules. A partir des résultats expérimentaux, nous discutons les limites d'application de la théorie.

1. Introduction

The interactions of a system of quasiparticles with an electromagnetic field can excite non-linear processes if the number of quasiparticles exceeds a threshold. Akhiezer has shown that dipolar energy in ferromagnetic crystals can induce transitions between spin waves [1]. From this theoretical analysis, it has been possible to explain the behaviour of the ferromagnetic resonance at high power levels [2, 3]. Indeed, Suhl has established that in the steady state [4] three-magnon transitions and four-magnon transitions give rise respectively to the subsidiary resonance and the saturation of the main resonance. In [4] the calculation of the non-linear instabilities has been derived phenomenologically by use of the Landau-Lifshits equation of motion of the magnetic moment and Maxwell's equations. The system of parametric equations thus defined has a cumbersome form and is not easy to obtain and to solve. In the present paper we have analysed the transient growth of spin-wave populations after an electromagnetic pulse is applied to a ferromagnetic crystal with perpendicular pumping. The calculations are derived on a quantum-mechanical basis and concern first-order processes assuming that the condition of coincidence is satisfied (three-magnon non-linear processes at resonance). With these conditions, second-order processes are neglected since their critical threshold is much higher than that for first-order processes. As previously outlined [5], we use the second quantization of the magnetic system and the electromagnetic field. It is assumed that the amplitude of the r.f. magnetic field is constant during the pulse. The advantage of this method lies in the fact that instead of the equations of motion of the magnetic moment we use the equation of motion of the creation and annihilation operators of magnons and photons. We can easily obtain and solve these equations. In addition, it is easier to perform transient analysis of the three-magnon non-linear transitions from a quantum-mechanical basis which well defines the thermal amplitude of the quasiparticles with the Bose distribution function. The formalism used in the present paper is similar, under some aspects, to that carried out recently by Savchenko and Tarasenko [6]. These authors have treated the general case of non-linear excitation of quasiparticles studying the evolution of the nonequilibrium density matrix of the magnetic system. With this method they have determined mainly the stationary amplitude of the parametric excitations.

To study the transient growth of spin-wave populations with three-magnon non-linear processes, we establish in Section 2 the basis of the problem in which the Hamiltonians which determine photon–magnon and magnon–magnon interactions are defined. In Sections 3 and 4 we derive the equations of motion of the creation and annihilation operators of the uniform and parametric modes. From the solution of these equations we can express the transient power absorbed by the ferromagnetic crystal from the electromagnetic field. The threshold values and the growth increment of the amplified magnons are also defined. In the last Section 5, we discuss, from experimental data, the validity of the present theory.

2. Hamiltonians of the Magnon–Magnon and the Photon–Magnon Interactions in Ferromagnetic Crystals

The quantum theory of the magnon–magnon interactions based on the second-quantization method of Holstein and Primakoff [7] has been developed by several authors [1, 8 to 10]. We represent the Hamiltonian of the system in the form

$$H = H_{\mathrm{m}} + H_{\mathrm{m\text{-}p}}\,, \tag{1}$$

where the terms on the right hand side are the respective Hamiltonians of the spin-wave system(H_{m}) and of the photon–magnon interaction ($H_{\mathrm{m\text{-}p}}$). H_{m} determines the behaviour of the spin system in the undeformed lattice. In what follows the magnon–phonon and electron–magnon interactions are neglected:

$$H_{\mathrm{m}} = \int\limits_V \mathrm{d}V \left[\frac{1}{2}\alpha_{ik}\left(\frac{\partial \boldsymbol{M}}{\partial x_i}\frac{\partial \boldsymbol{M}}{\partial x_k}\right) - \frac{1}{2}\beta\,(\boldsymbol{M}\,\boldsymbol{n})^2 - \frac{\boldsymbol{H}^2}{8\,\pi}\right]. \tag{2}$$

Here α_{ik} is the tensor of exchange constants, $\boldsymbol{M}$ is the magnetic moment density, β is the anisotropy constant, $\boldsymbol{n}$ is the unit vector directed along the easy axis

Fig. 1. Configuration of the magnetization in the non-equilibrium state

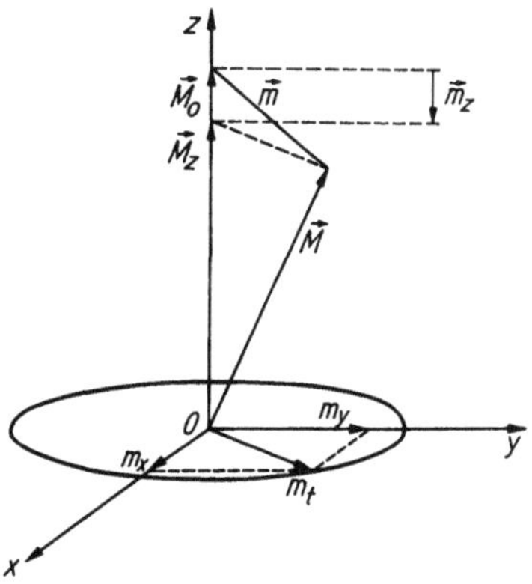

of magnetization, and V is the volume of the crystal. Let be $\boldsymbol{H} = \boldsymbol{H}_i + \boldsymbol{h}$ and $\boldsymbol{M} = \boldsymbol{M}_0 + \boldsymbol{m}$ where $\boldsymbol{H}_i$ and $\boldsymbol{M}_0$ are the internal dc magnetic field and the magnetization at the equilibrium, respectively; $\boldsymbol{h}$ and $\boldsymbol{m}$ are small deviations from $\boldsymbol{H}_i$ and $\boldsymbol{M}_0$ corresponding to the dipolar energy of the magnons (Fig. 1). We take the z-axis parallel to $\boldsymbol{M}_0$.

The magnetic system has a symmetry axis along $0z$, thus it is convenient to use the circular components of the magnetization:

$$M^{\pm} = M_x \pm i\, M_y \,. \tag{3}$$

If we consider states near saturation, we can, as Holstein and Primakoff [7], expand these quantities in terms of localized operators of creation and annihilation of spin deviation at the i-th site $a^+(\boldsymbol{r}_i)$ and $a(\boldsymbol{r}_i)$:

$$\left.\begin{aligned} M^+ \equiv m^+ &= (2\, g\, \mu_B\, M_0)^{1/2} \left(a - \frac{g\,\mu_B}{4\,M_0}\, a^+\, a\, a\right), \\ M^- \equiv m^- &= (2\, g\, \mu_B\, M_0)^{1/2} \left(a^+ - \frac{g\,\mu_B}{4\,M_0}\, a^+\, a^+\, a\right), \\ m_z &= M_z - M_0 = -g\,\mu_B\, a^+\, a, \end{aligned}\right\} \tag{4}$$

g and μ_B being the Landé factor and the Bohr magneton. a^+ and a satisfy the commutation relations

$$[a(\boldsymbol{r}, t), a^+(\boldsymbol{r}', t)] = \delta\, (\boldsymbol{r} - \boldsymbol{r}') \,. \tag{5}$$

The Hamiltonian (2) cannot be diagonalized with the use of the operators a^+ and a; thus, to define the eigen states of the ferromagnetic crystal, we must express these variables in terms of their Fourier components:

$$a(\boldsymbol{r}, t) = V^{-1/2} \sum_k a_k(t)\, e^{i\boldsymbol{k}\boldsymbol{r}} \,, \quad a^+(\boldsymbol{r}, t) = V^{-1/2} \sum_k a_k^+(t)\, e^{-i\boldsymbol{k}\boldsymbol{r}} \tag{6}$$

with the following commutation relations for the new operators:

$$[a_k(t), a_{k'}^+(t)] = \Delta\, (\boldsymbol{k} - \boldsymbol{k}') \,. \tag{7}$$

Then the components of $\boldsymbol{m}(\boldsymbol{r}, t)$ become

$$\left.\begin{aligned} m^+(\boldsymbol{r}_i) &= \left(\frac{2\, g\, \mu_B\, M_0}{V}\right)^{1/2} \left[\sum_k \mathrm{e}^{i \boldsymbol{k} \boldsymbol{r}_i}\, a_k - \frac{g\, \mu_B}{4\, M_0\, V} \sum_{1,2,3} \mathrm{e}^{i(-\boldsymbol{k}_1 + \boldsymbol{k}_2 + \boldsymbol{k}_3) \boldsymbol{r}_i}\, a_1^+\, a_2\, a_3\right], \\ m^-(\boldsymbol{r}_i) &= \left(\frac{2\, g\, \mu_B\, M_0}{V}\right)^{1/2} \left[\sum_k \mathrm{e}^{-i \boldsymbol{k} \boldsymbol{r}_i}\, a_k^+ - \frac{g\, \mu_B}{4\, M_0\, V} \sum_{1,2,3} \mathrm{e}^{i(-\boldsymbol{k}_1 - \boldsymbol{k}_2 + \boldsymbol{k}_3) \boldsymbol{r}_i}\, a_1^+\, a_2^+\, a_3\right], \\ m_z(\boldsymbol{r}_i) &= -\frac{g\, \mu_B}{V} \sum_{1,2} \mathrm{e}^{i(-\boldsymbol{k}_1 + \boldsymbol{k}_2) \boldsymbol{r}_i}\, a_1^+\, a_2\,. \end{aligned}\right\} \quad (8)$$

The indices 1, 2, 3 represent k_1, k_2, k_3 and the summation is made over the corresponding wave vectors. With relations (8) we can now develop Hamiltonian (2) in terms of a_k^+ and a_k. Neglecting terms upper to third order since they do not contribute to the three-magnon non-linear transitions, H_m becomes

$$H_m = H_2 + H_3\,, \quad (9)$$

where

$$H_2 = \sum_k A_k\, a_k^+\, a_k + \frac{1}{2}\,(B_k\, a_k\, a_{-k} + B_k^*\, a_k^+\, a_{-k}^+)\,, \quad (10)$$

$$H_3 = \sum_{1,2,3} \Phi\, a_1^+\, a_2^+\, a_3\, \Delta\, (\boldsymbol{k}_1 + \boldsymbol{k}_2 - \boldsymbol{k}_3) + \text{herm. conj.}; \quad (11)$$

$$A_k = g\, \mu_B\, (\alpha\, M_0\, k^2 + \beta\, M_0 + H_i + 2\, \pi\, M_0 \sin^2 \theta_k)\,, \quad (12)$$

$$B_k = 2\, \pi\, g\, \mu_B\, M_0 \sin^2 \theta_k\, \mathrm{e}^{-2\, i\, \varphi_k}\,, \quad (13)$$

$$\Phi = -\frac{1}{2}\, \pi\, g\, \mu_B \left(\frac{2\, g\, \mu_B\, M_0}{V}\right)^{1/2} \sin 2\, \theta_k\, \mathrm{e}^{-i \varphi_k}\,. \quad (14)$$

θ_k and φ_k are the polar and azimuthal angles of the propagation vector $\boldsymbol{k}$ measured from z-axis (direction of $\boldsymbol{H}_i$) and x-axis, respectively. Relations (10) and (11) define the dispersion law of the free quasiparticles and their interactions in three-boson processes. If $|B_k|$, which is defined only by the dipolar energy of the spin waves, is smaller than A_k, H_2 is diagonal and gives the spectrum of the ferromagnet. In the non-linear processes in which we are concerned, this approximation is no longer correct since parametric magnons are propagating with $\theta_k \approx \pi/4$ and have small propagation vectors. To diagonalize H_2 we must introduce new dynamical variables c_k^+ and c_k, using the canonical transformation of Bogolyubov and Tyablikov [11]:

$$a_k = u_k\, c_k + v_k^*\, c_{-k}^+\,, \qquad a_k^+ = u_k^*\, c_k^+ + v_k\, c_{-k} \quad (15)$$

with

$$\left.\begin{aligned} u_k &= \left(\frac{A_k + \hbar\, \omega_k}{2\, \hbar\, \omega_k}\right)^{1/2}, \qquad v_k = -\, \mathrm{e}^{-2\, i\, \varphi_k} \left(\frac{A_k - \hbar\, \omega_k}{2\, \hbar\, \omega_k}\right)^{1/2}, \\ \hbar\, \omega_k &= (A_k^2 - |B_k|^2)^{1/2} = g\, \mu_B\, [(H_i + D\, k^2)\, (H_i + D\, k^2 + 4\, \pi\, M_0 \sin^2 \theta_k)]\,. \end{aligned}\right\} \quad (16)$$

$D = g\, \gamma\, M_0\, \alpha$, $\gamma = e/2\, m\, c$ is the gyromagnetic factor and $\hbar$ is the Planck constant divided by $2\, \pi$. The new operators c_k^+ and c_k satisfy the commutation relation

$$[c_k, c_{k'}^+] = \Delta\, (\boldsymbol{k} - \boldsymbol{k}')\,. \quad (17)$$

Now, substituting (15) into (9), we get

$$H_m = \sum_k \hbar\, \omega_k\, c_k^+\, c_k + \sum_{k \neq 0} \hbar\, (f_k\, c_0^+\, c_k\, c_{-k} + f_k^*\, c_0\, c_k^+\, c_{-k}^+) \quad (18)$$

with

$$f_k = \pi g \gamma \left(\frac{2 g \mu_B M_0}{V}\right)^{1/2} \left(\frac{\omega_i + D k^2 + \omega_k}{4 \omega_k}\right) \sin 2\theta_k \, e^{-i\varphi_k} \tag{19}$$

and

$$\omega_i = g \gamma H_i .$$

f_k is a coupling parameter of the parametric state k with the uniform mode. In Hamiltonian (18) we have only retained third-order terms which are important for three-magnon non-linear transitions. We have for instance neglected four-magnon terms which determine second-order processes, since their critical threshold is much higher than that for first-order processes. In addition, we have not taken into account the two-magnon and three-magnon terms which contribute to the linear relaxation of the uniform mode and spin waves, respectively. The linear damping of these modes will be introduced later with the complex frequencies $\omega_k + i \eta_k$. Third-order terms in (18) are of two kinds. The first term $c_0 c_k^+ c_{-k}^+$ (splitting transition) is the direct process in which a $k = 0$ magnon is absorbed and two magnons ($\pm k$) are created. The second term $c_0^+ c_k c_{-k}$ (fusion transition) is the reverse process where two magnons with same energy and opposite momentum are destroyed and a uniform magnon is created. Taking energy from the uniform mode in the direct processes, the parametric spin waves keep its amplitude $\langle c_0^+ c_0 \rangle$ to a constant value. This property is the basis of the coincidence microwave limiters. Every photon absorbed by the crystal above the critical threshold is immediately converted into a $k \neq 0$ magnon. The population of the parametric magnons can reach very high values since, with these conditions, the probabilities of the direct processes exceed the probabilities of relaxation. When the electromagnetic field is suppressed the reverse terms $c_0^+ c_k c_{-k}$ play an important role. Indeed, the magnetic energy stored in the parametric magnon system vanishes in part in fusions with thermal magnons (linear relaxation) and in part by radiation of photons via the uniform mode. This reverse process is observed at the origin of the radiation peak for a short time when the electromagnetic field is switched off. We shall not consider this transient phenomenon in what follows. Finally, to obtain Hamiltonian (18) we have chosen the case of a cubic crystal ($\alpha_{ik} \to \alpha$) with a low anisotropy energy ($\beta \to 0$).

The Hamiltonian of the photon–magnon interactions can be represented in the form

$$H_m = -g \mu_B \sum_i h_x S_i = -\frac{1}{2} g \mu_B \sum_i (h^+ S_i^- + h^- S_i^+) . \tag{20}$$

The magnetic field operator is (see for example [11] or [12])

$$h_x = i \left(\frac{2 \pi \hbar \omega}{V_1}\right)^{1/2} [\boldsymbol{q}_0 \, \boldsymbol{e}_q] \left[q \, e^{-i\left(\frac{\omega}{c} y - \omega t\right)} - q^+ \, e^{i\left(\frac{\omega}{c} y - \omega t\right)}\right] \tag{21}$$

or

$$h_x = h^+ e^{i \omega t} + h^- e^{-i \omega t} . \tag{22}$$

c is the velocity of light and V_1 is the volume which defines, for n_q incident photons, an electromagnetic energy density $n_q \hbar \omega / V_1$. The wave vector of the electromagnetic field $\boldsymbol{q}$ ($|q| = \omega/c$) is parallel to the y-axis. $\boldsymbol{q}_0 = \boldsymbol{q}/|q|$, and $\boldsymbol{e}_q$ is the polarization vector of the r.f. magnetic field. The only non-vanishing matrix elements of the operators q^+ and q for creation and annihilation of photons are

$$\langle n_q + 1 \,|q^+|\, n_q\rangle = (n_q + 1)^{1/2}\,e^{-i\omega t} \quad \text{and} \quad \langle n_{q-1} \,|q|\, n_q\rangle = (n_q)^{1/2}\,e^{i\omega t}\,, \tag{23}$$

where n_q is the number of the incident photons. Expanding S_i^+ and S_i^- in the representation of the second quantization, $H_{\text{m-p}}$ becomes

$$H_{\text{m-p}} = -\frac{1}{2}\,g\,\mu_B(2\,N\,S)^{1/2}\left(h^+ \sum_k a_k^+ \sum_i e^{-i\boldsymbol{k}\boldsymbol{r}_i} + \text{herm. conj.}\right) \tag{24}$$

or

$$H_{\text{m-p}} = -i\,a\,(q\,a_0^+ - q^+\,a_0) \tag{25}$$

with

$$a = g\,\mu_B\,(N\,S\,2\,\pi\,\hbar\,\omega/2\,V_1)^{1/2}\,.$$

N and S are respectively the number of spins in the crystal of volume V and the spin quantum number related to M_0 with

$$M_0 = g\,\mu_B\,S/\delta^3\,, \tag{26}$$

where δ is the distance between two successive magnetic sites. The transformation of Bogolyubov and Tyablikov, which diagonalizes the second-order Hamiltonian of the magnetic system, is a simple identity for $H_{\text{m-p}}$. Thus, the total Hamiltonian becomes

$$H = \sum_k \hbar\,\omega_k\,c_k^+\,c_k + \sum_{k\neq 0} \hbar\,(f_k\,c_0^+\,c_k\,c_{-k} + f_k^*\,c_0\,c_k^+\,c_{-k}^+) - i\,a\,(q\,c_0^+ - q^+\,c_0)\,. \tag{27}$$

3. General Statement of the Three-Magnon Non-Linear Excitation

Since the second-order Hamiltonian of the crystal is diagonalized in terms of the dynamical variables c_k, we can write

$$c_0 = b_0\,e^{i\omega_0 t} \quad \text{and} \quad c_k = b_k\,e^{i\omega_k t}\,. \tag{28}$$

The reaction of the magnetic system to the external perturbation will be described by the evolution of the amplitudes b_0 and b_k. The operators $c(t)$ satisfy the following equation of motion:

$$i\,\hbar\,\dot{c} = [H, c]\,. \tag{29}$$

By use of (17), (27), and (29), the equations of motion for the operators c_0 and c_k can be obtained:

$$\dot{c}_0 = i\,(\omega_0\,c_0 + \sum_{k\neq 0} f_k\,c_k\,c_{-k} - i\,a_1\,q)\,, \tag{30}$$

$$\dot{c}_k = i\,(\omega_k\,c_k + f_k^*\,c_0\,c_{-k}^+) \tag{31}$$

with

$$a_1 = a/\hbar.$$

If we now introduce the amplitudes b_0 and b_k with (28) and the relaxation rates η_0 and η_k in the linear region of the uniform and spin-wave modes by the complex frequencies $\omega_0 + i\,\eta_0$ and $\omega_k + i\,\eta_k$, we have

$$\dot{b}_0 = -\,\eta_0\,b_0 + i\sum_{k\neq 0} f_k\,b_k\,b_{-k}\,e^{i(2\omega_k - \omega_0)t} + a_1\,q\,e^{i(\omega-\omega_0)t}\,, \tag{32}$$

$$\dot{b}_k = -\,\eta_k\,b_k + i\,f_k^*\,b_0\,b_{-k}^+\,e^{i(\omega_0 - 2\omega_k)t}\,. \tag{33}$$

We now suppose having realized the condition of coincidence in which the three magnon non-linear processes arise exactly at resonance. Thus, the energy of the first excited parametric states is equal to half the energy of the photons

($\omega = \omega_0 = 2\,\omega_k$). With these conditions, the transient non-linear motion for c_0 and c_k will be determined by equations (34), (35), and (36), and their hermitian conjugated equations for the quantum-dynamical variables c_0^+ and c_k^+:

$$\dot{b}_0 = -\eta_0\, b_0 + i \sum_{k \neq 0} f_k\, b_k\, b_{-k} + a_1\, q\,, \tag{34}$$

$$\dot{b}_k = -\eta_k\, b_k + i\, f_k^*\, b_0\, b_{-k}^+\,, \tag{35}$$

$$\dot{b}_{-k} = -\eta_k\, b_{-k} + i\, f_k^*\, b_0\, b_k^+\,. \tag{36}$$

At low power levels, equation (34) has the linear solution

$$b_0 = b_0^0\, \mathrm{e}^{-\eta_0 t} + \frac{a_1\, q}{\eta_0}(1 - \mathrm{e}^{-\eta_0 t})\,. \tag{37}$$

This linear solution is obtained neglecting third-order terms in the Hamiltonian (27). In the linear region, b_0 grows from the thermal level b_0^0 to a stationary amplitude $b_0^{\mathrm{m}} = a_1\, q/\eta_0$ with a time constant $1/\eta_0$. We can note that in the stationary linear region the occupation number of the uniform magnons is proportional to the number of the incident photons:

$$n_0 = \left(\frac{a_1}{\eta_0}\right)^2 n_q \gg n_0^0\,, \tag{38}$$

where n_0^0 is the Bose distribution function. In the non-linear region, equation (35) shows that a coupling exists between the mode $k = 0$ and the modes $\pm\, k$ by the coefficient f_k^*. With the elimination of b_{-k}^+ between (35) and the hermitian-conjugated equation of (36), we derive the stationary occupation number of the uniform mode when three-magnon transitions exist:

$$n_0 = \left(\frac{\eta_k}{|f_k|}\right)^2. \tag{39}$$

n_0 is now independent of the external excitation. From (38) and (39) we deduce the critical number of photons n_q^{crit} at the threshold of the non-linear processes:

$$n_q^{\mathrm{crit}} = \left(\frac{\eta_0\, \eta_k}{a_1\, |f_k|}\right)^2. \tag{40}$$

The elimination of b_{-k} between (34) and (36) gives

$$b_0 \left(\eta_0 + \sum_{k \neq 0} \frac{|f_k|^2}{\eta_k}\, b_k\, b_k^+\right) = a_1\, q\,. \tag{41}$$

From (41) and its hermitian-conjugated equation we obtain

$$n_0 = \left(\frac{a_1}{\eta_0 + \sum\limits_{k \neq 0} n_k\, |f_k|^2/\eta_k}\right)^2 \cdot n_q\,. \tag{42}$$

In the non-linear region, the excitation of n_k magnons results in an increase from η_0 to $\eta_0 + \sum\limits_k n_k\, |f_k|^2/\eta_k$ of the relaxation rate of the uniform mode. If $n_q > n_q^{\mathrm{crit}}$, n_k increases and fixes n_0 at an upper limit given by relation (39). Assuming the modes $\pm\, k$ to be only excited, we can define, from (39) and (42),

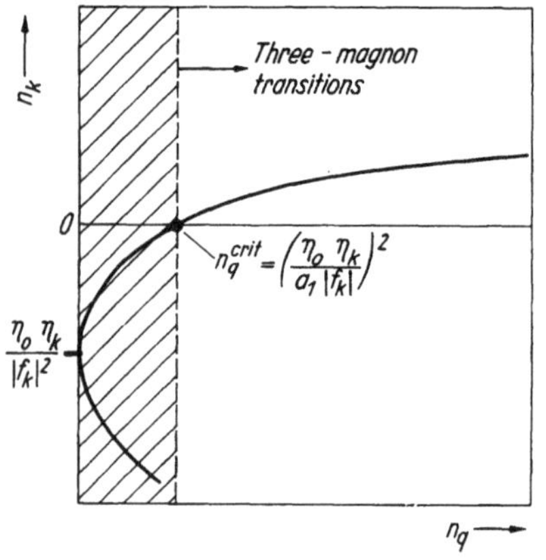

Fig. 2. Number of the parametric spin waves excited by three-magnon non-linear processes versus the number of the incident photons

the number of the parametric quasiparticles in terms of the number of the incident photons (Fig. 2):

$$n_k = \frac{a_1 (n_q)^{1/2}}{|f_k|} - \frac{\eta_0\, \eta_k}{|f_k|^2}\,. \tag{43}$$

4. Transient Evolution of the Uniform and Parametric Magnon Populations, with Three-Magnon Non-Linear Transitions

The transient evolution of the modes b_0, b_k, and b_{-k} is described by equations (34), (35), and (36), and their hermitian-conjugates for the variables b_0^+, b_k^+, and b_{-k}^+. First-order processes arise when b_0, during its transient growth, exceeds a critical value b_0^{crit} above which the interaction term $f_k^*\, b_0\, b_{-k}^+$ in (35), and hence b_k, becomes important (Fig. 3). The transient growth of b_k determines the decrease of b_0, with the coupling term $f_k\, b_k\, b_{-k}$ in (34) until the threshold value b_0^{crit}, which defines the stationary state, is reached. The linear solution (37) describes the transient evolution of b_0, unperturbed by spin waves ($t < t_1$, Fig. 3). In what follows we analyse theoretical evolution of the uniform mode at the beginning of the spin-wave growth as long as b_k does not become too

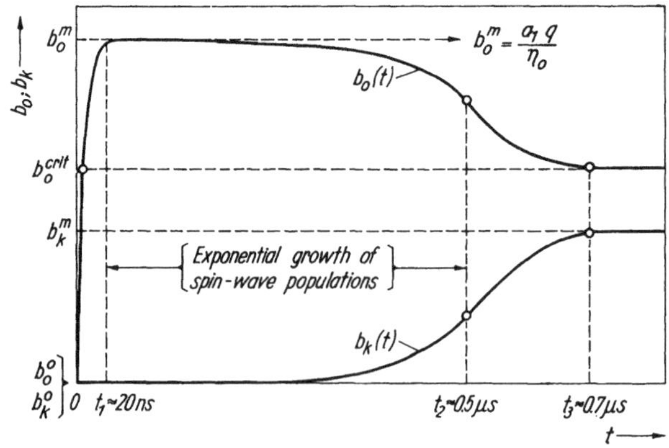

Fig. 3. Transient time dependence of the quantum-dynamical variables b_0 and b_k after the electromagnetic field is applied at $t = 0$

important ($t_1 < t < t_2$, Fig. 3). For this, we suppose that the maximum quasi-stationary amplitude b_0^m is constant at the beginning of the parametric excitation of the magnons (i.e. $b_0^m \gg b_k$ for $t_1 < t < t_2$). With this condition, equation (35) gives the time dependence of b_k:

$$b_k = e^{-\eta_k t} \, (b_k^0 \cosh \varrho_k t + i f_k^* \varrho_k^{-1} b_0^m b_{-k}^{+0} \sinh \varrho_k t) \,, \tag{44}$$

where we define an "excitation rate" with

$$\varrho_k = |f_k| \, (\langle b_0^{+m} b_0^m \rangle)^{1/2} = |f_k| \, (n_0^m)^{1/2} \,.$$

Since the mean values of $b_k^0 b_{-k}^0$ and $b_{-k}^{+0} b_k^{+0}$ are zero, we obtain immediately

$$b_k b_{-k} = e^{-2\eta_k t} \left[\frac{i}{2} f_k^* \varrho_k^{-1} b_0^m (b_{-k}^{+0} b_{-k}^0 + b_k^0 b_k^{+0} \sinh 2 \varrho_k t \right] . \tag{45}$$

From (34) and (45) we obtain the equation which describes the transient evolution of the uniform mode at the beginning of the spin-wave reaction:

$$\dot{b}_0 + \eta_0 b_0 + \sum_{k \neq 0} \frac{|f_k| \, b_0^m \, \eta_0}{a_1 \sqrt{n_q}} \left(b_k^{+0} b_k^0 + \frac{1}{2} \right) e^{-2\eta_k t} \sinh 2 \varrho_k t - a_1 q = 0 \,. \tag{46}$$

Integrating (46) for $t_1 < t < t_2$, we obtain the general analytic expression for b_0:

$$b_0 = b_0^m \left[1 - \sum_{k \neq 0} \frac{|f_k| \left(n_k^0 + \frac{1}{2} \right) \eta_0}{2 a_1 \sqrt{n_q}} \left(\frac{e^{2(\varrho_k - \eta_k)t} - e^{-\eta_0 t}}{\eta_0 + 2 \varrho_k - 2 \eta_k} - \frac{e^{-2(\varrho_k + \eta_k)t} - e^{-\eta_0 t}}{\eta_0 - 2 \varrho_k - 2 \eta_k} \right) \right] . \tag{47}$$

For the case $\varrho_k > \eta_k$, if t is higher than $1/\eta_0$ ($t > t_1 \approx 1/\eta_0$), the excitation of the parametric magnons determines an exponential decrease of the uniform mode from its maximal linear value $b_0^m = a_1 q/\eta_0$:

$$c_0 = b_0^m \left[1 - \sum_{k \neq 0} F_k \, e^{2(\varrho_k - \eta_k)t} \right] e^{i \omega t} \tag{48}$$

with

$$F_k = \frac{|f_k| \left(n_k^0 + \frac{1}{2} \right) \eta_0}{2 a_1 \sqrt{n_q} \, (\eta_0 + 2 \varrho_k - 2 \eta_k)} \,. \tag{49}$$

From the condition $\varrho_k > \eta_k$ we again find the critical threshold which we have calculated, as Suhl, in the stationary region (relation (40)).

In order to verify experimentally the evolution of c_0 and c_k, we derive the transient power P_a absorbed by the crystal at the beginning of the parametric excitation. For this, we start from

$$P_a = \frac{1}{2} \left\langle \frac{dm_0^+}{dt} h^- + \frac{dm_0^-}{dt} h^+ \right\rangle_{\text{time average}} . \tag{50}$$

From the second quantization of the magnetic system [1] we have

$$m_0^+ = (2 g \mu_B M_0 / V)^{1/2} c_0 \quad \text{and} \quad m_0^- = (2 g \mu_B M_0 / V)^{1/2} c_0^+ \,. \tag{51}$$

The transient power absorbed by the crystal of volume V is obtained from (21), (37), (50), and (51) in the linear region:

$$P_a = 2 \pi g \gamma M_0 V \frac{\omega_0}{\eta_0} \left(\frac{n_q \hbar \omega}{V_1} \right) [1 - \exp(-\eta_0 t)] \,. \tag{52}$$

With three-magnon non-linear processes the transient power absorbed by the magnetic system is obtained from (21), (48), (50), and (51):

$$P_a = 2\pi g \gamma M_0 V \frac{\omega_0}{\eta_0}\left(\frac{n_q \hbar \omega}{V_1}\right)\left[1 - \sum_{k\neq 0} F_k \exp\left(2(\varrho_k - \eta_k)\,t\right)\right]. \tag{53}$$

If the microwave magnetic field applied parallel to the x-axis is $h_x = h_0 \cos \omega t$, the electromagnetic energy density is

$$\frac{h_0^2}{8\pi} = \frac{n_q \hbar \omega}{V_1}. \tag{54}$$

Then we can express P_a in terms of the physical parameter h_0 which is experimentally accessible:

$$P_a = g \gamma M_0 V \omega \frac{h_0^2}{4\eta_0}\left[1 - \sum_{k\neq 0} \frac{\left(n_k^0 + \frac{1}{2}\right)}{V} \frac{\pi g \mu_B}{h_0}\left(\frac{\omega_i + D k^2 + \omega_k}{2\omega_k}\right)\sin 2\theta_k \times \right.$$
$$\left. \times \frac{\eta_0}{\eta_0 + 2(\varrho_k - \eta_k)} \cdot \exp\left(2(\varrho_k - \eta_k)\,t\right)\right]. \tag{55}$$

5. Discussion

It is important to develop the theoretical study of the transient processes at high power levels, since it allows us to understand better the three-magnon non-linear transitions. Relation (55) shows that the parametric modes which are coupled most strongly to the uniform mode are propagating in the direction $\theta_k \approx \pi/4$. From the growth increment measurement of the spin waves, it is possible to deduce the value of the relaxation rate and the coupling coefficient of the parametric magnons with the uniform mode. It would be interesting to compare these characteristic with those of the spin waves excited perpendicularly to the dc magnetic field ($\theta_k = \pi/2$) with the parallel pumping technique [14, 15]. These results are also of great interest in the study of the behaviour of microwave devices submitted to high power pulses, such as limiters, filters, and digital phase shifters. The coincidence limiters particularly, must absorb incident power above a critical level. This limitation is not effective at the beginning of the pulse as we can see from Fig. 3 and relation (55). From this relation it is possible to determine the limitation defect of the device by calculating the mean power unabsorbed by the crystal in the transient spike of the transmitted pulse.

The present theory has been developed assuming that the electromagnetic field rises in an infinitely short time and stays constant during the pulse. Thus, to test the theory it is preferable to use a wave-guide measurement technique. Indeed, in cavity experiments the time rise of the electromagnetic field is about between 100 ns and 1 μs following the cavity Q-factor. In addition, during the transient absorption the variation of the sample permeability produces a change of the field in the cavity. For these reasons, we have carried out accurate experiments at 5.3 GHz with a lithium ferrite single crystal. Three-magnon non-linear processes can take place at the resonance, since the critical frequency ($\omega_c = 2\,\omega_M/3$) is 6.6 GHz for this material. In fact, we observe, as described in Fig. 3, the growth of the uniform mode in a very short time t_1 ($t_1 \approx 50$ ns) until a maximum quasistationary value is reached ($b_0^m = a_1 q/\eta_0$, Fig. 3). Then the parametric excitation of the spin waves gives a quasi-exponential decrease of b_0 during a relatively long time ($t_2 - t_1 \approx 500$ ns) before the stationary state is reached at $t_3 \approx 1\ \mu$s.

Our experiments agree with the assumption that only the modes $\pm k$ are excited at least if the incident power is not too high. Thus we can limit the summation over k in (55) to a single exponential corresponding to the spin wave having the highest growth increment. From experimental data it is possible to conclude that the present theory is still applicable near the inflexion point of the transient absorbed power. Experimental results will be reported in a further publication.

Acknowledgements

In conclusion, the author expresses his gratitude to Dr. R. Vautier and Prof. J. Friedel for valuable discussions and for their interest in the work.

References

[1] A. I. Akhiezer, J. Phys. (SU) **10**, 217 (1946).
[2] R. W. Damon, Rev. mod. Phys. **25**, 239 (1953).
[3] N. Bloembergen and S. Wang, Phys. Rev. **93**, 72 (1954).
[4] H. Suhl, J. Phys. Chem. Solids **1**, 209 (1957).
[5] H. le Gall, E. Milot, and B. Dosermiere, Solid State Commun. **5**, 637 (1967).
[6] M. A. Savchenko and V. V. Tarasenko, Zh. eksper. teor. Fiz. **51**, 482 (1966); Soviet Phys. — J. exp. theor. Phys. **24**, 325 (1967).
[7] T. Holstein and H. Primakoff, Phys. Rev. **58**, 1098 (1940).
[8] A. I. Akhiezer, V. G. Baryakhtar, and S. V. Peletminskii, Zh. eksper. teor. Fiz. **36** (1959); Soviet Phys. — J. exp. theor. Phys. **9**, 146 (1959).
[9] M. I. Kaganov and V. M. Tsukernik, Zh. eksper. teor. Fiz. **37**, 823 (1959); Soviet Phys. — J. exp. theor. Phys. **10**, 587 (1960).
[10] M. Sparks, R. Loudon, and C. Kittel, Phys. Rev. **122**, 791 (1961).
[11] N. N. Bogolyubov and S. V. Tyablikov, Zh. eksper. teor. Fiz. **19**, 256 (1948).
[12] W. Heitler, The Quantum Theory of Radiation, Oxford University Press, 1954.
[13] L. I. Schiff, Quantum Mechanics, McGraw-Hill Publ. Comp., New York 1955.
[14] E. Schlomann, J. appl. Phys. **34**, 1998 (1963).
[15] H. le Gall, C. R. Acad. Sci. (France) **258**, 3986 (1964).

(Received December 29, 1967)

phys. stat. sol. 28, 507 (1968)

Subject classification: 18.2 and 18.4; 21.1

Laboratoire d'Electrostatique et de Physique du Metal, Grenoble

Couplage ferro-antiferromagnétique et traînage magnétique dans des couches minces multiples Co–CoO et Ni–NiO

Par

C. SCHLENKER

On a étudié le couplage ferro-antiferromagnétique sur des couches minces polycristallines, à anisotropie uniaxiale, oxydées superficiellement, du type Co–CoO ou Ni–NiO, entre −190 et 20 °C. L'anisotropie unidirectionnelle s'annule à une température de transition toujours inférieure au point de Néel et liée à l'épaisseur moyenne de la couche antiferromagnétique. L'anisotropie uniaxiale n'est pas modifiée par la présence du couplage. On explique le maximum du champ coercitif et la disparition de l'hystérésis de rotation au voisinage de la température de transition en supposant que la couche antiferromagnétique "déformable" du point de vue magnétique aux basses températures, devient "rigide" aux plus hautes températures. On observe un traînage magnétique de très grande amplitude qui provient de réorientations des moments dans certains grains antiferromagnétiques et suit une loi logarithmique dans des intervalles de temps de plusieurs heures, par suite de la dispersion de la taille des grains. Les ordres de grandeur de l'anisotropie antiferromagnétique déduits de l'étude détaillée du traînage sont supérieurs à l'anisotropie magnétocristalline des oxydes à l'état massif.

The temperature dependence of the ferromagnetic–antiferromagnetic coupling is studied between −190 and 20 °C in polycrystalline thin films of cobalt or nickel oxidized superficially. The unidirectional anisotropy vanishes at a transition temperature below the Néel temperature and is related to the average thickness of the antiferromagnetic film. The uniaxial anisotropy is found not to be dependent on the coupling. The observed maximum in the coercive force and the reduction in hysteresis loss at temperatures near the transition temperature are in qualitative agreement with a model in which the moments in the antiferromagnetic atomic planes are "distorted" by the coupling at low temperatures, the film being "rigid" at higher temperatures only. The magnetic aftereffect related to spin reorientations in some antiferromagnetic regions is found to be considerable and follows a logarithmic law for times of several hours, due to the size distribution of the antiferromagnetic crystallites. Finally, values greater than the magnetocrystalline anisotropy of the bulk materials are deduced for the anisotropy in the antiferromagnetic oxides.

1. Introduction

Les propriétés des couches minces de cobalt ou de nickel oxydées superficiellement s'interprètent par l'existence d'un couplage entre le métal ferromagnétique et l'oxyde antiferromagnétique [1]. Une étude détaillée de l'hystérésis dans de telles couches [2 à 6] a montré qu'elles contiennent deux sortes de régions, les unes à l'origine de l'anisotropie unidirectionnelle, les autres des pertes par hystérésis de rotation. Dans un cas, les moments antiferromagnétiques restent bloqués dans leur direction d'anisotropie ou effectuent des oscillations réversibles de faible amplitude autour de celle-ci lorsque les moments ferromagnétiques tournent. Dans l'autre, ils suivent les moments ferromagnétiques irréversiblement et l'on peut considérer que les directions faciles antiferromagnétiques sont distribuées d'une façon isotrope dans le plan de la couche.

D'autre part, les systèmes à couplage ferro-antiferromagnétique présentent généralement des phénomènes de traînage magnétique [7] qu'on attribue à des réorientations dans les grains antiferromagnétiques sous l'effet des fluctuations thermiques [5, 6]. Une étude quantitative de ce traînage, en particulier sa variation avec la température, comparée à la variation de l'anisotropie unidirectionnelle et de l'hystérésis de rotation, doit donner de nouveaux renseignements sur le matériau antiferromagnétique et sur le mécanisme du couplage.

Le présent article rend compte dans une première partie de la variation des cycles d'hystérésis et courbes de couple magnétique entre −190 et 20 °C et de son origine et dans une deuxième partie des résultats relatifs au traînage magnétique dans la même gamme de température et de leur interprétation possible.

Les conditions de préparation des échantillons et les techniques de mesure sont données dans [2]. Les couches ferromagnétiques sont polycristallines, à anisotropie uniaxiale, d'épaisseur comprise entre 500 et 1000 Å. Les matériaux antiferromagnétiques sont les oxydes CoO et NiO, à point de Néel de 18 et 250 °C respectivement. Les mesures d'aimantation à saturation montrent que les couches antiferromagnétiques sont d'épaisseur inférieure à 50 Å. Dans tous les cas, les échantillons sont refroidis de la température ambiante à −190 °C en présence d'un champ magnétique saturant l'aimantation ferromagnétique dans la direction facile uniaxiale et les mesures en fonction de la température sont faites lors du réchauffement.

2. Variation avec la température de l'anisotropie unidirectionnelle et des pertes par hystérésis de rotation

2.1 Résultats expérimentaux

Les Fig. 1 et 2 correspondent aux cycles d'hystérésis mesurés sur une couche Co–CoO et Ni–NiO respectivement, dans la direction d'anisotropie uniaxiale, avec un hystérésiscope fonctionnant à la fréquence de 1000 Hz. On n'étudie donc pas la contraction des cycles d'hystérésis mais seulement le cycle limite stabilisé [2]. H_c est la demi-largeur du cycle et H_e le décalage. Les résultats de l'analyse de Fourier des courbes de couple magnétique obtenues au moyen d'une balance de torsion en champ tournant à la fréquence de 5×10^{-3} Hz sont donnés sur la Fig. 3 pour Co–CoO et 4 pour Ni–NiO. On ne s'intéresse pas aux

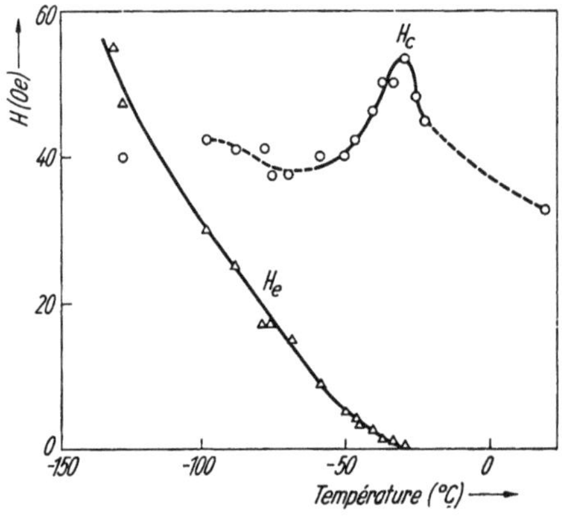

Fig. 1. Variation avec la température du champ coercitif H_c et du décalage de cycle H_e pour une couche Co–CoO

Fig. 2. Variation avec la température du champ coercitif H_c et du décalage de cycle H_e pour une couche Ni–NiO. L'échantillon utilisé est le même que pour les Fig. 4 à 7 et 9, 10

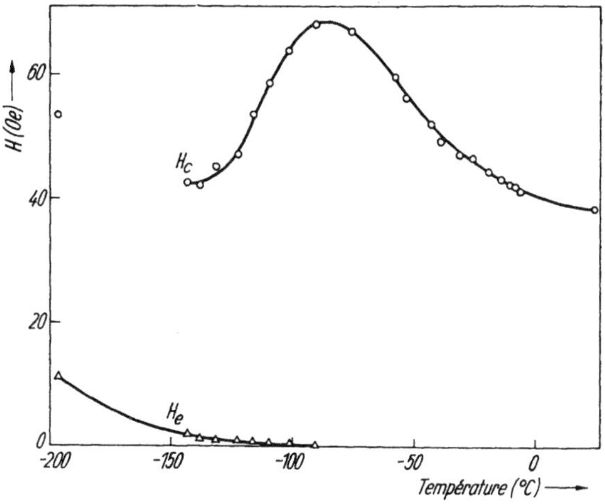

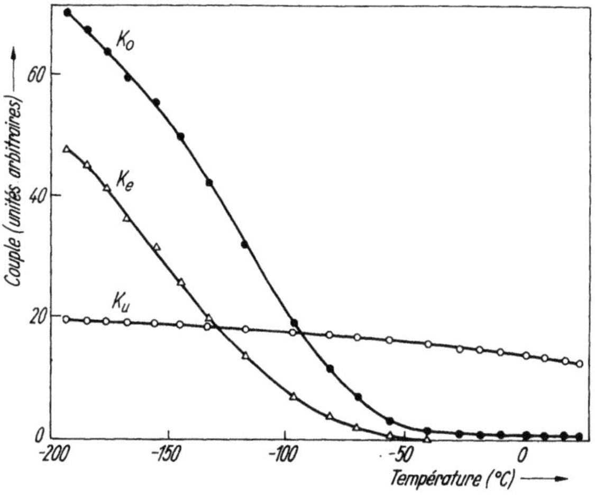

Fig. 3. Variation avec la température des composantes de Fourier de la courbe de couple magnétique pour une couche Co–CoO. K_0 est proportionnel aux pertes par hytérésis de rotation, K_e la composante unidirectionnelle et K_u la composante uniaxiale

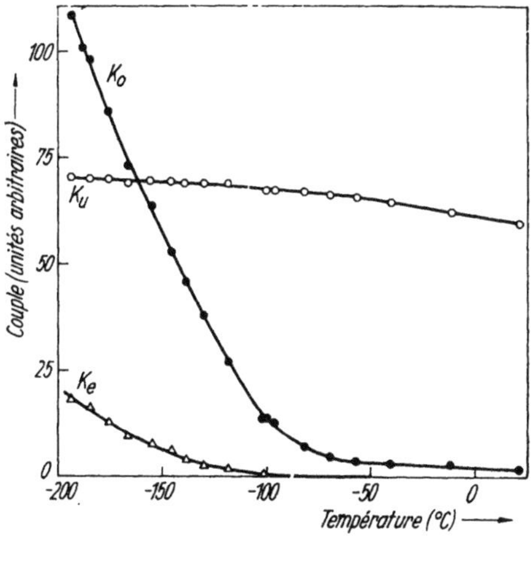

Fig. 4. Variation avec la température des composantes de Fourier de la courbe de couple magnétique pour une couche Ni–NiO

phénomènes transitoires et les valeurs sont relevées sur une courbe obtenue après plusieurs rotations du champ de 360° dans les deux sens. Le couple mesuré Γ est de la forme:

$$\Gamma = K_0 + K_e \sin(\theta - \Phi_e) + K_u \sin 2(\theta - \Phi_u) .$$

θ est l'angle entre la direction d'anisotropie uniaxiale et le champ appliqué. K_e et K_u sont les constantes d'anisotropie unidirectionnelle et uniaxiale. La composante constante K_0 est reliée aux pertes par hystérésis de rotation W_r par $W_r = 2\pi K_0$. On a mis en évidence des déphasages Φ_e et Φ_u pour les deux composantes unidirectionnelle et uniaxiale parfois considérables.

L'ensemble de ces résultats permet de définir une température de transition T_t, pour laquelle les quantités K_e et H_e, donc l'anisotropie unidirectionnelle s'annulent. Cette température est toujours inférieure au point de Néel. Il semble qu'elle soit d'autant plus élevée que l'épaisseur moyenne d'oxyde est plus grande: pour des couches de cobalt oxydées pendant 1 h à 200 °C, elle est comprise entre −30 et −50 °C, tandis que pour des couches oxydées pendant 2 h à 150 °C, elle est voisine de −100 °C; or les couches d'oxyde sont plus épaisses dans le premier cas que dans le second. Ce résultat confirme l'observation faite sur des couches de FeNi oxydées [8]. D'autre part, pour des conditions d'oxydation identiques, donc vraisemblablement pour des épaisseurs d'oxyde comparables, T_t est plus bas pour Ni–NiO que pour Co–CoO; ainsi pour des couches de nickel oxydées pendant 1 h à 200 °C, T_t est de l'ordre de −100 °C. On note aussi que la valeur relative de K_e par rapport à K_0 est toujours beaucoup plus faible dans le cas du nickel que dans le cas du cobalt.

L'hystérésis de rotation est toujours faible lorsqu'on atteint la température de transition, mais elle persiste au-delà de T_t. Sur les Fig. 1 et 2, on remarque au contraire que le champ coercitif, donc l'hystérésis alternative, passe par un maximum au voisinage de cette même température. Ce dernier résultat est caractéristique des systèmes à couplage ferro-antiferromagnétique [8]. Par contre, l'anisotropie uniaxiale K_u n'est pas modifiée par l'existence de la couche antiferromagnétique (Fig. 3 et 4); sa variation entre −190 et 20 °C est la même que pour une couche non oxydée.

La variation des déphasages Φ_e et Φ_u est donnée en fonction de la température sur la Fig. 5 et en fonction du champ magnétique sur la Fig. 6, où elle est comparée à la variation de K_0, K_e et K_u. Φ_e et Φ_u diminuent lorsque la température

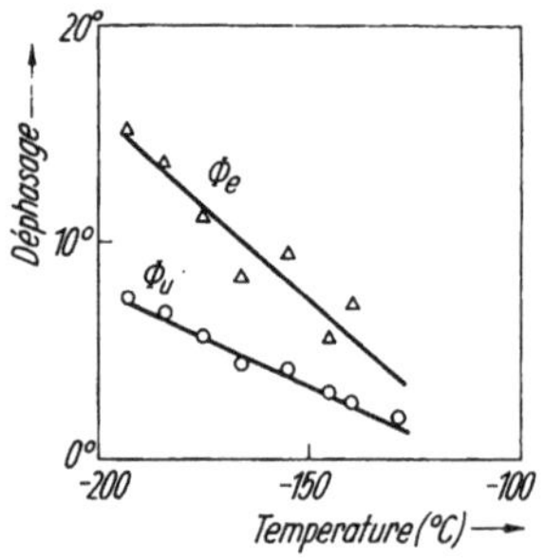

Fig. 5. Variation avec la température des déphasages de la composante unidirectionnelle Φ_e et de la composante uniaxiale Φ_u de la courbe de couple. $H = 360$ Oe

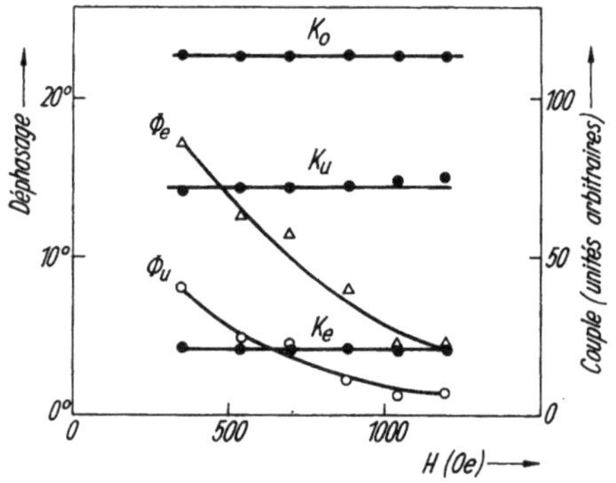

Fig. 6. Variation avec le champ magnétique des composantes de Fourier de la courbe de couple et de leurs déphasages. Température: −190 °C

augmente et disparaissent au voisinage de la température de transition; ils sont donc reliés au couplage ferro-antiferromagnétique. D'autre part, ils diminuent rapidement lorsqu'on augmente le champ, alors que pour les mêmes valeurs de H, les quantités K_0, K_e et K_u en sont indépendantes. On peut penser que Φ_e mesure l'angle moyen entre l'aimantation ferromagnétique et le champ. Mais l'on s'attend alors à ce que Φ_u soit égal à Φ_e, ce qui n'est pas vrai pour tous les échantillons . Φ_e est souvent plus grand que Φ_u. Ce résultat suggère, en première approximation, que lorsque le champ n'est pas assez fort, l'aimantation dans le plan de la couche ferromagnétique n'est pas toujours homogène. Dans certaines régions où le couplage est faible ou bien sans effet, l'aimantation est pratiquement alignée avec le champ, tandis que dans d'autres, l'angle moyen entre M et H peut atteindre Φ_e.

2.2 *Interprétation*

On peut considérer que la couche antiferromagnétique est constituée de cristallites dont la grosseur est déterminée par la taille des cristallites ferromagnétiques sous-jacents et par l'épaisseur moyenne de la couche d'oxyde. Nous ferons l'hypothèse que l'anisotropie antiferromagnétique K est uniaxiale et constante sur toute la surface de la couche tandis que l'intensité du couplage ferro-antiferromagnétique, caractérisé par une constante A, varie d'une manière considerable d'un cristallite à l'autre [5]. K et A sont rapportés à une unité d'interface. A l'aide de l'étude théorique développée dans [5], on peut distinguer deux comportements possibles de la couche antiferromagnétique. Une couche d'anisotropie faible reste "rigide" du point de vue magnétique sous l'effet du couple exercé à l'interface, c'est-à-dire que grâce à l'influence prépondérante des forces d'échange, les moments magnétiques restent parallèles entre eux d'un plan atomique au suivant. Au contraire, une couche de forte anisotropie se déforme, les moments de l'interface étant par exemple beaucoup plus écartés de la direction d'anisotropie que ceux de la surface extérieure. Elle se déforme d'autant plus qu'elle est plus épaisse. Nous admettrons que la couche ferromagnétique reste dans tous les cas "rigide".

Intéressons-nous au cas où la couche multiple est soumise à un champ tournant. Dans une couche antiferromagnétique "déformable", il y a deux possibilités suivant les valeurs relatives de K et A : ou bien les moments sont entraînés irréversiblement par l'aimantation ferromagnétique si la constante A est grande et le système présente de l'hystérésis de rotation; ou bien, lorsque A est faible, ils restent bloqués dans la direction d'anisotropie ou effectuent de petites oscillations autour de cette direction et on observe sur la couche multiple de l'anisotropie undirectionnelle.

Dans une couche antiferromagnétique rigide, qui correspond au modèle théorique donné dans [1], les moments antiferromagnétiques sont bloqués lorsque A est petit, suivent l'aimantation M irréversiblement pour des valeurs intermédiaires de A et réversiblement pour les grandes valeurs de A.

Aux basses températures, l'anisotropie K étant grande, les couches sont très probablement du type déformable. Par suite de la large distribution des valeurs de A, il y a simultanément anisotropie unidirectionnelle et hystérésis de rotation. Mais aucune région antiferromagnétique ne suit M réversiblement.

Lorsque la température augmente, les quantités K et A diminuent et s'annulent au point de Néel. Mais K diminue beaucoup plus rapidement que A, si bien

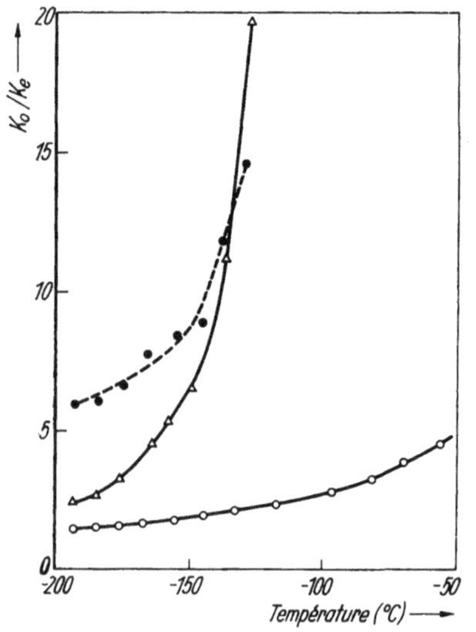

Fig. 7. Variation avec la température du rapport K_0/K_e proportionnel au rapport des pertes par hystérésis de rotation à l'anisotropie unidirectionnelle.
○ CoucheCo–CoO, température de transition: −45 °C.
△ Couche Co–CoO, température de transition: −90 °C.
● Couche Ni–NiO, température de transition: −95 °C

que, la proportion des régions antiferromagnétiques qui donnent lieu à l'anisotropie unidirectionnelle diminuant, K_e et H_e diminuent. Les régions à l'origine de l'hystéresis de rotation deviennent plus nombreuses, mais pour un cristallite donné, les pertes diminuent lorsque K diminue. Ce dernier effet l'emportant sur l'autre, K_0 diminue. On remarque d'ailleurs sur la Fig. 7 que le rapport K_0/K_e augmente considérablement entre les basses températures et T_t.

L'hystérésis de rotation étant très faible au-delà de T_t, on en déduit que la couche antiferromagnétique est devenue rigide. Dans un grand nombre de régions antiferromagnétiques, les moments suivent M réversiblement en champ tournant et sont sans effet sur l'allure de la courbe de couple. Par contre, en champ alternatif, ces mêmes régions gênent le basculement de l'aimantation et augmentent donc le champ coercitif. L'existence du maximum de H_c au voisinage de T_t semble indiquer que c'est au voisinage de la température de transition, probablement un peu au-dessous, que la couche antiferromagnétique déformable devient rigide. Lorsque la couche est rigide, il est d'ailleurs vraisemblable qu'une faible diminution relative de K par rapport à A, donc une faible augmentation de température fasse disparaître l'anisotropie unidirectionnelle résiduelle et l'hystérésis de rotation.

On comprend alors la relation qui existe entre la valeur de T_t et l'épaisseur de la couche antiferromagnétique. Plus cette couche est épaisse, plus l'anisotropie doit être faible et par conséquent plus la température doit être élevée pour que le passage de la couche déformable à la couche rigide ait lieu. On remarque en particulier que la présence d'une couche antiferromagnétique rigide de faible anisotropie peut n'avoir aucun effet sur le comportement de la couche multiple, même pour des températures bien inférieures au point de Néel. C'est le cas des couches Ni–NiO à température ambiante. La comparaison des couches Co–CoO et Ni–NiO indique que pour des épaisseurs d'oxyde voisines, une couche de NiO devient rigide pour des températures beaucoup plus basses qu'une couche de CoO. Cela tient à la différence d'intensité relative des forces d'échange et d'anisotropie dans les deux matériaux.

Cet ensemble de considérations interprète qualitativement les résultats expérimentaux. Il serait possible de faire une étude plus quantitative des phénomènes, mais la nature polycristalline des couches en réduit considérablement l'intérêt et la validité.

3. Traînage magnétique

3.1 Résultats expérimentaux

Le traînage magnétique a été étudié au moyen de la balance de torsion, dans les conditions du relevé de la courbe de couple. Après plusieurs rotations de 360° dans les deux sens, le champ magnétique est maintenu constant en direction et en intensité et on mesure le couple Γ en fonction du temps t. Γ_0 étant le couple enregistré à l'instant initial d'arrêt du champ et $\Gamma(t)$ à l'instant t, on définit $\Delta\Gamma(t)$ par: $\Delta\Gamma(t) = \Gamma_0 - \Gamma(t)$. Sur les Fig. 8 et 9, on a porté $\Delta\Gamma(t)$ en fonction de log t, pour des échantillons Co–CoO et Ni–NiO, à différentes températures. La direction et le sens du champ sont ceux de l'anisotropie unidirectionnelle des couches. L'étude de $\Delta\Gamma(t)$ en fonction de cette direction indique d'ailleurs que le rapport $\Delta\Gamma(t)/\Gamma_0$ en est sensiblement indépendant.

On remarque que la variation de Γ se poursuit pendant des temps supérieurs à plusieurs heures et que les quantités $\Delta\Gamma/\Gamma_0$ peuvent atteindre 30 à 60% selon les échantillons et selon la température. Le traînage magnétique est donc un phénomène important.

On note que la variation de Γ en fonction du temps est logarithmique dans un intervalle de temps d'autant plus grand que la température est plus basse. Les réorientations dans les grains antiferromagnétiques se font donc avec des constantes de temps variables d'une région à l'autre et il doit exister un spectre de constantes de temps.

3.2 Modèle et interprétation

Remarquons d'abord que le traînage magnétique ne doit avoir lieu que dans les régions antiferromagnétiques qui sont à l'origine des pertes par hystérésis, donc pour lesquelles la constante de couplage A est relativement grande. On peut considérer, en première approximation, que la hauteur de la barrière de

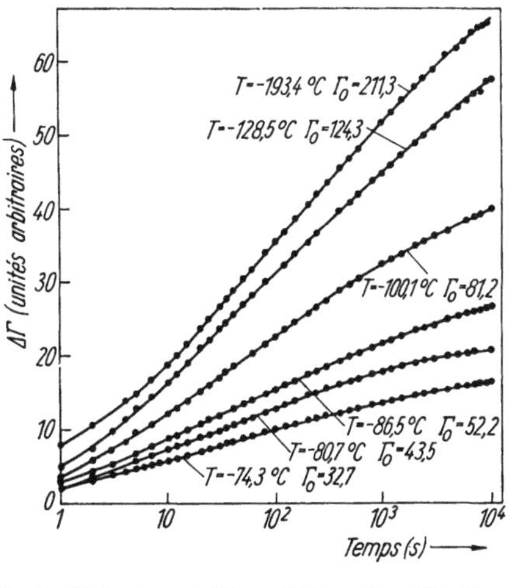

Fig. 8. Traînage magnétique $\Delta\Gamma(t) = \Gamma_0 - \Gamma(t)$. Même échantillon Co–CoO que pour la Fig. 3. Γ_0 est exprimé dans les mêmes unités que $\Delta\Gamma$

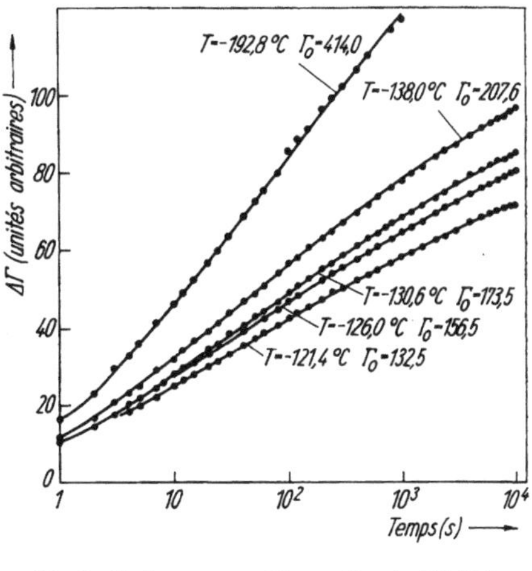

Fig. 9. Traînage magnétique. Couche Ni–NiO

potentiel qui sépare les deux états d'un grain antiferromagnétique avant et après réorientation est de l'ordre de grandeur de KV, K étant l'énergie d'anisotropie par unité de volume antiferromagnétique et V le volume du grain. Les effets de traînage sont prépondérants parce que l'énergie d'agitation thermique kT est du même ordre que la quantité $W = K\,V$, donc parce que les grains antiferromagnétiques sont de petit volume. La constante de temps θ du phénomène vérifie la relation $\theta = \theta_\infty \exp(W/kT)$, où θ_∞ est supposé indépendant de W et de T. Dans l'hypothèse où la constante d'anisotropie K ne varie pas d'un grain à l'autre, l'étalement des constantes de temps est due à la dispersion des volumes des grains. L'étude quantitative du traînage doit donc donner des renseignements sur les valeurs de $W = K\,V$. Nous admettrons que la répartition des constantes de temps est logarithmique, c'est-à-dire que les énergies W sont distribuées suivant un spectre rectangulaire de bornes inférieure W_1 et supérieure W_2 [9]. θ_1 et θ_2 étant les valeurs de θ associées à W_1 et W_2, on peut définir la constante de temps moyenne θ_0 et la largeur du spectre χ [10] par

$$\theta_0 = \sqrt{\theta_1\theta_2} = \theta_\infty \exp(W_0/kT)\,, \qquad \chi = \sqrt{\theta_2/\theta_1} = \exp(w/kT)$$

$$W_0 = (W_1 + W_2)/2\,, \qquad w = (W_2 - W_1)/2\,.$$

La loi de variation de $\Delta\Gamma$ est alors

$$\frac{\Delta\Gamma}{\Delta\Gamma_\infty} = \frac{\ln t + 0{,}577 - \ln \theta_1}{2 \ln \chi} \quad \text{pour} \quad \theta_1 < t < \theta_2\,.$$

$\Delta\Gamma_\infty$ est la limite de $\Delta\Gamma$ pour $t \to \infty$.

L'interprétation des courbes expérimentales selon le modèle ci-dessus nécessite la connaissance de $\Delta\Gamma_\infty$. Nous admettrons que $\Delta\Gamma_\infty$ est égal à Γ_0. Cette hypothèse est justifiée par les ordres de grandeur de $\Delta\Gamma(t)/\Gamma_0$ pour $t = 10000$ s et parce que les mesures sont faites dans la direction facile des couches. On peut alors déduire par comparaison de l'expression théorique de $\Delta\Gamma/\Delta\Gamma_\infty$ et de la partie linéaire des courbes expérimentales les valeurs de θ_0, χ et par suite w. Les tableaux 1 et 2 donnent les valeurs de θ_0 et χ pour différentes températures pour une couche Co–CoO et Ni–NiO. On remarque que θ_0 et χ sont du même ordre de grandeur quelle que soit la température T. La constante de temps θ_1

Tableau 1
Constante de temps moyenne θ_0 et largeur du spectre χ pour une couche Co–CoO

Température (°C)	θ_0	χ
−193,4	$2{,}75\times10^6$	$2{,}24\times10^6$
−184,6	$8{,}50\times10^5$	$7{,}41\times10^5$
−171,0	$3{,}42\times10^5$	$1{,}95\times10^5$
−154,5	$9{,}35\times10^4$	$6{,}60\times10^4$
−128,5	$1{,}95\times10^4$	$1{,}66\times10^4$
−100,1	$7{,}67\times10^3$	$6{,}31\times10^3$
− 86,5	$6{,}50\times10^3$	$7{,}41\times10^3$
− 80,7	$6{,}00\times10^3$	$6{,}92\times10^3$
− 74,3	$5{,}67\times10^3$	$6{,}03\times10^3$

Tableau 2
Constante de temps moyenne θ_0 et largeur du spectre χ pour une couche Ni–NiO

Température (°C)	θ_0	χ
−192,8	$3{,}00\times10^5$	$2{,}82\times10^5$
−138,0	$1{,}52\times10^4$	$1{,}82\times10^4$
−130,6	$1{,}08\times10^4$	$1{,}45\times10^4$
−126,0	$0{,}83\times10^4$	$1{,}10\times10^4$
−121,4	$3{,}70\times10^3$	$6{,}16\times10^3$

Fig. 10. Variation avec la température de w, peu différent de la valeur moyenne de KV.
○ Couche Co–CoO, température de transition: −45 °C.
△ Couche Co–CoO, température de transition: −90 °C.
● Couche Ni–NiO, température de transition: −95 °C

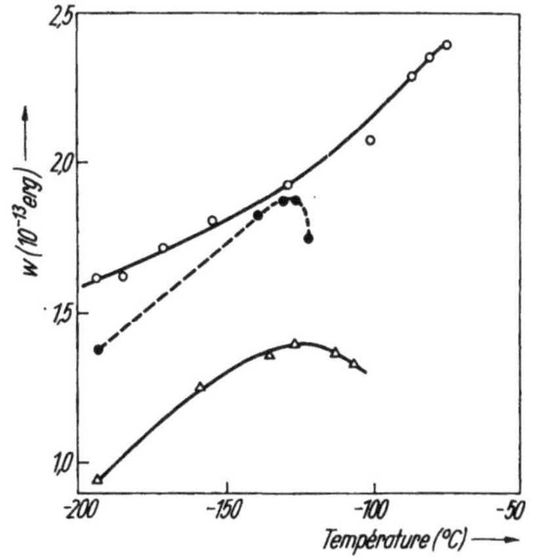

est donc voisine de 1 s et varie peu avec T, ce qui signifie que W_1 est plus petit que kT. θ_2 variant au contraire beaucoup avec T, on en conclut que W_2 est assez grand devant W_1 et l'on peut considérer que: $w \approx W_0 \approx W_2/2$. La Fig. 10 montre la variation de w avec T pour trois échantillons. w augmente avec T et passe par un maximum pour une température T_m inférieure à T_t. w étant approximativement la valeur moyenne de $W = KV$ et K diminuant lorsque T augmente, on en déduit que les grains donnant lieu au traînage sont de taille en moyenne plus faible à basse température qu'aux plus hautes températures. D'après ce qu'on a vu dans II-2, c'est dans les grains où la constante de couplage A est relativement faible que les moments sont bloqués à basse température et ne le sont plus aux températures plus élevées. On peut donc conclure que la constante de couplage est en moyenne d'autant plus faible que le grain est plus gros. Dans l'hypothèse de grains cylindriques ayant à peu près tous la même épaisseur, A est donc d'autant plus petit que la surface de base du grain est plus grande. Ce résultat n'est pas surprenant si l'on considère que la valeur de A est déterminée en particulier par l'existence de marches dans la structure de l'interface, le couplage changeant par exemple de sens d'une marche à la suivante [5] et le nombre moyen de marches par grain augmentant avec la surface.

Le maximum de $w(T)$ apparaît lorsque la diminution de K avec la température devient prépondérante par rapport à la variation de V. Pour $T \approx T_m$, l'anisotropie unidirectionnelle est très faible, peu de grains ont leurs moments encore bloqués et le volume moyen des grains donnant lieu au traînage varie peu pour $T > T_m$.

On peut obtenir à partir de la valeur de w un ordre de grandeur de la constante d'anisotropie K du matériau antiferromagnétique. Des études en microscopie électronique ont montré que le diamètre des cristallites de la couche ferromagnétique est inférieur à 50 Å. Il en est de même pour les grains antiferromagnétiques et l'on peut admettre que ceux qui donnent lieu au traînage à −190 °C ont un diamètre moyen compris entre 20 et 40 Å, tandis que l'épaisseur moyenne de la couche antiferromagnétique doit varier selon les échantillons entre 10 et 30 Å. La constante K est par conséquent à −190 °C de l'ordre de 5×10^6 à 5×10^7 erg/cm^3 aussi bien pour des couches de CoO que de NiO.

Il est intéressant de comparer ces valeurs à l'anisotropie magnétocristalline des deux oxydes à l'état massif. En ce qui concerne NiO, l'énergie d'anisotropie est de l'ordre de 2 à 3×10^6 erg/cm^3, le plan (111) étant un plan facile [11]. Mais l'anisotropie dans le plan (111) est beaucoup plus faible. Or le couplage ferro-antiferromagnétique doit provenir en très grande partie de grains dont le plan d'interface est un plan (111), plan d'alignement ferromagnétique des moments. Il est donc impossible d'attribuer la forte anisotropie déterminée expérimentalement à l'anisotropie magnétocristalline de NiO massif. On pourrait invoquer la non stoechiomètrie de NiO [12], mais il ne semble pas non plus qu'on puisse justifier ainsi des valeurs aussi fortes.

L'anisotropie magnétocristalline de CoO serait de l'ordre de 10^8 erg/cm^3, la direction facile étant voisine de [001] et l'anisotropie dans le plan (111) étant très faible [13]. De même que dans NiO, les plans (111) sont des plans d'alignement ferromagnétique des moments. Il est encore peu vraisemblable que l'anisotropie déduite de nos mesures soit l'anisotropie magnétocristalline de CoO massif.

Par contre, la très faible épaisseur des couches antiferromagnétiques suggère que l'anisotropie de surface [14] doit être prépondérante [15]. Les valeurs de K données précédemment correspondent à des anisotropies superficielles de 0,1 à 1 erg/cm^2, ce qui semble raisonnable.

4. Discussion et conclusion

Dans le modèle choisi pour interpréter les propriétés du traînage magnétique, nous avons supposé que la hauteur de la barrière de potentiel W séparant deux états d'un grain antiferromagnétique avant et après réorientation ne dépend que de l'anisotropie et du volume de ce dernier. En fait, W doit dépendre aussi de la constante de couplage A et de l'angle entre la direction facile du grain et celle de l'aimantation ferromagnétique. Mais nous pensons que les ordres de grandeur obtenus pour K ne seraient pas beaucoup modifiés par une amélioration du modèle.

D'autre part, nous avons fait l'hypothèse dans tout ce travail que la constante d'anisotropie K était la même pour tous les grains. Il est probable que K varie d'un grain à l'autre, d'autant plus qu'il s'agit vraisemblablement d'une anisotropie de surface qui dépend fortement des plans cristallographiques. Mais l'influence des plans d'interface (111) étant prépondérante pour le couplage ferro-antiferromagnétique, l'approximation qui consiste à prendre une seule valeur pour K est acceptable.

Il semble qu'il soit difficile de faire une théorie quantitative du couplage ferro-antiferromagnétique dans nos échantillons, surtout parce qu'on ne connait pas les fonctions de distribution de la constante de couplage et du volume des grains. Cependant, nous pensons avoir mis en évidence l'importance du traînage magnétique et avoir interprété qualitativement de manière satisfaisante les résultats expérimentaux concernant la variation avec la température des propriétés du couplage, et en particulier du traînage.

Bibliographie

[1] W. H. Meiklejohn, J. appl. Phys. **33**, 1328 (1962).
[2] D. Paccard, C. Schlenker, O. Massenet, R. Montmory et A. Yelon, phys. stat. sol. **16**, 301 (1966).
[3] C. Schlenker et D. Paccard, J. Physique **28**, 611 (1967).

[4] L. Néel, C.R. Acad. Sci. (France) **B264**, 1002 (1967).
[5] L. Néel, Ann. Phys. (France) **2**, 61 (1967).
[6] C. Schlenker, J. Physique **29**, C2 — 157 (1968).
[7] A. A. Glazer, A. P. Potapov, R. I. Tagirov et Ya. S. Shur, phys. stat. sol. **16**, 745 (1966).
[8] F. B. Hagedorn, J. appl. Phys. **38**, 3641 (1967).
[9] L. Néel, J. Phys. Radium **12**, 339 (1951); **13**, 249 (1952).
[10] P. Brissonneau, C.R. Acad. Sci. (France) **244**, 1174 (1957).
[11] G. T. Rado et H. Suhl, Magnetism, Vol. 1, Academic Press 1963 (p. 187).
[12] G. T. Rado et H. Suhl, Magnetism, Vol. 3, Academic Press 1963 (p. 333).
[13] G. T. Rado et H. Suhl, Magnetism, Vol. 1, Academic Press 1963 (p. 198).
[14] L. Néel, J. Phys. Radium **15**, 225 (1954).
[15] L. Néel, J. Phys. Soc. Japan **17**, Suppl. BI, 676 (1962), et communication personnelle.

(Received May 3, 1968)

phys. stat. sol. 28, 519 (1968)

Subject classification: 14.4.1; 22

Department of Physics, State University Moscow

On a Certain Type of Instability of Electrical Domains

By

R. ENDERLEIN[1])

The stability of the Riemann solution in the theory of electrical domains with respect to variations of the boundary conditions is investigated. Both the drift and the recombination non-linearities are considered. The Riemann solution is found to be stable under certain conditions of experimental interest.

Die Riemannsche Lösung in der Theorie der elektrischen Domänen wird auf ihre Stabilität gegenüber Variationen der Randbedingungen an den Kontakten untersucht. Sowohl der Drift- als auch der Rekombinations-Nichtlinearitäts-Fall wird betrachtet. Es zeigt sich, daß die Riemannsche Lösung unter bestimmten experimentell interessanten Bedingungen stabil sein kann.

1. Introduction

In the theory of electrical domains in semiconductors it is assumed that the stationary domain motion is described by the Riemann-type solution [1]. For the validity of this assumption the Riemann solution must have the necessary properties of stability. Its stability with respect to local fluctuations was studied by various authors [2 to 5]. In this paper the stability of the Riemann solution with respect to the variations of the boundary conditions is considered. The reason to study this question is obvious. The actual problem in the theory of electrical domains in semiconductors is to calculate the time and space behaviour of the electric quantities in the circuit of Fig. 1, consisting of the semiconductor sample and the load resistance. The stable solution of this problem for the complete circuit is uniquely determined by the battery e.m.f. or the total current. The relevant simultaneous differential equations must be supplemented by conditions describing the physical situation at the contacts. In the domain theory, however, the solutions are studied only in the sample. In this case the battery e.m.f. or the total current and the conditions of stability are not sufficient

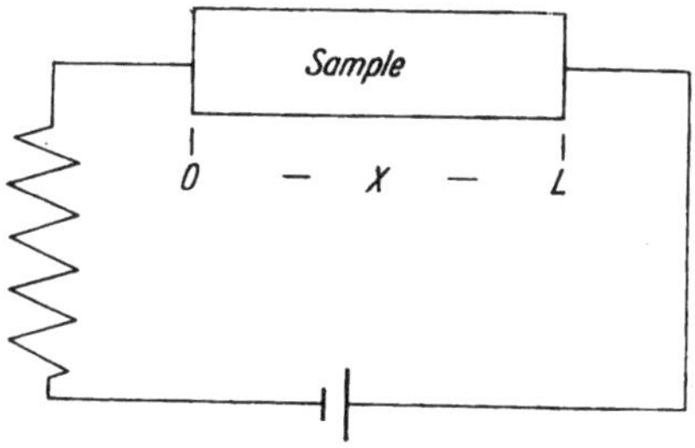

Fig. 1. Current circuit for observing domain instabilities

[1]) Permanent address: Bereich Theoretische Physik der Sektion Physik der Humboldt-Universität zu Berlin.

to determine the solution uniquely. Formulation of boundary conditions, e.g. for the field strengths E, becomes necessary [1, 6]. The boundary conditions postulate a priori some properties of the unknown solution $E(x, t)$ of the total circuit problem and are arbitrary from this point of view. Therefore, the Riemann solution $E_R(x - u_0 t)$, corresponding to a special choice of the boundary values, can be regarded as a good approximation of $E(x, t)$ only if it is stable with respect to variations of the boundary conditions.

To investigate this question we assume that the electric field $E(x, t)$ at the contacts is not given by the Riemann solution, but by the values

$$E(0, t) = E_R(-u_0 t) + f(t) , \tag{1}$$

$$E(L, t) = E_R(L - u_0 t) + \xi(t) f(t) , \tag{2}$$

where $f(t)$ and $\xi(t) f(t)$ are small disturbations which, in a phenomenological way, describe the influence of the contacts. In particular, the deviation of the parameter $\xi(t)$ from 1 is a measure of the disturbance of the total charge neutrality of the sample. The time dependence of $f(t)$ and $\xi(t)$ is periodic with the period T of the domain motion. Inside the sample the field $E(x, t)$ is given by

$$E(x, t) = E_R(x - u_0 t) + \delta E(x, t) , \tag{3}$$

where $\delta E(x, t)$ describes the deviation from the Riemann solution induced by the variation of the boundary conditions. In terms of $\delta E(x, t)$ the stability or instability of the Riemann solution means decreasing or increasing of $\delta E(x, t)$ if x moves away from the contacts into the sample.

2. Drift Non-Linearity

In case of the drift non-linearity the electrical state of the sample is described besides $E(x, t)$ by the current density $j(x, t)$ and the charge density $\varrho_0 - \varrho(x, t)$, where ϱ_0 is the positive compensating charge density and $\varrho(x, t)$ is the absolute value of the negative charge density. We have the equations

$$\frac{\partial j}{\partial x} - \frac{\partial \varrho}{\partial t} = 0 , \tag{4}$$

$$\frac{\partial E}{\partial x} = \frac{4\pi}{\varepsilon} (\varrho_0 - \varrho) , \tag{5}$$

$$j = \varrho \mu E + D \frac{\partial \varrho}{\partial x} . \tag{6}$$

In (6) the mobility μ and the diffusion coefficient D are functions of the field strengths $E(x, t)$. Relation (6) holds as long as the mean electron energy is not too high [7]. From (4) to (6) for the deviation $\delta E(x, t)$ follows in linear approximation

$$\left[D \frac{\partial^2}{\partial x^2} + v \frac{\partial}{\partial x} - \frac{4\pi\tilde{\sigma}}{\varepsilon} - \frac{\partial}{\partial t}\right] \delta E(x, t) = -\frac{4\pi}{\varepsilon} \delta j_e(t) , \tag{7}$$

$$v = \mu E , \qquad \tilde{\sigma} = \sigma + \sigma_3 , \qquad \sigma_3 = -\frac{\varepsilon}{4\pi} \frac{\partial^2 E}{\partial x^2} \frac{dD}{dE} \tag{8}$$

with $E = E_R(x - u_0 t)$. Here σ is the differential conductivity and $\delta j_e(t)$ the change of the total current density

$$j_e(t) = j(x, t) + \frac{\varepsilon}{4\pi} \frac{\partial E}{\partial t} . \tag{9}$$

First we consider the current-controlled case, i.e. we put

$$\delta j_e(t) = 0 . \tag{10}$$

The voltage-controlled regime is investigated in Appendix B. In (7) the coefficients D, v and $\tilde{\sigma}$ depend on $E = E_R(x - u_0 t)$ and therefore are functions of x and t. Let l be the characteristic length for the spatial variation of D, v, $\tilde{\sigma}$ and λ the characteristic length for the spatial variation of $\delta E(x, t)$. If λ is small compared with l, i.e.

$$\left(\frac{\lambda}{l}\right) \ll 1 , \tag{11}$$

then in order to solve equation (7) we can use the WJKB-approximation. Condition (11) is not fulfilled for sufficiently steep domain walls. This case is considered in Appendix A. Further, we suppose, that the time variation of D, v, $\tilde{\sigma}$ and of the boundary conditions (1), (2) is adiabatically slow compared to the other processes determining $\delta E(x, t)$ according to equation (7). For this purpose the characteristic time τ for these processes must be small compared to the period T of the domain motion, i.e.

$$\left(\frac{\tau}{T}\right) \ll 1 . \tag{12}$$

In particular τ may be the dielectric relaxation time. Under the conditions (11) and (12) the solution of (7) is approximately given by

$$\delta E(x,t) = \delta E_1(t) \exp\left(\int_0^x \mathrm{d}x'\, k_1(x' - u_0 t)\right) + \delta E_2(t) \exp\left(\int_0^x \mathrm{d}x'\, k_2(x' - u_0 t)\right), \tag{13}$$

$$k_{1,2} = -\frac{v}{2D} \pm \sqrt{\left(\frac{v}{2D}\right)^2 + \frac{4\pi\tilde{\sigma}}{\varepsilon D}} . \tag{14}$$

The boundary conditions (1), (2) for $\delta E(x, t)$ give

$$\delta E_2(t) = -\delta E_1(t) \frac{1 - \xi(t) \exp\left(\int_0^L \mathrm{d}x'\, k_1(x')\right)}{1 - \xi(t) \exp\left(\int_0^L \mathrm{d}x'\, k_2(x')\right)} , \tag{15}$$

$$\delta E_1(t) = \frac{f(t)}{\xi(t)} \frac{1 - \xi(t) \exp\left(\int_0^L \mathrm{d}x'\, k_2(x')\right)}{\exp\left(\int_0^L \mathrm{d}x'\, k_1(x')\right) - \exp\left(\int_0^L \mathrm{d}x'\, k_2(x')\right)} . \tag{16}$$

To study the spatial behaviour of $\delta E(x, t)$ we distinguish the three cases:

(i) Re $k_1(x, t)$ and Re $k_2(x, t)$ have different signs for all x, t. Then $\delta E(x, t)$ decreases if x is moved away from the contacts into the sample (Fig. 2). The Riemann solution is stable with respect to variations of the boundary conditions.

(ii) Re $k_1(x, t)$ and Re $k_2(x, t)$ have the same sign for all x, t. In this case $\delta E(x, t)$ increases if x is moved away from the contacts (Fig. 3). The Riemann solution is instable with respect to variations of the boundary conditions.

(iii) For various x, t case (i) holds and for other x, t case (ii). General conclusions about the spatial behaviour of $\delta E(x, t)$ cannot be made. With the dependence of the particular form of the functions Re $k_{1,2}(x, t)$ the Riemann

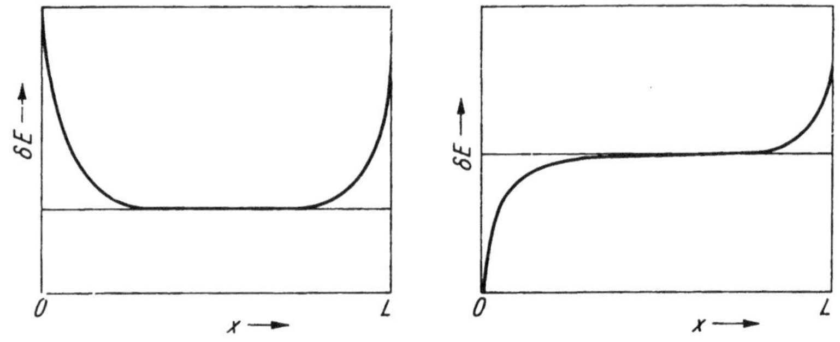

Fig. 2. Field deviation δE from the Riemann solution over x for $\tilde{\sigma} > 0$ (schematic)

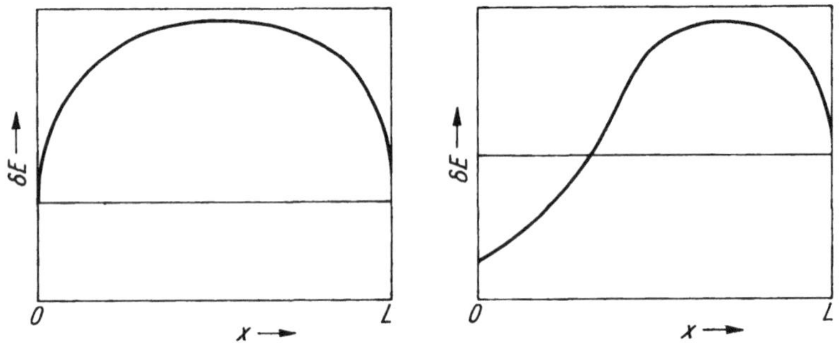

Fig. 3. Field deviation δE from the Riemann solution over x for $\tilde{\sigma} < 0$ (schematic)

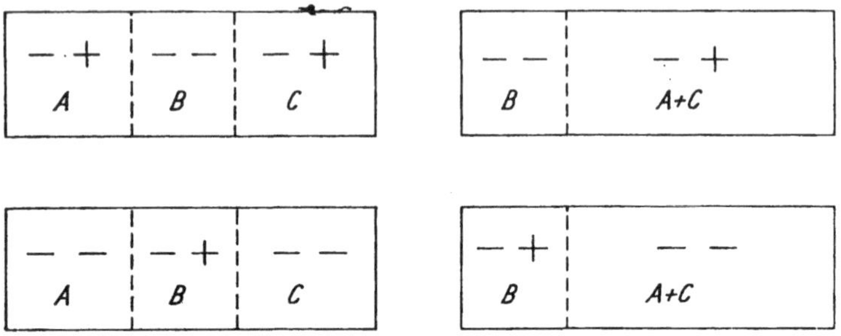

Fig. 4. Characteristic distribution of sign Re k_1 and sign Re k_2 over the sample

solution may be stable or instable. For instance, let the distribution of signs of Re $k_{1,2}(x, t)$ over the sample be as shown in the Fig. 4. The lengths l_B of section B may be not greater than the lengths l_A and l_C of A and C, respectively, and the values of $|\text{Re}\, k_{1,2}|$ may be not greater in B than those in A and C. Then the Riemann solution is stable for the distribution of Fig. 4a and instable for the distribution of Fig. 4b.

As shown in Appendices A and B, these conclusions about the stability or instability of the Riemann solution are valid in the voltage-controlled case as well as in the case of very steep domain walls.

Table 1

Characteristic quantities determining the stability of the Riemann solution in the case of the Gunn-effect in GaAs inside and outside the domain (T^* mean electron energy)

	T^* (eV)	σ (s^{-1})	σ_3 (s^{-1})	k_1 (cm^{-1})	k_2 (cm^{-1})
inside	0.2	-0.6×10^{10}	-2×10^{10}	-3×10^{3}	-1×10^{5}
outside	0.02	3×10^{11}	0	0.3×10^{5}	-1×10^{5}

It follows from equation (14), that the case (i) takes place then and only then, if for all x, t one has

$$\sigma(x, t) + \sigma_3(x, t) > 0 . \tag{17}$$

For the domain solution we can assume that $\sigma > 0$ and $\sigma_3 \approx 0$ outside the domain. Consequently, condition (17) is fulfilled there. Inside the domain σ may be negative. Then $\sigma_3 > |\sigma|$ must hold to satisfy (17). In particular it means that for $\mathrm{d}D/\mathrm{d}E > 0$ the domain must be a high field domain and for $\mathrm{d}D/\mathrm{d}E < 0$ a low field domain. However, for the stability of the Riemann solution it is not necessary to satisfy condition (17) for all x, t. If $\sigma < 0$ and $\sigma_3 < |\sigma|$ inside of the domain, case (iii), which does not exclude the stability of the Riemann solution, holds. The results of a numerical estimate for the Gunn-effect in GaAs, using the Einstein relation for D[2]) and the numerical data of the paper [8], are shown in Table 1. The domain length is one tenth of the sample length. Therefore, the conditions of Fig. 4a are fulfilled and the Riemann solution is stable.

The physical reason for the stability condition (17) becomes clear through the discussion of equation (7). According to (7) the deviation $\delta E(x, t)$ from the Riemann solution is determined by three processes: diffusion, drift, and dielectric relaxation. Because of diffusion and drift the boundary values of $\delta E(x, t)$ propagate from the contacts into the sample. On account of the dielectric relaxation the δE-values there increase or decrease in time depending upon the sign of $\tilde{\sigma}$. Analogous to this in the stationary case the δE inside the sample increases compared to its boundary value if $\tilde{\sigma} < 0$ and decreases, if $\tilde{\sigma} > 0$.

3. Recombination Non-Linearity

In case of recombination non-linearity the density ϱ_e of the free charge carriers differs from ϱ by the density ϱ_t of the trapped carriers. It holds that

$$\varrho = \varrho_e + \varrho_t \tag{18}$$

$$\frac{\partial \varrho_t}{\partial t} = c_n \left[\varrho_e \left(N - \frac{\varrho_t}{e}\right) f(E) - n_1 \varrho_t\right] . \tag{19}$$

Here n_1 is the Shockley-Reed-Hall-factor, $c_n f(E)$ is the capture coefficient in the presence of the electric field E and c_n its value for vanishing field E. For the sake of simplicity we assume that μ and D are field-independent. In addition

[2]) The Einstein relation is valid for small deviations from the thermodynamic equilibrium. In our case this condition is not fulfilled. In particular, the electron temperature, which appears in the Einstein relation, does not exist and is replaced by the mean electron energy. Therefore, the numerical results in Table 1 have, at the best, qualitative meaning.

to (19) for the electric quantities of interest we have the equations (4) to (6), where in (6) ϱ must be substituted by the density ϱ_e of the free charge carriers. With the help of these equations for the deviations $\delta E(x, t)$ and $\delta\varrho_t(x, t)$ from the Riemann solution $E_R(x - u_0 t)$, $\varrho_{tR}(x - u_0 t)$ we obtain the following system of differential equations:

$$\left[D \frac{\partial^2}{\partial x^2} + \tilde{v} \frac{\partial}{\partial x} - \frac{4\pi\sigma}{\varepsilon} - \frac{\partial}{\partial t}\right] \delta E(x, t) = -\frac{4\pi}{\varepsilon}\left[D \frac{\partial}{\partial x} + v\right] \delta\varrho_t(x, t)\,, \tag{20}$$

$$\left[\frac{\partial}{\partial t} + C\right] \delta\varrho_t(x, t) = \left[A \frac{\partial}{\partial x} + B\right] \delta E(x, t)\,, \tag{21}$$

$$\tilde{v} = v - \frac{4\pi}{\varepsilon} D \frac{d\varrho_{tR}}{dE}\,, \tag{22}$$

$$A = -c_n \frac{\varepsilon}{4\pi}\left(N - \frac{\varrho_{tR}}{e}\right) f(E_R)\,, \tag{23}$$

$$B = c_n \varrho_{eR}\left(N - \frac{\varrho_{tR}}{e}\right) f'(E_R)\,, \tag{24}$$

$$C = c_n\left[\left(N - \frac{\varrho_{tR}}{e} + \frac{\varrho_{eR}}{e}\right) f(E_R) + n_1\right] \tag{25}$$

with $f' = df/dE$. In (20) the current-controlled case is assumed. To determine the solution of (20), (21) uniquely, in addition to the boundary conditions for δE the initial value of $\delta\varrho_t(x, t)$ must be given. The choice of the initial value is restricted by the condition that $\delta\varrho_t(x, t)$ vanishes, if the boundary values of δE tend to zero. As in the drift non-linearity case we can neglect the time derivative in (20), because the process of domain motion can be considered as adiabatically slow compared to the other processes in equation (20). In order to neglect the time derivative in equation (21) it would also be necessary that the recombination time $\tau_R = 1/C$ should be small compared to the period T of the domain motion. However, in practical situations the ratio (τ_R/T) may be of the order 1. In this case the recombination process is enough to give an essential effect in the interesting time interval $t \approx T$, but not quick enough to prevent $\delta\varrho_t$ from becoming markedly dependent also on $\delta E(x, t')$ for $t' < t$. Because of this the WJKB-approximation no longer is applicable. Therefore, we consider only the limiting cases $(\tau_R/T) \gg 1$ and $(\tau_R/T) \ll 1$.

The stationary solution of (21), tending to zero for vanishing boundary values of δE, has the form

$$\delta\varrho_t(x, t) = \int\limits_{-\infty}^{t} dt' \exp\left(-\int\limits_{t'}^{t} dt''\, C(t'')\right)\left[A(t') \frac{\partial}{\partial x} + B(t')\right] \delta E(x, t')\,. \tag{26}$$

First of all we consider the case $(\tau_R/T) \ll 1$. From (26) follows approximately

$$\delta\varrho_t(x, t) = \frac{1}{C(t)}\left[A(t) \frac{\partial}{\partial x} + B(t)\right] \delta E(x, t)\,. \tag{27}$$

Substituting $\delta\varrho_t$ in (21) by (27) we obtain an equation which differs from equation (7) only by the values of the coefficients. Therefore, the conclusions concerning the stability of the Riemann solution in the drift case are valid in this case as well, if we replace σ_3 by $\sigma_4 = -(v\,B)/C$. It follows from (24) and (25)

$$\sigma_4 = -v\,\varrho_{eR}\,f'(E_R)\,\frac{\left(N - \frac{\varrho_{tR}}{e}\right)}{\left(N - \frac{\varrho_{tR}}{e} + \frac{\varrho_{eR}}{e}\right) f(E_R) + n_1}\,. \tag{28}$$

In the limiting case $(\tau_R/T) \gg 1$ equation (26) gives

$$\delta\varrho_t = \frac{1}{\int\limits_0^T dt\,C(t)} \int\limits_0^T dt' \left[A\,(t-t')\,\frac{\partial}{\partial x} + B\,(t-t')\right] \delta E\,(x, t-t')\,. \tag{29}$$

The quantity σ_4 from (28) for $(\tau_R/T) \gg 1$ must be replaced by an another function which in any case has the order of σ_4, as is shown by (29).

It may be expected that an approximate stability condition for $(\tau_R/T) \approx 1$ can be obtained by means of interpolation between both these limiting cases. That means replacing σ_4 corresponding to

$$\sigma_4 \to \varphi\left(\frac{\tau_R}{T}\right)\sigma_4$$

with $\varphi(0) = 1$ and $\varphi(\infty) \approx 1$. Regarding the results in Section 2 we conclude that under practical conditions the Riemann solution can be stable with respect to variations of the boundary conditions in the case of the recombination nonlinearity also.

Acknowledgement

The author is greatly indebted to Prof. V. L. Bonch-Bruevich for suggesting the problem and for continuous help in preparing this paper.

Appendix A

We consider the case of sufficiently steep domain walls, which is defined as follows: The interval for increase or decrease of $E_R\,(x - u_0\,t)$ to the value inside the domain is small enough that the variations of $\delta E(x, t)$ and its derivative $\frac{\partial}{\partial x}\,\delta E(x, t)$ in this interval can be neglected. This means that the domain walls can be approximated by perpendicular walls, at which $\delta E(x, t)$ and $\frac{\partial}{\partial x}\,\delta E(x, t)$ are continuous functions of x.

First we remark that the stability properties of the Riemann solution are the same as in the case $(\lambda/l) \gg 1$, if the distance between the domain and the contacts is long compared with λ. A new calculation becomes necessary only for such times, for which the domain is placed at one of the contacts. The deviations $\delta E_A(x, t)$ and $\delta E_B(x, t)$ from the Riemann solution inside and outside the domain respectively have the form (13), if the indices A or B are added:

$$\delta E_A(x, t) = E_{A1}(t)\,e^{S_{A1}(x,t)} + E_{A2}(t)\,e^{S_{A2}(x,t)}\,, \tag{A 1}$$

$$\delta E_B(x, t) = E_{B1}(t)\,e^{S_{B1}(x,t)} + E_{B2}(t)\,e^{S_{B2}(x,t)}\,, \tag{A 2}$$

$$S_{\substack{A1,2\\B1,2}}(x, t) = \int\limits_0^x dx'\,k_{\substack{A1,2\\B1,2}}(x' - u_0\,t)\,. \tag{A 3}$$

The continuity conditions for $\delta E(x, t)$ and $\frac{\partial}{\partial x}\,\delta E(x, t)$ at the domain wall $x = x_0$ give the relations

$$E_{A2} = E_{A1}\,\frac{e^{S_{A1}(x_0)}}{e^{S_{A2}(x_0)}}\,\frac{k_{A1} - k_{B1}}{k_{B1} - k_{A2}} + E_{B2}\,\frac{e^{S_{B2}(x_0)}}{e^{S_{A2}(x_0)}}\,\frac{k_{B1} - k_{B2}}{k_{B1} - k_{A2}}, \tag{A 4}$$

$$E_{B1} = E_{A2}\,\frac{e^{S_{A1}(x_0)}}{e^{S_{B1}(x_0)}}\,\frac{k_{A1} - k_{A2}}{k_{B1} - k_{A2}} + E_{B2}\,\frac{e^{S_{B2}(x_0)}}{e^{S_{B1}(x_0)}}\,\frac{k_{B2} - k_{A2}}{k_{A2} - k_{B1}}. \tag{A 5}$$

The boundary conditions have the form

$$E_{A1} + E_{A2} = f\,, \tag{A 6}$$

$$E_{B1}\,e^{S_{B1}(L)} + E_{B2}\,e^{S_{B2}(L)} = \xi\, f\,. \tag{A 7}$$

Following these equations it is easy to verify that the stability of the Riemann solution is governed by the conditions given in section 2.

Appendix B

The voltage-controlled case is defined by the relation

$$\int_0^L \delta E(x, t)\,dx = 0\,. \tag{B 1}$$

The general solution of the inhomogeneous equation (7) in the WJKB-approximation has the form

$$\delta E(x, t) = E_1(t)\exp\left(\int_0^x dx'\,k_1\,(x' - u_0\,t)\right) + E_2(t)\exp\left(\int_0^x dx'\,k_2\,(x' - u_0\,t)\right) + $$
$$+\,\delta E_{sp}(x, t)\,, \tag{B 2}$$

$$\delta E_{sp}(x, t) =$$
$$= -\frac{4\pi\,\delta j_e(t)}{\varepsilon\,D\,k_1\,k_2}\left[1 + \frac{\exp\left(\int_0^x dx'\,k_1\,(x' - u_0\,t)\right) - \exp\left(\int_0^x dx'\,k_2\,(x' - u_0\,t)\right)}{k_1 - k_2}\right]. \tag{B 3}$$

For the validity of the WJKB-approximation we have in particular

$$\left|\frac{k_{1,2}}{k_1 - k_2}\right| \leqq 1\,. \tag{B 4}$$

The coefficients $E_1(t)$, $E_2(t)$ and $\delta j_e(t)$ are determined by the relations (2), (3) and (B 1). Estimating the order of magnitude for $\delta j_e(t)$ we obtain

$$\delta j_e \approx \begin{cases} \left(\dfrac{\lambda}{L}\right)\tilde{\sigma}\, f\,, & \tilde{\sigma} > 0\,,\\ \tilde{\sigma}\, f\,, & \tilde{\sigma} < 0\,. \end{cases} \tag{B 5}$$

The conclusions on the stability of the Riemann solution, obtained for the current-controlled regime are valid also in the voltage-controlled case. This statement is true also for sufficiently steep domain walls.

References

[1] V. L. Bonch-Bruevich, Fiz. tverd. Tela **8**, 1753 (1966).
[2] B. W. Knight and G. A. Peterson, Phys. Rev. **155**, 393 (1967).
[3] V. L. Bonch-Bruevich, On the Theory of Electrical Domains in Hot Electron Semiconductors, Preprint of the Institute of Radioengineering and Electronics, Academy of Sciences of the USSR, Moscow 1966; Fiz. Tekh. Poluprov. **1**, 239 (1967).
[4] V. M. Eleonskii, Zh. eksper. teor. Fiz. **53**, 592 (1967).
[5] I. P. Zviaguine, Fiz. Tekh. Poluprov. (in the press).
[6] V. L. Bonch-Bruevich, Fiz. tverd. Tela **9**, 355 (1967).
[7] V. L. Bonch-Bruevich and Yu. V. Gulyaev, Preprint of the Institute of Radioengineering and Electronics, Academy of Sciences of the USSR, Moscow.
[8] A. G. Chynoweth, W. L. Feldmann, and D. E. McCumber, Proc. Internat. Conf. Phys. Semicond. Kyoto 1966 (p. 514).

(Received May 3, 1968)

phys. stat. sol. 28, 529 (1968)

Subject classification: 10 and 12; 22.5.2

Department of Experimental Physics, Growth of Crystals Research Group, University of Technology, Budapest

Elastic Constants and Tensile Strength of NaCl-Type Alkali Halides

By

P. O. SZOMOR

The second-order elastic constants and tensile strength of NaCl-type alkali halides are calculated on the basis of the Born-Mayer model of the lattice energy of ionic solids. The repulsive parameters which include the contribution of the second nearest neighbours in the expression of the repulsive energy are obtained from the equation of state of static solids and its volume derivative. For determining the elastic constants and the tensile strength a procedure similar to Zwicky's procedure is chosen. The comparison of the calculated elastic constants and the available experimental results show that the basic radii obtained by the Huggins-Mayer procedure are not accurate enough in the case of certain ions.

Die elastischen Konstanten zweiter Ordnung und die Zerreißfestigkeit für Alkalihalogenide vom NaCl-Typ werden mit dem Born-Mayerschen Modell der Gitterenergie des Ionenfestkörpers berechnet. Die Abstoßungsparameter im Ausdruck für die den Beitrag der zweiten nächsten Nachbarn berücksichtigende Abstoßungsenergie werden mit der Zustandsgleichung des statischen Festkörpers und mit ihrer Ableitung nach dem Volumen bestimmt. Bei der Bestimmung der elastischen Konstanten und der Zerreißfestigkeit wird ein ähnliches Verfahren wie bei Zwicky angewandt. Der Vergleich der berechneten elastischen Konstanten mit den zur Verfügung stehenden experimentellen Ergebnissen zeigt, daß die nach dem Huggins-Mayerschen Verfahren bestimmten Grundradien im Falle von einigen Ionen nicht genau genug sind.

1. Introduction

Owing to the relative simplicity of alkali halides there are several papers concerning the theoretical determination of elastic constants [1 to 6] and tensile strength [7 to 9]. The elastic constants were mainly determined for 0 °K except the results of Karo [2] and Lincoln et al. [4]. The tensile strength was theoretically determined only for NaCl.

Considering that the elastic constants of solids together with other physical properties give valuable informations on the nature of forces between the atoms, the theoretical determination of elastic constants gives us an opportunity to subject the Born-Mayer [10] model of the lattice energy of ionic crystals to a test being more severe than that of determining the cohesive energy. These considerations allowed us to determine the second-order elastic constants together with the tensile strength of 17 NaCl-type alkali halides.

The lattice energy per molecule of an alkali halide crystal W_L as function of the interionic distance may be given as sum of the Coulomb energy, van der

Waals energy, and Born repulsive energy (cf. [11], equation (6.17)):

$$W_L = -\frac{\alpha_r e^2}{r} - \frac{C_r}{r^6} - \frac{D_r}{r^8} + M\,\beta_{+-}\, b \exp\left(\frac{r_+ + r_- - r}{\varrho}\right) +$$

$$+ M'\, b \left[\beta_{++} \exp\left(\frac{2\,r_+}{\varrho}\right) + \beta_{--} \exp\left(\frac{2\,r_-}{\varrho}\right)\right] \exp\left(\frac{-r\sqrt{2}}{\varrho}\right), \qquad (1)$$

where α_r is the Madelung constant, C_r, D_r, are the van der Waals coefficients, M, M' are coordinate numbers, β_{++}, β_{--}, β_{+-} are the Pauling coefficients, r_+, r_- are the basic radii, b and ϱ are adjustable parameters.

The method of determining the elastic constants and tensile strength, that agrees in part with Zwicky's method of determining the tensile strength [7], is as follows: The b- and ϱ-parameters are fixed by suitable equations. Then a crystal is placed in a coordinate system in such a manner that a chosen ion is placed in the origin and the chains of ions are parallel, or, if the chains contain the chosen ion, coincident with the coordinate axes. As all the crystals tested are of NaCl-type, this can easily be done. Let us suppose that the crystal is deformed by a stretching force acting along the X-axis while there are no forces in the directions of the Y- and Z-axes. Under these conditions the crystal is dilated in the direction of the X-axis while there is a diagonal contraction in the directions of Y- and Z-axes. The diagonal contraction results in turning the originally r-sided square into a y-sided square in the plane perpendicular to the X-axis providing that the diagonal contraction in the direction of the Z-axis is the same as that in the direction of the Y-axis.

The elastic strain energy density of an undeformed cubic lattice is

$$U = \frac{1}{2\,r^3} W_L\,. \qquad (2)$$

It is known that the differential quotient of the elastic strain energy density with respect to the deformation gives the corresponding value of the stress. Assume that the distance between two ions is increased by dx due to a stretching force along the X-axis; then the unit length edge of a cube is given by

$$ds = \frac{1}{r}\, dx\,.$$

Thus, the corresponding value of stress amounts to

$$\sigma = \frac{1}{2\,r^2} \frac{dW_L}{dx}\,. \qquad (3)$$

From the deformation and corresponding stress values obeying Hook's law the characteristic elastic constants (Young's modulus, Poisson's ratio) were determined. The value of the tensile strength was determined from the maximum of the function (3).

2. Determination of the Repulsive Parameters

There are two thermodynamic conditions which must be applied to the lattice energy of ionic solids which is the inner energy of solids. These conditions are the equation of state applied to the static cubic crystals and its volume derivative ([11], equations (1.2) and (1.3)):

$$\frac{\mathrm{d}W_{\mathrm{L}}}{\mathrm{d}V} = -P\,, \tag{4}$$

$$V\frac{\mathrm{d}^2 W_{\mathrm{L}}}{\mathrm{d}V^2} = \frac{1}{K}\,. \tag{5}$$

Here $V = k\,r^3$ is the molecular volume (the value of k is 2 in the case of NaCl-type crystals), K is the isothermal compressibility, and P is the hydrostatic pressure.

Inserting equation (1), the expressions (4) and (5) may be written in the following form providing the pressure to be negligible:

$$M\,\beta_{+-}\,b\,\frac{r}{\varrho}\exp\left(\frac{r_+ + r_- - r}{\varrho}\right) + \frac{1}{2}\,M'\,b\,\frac{r\sqrt{2}}{\varrho}\times$$
$$\times\left[\beta_{++}\exp\left(\frac{2\,r_+}{\varrho}\right) + \beta_{--}\exp\left(\frac{2\,r_-}{\varrho}\right)\right]\exp\left(\frac{-r\sqrt{2}}{\varrho}\right) = \tau\,, \tag{6}$$

$$M\,\beta_{+-}\,b\,\frac{r^2}{\varrho^2}\exp\left(\frac{r_+ + r_- - r}{\varrho}\right) + M'\,b\,\frac{r^2}{\varrho^2}\times$$
$$\times\left[\beta_{++}\exp\left(\frac{2\,r_+}{\varrho}\right) + \beta_{--}\exp\left(\frac{2\,r_-}{\varrho}\right)\right]\exp\left(\frac{-r\sqrt{2}}{\varrho}\right) = \psi\,, \tag{7}$$

where

$$\tau = \frac{\alpha_r\,e^2}{r} + 6\,\frac{C_r}{r^6} + 8\,\frac{D_r}{r^8}\,, \tag{8}$$

$$\psi = \frac{18\,r^3}{K} + 2\,\frac{\alpha_r\,e^2}{r} + 42\,\frac{C_r}{r^6} + 72\,\frac{D_r}{r^8}\,. \tag{9}$$

The solution of the system of equations made up of the independent equations (6) and (7) gives the fixed repulsive parameters.

3. Elastic Strain Energy Density of the Crystal

The elastic strain energy density of a crystal placed in a coordinate system as described earlier is defined by equation (2), where

$$W_{\mathrm{L}} = W_{\mathrm{L}}(x, y)$$

is the lattice energy of the crystal taking into consideration that $y = z$ according to the assumption made. It has to be given exactly now taking into account equation (1). First of all, it is evident that, when the crystal is deformed, the quantities α_r, C_r, and D_r occurring in equation (1) are not constants but depend on the degree of deformation of the crystal. Therefore, let us examine the first term of equation (1), the Coulomb energy term, first of all. Following the investigations of Fáthy and Bukovszky [12] we have for the Coulomb energy

$$-\frac{\alpha_r\,e^2}{r} = -\frac{e^2}{x}\sum_a\sum_{n=-\infty}^{\infty}(-1)^{a+n-1}\frac{1}{\sqrt{\left(\frac{y}{x}\right)^2 a + n^2}} \tag{10}$$

(the case $a = n = 0$ must be excluded in the sum), where

$$a = l^2 + m^2; \quad l, m \text{ integer}, \quad -\infty < l, m < \infty .$$

When converting the van der Waals energy term the definition of C_r and D_r must be taken into consideration (cf. [11], equations (4.6) and (4.7)):

$$C_r = c_{+-} S_6' + \frac{1}{2} (c_{++} + c_{--}) S_6'' ,$$

$$D_r = d_{+-} S_8' + \frac{1}{2} (d_{++} + d_{--}) S_8'' ,$$

where S_6' and S_8' are the sixth and eighth order odd lattice sums, respectively, counted from the point of reference; similarly S_6'' and S_8'' are even lattice sums, c_{++}, c_{--}, c_{+-}, d_{++}, d_{--}, and d_{+-} are the van der Waals coefficients as determined by Mayer [13]. For the van der Waals energies there were obtained, similarly to the above, the following expressions:

$$-\frac{C_r}{r^6} = -\frac{1}{x^6} \sum_a \sum_{n=-\infty}^{\infty} \frac{1}{\left[\left(\frac{y}{x}\right)^2 a + n^2\right]^3} \times \begin{cases} \frac{1}{2}(c_{++} + c_{--}), & \text{if} \quad a + n \text{ is even}, \\ c_{+-}, & \text{if} \quad a + n \text{ is odd}, \end{cases} \tag{11}$$

$$-\frac{D_r}{r^8} = -\frac{1}{x^8} \sum_a \sum_{n=-\infty}^{\infty} \frac{1}{\left[\left(\frac{y}{x}\right)^2 a + n^2\right]^4} \times \begin{cases} \frac{1}{2}(d_{++} + d_{--}), & \text{if} \quad a + n \text{ is even}, \\ d_{+-}, & \text{if} \quad a + n \text{ is odd}. \end{cases} \tag{12}$$

(The case $a = n = 0$ must be excluded in the sums of both equations.)

When considering the Born repulsive energy term, first of all the meaning of the M and M' coordinate numbers should be examined. M designates the number of nearest ions with opposite sign around the reference ion. For crystals of NaCl-type it is $M = 6$. It follows from the supposed positioning of the crystal lattice in the coordinate system that two of these ions are on the X-axis, two on the Y-axis, and two on the Z-axis. M' designates the number of nearest ions with the same sign around the reference ion. For crystals of NaCl-type it is $M' = 12$. It is easy to see that four of these are in the (y, z) plane and the expression describing this case is independent of x. Considering that in the repulsive energy term, we have

$$2 b \beta_{+-} \left[\exp\left(-\frac{x}{\varrho}\right) + 2 \exp\left(-\frac{y}{\varrho}\right)\right] \exp\left(\frac{r_+ + r_-}{\varrho}\right) +$$
$$+ 2 b \left[\beta_{++} \exp\left(\frac{2 r_+}{\varrho}\right) + \beta_{--} \exp\left(\frac{2 r_-}{\varrho}\right)\right] \times$$
$$\times \left[\exp\left(-\frac{y\sqrt{2}}{\varrho}\right) + 2 \exp\left(-\frac{x}{\varrho}\sqrt{\left(\frac{y}{x}\right)^2 + 1}\right)\right]. \tag{13}$$

Making use of equations (10), (11), (12), and (13) we obtain

$$W_L(x, y) = -\frac{e^2}{x} \sum_a \sum_{n=-\infty}^{\infty} (-1)^{a+n-1} \frac{1}{\sqrt{\left(\frac{y}{x}\right)^2 a + n^2}} -$$

$$-\frac{1}{x^6} \sum_a \sum_{n=-\infty}^{\infty} \frac{1}{\left[\left(\frac{y}{x}\right)^2 a + n^2\right]^3} \times \begin{Bmatrix} \frac{1}{2}(c_{++} + c_{--}), & \text{if } a+n \text{ is even} \\ c_{+-}, & \text{if } a+n \text{ is odd} \end{Bmatrix} -$$

$$-\frac{1}{x^8} \sum_a \sum_{n=-\infty}^{\infty} \frac{1}{\left[\left(\frac{y}{x}\right)^2 a + n^2\right]^4} \times \begin{Bmatrix} \frac{1}{2}(d_{++} + d_{--}), & \text{if } a+n \text{ is even} \\ d_{+-}, & \text{if } a+n \text{ is odd} \end{Bmatrix} +$$

$$+ 2b\beta_{+-}\left[\exp\left(-\frac{x}{\varrho}\right) + 2\exp\left(-\frac{y}{\varrho}\right)\right]\exp\left(\frac{r_+ + r_-}{\varrho}\right) +$$

$$+ 2b\left[\beta_{++}\exp\left(\frac{2r_+}{\varrho}\right) + \beta_{--}\exp\left(\frac{2r_-}{\varrho}\right)\right] \times$$

$$\times\left[\exp\left(-\frac{y\sqrt{2}}{\varrho}\right) + 2\exp\left(-\frac{x}{\varrho}\sqrt{\left(\frac{y}{x}\right)^2 + 1}\right)\right]. \tag{14}$$

With that and with use of equation (2) the elastic strain energy density of the crystal is easily obtained.

4. The Determination of Elastic Constants

As the function $W_L(x, y)$ is known, it must be taken into consideration when equation (3) is solved that according to our conditions there are no external forces acting in the directions of the Y- or Z-axes. This leads to the following condition

$$\frac{\partial W_L}{\partial y} = 0 . \tag{15}$$

It follows from this condition that the lattice energy is a function of x only, that is

$$W_L = W_L(x, y(x)) .$$

Because of equation (15) it is

$$\frac{dW_L}{dx} = \frac{\partial W_L}{\partial x} + \frac{\partial W_L}{\partial y}\frac{dy}{dx} = \frac{\partial W_L}{\partial x} .$$

Now with the knowledge of equation (14) we can determine $\partial W_L/\partial x$ and $\partial W_L/\partial y$ and solve then equations (15) and (3). From equation (15) one can calculate the corresponding values of y for any $x > r$. With the corresponding values of x and y thus calculated the value of the stress belonging to x may be obtained from equation (3). If the value of x was chosen in such a way that Hooke's law is valid Young's modulus and Poisson's ratio are easily determined taking into consideration their definition known from the theory of elasticity:

$$E = \sigma \frac{r}{x - r} \tag{16}$$

and

$$\mu = -\frac{y - r}{x - r}. \tag{17}$$

It is also known from the theory of elasticity [14] that Young's modulus, Poisson's ratio, and the second-order elastic constants are related in the following manner:

$$C_{12} = \frac{E\,\mu}{(1 + \mu)(1 - 2\,\mu)} \tag{18}$$

and

$$C_{11} = \frac{E\,(1 - \mu)}{(1 + \mu)(1 - 2\,\mu)}. \tag{19}$$

There are three elastic constants for an anisotropic cubic crystal. The determination of the third elastic constant C_{44}, however, may not be carried out by the method used so far.

5. Results and Discussion

The solution of the system of equations (6) and (7) and the solutions of equations (15) and (3) together with the determination of the extreme values of the function given in equation (3) were carried out on the computer Ural-2. The room temperature data necessary for the solution are taken from the paper of Tosi [11] (Table VII). The basic radii r_+ and r_- were taken from the paper of Tosi and Fumi [15] using the Huggins-Mayer treatment and checking it by calculating the cohesive energy. The values of the coefficients β_{++}, β_{--}, β_{+-} were determined by Pauling [16].

It follows from the fact that we had used the equation of state for static crystals and its volume derivation to determine the repulsive parameters, that the elastic constants thus obtained will be the isothermal elastic constants of the crystals taken at room temperature. As the experimentally determined elastic constants given in various papers are the adiabatic elastic ones, the isothermal values obtained were converted into adiabatic ones with the known thermodynamic correction [17]

$$C_{11}^{\mathrm{ad}} - C_{11}^{\mathrm{is}} = C_{12}^{\mathrm{ad}} - C_{12}^{\mathrm{is}} = \frac{V\,T\,\beta^2}{C_v\,K_{\mathrm{is}}^2}, \tag{20}$$

where V stands for the crystal molecular volume, β is the coefficient of the volume thermal expansion, K is the isothermal compressibility, and C_v the thermal capacity for constant volume. The values of C_v and β being necessary for the conversion are also taken from the paper of Tosi [11] (Table VII). The results obtained are listed in Tables 1 and 2. Table 1 contains the isothermal elastic constants of 17 NaCl-type alkali halides at room temperature in addition to Young's modulus E, Poisson's ratio μ, and the second-order elastic constants C_{11} and C_{12}. Table 2 contains the values of the position of cleavage and tensile strength. Table 3 gives the adiabatic elastic constants at room temperature that were obtained from the isothermal values by the thermodynamic correction (equation (20)). It also contains the experimental data of Haussühl [18, 19] and the theoretical data of Karo [2] valid for room temperature as well as those of Lincoln, Koliwad, and Ghate [4].

Table 1

Isothermal elastic constants of NaCl-type alkali halides at room temperature

	E (10^{11} dyn cm^{-2})	μ	C_{11} (10^{11} dyn cm^{-2})	C_{12} (10^{11} dyn cm^{-2})
LiF	4.00	0.398	8.48	5.62
LiCl	2.42	0.360	4.09	2.31
LiBr	1.96	0.359	3.28	1.83
LiI	1.30	0.368	2.29	1.33
NaF	6.81	0.2542	8.24	2.808
NaCl	4.21	0.2045	4.71	1.212
NaBr	3.44	0.2009	3.94	0.992
NaI	2.60	0.2091	2.92	0.773
KF	5.61	0.1908	6.17	1.455
KCl	4.06	0.1096	4.17	0.514
KBr	3.50	0.1033	3.59	0.413
KI	2.78	0.1000	2.85	0.317
RbF	5.06	0.1755	5.47	1.166
RbCl	3.94	0.0767	3.99	0.332
RbBr	3.31	0.0724	3.35	0.261
RbI	2.74	0.0634	2.76	0.187
CsF	5.08	0.1368	5.31	0.843

Table 2

Theoretical tensile strength of NaCl-type alkali halides

	X_t (10^{-8} cm)	Y_t (10^{-8} cm)	σ_t (10^{10} dyn cm^{-2})
LiF	2.719	1.900	4.733
LiCl	3.420	2.432	2.711
LiBr	3.661	2.607	2.267
LiI	4.150	2.836	1.824
NaF	2.827	2.236	5.004
NaCl	3.420	2.742	2.949
NaBr	3.609	2.911	2.478
NaI	3.942	3.149	1.894
KF	3.279	2.600	4.118
KCl	3.762	3.096	2.664
KBr	3.948	3.247	2.281
KI	4.233	3.481	1.808
RbF	3.475	2.742	3.829
RbCl	3.941	3.250	2.555
RbBr	4.145	3.403	2.152
RbI	4.381	3.633	1.755
CsF	3.674	2.945	3.647

Table 3

Adiabatic elastic constants of NaCl-type alkali halides at room temperature, in units of 10^{11} dyn cm^{-2}

	C_{11}				C_{12}			
	Ours (theor.)	Haussühl (exp.)	Karo (theor.)	Lincoln (theor.)	Ours (theor.)	Haussühl (exp.)	Karo (theor.)	Lincoln (theor.)
LiF	8.76	11.355	9.60	—	5.90	4.76	4.94	—
LiCl	4.27	4.943	5.08	—	2.49	2.28	1.86	—
LiBr	3.59	3.937	4.13	—	2.01	1.87	1.42	—
LiI	2.46	2.85	3.01	—	1.50	1.40	0.99	—
NaF	8.44	9.70	8.57	—	3.012	2.43	2.82	—
NaCl	4.83	4.936	4.48	4.05	1.328	1.29	1.28	1.25
NaBr	4.05	4.012	3.87	3.06	1.102	1.09	1.02	1.00
NaI	3.03	3.025	2.77	—	0.875	0.88	0.74	—
KF	6.31	6.56	5.90	2.45	1.590	1.46	1.59	1.98
KCl	4.24	4.078	3.67	2.62	0.584	0.69	0.83	0.75
KBr	3.66	3.476	3.11	2.56	0.482	0.57	0.68	0.58
KI	2.92	2.76	2.47	—	0.385	0.45	0.52	—
RbF	5.57	5.525	4.75	—	1.269	1.395	1.28	—
RbCl	4.05	3.634	3.10	—	0.393	0.615	0.70	—
RbBr	3.40	3.157	2.61	2.30	0.315	0.495	0.58	0.47
RbI	2.82	2.583	2.24	—	0.243	0.37	0.45	—
CsF	5.41	—	5.08	—	0.944	—	0.99	—

The comparison shows that in the case of NaBr and NaI there is good agreement while in the case of the other alkali-halide crystals the agreement is less satisfactory. It may be observed in any case that the deviation from the experimental data is systematic for each group of crystals. For instance, for fluorides it is C_{11} (theor.) $<$ C_{11} (exp.) in the case of the lithium salts, or C_{11} (theor.) $>$ $>$ C_{11} (exp.) in the case of the potassium or rubidium salts. There are similar systematic deviations in the data calculated by Karo and determined experimentally but in the reverse direction, i.e. C_{11} (theor.) $>$ C_{11} (exp.) in the case of lithium salts, while C_{11} (theor.) $<$ C_{11} (exp.) in the case of the other salts. He found also that the behaviour of the fluorides differs from that of the other salts.

According to Karo the deviation of the calculated values from the experimental ones is due to the uncertainty in the compressibility value. In our opinion the systematic deviation from the experimental data may be interpreted as follows: Tosi and Fumi [15] when determining the basic radii r_+ and r_- checked the results by calculating the cohesive energy of the crystal. They found that there is good agreement with the experimental data obtained from the Born-Haber cycle. The determination of the elastic constants, on the other hand, subjects the expression of lattice energy (1) and the constants in it to a more severe test. The data obtained for the elastic constants show that the structure of the expression for the lattice energy is suitable, too, for the description of elastic characteristics of crystals. The systematic deviations from the experimental data are probably due to the fact that the basic radii of certain ions are not accurate enough. It may be expected that a more accurate determination of the basic radii will result in a good agreement between the experimental and theoretical values.

The comparison between the theoretical and experimental values of the tensile strength may only be done on NaCl. Zwicky's [7] value for the tensile strength of NaCl is 2×10^{10} dyn cm^{-2}, Tyson's one [9] is 2.66×10^{10} dyn cm^{-2}. The agreement with our data (Table 2) is quite good considering that both Zwicky and Tyson applied an expression for the repulsive energy term that is different from that used by us. The value for the tensile strength of NaCl obtained experimentally [20] is 2.13×10^{7} dyn cm^{-2} though it is worth mentioning that according to Gyulai's measurements [21] the tensile strength of certain NaCl needles may be as high as 10^{10} dyn cm^{-2}. This is of the same order as the theoretical value.

Acknowledgements

We would like to express our gratitude to the Computation Centre of the Hungarian Academy of Sciences for their help in programming and for placing the computer Ural-2 at our disposal.

We would like to thank Prof. Dr. Z. Gyulai for his constant interest in this paper and Mr. E. Hartmann, Research Fellow at the University of Technology, Budapest, for his many useful advices.

References

[1] K. S. Krishnan and S. K. Roy, Proc. Roy. Soc. **210**, 481 (1952).
[2] A. M. Karo, J. chem. Phys. **31**, 1489 (1959); **33**, 7 (1960).
[3] V. V. Mitskevich, Fiz. tverd. Tela **3**, 3022 (1961).
[4] R. C. Lincoln, K. M. Koliwad, and P. B. Ghate, phys. stat. sol. **18**, 265 (1966).
[5] P. O. Löwdin, Adv. Phys. **5**, 1 (1956).
[6] B. G. Dick, Phys. Rev. **129**, 1583 (1963).
[7] F. Zwicky, Phys. Z. **24**, 131 (1923).
[8] J. H. de Boer, Trans. Faraday Soc. **32**, 10 (1936).
[9] W. R. Tyson, Phil. Mag. **14**, 925 (1966).
[10] M. Born and J. E. Mayer, Z. Phys. **75**, 1 (1932).
[11] M. P. Tosi, Solid State Phys. **16**, 1 (1964).
[12] F. Fáthy and F. Bukovszky, Acta phys. Hungar. **8**, 89 (1957); **9**, 275 (1959).
[13] J. E. Mayer, J. chem. Phys. **1**, 270 (1933).
[14] L. D. Landau and E. M. Lifshits, Theory of Elasticity, Pergamon Press, London 1959.
[15] M. T. Tosi and F. G. Fumi, J. Phys. Chem. Solids **25**, 45 (1964).
[16] L. Pauling, Z. Krist. **67**, 377 (1928).
[17] G. Leibfried and W. Ludwig, Solid State Phys. **12**, 275 (1961).
[18] S. Haussühl, Z. Krist. **110**, 67 (1958).
[19] S. Haussühl, Z. Phys. **159**, 223 (1960).
[20] W. Burgmüller, Z. Phys. **83**, 317 (1933).
[21] Z. Gyulai, Z. Phys. **138**, 317 (1954).

(Received May 9, 1968)

phys. stat. sol. 28, 539 (1968)

Subject classification: 5; 21

Department of Physics, University of Helsinki

Absolute Atomic Scattering Factors of Tungsten

By

T. PAAKKARI and P. SUORTTI

The atomic scattering factors f of tungsten are determined in the region $\sin\theta/\lambda \leqq 1.1$ Å^{-1} from the integrated intensities of Bragg reflections. The reflections from a powder sample were measured at 300 °K with monochromatic MoKα radiation. The intensities are corrected for the effects of porosity and surface roughness, and for the contribution of thermal diffuse scattering. The intensities are put on an absolute scale with respect to the absolute value $f(110)$ of iron. The mass absorption coefficient of tungsten for MoKα radiation is measured to be $\mu_\varrho = (101.8 \pm 0.4)$ cm^2/g. A value for the temperature parameter of $B_W = (0.183 \pm 0.008)$ Å^2 is obtained by a measurement of the integrated intensity of a high-order reflection both at 300 and 78 °K. This corresponds to the Debye temperature $\Theta_M = (325 \mp 7)$ °K. The experimental f-values are compared with various theoretical calculations. It is found out that the Dirac-Slater values are in excellent agreement with present results.

Die Atomformfaktoren f des Wolframs im Intervall $\sin\theta/\lambda \leqq 1{,}1$ Å^{-1} wurden durch die Messung der Intensitäten der Bragg-Reflexe bestimmt. Die Reflexe an einem Pulverpräparat wurden bei 300 °K mit monochromatischer MoKα-Strahlung ausgemessen. Die Intensitäten wurden sowohl auf die Effekte der Porosität und der Rauhigkeit der Präparatoberfläche als auch auf die Beiträge der thermisch diffusen Streustrahlung korrigiert. Das absolute Niveau der Intensitäten wird im Verhältnis zum Absolutwert $f(110)$ des Eisens bestimmt. Für den Massenabsorptionskoeffizient wird der Wert $\mu_\varrho = (101{,}8 \pm 0{,}4)$ cm^2/g gemessen. Durch die Messung der Intensität eines Reflexes hoher Ordnung bei 300 und 78 °K wird für den Temperaturfaktor der Wert $B_W = (0{,}183 \pm 0{,}008)$ Å^2 erhalten; dieser entpsricht einer Debye-Temperatur von $\Theta_M = (325 \mp 7)$ °K. Die experimentellen f-Werte werden mit verschiedenen theoretischen Berechnungen verglichen. Es wird festgestellt, daß die Dirac-Slater-Werte sehr gut mit den Ergebnissen der vorliegenden Arbeit übereinstimmen.

1. Introduction

Crystallographers are nowadays overwhelmed with calculated atomic scattering factors of free atoms and one might well ask which of these one should use. For the lighter elements the values computed from Hartree-Fock (HF) wave functions and in the case of the lightest elements from even more accurate wave functions are found to be consistent to one per cent with experimental values [1]. Except recent HF calculations [2] there are today four essentially different atomic models for heavier atoms to calculate scattering factors: the Thomas-Fermi-Dirac (TFD), which is the statistical model including exchange [3], the non-relativistic Hartree (H) without exchange [4], the Hartree-Fock-Slater (HFS), similar to Hartree but including Slater's approximate exchange correction [5], and the Dirac-Slater (DS), the relativistic equivalent of the HFS model [6]. It has been suggested [7] that the Dirac-Slater scattering factors are to be preferred for heavier elements, because relativistic effects are accounted for and at least an approximation for exchange has been made.

Purpose of the experimental work presented in this paper is to study the validity of different atomic models by means of the X-ray diffraction method. Because the effect of crystalline environment is limited to the outer parts of the electron distribution of an atom in solid state, an X-ray determination of scattering factors is able to bring this kind of information if the measurements are extended to high values of $\sin \theta/\lambda$. However, these measurements have little confidence if the following subjects are not carefully considered: 1. The sample must fulfill the conditions of the diffraction theory. 2. The background scattering inclusing thermal diffuse scattering (TDS) has to be subtracted reliably. 3. The thermal motion of atoms must be well known. 4. An absolute measurement including measurement of the absorption coefficient should be made.

To approach the first requirement we have chosen the X-ray powder method. Tungsten was selected as the object of this measurement, because it is heavy enough ($Z = 74$), and suitable powder is available. Further the crystal structure of tungsten (b.c.c., $a = 3.1647$ Å at 300 °K) is favourable due to the absence of overlapping reflections, thermal motion of atoms is small and well known, and the correction for anomalous dispersion is relatively small for MoKα radiation.

2. Experimental

The tungsten powder used in the measurements was manufactured by Sylvania Electric Products Inc. through hydrogen reduction. The average particle size was 2.4 μm obtained by Fisher sub-sieve sizer, and the purity 99.95%. The samples were pressed against a polished steel plate. All attempts to remove this original surface were followed by a reduction of 10 to 20% in the absolute intensity of 110 reflection. To reveal effects of preferred orientation, the ratio of integrated intensities, $I(200)/I(110)$, was measured against moulding pressure, which varied from 1500 to 5000 kp/cm^2. When the moulding pressure was 5000 kp/cm^2 this ratio increased by 4% compared with those obtained from the samples pressed with 1500 to 3000 kp/cm^2. For these reasons the final measurements were made on a sample consisting of an undisturbed surface of a block pressed with 2500 kp/cm^2.

When absorption of X-rays is large there will exist considerable reduction in the diffracted intensity due to porosity and surface roughness of the sample. This is likely although the powder is very fine-grained and the surface seems to be perfectly smooth. There have been put forth several theoretical calculations (see [8]) on these corrections, but because of the complexity of this problem the results have been only qualitative. It is obvious that an actual powder sample cannot be characterized so completely that any theoretical model could be applied reliably. Accordingly, it is necessary to develop some experimental techniques to obtain these inevitable corrections to the measured Bragg intensities.

The main problem is to separate the effects of particle size from the effects of preferred orientation. This is not possible by the commonly used comparison between the Bragg intensities, measured as a function of packing density, due to the overlapping of these two effects. A measurement of the fluorescent intensity from a sample solves this problem, because fluorescent intensity is not affected by the preferred orientation and a comparison with a very well polished bulk sample can be made. If the absorption coefficient μ^* for the fluorescent radiation is close to that, μ, of the incident radiation, the fluorescent intensity from a powder sample relative to that from the bulk sample, measured at the same angle as the Bragg reflection, gives immediately the granularity correction. This method has been suggested by de Wolff [9]. DeMarco and Weiss [10] have considered this method in the case that μ^* differs from μ by means of a simplified model. The relation they have obtained between Bragg scattering and fluorescence is evidently insensitive to the choice of granularity model. Therefore, it can be reliably used if the correction is small and μ^* does not considerably differ from μ. In this case the integrated intensity of a Bragg reflection is reduced by a factor R given by

$$R = 1 - \frac{(\mu + \mu^*)\,(1 - R^*)}{2\,\mu^*}, \tag{1}$$

where R^* is the ratio of the fluorescent intensity from a powder sample to that from a polished bulk sample.

For tungsten μ is about 2000 cm^{-1} for MoKα and μ^* about 2500 cm^{-1} averaged over the L-emission spectrum of tungsten. The situation is thus very favourable for a reliable determination of the granularity correction. Fluorescence was exited by monochromatized MoKα radiation and detected by a scintillation counter. Bragg scattering was eliminated by pulse height discrimination and by selecting the points of measurement far away from the Bragg reflections. The results of this measurement are given in Fig. 1. The S-shape reported by de Wolff [9] is observable in the curve, which has been extrapolated to $R^* = 1$ at $2\,\theta = 180°$ according to the boundary condition. The over-all reduction of 3 to 4% arises apparently from porosity and the steep descent at small angles from surface roughness absorption.

The integrated intensities of Bragg reflections were measured by the same apparatus as the fluorescent intensity except the scattered radiation was monochromatized. The polarization factor of the monochromator was experimentally found to be consistent with $\cos^2 2\theta_M$, where θ_M is the Bragg angle of the monochromator crystal. Counting rate was kept low enough to make dead-time corrections negligible. The background scattering was scanned thoroughly over the whole measuring range with intervals of 0.05° in 2θ. Since the background intensity was mainly due to the dark current of the counter, it could be easily subtracted by using points far away from the reflections and assuming linearity of the background between these points. The resulting relative integrated intensities, I_{meas}, which have been corrected for granularity effects after Fig. 1 and equation (1), are given in Table 1. The intensity distribution of TDS was

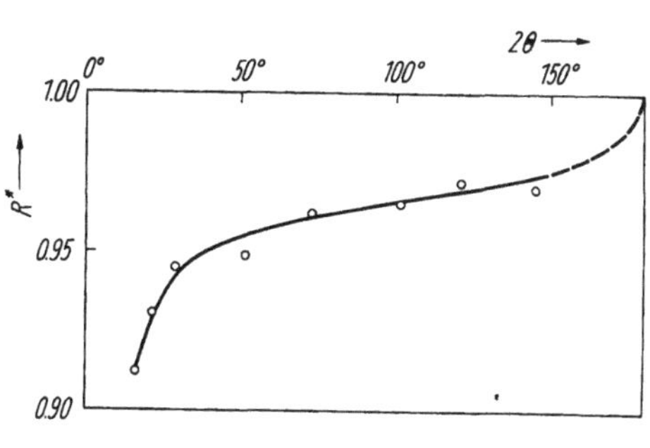

Fig. 1. The ratio R^* of the fluorescent intensity of a sample pressed with 2500 kp/cm² to that of a polished bulk sample as a function of 2θ

Table 1

Experimental integrated intensities I_{meas} of tungsten at 300 °K. The values I_{Bragg} have been corrected for TDS. Last two columns give experimental absolute scattering factors f_0 for a tungsten atom and their estimated errors, δf_0, due to the intensity measurements on a relative scale

$h\,k\,l$	$\sin\theta/\lambda$	I_{meas}	I_{Bragg}	f_0	δf_0
110	0.2234	573.18	572.03	61.15	0.06
200	0.3159	105.09	104.69	53.79	0.14
211	0.3869	241.14	239.86	51.22	0.05
220	0.4468	70.85	70.35	46.53	0.14
310	0.4995	95.22	94.20	43.70	0.11
222	0.5472	22.99	22.74	41.78	0.22
321	0.5911	104.10	101.78	39.93	0.11
330, 411	0.6702	45.68	44.80	36.33	0.12
420	0.7065	23.97	23.68	34.82	0.20
332	0.7409	19.84	19.38	33.68	0.24
422	0.7739	16.87	16.15	32.69	0.23
431, 510	0.8055	41.57	39.92	31.39	0.12
521	0.8652	19.81	19.40	29.53	0.18
433, 530	0.9211	15.24	14.84	27.92	0.22
532, 611	0.9738	18.80	18.10	26.73	0.19
550, 543, 710	1.1170	14.60	13.58	23.15	0.25
552, 633, 721	1.1608	16.38	14.95	22.56	0.21

calculated point by point according to Suortti [11] beneath the measured reflections, after which its contribution was subtracted. This method to correct for TDS is as reliable as the one-phonon approximation, which is supposed to be rather good because of the small temperature factor of tungsten. The resulting values, I_{Bragg}, are given in the fourth column of Table 1. The atomic scattering factor, f_0, is related to the integrated Bragg intensity by the well-known formula

$$I_{Bragg} = \frac{I_{meas}}{1+\sigma} = C\,\frac{p\,(f_0 + \Delta f' + i\,\Delta f'')^2\,e^{-2M}}{K(\theta,\theta_M)}\,, \tag{2}$$

where C is the scale factor, p the multiplicity, $\Delta f'$ and $\Delta f''$ the dispersion terms, $\exp(-2M) = \exp[-2B(\sin\theta/\lambda)^2]$ the Debye-Waller factor, σ the correction for TDS, and $K(\theta, \theta_M)$ the Lorentz-polarization factor for monochromatized radiation.

The temperature factor B at room temperature was determined by measuring the integrated intensity of a high-angle reflection at two temperatures, T_1 (liquid nitrogen) and T_2 (room temperature). Because changes in f_0 and $K(\theta, \theta_M)$ are negligible, one obtains

$$B_2 - B_1 = \frac{1}{2\left(\frac{\sin\theta}{\lambda}\right)^2}\ln\frac{I_{1,\,Bragg}}{I_{2,\,Bragg}}\,. \tag{3}$$

According to several measurements [12, 13], the Debye characteristic temperature Θ_M has practically the same value at liquid nitrogen temperature and at room temperature. Thus one obtains

$$\frac{B_2}{B_1} = \frac{T_2}{T_1} \frac{\frac{\Theta_M}{4 T_2} + \varphi\left(\frac{\Theta_M}{T_2}\right)}{\frac{\Theta_M}{4 T_1} + \varphi\left(\frac{\Theta_M}{T_1}\right)}, \quad (4)$$

where $\varphi(\Theta_M/T)$ is the Debye function. The right-hand side of (4) varies very slowly with Θ_M, and therefore only a crude approximation for Θ_M is needed. By substituting B_1 and the measured values to (3) it is found

$$B_2 = 0.183 \pm 0.008 \text{ Å}^2 ,$$

which corresponds to $\Theta_M = 325 \mp 7$ °K. This is in excellent agreement with the value $\Theta_M = 321$ °K derived from the value $\Theta_D = 312$ °K, which has been measured by means of inelastic scattering of neutrons [13].

The scale factor was determined relative to the recent absolute measurement of the 110 reflection of iron [14]. Because of the very high absorption coefficient of tungsten for MoKα radiation, extremely thin foils are needed for the measurement of absorption coefficient. We prepared half a dozen of foils having thicknesses from 8 to 17 μm utilizing an electrolytic film stripping method [15]. The masses per unit area of foils were determined by weighing rectangular sheets of known area. Changes of the thickness were compensated by measuring the attenuation in about 40 points, which were evenly distributed over the foil area. The resulting value of the mass absorption coefficient of tungsten for MoKα radiation was

$$\mu_\varrho = 101.8 \pm 0.4 \text{ cm}^2/\text{g} .$$

The experimental values of the absolute atomic scattering factors of tungsten atom at rest are given in Table 1. These values have been corrected for anomalous dispersion using the corrections given in International Tables for X-Ray Crystallography [16], which are considered to be the most reliable ones.

3. Discussion

Figs. 2a and 2b give the comparison between experimental atomic scattering factors and various theoretical ones. The broken lines are drawn according to the estimated errors in scale factor and temperature factor. If other experimental errors (given in Table 1) are considered to be statistical, all deviations in the average behaviour of the differences $f_0 - f_{th}$ (solid lines) exceeding the magnitude given by broken lines are significant. With this criterion only the DS values (Fig. 2a) are in agreement with the experimental ones. This proves that Slater's method to incorporate exchange is valid when average electron density is large. It is remarkable that the TFD model completely fails (Fig. 2a) although it has been generally applied (see [16]). Hartree values (Fig. 2b) are about 1.2 electron units too low. With Slater's exchange correction the fit becomes considerably better although the lack of relativistic correction is obvious in the general course of HFS values at high values of sin θ/λ (Fig. 2b). Comparison with recently calculated HF values of tungsten [2] shows that these are about 0.3 electron units smaller than HFS values giving thus worse fit compared with present experimental results.

Some f_0-values obtained from low-order reflections deviate more from the general course than expected within the limits of experimental errors given in

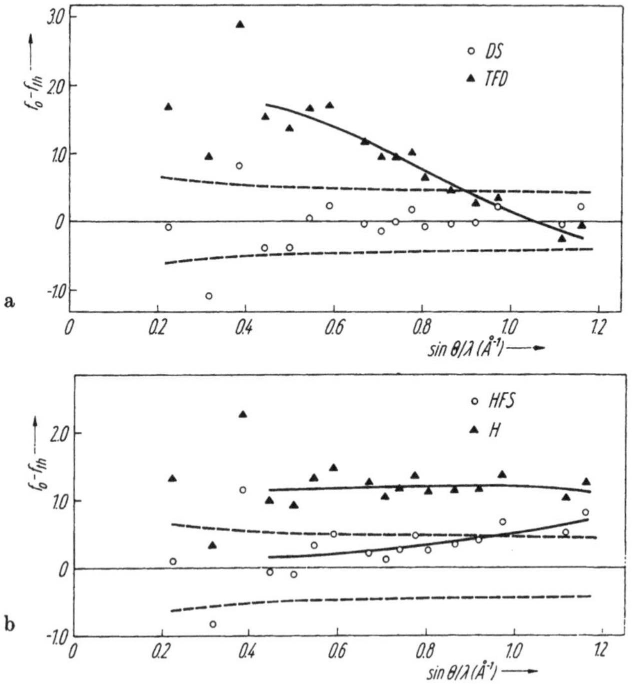

Fig. 2. Differences between experimental and theoretical atomic scattering factors of tungsten. Average behaviour of the results are shown with solid lines. The broken lines give the estimated uncertainty of the determination of scale factor (about 1%) together with the estimated uncertainty of temperature parameter

Table 1. These deviations may be due to non-spherical effects in the outer electron distribution of a tungsten atom in solid state. A quantitative study of this would require a more sophisticated investigation of preferred orientation. However, it is likely that in this case the effect of preferred orientation cannot be completely avoided with the powder method, because the uncertainties introduced by granularity effects are more dominant if the measurements are made on lightly pressed or surface-treated samples.

Acknowledgements

We wish to thank Prof. P. E. Tahvonen for his interest in our work and Dr. O. Inkinen for valuable discussions.

This work is made possible by a Finnish Government Grant for Young Scientists (P.S.) and by a fellowship of the Alfred Kordelin Foundation (T.P.).

References

[1] D. R. CHIPMAN and L. D. JENNINGS, Phys. Rev. **132**, 728 (1963).
[2] D. T. CROMER and J. B. MANN, Acta cryst. **A24**, 321 (1968).
[3] L. H. THOMAS and K. UMEDA, J. chem. Phys. **26**, 239 (1957).
[4] D. T. CROMER, A. C. LARSON, and J. T. WABER, Acta cryst. **17**, 1044 (1964).
[5] H. P. HANSON, F. HERMAN, J. D. LEA, and S. SKILLMAN, Acta cryst. **17**, 1040 (1964).
[6] D. T. CROMER and J. T. WABER, Acta cryst. **18**, 104 (1965).
[7] D. T. CROMER, Acta cryst. **19**, 224 (1965).
[8] R. J. HARRISON and A. PASKIN, Acta cryst **17**, 325 (1964).
[9] P. M. DE WOLFF, Acta cryst. **9**, 682 (1956).

[10] J. J. DeMarco and R. J. Weiss, U.S. Army Materials Res. Lab. Rep. 114, Watertown (Mass.) 1962.
[11] P. Suortti, Ann. Acad. Sci. Fenn., Ser. A VI, No. 240 (1967).
[12] A. Myers, Phil. Mag. **5**, 927 (1960).
[13] S. H. Chen and B. N. Brockhouse, Solid State Commun. **2**, 73 (1964).
[14] T. Paakkari and P. Suortti, Acta cryst. **22**, 755 (1967).
[15] M. McCargo, J. A. Davies, and F. Brown, Canad. J. Phys. **41**, 1231 (1963).
[16] International Tables for X-Ray Crystallography, Vol. III, Kynoch Press, Birmingham 1962.

(Received March 29, 1968)

phys. stat. sol. 28, 547 (1968)

Subject classification: 16; 22.2.2

Institute of Technology, Department of Solid State Physics, Lund

Storage Effects in Photoconducting GaP–Cu

By

H. G. GRIMMEISS and G. OLOFSSON

Experimental evidence is given that charge storage exists in photoconducting GaP–Cu after excitation with both intrinsic and extrinsic light. Appreciable storage is still observed one hour after excitation even at room temperature. Measurements of stored charge as a function of ir photon energy, external voltage, and excitation density confirm that such effects as supralinear photoconductivity, etc., are governed by optical charge exchange in agreement with previous investigations.

Es wird experimentell gezeigt, daß in photoleitendem GaP–Cu nach der Anregung mit Licht größerer oder kleinerer Energie als die Bandkante eine Ladungsspeicherung auftreten kann. Eine beträchtliche Speicherung ist eine Stunde nach der Anregung selbst bei Zimmertemperatur noch nachzuweisen. Die gespeicherte Ladungsmenge wurde in Abhängigkeit von der IR-Photonenenergie, äußerer Spannung und Anregungsdichte gemessen. Die erhaltenen Ergebnisse bestätigen in Übereinstimmung mit früheren Messungen, daß Effekte wie superlineare Photoleitung usw. durch einen optischen Ladungsaustausch bestimmt werden.

1. Introduction

Copper doping has been shown to provide a remarkable photosensitivity in GaP in both the visible and ir regions of the spectrum [1 to 5]. Under certain circumstances, a supralinear dependence of the photoconductivity on excitation density and a strong ir quenching of the photoconductivity have been observed [3]. These effects have been attributed [3] to optical charge exchange and Bowman [6] has given additional evidence for this explanation by photo-Hall measurements. In this paper, we will report and discuss several experiments carried out to obtain further information on the optical charge exchange mechanism [7] in Cu-doped GaP, since we believe that this mechanism is also responsible for some other interesting effects.

Using the same type of material as in [3], we have obtained evidence suggesting that in our samples at least two types of centres are involved in the charge exchange: acceptor levels "a" having low electron-capture cross-sections and deep-lying donor centres "k" having high electron-capture cross-sections. The acceptor centres are present after doping the sample with Cu, while the origin of the donor centres is unknown.[1]) We assume that the donor centres act as recombination centres and that the Fermi level lies between the acceptor and donor levels in our high-ohmic samples. The supralinearity of photoconductivity in highly-compensated material has thus been explained [3] as originating from the filling-up of donor centres by electrons from the occupied acceptor centres (due to hole capture at the acceptor centres). The ir quenching of the photo-

[1]) ESR investigations indicate, however, that the donor centres may originate from Fe.

conductivity is the reverse of this process, namely the emptying of filled donor centres by means of free holes created by ir ionisation of unoccupied acceptor levels. Therefore appreciable ir quenching is only observed after the onset of supralinear photoconductivity [3]. From this explanation, it should be expected that those electrons which occupy donor centres by excitation from the acceptor centres should remain in the donor centres even after the source of excitation is removed when no direct recombination between charge carriers in the acceptor and donor levels is possible and both types of centres are assumed to have large binding energies i.e. thermal release can be neglected. In the next chapter we give experimental evidence that charge storage indeed occurs and that appreciable storage is still observed one hour after excitation even at room temperature.

2. Experimental Observations and Discussion

From previous investigations [1, 3, 5], it is known that electrons can be excited from the valence band into the acceptor levels by optical means, if not all of these levels are occupied by electrons. This process gives rise to an ir sensitive photoconductivity in (weakly) p-type GaP–Cu. In our n-type material, no change in conductivity is observed if the samples are illuminated only with ir light. These results offer a possibility of investigating the storage effects mentioned above. We therefore selected a highly compensated sample with the Fermi level lying just beneath the donor levels. For low voltages, the change of conductivity with ir illumination could be neglected, because nearly all acceptor levels were occupied with electrons. On the other hand, in thermal equilibrium many of the donor levels were empty. It was therefore possible to fill these donor centres with electrons from the acceptor centres by illuminating the sample, e.g. with intrinsic light (5000 Å) of reasonable intensity. At the same time the acceptor centres were emptied. If this situation remained even after irradiation, the concentration of empty acceptor levels and filled donor levels in darkness should have increased considerably. Consequently, the sample should now show a change in conductivity when illuminated with ir light alone because free holes can then be created by the ir light, due to ionisation of the empty acceptor levels. When all the acceptor levels are filled by this process, the supply of free holes ceases and hence the change in conductivity due to the recombination of free holes with electrons from the donor levels. To prove this, the sample was first illuminated with ir light and then kept in darkness for some time. After being biased with a p.d. of 4 V, it was again illuminated with ir light (0.97 eV) for a time $t_s = 0.5$ s. Little change in conductivity could be observed (Fig. 1, above). The sample was then illuminated with intrinsic light (2.5 eV) for a time $t_p = 15$ s, kept in darkness for a time $t_w = 15$ s and again illuminated with ir light (0.97 eV). This time a transient increase σ_{st} in conductivity was clearly observed (Fig. 1, below).

Such a pronounced storage effect can only be observed when the material is properly compensated. If the Fermi level is situated too low, the above-mentioned ir photosensitivity masks any storage effect present. If the Fermi level lies too high, many of the k-levels are already occupied in thermal equilibrium and no appreciable optical charge exchange can be achieved. Furthermore, in less compensated samples the dark current can also give rise to difficulties. From seven samples with different degrees of compensation which all showed

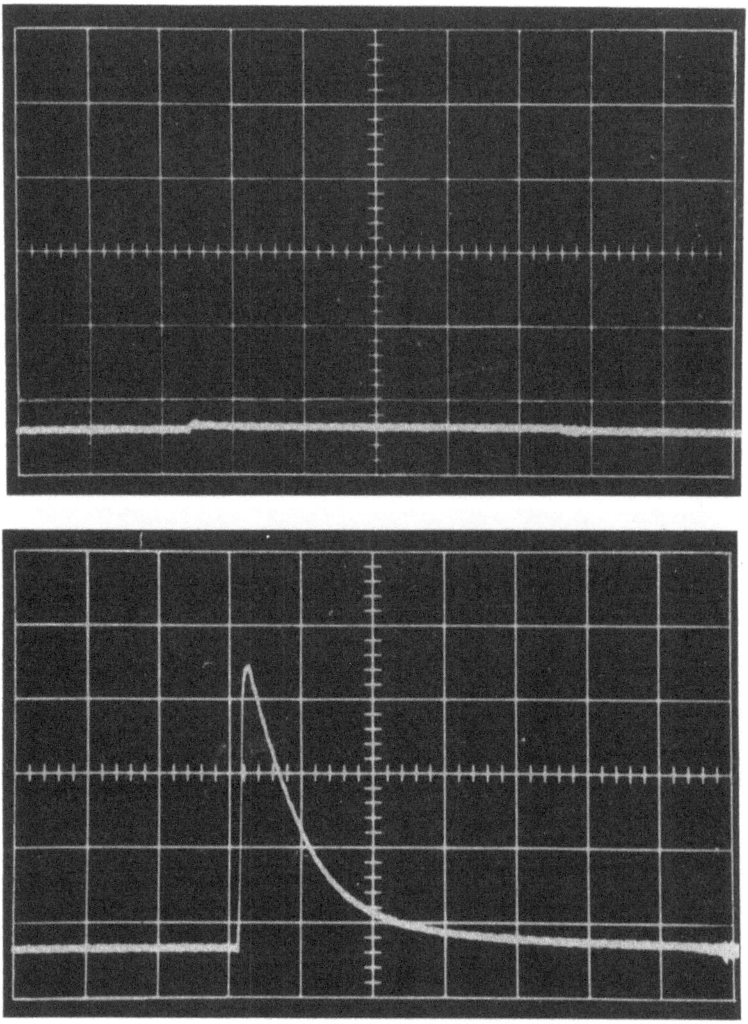

Fig. 1. Change in conductivity due to illumination with ir light (1.2 μm) before (above) and after pumping. Vertical scale: 1.7 nA/div., horizontal scale: 0.1 s/div.; t_p = 15 s, t_w = 15 s, U = 4 V (see text)

a more or less pronounced storage effect, we therefore selected one sample with the most suitable properties for our measurements. The sample had two alloyed Sn-contacts and a dark resistivity of about 10^{10} Ωcm. Below 400 V essentially no rectification could be observed and the photovoltaic effects neglected. The results on the storage effect obtained were not changed when the sample was short-circuited during the pumping time t_p and/or during the waiting time t_w.

The time-dependence of the transient increase σ_{st} in conductivity is strongly influenced by the intensity of the ir light (Fig. 2), but the charge Q carried by this induced photocurrent is independent of the ir intensity. The charge Q induced in the sample by optical charge exchange has been measured using an

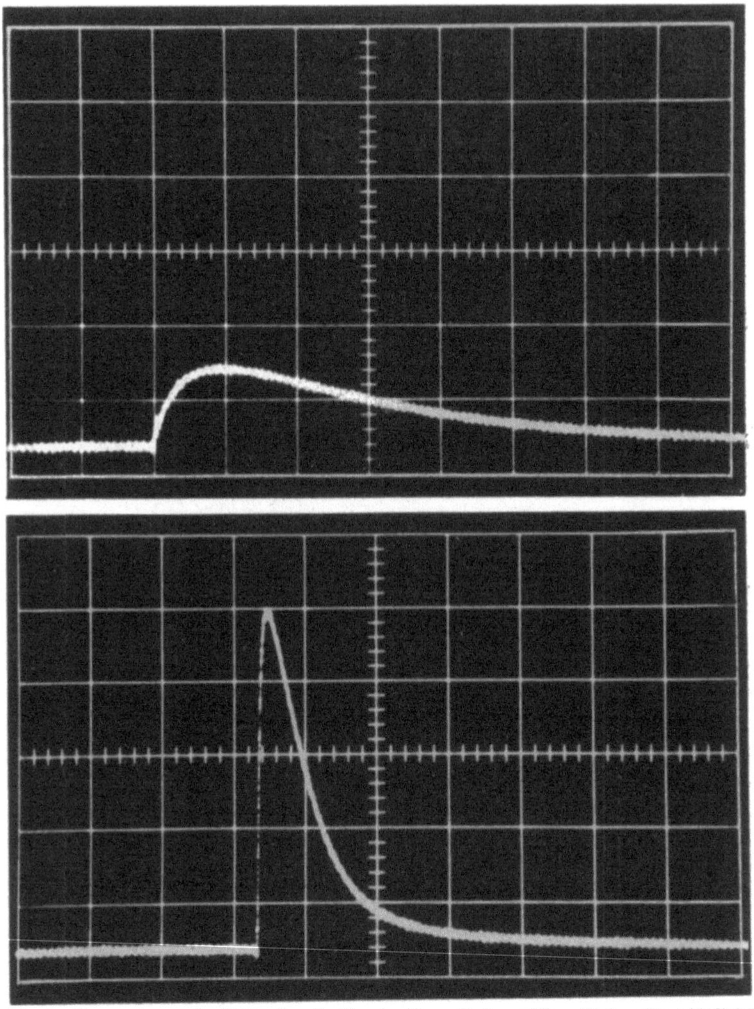

Fig. 2. Change in conductivity due to illumination with two different intensities of ir light after pumping. Upper curve with lower intensity. Vertical scale: 1 nA/div., horizontal scale: 0.2 s/div.; $t_p = 15$ s, $t_w = 15$ s

operational amplifier and a storage oscilloscope.[2]) At higher voltages, Q is defined as the difference of the charge liberated in the circuit by illuminating the sample with ir light both with and without preceding illumination with intrinsic light. A change in intensity of the ir light by nearly 2 orders of magnitude gave no indication of any change in Q, in agreement with our explanation that Q is primarily caused by ionising the unoccupied acceptors originating from optical charge exchange. After complete pumping, Q is a constant and independent of the rate of emptying.

It may be worthwhile mentioning that in GaP optical charge exchange can be produced not only at intrinsic but also at extrinsic light energies. However,

[2]) Tektronix type 549.

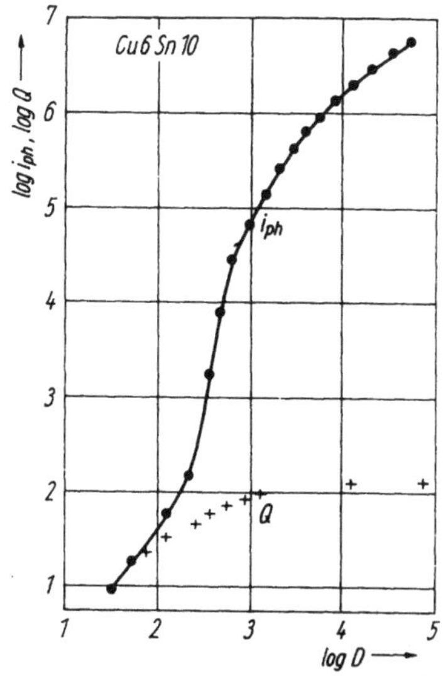

Fig. 3. Photocurrent i_{ph} (●) and stored charge Q (+) as a function of excitation density D (6000 Å), expressed in arbitrary units

by this kind of excitation, not only free electrons but apparently also free holes are created by two-step processes [8]. It has been found by a variation of the extrinsic excitation densities that a supralinear increase of photoconductivity can also be achieved (Fig. 3). It is therefore not unreasonable that the excitation spectra for the storage effect are in correspondence with the spectral sensitivity of the estrinsic photovoltaic effect observed in GaP–Cu photocells [4]. The low-energy cut-off for the creation of optical charge exchange is consequently determined by the energy difference between the conduction band and the acceptor levels responsible for the two-step excitation process. This has not necessarily to be a Cu level. The use of extrinsic light for charge exchange has the advantage of more homogeneous excitation and avoids surface effects.

On the other hand, it must be expected that the lowest ir photon energy capable of causing a charge Q is given by the binding energy of the acceptors and must be the same as for ir quenching of the photoconductivity according to our explanation, because both processes are caused by the same type of excitation. A comparison between the "relative quenching" of the photoconductivity [3] (Fig. 4) and Q (Fig. 5) as a function of the photon energy of ir light shows excellent agreement. Both sensitivities start between 0.6 and 0.65 eV (see also Fig. 8 in [3]). It is interesting to note that the decay in sensitivity in Fig. 4 is much more abrupt for extrinsic than for intrinsic light. Although the spectral sensitivity of Q is more or less constant for photon energies larger than 0.85 eV, the time dependence of the transient conductivity increase σ_{st} is rather

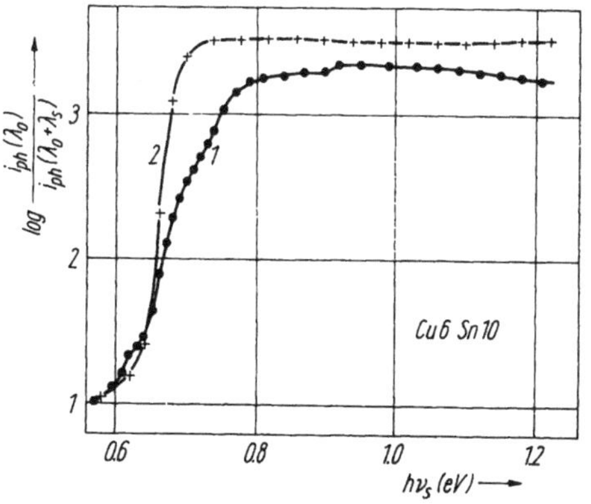

Fig. 4. Spectral distribution of "relative quenching" (see text). The exciting radiation, which was of constant intensity, has a wavelength of 5030 Å (curve 1) and 6360 Å (curve 2), respectively

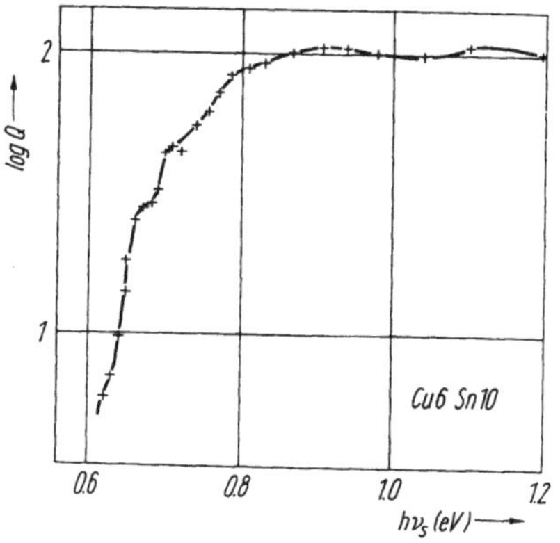

Fig. 5. Stored charge Q as a function of ir photon energy. $t_p = 15$ s, $t_w = 60$ s, $U = 4$ V

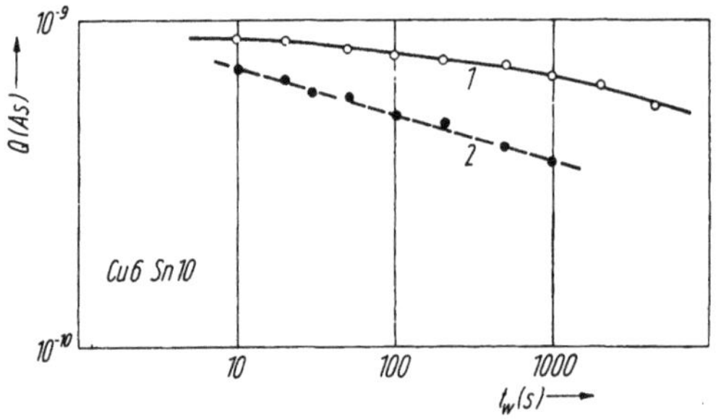

Fig. 6. Stored charge Q as a function of waiting time t_w. Curve 1: no external voltage during t_w; curve 2: with external voltage during t_w. Pumping time $t_p = 15$ s

dependent on the photon energy. Typically the decay time of the storage peak (Fig. 1, below) decreases between 0.85 and 1 eV and increases for larger photon energies.

We have already mentioned that both centres involved in optical charge exchange are assumed to have large binding energies. This means that large storage times can be expected if the interactions between these two centres are small. In fact the decrease in Q with increasing waiting time t_w between pumping and ir illumination is rather small during the first hour if the sample is not biased during t_w (Fig. 6, curve 1). This apparently indicates that any interaction between these two levels is indeed small. A somewhat larger decrease in Q with increasing t_w could be observed if the sample was biased immediately after pumping, i.e. also during t_w. It has been found that the decrease in Q at constant t_w was continuous and depended on the fraction of time the sample was biased during t_w. This shows that the application of a voltage to the sample acts as a kind of de-ionisation of the optical charge exchange.

A similar behaviour could be observed when the charge Q was measured as a function of the applied voltage U. According to our explanation, Q can be expected to increase with external voltage U. If we assume a simple relationship

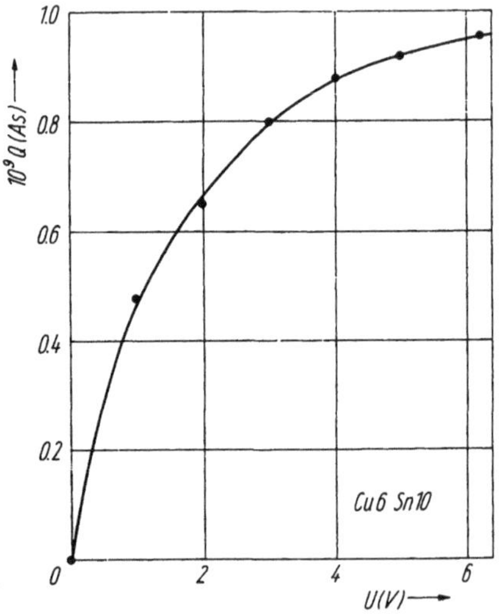

Fig. 7. Stored charge Q as a function of the external voltage U. $t_p = 15$ s, $t_w = 30$ s

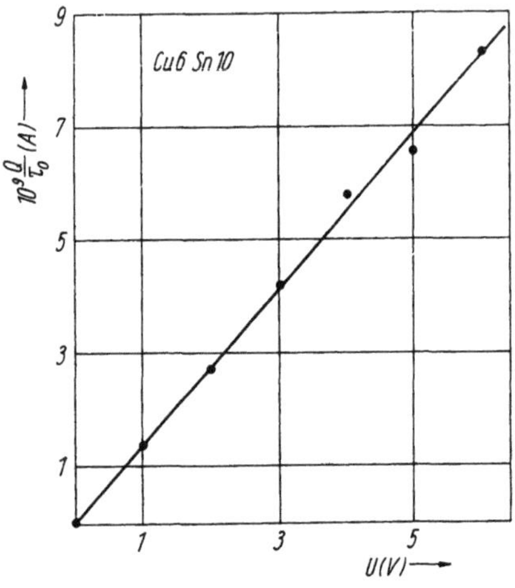

Fig. 8. Ratio of Q and decay time τ_0 of the transient photocurrent as a function of the external voltage U. $t_p = 15$ s, $t_w = 30$ s

such as $L_d = \mu \tau E$ between the drift-length L_d and the electric field E, then the dependence of L_d on E should be linear if the mobility μ and lifetime τ are assumed to be independent of E. A plot of Q against U (Fig. 7) showed no such linearity. If the electric field has really de-ionising properties, a simple linearity between L_d and E can no longer be expected because then a field-independent lifetime is rather unlikely. It has been found that the decay time τ_0 of the transient conductivity change σ_{st} decreases with increasing voltage U. Assuming the proportionality of τ_0 and τ, we have divided Q by the corresponding decay time and plotted Q/τ_0 against U (Fig. 8). The result is a straight line. Note that for $U = 0$, Q is zero. This was valid for all samples and indicates that we were not dealing with space-charge regions. In space-charge regions, we have found that storage currents can be achieved even without external voltages [9].

All these results can also be explained by the assumption that the donor levels are not recombination centres but traps with the same binding energies as the acceptor levels. One possibility of distinguishing between these possibilities is the determination of the type of free carriers created by ir light. So far such experiments have not yielded definite results with our sample because of experimental difficulties.

Another possibility for obtaining information about whether or not the donor levels are recombination centres is offered by the intensity dependence of the concentration of occupied recombination centres. Using a simple model, previous investigations have shown [3] that the number of occupied donor levels acting as recombination centres increases with the root of the excitation density until superlinearity is achieved. For higher excitation density the number of occupied donor levels is constant. If, therefore, the storage effect is caused by donor levels acting as recombination centres, the induced charge Q should vary with excitation density in the same way as the number of occupied donor levels. This has been found (Fig. 3). In these measurements Q has been determined with so long pumping times t_p that any further increase in t_p had no influence on Q. It is rather unlikely that the donor levels should show the same properties when acting as traps. Ryvkin [7], for example, found with Ge at low illumination intensity that the concentration of electrons in the donor centres acting as traps is independent of the intensity. Nevertheless further investigations are under way to clarify this question.

In agreement with previous experiments [4], additional measurements showed that at least two acceptor levels are present in GaP–Cu with binding energies of about 0.6 and 0.8 eV. As will be shown elsewhere, we have also reason to believe that each of the centres can be active in optical charge exchange depending on temperature and compensation. This might be the reason why the binding energy of the Cu level in GaP is given in the literature with different values. Whereas the 0.8 eV level is only observed in GaP–Cu, the 0.6 eV level has also been found in GaP with other dopings; it is therefore still an open question whether both levels are caused by Cu or not. Furthermore, it is not clear whether the acceptor levels are "pure" acceptors or complex.

3. Conclusion

The measurements of stored charges in photoconducting GaP–Cu as a function of time, ir photon energy, external voltage, and excitation density above described represent independent evidence that optical charge exchange is possible in this material, and further support the assumption that effects such as supralinear photoconductivity, etc., are governed by these processes.

Acknowledgement

The authors wish to thank the Philips Zentrallaboratorium GmbH, Aachen, for supplying the material used in this investigation.

References

[1] J. W. Allen and R. J. Cherry, J. Phys. Chem. Solids **23**, 509 (1962).
[2] D. N. Nasledov and S. V. Slobodshikov, Soviet Phys. — Solid State **4**, 2315 (1963).
[3] H. G. Grimmeiss and H. Scholz, Philips Res. Rep. **20**, 107 (1965).
[4] H. G. Grimmeiss and H. Scholz, Philips Res. Rep. **21**, 246 (1966).
[5] B. Goldstein and S. S. Perlman, Phys. Rev. **148**, 715 (1966).
[6] D. L. Bowman, J. appl. Phys. **38**, 568 (1967).
[7] See, for instance, S. M. Ryvkin, Photoelectric Effects in Semiconductors, Consultants Bureau, New York 1964.
[8] H. G. Grimmeiss and H. Koelmans, Philips Res. Rep. **15**, 290 (1960).
[9] H. G. Grimmeiss and G. Olofsson, to be published.

(Received April 2, 1968)

phys. stat. sol. 28, 555 (1968)

Subject classification: 13.1; 22.1.2

Haldor Topsøe Research Laboratory, Vedbæk

Empirical Third Neighbour LCAO Energy Bands of Silicon

By

I. ALSTRUP and K. JOHANSEN

A third nearest neighbour LCAO description, including overlap integrals, is constructed for the energy bands of silicon. The overlap integrals are calculated from Slater-type atomic orbitals, the interaction integrals and the screening parameter are determined empirically by fitting the bands to recent adjusted OPW calculations. Good overall agreement is obtained. The second nearest neighbour interaction integrals obtained cannot be derived from two-center integrals only. Previous failures to obtain reasonable bands for diamond and silicon by the empirical LCAO scheme are briefly discussed.

Eine LCAO-Beschreibung, die übernächste Nachbarn und Überlappungsintegrale einschließt, wird für die Energiebänder des Siliziums aufgestellt. Die Überlappungsintegrale werden auf der Grundlage der Slaterschen Atomfunktionen berechnet. Die Wechselwirkungsintegrale und der Abschirmparameter werden durch Anpassung der Bänder an neue, angepaßte OPW-Berechnungen empirisch bestimmt. Gute Gesamtübereinstimmung wird erreicht. Die empirisch bestimmten Nächste-Nachbarn-Wechselwirkungsintegrale können von Zweizentrenintegralen allein nicht abgeleitet werden. Frühere, verfehlte Versuche vernünftige Bänder für Diamant und Silizium durch die empirische LCAO-Methode zu erhalten, werden kurz diskutiert.

1. Introduction

Since Slater and Koster [1] (hereafter referred to as S-K) suggested to use the LCAO method as an interpolation scheme very few attempts have been made to apply this method to covalent semiconductors. This lack of favour is probably due to the unrealistic bands obtained by S-K for diamond [1] and by Bassani [2] for silicon. It seems as if the method has since then been regarded as not being adequate for the calculation of the energy bands of covalent semiconductors [3, 4].

Cohan et al. [5] made an *a priori* calculation of diamond bands by using combinations of hybridized Slater orbitals and neglecting overlap integrals but taking interaction integrals up to and including third parallel nearest neighbour bonds into account. Recently Dresselhaus and Dresselhaus [6] have used the S-K approximation, i.e. taking into account next nearest neighbour interaction integrals and disposing of the overlap integrals (OI) by assuming an orthogonal basis of localized orbitals (AO). S-K and Bassani used OPW energies at symmetry points while in [6] available experimental information was used for the determination of the interaction integrals (INI).

We have applied the LCAO method to silicon taking explicitly OI's into account and including third nearest neighbour OI's and INI's.

The OI's are calculated by assuming Slater type AO's. The twenty INI's and the screening constant of the AO's are adjusted to give bands in accordance with the recent slightly adjusted *a priori* OPW calculations of Herman et al. [7].

The OI's are explicitly included in order to be able to use modifications of the LCAO scheme for future studies of the effect on the band structure of different types of defects such as vacancies and surfaces. The inclusion of OI's makes it further possible to discuss qualitatively the INI values obtained.

2. LCAO Equations for the Diamond Structure

Silicon crystals have the diamond structure with space group O_h^7 (Fd3m). The cubic unit cell contains eight atoms, each held to four nearest neighbours by covalent bonds. The lattice can be regarded as being composed of two f.c.c. lattices, the one shifted $\bar{\tau} = a/4\,(1, 1, 1)$ from the other, where a is the side length of the unit cell.

We assume that the Bloch orbitals of the valence electrons can be constructed with sufficient accuracy from four different orbitals Φ_1, Φ_2, Φ_3, Φ_4 localized on each atom and with the same symmetry as s-, p_x-, p_y-, p_z-atomic orbitals, respectively. From these AO's eight Bloch orbitals (four per f.c.c.-lattice) are constructed in the usual way:

$$\psi_{\boldsymbol{k}, n} = \sum_j \exp\,(i\,\boldsymbol{k}\cdot\boldsymbol{R}_j)\,\Phi_n\,(\boldsymbol{r}-\boldsymbol{R}_j)\ ; \qquad n = 1, 2, \ldots, 8\,, \tag{1}$$

where Φ_n for $n = 1, 2, 3, 4$ are localized on the atoms of one f.c.c. lattice and $\Phi_{n+4} = \Phi_n$ for $n = 1, 2, 3, 4$ are localized on the atoms of the other f.c.c. lattice. $\boldsymbol{R}_j$ and $\boldsymbol{r}$ are the position vectors of the atom cores and the valence electrons, respectively, and $\boldsymbol{k}$ is the wave vector of the Bloch orbital.

We now further assume, as usually in band calculations, that the bands can be derived from a $\boldsymbol{k}$-independent one-electron crystal potential V common to all the valence electrons.

It is then straightforward to express the matrix elements of the crystal one-electron Hamiltonian $H = -\hbar^2/(2\,m)\,\nabla^2 + V$ between Bloch states (1) by matrix elements of H between the localized orbitals Φ_1, Φ_2, Φ_3, Φ_4, and the same relations exist between Bloch orbital overlap integrals and the overlap integrals for the localized orbitals.

S-K have given expressions for the case where next nearest neighbour interactions are included. A few misprints in the expressions in [1] are corrected in [6], in which expressions up to and including fourth nearest neighbour contributions can be found, but expressed in a representation of the Hamiltonian different from the one used by S-K, but related to this by a unitary transformation. Also the set of basic parameters in [6] is different from the set of matrix elements between the localized orbitals Φ_1, Φ_2, Φ_3, Φ_4 used by S-K.

As we want to relate the energy band structure to atomic properties, we use like S-K the Bloch functions (1) as basis by the derivation of the elements $H_{m,n}$ of the Hamiltonian expressed by the matrix elements $E_{m,n}(u, v, w)$ between the localized orbitals:

$$H_{m,n} = \int \psi_m^*\,H\,\psi_n\,\mathrm{d}v = \sum_j \exp\,\big(i\,\boldsymbol{k}\cdot(\boldsymbol{R}_j - \boldsymbol{R}_i)\big)\int \Phi_m^*\,(\boldsymbol{r}-\boldsymbol{R}_i)\,H\,\Phi_n\,(\boldsymbol{r}-\boldsymbol{R}_j)\,\mathrm{d}v =$$

$$= \sum_{uvw} \exp\,\left(i\,\frac{a}{4}\,(k_x\,u + k_y\,v + k_z\,w)\big)E_{m,n}(u, v, w)\right) \tag{2}$$

and similar for the overlap integrals:

$$S_{m,n} = \int \psi_m^* \psi_n \, dv = \sum_{uvw} \exp\left(i \frac{a}{4} (k_x u + k_y v + k_z w)\right) s_{m,n}(u, v, w) , \quad (3)$$

where a is the lattice constant and $a/4(u, v, w)$ are atom coordinates in the diamond lattice.[1])

3. Determination of the Interaction Integrals

With a given set of OI's, the twenty INI's, $E_{m,n}(u, v, w)$, in the matrix elements $H_{m,n}$ are left to be determined.

For this purpose we need the fourteen energies at the symmetry points Γ, X, and L of the eight bands involved. The nine energies nearest to the energy gap can be taken directly from [7], but the other five energies have been chosen to be in accordance with reasonable extrapolations of Fig. 4 of [7] and with the energy relation (4) below. Because the latter energies are relatively far from the energy gap, we believe that introducing the correct OPW values will near the band gap result in but small corrections to the bands we have obtained.

In the third nearest neighbour approximation the following relation can be derived by assuming the AO's have been orthogonalized:

$$E(\Gamma_1) + E(\Gamma_{2'}) + 3\,[E(\Gamma_{15}) + E(\Gamma_{25'})] + 6\,[E(X_1^{\mathrm{l}}) + E(X_1^{\mathrm{u}}) + E(X_3) + E(X_4)] =$$
$$= 4\,[E(L_1^{\mathrm{l}}) + E(L_1^{\mathrm{u}}) + E(L_{2'}^{\mathrm{l}}) + E(L_{2'}^{\mathrm{u}})] + 8\,[E(L_3) + E(L_{3'})] .\text{[2])} \quad (4)$$

In the second nearest neighbour approximation an additional relation can be derived:

$$E(\Gamma_{15}) - E(\Gamma_{25'}) + E(X_3) - E(X_4) = 2\,[E(L_3) - E(L_{3'})] . \quad (5)$$

Only $E(X_3)$ in (5) is not directly available from Fig. 4 of [7] and in order to

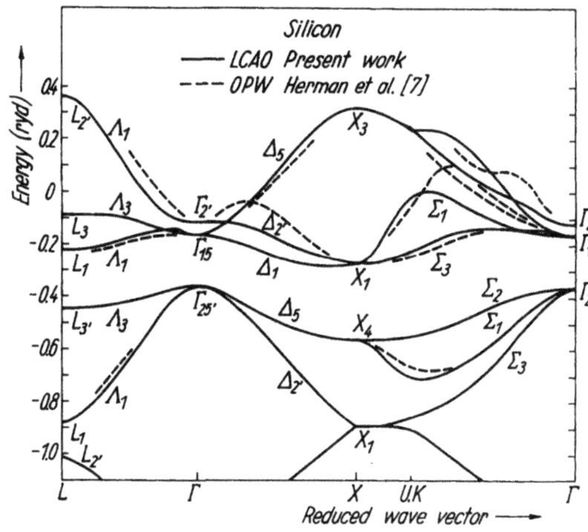

Fig. 1. LCAO energy bands for silicon compared with the OPW bands of Herman et al. [7] (spin-orbit splitting neglected). Where the two models nearly coincide, the dashed line (OPW bands) is not drawn

[1]) There are small differences between the S–K and our expressions: our u-, v-, w-values are twice the S–K values because we express atom coordinates in units of $a/4$ and S–K in units of $a/2$ (which is called a in [1] and [6]).

[2]) In order to distinguish between different energies belonging to the same representation we use the superscripts l and u (lower and upper, respectively).

Table 1

Energies (in Ry) and parameters of Δ_1 conduction band minimum

$E(\Gamma_1)$	-1.23	$E(L_1^l)$	-0.88
$E(\Gamma_{2'})$	-0.116	$E(L_1^u)$	-0.226
$E(\Gamma_{15})$	-0.162	$E(L_{2'}^l)$	-1.012
$E(\Gamma_{25'})$	-0.362	$E(L_{2'}^u)$	0.36
$E(X_1^l)$	-0.89	$E(L_3)$	-0.089
$E(X_1^u)$	-0.27	$E(L_{3'})$	-0.447
$E(X_3)$	0.32	$E(\Delta_{1m})$	-0.28
$E(X_4)$	-0.56		

k_0	0.496 a.u.
m_l/m	0.98 (1.1)
m_t/m	0.2

The m_l- and m_t-values are from [6]. The figure in parentheses is the m_l-value obtained.

fulfil (5) $E(X_3) - E(\Gamma_{25'})$ should be 0.32 Ry, but according to the bands in [7] it is larger than the highest energy in the figure, i.e. 0.55 Ry and reasonable extrapolation of the bands gives 0.65 to 0.67 Ry.

We conclude that a LCAO scheme which does not include more distant neighbours than second nearest cannot adequately reproduce the silicon bands of Herman et al. [7].

One of the difficulties when working with non-orthogonal orbitals is that the band structure depends on the mean value of the one-electron potential V, so it is necessary to specify absolute values of the energies. In the case of silicon precise values for the threshold for photoemission of electrons are available [8], and as the surface dipol is expected to be small [9] we take the absolute value of the energy at the top of the valence band, $E(\Gamma_{25'})$, to be identical with the photo-threshold of the clean silicon surface.

In the Appendix expressions for the elements (2) of the 8×8 matrix, the eigenvalues of which give the energies at a general point of the Brillouin zone of the diamond structure, and expressions appropriate for the Δ- and Λ-directions are given.

The factorization of the 8×8 matrix at the symmetry points Γ, X, and L gives 8 first order and 6 second order equations relating the energies in Table 1 to the interaction integrals. To these equations we add four which relate interaction integrals to the four parameters of the minimum of the Δ_1-conduction band, i.e. the position k_0, the energy $E(\Delta_{1m})$, and the longitudinal and transverse effective masses m_l and m_t, respectively. The values used for these parameters are also shown in Table 1.

Now we have a sufficient number of equations to determine 18 of the INI's. From the values of the OI's (Table 3) we judge that $E_{x,x}(\bar{1}13)$[3]) and $E_{x,y}(\bar{1}13)$ should be small and have only little influence on the bands. So we start by putting these integrals equal to zero and eventually adjust them to give a good overall fit to the bands of [7] determining the other integrals from the 18 equations.

[3]) When referring to specific interaction or overlap integrals we use (like S–K) the orbital letters s, x, or y as subscripts instead of the numbers implied in (2) and (3).

The second order eigenvalue equations are rewritten into an equation, first order in the INI's, corresponding to the sum of the two energies involved, and one, second order in the INI's, corresponding to the product of the two energies. In the following, these two equations are called the sum and the product equation, respectively.

We start by guessing a value for $E_{x,x}(000)$ and calculating $E_{s,s}(000)$ from the X_1 product equation.

In step 2 $E_{s,s}(111)$ and $E_{x,y}(\bar{3}11)$ are determined from two coupled second order equations: the L_1 and the $L_{2'}$ product equation.

In step 3 $E_{x,x}(\bar{3}11)$ is determined from a first order equation constructed from the eigenvalue equations giving the energies $E(\Gamma_{15})$ and $E(\Gamma_{25'})$.

In step 4 $E_{s,x}(220)$ and $E_{s,x}(\bar{3}11)$ are determined from the two coupled equations expressing the energy and the position of the minimum of the Δ_1 conduction band with $E_{s,x}(111)$, $E_{s,x}(022)$, and $E_{s,x}(\bar{3}11)$ eliminated by means of the sum equations for the X_1-, L_1-, and $L_{2'}$-energies.

In step 5 the equation giving the longitudinal component m_l of the effective mass tensor at this minimum is solved, and if the prescribed value is not obtained $E_{x,x}(000)$ is changed and the steps 1 to 5 are repeated until either we end up with the prescribed value of m_l or we cannot come nearer because a solution corresponding to a set of real integrals cannot be found.

When this procedure has been finished, the remaining ten integrals are easily calculated from equations constructed by means of the first order eigenvalue equations: $E_{s,s}(220)$ from a $(\Gamma_1, \Gamma_{2'})$ equation, $E_{x,x}(022)$ from a (X_3, X_4) equation, $E_{x,x}(220)$ from a $(\Gamma_{15}, \Gamma_{25'})$ equation, $E_{x,y}(220)$ from a $(L_3, L_{3'})$ equation, $E_{s,s}(\bar{3}11)$ from a $(\Gamma_1, \Gamma_{2'})$ equation, $E_{x,x}(111)$ from a $(\Gamma_{15}, \Gamma_{25'})$ equation, and $E_{x,y}(111)$ from a (X_3, X_4) equation.

Finally, $E_{x,y}(022)$ is computed from the equation for the transverse component m_t of the effective mass tensor at the minimum of the Δ_1 conduction band.

The calculations described above have been carried out for a large number of different sets of energies and overlap integrals and in some cases it was found that the set of equations could be satisfied by four different sets of real values for the INI's, but in all cases at most one of the solutions gave physically sensible bands.

4. Overlap Integrals

We assume that the OI's can be calculated with sufficient accuracy by choosing orbitals of the form suggested by Slater [10]:

$$\left.\begin{aligned}
\Phi_1 &= \Phi_5 = R(r) \cdot (1/4\,\pi)^{1/2} && \text{(s-orbital)}\,,\\
\Phi_2 &= \Phi_6 = R(r) \cdot (3/4\,\pi)^{1/2} \sin\theta \cos\varphi && (\text{p}_x\text{-orbital})\,,\\
\Phi_3 &= \Phi_7 = R(r) \cdot (3/4\,\pi)^{1/2} \sin\theta \sin\varphi && (p_y\text{-orbital})\,,\\
\Phi_4 &= \Phi_6 = R(r) \cdot (3/4\,\pi)^{1/2} \cos\theta && (\text{p}_z\text{-orbital})\,,
\end{aligned}\right\} \tag{6}$$

where r, θ, φ are the coordinates in a spherical coordinate system. We have assumed that the same $R(r)$ function can be used for s- and p-orbitals:

$$R(r) = N_r \cdot r^2 \exp(-\mu\, r)\,, \tag{7}$$

where N_r is determined by

$$1/N_r^2 = \int_0^\infty r^6 \exp(-2\,\mu\, r)\,\mathrm{d}r\,. \tag{8}$$

Table 2

Atomic orbital overlap integrals ($\mu = 1.54$)

Integral	Neighbour				
	1	2	3	4	5
$s\,s\,\sigma$	0.28027	0.03761	0.01245	0.00227	0.00088
$s\,p\,\sigma$	0.32879	0.05358	0.01843	0.00348	0.00136
$p\,p\,\sigma$	0.35578	0.07554	0.02715	0.00533	0.00212
$p\,p\,\pi$	0.14852	0.01226	0.00342	0.00051	0.00018

Because of the nodeless character of $R(r)$, the expressions (6) give a poor description of the AO's near the centre of the atom, but a fairly good one of the outer parts of the AO's, and as shown by Mulliken et al. [11] only small changes occur in the overlap integrals when the Slater AO's belonging to the same atom are orthogonalized.

In Table 2 the σ and π overlap integrals corresponding to $\mu = 1.54$ are presented. We have included fourth and fifth nearest neighbour atomic OI's, and from the small value of these integrals we judge that it is sufficient to include third nearest neighbour OI's in order to obtain an accuracy consistent with other uncertainties of our scheme.

Determination of INI's and energy bands has been carried out for a number of different μ-values and as expected much better bands were obtained with a μ-value (1.54) larger than the value (1.383) appropriate to free silicon atoms.

5. Results and Discussion

The Δ- and Λ_3-bands can in the LCAO scheme very easily be calculated by solving second order equations, but generally we have to solve determinantal equations of the form

$$\det |H_{r,s} - \lambda\, S_{r,s}| = 0\,, \tag{10}$$

where $H_{r,s}$ and $S_{r,s}$ are Hermitian matrices of order higher than two. We do this by using algorithms which are generalizations of the ones recently published [12] for real, symmetric matrices.

The results obtained for the Δ- and Λ-directions and for the bands along X–U and K–Γ lines are shown in Fig. 1 together with those parts of the silicon bands in [7] which deviate significantly from ours. Small changes of some of the starting parameters could probably improve the accordance further, but Fig. 1 clearly demonstrates the ability of the scheme to give bands in good agreement with the OPW results.

In Table 3 the corresponding INI's and OI's are shown. We notice that with four exceptions the interaction and corresponding overlap integral have opposite sign and their proportion vary between approximately -0.5 and -2.5.

The OI's of the exceptions s, $x(022)$ and x, $y(022)$ are zero by symmetry so the corresponding INI's are determined by three-centre integrals. The signs of $E_{s,x}(022)$ and $E_{x,y}(022)$ are in accordance with what should be expected for the dominating three-centre integral, i.e. the one with atom potential localized at the $a/4(1, 1, 1)$ atom.

Table 3

Interaction and overlap integrals for silicon

$m, n(u, v, w)$	Notation in Appendix	$E_{m, n}(u, v, w)$ (Ry)	$s_{m, n}(u, v, w)$
$s, s(000)$	h_{01}	−0.9406	1.000
$x, x(000)$	h_{02}	−0.5028	1.000
$s, s(111)$	h_{11}	−0.3481	0.280
$s, x(111)$	h_{12}	0.1974	−0.190
$x, x(111)$	h_{13}	−0.0030	−0.020
$x, y(111)$	h_{14}	0.1410	−0.168
$s, s(220)$	h_{21}	−0.0619	0.038
$s, x(220)$	h_{22}	0.0530	−0.038
$x, x(220)$	h_{23}	0.0494	−0.032
$x, y(220)$	h_{24}	0.0543	−0.044
$s, x(022)$	h'_{22}	−0.0009	0.000
$x, x(022)$	h'_{23}	−0.0290	0.012
$x, y(022)$	h'_{24}	−0.0084	0.000
$s, s(\bar{3}11)$	h_{31}	−0.0224	0.012
$s, x(\bar{3}11)$	h_{32}	−0.0218	0.017
$x, x(\bar{3}11)$	h_{33}	0.0253	−0.022
$x, y(\bar{3}11)$	h_{34}	−0.0103	0.008
$s, x(\bar{1}13)$	h'_{32}	−0.0064	0.006
$x, x(\bar{1}13)$	h'_{33}	0.004	0.001
$x, y(\bar{1}13)$	h'_{34}	−0.003	0.003

The overlap integral $s_{x,x}(111)$ can be written

$$s_{x,x}(111) = \frac{1}{3}(2\,p\,p\,\pi - p\,p\,\sigma)\,, \tag{10}$$

where $p\,p\,\pi$ and $p\,p\,\sigma$ are nearest neighbour integrals from Table 2. $s_{x,x}(111)$ is unusually small because the components nearly cancel (they do exactly at $\mu \approx 1.43$). If the two-centre approximation is valid for nearest neighbours, the INI's corresponding to π- and σ-OI can be determined by expressions analogous to (11). We find $E_{pp\pi} = -0.141$ and $E_{pp\sigma} = -0.279$ while the corresponding OI's are 0.149 and 0.356, respectively (Table 2) in accordance with the general rule.

The fourth exception $(x, x(\bar{1}13))$ is probably not of real significance as this INI may be strongly dependent on the starting parameters, and relatively large changes of this integral have little influence on the bands.

In the S-K [1] and the Bassani [2] case we find that sensible Δ-bands can be obtained simply by changing $E_{s,x}(220)$, which cannot be determined from the Γ, X, and L-energies. In the S-K case the agreement with the L-energies is then spoiled, but the resulting Λ-bands are in better accordance with recent OPW calculations [7].

Acknowledgement

We wish to thank Mr. Haldor Topsøe for encouragement and for permission to carry out and publish this work.

Appendix

LCAO Matrix Elements

Electron energies E and state vectors $\boldsymbol{v}$ are determined from eigenvalue problems of the form

$$(H_{r,s} - E \cdot S_{r,s})\, \boldsymbol{v} = 0\,, \tag{A1}$$

where $H_{r,s}$ and $S_{r,s}$ are Hermitian matrices the elements of which can be calculated from the expressions (A3) to (A7) below by inserting interaction or overlap integral values, respectively. A simplified notation ($h_{01}, h_{02}, \ldots, h_{11}, \ldots, h'_{34}$) defined in Table 3 is used here for the AO matrix elements instead of $E_{m,n}(u, v, w)$ or $s_{m,n}(u, v, w)$ elsewhere in the paper.

As arguments for the trigonometric functions are used

$$\alpha = (1/4)\, k_x a\,, \quad \beta = (1/4)\, k_y a\,, \quad \gamma = (1/4)\, k_z a\,, \tag{A2}$$

where $\boldsymbol{k} = (k_x, k_y, k_z)$ is the wave vector of the Bloch state and a is the unit length in the diamond lattice.

Δ_1- and $\Delta_{2'}$-matrix:

$$\left.\begin{aligned}
H_{1,1} &= H_{2,2} = A + B\,,\\
H_{1,2} &= A - B + i\,C\,;\\
A &= h_{01} + 4\,h_{21} \pm 4\,(h_{11} + 2\,h_{31}) \cos\alpha + 8\,h_{21} \cos 2\,\alpha \pm 4\,h_{31} \cos 3\,\alpha\,,\\
B &= h_{02} + 4\,h'_{23} \pm 4\,(h_{13} + 2\,h'_{33}) \cos\alpha + 8\,h_{23} \cos 2\,\alpha \pm 4\,h_{33} \cos 3\,\alpha\,,\\
C &= 8\,[\mp(h_{12} - 2\,h'_{32}) \sin\alpha - 2\,h_{22} \sin 2\,\alpha \pm h_{32} \sin 3\,\alpha]\,.
\end{aligned}\right\} \tag{A3}$$

Upper sign: Δ_1,
lower sign: $\Delta_{2'}$.

Δ_5-matrix:

$$\left.\begin{aligned}
H_{1,1} &= H_{2,2} = h_{02} + 4\,h_{23} + 4\,(h_{23} + h'_{23}) \cos 2\,\alpha\,,\\
H_{1,2} &= -4\,(h_{13} + h_{33} + h'_{33}) \cos\alpha - 4\,h'_{33} \cos 3\,\alpha -\\
&\quad - 4\,i\,[(h_{14} + 2\,h_{34}) \sin\alpha + h'_{34} \sin 3\,\alpha]\,.
\end{aligned}\right\} \tag{A4}$$

Δ_3-matrix:

$$\left.\begin{aligned}
H_{1,1} &= A + B,\\
H_{2,2} &= A - B;\\
A &= h_{02} + 4\,(2\,h_{23} + h'_{23}) \cos^2 2\,\alpha + 4\,h_{24} \sin^2 2\,\alpha,\\
B &= 4\,[h_{13} \cos^3\alpha + (h_{33} + 2\,h'_{33}) \cos^2\alpha \cos 3\,\alpha +\\
&\quad + (h_{14} \cos\alpha - h'_{34} \cos 3\,\alpha) \sin^2\alpha - 2\,h_{34} \cos\alpha \sin\alpha \sin 3\,\alpha];\\
H_{1,2} &= 4\,i\,[h_{13} \sin^3\alpha - (h_{33} + 2\,h'_{33}) \sin^2\alpha \sin 3\,\alpha +\\
&\quad + (h_{14} \sin\alpha + h'_{34} \sin 3\,\alpha) \cos^2\alpha + 2\,h_{34} \sin\alpha \cos\alpha \cos 3\,\alpha]\,.
\end{aligned}\right\} \tag{A5}$$

Λ_1-*matrix:*

$$\left.\begin{aligned}
H_{1,1} &= h_{01} + 12\, h_{21} \cos^2 2\,\alpha\,, \\
H_{1,2} &= 12\,(-h'_{22} \sin^2 2\,\alpha + 2\,i\,h_{22} \cos 2\,\alpha \sin 2\,\alpha)\,, \\
H_{1,3} &= 4\,[(h_{11} \cos \alpha + 3\,h_{31} \cos 3\,\alpha) \cos^2 \alpha - \\
&\quad - i\,(h_{11} \sin \alpha - 3\,h_{31} \sin 3\,\alpha) \sin^2 \alpha]\,, \\
H_{1,4} &= -12\,\{(h_{12} \cos \alpha + h_{32} \cos 3\,\alpha) \sin^2 \alpha + 2\,h'_{32} \cos \alpha \sin \alpha \sin 3\,\alpha + \\
&\quad + i\,[(-h_{12} \sin \alpha + h_{32} \sin 3\,\alpha) \cos^2 \alpha + 2\,h'_{32} \sin \alpha \cos \alpha \cos 3\,\alpha]\}\,, \\
H_{2,2} &= 3\,h_{02} + 12\,(2\,h_{23} + h'_{23}) \cos^2 2\,\alpha - 24\,h_{24} \sin^2 2\,\alpha\,, \\
H_{2,3} &= -H_{1,4}\,, \\
H_{2,4} &= 12\,\{h_{13} \cos^3 \alpha - 2\,(h_{14} \cos \alpha - h'_{34} \cos 3\,\alpha) \sin^2 \alpha + \\
&\quad + (h_{33} + 2\,h'_{33}) \cos^2 \alpha \cos 3\,\alpha + 4\,h_{34} \cos \alpha \sin \alpha \sin 3\,\alpha + \\
&\quad + i\,[-h_{13} \sin^3 \alpha + 2\,(h_{14} \sin \alpha + h'_{34} \sin 3\,\alpha) \cos^2 \alpha + \\
&\quad + (h_{33} + 2\,h'_{33}) \sin^2 \alpha \sin 3\,\alpha + 4\,h_{34} \sin \alpha \cos \alpha \cos 3\,\alpha]\}\,, \\
H_{3,3} &= H_{1,1}\,, \\
H_{3,4} &= 12\,(h'_{22} \sin^2 2\,\alpha + 2\,i\,h_{22} \cos 2\,\alpha \sin 2\,\alpha)\,, \\
H_{4,4} &= H_{2,2}\,.
\end{aligned}\right\} \quad \text{(A6)}$$

General point matrix:

$$\left.\begin{aligned}
H_{1,1} &= h_{01} + 4\,h_{21} (\cos 2\,\alpha \cos 2\,\beta + \cos 2\,\alpha \cos 2\,\gamma + \cos 2\,\beta \cos 2\,\gamma)\,, \\
H_{1,2} &= 4\,[-h'_{22} \sin 2\,\beta \sin 2\,\gamma + i\,h_{22} \sin 2\,\alpha\,(\cos 2\,\beta + \cos 2\,\gamma)]\,, \\
H_{1,3} &= H_{1,2}\,(\alpha \to \beta \to \gamma \to \alpha)^{4)}\,, \\
H_{1,4} &= H_{1,2}\,(\alpha \to \gamma \to \beta \to \alpha)\,, \\
H_{1,5} &= 4\,[h_{11} \cos \alpha \cos \beta \cos \gamma + h_{31} (\cos 3\,\alpha \cos \beta \cos \gamma + \\
&\quad + \cos \alpha \cos 3\,\beta \cos \gamma + \cos \alpha \cos \beta \cos 3\,\gamma)] + \\
&\quad + 4\,i\,[-h_{11} \sin \alpha \sin \beta \sin \gamma + h_{31} (\sin 3\,\alpha \sin \beta \sin \gamma + \\
&\quad + \sin \alpha \sin 3\,\beta \sin \gamma + \sin \alpha \sin \beta \sin 3\,\gamma)]\,, \\
H_{1,6} &= -4\,[h_{12} \cos \alpha \sin \beta \sin \gamma + h_{32} \cos 3\,\alpha \sin \beta \sin \gamma + \\
&\quad + h'_{32} \cos \alpha\,(\sin \beta \sin 3\,\gamma + \sin 3\,\beta \sin \gamma)] - \\
&\quad - 4\,i\,[-h_{12} \sin \alpha \cos \beta \cos \gamma + h_{32} \sin 3\,\alpha \cos \beta \cos \gamma + \\
&\quad + h'_{32} \sin \alpha\,(\cos \beta \cos 3\,\gamma + \cos 3\,\beta \cos \gamma)]\,, \\
H_{1,7} &= H_{1,6}\,(\alpha \to \beta \to \gamma \to \alpha)\,, \\
H_{1,8} &= H_{1,6}\,(\alpha \to \gamma \to \beta \to \alpha)\,, \\
H_{2,2} &= h_{02} + 4\,[h_{23} \cos 2\,\alpha\,(\cos 2\,\beta + \cos 2\,\gamma) + h'_{23} \cos 2\,\beta \cos 2\,\gamma]\,, \\
H_{2,3} &= -4\,h_{24} \sin 2\,\alpha \sin 2\,\beta - 4\,i\,h'_{24} \sin 2\,\gamma\,(\cos 2\,\alpha - \cos 2\,\beta)\,, \\
H_{2,4} &= H_{2,3}\,(\alpha \to \gamma \to \beta \to \alpha)\,, \\
H_{2,5} &= -H_{1,6}\,,
\end{aligned}\right\} \quad \text{(A7)}$$

[4]) This notation means that $H_{1,3}$ can be generated from $H_{1,2}$ by the changes indicated in the paranthesis: α is replaced by β, β by γ etc.

$$H_{2,6} = 4\,[h_{13} \cos\alpha \cos\beta \cos\gamma + h_{33} \cos 3\alpha \cos\beta \cos\gamma + h'_{33} \cos\alpha\,(\cos\beta \cos 3\gamma + \cos 3\beta \cos\gamma)] + 4\,i\,[-h_{13} \sin\alpha \sin\beta \sin\gamma + h_{33} \sin 3\alpha \sin\beta \sin\gamma + h'_{33} \sin\alpha\,(\sin\beta \sin 3\gamma + \sin 3\beta \sin\gamma)]\,,$$

$$H_{2,7} = 4\,[-h_{14} \sin\alpha \sin\beta \cos\gamma + h_{34} \cos\gamma\,(\sin 3\alpha \sin\beta + \sin\alpha \sin 3\beta) + h'_{34} \sin\alpha \sin\beta \cos 3\gamma] + 4\,i\,[h_{14} \cos\alpha \cos\beta \sin\gamma + h_{34} \sin\gamma\,(\cos 3\alpha \cos\beta + \cos\alpha \cos 3\beta) + h'_{34} \cos\alpha \cos\beta \sin 3\gamma]\,,$$

$$H_{2,8} = H_{2,7}\,(\alpha \to \gamma \to \beta \to \alpha)\,,$$
$$H_{3,3} = H_{2,2}\,(\alpha \to \beta \to \gamma \to \alpha)\,,$$
$$H_{3,4} = H^*_{2,3}\,(\alpha \to \beta \to \gamma \to \alpha)\,,$$
$$H_{3,5} = -\,H_{1,6}\,(\alpha \to \beta \to \gamma \to \alpha)\,,$$
$$H_{3,6} = H_{2,7}\,,$$
$$H_{3,7} = H_{2,6}\,(\alpha \to \beta \to \gamma \to \alpha)\,,$$
$$H_{3,8} = H_{2,7}\,(\alpha \to \beta \to \gamma \to \alpha)\,,$$
$$H_{4,4} = H_{2,2}\,(\alpha \to \gamma \to \beta \to \alpha)\,,$$
$$H_{4,5} = -\,H_{1,6}\,(\alpha \to \gamma \to \beta \to \alpha)\,,$$
$$H_{4,6} = H_{2,7}\,(\alpha \to \gamma \to \beta \to \alpha)\,,$$
$$H_{4,7} = H_{2,7}\,(\alpha \to \beta \to \gamma \to \alpha)\,,$$
$$H_{4,8} = H_{2,6}\,(\alpha \to \gamma \to \beta \to \alpha)\,,$$
$$H_{5,5} = H_{1,1}\,,\quad H_{5,6} = -\,H^*_{1,2}\,,\quad H_{5,7} = -\,H^*_{1,3}\,,\quad H_{5,8} = -\,H^*_{1,4}\,,$$
$$H_{6,6} = H_{2,2}\,,\quad H_{6,7} = H^*_{2,3}\,,\quad H_{6,8} = H^*_{2,4}\,,$$
$$H_{7,7} = H_{3,3}\,,\quad H_{7,8} = H^*_{3,4}\,,$$
$$H_{8,8} = H_{4,4}\,.$$

References

[1] J. C. SLATER and G. F. KOSTER, Phys. Rev. **94**, 1498 (1954).
[2] F. BASSANI, Phys. Rev. **108**, 263 (1957).
[3] J. C. PHILLIPS, Phys. Rev. **112**, 685 (1958).
[4] J. CALLAWAY, Energy Band Theory, Academic Press, 1964.
[5] N. V. COHAN, D. PUGH, and R. H. TREDGOLD, Proc. Phys. Soc. **82**, 65 (1963).
[6] G. DRESSELHAUS and M. S. DRESSELHAUS, Phys. Rev. **160**, 649 (1967).
[7] F. HERMAN, R. L. KORTUM, C. D. KUGLIN, and R. A. SHORT in: Quantum Theory of Atoms, Molecules, and the Solid State (a tribute to J. C. Slater), Ed. P.-O. Löwdin, Academic Press, 1966 (p. 381).
F. HERMAN, R. L. KORTUM, C. D. KUGLIN, and R. A. SHORT, Proc. Internat. Conf. Phys. Semicond., Kyoto 1966.
[8] F. G. ALLEN and G. W. GOBELI, J. appl. Phys. **35**, 597 (1964).
[9] J. BARDEEN, Surface Sci. **2**, 381 (1964).
[10] J. C. SLATER, Phys. Rev. **36**, 57 (1930).
[11] R. S. MULLIKEN, C. A. RIEKE, D. ORLOFF, and H. ORLOFF, J. chem. Phys. **17**, 1248 (1949).
[12] J. BOOTHROYD, Commun. ACM **10**, 181 (1967).

(Received May 9, 1968)

phys. stat. sol. 28, 565 (1968)

Subject classification: 8; 22

Laboratorium voor Metaalkunde, Technische Hogeschool Delft

Specific Heat of Ice near the "Ferroelectric" Transition Temperature

By

A. VAN DEN BEUKEL

According to Dengel et al. [1] a transition (believed to be ferroelectric) occurs in ice near 100 °K. Specific heat measurements in this temperature range show the existence of two peaks, at 104 and 132 °K, respectively. The excess heat and the transition entropy corresponding to these peaks amount to (4.3 ± 0.8) cal/mol and to about 0.04 cal/mol °K, respectively.

Nach Dengel et al. [1] findet in Eis bei 100 °K ein Übergang statt, von dem angenommen wird, daß er ferroelektrisch ist. Messungen der spezifischen Wärme in diesem Temperaturbereich ergeben zwei Maxima bei 104 und 132 °K. Die Energie und Übergangsentropie, die diesen beiden Maxima entspricht, beträgt (4,3 ± 0,8) cal/Mol bzw. 0,04 cal/Mol °K.

1. Introduction

The dielectric constant of ice, as observed by Dengel, Eckener, Plitz, and Riehl [1] shows a sharp maximum near 100 °K. The authors suggest that this effect is due to a ferroelectric transition in ice near 100 °K. A second set of experiments where the charging currents to the ice sample were measured during a warming up from 77 to 175 °K indicated the existence of a second Curie point between 125 and 150 °K. This behaviour was observed only if the ice specimen was "not very pure".

According to Onsager [2] acid impurities may play an essential part in the molecular rearrangements at low temperatures, leading to the formation of a ferroelectric order. Onsager emphasizes the need of calorimetric measurements in this temperature range, in order to find the heat of transition. He recommends addition of acid impurity to the sample which is then stored for several days or weeks at 95 °K, in order that equilibrium should be established. Measurements of this kind have been carried out and the results will be described below.

2. Experimental Procedure

Specific heat measurements were carried out on an ice specimen of about 0.275 g. To the water for the preparation HF has been added to a p_H of 3.

The calorimeter has been described elsewhere [3]. It has been constructed as a differential calorimeter; for the present purpose, however, only one set of thermocouples is used. The calorimeter was warmed up continuously at a rate of about 5 °K/h. The heat flow j to the ice specimen passes through a set of silver–constantan thermocouples, giving rise to a thermoelectric voltage

$$V = s\,j\,, \qquad (1)$$

where s is a temperature-dependent proportionality factor, which we call the sensitivity (unit: V/W).

For the heat flow j to the specimen we can write

$$j = m\,c\,\frac{dT}{dt}\,, \tag{2}$$

where m and c are the mass and the specific heat of the specimen, respectively, and T and t have their usual meaning. In order to determine s as a function of temperature prior to the measurements on ice, a run with a silver specimen of known mass (1.584 g) and specific heat (taken from Eucken, Clusius, and Woitinek [4]) was carried out. By measuring V and dT/dt during the warming up, from equations (1) and (2) s can be determined. In the next run the same procedure is repeated with an ice specimen in the calorimeter. When the mass of the sample is known, the values of V and dT/dt substituted in (1) and (2) give the specific heat of ice versus the temperature. So in fact this measurement is a relative one: the specific heat values of ice are found relative to those of silver at the same temperatures.

The calorimeter temperature was measured by a copper-constantan thermocouple close to the specimen. The thermoelectric voltage of this couple was compensated by a bridge circuit, which was in equilibrium at the beginning. The output of the bridge caused by the rise of the temperature and thus of the thermoelectric voltage was written on a Kipp BD1 recorder. The full scale of the recorder corresponded with a temperature change of about 5 °K. At the same time the voltage V (equation (1)) was recorded. In this way a continuous set of V and dT/dt, and thus of c-values, is obtained as a function of temperature. The main error comes from the determination of dT/dt from the recorded T–t curves. From these curves the dT/dt values were determined at intervals of about 1 °K and a smooth curve was drawn through the resulting points. The points deviate from the smoothed curve by 0.5% on the average.

3. Results and Discussion

Fig. 1 shows the result for a specimen which had been stored previously at 95 °K during 8 h. The straight line shown was obtained by connecting the straight parts of the curve in the 90 and 140 °K temperature ranges. The abso-

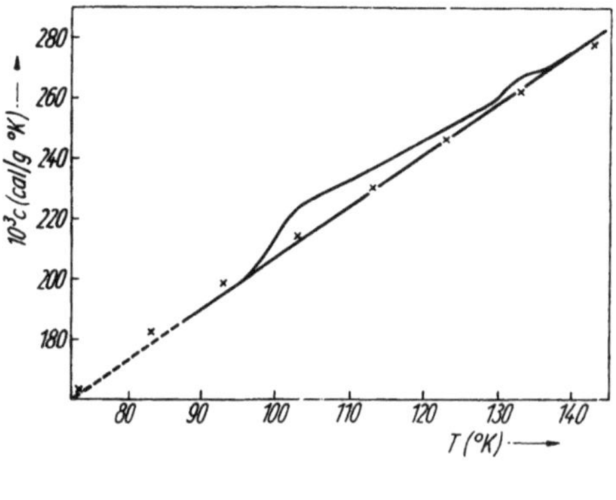

Fig. 1. Specific heat of ice versus temperature. Crosses: data of Nernst [5]

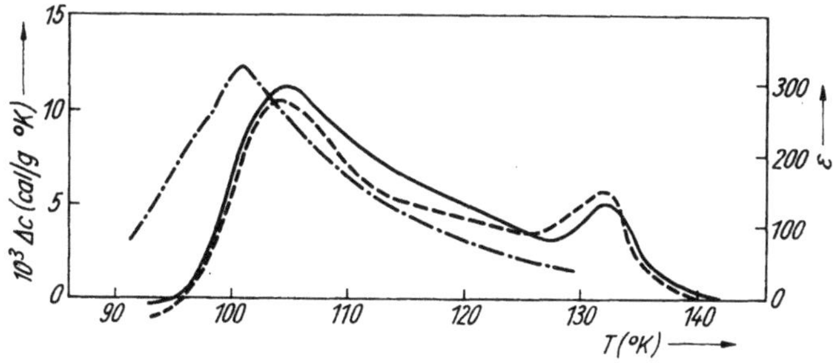

Fig. 2. Excess specific heat relative to the straight line in Fig. 1. ——— and - - - - -: measurements on the same specimen, preannealed 8 h at 95 °K; — · — ·: dielectric constant, according to [1]

lute level of the curve is not known with a better accuracy than 2%, this being due to the uncertainty in the mass of the specimen. For this reason the position of the straight line in Fig. 1 has been chosen in such a way as to fit reasonably the specific heat data of Nernst [5], which are indicated by crosses.

The excess specific heat Δc, as found by subtracting the observed curve from the straight line in Fig. 1, is shown in Fig. 2 (full line). An impression of the reproducibility is obtained by comparing the full line with the dashed line representing two different measurements on the same specimen, preannealed at 95 °K for 8 h in both cases. Two peaks appear in both curves, the larger one being situated at 104 °K, the smaller one at 132 °K. For comparison the ε vs. T curve as observed by Dengel et al. [1] is also shown in the figure, making clear why the second peak has not been found in this experiment. (As mentioned in the introduction the existence of the second peak was indicated by another set of experiments.) The total excess heat, obtained from the area between the T-axis and the full curve, amounts to 4.3 cal/mol. The estimated accuracy of this value is about 20%. The corresponding entropy of transition amounts to about 0.04 cal/mol °K. This value is rather low compared with other well-known ferroelectrics like $BaTiO_3$ ($\Delta S = 0.12$) and KH_2PO_4 ($\Delta S = 0.74$) (see, e.g., the review by Känzig [6]).

In a subsequent experiment the influence of the annealing time at 95 °K on the excess specific heat was investigated on the same specimen. An increase of this time to 16 h did not change the results. On the other hand in Fig. 3 (dashed line) the result is given for a run without preannealing at 95 °K. (The

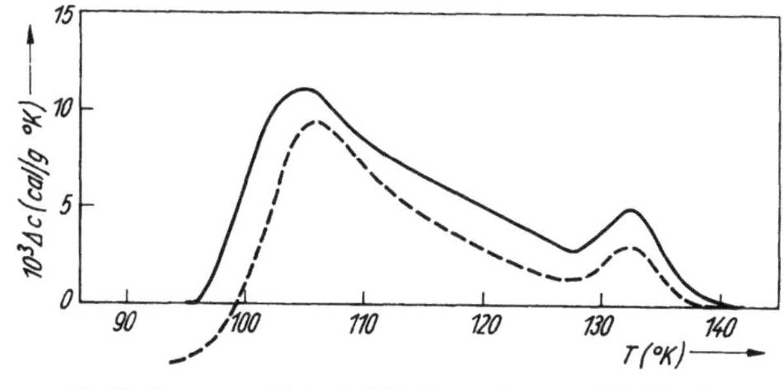

Fig. 3. Excess specific heat of the ice specimen used for Fig. 2.
——— Preannealed 8 hours at 95 °K;
- - - - no preannealing

full curve is redrawn from figure 2 for comparison.) In this case Δc is negative between 93 and 99 °K, indicating that now ordering takes place during the warming up. These results are fully compatible with the observation of Dengel et al. [1] that the relaxation time for the establishment of equilibrium is of the order of hours.

Concluding we can say that the peaks observed in the specific heat of ice are in accordance with the expectations based on measurements of the dielectric constant in the same temperature range [1], which were interpreted as caused by a ferroelectric transition. The observed transition entropy is rather small, however, compared to the entropy observed in other ferroelectrics.

References

[1] O. DENGEL, V. ECKENER, H. PLITZ, and N. RIEHL, Phys. Letters (Netherlands) **9**, 291 (1964).
[2] L. ONSAGER, in: Ferroelectricity, Ed. E. F. WELLER, Elsevier, Amsterdam 1967 (p. 16).
[3] A. VAN DEN BEUKEL, Acta metall. **11**, 97 (1963).
[4] A. EUCKEN, K. CLUSIUS, and K. WOITINEK, Z. anorg. Chem. **203**, 39 (1932).
[5] W. NERNST, Ann. Phys. (Germany) **36**, 395 (1911).
[6] W. KÄNZIG, Ferroelectrics and Antiferroelectrics, Academic Press, New York 1957.

(Received March 14, 1968)

phys. stat. sol. 28, 569 (1968)

Subject classification: 9; 21.1

Natuurkundig Laboratorium van de Universiteit van Amsterdam

A Curvature in the ln D versus $1/T$ Plot for Self-Diffusion in Nickel at Temperatures from 980 to 1400 °C

By

H. BAKKER

A rather gradual curvature was found for the ln D versus $1/T$ plot for tracer-self-diffusion in nickel single crystals at temperatures from 980 to 1400 °C. The activation energies and frequency factors at the highest and lowest temperatures were in agreement with values measured by other investigators in a similar temperature region. It proved to be possible to analyse the curve in terms of contributions of single and multiple vacancies. Especially accurate values of the characteristic quantities for the diffusion via single vacancies were obtained. Other possible explanations of the curvature were investigated, but rejected.

Eine ziemlich gleichmäßig verlaufende Krümmung wurde in der graphischen Darstellung von ln D gegen $1/T$ für Tracerselbstdiffusion in einkristallinem Nickel bei Temperaturen von 980 bis 1400 °C gefunden. Die Aktivierungsenergien und Frequenzfaktoren bei den höchsten und niedrigsten Temperaturen waren in Übereinstimmung mit den Werten anderer Experimentatoren im gleichen Temperaturgebiet. Es erwies sich als möglich, die Kurve in Beiträge von einfachen und mehrfachen Leerstellen zu zerlegen. Insbesondere wurden genaue Werte für die charakteristischen Größen der Diffusion über einfache Leerstellen gefunden. Andere mögliche Erklärungen für die Krümmung wurden untersucht, aber verworfen.

1. Introduction

Nowadays it is generally accepted that in f.c.c. metals self-diffusion occurs by a vacancy mechanism. If a pure single vacancy mechanism is operative and if vacancy enthalpies and entropies are not dependent on temperature, the self-diffusion coefficients are represented by the Arrhenius relation

$$D = D_0 \exp\left(-\frac{Q}{kT}\right), \tag{1}$$

the frequency factor D_0 and the activation energy Q being constants. However, it is quite possible that multiple vacancies contribute to the self-diffusion coefficient. Under certain conditions this might result in a measurable curvature of the ln D vs. $1/T$ plot at higher temperatures. Seeger et al. [1] have given three arguments why this may be the case for self-diffusion in nickel:

1. When we consider the measurements of different authors, there is a tendency for D_0 and Q to increase with temperature.

2. From what is known about the behaviour of divacancies in nickel (especially the relatively high binding energy), they are expected to contribute considerably to the self-diffusion coefficient at higher temperatures.

3. The values of D_0 obtained in the higher-temperature measurements are too high to be consistent with reasonable single vacancy properties.

Seeger et al. [1] have analysed some lower- and higher-temperature measurements in terms of single and multiple vacancies. Such an analysis is complicated by the rather large experimental errors and by the fact that the measurements were performed by different experimental techniques. Therefore it seemed worthwhile to carry out one single, accurate measurement in a large temperature region. The upper limit of this region was determined by the melting point of the metal, the lower one by the experimental technique.

2. Experimental

A nickel single crystal was grown from Johnson & Matthey nickel with a nominal purity of 99.999%, following the Bridgman method. Four samples of 4 mm thickness and 11 mm diameter were machined and spark-cut from the ingot, stresses being removed by polishing and etching.

To detect eventual recrystallisation the samples were annealed for some hours at 1400 °C. A thin layer of radio-active Ni-63 was plated from the solution on one of the flat surfaces of the sample. Diffusion anneals were performed in a slow stream of pure argon at the higher temperatures and in evacuated ampullae at the lower ones. To prevent evaporation of the tracer an alumina flat was placed against the end of the specimen. Both the controlling thermocouple of the furnace and the measuring thermocouple were located in the furnace tube against the sample. Pt–Pt 10% Rh thermocouples were used. The temperature was controlled within ± 0.2 °C by means of a set of temperature controlling units and measured by compensating the signal from the measuring thermocouple.

Because during installation a thermocouple may be subjected to a certain amount of bending and distortion, which results in some uncertainty in the thermal e.m.f., calibration was done "in situ" by melting a 99.99% pure golden and silver wire in the same crucible and under exactly the same conditions as the diffusion anneals were performed. The gold point proved to be reproducible within ± 1 μV, the silver point within ± 2 μV. The melting of the wire was concluded from the fact that its resistance became suddenly infinite. A correction curve was drawn by interpolating and extrapolating linearly. This curve proved to be exactly the same as the one obtained for a thermocouple of the same manufacture, calibrated at the freezing points of Sn, Zn, Al, and Cu.

After the high temperature diffusion anneals the thermocouple was recalibrated, the decrease of its silver point being only 3 μV. If necessary, the diffusion time was corrected for the heating and cooling of the sample.

Because of its outstanding accuracy a sectioning method was employed. Sectioning was done with the aid of a precision lapping device, described by Leblans and Verheijke [2]. In our version the sample was held by vacuum on the holder, which secured a reproducible mounting of the sample. The layers taken off were uniform in thickness much better than 1 μm. About 8 sections some microns in thickness, were lapped off each sample until its activity had been reduced to one tenth of the initial activity.

Since Ni-63 emits only β-radiation of extreme low energy, it was possible to apply Gruzin's method [3], i.e. counting the surface activity of the specimen each time after taking off one section. This was done with the aid of a window-

Fig. 1. The logarithm of the concentration of radioactive isotopes (C in arbitrary units) versus the second power of the distance taken off, for the lowest and for the highest temperature employed

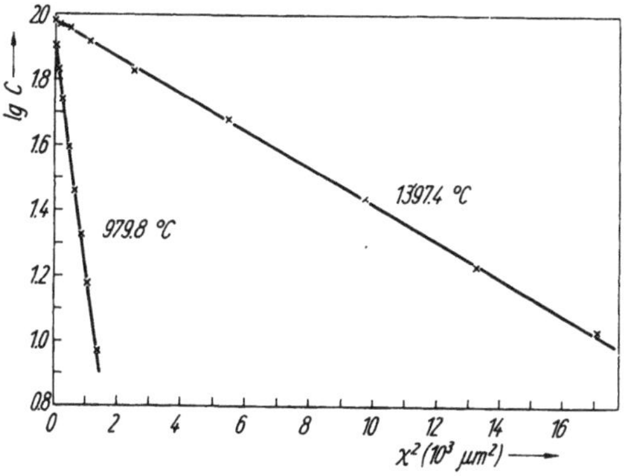

less proportional gas-flow counter. Before each counting, the detector was calibrated by means of a Ni-63 calibration source. The sample was covered with an aluminium diafragm with a hole 1 mm smaller than the sample diameter. This prevented activity from its cylindrical surface (by surface diffusion) from being detected. As many as 100000 counts were collected. Background was measured with this diafragm covered by a 20 μm thick aluminium absorber. In this foil the β-radiation from Ni-63 was completely absorbed, so that the contribution of radiochemical impurities could be subtracted. By least square fit the diffusion coefficient was calculated.

The penetration curves for the highest and the lowest temperature are plotted in Fig. 1.

The total error in $D(T)$, including possible inaccuracy of temperature, was estimated to be 2%.

3. Experimental Results

The values of the measured diffusion coefficients are represented by Table 1. The value at 1121.5 °C is the mean of two measurements. If several diffusion mechanisms are operative we must write

$$D = \sum_{i=1}^{N} D_i \tag{2}$$

with

$$D_i = D_{i0} \exp\left(-\frac{Q_i}{kT}\right). \tag{3}$$

Following Seeger et al. [1] we shall make a least square approximation of the measured D's by the function

$$\ln D - \ln\left\{1 + \exp\left(\ln\frac{D_{20}}{D_{10}} - \frac{Q_2 - Q_1}{kT}\right) + \exp\left(\ln\frac{D_{30}}{D_{10}} - \frac{Q_3 - Q_1}{kT}\right)\right\} = \ln D_{10} - \frac{Q_1}{kT}. \tag{4}$$

By choosing values for D_{i0}/D_{10} and $(Q_i - Q_1)$ we shall minimize the function χ^2, defined as follows:

$$\chi^2 = \sum_{j=1}^{n} \frac{d_j^2}{(n-q)\,\Delta_j^2}. \tag{5}$$

Table 1

Experimental values of the self-diffusion coefficient of nickel

Temperature (°C)	Diffusion coefficient (cm²/s)
1397.4	2.264×10^{-9}
1374.2	1.614×10^{-9}
1351.6	1.213×10^{-9}
1332.8	9.429×10^{-10}
1326.2	8.773×10^{-10}
1295.8	5.756×10^{-10}
1272.6	4.027×10^{-10}
1246.0	2.687×10^{-10}
1226.1	2.037×10^{-10}
1200.1	1.370×10^{-10}
1163.6	7.674×10^{-11}
1139.0	4.920×10^{-11}
1121.5	3.545×10^{-11}
1099.4	2.506×10^{-11}
1080.0	1.771×10^{-11}
1058.4	1.139×10^{-11}
1040.5	8.190×10^{-12}
1020.1	5.391×10^{-12}
1000.3	3.557×10^{-12}
979.8	2.428×10^{-12}

q is the number of unknown parameters to be found, i.e. 2, n is the number of measured D-values, i.e. 20; the d_j's are the residuals when the most probable $\ln D_{10}$ and Q_1-values are substituted in the given equations, Δ_j is the estimated experimental error in $\ln D$: $\Delta_j = 0.02$.

We shall now examine the several possibilities:

1. *A pure single vacancy mechanism.* — When we choose $N = 1$ in formula (2), we obtain for $T = 980$ to 1400 °C

$$Q = 2.955 \pm 0.008 \text{ eV},$$
$$D_0 = 1.77 \pm 0.11 \text{ cm}^2/\text{s},$$
$$\chi^2 = 1.5 .$$

The Q- and D_0-values are just between the higher values, obtained by some authors, and the lower values, measured by other ones (see Table 1 from [1]). However, for 8 points at our lower temperatures, we find for $T = 980$ to 1120 °C

$$Q = 2.89 \text{ eV},$$
$$D_0 = 1.16 \text{ cm}^2/\text{s}$$

(cf. Hoffman et al. [4]: for $T = 870$ to 1250 °C, $Q = 2.88$ eV and $D_0 = 1.27$ cm²/s).

Carrying out the same procedure at the higher temperatures we find for $T = 1250$ to 1400 °C

$$Q = 3.04 \text{ eV},$$
$$D_0 = 3.34 \text{ cm}^2/\text{s}$$

(cf. MacEwan et al. [5]: for $T = 1150$ to 1400 °C, $Q = 3.01$ eV and $D_0 = 3.36$ cm²/s).

When we take each time 8 successive points, the activation energy and frequency factor increase rather gradually with temperature. So it appears that there exists some curvature in the ln D vs. $1/T$ plot.

2. *A single-plus divacancy mechanism.* — We find a better approximation ($\chi^2 = 0.80$) if we choose $N = 2$ in formula (2). The values we then obtain for Q_1 and D_{10} are reasonable (about 2.80 eV and 0.36 cm²/s respectively). However, in the rather broad minimum of χ^2, we find values of $(Q_2 - Q_1)$ and D_{20}/D_{10} that are not consistent with known divacancy properties (about 0.65 eV and 65, respectively).

Consistent values for the last two quantities are found at $\chi^2 \approx 1$, the values for Q_1 and D_{10} changing only slightly.

3. *A single-plus divacancy-plus trivacancy mechanism.* — So let us try $N = 3$. Again a rather broad minimum of χ^2 appeared. However, the value of Q_1 was about 2.80 eV in all these approximations. In this minimum we have chosen that value for $(Q_2 - Q_1)$ that gave the best consistency with known divacancy enthalpies. So we have calculated (with $\chi^2 = 0.80$):

$$Q_1 = 2.81 \text{ eV}, \qquad D_{10} = 0.38 \text{ cm}^2/\text{s};$$

$$Q_2 = 3.21 \text{ eV} \quad \text{(chosen consistent)} \quad D_{20} = 3.07 \text{ cm}^2/\text{s};$$

$$Q_3 \approx 3.91 \text{ eV}, \qquad D_{30} \approx 170 \text{ cm}^2/\text{s}.$$

4. Discussion

Reimers [6] has proved that evaporation of the tracer during the diffusion anneal results in a curvature downwards of the beginning of the penetration plot, causing D-values that are somewhat too high. Fig. 1 shows that even at 1400 °C such an effect was not present in our experiment. Apparently it was avoided by placing an alumina flat against the end of the specimen.

Furthermore we have considered the possibility that the measured curvature in the temperature plot of ln D might be due to diffusion through dislocations following Hart's model [7]. We have used an activation energy for this process of 1.6 eV as measured by Wuttig and Birnbaum [8]. The best fit to our curve was then ($\chi^2 = 0.92$)

$$D(T) = 2.53 \exp\left(-\frac{3.01}{kT}\right) + 0.9 \times 10^{-6} \exp\left(-\frac{1.6}{kT}\right). \tag{6}$$

The first term corresponds to the diffusion by monovacancies, the second one to diffusion through dislocations, the dislocation density then — following [7, 8] — being 3.5×10^5 cm⁻².

However such an analysis would yield surprisingly high values for the monovacancy frequency factor and activation energy. Moreover, from what is known about multiple vacancies from other sources and techniques, one would expect these to play an important part in diffusion at higher temperatures.

A further argument is found in the extrapolation to lower temperatures of both possibilities, which yields the values of Table 2.

It is seen that an explanation in terms of multiple vacancies is, within experimental error, compatible with nearly all lower temperature measurements. This is not the case for an extrapolation by (6). Usually a dislocation density of 10^5 to 10^6 cm^{-2} is assumed for f.c.c. metals. But even the assumption of a dislocation density of 10^5 cm^{-2} is not consistent with the measured low-temperature values.

Finally, that dislocations are not very effective for volume diffusion in nickel is in agreement with experiments performed by Wazzan and Dorn [14], who did not find any effect of prestraining on self-diffusion in nickel.

X-ray examination of our crystal showed subgrains about 1 mm in diameter, orientation differences being some tenths of a degree. Most of the dislocations will be in subboundaries, the dislocation density within the subgrains being relatively small. So the condition for the application of Hart's theory may be not fulfilled if tracer penetration depths are only small, because then most of the tracers will not meet dislocations. Especially this will be the case when the surface decrease technique is used, which requires — using Ni-63 as a tracer — a penetration depth of only some microns.

This may be the reason, why, in measuring the volume self-diffusion coefficient in nickel, dislocations turn out to be not very effective.

Nowick and Dienes [15] have shown that the inherent curvature in the $\ln D$ vs. $1/T$ plot alone is too small to be responsible for the measured curvature.

So it seemed to be justified to make an analysis in terms of contributions of multiple vacancies. Let us consider the associated activation enthalpies as a function of formation, jump, and binding enthalpies. In usual notation

$$Q_1 = E_1^F + E_1^M , \tag{7}$$

$$Q_2 = 2\, E_1^F - B_2 + E_2^M , \tag{8}$$

$$Q_3 = 3\, E_1^F - B_3 + E_3^M . \tag{9}$$

At present we have quite a good knowledge of the numerical value of E_1^M, whereas E_1^F has not been accessible to a reliable determination as yet. When we take $E_1^M = 1.48$ eV [16], then we get from our analysis $E_1^F = 1.33$ eV. Moreover, $E_2^M - B_2 = 0.54$ eV seems to be quite reliable for this quantity, so that consistency requires that analysis of the experiment yields $Q_2 \approx 3.20$ eV. Concerning trivacancies, the accuracy of our analysis is only small, but it is reasonable to assume $B_3 \approx 3\, B_2$, and E_3^M somewhat larger than E_2^M. With $B_2 = 0.3$ eV, $E_2^M = 0.8$ eV and $E_3^M - B_3 \approx 0$ eV; the value calculated for Q_3 by (9) is compatible with the measured value of $Q_3 \approx 3.90$ eV. A similar discussion of frequency factors in terms of entropies has been given by Seeger et al. [1, 16].

Recently Seeger and Schumacher [16] have reanalysed measurements by Hoffman et al. and by Hirano et al. in terms of mono- and divacancies, choosing the activation enthalpies linearly dependent on temperature:

$$H(T) = H(T_0) + \alpha\,(T - T_0) \tag{10}$$

and consequently for the entropies:

$$S(T) = S(T_0) + \alpha \ln\left(\frac{T}{T_0}\right). \tag{11}$$

Table 2

Extrapolation of the various possibilities to lower temperatures and comparison with values obtained by other authors

Temperature range (°C)	Extrapolation by $0.38 \exp(-2.81/kT) + 3.07 \exp(-3.21/kT) + 170 \exp(-3.91/kT)$		Extrapolation by $2.53 \exp(-3.01/kT) + 0.9 \times 10^{-6} \exp(-1.6/kT)$ corresponding to a dislocation density of 3.5×10^5 cm^{-2}		Extrapolation by $2.53 \exp(-3.01/kT) + 2.5 \times 10^{-7} \exp(-1.6/kT)$ corresponding to a dislocation density of 10^5 cm^{-2}		Experimental values and references		
	Q (eV)	D_0 (cm^2/s)	Q (eV)	D_0 (cm^2/s)	Q (eV)	D_0 (cm^2/s)	Q (eV)	D_0 (cm^2/s)	Ref.
250–1250	2.83	0.60	1.86	7.00×10^{-5}	1.98	1.50×10^{-4}	2.63 to 2.80		[9]
850–1100	2.88	0.86	2.67	0.12	2.89	0.89	2.84	0.48	[10]
870–1250	2.91	1.20	2.87	0.81	2.96	1.77	2.88 ± 0.04	1.27	[4]
675– 750	2.84	0.58	1.83	1.87×10^{-5}	2.17	3.17×10^{-4}	2.88	1.9	[11]
700–1100	2.87	0.81	2.48	2.21×10^{-2}	2.76	0.27	2.94	1.7	[12]
680– 830	2.84	0.61	1.97	8.31×10^{-5}	2.34	3.40×10^{-3}	3.01	5.8 ± 2.0	[13]
950–1020	2.89	1.00	2.82	0.53	2.95	1.50	2.90	1.9	[14]

However, such an analysis fails to improve the fit to our measurements. When trivacancies are not considered and if we require consistency with known divacancy properties, only a small temperature dependence ($\alpha \approx 1.1 \times \times 10^{-5}$ eV/deg) is acceptable, leading to a relatively high value of the χ^2-function ($\chi^2 \approx 1$); a temperature dependence that is an order of magnitude greater reduces the χ^2-function to 0.80, but then no longer consistency with known divancancy properties is maintained.

The reason for this may be that our measurements extend to such high temperatures that a contribution of trivacancies is no longer negligible and overshadows some possible temperature dependence of activation enthalpies and entropies.

5. Conclusion

In this paper we showed that a curvature in the ln D vs. $1/T$ plot is observed in a f.c.c. metal like nickel. By calling upon known divacancy properties we were able to attribute this curvature to the influence of di- and trivacancies. Though some extra knowledge is gained about multiple vacancies, the most reliable numerical values from the analysis are those for monovacancy properties. Of special importance is the value calculated for the formation enthalpy of a monovacancy, $E_1^F = 1.33$ eV.

Acknowledgements

I thank Prof. Dr. G. de Vries for helpful discussions, Mr. H. V. M. Mirani and Mr. J. G. E. M. Backus for help in the experiment. I am obliged to the N. V. Philips Gloeilampenfabrieken for placing at our disposal the design of the lapping device, and to Mr. H. Witteveld and Mr. C. J. Pauël for constructing it.

References

[1] A. SEEGER, G. SCHOTTKY, and D. SCHUMACHER, phys. stat. sol. **11**, 363 (1965).
[2] L. M. L. J. LEBLANS and M. L. VERHEIJKE, Philips techn. Rev. **25**, 191 (1963/64).
[3] C. LEYMONIE, Radioactive Tracers in Physical Metallurgy, Chapman & Hall, London 1963 (p. 55).
[4] R. E. HOFFMAN, F. W. PIKUS, and R. A. WARD, Trans. AIME **206**, 483 (1956).
[5] J. R. MACEWAN, J. W. MACEWAN, and L. YAFFE, Canad. J. Chem. **37**, 1623 (1959).
[6] P. REIMERS, phys. stat. sol. **22**, K27 (1967).
[7] E. HART, Acta metall. **5**, 597 (1957).
[8] M. WUTTIG and H. K. BIRNBAUM, Phys. Rev. **147**, 495 (1966).
[9] H. BURGESS and R. SMOLUCHOWSKY, J. appl. Phys. **26**, 491 (1955).
[10] W. R. UPTHEGROVE and M. J. SINNOTT, Trans. ASM **50**, 1031 (1958).
[11] A. R. WAZZAN, Thesis, University of California, Berkeley 1962.
[12] K. HIRANO, M. COHEN, and B. L. AVERBACH, Trans. ASM **53**, 910 (1961).
[13] A. MESSNER, R. BENSON, and J. E. DORN, Trans. ASM **53**, 227 (1961).
[14] A. R. WAZZAN and J. E. DORN, J. appl. Phys. **36**, 222 (1965).
[15] A. S. NOWICK and G. J. DIENES, phys. stat. sol. **24**, 461 (1967).
[16] A. SEEGER and D. SCHUMACHER, Mater. Sci. Engng. **2**, 31 (1967).

(Received May 24, 1968)

phys. stat. sol. 28, 577 (1968)

Subject classification: 13.1; 22.4

Institut für Theoretische Physik II, Universität Marburg/Lahn

The Electronic Structure of CdO

I. The Energy-Band Structure (APW Method)

By

K. MASCHKE and U. RÖSSLER

The energy-band structure of CdO was calculated along the high symmetrical axes of the Brillouin zone by use of the APW method. This method seems to be more suitable for the rocksalt lattice than for the zincblende structure. The resulting energy bands are similar to the bands of silver halides.

Es wurde die Energiebandstruktur für CdO längs der Symmetrieachsen der Brillouinzone mit Hilfe der APW-Methode berechnet. Diese Methode scheint für die Natriumchloridstruktur besser geeignet zu sein als für die Zinkblendestruktur. Die erhaltenen Energiebänder sind denjenigen der Silberhalogenide ähnlich.

1. Introduction

The IIb–VI compounds usually crystallize in either the zincblence or the wurtzite structure, CdO, however, forms only the rocksalt lattice.

We have calculated the energy bands of CdO by use of the APW method, since there was a great interest in a calculated band model in order to understand the known features of this material and to get some hints for further experimental work. When the calculation was started, little was known about the higher optical transitions in CdO (see references given in part II), the experiments published in the following paper[1]) were done simultaneously to our investigations. Thus the calculation had to be undertaken with the only knowledge of the quantity of the direct optical energy gap (2.35 eV). In this case either the APW method [1, 2, 3] or the KKR method [4, 5] are suitable. Both methods involve the use of a muffin-tin potential, i.e. the crystal potential is approximated by spherically symmetric atomic potentials within nonoverlapping spheres around each lattice site and a constant potential between these spheres. The constant potential value V_0 outside the muffin-tin spheres serves as an adjustable parameter, to fit the calculation to an experimental value, usually to the smallest direct optical gap. As shown by Johnson [3], both methods have the same theoretical base, since both result from solving the Schroedinger equation of a crystal electron by means of a Green's function. Therefore, the energy eigenvalues and eigenfunctions of both methods converge towards the exact solutions of the muffin-tin problem, the eigenfunctions of which are continuous and have continuous derivatives throughout the crystal. We have decided to use the APW method, since the computational time of the KKR method is much higher than that of the APW method if only the band structure of one compound with a given crystal lattice is to be calculated.

[1]) Part II will be published in phys. stat. sol. 29, No. 1, by the authors: M. Altwein, H. Finkenrath, Č. Koňák, J. Stuke, and G. Zimmerer.

We have also been interested in comparing energy eigenvalues, which are calculated by use of the different modifications of the APW method [1, 2, 3]. These calculations for cubic ZnS [6, 7] had the result that the original APW formulation of Slater [1] leads to the most rapidly convergent results. The rapid convergence which Johnson [3] supposed for his modified APW method (M-APW) was not found. But even the most rapidly convergent APW eigenvalues, calculated by use of 65 trial functions, are at somewhat higher energies than the respective KKR results [8] for cubic ZnS. The discrepancy is about 0.4 eV for valence-band states, while the conduction bands differ by only 0.1 eV. Since by use of 65 trial functions in an APW expansion the capacity of the TR4 computer in Marburg was almost exhausted, this result does not seem encouraging in order to undertake an APW calculation for CdO. As shown in Section 3, however, we can expect a better convergence in a rocksalt lattice than for a zincblence structure.

2. The Muffin-Tin Potential

As a spherically symmetric potential within the muffin-tin spheres we have chosen a superposition of self-consistent atomic potentials [9] for Cd and O along the nearest neighbour distance of the rocksalt lattice ((100)-direction). The radii of the muffin-tin spheres are then determined by the maximum of the superimposed potential V_{sup} (Fig. 1). From the lattice constant $a = 8.873$ a.u. (atomic units) we have found the radii $\varrho_{\text{Cd}} = 2.499$ a.u. and $\varrho_{\text{O}} = 1.939$ a.u.

In the rocksalt structure the two interpenetrating f.c.c. lattices are separated by one half of the body diagonal of the f.c.c. cube, while the separation in the zincblende structure is only one quarter of the body diagonal. Thus the nearest neighbour distance in CdO is larger and the muffin-tin spheres can be enlarged, as a consequence of which the filling of the unit cell by the muffin-tin spheres is 55% for CdO, compared with only 35% in cubic ZnS. Furthermore, the muffin-tin approximation for the crystal potential seems to be better in the more ionic bound CdO crystal than in the more covalent zincblende crystal. From these reasons we expect better results for CdO than we have got for ZnS.

The choice of the superimposed potential within the muffin-tin spheres seems to be better justified for the rocksalt structure, which has coordination number 6, than for the zincblende lattice, where each atom has only 4 nearest neighbours. But as is known from the KKR calculations [10, 11], the band structure is very insensitive to the superposition of the atomic potentials. The only effect is the shifting of the valence and conduction bands relative to the d-bands. An

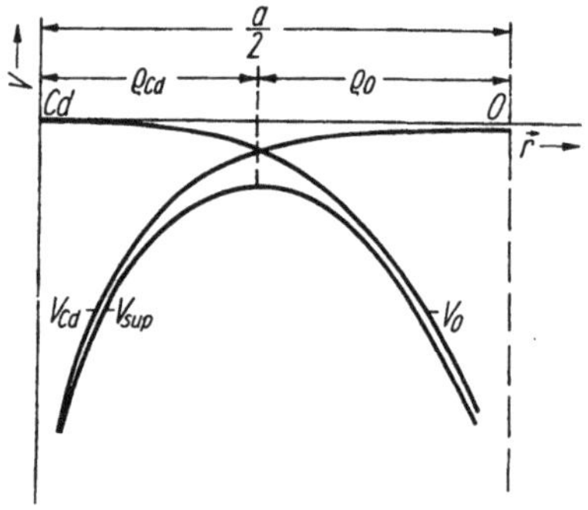

Fig. 1. Superposition of the atomic potentials (qualitatively)

exact examination of interband transitions from the d-bands could give information whether the superimposed or the simple atomic potential (or an intermediate case) is the best possible choice in the frame of the muffin-tin approximation.

The constant potential value V_0 outside the muffin-tin spheres was used to fit the calculated direct gap, which was supposed to correspond to a $\Gamma_{15} - \Gamma_1$ transition, to the experimental value. This assumption was verified by the resulting band structure. By use of $V_0 = -1.86$ ry (in the energy scale of free atoms) the $\Gamma_{15} - \Gamma_1$ transition is 2.38 eV.

3. Results

The energy bands of CdO have been calculated by an expansion of the one-electron wave function in 65 trial functions, corresponding to plane waves with $|\boldsymbol{K}_n| \leqq \left|\frac{2\pi}{a}(4, 0, 0)\right|$ outside the muffin-tin spheres. The convergence of the eigenvalues in Γ, L, and X is demonstrated in Table 1, where 51, 59, and 65 trial functions, according to $|\boldsymbol{K}_n| \leqq \left|\frac{2\pi}{a}(3, 1, 1)\right|$, $|\boldsymbol{K}_n| \leqq \left|\frac{2\pi}{a}(2, 2, 2)\right|$, and $|\boldsymbol{K}_n \leqq \left|\frac{2\pi}{a}(4, 0, 0)\right|$, respectively, have been taken into account. The convergence seems to be sufficient, i.e., the convergence is much better for CdO than for ZnS: the energy differences $E_{311} - E_{400}$ are about 50% less than in the case of ZnS.

The energy bands have been calculated along the high symmetrical 100–(Δ), 111–(Λ), and 110–(Σ) axes of the Brillouin zone (Fig. 2). These bands are characterized by the small energy separation between d-bands and upper valence band, the valence band maxima in Σ and L off $\boldsymbol{k} = 0$, and a simple lowest conduction band, which is separated from the higher conduction bands by a small gap. These general features of the CdO band structure are also found in the band models for the silver halides [12, 13]. The valence band maxima off $\boldsymbol{k} = 0$ result from the small energy separation between the d-bands and the upper valence band and the repulsion between states of equal parity [13]. This repulsion is due to an interaction between p- and d-states via the crystal potential.

Table 1

Test of convergence $\delta_1 E = E_{311} - E_{222}$, $\delta_2 E = E_{222} - E_{400}$ (energy scale of free atoms)

	E_{311} (ry)	E_{222} (ry)	$\delta_1 E$ (ry)	E_{400} (ry)	$\delta_2 E$ (ry)
Γ_{15}	−1.676	−1.680	0.004	−1.682	0.002
Γ_1	−1.507	−1.507	0.000	−1.507	0.000
$\Lambda_1(L)$	−1.773	−1.800	0.027	−1.805	0.005
$\Lambda_3(L)$	−1.563	−1.566	0.003	−1.567	0.001
$\Lambda_1(L)$	−1.181	−1.182	0.001	−1.182	0.000
$\Lambda_1(L)$	−0.964	−0.974	0.010	−0.974	0.000
$\Delta_1(X)$	−1.854	−1.854	0.000	−1.856	0.002
$\Delta_5(X)$	−1.708	−1.723	0.015	−1.725	0.002
$\Delta_1(X)$	−1.095	−1.095	0.000	−1.095	0.000
$\Delta_1(X)$	−0.941	−0.942	0.001	−0.943	0.001

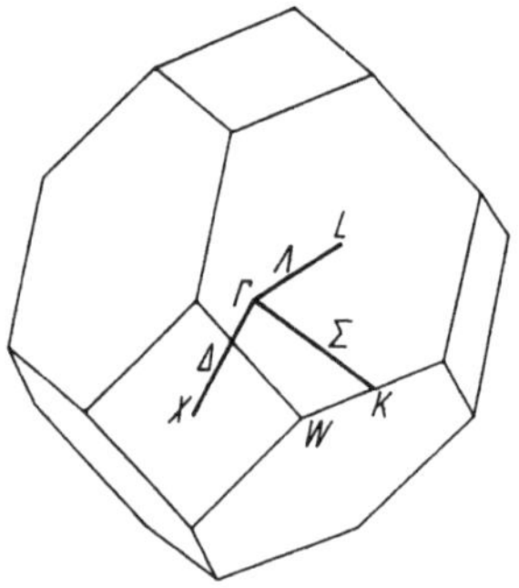

Fig. 2. Brillouin zone of the f.c.c. lattice

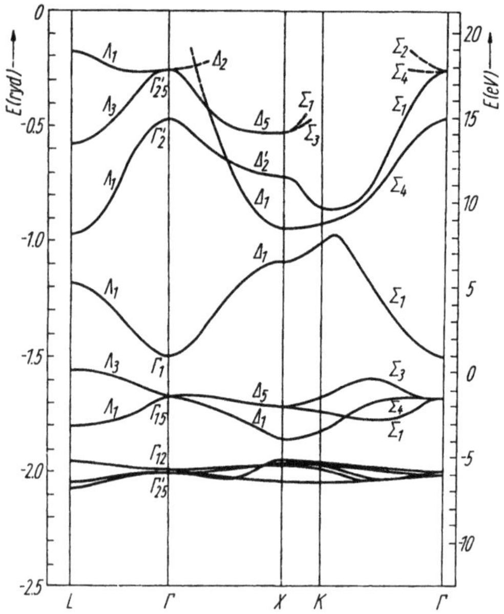

Fig. 3. Band structure of CdO. The xy- energy scale corresponds to the scale of free atoms, the eV- energy scale is fitted to $E = 0$ at the valence-band maximum in L

The matrix element describing this interaction is zero at $\boldsymbol{k} = 0$, where the p(Γ_{15}) and d(Γ'_{25}, Γ_{12}) eigenfunctions have opposite parity, but has a large value for instance at L, where p- and d-states have equal parity. Also the broadening of the d-bands and the valence-band maximum in Σ can be explained by means of the repulsion of equal parity states.

The values for the indirect gaps of CdO are 0.8 eV for the Γ–L gap and 1.2 eV for the Γ–Σ gap. Effective masses have been estimated by comparing the free-electron parabola with the calculated bands in the vicinity of the valence-band maxima in L ($m = -1.4\ m_0$) and Σ ($m = -0.4\ m_0$), and the conduction-band minimum in Γ ($m = 0.3\ m_0$). The accuracy of all these values is of course limited by the muffin-tin approximation. The existence of an indirect gap of about 1 eV, however, is felt to be proven.

A discussion of the most recent experimental results for CdO and their comparison with the calculated band structure will be given in part II.

Acknowledgements

We wish to express our gratitude to Prof. Dr. O. Madelung for his support and interest in this work, to Prof. Dr. M. Cardona for a critical reading of the manuscript, and to Dr. J. Treusch for helpful discussions. We are indebted to the staff of the "Zentrale Rechenanlage, Marburg" for technical assistance.

References

[1] J. C. SLATER, Phys. Rev. **51**, 846 (1937).
[2] M. M. SAFFREN and J. C. SLATER, Phys. Rev. **92**, 603 (1953).
[3] K. H. JOHNSON, Phys. Rev. **150**, 429 (1966).
[4] W. KOHN and N. ROSTOKER, Phys. Rev. **94**, 1111 (1954).

[5] B. Segall, Phys. Rev. **105**, 108 (1957).
[6] K. Maschke, Diplomarbeit, Marburg 1968 (unpublished).
[7] U. Roessler and M. Lietz, phys. stat. sol. **17**, 597 (1966).
[8] P. Eckelt, O. Madelung, and J. Treusch, Phys. Rev. Letters **18**, 656 (1967).
[9] F. Herman and S. Skillman, Atomic Structure Calculations, Prentice Hall, Inc., Englewood Cliffs (New Jersey) 1963.
[10] J. Treusch and R. Sandrock, phys. stat. sol. **16**, 487 (1966).
[11] P. Eckelt, phys. stat. sol. **23**, 307 (1967).
[12] F. Bassani, R. S. Knox, and W. B. Fowler, Phys. Rev. **137**, A1217 (1965).
[13] P. M. Scop, Phys. Rev. **139**, A934 (1965).

(Received May 24, 1968)

phys. stat. sol. 28, 583 (1968)

Subject classification: 11; 10.2; 19; 20.3; 22

I. Physikalisches Institut der Justus-Liebig-Universität Giessen

EPR and Thermoluminescence Measurements on $CaWO_4$ Crystals

By

G. K. BORN, R. J. GRASSER, and A. O. SCHARMANN

After irradiation with X-rays in the energy range 50 to 150 keV at the temperature of liquid air paramagnetic centres are created in single crystals of $CaWO_4$. The centre with a spectroscopic splitting factor slightly higher than $g = 2$ originates from a hole stably trapped between two tungstate tetrahedrons. The signal of the corresponding electron is broad and rather difficult to detect. The concentration of these centres depends linearly on the dose of irradiation. The decrease of the EPR signal of the hole and electron centres with temperature is correlated with the thermoluminescence intensity.

Nach Bestrahlung mit Röntgenquanten im Energiebereich von 50 bis 150 keV bei der Temperatur der flüssigen Luft werden in $CaWO_4$-Einkristallen paramagnetische Zentren nachgewiesen. Das Zentrum mit einem g-Faktor etwas über 2 wird einem Defektelektron zugeschrieben, das stabil zwischen zwei Wolframatkomplexen eingefangen ist. Das Signal des zugehörigen Elektrons ist breit und schwer nachweisbar. Die Zentrendichte im Kristall hängt linear von der Bestrahlungsdosis ab. Die Abnahme des EPR-Signals dieser Zentren mit der Temperatur ist mit der vom Kristall emittierten Thermolumineszenz korreliert.

1. Introduction

The luminescence properties of single crystals of $CaWO_4$ have since long been subjects of intensive studies. Emission characteristics, phosphorescence, and trapping phenomena [1] have been investigated. Excitation by UV light, energetic electrons, X-rays, and γ-rays was employed. These investigations throw some light on the nature of the electronic states participating in luminescence.

The luminescence decays exponentially, indicating kinetics of first order [2 to 5]. The earlier experiments show that charge transport is normally not involved in the luminescence mechanisms of $CaWO_4$. Thus, the energy level relationships and interaction of the centre with the host lattice are considered on the basis of energy versus configuration coordinate diagrams.

More recent studies of thermoluminescence and above all of thermally stimulated conductivity and photoconductivity present evidence that a pure centre model is incorrect in the case of $CaWO_4$ [6, 7]. The transport properties can be explained in the picture of a "small polaron" [8].

Self-activated $CaWO_4$ crystals emit thermoluminescence after irradiation with X-rays or fast electrons, while no thermoluminescence is observed after UV excitation (253 nm). An irradiated and afterwards thermally bleached crystal does not show thermoluminescence by UV illumination at liquid nitrogen temperature. These results indicate an irreversible recombination process stimulated by temperature.

In the case of X-ray irradiation correlation between thermally stimulated current and thermoluminescence maxima is observed. There is a marked difference between the luminescence and thermoluminescence spectra.

After excitation with fast electrons a saturation of the thermoluminescence is found. Comparison of the light sums after X-ray irradiation shows saturation effects, too.

From all these measurements on pure unactivated calcium tungstate crystals by high-energy excitation it is not possible to decide if new trapping states are created or if already existing defects are ionized or filled.

The purpose of this paper is to report on EPR investigations of paramagnetic defects in $CaWO_4$ crystals caused by irradiation with X-rays. We observe a correlation betwen paramagnetic resonance and thermoluminescence.

2. Experimental

The used calcium tungstate single crystals were obtained from Semi Elements Inc. and Philips. The size of the nominally pure samples was about $1.5 \times 1.5 \times 4.0$ mm³. Calcium tungstate crystallizes in the scheelite structure and has space group C_{4h}^6. There are four molecules in the unit cell. The dimensions of this cell are $a = b = 5.24$ and $c = 11.37$ Å [9, 10].

EPR measurements were made using a AEG 20X spectrometer with 125 kHz modulation. The H_{103}-cavity permitted optical illumination of the samples and observation of the thermoluminescence light emitted by the crystal. The magnetic field strength was measured via proton resonance with an electronic counter (Hewlett Packard, type 5246 L).

Irradiation with X-rays was done at liquid air temperature. The tube voltage could be varied from 50 to 150 kV. The currents were between 2 and 20 mA at all voltages. Without warming the irradiated specimen were brought into the cooled cavity.

Excitation by UV light was performed with an Osram high-pressure mercury-vapour lamp (HBO 100 W/2) and a set of suitable filters. Thermoluminescence emission was measured with a RCA 1P28 photomultiplier simultaneously with the EPR signal.

Orientation of the samples with regard to the dc magnetic field was done by means of the Mn^{2+} impurities of our crystals. EPR measurements have shown that Mn^{2+} substitutes Ca^{2+} and is found only on this lattice site [11]. In $CaWO_4$: Mn crystals a nuclear quadrupole interaction of the Mn^{55} nucleus is found by means of forbidden hyperfine transitions [12]. This effect is strongly anisotropic. For certain orientations of the crystalline c-axis to the applied static magnetic field "forbidden" $\Delta m = \pm 1, \pm 2$ transitions disappear. By measuring the angular dependence of the $CaWO_4$:Mn spectrum the crystallographic axes could be found.

3. Experimental Results

If calcium tungstate crystals are exposed to X-ray irradiation at low temperatures, they acquire a brown colouration which becomes black with increasing dose of absorbed radiation. At liquid air temperature X-rays produce stable colour centres in $CaWO_4$. An unirradiated crystal does not absorb in the visible region and appears clear.

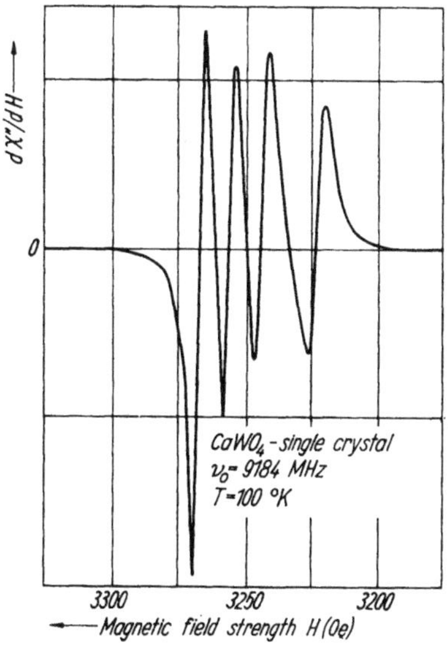

Fig. 1. EPR spectrum of the hole centre in $CaWO_4$. Arbitrary orientation of the crystal axes to the magnetic field

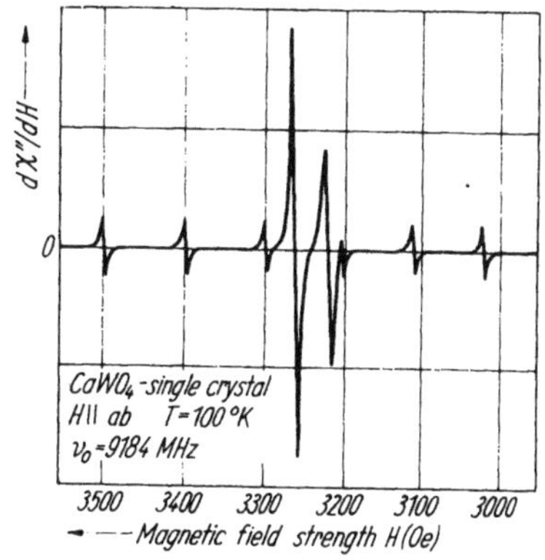

Fig. 2. EPR spectrum of the hole centre in $CaWO_4$ together with the Mn^{2+} hyperfine spectrum ($\boldsymbol{H} \parallel a\,b$)

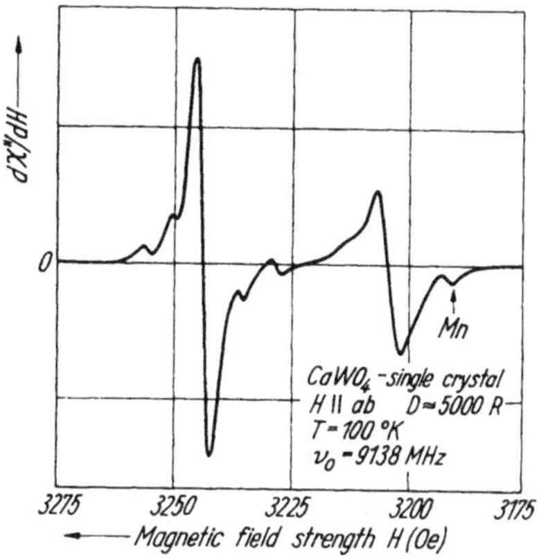

Fig. 3. EPR spectrum of the hole centre in $CaWO_4$. Partially resolved hyperfine structure of the W^{183} nuclei ($\boldsymbol{H} \parallel a\,b$)

After irradiation with X-rays of 50 to 150 keV at liquid air temperature paramagnetic centres are created in $CaWO_4$ crystals. We observe two groups of lines. One group has a spectroscopic splitting factor g slightly higher than the free spin value $g_e = 2.002319$, the other group of relatively broad lines is shifted towards smaller g-values ($g \approx 1.6$, $g \approx 1.8$).

At arbitrary orientations of the crystallographic axes to the static magnetic field a spectrum appears consisting of four lines (Fig. 1). All these lines are nearly equal in shape. The shoulder on the left line can be interpreted as resulting from unresolved hyperfine interaction of the paramagnetic centre with the nuclear magnetic moment of the W^{183} nucleus ($I = 1/2$).

If the static magnetic field has the direction of the $a\,b$ plane of the tungstate crystal a two-line spectrum is measured. Always two lines of the four lines collapse and form the new spectrum. The position of the resonance of the centre can easily be seen in comparison to the six hyperfine lines of the Mn^{2+} impurities (Fig. 2).

The resonance of the created centre with partially resolved hyperfine structure is shown in Fig. 3 for the orientation $\boldsymbol{H} \parallel a\,b$ plane. If the magnetic field has the direction of the c-axis of the single crystal only one line with accompanying hyperfine lines appears. The position of the resonance lines indicates that the spectrum arises from an electron deficiency (hole). We find a small positive deviation of the spectroscopic splitting factor from the free spin value:

$$g = 2.024\ .$$

The averaged difference Δg, the g-shift, has the value $\Delta g = 0.0216$ and measures the admixture of higher excited states into the ground state.

Hyperfine effects are best interpreted by means of Fig. 3. The observed hyperfine splitting arises from interaction of the hole with the magnetic moments of the present W^{183} nuclei. The abundance of the isotope W^{183} in the natural composition is 14.40%. From the intensity of the hyperfine components an interaction of the hole with two equivalent tungsten nuclei can be deduced. The hyperfine splitting consists of an isotropic and an anisotropic part.

Irradiating calcium tungstate with Co^{60} γ-quants ($E = 1.17$ and 1.33 MeV) paramagnetic centres are found [13]. In this energy range γ-radiation produces defects by direct impact in contrast to our measurements. The so formed centres will probably correspond to the centres produced by relatively low energetic X-rays. The authors report on an isotropic hyperfine interaction of the centre with the surrounding tungsten nuclei.

From the magnitude of the hyperfine splitting the probability of the hole being at the tungsten nucleus can be estimated to be small. In the related V-centre in the alkali halides these splittings are much greater [14]. That indicates that the electron deficiency is essentially localized at oxygen between two tungsten nuclei. This is supported by the anisotropic part of the hyperfine splitting.

The resonances at lower g-values are interpreted as resonances of trapped electrons [13]. These lines were rather broad in the investigated temperature range; therefore no further measurements were made.

In the observed temperature region the resonance lines of the hole centres show nearly lorentzian shape indicating a linewidth given by relaxation processes ($\tau \approx 10^{-8}$ s).

For the orientation $\boldsymbol{H} \parallel a\,b$ plane the lines show an anomaly in linewidth. For an arbitrary orientation of the crystallographic axes to the applied magnetic

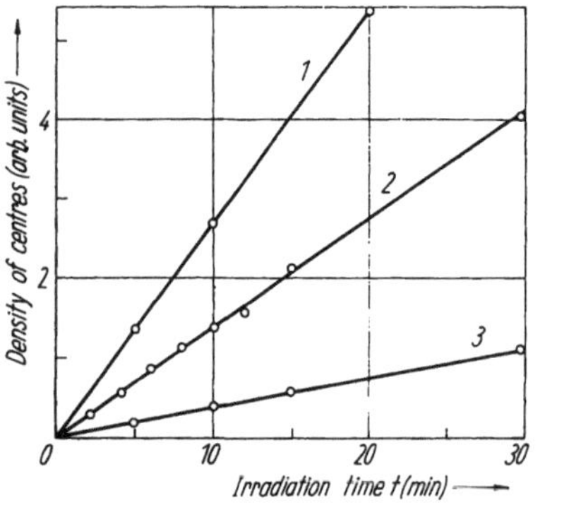

Fig. 4. Density of hole centres in $CaWO_4$ after X-ray irradiation as function of irradiation time.
Curve 1: 150 kV, 20 mA;
curve 2: 150 kV, 14 mA;
curve 3: 100 kV, 14 mA

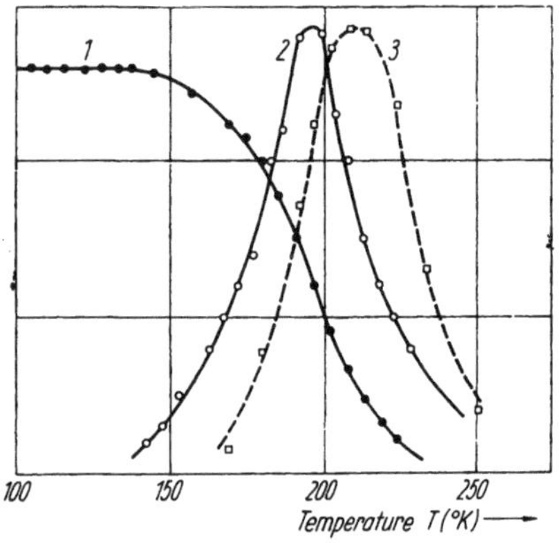

Fig. 5. EPR and thermoluminescence. Linear heating rate ≈ 12 deg/min. Curve 1: intensity I of the EPR signals (population changes corrected), curve 2: negative first derivative of curve 1, $-dI/dT$; curve 3: intensity of the emitted glow light

field each of the four lines has approximately the same width, while for $\boldsymbol{H} \parallel a\,b$ plane a high narrow line and a flat broad one arise. Twofold integration of the lines shows that equal numbers of paramagnetic centres contribute to both lines. From the symmetry properties of the scheelite-structure two groups of two coupled WO_4^{2-} tetrahedrons for this special orientation ($\boldsymbol{H} \parallel a\,b$ plane) can be deduced which are magnetically equivalent. We suppose that by the influence of the static magnetic field the states of these two with respect to symmetry nonequivalent species of holes are affected in a different manner. Thus different relaxation times and different line shapes could arise. A strain-broadening could also be possible.

The density of paramagnetic centres as a function of the irradiation time is shown in Fig. 4 with various tube voltages and currents as parameters. The cross-section for centre production decreases rapidly as a function of the energy of the incident radiation. Fig. 4 is only valid in the range where no saturation effects occur. The saturation concentrations have different values for different crystals and are about 10^{16} to 10^{17} centres/cm^3.

4. Correlations between EPR and Thermoluminescence

After X-ray irradiation at the temperature of liquid air $CaWO_4$ crystals have one broad thermoluminescence maximum when heated up to room temperature. The simultaneously measured thermally stimulated conductivity shows parallel behaviour with a small shift in the position of the maxima. We have investigated the behaviour of the paramagnetic centres when raising the temperature. The centres disappear with increasing temperature. The density of centres was measured as a function of temperature; simultaneously the emitted thermoluminescence light was recorded. Fig. 5 shows the observed correlations. Curve 1 (full circles) gives the normalized density of paramagnetic centres as a function of temperature corrected with the changes in the population difference (Boltzmann) during heating-up. Curve 2 (open circles) is the negative first derivative of curve 1 and should measure the number of decaying centres per temperature interval. The dotted curve 3 shows the intensity of the glow light emitted by the sample. The lineshapes and halfwidths of the curves 2 and 3 are rather similar. The shift of their maxima is beyond the experimental error. The EPR signal of holes and electrons seem to disappear simultaneously and in the same proportion.

5. Discussion

Irradiation of calcium tungstate single crystals at low temperatures with low energetic X-rays creates stable paramagnetic centres. The experimental results strongly support the assumption that there is a direct correspondence between the produced paramagnetic centres and the centres responsible for thermoluminescence. Because of the simultaneous disappearance of electrons and holes we suppose that the back-reaction of the two species is radiative. This model explains qualitatively all investigated thermoluminescence properties of unactivated $CaWO_4$ crystals. The proposed mechanism would simply explain the parallelity of the thermo-luminescence and the thermally stimulated conductivity glow curves, too.

A quantitative treatment of the hole centre is connected with great difficulties because of lack of higher symmetry.

Acknowledgements

We wish to thank Prof. Dr. W. Hanle for his stimulating interest, Mr. A. Hofstaetter for valuable assistance during our experiments, and the Deutsche Forschungsgemeinschaft for financial support.

References

[1] D. Hahn and K. Lertes, Z. Phys. **183**, 431 (1965).
[2] V. L. Levshin, V. B. Guman, and E. N. Karzhavina, Optika i Spektroskopiya **6**, 234 (1959).
[3] A. Scharmann, Z. Phys. **157**, 301 (1959).
[4] M. Sayer and W. R. Hardy, Canad. J. Phys. **43**, 1925 (1965).
[5] G. B. Beard, W. H. Kelly, and M. L. Mallory, J. appl. Phys. **33**, 144 (1962).
[6] P. Bräunlich, K. Reiber, and A. Scharmann, Z. Phys. **183**, 431 (1965).
[7] M. Sayer and A. Souder, Phys. Letters (Netherlands) **24A**, 246 (1967).
[8] K. Reiber and A. Scharmann, Z. Phys. **191**, 480 (1966).
[9] A. Zalkin and D. H. Templeton, J. chem. Phys. **40**, 501 (1964).
[10] M. I. Kay, B. C. Frazer, and I. Almodovar, J. chem. Phys. **40**, 504 (1964).
[11] R. W. Kedzie and M. Kestigan, Appl. Phys. Letters **3**, 86 (1963).
[12] D. H. Lyons and R. W. Kedzie, Phys. Rev. **145**, 148 (1966).
[13] H. Zeldes and R. Livingston, J. chem. Phys. **34**, 247 (1961).
[14] T. G. Castner and W. Känzig, J. Phys. Chem. Solids **3**, 178 (1957).

(Received April 30, 1968)

phys. stat. sol. 28, 589 (1968)

Subject classification: 14.1; 21.1

Max-Planck-Institut für Metallforschung, Institut für Metallkunde, Stuttgart

Hall Effect and Magnetoresistance in Copper Single Crystals at Low Temperatures

By

K. E. SAEGER[1])

The Hall coefficient and transverse magnetoresistance of copper single crystals of different purity are measured in dependence on the magnetic field strength, temperature, and crystal orientation. For high magnetic fields and low temperatures, Kohler's rule is valid for both effects in pure samples for all crystal orientations investigated. Departures from Kohler's rule occur for increasing temperatures and impurity contents. The investigation of the temperature dependence of the Hall coefficient shows that the maximum in the Hall coefficient versus temperature curves recently found by several authors is due to anisotropic electron-phonon scattering. The results of the present investigation indicate that the temperature dependence of the low field Hall coefficient differs significantly from that previously assumed to be valid.

An Kupfer-Einkristallen verschiedener Reinheit wurden Messungen der Feld-, Temperatur- und Orientierungsabhängigkeit von Hallkoeffizient und Widerstandsänderung im transversalen Magnetfeld durchgeführt. Im Falle hoher Magnetfelder und tiefer Temperaturen ist für beide Größen die Kohler-Regel bei reinen Proben für alle untersuchten Kristallorientierungen erfüllt. Durch Erhöhung der Temperatur oder Verunreinigung der Proben werden Abweichungen von der Kohler-Regel hervorgerufen. Die Untersuchung der Temperaturabhängigkeit des Hallkoeffizienten ergab, daß das in verschiedenen neueren Arbeiten gefundene Maximum von der Streuung der Elektronen an den thermischen Gitterschwingungen herrührt. Weiterhin läßt sich aus den in der vorliegenden Untersuchung gewonnenen Ergebnissen schließen, daß der Temperaturverlauf des Hallkoeffizienten im Niedrigfeldfall bei sehr reinen Proben sich stark vom bisher angenommenen unterscheidet.

1. Introduction

The last ten years have seen considerable progress in the understanding of the field dependence of the galvanomagnetic properties of metals at low temperatures. For the case of high magnetic fields the problem was solved semiquantitatively by Lifshits et al. [1, 2]. The results of their theory are in good agreement with the experimental work on noble metals [3 to 10]. Recently Seeger, Mann, and Clausecker [11] approached the problem on a somewhat different basis in order to obtain quantitative results. They were able to calculate the complete field dependence of the galvanomagnetic effects for a copper single crystal with the magnetic field in the [100] direction.

The state of knowledge concerning the temperature dependence of these effects, however, appears to be less advanced. The results of the calculations

[1]) Reported in part at the Bad Nauheim meeting of the Arbeitsgemeinschaft Metallphysik, April 17 to 22, 1967.

for very small magnetic fields carried out by Bross [12] do not even qualitatively agree with the experimental results of Fukuroi and Ikeda [13], which have been considered to be typical for the low-field region. The temperature dependence of the Hall coefficient of the noble metals calculated by Gabriel [14] agrees rather well with the work of Fukuroi and Ikeda [13]; however, it does not show the maximum of the Hall coefficient at approximately 35 °K which was found in several more recent investigations [15 to 18]. The different Hall effect results, of which only one set was obtained from a single crystal, are not in close agreement with respect to the temperature and height of the maximum. Thus it is very difficult to state which kind of a mechanism might be responsible for the occurrence of this maximum.

Neither Bross [12] nor Gabriel [14] fully take into account the influence of both the shape of the Fermi surface and the details of electron–phonon interaction. For this reason one is tempted to assume that the maximum is due to the combined action of both influences. On the other hand the "hump" in the Hall coefficient versus temperature curves might be due to open electron orbits. This line of reasoning is supported to a certain extent by the work of Klauder et al. [9, 10], who measured the orientation dependence of the Hall effect of copper at 4.2 °K. These authors found minima for magnetic field directions normal to planes containing open orbits. Influences of this type should, of course, become noticeable only for relatively high magnetic fields. This means that the low-field theories [12, 14] could not be expected to provide an explanation for the occurrence of the "hump", if the decrease of the Hall coefficient below approximately 35 °K is caused by the influence of open orbits. The experimental curves mentioned above [15 to 18] were found with magnetic fields ranging from 9 to 32.4 kG. Thus, it seems quite possible that they do not represent the true low-field behaviour of the Hall coefficient, and therefore the differences between the curve calculated by Gabriel [14] for very small fields and those found experimentally might indeed be explained as arising from the influence of open orbits. It will be shown, however, in Section 3 that this latter explanation, reasonable at is appears, is not true. The information drawn from the experimental results presented in this paper indicates rather convincingly that the "hump" is caused by anisotropy in the thermal scattering.

2. Experimental Procedure

The measurements reported in this paper were carried out in a cryomagnet designed by Ricker and Lück [19], which is completely submerged in liquid nitrogen. This magnet produces fields up to 14.7 kOe. The cryostat consists of a commercial glass dewar, which fits exactly into the air gap of the cryomagnet, and a brass head assembly which is clamped to the surface-ground flange of the dewar. A sample holder of the type described by Lück [20] is connected to the rotatable cryostat cover by means of a thin-walled section of Contracid tube. The rotatable cover also carries the feed-through for the current and potential leads. This set-up allows rotation of sample and holder about the current direction in a manner similar to that described by Borovik [21] and Klauder et al. [10]. By this means it is possible to measure Hall coefficient and magnetoresistance on each sample in many different orientations relative to the direction of the magnetic field. As the Hall voltage varies with φ the Hall coefficient is calculated from

$$R = \frac{U_H\, d}{I\, B \cos\varphi} \tag{1}$$

(cf. Lück and Saeger [22]). The uncertainty in the rotation angle φ is $\pm\ 1/2°$. The measuring circuit corresponds to that used by Lück and Saeger [22]. For some of the measurements, however, a Zernicke galvanometer of the type Zc (Kipp & Zonen, Delft) was used as a null detector instead of the Keithley nanovoltmeter. Using the nanovoltmeter the sensitivity of the measuring system amounted to $\pm\ 5 \times 10^{-9}$ V, with the Zernicke galvanometer it was approximately $\pm\ 10^{-8}$ V. Measurements of the temperature dependence of both Hall coefficient and magnetoresistance between the boiling point of helium and that of nitrogen were carried out during warm-up of the sample holder. The tempera-

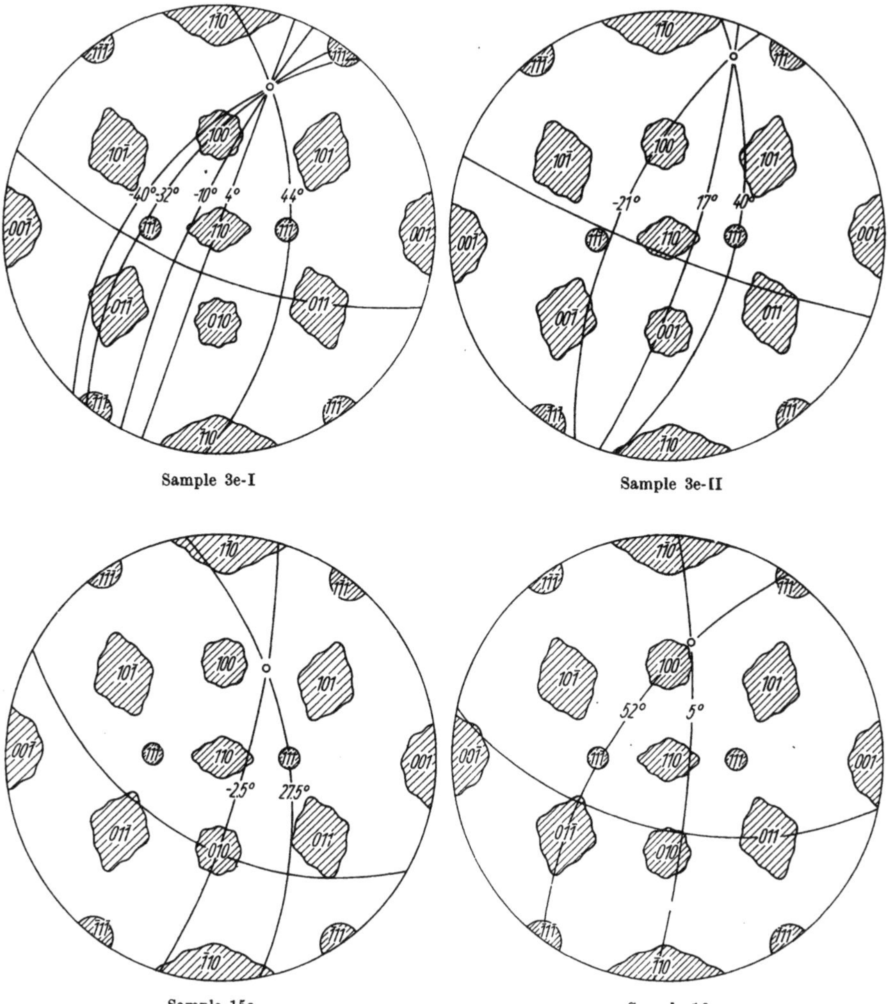

a Sample 3e-I

b Sample 3e-II

c Sample 15a

d Sample 16a

Fig. 1. Orientation diagrams for four different single crystal samples

ture was determined with the aid of a 100 Ω Allen-Bradley carbon resistor mounted inside the sample holder directly below the sample. The single crystal samples were produced from 99.999% pure copper (Johnson & Matthey) by the recrystallisation technique. Sample dimensions were 60 mm long by 7 mm wide by 0.18 mm thick. The residual resistivity ratio of sample 3e-I was $\varrho_{273}/\varrho_{4.2} \approx 1200$. After finishing the measurements on this sample a new one with a different current orientation, 3e-II, was cut from it. The residual resistivity ratio deteriorated to $\varrho_{273}/\varrho_{4.2} \approx 1000$ during this procedure. Two more samples of lower purity were also investigated. Their residual resistivity ratios were $\varrho_{273}/\varrho_{4.2} \approx 126$ for sample 15a and $\varrho_{273}/\varrho_{4.2} \approx 168$ for sample 16a.

Crystal orientation of the samples was determined from Laue back-reflection patterns. Current direction for 3e-I was approximately [8$\bar{3}$3], for 3e-II it was approximately [643]. The normal to the sample surface was close to [$\bar{2}$38] in both cases. For samples 15a and 16a the current and surface normal directions were approximately [602] and [$\bar{2}$06], and [6$\bar{1}$1] and [11$\bar{5}$], respectively. The current directions are marked as poles in Fig. 1, while the meridians represent the possible planes containing all directions normal to a given magnetic field direction. This representation of the crystallographic data corresponds to that given by Lück and Saeger [22]. The shaded areas represent the regions of open orbits as taken from Klauder et al. [10].

As the samples were extremely soft, it was impossible to determine their thickness by mechanical means with an accuracy of better than ± 2%. On the other hand, it was found that the anisotropy of the Hall coefficient measured at 78 °K does not exceed 1%. For this reason it was decided to take the value of R measured at 78 °K by Köster, Hagmann and Saeger [23] as a standard value. This appears to be justified as the sample used by these authors was produced from the same material and by the same technique. The measurements are not corrected for thermal expansion. Sample thickness should, according to Nix and MacNair [24] and Frazer and Hollis Hallett [25], not decrease by more than approximately 0.3% upon cooling from room temperature to 4.2 °K.

3. Experimental Results and Discussion

Information obtained from the combined measurement of Hall coefficient and transverse magnetoresistance versus angle of rotation about the current axis should, in the high-field region, be sufficient for the determination of the type of electron behaviour which is prevalent for different directions of magnetic field (cf. Fawcett [26]). Thus, the first step in the investigation was the determination of the dependence of both effects on field orientation. An estimate of $\omega\tau$, where ω is the cyclotron frequency and τ the relaxation time, from the applied field strength and the residual resistivity ratio of the two purer samples (3e-I and 3e-II) gives the values $\omega\tau \approx 7.8$ and $\omega\tau \approx 6.5$. The results of the high-field theory of Lifshits et al. [1, 2] are therefore applicable for the interpretation of the rotation diagrams shown in Figs. 2 and 3.

The dependence of the Hall coefficient, R, on field orientation (Figs. 2a and 3a) agrees quite well with the experimental curves obtained by Klauder et al. [9, 10], if account is taken of the fact that their curves are not corrected for the factor $\cos\varphi$ appearing in equation (1). The pronounced maximum of R at $-32°$ in Fig. 2a and the minimum of $\Delta\varrho/\varrho_0$ for the same orientation (Fig. 2b) point to the simultaneous existence of electron and hole orbits with negligible influence

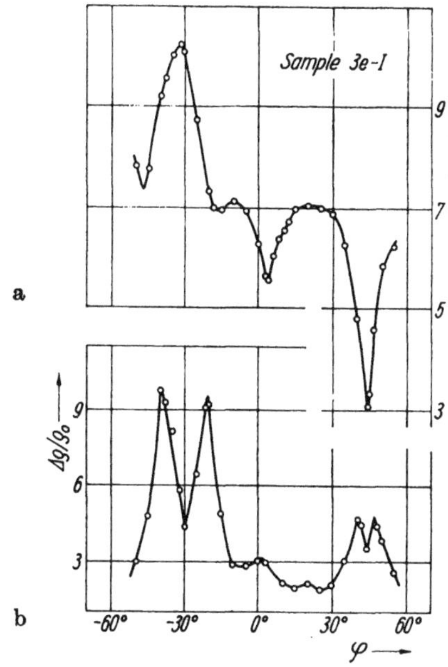

Fig. 2

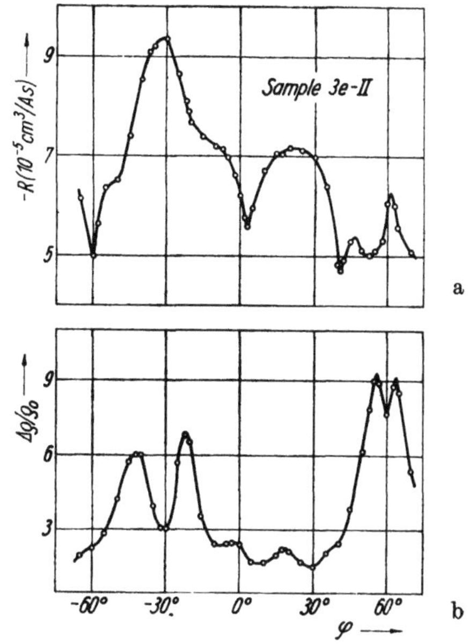

Fig. 3

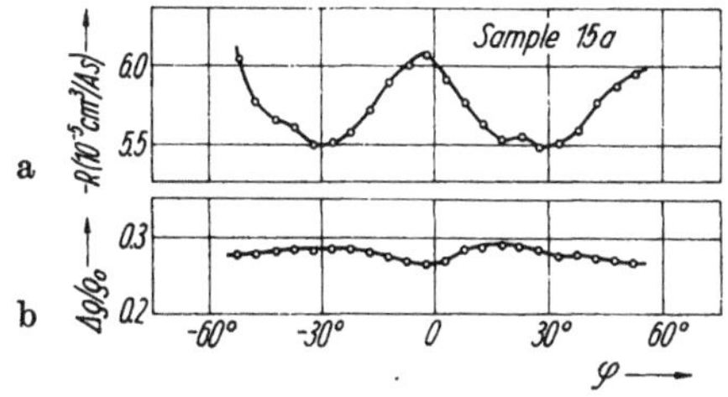

Fig. 4

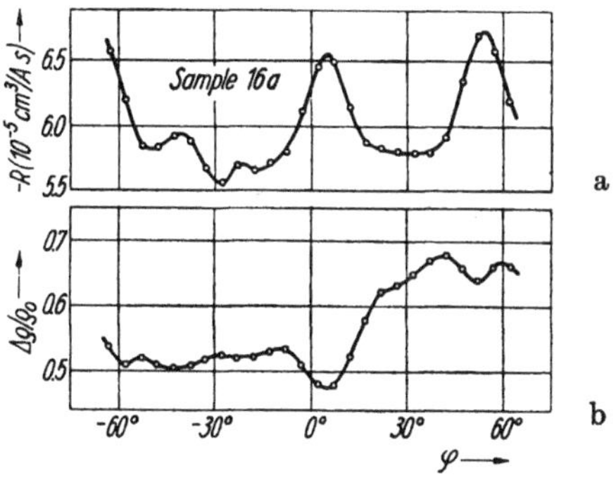

Fig. 5

Figs. 2 to 5. Rotation diagrams of Hall coefficient, R, and transverse magnetoresistance, as measured on four different single crystal samples

of open orbits. For this direction, which is roughly equivalent to $\boldsymbol{B} \parallel [011]$, the Hall coefficient is found to be $R_{011} = -10.3 \times 10^{-5}$ cm³/As. By extrapolation for $B \to \infty$ with the aid of equation (3) (see below) this gives $R_{011} = -10.7 \times 10^{-5}$ cm³/As while the values given by Klauder et al. [9, 10] are $R_{011} = -11.4 \times 10^{-5}$ cm³/As and -11.9×10^{-5} cm³/As, respectively. The value calculated by Klauder [27] on the basis of Pippard's [28] Fermi surface of copper is $R_{011} = -10.2 \times 10^{-5}$ cm³/As. The deep minimum of R at $+44°$ in Fig. 2a together with the twin maxima of $\Delta\varrho/\varrho_0$ between $+38$ and $+50°$ in Fig. 2b indicates the existence of two non-intersecting open orbits.

As predicted in Lifshits's theory [1, 2] and confirmed by several experimental investigations [3 to 8], the maxima in the rotation diagrams of $\Delta\varrho/\varrho_0$ are characteristic of field directions normal to the orientations of open orbits. With the

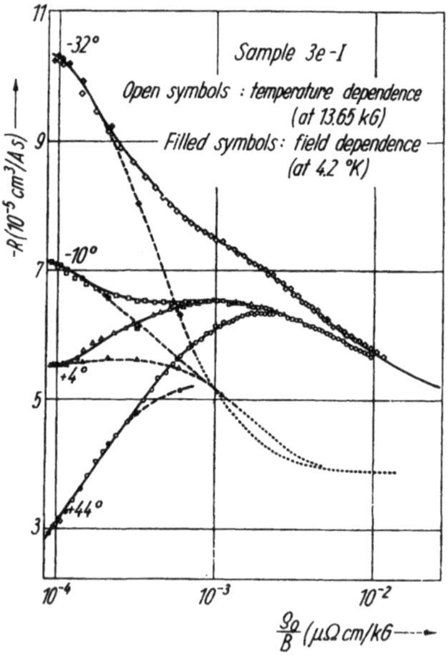

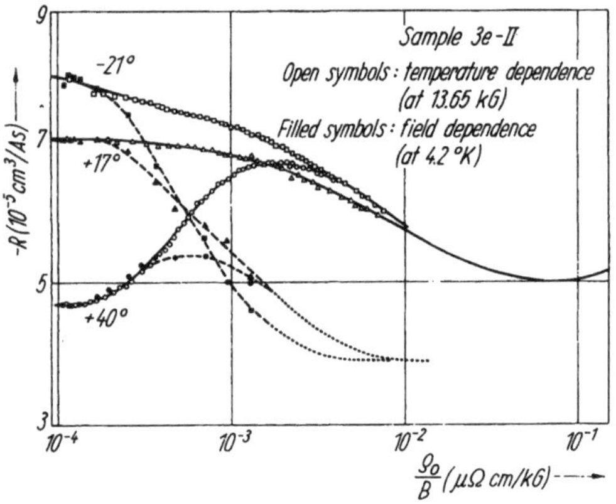

◀ Fig. 6 ▲ Fig. 7

Figs. 6 and 7. Field and temperature dependence of the Hall coefficient for different crystal orientations

aid of Figs. 1a and 1b it is possible to identify the open orbit direction responsible for each maximum in Figs. 2b and 3b. It is to be seen that the higher maxima belong to open orbit regions with short angular distance from the pole, i.e., the current direction, while the lower ones are due to open orbit regions farther distant from the pole. Thus, qualitative agreement is found with the theoretical predictions (cf. Klauder et al. [10] or Fawcett [26]). These state that $\varrho(B)$ in the high-field region varies with the angle ψ between the directions of current and open orbit in the following manner:

$$\varrho(B) = c_1 + c_2\, B^2 \cos^2 \psi\,, \tag{2}$$

where B is the magnetic induction and c_1 and c_2 are constants with regard to changes in B. A closer check is possible if measurements of the transverse-even voltage [22, 29] are also obtained. For the two samples in question these have been obtained by Lück and Saeger [22]. Figs. 4 and 5 show the rotation diagrams for the two samples of lower purity. For these curves the details are somewhat less distinct but it is still possible to relate their shape to the different types of electronic behaviour dictated by Fermi surface topology and magnetic field orientation. Thus meridians to be used with Fig. 1 may be obtained from Figs. 4 and 5; these indicate that the selected orientations correspond to the interesting orientations in Figs. 2 and 3.

Figs. 6 and 7 give the temperature dependence of R for samples 3e-I and 3e-II at 13.65 kG in the range between 4.2 and 77.8 °K as well as its field dependence at 4.2 °K in a Kohler plot. Corresponding curves of $\Delta\varrho/\varrho_0$ are shown in Figs. 8 and 9. The field dependence of the Hall coefficient (Figs. 6 and 7) shows a tendency towards saturation with increasing fields for all orientations which are not influenced by open orbits or which are influenced by only one open orbit. This is in good agreement with theory [1, 2] and with the findings of Klauder et al. [10]. For magnetic field direction parallel to [112], however, R disappears with increasing field strength (Fig. 5, 44°). The $1/B^2$-dependence predicted by theory (cf. Fawcett [26]) for this direction, which contains two non-intersecting open orbits, is only approximately verified. This may be due to the fact that the field direction is not exactly parallel to [112]. The field dependence of $\Delta\varrho/\varrho_0$ for orientations with high maxima (cf. Fig. 2b, $-40°$, and Fig. 3b, $-21°$) is in agreement with theory and the experimental results of Funes and Coleman [7] (cf. equation (2)). For orientations with less pronounced maxima (4° in Fig. 2b, 17° in Fig. 3b) an approximately linear field dependence is observed. For these directions the saturating part of magnetoresistance is, due to the small values of $\cos^2\psi$, comparable in magnitude to the part with quadratic field dependence for the fields applied here. The tendency towards saturation with increasing field strength for orientations without open orbits and for regions with two intersecting open orbits (Fig. 8, -10 and $+44°$, respectively) is quite obvious. This also agrees with Lifshits's theory [1, 2] and the experimental work of Alekseevskii and Gaidukov [3, 4]. Not quite clear is the behaviour of $\Delta\varrho/\varrho_0$ for $-32°$ in Fig. 8, where a linear field dependence is found. This could be due to the narrow band of non-vanishing open orbits for magnetic field in [110] direction (cf. Fawcett [26]). In this case a quadratic field dependence should be found at still higher fields as seen from equation (2). However, as the field direction for this orientation does not coincide exactly with the [110] direction it is also possible that only highly extended orbits, for which the conditions $\omega\tau \gg 1$ is not fulfilled due to their low values of ω (cf. Ziman [30],), exist.

It is seen from Figs. 6, 7, 8, and 9 that Kohler's rule is valid for the lower values of ϱ_0/B for both magnetoresistance and Hall effect at all orientations investigated. For values of ϱ_0/B above 2.5×10^{-4} $\mu\Omega$ cm/kG, however, the curve

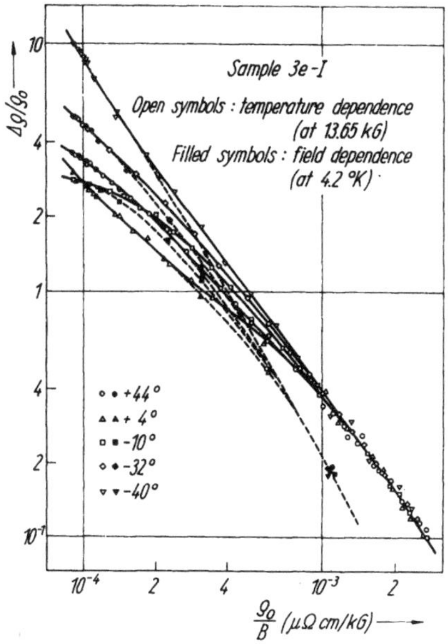

Fig. 8

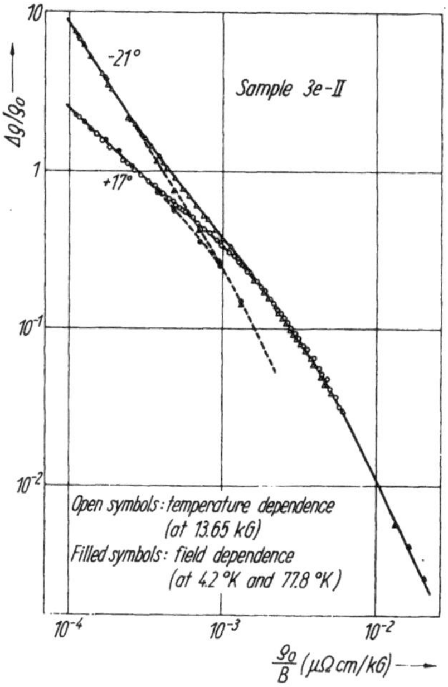

Fig. 9

Figs. 8 and 9. Field and temperature dependence of transverse magnetoresistance for different crystal orientations

of the field dependence at 4.2 °K and that of the temperature dependence at 13.65 kG no longer coincide. The various curves of the field dependence of R appear to approach a constant value, which is extrapolated with the aid of a formula given by Pippard [31] (see below) while the corresponding curves of $\Delta\varrho/\varrho_0$ combine to form a single one with quadratic field dependence. For both effects the curves of the temperature dependence also converge above the curve of the field dependence. The departure from Kohler's rule is considerable for both effects. In the case of magnetoresistance, the values of the temperature dependence are as much as three times the corresponding values of the field dependence, i.e., those at the same values of ϱ_0/B.

This behaviour finds a qualitative explanation in the changes in the type of scattering of the electrons with changing temperature. From the theory of Seeger et al. [11] it follows that Kohler's rule is valid only when one type of scattering mechanism is present . It is no longer valid if a second mechanism appears, e.g., scattering at interstitial sites in addition to scattering at vacancies [32]. Now, with thermal scattering becoming noticeable, such a second mechanism is introduced and therefore, the prerequisites for the validity of Kohler's rule are no longer met. It has to be assumed that these deviations from Kohler's rule are caused by an increase in the anisotropy of scattering of the electrons as one expects higher values of $\Delta\varrho/\varrho_0$ with an inreasec in the anisotropy of the scattering [12].

The question of the validity of Kohler's rule is not only interesting in connection with changes in resistivity due to the influence of temperature, but also in the case of variation of resistivity with impurity content. Comparing the field dependence of the Hall coefficient measured on the purer samples with that found for the corresponding orientations of the less pure samples, one finds that the departures from Kohler's rule are not very large (cf. Fig. 10). Therefore Gabriel's [14] assumption that the data given by Fukuroi and Ikeda [13], which were obtained at a magnetic field of 20 kG, are to be regarded as "low-field" ($\omega\tau \ll 1$) measurements because of the low residual resistivity ratio of $\varrho_{273}/\varrho_{4.2} \approx 26$, could be justified, at least in principle.

The comparison of the magnetoresistance measurements on two crystals of markedly different purity (3e-II: $\varrho_{273}/\varrho_{4.2} \approx 1000$, 15a: $\varrho_{273}/\varrho_{4.2} \approx 126$) carried out with the data given in Figs. 9 and 11 shows that Kohler's rule is violated even at 4.2 °K (cf. Fig. 12). For higher fields the curve of the field dependence of sample 15a comes very close to those of sample 3e-II; for lower fields, however, the curve for sample 15a reaches values of $\Delta\varrho/\varrho_0$ up to two times as high as the corresponding values for sample 3e-II. That means that the departures from Kohler's rule caused by foreign atoms at 4.2 °K may nearly reach the magnitude of those due to a considerable increase in temperature. This indicates that the change in anisotropy of scattering by impurity is comparable to that caused by thermal vibrations of the lattice. As found from Fig. 12 the increment in magnetoresistance is not superimposed by addition as one might expect. It rather appears as if a certain amount of increase over the low-field values of the purer samples is not exceeded. At temperatures in the vicinity of 77.8 °K field and temperature dependence of magnetoresistance coincide again, i.e., Kohler's rule is valid. This statement, however, must not be generalized as shown by the measurements of de Launey et al. [33]. The conclusion to be drawn is rather that the cases of the validity of Kohler's rule seem to constitute exceptions to the behaviour more generally found in experimental work.

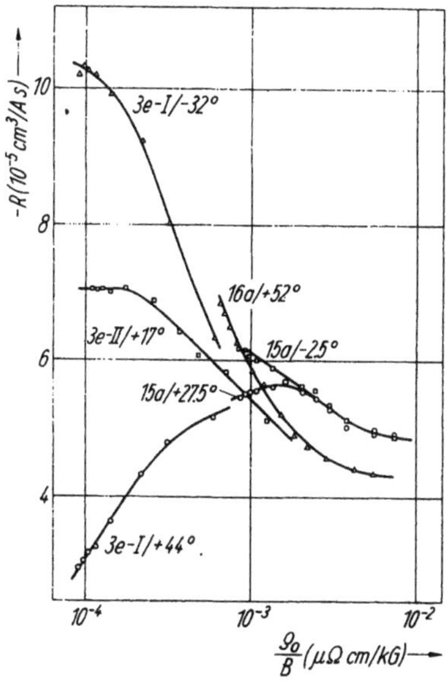

Fig. 10. Field dependence of the Hall coefficient at 4.2 °K for different crystal orientations in specimens of various purity

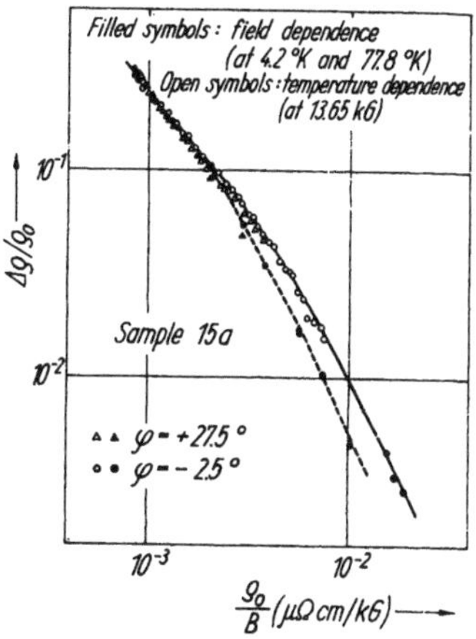

Fig. 11. Field and temperature dependence of transverse magnetoresistance for different crystal orientations

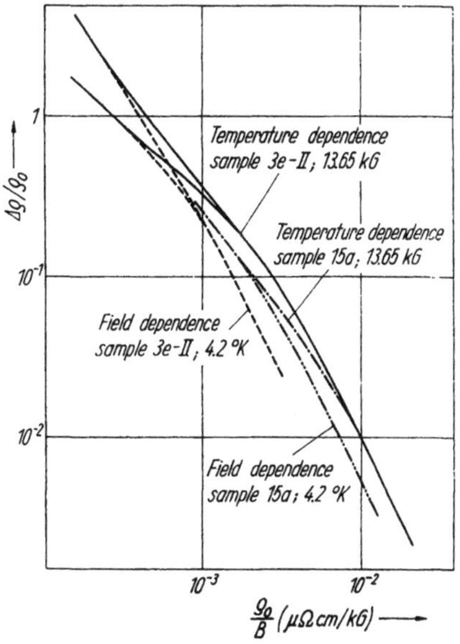

Fig. 12. Field and temperature dependence of transverse magnetoresistance for different sample purities (compiled from the measured values given in Figs. 9 and 11)

Figs. 8, 9, and 12 make it possible to explain the approximately linear increase of $\Delta\varrho/\varrho_0$ with magnetic field which was found by Kapitza [34] for polycrystalline copper samples at very high magnetic fields. These measurements were taken in boiling nitrogen; therefore the results for low magnetic fields correspond to the curve of the temperature dependence in Fig. 9. From the data given by Kapitza [34] a value of $\varrho_0/B \approx 8 \times 10^{-4}\,\mu\Omega\,\mathrm{cm/kG}$ is calculated for the highest fields applied. This means that the true high-field region is not yet attained. The values calculated for 300 and 200 kG rather belong to the region of transition between low-field behaviour at 77.8 °K and true high-field behaviour as described by Lifshits's [1, 2] theory. As seen in Figs. 8, 9, and 12 the field dependence of $\Delta\varrho/\varrho_0$ is fairly close to being linear in this transition region for all crystal orientations. Ziman's [35] explanation that the linear field dependence of $\Delta\varrho/\varrho_0$ found by Kapitza [34] is due to an averaging process over the different types of high-field behaviour is therefore not applicable in this case.

The comparison of temperature and field dependence of the Hall coefficient for the low-purity sample 15a (Figs. 10 and 13) leads to the conclusion that, for this sample, Kohler's rule is not valid at 4.2 °K. The same appears to be true at 77.8 °K where no appreciable field dependence was found, while the temperature dependence is quite strong as seen, for instance, in Figs. 13 and 14 between 40 and 200 °K.

A point of particular interest in this investigation was the maximum of R between 30 and 40 °K. For some of the field orientations of the two purer samples that support open orbits a maximum occurs. For $\boldsymbol{B} \parallel [112]$ and its vicinity this maximum is particularly strong, while for other orientations, especially those supporting electron and hole orbits with negligible influence of open orbits, no such maximum is found. From this an explanation for the behaviour of the polycrystalline samples analogous to that proposed by Ziman [35] for Kapitza's [34] measurements of $\Delta\varrho/\varrho_0$ could be constructed. However, such an explanation cannot be upheld in the light of the information given by Fig. 14 and a closer examination of Figs. 6 and 7. Fig. 14 gives the temperature dependence of R as measured on one of the less pure samples for an orientation that shows a maximum in the rotation diagrams, i.e., a direction for which electron and hole orbits occur and no open orbit influence is expected. For this orientation the "hump" is seen to be quite pronounced. On the other hand it is found that the maxima in the R–T curves, that may be drawn from Figs. 6 and 7 with the aid of Fig. 15, though slightly shifted with crystal orientation, occur in the same temperature range as those in Figs. 13 and 14 and not at similar $\omega\,\tau$ values. Thus it must be concluded that this "hump" can only be caused by a directly temperature-dependent mechanism. This view is also supported by the results of Raetz [17], who finds that the temperature of the Hall coefficient maximum for his polycrystalline copper, silver, and gold samples varies in a way similar to the variation of the Debye temperature for these metals.

From the measurements of the field dependence of R (Fig. 10) it is found that a considerable amount of impurities has to be introduced in order to make possible a "low-field" ($\omega\,\tau \ll 1$) measurement of the temperature dependence of the Hall coefficient at fields larger than 5 kG. However, as the magnetoresistance

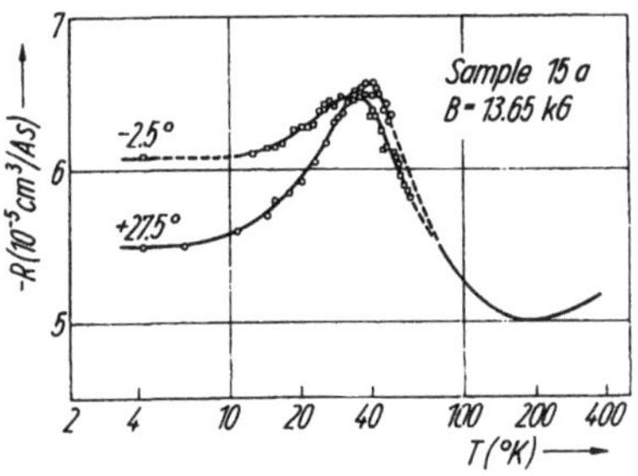

Fig. 13. Temperature dependence of the Hall coefficients for different crystal orientations

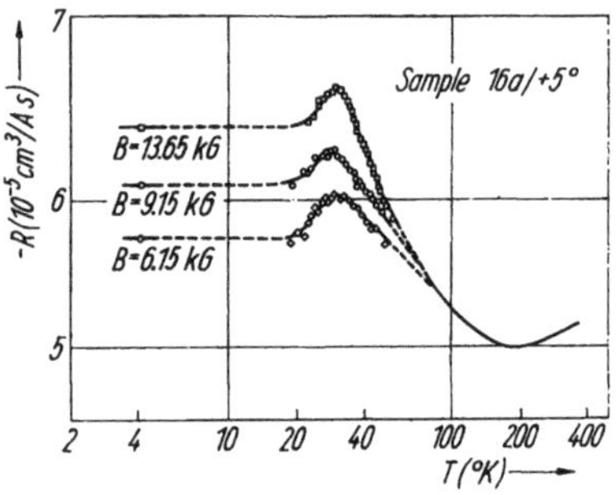

Fig. 14. Temperature dependence of the Hall coefficient at different field strengths

Fig. 15. Temperature dependence of resistivity for four different single crystal samples

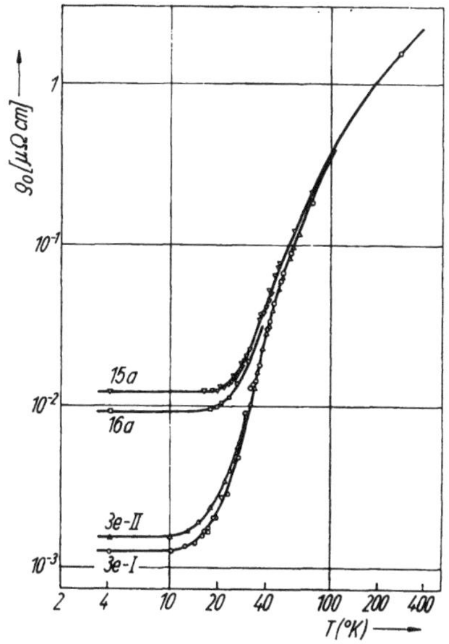

measurements show, this addition of impurities may produce a strong scattering anisotropy. Therefore, it is quite possible that at 4.2 °K the low-field value of R is increased by impurities as well. The departures from Kohler's rule found in Fig. 10 may well be interpreted to support such a possibility. In this case "low-field" measurements of the Hall coefficient at field strengths of several kG become rather problematic. Since, on the other hand, it is extremely difficult to measure pure samples in the vicinity of $\omega\tau \approx 0.1$ with a satisfactory accuracy, it seems reasonable to try an extrapolation of the low-field value of R at liquid helium temperature from the high and intermediate field measurements carried out on one of the purer samples. This extrapolation is tried with the aid of Pippard's [31] two-band formula for the Hall coefficient:

$$R = \frac{(N_- - N_+)\,(1 + \omega^2\tau^2)}{e\,[(N_- + N_+)^2 + (N_- - N_+)^2\,\omega^2\tau^2]}\,. \tag{3}$$

The assumptions made in deriving this equation are, of course, somewhat problematic for the case of the noble metals. Still, because the high-field value found from this formula is identical with that given by Lifshits et al. [1, 2] for the case of field orientations with electron and hole orbits and negligible influence of open orbits, and because equation (3) gives the type of curve one expects in this case for the low-field region, its use for extrapolation in this case appears to be reasonable. It must be kept in mind, however, that the parameters $(N_- + N_+)$ and $(N_- - N_+)$ should not be given any special significance for the low-field region. Use of equation (3) with the parameters $(N_- - N_+)$ and $(N_- + N_+)$ chosen so as to give the best fit to the measured values of orientation $-32°$ of sample 3e-I yields the low-field value $R = 3.9\times10^{-5}$ cm³/As. The field dependence of the Hall coefficient as measured for the corresponding orientation on the less pure sample 16a is given in Fig. 10 (16a / 52°). For $\omega\tau \approx 0.1$ a field-independent value of R is almost reached for this sample; in this case it amounts to approximately -4.3×10^{-5} cm³/As. This and the low-field value of approximately -4.85×10^{-5} cm³/As found for the least pure sample 15a indicates that the "low-field" value of the Hall coefficient is indeed influenced by impurity content. As the low-field value of R is expected to be isotropic (see, for instance, Bross [12]) it is possible to compare these values obtained by experiment and extrapolation with the theoretical results for the [100] direction as given by Seeger et al. [11]. Their result of approximately -4.5×10^{-5} cm³/As is somewhat higher than the value found by extrapolation from the measurements on sample 3e-I; in fact, it appears to agree better with

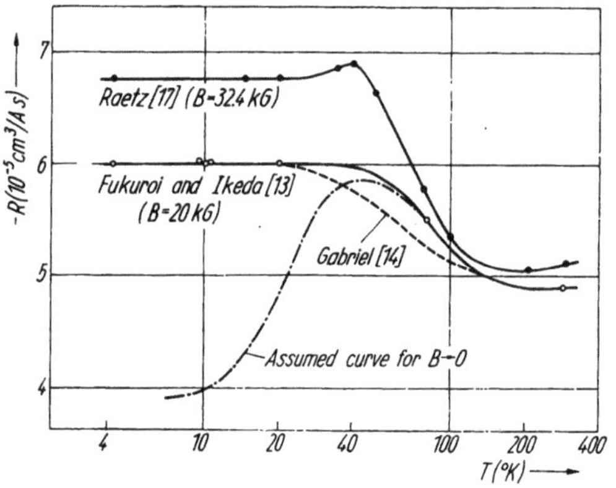

Fig. 16. Assumed temperature dependence of the Hall coefficient at vanishingly small magnetic field; for comparison several measured and calculated curves taken from literature

the value measured on the less pure sample 16a. Considering the fact that the previous theoretical results [12, 14] always contained at least one free parameter the agreement with the results of Seeger et al. [11] is extremely satisfactory.

The low-field value of R extrapolated as described above is drawn in Figs. 6 and 7. With this and the field dependence curves measured for high and intermediate fields it is possible to construct curves of the field dependence of R for values of $\varrho_0/B > 10^{-3}$ cm/kG for the purer samples. With the aid of these curves, which are given as dotted lines, one can see that not only the temperature at which the departure from Kohler's rule begins, varies with crystal orientation, but also that the same is true for the amount of deviation. This gives a phenomenological explanation for the considerable difference in relative height of the "hump" for the two different crystal orientations shown in Fig. 13.

If it is assumed that for temperatures above 80 °K a noticeable field dependence of R no longer exists — Fig. 14 supports this assumption — and that the amount of deviation from Kohler's rule as taken from Figs. 6 and 7 does not vary too strongly with the applied magnetic field, then it is possible to estimate the shape of the R–T curve for a pure copper sample as would be measured at vanishingly small fields. Such a curve assembled from the data produced in this paper is given in Fig. 16 together with the results of several previous investigations.

It is seen that the shape of this estimated curve differs considerably from the curves so far regarded as typical low-field measurements. In the case of Raetz's [17] experiments the reason for this is in all probability to be found in the fact that they are by no means to be taken as low-field measurements, as from his data values of $\omega\,\tau$ between 5 and 2 are found for 4.2 °K. In this region of $\omega\,\tau$ there should still be a considerable field dependence. The measurements of Fukuroi and Ikeda [13] were taken on cold-rolled samples with a relatively high impurity content. For this reason a higher value of R had to be expected as shown above. It is quite probable that the high concentration of dislocations introduced by cold-rolling exert a particularly strong influence in this direction. Evidence for this was found an cold-rolled silver samples at 77.8 °K and even at room temperature by Köster et al. [23, 36]. All things considered, Gabriel's [14] theory cannot be said to be any longer very satisfactory, even if one makes allowances for the fact that our constructed curve cannot give exact values. On the other hand the basic objection to Bross's [12] results, i.e., the argument that at low temperatures decreasing values of R are not found experimentally [14], is not valid any longer. It appears necessary, however, to base a calculation of the temperature dependence of R in the low-field region not only on the true lattice dynamics as done by Bross [12], but also on the true Fermi surface instead of introducing a plausible value of the effective mass.

Acknowledgements

The author wishes to thank Dr. R. Lück for many helpful discussions and Prof. Dr. A. Seeger for reading the manuscript.

References

[1] I. M. Lifshits, M. Ya. Azbel, and M. I. Kaganov, Soviet Phys. — J. exper. theor. Phys. **3**, 143 (1956); **4**, 41 (1967).
[2] I. M. Lifshits and V. G. Peshanskii, Soviet Phys. — J. exper. theor. Phys. **8**, 875 (1959); **11**, 137 (1960).
[3] N. E. Alekseevskii and Yu. P. Gaidukov, Soviet Phys. — J. exper. theor. Phys. **10**, 481 (1960).
[4] Yu. P. Gaidukov, Soviet Phys. — J. exper. theor. Phys. **10**, 913 (1960).
[5] N. E. Alekseevskii and Yu. P. Gaidukov, Soviet Phys. — J. exper. theor. Phys. **15**, 49 (1962).
[6] J. R. Klauder and J. E. Kunzler, in: The Fermi Surface, John Wiley & Sons, New York 1960.
[7] A. J. Funes and R. V. Coleman, Phys. Rev. **131**, 2084 (1963).
[8] R. V. Coleman, A. J. Funes, J. S. Plaskett, and C. M. Tapp, Phys. Rev. **133**, A521 (1964).
[9] J. E. Kunzler and J. R. Klauder, Phil. Mag. **6**, 1054 (1961).
[10] J. R. Klauder, W. A. Reed, G. F. Brennert, and J. E. Kunzler, Phys. Rev. **141**, 592 (1966).
[11] A. Seeger, E. Mann, and K. Clausecker, phys. stat. sol. **24**, 721 (1967). K. Clausecker, Diploma Work, Stuttgart 1967.
[12] H. Bross, Z. Naturf. **14a**, 560 (1959).
[13] T. Fukuroi and T. Ikeda, Sci. Rep. Res. Inst. Tôhoku Univ. 8, 205 (1956).
[14] H. Gabriel, Thesis, Braunschweig 1960.
[15] T. G. Berlincourt, Phys. Rev. **112**, 381 (1958).
[16] W. F. Love, J. Phys. Chem. Solids **9**, 281 (1959).
[17] K. Raetz, Thesis, Gießen 1968.
[18] K. E. Saeger, Thesis, Stuttgart 1967.
[19] Th. Ricker and R. Lück, Z. Metallk. **56**, 799 (1965).
[20] R. Lück, phys. stat. sol. **18**, 59 (1966).
[21] E. S. Borovik, Soviet Phys. — J. exper. theor. Phys. **3**, 303 (1956).
[22] R. Lück and K. E. Saeger, phys. stat. sol. **21**, 671 (1967).
[23] W. Köster, D. Hagmann, and K. E. Saeger, Z. Metallk. **54**, 619 (1963).
[24] F. C. Nix and D. MacNair, Phys. Rev. **60**, 587 (1941); **61**, 74 (1942).
[25] D. B. Frazer and A. C. Hollis Hallet, Proc. VII. Internat. Conf. Low Temperature Physics, Toronto 1960.
[26] E. Fawcett, Adv. Phys. **13**, 139 (1964).
[27] J. R. Klauder, Bell Syst. techn. J. **40**, 1349 (1961).
[28] A. B. Pippard, Phil. Trans. Roy. Soc. (London) **A250**, 325 (1957).
[29] J. R. Klauder and J. E. Kunzler, Phys. Rev. Letters **6**, 179 (1961).
[30] J. M. Ziman, Electrons and Phonons, Oxford University Press, Oxford 1960 (p. 520).

[31] A. B. PIPPARD, Rep. Progr. Phys. **23**, 176 (1960).
[32] K. CLAUSECKER, private communication.
[33] J. DE LAUNEY, R. L. DOLECEK, and R. T. WEBBER, J. Phys. Chem. Solids **11**, 37 (1959).
[34] P. KAPITZA, Proc. Roy. Soc. **A213**, 292 (1929).
[35] J. M. ZIMAN, Phil. Mag. **3**, 1117 (1958).
[36] W. KÖSTER and W. SCHÜLE, Z. Metallk. **48**, 634 (1957).

(Received May 16, 1968)

phys. stat. sol. 28, 603 (1968)
Subject classification: 10.1; 21; 21.1; 21.1.1

Institut für Metallphysik und Reinstmetalle der Deutschen Akademie der Wissenschaften zu Berlin, Dresden

Stacking Faults on {110} Planes in the B.C.C. Lattice

By
B. PEGEL

Maximum stacking fault energies for α-type stacking faults in the b.c.c. lattice are estimated for the group Va and VIa transition metals and iron. The half widths of glissile type split screw dislocations should be in all cases at least one total Burgers vector.

Es wird eine Abschätzung der oberen Grenze der Stapelfehlerenergien von α-Typ Stapelfehlern für die k.r.z. Übergangsmetalle der fünften und sechsten Nebengruppe sowie für Eisen gegeben. Die Versetzungsweiten aufgespaltener, gleitfähiger Schraubenversetzungen sollten in allen Fällen mindestens vom Betrag eines Burgersvektors sein.

1. Introduction

The concept of splitting of dislocations on {110} planes in b.c.c. crystals was first introduced by Crussard [1] and Cohen et al. [2] and has since been used in various theoretical models [3 to 7]. However, the possibility of stacking faults on {110} planes is at present only a theoretical proposal based on hard-sphere model considerations which are unrealistic for b.c.c. crystals.

In this paper we presuppose from the outset that stable stacking faults on {110} planes do exist and then we calculate their energies. The calculations are based on two-body central forces between nearest and next nearest neighbours in the b.c.c. lattice. From the very existence of stacking faults some information on the two-body interaction potentials can be drawn. It is well known that central forces imply the Cauchy relation $C_{12} = C_{44}$ for the elastic constants which, however, is not satisfied for the metals considered. In our model this difficulty is by-passed by using only two linear combinations of the three elastic constants. If any further linear combination is either fully determined (due to the Cauchy relation) or not is then quite immaterial for the problem.

2. Stability Considerations

In Fig. 1 the atomic structure of a $(01\bar{1})$ plane (called plane 0 in the following) is shown . All nearest and next nearest neighbours of an arbitrary atom in plane 0 are situated either in its own plane or in the planes immediately above (plane I) and below (plane II) it. The distance of neighbouring planes is $d = a/\sqrt{2}$ where a is the lattice constant. Imagine a rigid displacement of all crystal planes above the plane 0 shown in Fig. 1, i.e. a rigid displacement of the upper half crystal with respect to the lower one. With this displacement all nearest and next nearest atomic distances except those between plane 0 and plane I remain unchanged. In Fig. 1 the point A is a representative atom of plane I projected into plane 0. Its nearest and next nearest neighbours in plane 0 are numbered 1, 2 and 3, 4, respectively. The rigid displacements mentioned are given by a set of vectors, **s**, which, for reasons of symmetry, will be assumed to lie in a plane piercing verti-

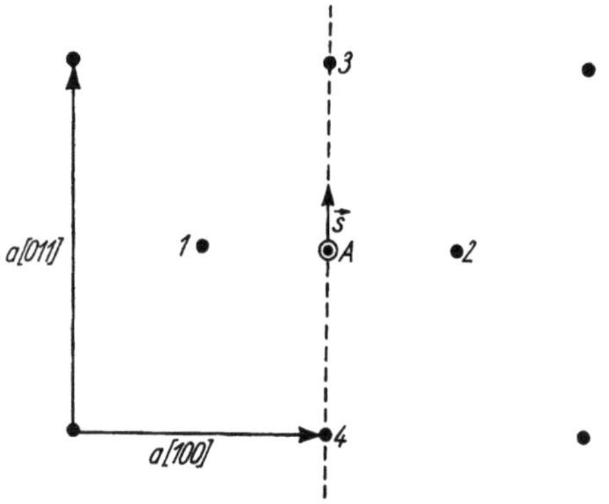

Fig. 1. Atomic structure of a $(01\bar{1})$ plane in the b.c.c. lattice

cally through the $(01\bar{1})$ crystal planes. Its trace in plane 0 is indicated by a dotted line in Fig. 1. The components of **s** in the directions [011] and $[01\bar{1}]$ are designated by $s_1 = d\,\varepsilon$ and $s_2 = d\,\mu$ where ε and μ are the corresponding components in units of the distance d. Let R_i ($i = 1, 2, 3, 4$) the distances between A and its neighbours in plane 0. These are given by

$$\left.\begin{aligned} R_1^2 = R_2^2 &= \frac{a^2}{4}\,[1 + 2\,\varepsilon^2 + 2\,(1+\mu)^2]\,, \\ R_3^2 &= \frac{a^2}{2}\,[(1-\varepsilon)^2 + (1+\mu)^2]\,, \\ R_4^2 &= \frac{a^2}{2}\,[(1+\varepsilon)^2 + (1+\mu)^2]\,. \end{aligned}\right\} \tag{1}$$

From these equations we find the relation

$$R_3^2 + R_4^2 = 2\,R_1^2 + \frac{a^2}{2}\,. \tag{2}$$

For $\varepsilon = \mu = 0$ we have $R_1 = R_2 = r_1 = a\,\sqrt{3}/2$ and $R_3 = R_4 = r_2 = a$ where r_1 and r_2 are the nearest and next nearest neighbour distances in the ideal lattice, respectively.

Let $V(r)$ the interatomic potential which will be assumed to vanish somewhere between second and third nearest neighbour distances. Because the energy change of the deformed crystal is only due to the change of atomic distances between plane 0 and plane I the contribution of each atom in plane I (or plane 0) to this energy change is

$$W(\varepsilon, \mu) = 2\,V(R_1) + V(R_3) + V(R_4)\,. \tag{3}$$

W has, of course, a minimum at $\varepsilon = \mu = 0$. If stable α-type stacking faults exist there must be other values of ε and μ for which the minimum conditions

$$\frac{\partial W}{\partial \varepsilon} = \frac{\partial W}{\partial \mu} = 0, \quad \frac{\partial^2 W}{\partial \varepsilon^2} > 0\,, \quad \frac{\partial^2 W}{\partial \mu^2} > 0; \quad \frac{\partial^2 W}{\partial \varepsilon^2}\,\frac{\partial^2 W}{\partial \mu^2} > \left(\frac{\partial^2 W}{\partial \varepsilon\,\partial \mu}\right)^2 \tag{4}$$

are satisfied. For shortness we put $V'(R_i)/R_i \equiv g_i$. Then, using equations (1),

Fig. 2. Qualitative shape of the two-body interaction potential

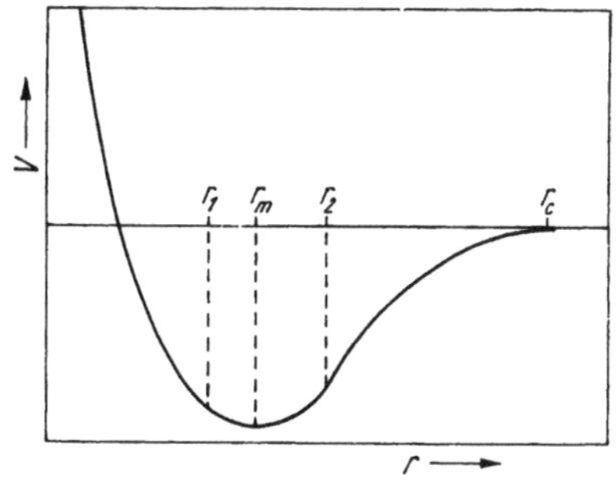

we find

$$\left.\begin{aligned}\frac{\partial W}{\partial \mu} &= \frac{a^2}{2}(1+\mu)(2\,g_1+g_3+g_4)\,,\\ \frac{\partial W}{\partial \varepsilon} &= \frac{a^2}{2}\,\varepsilon\,(2\,g_1+g_3+g_4)+\frac{a^2}{2}(g_4-g_3)\,.\end{aligned}\right\} \tag{5}$$

Excluding the formal possibility $1 + \mu = 0$ (i.e. pressing two neighbouring lattice planes together into one plane) the conditions $\partial W/\partial\mu = \partial W/\partial\varepsilon = 0$ become

$$g_3 = g_4 \quad \text{and} \quad g_1 + g_3 = 0\,. \tag{6}$$

For $\varepsilon = \mu = 0$ the first of these conditions is fulfilled trivially ($R_3 = R_4$) and the second one reads

$$\frac{V'(r_1)}{r_1} + \frac{V'(r_2)}{r_2} = 0\,. \tag{7}$$

The interatomic potential must look qualitatively as shown in Fig. 2 [8, 9]. r_c is the range of the interatomic forces. Evidently, a stacking fault is possible for the configuration

$$R_1 = R_3 = r_m \quad \text{and} \quad R_4 \geq r_c\,. \tag{8}$$

In this case we simply have $g_1 = g_3 = g_4 = 0$, satisfying equation (6). Further it is obvious from Fig. 2 and equation (3) that this configuration is a stable one. Using the equations (1) and (2) the conditions (8) can be written as

$$\varepsilon = \frac{1}{4}\,, \quad \mu = 1 - \sqrt{2\,(r_m/a)^2 - 9/16}\,, \quad R_4 = \sqrt{r_m^2 + a^2/2} \equiv R_4^0 \geq r_c\,. \tag{9}$$

The value $\varepsilon = 1/4$ is exactly that predicted from the hard-sphere model (e.g. [5]). The position r_m of the potential minimum will be roughly halfway between r_1 and r_2 (see Table 2), i.e. $\mu \approx -0.1$. This means that the faulted planes are compressed by approximately 10 per cent.

3. Stacking Fault Energies

Let us first consider the interatomic potentials for iron used by Johnson [8] and Chang and Graham [9]. Both do not satisfy the condition $R_4^0 \geq r_c$. Johnson assumed r_c to be exactly halfway between second and third nearest neighbour distances whereas the cut-off in the model of Chang and Graham is at a slightly smaller distance. Moreover, in the Chang model the stability condition (7) does not hold. Since the cut-off distance r_c is chosen rather arbitrarily in these models and a slight change of this parameter is sufficient to guarantee a stable stacking fault, we may calculate the stacking fault energy by changing r_c a little bit but leaving invariant the potential curve between r_1 and r_2. Proceeding in this way we obtain for the stacking fault energy per atom, $\Delta W = 3\,V(r_m) - 2\,(V(r_1) + V(r_2))$, $\Delta W = 0.1158$ eV (Johnson) and $\Delta W = 0.0277$ eV (Chang, Graham), respectively. This corresponds to stacking fault energies of 320 erg/cm² (Johnson) and 77 erg/cm² (Chang, Graham), respectively.

In the following we assume the condition $R_4^0 \geq r_c$ to hold, i.e. stable stacking faults to exist. Then we need, for a calculation of ΔW, the shape of the potential curve between r_1 and r_c. At first we try to find out the shape of the potential curve between r_1 and r_2. For this purpose we expand the potential curve around its minimum

$$V(r) = V_0 + \alpha\,(r - r_m) + \beta\,(r - r_m)^3 \tag{10}$$

assuming that a cubic parabola is a sufficient good approximation. The constants α, β, and r_m can be determined from the elastic constants of the solid as follows. A small rigid displacement of the type discussed in section 2 is, in linear elasticity theory, described by an elastic displacement field $\boldsymbol{u}(\boldsymbol{r}) = \boldsymbol{s}\,f(\boldsymbol{n}\,\boldsymbol{r})$. Here $\boldsymbol{n}$ is the normal of the set of lattice planes considered and $f(x)$ is a smooth step function rising from 0 to 1 in a small interval around $x = 0$. The elastic energy density η corresponding to such a displacement is

$$\eta = \frac{1}{4} f'^2\,\{s_1^2\,(C_{11} - C_{12}) + s_2^2\,(2\,C_{44} + C_{11} + C_{12})\}\,. \tag{11}$$

Therefore, for rigid displacements of the type considered only those two linear combinations of elastic constants appearing in (11) are effective. Using equation (7) the interesting combinations of elastic constants can be expressed through V as follows (e.g. [10]):

$$\left.\begin{aligned} \frac{3}{2} a\,[2\,C_{44} + C_{11} + C_{12}] &= 4\,V''(r_1) + 3\,V''(r_2) + V'(r_2)/r_2\,,\\ \frac{a}{2}\,[C_{11} - C_{12}] &= V''(r_2) + V'(r_1)/r_1\,. \end{aligned}\right\} \tag{12}$$

The two equations (12) together with equation (7) determine the constants α, β, and r_m of equation (10).

Table 1

Adiabatic elastic constants at room temperature in units 10^{11} N m^{-2}

	C_{11}	C_{12}	C_{44}	Ref.
V	2.28	1.19	0.426	[11]
Nb	2.46	1.34	0.287	[11]
Ta	2.61	1.57	0.818	[12]
Cr	3.50	0.578	1.008	[13]
Mo	4.408	1.724	1.217	[12]
W	5.23	2.05	1.607	[12]
Fe	2.28	1.32	1.165	[14]

Using the elastic constants given in Table 1 (a correction for isothermal elastic constants would only concern not significant decimals and is not justified in view of the whole estimation) the results are presented in Table 2.

Tabel 2

Parameters of equation (10), describing the shape of the interatomic potential between r_1 and r_2

	r_m/r_1	αr_1^2 [eV]	βr_1^3 [eV]
V	1.0605	6.067	$-$ 7.530
Nb	1.0606	7.958	$-$ 9.590
Ta	1.0549	11.097	$-$20.062
Cr	1.0784	7.373	$+$ 5.414
Mo	1.0654	13.023	$-$ 9.262
W	1.0655	16.476	$-$13.196
Fe	1.0534	7.463	$-$14.570

Except for Cr the positions r_m of the potential minima are somewhat left-hand from the midpoint of r_1 and r_2 $\left(\frac{1}{2}\frac{r_1 + r_2}{r_1} = 1.0774\right)$.

$\Delta W_1 = 2\,V(r_1) + V(r_2) - 3\,V(r_m)$ is the energy gain per atom due to the change of the distances R_1, R_2 and R_3 (when the atoms originally situated at distances r_1 and r_2 in Fig. 2 slide into the minimum). In Table 3 ΔW_1 is presented in units eV.

Table 3

Maximum stacking fault energies and minimum half widths of a glissile type split screw dislocation

	W_1 (eV)	$W_{2\,max}$ (eV)	γ_{max} (erg cm^{-2})	$(r/b)_{min}$
V	0.095	0.160	160	0.8
Nb	0.125	0.211	190	0.9
Ta	0.164	0.272	220	1.4
Cr	0.131	0.273	390	1.2
Mo	0.214	0.391	410	1.3
W	0.269	0.481	480	1.4
Fe	0.198	0.162	150	2.3

By $\Delta W_2 = -V(r_2)$ we denote the energy necessary to enlarge the distance R_4 to values larger than r_c, i.e. to let the atom 4 in Fig. 1 escape from the attraction exerted on it from atom A. ΔW_2 depends, of course, on the potential depth

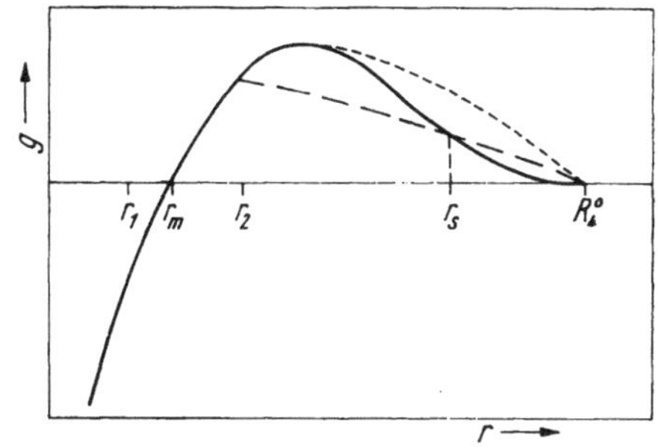

Fig. 3. Qualitative shape of the curve $g(r) = V'(r)/r$; dashed line: fictitious $g(r)$ curve; dotted line: calculated from equation (13)

V_0. The stacking fault energy per atom is then $\Delta W = \Delta W_2 - \Delta W_1$. Unfortunately, no reliable values of V_0 are available. But we can draw some conclusions about the shape of the potential curve between r_2 and r_c provided that stable stacking faults do exist. For this purpose we consider the equilibrium conditions (6). In Fig. 3 the qualitative shape of the curve $g(r)$ is shown which is representative for all special cases. ΔW_2 is geometrically represented by the integral $\Delta W_2 = \int_{r_2}^{r_c} r\, g(r)\, dr$. If there existed a stable stacking fault there must be an unstable saddle point configuration somewhere between the ideal lattice configuration and the stacking fault configuration. This saddle point configuration is characterized by $r_1 < R_1 < r_m$, $r_m < R_3 < r_2$, $r_2 < R_4 < r_c$ whereby equations (2) and (6) must be satisfied. The dashed line between r_2 and R_4^0 in Fig. 3 represents a fictitious $g(r)$ curve for which every atomic configuration satisfying equation (2) is an equilibrium configuration, i.e. for which the equations (6) are fulfilled simultaneously. The real $g(r)$ curve must intersect the fictitious $g(r)$ curve at some point r_s between r_2 and R_4^0 if there is a stable stacking fault separated from the ideal lattice by a saddle point. r_s represents the distance R_4 corresponding to the saddle point configuration. Let us assume that the potential curve between r_2 and R_4^0 can be represented by a cubic parabola

$$V(r) = A\,(r_c - r)^2 + B\,(r_c - r)^3 \tag{13}$$

vanishing with zero slope at $r = r_c$. Given r_c, the constants A and B can be calculated by adapting the slope and curvature at the point r_2 to the values known from equation (10). Putting $r_c = R_4^0$ and calculating $g(r) = V'(r)/r$ from (13) we obtain the dotted curve of Fig. 3. Its slope at $r = R_4^0$ is always larger than the slope of the fictitious $g(r)$ curve, hence it runs always above the fictitious curve, joining $g(r)$ smoothly at $r = r_2$. Thus, if we calculate ΔW_2 from the potential (13) $(r_c = R_4^0)$ we get an upper boundary value for ΔW_2. In the second column of Table 3 the maximum values $\Delta W_{2\,max}$ obtained in this way are given. From the maximum stacking fault energies per atom, $\Delta W_{max} = \Delta W_{2\,max} - \Delta W_1$, the maximum stacking fault energies in usual units are easily calculated. They are also presented in Table 3.

4. Conclusion

Kroupa and Vitek [3] have calculated the equilibirum widths of split dislocations using linear, isotropic elasticity theory. For a rough estimation of the widths of split dislocations we consider a glissile type splitting of a pure screw dislocation on a {110} plane. Its half width, r, is given by [3]

$$\frac{r}{b} = \frac{b\,\mu\,k}{\gamma}\,; \quad k = \frac{5 - 6\,\nu}{96\,\pi\,(1 - \nu)}\,,$$

where μ is the shear modulus, b the absolute value of the total Burgers vector, and ν the Poisson ratio. Deriving isotropic elastic constants μ and ν from Table 1 by means of the procedure of Leibfried [15] and putting $b = \sqrt{3}\,a/2$ we can calculate the minimum values of r/b using the maximum values of the stacking fault energies derived in section 3. The results are also given in Table 3. Since the tabulated r/b values are lower boundary values we expect the half widths of split dislocations to be at least one Burgers vector for all metals considered.

Concluding we may state that (provided that stable stacking faults do exist) the concept of dislocation splitting on {110} planes is reasonable for all metals considered. In most cases (except, perhaps, iron) splitting is more adequately described as a state of the dislocation core which is not covered by linear elasticity theory. There is at present no reasonable idea how to estimate minimum stacking fault energies for the transition metals. Probably, electronic relaxation will play an important role with regard to this problem.

Acknowledgement

The author is indepted to Mr. H. Eichler for valuable discussions as well as for checking some calculations.

References

[1] C. Crussard, C. R. Acad. Sci. (France) **252**, 273 (1961).
[2] J. B. Cohen, R. Hinton, K. Lay, and S. Sass, Acta metall. **10**, 894 (1962).
[3] F. Kroupa and V. Vitek, Czech. J. Phys. **B14**, 337 (1964).
[4] F. Kroupa, phys. stat. sol. **3**, K391 (1963).
[5] F. Kroupa and V. Vitek, Canad. J. Phys. **45**, 945 (1967).
[6] R. J. Wasilewski, Acta metall. **13**, 40 (1965).
[7] J. P. Hirth and J. Lothe, phys. stat. sol. **15**, 487 (1966).
[8] P. A. Johnson, Phys. Rev. **134**, A1329 (1964).
[9] R. Chang and L. J. Graham, phys. stat. sol. **18**, 99 (1966).
[10] G. Leibfried, Encyclopedia of Physics, Vol. 7, Chap. 1, Springer-Verlag 1955 (p. 201).
[11] D. J. Bolef, J. appl. Phys. **32**, 100 (1961).
[12] F. H. Featherstone and J. R. Neighbours, Phys. Rev. **130**, 1324 (1963).
[13] D. J. Bolef and J. De Klerk, Phys. Rev. **129**, 1063 (1963).
[14] A. E. Lord and D. N. Beshers, J. appl. Phys. **36**, 1620 (1965).
[15] G. Leibfried, Z. Phys. **135**, 23 (1953).

(Received May 13, 1968)

phys. stat. sol. 28, 611 (1968)

Subject classification: 6 and 10; 22.5.2

Institut für Kristallphysik der Deutschen Akademie der Wissenschaften zu Berlin

Multiphononabsorption von Kristallen mit Punktdefekten im starken Strahlungsfeld

Von

H. BÖTTGER

Es wird die Gitterabsorption durch Summenprozesse in Kristallen mit einer lokalisierten Schwingung im starken Strahlungsfeld unter Verwendung einer Diagrammtechnik untersucht. Für die hochfrequenten Seitenbanden der lokalisierten Schwingung von H^--Defekten in Alkalihalogeniden wird die kritische Feldstärke, in deren Nähe die betrachtete Absorption rapide anwächst, zu ca. 10^6 V/cm abgeschätzt.

An investigation is made of the lattice absorption by summing processes in crystals with a localized oscillation in a strong radiation field using a diagram technique. For the high frequency sidebands of the localized mode of H^--defects in alkali halides, the critical field strength, at which the mentioned absorption rapidely increases, is estimated to $\approx 10^6$ V/cm.

1. Einleitung

Die optische Absorption nichtmetallischer Kristalle kann im mittleren und fernen Ultrarot als reine Gitterabsorption beschrieben werden, die nicht mit einer Elektronenanregung im Kristall verbunden ist. Die Gitterabsorptionsprozesse werden in Ein- und Multiphononprozesse eingeteilt, je nachdem ob an einem Absorptionsakt ein oder mehrere Phononen teilnehmen.

Die Einphononabsorption wird durch das in den Auslenkungen der Gitterbausteine aus ihrer Gleichgewichtslage lineare elektrische Dipolmoment bestimmt, die Multiphononabsorption durch das lineare Dipolmoment und die Anharmonizität des Gitterpotentials bzw. durch das nichtlineare Dipolmoment. In einem Kristall können z. B. durch den Einbau sehr leichter Punktdefekte lokalisierte Schwingungen entstehen, deren Frequenzen oberhalb der maximalen Bandfrequenz des Kristalls liegen. Liefern diese lokalisierten Schwingungen einen genügend großen Beitrag zum Dipolmoment des Kristalls, so ist eine Absorption durch Ein- bzw. Multiphononprozesse oberhalb der maximalen Bandfrequenz zu erwarten.

In einigen Alkalihalogeniden [1, 2] und Erdalkalihalogeniden [3] mit H^--Defekten wird oberhalb der maximalen Bandfrequenz eine intensive symmetrisch von Seitenbanden begleitete Absorptionsbande gefunden, die als Einphononabsorptionsbande der lokalisierten Schwingung des H^- gedeutet wird. Die hochfrequenten Seitenbanden werden durch Summenprozesse gedeutet, d. h. durch die Erzeugung eines Phonons einer Bandschwingung und eines Phonons der lokalisierten Schwingung, und die niederfrequenten Seitenbanden durch Differenzprozesse, d. h. durch die Vernichtung eines Phonons einer Bandschwingung und Erzeugung eines Phonons der lokalisierten Schwingung.

In einigen Erdalkalihalogeniden mit H^--Defekten [3] werden auch Absorptionsbanden der zweiten und dritten Harmonischen der lokalisierten Schwingung des H^- gefunden, die der Erzeugung von zwei bzw. drei Phononen der lokalisierten Schwingung bei einem Absorptionsakt entsprechen. In Si mit C^{12} und C^{13} [4] wird ebenfalls eine Absorptionsbande der zweiten Harmonischen der lokalisierten Schwingung gefunden.

Die bisher durchgeführten experimentellen [1 bis 4] und theoretischen ([5, 6], dort weitere Literatur) Untersuchungen der mit einer lokalisierten Schwingung verbundenen Gitterabsorption wurden stets für ein schwaches Strahlungsfeld durchgeführt. Das Absorptionsproblem ist dann ein Spezialfall einer linearen „Response" eines Systems auf ein äußeres Feld, d. h. die Suszeptibilität ist von der Stärke des Strahlungsfeldes unabhängig.

Die Multiphononabsorption kann jedoch in genügend starkem Strahlungsfeld empfindlich von der Feldstärke des Strahlungsfeldes abhängen [7]. Die Absorption durch Differenzprozesse wird mit wachsender Feldstärke geringer, während die Absorption durch Summenprozesse rapide anwächst, wenn die Feldstärke in die Nähe einer kritischen Feldstärke E_{krit} kommt.

In der vorliegenden Arbeit wird eine Theorie für die mit einer lokalisierten Schwingung verbundene Multiphononabsorption im starken Strahlungsfeld angegeben, die es u. a. ermöglicht E_{krit} abzuschätzen. Es wird speziell die Feldstärkeabhängigkeit der Absorption durch Summenprozesse im Gebiet der hochfrequenten Seitenbanden und der zweiten Harmonischen einer lokalisierten Schwingung bei tiefen Temperaturen untersucht. Dabei werden die beiden Grenzfälle betrachtet, daß die Multiphononabsorption entweder durch das nichtlineare Dipolmoment ermöglicht wird oder durch das lineare Dipolmoment und die Anharmonizität des Potentials.

Zur Berechnung der Absorptionskonstante wird eine Diagrammtechnik von Konstantinov und Perel [8] verwendet.

2. Zur Berechnung der Gitterabsorption im starken Strahlungsfeld

Der Hamiltonoperator der Gitterschwingungen ist gegeben durch

$$H = H_0 + H_A \tag{1}$$

mit

$$H_0 = \sum_{\lambda} \hbar\, \omega_\lambda \left(a_\lambda^+ a_\lambda + \frac{1}{2} \right),$$

$$H_A = \sum_{n \geqq 3} \sum_{\lambda_1 \ldots \lambda_n} V_{\lambda_1 \ldots \lambda_n} A_{\lambda_1} \ldots A_{\lambda_n},$$

$$A_\lambda = a_\lambda^+ + a_\lambda, \quad [a_\lambda, a_{\lambda'}^+] = \delta_{\lambda\lambda'}, \quad [a_\lambda, a_{\lambda'}] = [a_\lambda^+, a_{\lambda'}^+] = 0 .$$

Dabei kennzeichnen die Indizes λ_i sowohl die lokalisierte Schwingung als auch die Bandschwingungen.

Wird der betrachtete Kristall in ein Strahlungsfeld gebracht, dessen elektrische Feldstärke $\boldsymbol{E}$ in x-Richtung zeigen soll, so wird die Dipolwechselwirkung des Kristalls mit dem Strahlungsfeld beschrieben durch

$$H'(t) = -M\, E \cos \omega t\, e^{\varepsilon t} \quad \text{mit} \quad \varepsilon \to +0 . \tag{2}$$

Dabei hat die x-Komponente M des Dipolmoments des Kristalls, entwickelt nach A_{λ_i}, folgende Form:

$$M = \sum_{n \geqq 1} \sum_{\lambda_1 \ldots \lambda_n} M_{\lambda_1 \ldots \lambda_n} A_{\lambda_1} \ldots A_{\lambda_n} . \tag{3}$$

Durch die zeitabhängige Störung $H'(t)$ wird der Dichteoperator ϱ und damit auch der Erwartungswert von M zeitabhängig. Unter der Annahme, daß durch $H'(t)$ das thermodynamische Gleichgewicht des Systems nur wenig gestört wird, kann für $\varrho(t)$ der Ansatz gemacht werden [9]

$$\varrho(t) = \varrho_0 + \sum_{k=1}^{\infty} \Delta_k \varrho(t) \quad \text{mit} \quad \varrho_0 = \mathrm{e}^{-\beta H} / \mathrm{Sp}\,\{\mathrm{e}^{-\beta H}\}\,, \qquad \beta = \frac{1}{kT}\,, \tag{4}$$

und $\Delta_k \varrho(t)$ ergibt sich aus

$$\frac{\mathrm{d}\varrho}{\mathrm{d}t} = \frac{i}{\hbar} [\varrho, H + H'(t)] \tag{5}$$

zu

$$\Delta_k \varrho(t) = \left[\frac{i}{\hbar}\right]^k \int_{-\infty}^{t} \mathrm{d}t_1 \int_{-\infty}^{t_1} \cdots \int_{-\infty}^{t_{k-1}} \mathrm{d}t_k \, \mathrm{e}^{-i/\hbar H t} \times$$

$$\times \, [\cdots [\,[\varrho_0, H'_I(t_k)], H'_I(t_{k-1})], \cdots, H'_I(t)]\, \mathrm{e}^{i/\hbar H t} \tag{6}$$

mit

$$H'_I(t) = \mathrm{e}^{i/\hbar H t}\, H'(t)\, \mathrm{e}^{-i/\hbar H t} .$$

Mit dem Dichteoperator (4), (6) ergibt sich für die durch $H'(t)$ bewirkte Änderung des Erwartungswertes von M [8, 10]

$$\Delta M(t) = \sum_{k=1}^{\infty} \Delta_k M(t) \tag{7}$$

mit

$$\Delta_k M(t) = \mathrm{Sp}\,\{\Delta_k \varrho(t)\, M\} = \sum_{\mathrm{C}} (-1)^s \left[-\frac{i}{\hbar}\right]^k [\mathrm{Sp}\,(\mathrm{e}^{-\beta H})]^{-1} \times$$

$$\times \int_{-\infty}^{t} \cdots \int_{-\infty}^{t_{k-1}} \mathrm{d}t_1 \ldots \mathrm{d}t_k \times$$

$$\times \, \mathrm{Sp}\left\{\mathrm{e}^{-\beta H_0}\, T_{\mathrm{C}}\left[\mathrm{e}^{-i/\hbar \int_{\mathrm{C}} H^A_{\mathrm{w}}(z)\, dz} H'^{\mathrm{w}}(t_1) \ldots H'^{\mathrm{w}}(t_k)\, M^{\mathrm{w}}(t)\right]\right\} + \text{k. k.}$$

mit

$$H^{\mathrm{w}}_A(z) = \mathrm{e}^{i/\hbar H_0 z}\, H_A\, \mathrm{e}^{-i/\hbar H_0 z} \qquad \left(H'^{\mathrm{w}}(t_i),\, M^{\mathrm{w}}(t) - \text{analog}\right) .$$

Die Konturen C in (7) werden in der komplexen Ebene folgendermaßen gewählt [10]: Jede Kontur C besteht aus einem vertikalen Teil $t_i + i\,\hbar\,\beta \to t_i$ (mit $t_i = t, t_1 \ldots t_{k-1}$), der den Anfang der Kontur bildet, zwei oberen horizontalen Teilen, von denen der linke die Fortsetzung und der rechte das Ende der Kontur bildet und einem unterem horizontalen Teil. Die „Klemmen“ $t, t_1 \ldots t_k$ werden auf der Kontur so verteilt, daß sich auf dem vertikalen und dem linken oberen horizontalen keine Klemmen befinden, und t_i das Ende der Kontur bildet. Die restlichen Klemmen werden auf dem unteren horizontalen und dem rechten oberen horizontalen Teil so verteilt, daß bezüglich der reellen Achse gilt: $t_k \leqq t_{k-1} \leqq \cdots \leqq t_1 \leqq t$. z. B. die vier Konturen C für $k = 3$ sind in ([10] Fig. 1) angegeben.

T_C ordnet in (7) die rechts neben ihm stehenden Operatoren längs der Kontur C so, daß die zu kleineren „Zeiten" gehörenden Operatoren stets rechts von den zu größeren „Zeiten" gehörenden Operatoren stehen, und s ist die Zahl der Klemmen auf dem oberen horizontalen Teil ($t-$ zählt nicht mit).

Wird für $H'(t)$ geschrieben

$$H'(t) = -M\,E(t) \tag{8}$$

mit

$$E(t) = \int_{-\infty}^{\infty} \mathrm{d}\omega\, e^{st}\, E(\omega)\,, \qquad s = i\,\omega + \varepsilon\,, \quad \varepsilon \to +0$$

und in (7) die S-Matrix nach Potenzen von H_A entwickelt, so ergibt sich für $\Delta_k M(t)$

$$\Delta_k M(t) = \sum_C \sum_{n=0}^{\infty} \int_{-\infty}^{\infty} \cdots \int_{-\infty}^{\infty} \mathrm{d}\omega_1 \ldots \mathrm{d}\omega_k\, E(\omega_1) \ldots E(\omega_k)\, G_C^n(\omega_1 \ldots \omega_k, t) + \text{k.k.} \tag{9}$$

mit

$$G_C^n(\omega_1 \ldots \omega_k, t) = (-1)^{s+k} \left(-\frac{i}{\hbar}\right)^{k+n} \frac{1}{n!} [\mathrm{Sp}\,(e^{-\beta H})]^{-1} \times$$
$$\times \int_{-\infty}^{t} \cdots \int_{-\infty}^{t_{k-1}} \mathrm{d}t_1 \ldots \mathrm{d}t_k\, e^{s_1 t_1} \ldots e^{s_k t_k} \int_C \cdots \int_C \mathrm{d}z_1 \ldots \mathrm{d}z_n \times$$
$$\times \mathrm{Sp}\,\{e^{-\beta H_0}\, T_C\, [H_A^w(z_1) \ldots H_A^w(z_n)\, M^w(t_1) \ldots M^w(t_k)\, M^w(t)]\}\,.$$

Zur Berechnung von G_C^n werden H_A und M in (9) nach (1), (3) durch $a_{\lambda_i}^+$, a_{λ_i} ausgedrückt und die T_C-Produkte mit dem Wickschen Theorem ausgewertet. Die Integrale über die Punkte $z_1 \ldots z_n$ werden in eine Summe von Teilintegralen zerlegt, in denen jeweils diese Punkte eine bestimmte Lage zu den Klemmen und zueinander haben. Es ergibt sich [8], daß G_C^n als Summe aller Diagramme dargestellt werden kann, die $k + 1$ Klemmen und n Punkte auf C haben, deren Klemmen und Punkte paarweise mit Linien verbunden sind, die jeweils mit einer Pfeilrichtung und einem Phononenindex gekennzeichnet sind, und bei denen jeweils benachbarte Punkte und Klemmen auf den horizontalen Teilen von C durch vertikale Schnitte und benachbarte Punkte auf dem vertikalen Teil von C durch horizontale Schnitte getrennt werden. Linien, deren Ende auf einem späteren Teil von C liegt als deren Anfang, sollen als richtige Linien bezeichnet werden, die anderen als nichtrichtige.

Jedem dieser Diagramme wird nach folgenden Regeln ein analytischer Ausdruck zugeordnet:

1. Jeder Klemme entspricht ein Faktor $M_{\lambda_1 \ldots \lambda_p}$, d. h. in jeder Klemme können $p \geqq 1$ Linien ein- bzw. auslaufen.

2. Jedem Punkt entspricht ein Faktor $V_{\lambda_1 \ldots \lambda_m}$, d. h. in jedem Punkt können $m \geqq 3$ Linien ein- bzw. auslaufen.

3. Der Klemme t entspricht der Faktor $\exp\{(s_1 + \cdots + s_k)\,t\}$.

4. Jeder richtigen Linie entspricht der Faktor $(1 + n_\lambda)$, jeder nichtrichtigen Linie der Faktor n_λ mit $n_\lambda = [e^{\hbar\omega_\lambda/kT} - 1]^{-1}$.

5. Jedem Punkt auf dem unteren horizontalen, dem oberen horizontalen und dem senkrechten Teil von C entspricht der Faktor $[-i/\hbar]$, $[+i/\hbar]$, -1.

6. Jeder Klemme auf dem unteren horizontalen, dem oberen horizontalen Teil von C entspricht der Faktor $[+i/\hbar]$, $[-i/\hbar]$.

7. Jedem vertikalen Schnitt entspricht der Faktor $\left(\sum^{l} s_i + i\,\omega_{ab}\right)^{-1}$ mit $\omega_{ab} = \omega_a - \omega_b$. Dabei ist $\omega_a(\omega_b)$ die Summe der Frequenzen der nach (von)

rechts durch den Schnitt gehenden Linien, und $\sum^{l}$ geht über alle Klemmen, die links vom Schnitt liegen.

8. Jedem horizontalen Schnitt entspricht der Faktor $(\sigma + \omega_{ab})$. Dabei ist $\omega_a(\omega_b)$ die Summe der Frequenzen der durch den Schnitt nach unten (oben) laufenden Linien.

Weiterhin ist über alle Phononenindizes zu summieren und von σ zu β durch Umkehrung der Laplacetransformation überzugehen. Wird die durch $H'(t)$ (2) bewirkte Änderung des Erwartungswertes des Dipolmoments des Kristalls in folgende Form gebracht:

$$\Delta M(t) = \chi''(\omega)\, E \sin \omega t + \cdots, \tag{10}$$

so wird das Absorptionsspektrum durch $\chi''(\omega)$ beschrieben. Ein Vergleich von (9) mit $E(\omega_i) = E/2\, [\delta\, (\omega_i - \omega) + \delta\, (\omega_i + \omega)]$ und (10) zeigt, daß zu $\chi''(\omega)$ nur die $G^n_{\mathrm{C}}(\omega_1 \ldots \omega_k, t)$ mit $(\omega_1 + \cdots + \omega_k) = \pm\, \omega$ beitragen. Mit

$$G^n_{\mathrm{C}}(\omega_1 \ldots \omega_k, t) = \begin{cases} \mathrm{e}^{+i\omega t}\, G^n_{\mathrm{C}}(\omega) & \text{für} \quad \omega_1 + \cdots + \omega_k = +\, \omega \\ \mathrm{e}^{-i\omega t}\, G^n_{\mathrm{C}}(-\, \omega) & \text{für} \quad \omega_1 + \cdots + \omega_k = -\, \omega \end{cases} \tag{11}$$

ergibt sich für $\chi''(\omega)$

$$\chi''(\omega) = \sum_{k\text{-ungerade}} \chi''_k(\omega) \tag{12}$$

mit

$$\chi''_k(\omega) = \left(\frac{E}{2}\right)^{k-1} \mathrm{Im} \left\{ \sum_n \sum_{\mathrm{C}} G^n_{\mathrm{C}}(-\, \omega) - \sum_n \sum_{\mathrm{C}} G^n_{\mathrm{C}}(+\, \omega) \right\}.$$

Zur Berechnung von $G^n_{\mathrm{C}}(+\, \omega)$ und $G^n_{\mathrm{C}}(-\, \omega)$ gelten die oben angegebenen Regeln außer der dritten unter der Bedingung, daß $(\omega_1 + \cdots + \omega_k) = \pm\, \omega$ ist. Folgende Näherungen sollen bei der Untersuchung der betrachteten Gitterabsorption verwendet werden:

1. Unter der Annahme, daß $|H_A|$ klein gegen $|H_0|$ und klein gegen $1/\beta$ ist, wird $\mathrm{e}^{-\beta H}/\mathrm{Sp}\, \{\mathrm{e}^{-\beta H}\}$ durch $\mathrm{e}^{-\beta H_0}/\mathrm{Sp}\, \{\mathrm{e}^{-\beta H_0}\}$ ersetzt, d. h. es werden nur Diagramme betrachtet, die keine Punkte auf dem senkrechten Teil von C haben.

2. Da die Absorption nur für tiefe Temperaturen untersucht werden soll, wird die mittlere Besetzungszahl n_λ der Gitteroszillatoren gleich Null gesetzt, d. h. es werden nur Diagramme mit richtigen Linien berücksichtigt.

3. Von Diagrammen mit einer bestimmten Zahl vertikaler Schnitte werden nur die mit einer möglichst großen Zahl von Resonanzschnitten [8, 10] berücksichtigt.

3. Summenprozesse im starken Strahlungsfeld

3.1 Hochfrequente Seitenbanden durch das Dipolmoment zweiter Ordnung

Es wird der Grenzfall betrachtet, daß die hochfrequenten Seitenbanden der Absorptionsbande einer lokalisierten Schwingung allein durch das Dipolmoment zweiter Ordnung ermöglicht werden. Folgende Terme des Dipolmoments werden berücksichtigt:

$$M = 2 \sum_{n \neq l} M_{ln}\, A_l\, A_n\,. \tag{13}$$

Dabei bezeichnet l die lokalisierte Schwingung und n die Bandschwingungen. Das zu $\chi''_1(\omega)$ beitragende Diagramm ist in Fig. 1a eingetragen. Der ent-

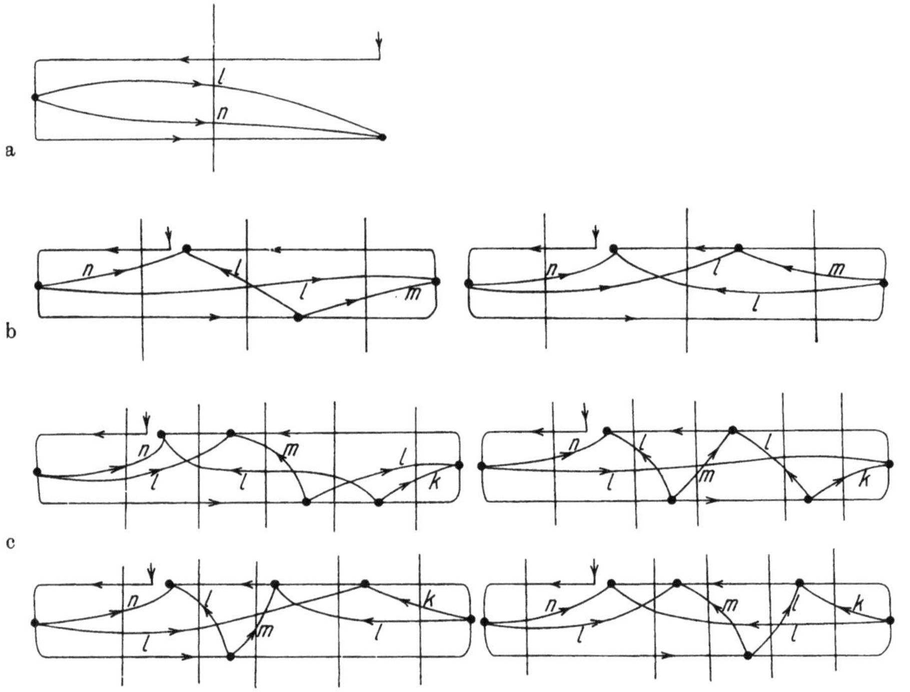

Fig. 1. Diagramme zu $\chi_1''(\omega)$ (a), $\chi_3''(\omega)$ (b), $\chi_5''(\omega)$ (c) bei Berücksichtigung des Dipolmoments zweiter Ordnung

sprechende analytische Ausdruck ist

$$\chi_1''(\omega) = \frac{4\pi}{\hbar} \sum_n M_{ln}^2 \, \delta(\omega_l + \omega_n - \omega) \,. \tag{14}$$

Die Diagramme zu $\chi_3''(\omega)$ sind in Fig. 1b dargestellt. In jedem dieser beiden Diagramme liefert der mittlere vertikale Schnitt nach den in Abschnitt 2 angegebenen Regeln einen Faktor $1/\varepsilon$, der mit $\varepsilon \to +0$ unendlich groß wird. Treten solche divergierenden Terme auf, so ist eine Beschränkung auf endlich viele Glieder der Störungstheorie nicht mehr möglich und es müssen unendlich viele Glieder aufsummiert werden [8].

Ein vertikaler Schnitt, der einen divergierenden Ausdruck liefert, wird „freier Schnitt" genannt, und als irreduzibler Teil wird der Teil eines Diagramms bezeichnet, der von zwei freien Schnitten begrenzt ist, aber keinen freien Schnitt enthält [8].

Werden unendlich viele Diagramme aufsummiert, die sich in der Anzahl ihrer irreduziblen Teile unterscheiden, so ergibt sich, daß die betrachteten Schnitte in Fig. 1b durch $[-\Gamma_l^l]^{-1}$ zu ersetzen sind. Die Diagramme des irreduziblen Teils Γ_l^l sind in Fig. 2 eingetragen für den Fall, daß zu H_A nur Terme dritter Ordnung beitragen. (An dem Symbol für einen irreduziblen Teil kennzeichnet der untere Index jeweils die von links in den irreduziblen Teil einlaufenden und nach rechts auslaufenden Linien und der obere Index die von rechts einlaufenden und nach links auslaufenden Linien). Der analytische Aus-

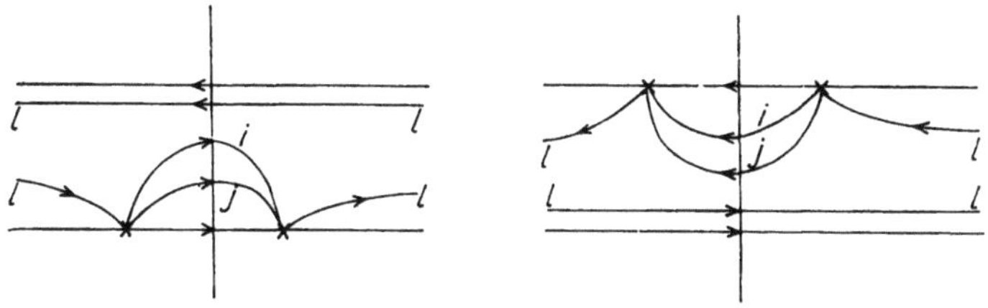

Fig. 2. Diagramme des irreduziblen Teils Γ_l^l

druck für Γ_l^l ist dann

$$\Gamma_l^l = -\frac{36\pi}{\hbar^2} \sum_{ij} V_{lij}^2 \,\delta\,(\omega_l - \omega_i - \omega_j), \tag{15}$$

und $\chi_3''(\omega)$ ist gegeben durch

$$\chi_3''(\omega) = \frac{4\pi}{\hbar} \sum_n M_{ln}^2 \,\delta\,(\omega_l + \omega_n - \omega) \cdot \frac{4\pi E^2}{2\hbar^2} [-\Gamma_l^l]^{-1} \sum_m M_{lm}^2 \,\delta\,(\omega_l + \omega_m - \omega)\,. \tag{16}$$

In analoger Weise wird in jedem zu $\chi_5''(\omega)$ gehörenden Diagramm in Fig. 1c jeweils der zweite und vierte Schnitt durch $[-\Gamma_l^l]^{-1}$ ersetzt, so daß sich für $\chi_5''(\omega)$ ergibt

$$\chi_5''(\omega) = \frac{4\pi}{\hbar} \sum_n M_{ln}^2 \,\delta\,(\omega_l + \omega_n - \omega) \times$$
$$\times \left\{\frac{4\pi E^2}{2\hbar^2} [\Gamma_l^l]^{-1} \sum_m M_{lm}^2 \,\delta\,(\omega_l + \omega_m - \omega)\right\}^2. \tag{17}$$

Eine Fortsetzung dieses Verfahrens liefert eine unendliche geometrische Folge für $\chi_1''(\omega), \chi_3''(\omega), \chi_5''(\omega)\ldots$ deren Summe (12) für $\chi''(\omega)$ folgenden Ausdruck liefert:

$$\chi''(\omega) = \chi_1''(\omega) \frac{1}{1 - \frac{E^2}{2\hbar} [-\Gamma_l^l]^{-1} \chi_1''(\omega)} \tag{18}$$

mit $\chi_1''(\omega)$ aus (14). Da nach (15) $[-\Gamma_l^l]^{-1}$ positiv ist, wird mit wachsender Feldstärke E auch die durch $\chi''(\omega)$ beschriebene Absorption wachsen, besonders stark dann, wenn der zweite Summand im Nenner von (18) vergleichbar mit 1 wird.

3.2 Zweite Harmonische durch das Dipolmoment zweiter Ordnung

Mit dem Dipolmoment

$$M = M_{ll}\, A_l\, A_l \tag{19}$$

wird der Grenzfall untersucht, daß die Absorptionsbande der zweiten Harmonischen einer lokalisierten Schwingung durch das Dipolmoment zweiter Ordnung erzeugt wird. Zur Beschreibung von $\chi_1''(\omega)$, $\chi_3''(\omega)$, $\chi_5''(\omega)$ können wieder die Diagramme in Fig. 1 verwendet werden, wenn in diesen Diagrammen alle Linien durch den Index l der lokalisierten Schwingung gekennzeichnet werden. Da jetzt aber nicht mehr über die Bandschwingungen summiert wird, liefern alle

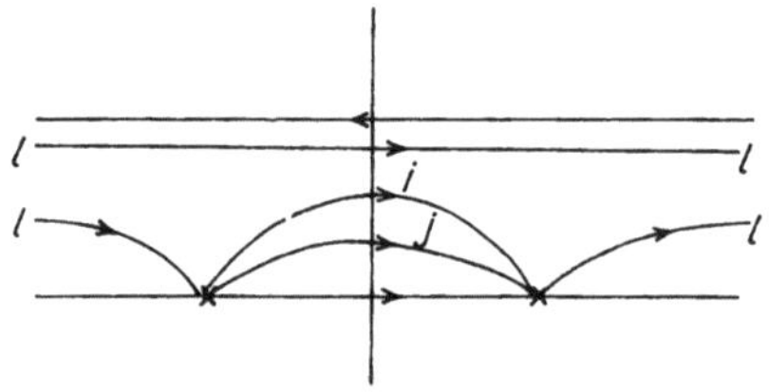

Fig. 3. Diagramm des irreduziblen Teils Γ_{ll}

vertikalen Schnitte in Fig. 1 divergierende Faktoren, d. h. es ist für jeden Schnitt eine Aufsummation unendlich vieler Terme der Störungstheorie nötig. So ist z. B. der erste Schnitt (von links) in einem Diagramm in Fig. 1b durch $[i\,(2\,\omega_l - \omega) - \Gamma_{ll}\,(-\,i\,\omega + \varepsilon)]^{-1}$ und der dritte Schnitt durch $[-\,i\,(2\,\omega_l - \omega) - \Gamma^{ll}\,(+\,i\,\omega + \varepsilon)]^{-1}$ zu ersetzen. Unter Ausnutzung der Beziehung

$$\Gamma^{ll}\,(+\,i\,\omega + \varepsilon) = \Gamma_{ll}^{*}\,(-\,i\,\omega + \varepsilon)\,, \tag{20}$$

die sich leicht verifizieren läßt, können $\chi_1''(\omega)$, $\chi_3''(\omega)$, $\chi_5''(\omega)$ berechnet werden:

$$\chi_1''(\omega) = \frac{2\,M_{ll}^2}{\hbar}\,\mathrm{Re}\,[i\,(2\,\omega_l - \omega - \mathrm{Im}\,\Gamma_{ll}) - \mathrm{Re}\,\Gamma_{ll}]^{-1}\,, \tag{21}$$

$$\chi_3''(\omega) = \frac{2\,M_{ll}^2}{\hbar}\,\mathrm{Re}\,[i\,(2\,\omega_l - \omega - \mathrm{Im}\,\Gamma_{ll}) - \mathrm{Re}\,\Gamma_{ll}]^{-1} \times$$
$$\times \frac{E^2\,M_{ll}^2}{\hbar^3}\,[-\,\Gamma_l^l]^{-1}\,\mathrm{Re}\,[i\,(2\,\omega_l - \omega - \mathrm{Im}\,\Gamma_{ll}) - \mathrm{Re}\,\Gamma_{ll}]^{-1}\,, \tag{22}$$

$$\chi_5''(\omega) = \frac{2\,M_{ll}^2}{\hbar}\,\mathrm{Re}\,[i\,(2\,\omega_l - \omega - \mathrm{Im}\,\Gamma_{ll}) - \mathrm{Re}\,\Gamma_{ll}]^{-1} \times$$
$$\times \frac{E^2\,M_{ll}^2}{\hbar^3}\,[-\,\Gamma_l^l]^{-1}\,\mathrm{Re}\,[i\,(2\,\omega_l - \omega - \mathrm{Im}\,\Gamma_{ll}) - \mathrm{Re}\,\Gamma_{ll}]^{-1}\,. \tag{23}$$

$\chi_1''(\omega)$, $\chi_3''(\omega)$, $\chi_5''(\omega)$ aufsummiert, liefert wieder $\chi''(\omega)$ (18) mit $\chi_1''(\omega)$ aus (21). Bei Berücksichtigung der Anharmonizität dritter Ordnung wird der für die kritische Feldstärke wichtige $\mathrm{Re}\,\Gamma_{ll}\,(-\,i\,\omega + \varepsilon)$ durch das Diagramm in Fig. 3 beschrieben, dessen analytischer Ausdruck gegeben ist durch

$$\mathrm{Re}\,\Gamma_{ll} = -\,\frac{18\,\pi}{\hbar^2}\sum_{ij} V_{lij}^2\,\delta\,(\omega_l + \omega_i + \omega_j - \omega)\,. \tag{24}$$

3.3 Hochfrequente Seitenbanden und zweite Harmonische durch das lineare Dipolmoment und die Anharmonizität des Potentials

Mit dem Dipolmoment

$$M = M_l\,A_l \tag{25}$$

wird der Grenzfall untersucht, daß die hochfrequenten Seitenbanden und die zweite Harmonische einer lokalisierten Schwingung durch das lineare Dipolmoment und die Anharmonizität des Gitterpotentials beschrieben werden.

Für die hochfrequenten Seitenbanden sind in Fig. 4a die zu $\chi_1''(\omega)$ beitragenden Diagramme eingetragen und von den zu $\chi_3''(\omega)$ bzw. $\chi_5''(\omega)$ beitragenden

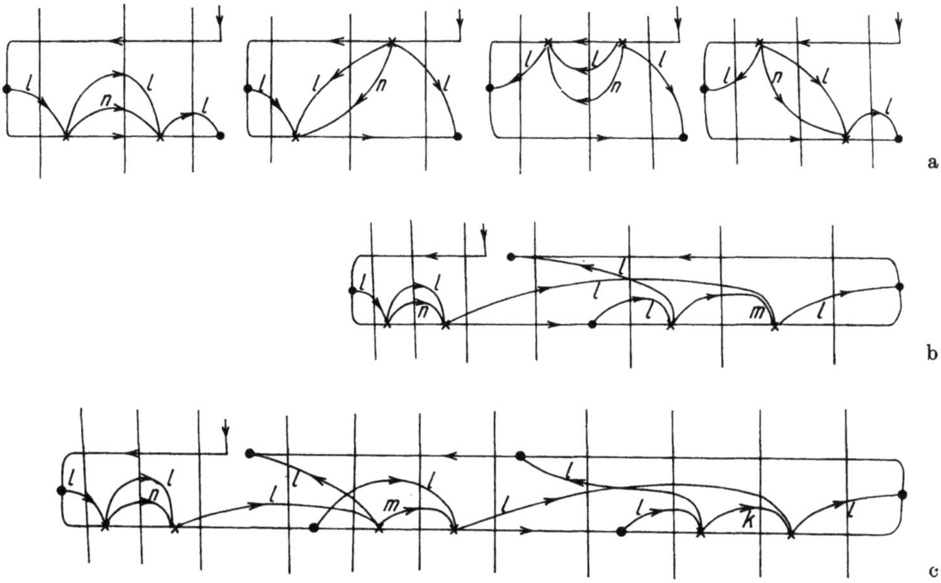

Fig. 4. Diagramme zu $\chi_1''(\omega)$ (a), $\chi_3''(\omega)$ (b), $\chi_5''(\omega)$ (c) bei Berücksichtigung des linearen Dipolmoments und der Anharmonizität des Potentials

Diagrammen in Fig. 4b bzw. Fig. 4c der Einfachheit halber jeweils nur ein einziges als Beispiel. Für $\chi_1''(\omega)$, $\chi_3''(\omega)$, $\chi_5''(\omega)$ ergibt sich

$$\chi_1''(\omega) = \frac{36\,\pi}{\hbar^3} M_l^2 \left[\frac{2\,\omega_l}{(\omega_l^2 - \omega^2)}\right]^2 \sum_n V_{lln}^2\, \delta\,(\omega_l + \omega_n - \omega)\,, \tag{26}$$

$$\chi_3''(\omega) = \frac{36\,\pi}{\hbar^3} M_l^2 \left[\frac{2\,\omega_l}{(\omega_l^2 - \omega^2)}\right]^2 \sum_n V_{lln}^2\, \delta\,(\omega_l + \omega_n - \omega) \times$$
$$\times \frac{E^2}{\hbar} [-\Gamma_l^l]^{-1} \frac{(3\,\omega_l\,\omega + \omega^2)}{4\,\omega_l^2} \frac{36\,\pi}{\hbar^3} M_l^2 \left[\frac{2\,\omega_l}{(\omega_l^2 - \omega^2)}\right]^2 \sum_m V_{llm}^2\, \delta\,(\omega_l + \omega_m - \omega)\,, \tag{27}$$

$$\chi_5''(\omega) = \frac{36\,\pi}{\hbar^3} M_l^2 \left[\frac{2\,\omega_l}{(\omega_l^2 - \omega^2)}\right]^2 \sum_n V_{lln}^2\, \delta\,(\omega_l + \omega_n - \omega) \times$$
$$\times \left\{\frac{E^2}{\hbar} [-\Gamma_l^l]^{-1} \frac{(3\,\omega_l\,\omega + \omega^2)}{4\,\omega_l^2} \frac{36\,\pi}{\hbar^3} M_l^2 \left[\frac{2\,\omega_l}{(\omega_l^2 - \omega^2)}\right]^2 \sum_m V_{llm}^2\, \delta\,(\omega_l + \omega_m - \omega)\right\}^2, \tag{28}$$

und die Aufsummation von $\chi_1''(\omega)$, $\chi_3''(\omega)$, $\chi_5''(\omega)$ ... liefert für $\chi''(\omega)$ im Bereich der hochfrequenten Seitenbanden

$$\chi''(\omega) = \chi_1''(\omega) \frac{1}{1 - \frac{E^2}{\hbar}[-\Gamma_l^l]^{-1} \frac{(3\,\omega_l\,\omega + \omega^2)}{4\,\omega_l^2}\,\chi_1''(\omega)} \tag{29}$$

mit $\chi_1''(\omega)$ aus (26).

Die Diagramme der zweiten Harmonischen ergeben sich aus den Diagrammen in Fig. 4, wenn in diesen alle Linien mit dem Index l der lokalisierten Schwingung bezeichnet werden. Die divergierenden Terme werden wie in Abschnitt 3.2 durch Aufsummation von unendlich vielen Termen beseitigt. Es ergibt sich für $\chi''(\omega)$ der zweiten Harmonischen der Ausdruck (29) mit

$$\chi_{\text{I}}''(\omega) = \frac{18}{\hbar^3} M_l^2\, V_{lll}^2 \left[\frac{2\,\omega_l}{(\omega_l^2 - \omega^2)}\right]^2 \operatorname{Re}\,[i\,(2\,\omega_l - \omega_l - \operatorname{Im} \Gamma_{ll}) - \operatorname{Re} \Gamma_{ll}]^{-1}\,. \quad (30)$$

4. Abschätzung der kritischen Feldstärke

Als kritische Feldstärke wird die Feldstärke bezeichnet, für die der Nenner von (18) bzw. (29) verschwindet. D. h. für den Fall der Summenprozesse durch das Dipolmoment zweiter Ordnung ergibt sich

$$E_{\text{krit}}^2 = -\,2\,\hbar\, \Gamma_l^l / \chi_{\text{I}}''(\omega) \quad (31)$$

und für den Fall der Summenprozesse durch das lineare Dipolmoment und die Anharmonizität des Potentials

$$E_{\text{krit}}^2 = -\,4\,\omega_l^2\,\hbar\,\Gamma_l^l/(3\,\omega_l\,\omega + \omega^2)\,\chi_{\text{I}}''(\omega)\,. \quad (32)$$

Die in $\chi_{\text{I}}''(\omega)$ eingehenden Parameter des Potentials und des Diplomoments sind praktisch unbekannt. Es soll auch nicht versucht werden, diese Parameter mit Hilfe irgendwelcher Modelle abzuschätzen, um so E_{krit} berechnen zu können. E_{krit} kann aus Daten abgeschätzt werden, die sich aus dem Absorptionsspektrum im schwachen Strahlungsfeld ergeben.

Beschreibt $\varkappa(\omega)$ die Seitenbandenabsorption im schwachen Strahlungsfeld, so ergibt sich $\chi_{\text{I}}''(\omega)$ aus

$$\chi_{\text{I}}''(\omega) = \frac{\varepsilon_0\, c}{\omega\, N}\varkappa(\omega) \quad (33)$$

(N-Zahl der Defekte pro cm³). Γ_l^l kann aus der Halbwertsbreite ΔH der Einphononabsorptionsbande der lokalisierten Schwingung bestimmt werden, denn diese ergibt sich, wenn nur die Anharmonizität dritter Ordnung berücksichtigt wird, zu

$$\Delta H = \frac{18}{\hbar^2} \sum_{ij} V_{lij}^2\, \delta\,(\omega_i + \omega_j - \omega)\,, \qquad (\omega \approx \omega_l)\,. \quad (34)$$

Ein Vergleich von (34) und (15) zeigt, daß $-\Gamma_l^l = 2\,\pi\,\Delta H$ gilt. Diese Beziehung gilt auch dann, wenn ω_l oberhalb der doppelten maximalen Bandfrequenz liegt, und Γ_l^l und ΔH durch die Anharmonizität vierter Ordnung bestimmt werden.

Für Alkalihalogenide mit H^--Defekten ergibt sich für E_{krit} im Bereich der hochfrequenten Seitenbanden der lokalisierten Schwingung mit $N \approx 4 \times 10^{17}$ cm^{-3}, $\omega \approx \omega_l \approx 2\,\pi \times 500$ cm^{-1}, $\varkappa \approx 1$ cm^{-1} und $\Delta H \approx 1$ cm^{-1} aus (32) $E_{\text{krit}} \approx 5 \times 10^5$ V/cm und aus (31) $E_{\text{krit}} \approx 7 \times 10^5$ V/cm.

5. Einige abschließende Bemerkungen

Es wurde die Abhängigkeit der Absorption durch Summenprozesse im Bereich der hochfrequenten Seitenbanden und der zweiten Harmonischen einer lokalisierten Schwingung untersucht und gezeigt, daß im starken Strahlungsfeld in der Nähe einer kritischen Feldstärke E_{krit} diese Absorption rapide mit der Feldstärke des Strahlungsfeldes anwächst. Für H^--Defekte in Alkalihalogeniden wurde für die hochfrequenten Seitenbanden $E_{\text{krit}} \approx 10^6$ V/cm abgeschätzt. Feldstärken dieser Größenordnung sind mit fokussierten Laserstrahlen durchaus erreichbar, so daß die gefundene Abhängigkeit der betrachteten Absorption von der Feldstärke (18), (29) prinzipiell experimentell nachprüfbar sein müßte. Ein wesentliches Problem einer experimentellen Nachprüfung wird jedoch darin bestehen, einen geeigneten Laser in dem jeweils interessierenden Spektralbereich zu finden, der möglichst auch abstimmbar ist.

Da nach (18), (29) die kritische Feldstärke bzw. eine Feldstärke, die die im schwachen Strahlungsfeld gemessene Absorption $\chi_1''(\omega)$ um einen beliebigen Faktor vergrößert, proportional zu $1/\sqrt{\chi_1''(\omega)}$ ist, lassen sich aus einer Untersuchung der Abhängigkeit der Absorption von der Feldstärke analoge Informationen gewinnen wie aus der Untersuchung von $\chi_1''(\omega)$ [5, 6].

In den Alkalihalogeniden und Erdalkalihalogeniden, in denen die Frequenz der lokalisierten Schwingung der H^--Defekte oberhalb und der D^--Defekte unterhalb der doppelten maximalen Bandfrequenz liegt, müßten sich für H^-- und D^--Defekte unterschiedliche Werte für E_{krit} ergeben, da E_{krit} nach (31), (32) proportional zu $\sqrt{-\Gamma_l^l}$ ist, und Γ_l^l dann für H^--Defekte durch die Anharmonizität vierter Ordnung des Potentials und für D^--Defekte durch die Anharmonizität dritter Ordnung bestimmt wird.

Ein Mangel der hier durchgeführten Untersuchung der Multiphononabsorption im starken Strahlungsfeld besteht darin, daß die erhaltenen Resultate nur für $E < E_{\text{krit}}$ gelten, also nicht im Bereich der Feldstärke, wo eine parametrische Erzeugung von Phononen stattfindet.

Herrn Prof. Dr. O. Stasiw danke ich für zahlreiche anregende Diskussionen.

Literatur

[1] G. Schäfer, J. Phys. Chem. Solids **12**, 233 (1960).
[2] B. Fritz, U. Gross und D. Bäuerle, phys. stat. sol. **11**, 231 (1965).
[3] W. Hayes, G. D. Jones, H. F. MacDonald, C. T. Sennett und R. J. Elliott, Proc. Roy. Soc. **A289**, 1 (1965).
[4] R. C. Newman und J. B. Willis, Phys. Chem. Solids **26**, 373 (1965).
[5] H. Bilz, D. Strauch und B. Fritz, J. Phys. Radium (Suppl.) **C 2**, 3 (1966).
[6] J. B. Page und B. G. Dick, Phys. Rev. **163**, 910 (1967).
[7] V. M. Fain, Zh. eksper. teor. Fiz. **53**, 2070 (1967).
[8] O. V. Konstantinov und V. I. Perel, Zh. eksper. teor. Fiz. **39**, 197 (1960).
[9] R. Kubo, J. Phys. Soc. Japan **12**, 579 (1957).
[10] L. K. Aminov, Zh. eksper. teor. Fiz. **48**, 1399 (1965).

(Received June 7, 1968)

phys. stat. sol. 28, 623 (1968)

Subject classification: 6 and 13; 14.3; 22

Institut für Theoretische Physik der Universität zu Köln

Polaronic Impurity Hopping Conduction

By

J. SCHNAKENBERG

Impurity hopping conduction in polar lattices and amorphous substances is investigated by application of the canonical transformation of the small polaron theory. In the case of acoustic coupling only, the results of Miller and Abrahams and Gosar for one-phonon processes are recovered and compared with the contributions of multi-phonon processes. In the polar case the assisting roles of acoustic and polar phonons are discussed seperately at temperatures $T > \theta/2$ and $T < \theta/2$ (θ being the Debye temperature). The type of the temperature dependence of the conductivity is determined.

Es wird die Störstellen-Hoppingleitfähigkeit in polaren Gittern und amorphen Substanzen mit der Anwendung der kanonischen Transformation aus der Theorie des kleinen Polarons untersucht. Im Falle der rein akustischen Kopplung werden die Ergebnisse von Miller und Abrahams und von Gosar für Ein-Phonon-Prozesse verifiziert und mit den Beiträgen von Mehr-Phononen-Prozessen verglichen. Im Fall des polaren Gitters wird die assistierende Rolle der akustischen und polaren Phononen getrennt für die Temperaturen $T > \theta/2$ und $T < \theta/2$ (θ Debye Temperatur) diskutiert. Die Temperaturabhängigkeit der Leitfähigkeit wird bestimmt.

1. Introduction

By the term "impurity conduction" one denotes charge carrier transport processes between impurity states in an ordered or random lattice as to be distinguished from intrinsic conduction of free carriers which are thermally excited into the conduction or valence band. In a semiconductor containing N_d donor and N_a acceptor states with energies ε_d and ε_a such that $E_v < \varepsilon_a < \varepsilon_d < E_c$ (E_v and E_c being the upper valence and the lower conduction band edges) a fraction $K = N_a/N_d < 1$ of the donor states will be compensated or vice versa if $K > 1$ allowing for finite transition probabilities of the carriers between occupied and vacant impurity sites. In this case, we therefore expect impurity conduction at temperatures $kT < E_c - \varepsilon_d$ if $K < 1$ (or $kT < \varepsilon_a - E_v$ if $K > 1$) whereas intrinsic conduction becomes appreciable only if kT is of the order of $E_c - \varepsilon_d$ (or $\varepsilon_a - E_v$). For the sake of simplicity let us discuss in the following considerations only the case $0 < K < 1$; $K > 1$ may be treated quite analogously.

Due to the compensation the acceptor states become ionized and thus produce local varying fields which will remove the degeneracy of the donor states and lead to a statistical distribution of the donor energies ε_n with a mean energy difference Δ between neighbouring sites. As pointed out by Anderson [1] in detail we have to compare Δ with the mean energy overlap I (resonance integral) between neighbouring donor sites to distinguish between two possible mechanisms of impurity conduction. As one may expect and as confirmed by Andersons treatment there exists a critical value $(\Delta/I)_c$ of order of magnitude unity such that if $\Delta/I > (\Delta/I)_c$ the wave functions of the donor states remain localized whereas if $\Delta/I < (\Delta/I)_c$ the donor states will form a so-called "impurity band". Since I is exponentially decreasing with the mean distance between the donor states we expect localized donor states at low concentrations and impurity bands at high concentrations. In this paper, we shall restrict ourselves to the case of localized donor states. The transport processes may now be described as hopping transitions between donor states with different energies ε_n. In order to conserve energy these transitions have to be accompanied by emission and absorption of phonons. Therefore, we expect the conductivity σ to be a strongly increasing function of temperature and $\sigma = 0$ at $T = 0$. This type of hopping conduction including acoustic one-phonon processes only has been calculated and discussed in detail by Miller and Abrahams [2] and Gosar [3]. We shall refer to these results in our following considerations.

To generalize our model to polar lattices and particularly to amorphous polar substances like glasses and glassy semiconductors we first of all notice that the energy band spectrum of an amorphous solid is very similar to that of a crystal, the only difference being the existence of small tails of the density of states instead of sharp band edges. Since we are concerned only with impurity processes we may neglect the tails of the density of states if the donor state energies are separated far enough from the band edge region.

A very important feature of polar substances, however, is the presence of polar vibrations of the "lattice" which have to be taken into account in the phonon emission and absorption processes during a carrier transition in addition to the acoustic phonons. Moreover, the carriers may form small polaron states by polarizing the surrounding medium and thus get self-trapped. As well known from the small polaron theory [4] this self-trapping causes a decrease of the resonance integral I by at least an order of magnitude. Hence, the case of localized donor states may apply even to high concentrations of impurities in polar lattices.

A possible realization of the polar case is a P_2O_5-glass containing V_2O_5 and VO_2 as V^{5+}, V^{4+}, and O^{2-}-ions. Local varying fields are produced by O^{2-} ions in addition to the local fields due to the amorphous structure itself and hopping of carriers may take place between V^{4+} and V^{5+} ions.

In sections 2, 3, and 4 of this paper we shall try to give a general formulation of the problem including both the normal and the polar case. After having obtained general results we shall discuss the electric conductivity separately for the two cases in sections 5 and 6.

2. Model Hamiltonian

Let us start with the Hamiltonian for charge carriers in impurity states with non-equal energy levels ε_n which may be written as

$$H_e = \sum_n (\varepsilon_n - \mu)\, a_n^+ a_n + \sum_{n,n'} I_{nn'}\, a_n^+ a_{n'} \,. \tag{2.1}$$

a_n, a_n^+ are annihilation and creation operators for carriers in a donor state n with position vector $\boldsymbol{r}_n$, $I_{nn'}$ is the resonance integral between donor states n and n', $I_{nn'} = 0$ if $n = n'$, μ is the chemical potential. The phonon part of the

Hamiltonian reads[1])

$$H_p = \sum_{j=1}^{2} \sum_{\boldsymbol{q}} \omega_{\boldsymbol{q}j} b^+_{\boldsymbol{q}j} b_{\boldsymbol{q}j} \tag{2.2}$$

$j = 1, 2$ referring to acoustic and polar phonons, respectively; $\omega_{\boldsymbol{q}j}$ is the frequency of phonons of type j with wave vector $\boldsymbol{q}$. The electron–phonon interaction part of the Hamiltonian may be written as

$$H_{ep} = -\sum_{j=1}^{2} \sum_{n,\boldsymbol{q}} u^{1/2}_{\boldsymbol{q}j} \omega_{\boldsymbol{q}j} (e^{i\boldsymbol{q}\boldsymbol{r}_n} b_{\boldsymbol{q}j} + e^{-i\boldsymbol{q}\boldsymbol{r}_n} b^+_{\boldsymbol{q}j}) a^+_n a_n , \tag{2.3}$$

where

$$u_{\boldsymbol{q}1} = \frac{g^2 \boldsymbol{q}^2}{2 M N \omega^3_{\boldsymbol{q}1}} , \qquad u_{\boldsymbol{q}2} = \frac{\gamma^2}{2 M N \omega^3_{\boldsymbol{q}2}} . \tag{2.4a, b}$$

N is the total number of atoms or molecules with mass M; g, γ are coupling constants. (2.4a) corresponds to the usual deformation potential coupling, (2.4b) is identical with the dipole moment coupling of the small polaron theory.

In (2.3) we restrict ourselves to interaction processes $\sim a^+_n a_{n'}$ with $n = n'$. The $n \neq n'$ processes contain a further factor of order of magnitude $I_{nn'}$ and may be neglected as $I_{nn'}$ is assumed to be small.
The total Hamiltonian is

$$H = H_e + H_p + H_{ep} . \tag{2.5}$$

The operator of the electical current reads

$$\boldsymbol{j} = i \sum_{n,n'} I_{nn'} \boldsymbol{r}_{n'n} a^+_n a_{n'} , \tag{2.6}$$

where $\boldsymbol{r}_{n'n} = \boldsymbol{r}_{n'} - \boldsymbol{r}_n$.

3. Canonical Transformation

Since the typical coupling energies of H_{ep} may be greater than $I_{nn'}$ even in the absence of polar phonons we shall carry out a perturbation expansion in powers of $I_{nn'}$ instead of $u_{\boldsymbol{q}j}$ as in the usual theory of electron–phonon interaction. For this purpose we apply Holstein's canonical transformation [4, 5] to H of (2.5) including now the interaction with both acoustic and polar phonons to eliminate the coupling term H_{ep}:

$$\bar{H} = e^{-S} H e^{S} , \tag{3.1}$$

$$S = \sum_{j=1}^{2} \sum_{n,\boldsymbol{q}} u^{1/2}_{\boldsymbol{q}j} (e^{-i\boldsymbol{q}\boldsymbol{r}_n} b^+_{\boldsymbol{q}j} - e^{i\boldsymbol{q}\boldsymbol{r}_n} b_{\boldsymbol{q}j}) a^+_n a_n . \tag{3.2}$$

---

[1]) For the sake of simplicity we use atomic units by setting $\hbar = 1$, e (electronic charge) $= 1$, m (electronic mass) $= 1$.

The result of this transformation without any approximations up to now is

$$\overline{H} = \overline{H}_0 + \overline{H}_1 + \overline{H}_2 , \tag{3.3}$$

$$\overline{H}_0 = \sum_n \eta_n a_n^+ a_n + \sum_{j=1}^{2} \sum_{\boldsymbol{q}} \omega_{\boldsymbol{q}j} b_{\boldsymbol{q}j}^+ b_{\boldsymbol{q}j} , \tag{3.4}$$

$$\overline{H}_1 = \sum_{n,n'} I_{nn'} a_n^+ a_{n'} B_{n'n} , \tag{3.5}$$

$$\overline{H}_2 = \sum_{n \neq n'} v(\boldsymbol{r}_{n'n}) a_n^+ a_n a_{n'}^+ a_{n'} , \tag{3.6}$$

$$\overline{\boldsymbol{j}} = i \sum_{n,n'} I_{nn'} \boldsymbol{r}_{n'n} a_n^+ a_{n'} B_{n'n} , \tag{3.7}$$

where

$$\eta_n = \varepsilon_n - \sum_{j=1}^{2} \sum_{\boldsymbol{q}} u_{\boldsymbol{q}j} \omega_{\boldsymbol{q}j} - \mu , \tag{3.8}$$

$$v(\boldsymbol{r}) = \sum_{j=1}^{2} \sum_{\boldsymbol{q}} u_{\boldsymbol{q}j} \omega_{\boldsymbol{q}j} e^{i\boldsymbol{q}\boldsymbol{r}} , \tag{3.9}$$

$$B_{n'n} = \exp\left\{\sum_{j=1}^{2} \sum_{\boldsymbol{q}} u_{\boldsymbol{q}j}^{1/2} [(e^{-i\boldsymbol{q}\boldsymbol{r}_{n'}} - e^{-i\boldsymbol{q}\boldsymbol{r}_n}) b_{\boldsymbol{q}j}^+ - \text{c.c.}]\right\} . \tag{3.10}$$

We see that the electron–phonon interaction involving multi-phonon processes of arbitrary high order is showing up now in the operators of kinetic energy and electric current. This means that we have changed to the polaron picture where the particles are carrying along with the polarization of the surrounding medium. To elucidate the meaning of $\overline{H}_1$ we split it into two parts:

$$\overline{H}_1 = \overline{H}_{\mathrm{B}} + \overline{H}_{\mathrm{ep}} , \tag{3.11}$$

$$\overline{H}_{\mathrm{B}} = \sum_{n,n'} \tilde{I}_{nn'} a_n^+ a_{n'} , \tag{3.12}$$

$$\overline{H}_{\mathrm{ep}} = \sum_{n,n'} I_{nn'} a_n^+ a_{n'} (B_{n'n} - \overline{B}_{n'n}) , \tag{3.13}$$

$$\tilde{I}_{nn'} = I_{nn'} \overline{B}_{n'n} , \tag{3.14}$$

where $\overline{B}_{n'n}$ means the expectation value of $B_{n'n}$ with respect to $\overline{H}_1 = 0$. $\overline{H}_{\mathrm{B}}$ now describes the mean kinetic energy of the polaron in the impurity band and it is the quantity $\tilde{I}_{nn'}$ rather than $I_{nn'}$ which has to be compared with the mean spacing Δ of the levels ε_n or η_n to decide whether impurity hopping or impurity band processes will take place. Since

$$\overline{B}_{n'n} = \exp\left\{- 2 \sum_{j=1} \sum_{\boldsymbol{q}} u_{\boldsymbol{q}j} \sin^2 \frac{1}{2} \boldsymbol{q} \boldsymbol{r}_{n'n} \coth \frac{1}{2} \beta \omega_{\boldsymbol{q}j}\right\} , \tag{3.15}$$

$\tilde{I}_{nn'}$ will always be smaller than $I_{nn'}$ such that Anderson's criterion

$$\frac{\Delta}{\tilde{I}} > \left(\frac{\Delta}{I}\right)_{\mathrm{c}} \tag{3.16}$$

for the hopping case may be satisfied even at rather high concentrations. $\overline{H}_{\mathrm{ep}}$ describes the carrier–phonon interaction in arbitrary high order. Clearly, $\overline{H}_{\mathrm{ep}} = 0$ if $u_{\boldsymbol{q}j} = 0$.

In the same way we have

$$\bar{j} = \bar{j}_B + \bar{j}_H , \tag{3.17}$$

$$\bar{j}_B = i \sum_{n,n'} \tilde{I}_{nn'} r_{n'n} a_n^+ a_{n'} , \tag{3.18}$$

$$\bar{j}_H = i \sum_{n,n'} I_{nn'} r_{n'n} a_n^+ a_{n'} (B_{n'n} - \bar{B}_{n'n}) , \tag{3.19}$$

where $\bar{j}_B$ and $\bar{j}_H$ describe the current of impurity band and phonon-assisted hopping transitions, respectively.

Since we are dealing with the hopping case in this paper we drop $\bar{H}_B$ and $\bar{j}_B$ assuming that Anderson's criterion (3.16) is satisfied.

Apart from the carrier–phonon interaction term $\bar{H}_{ep}$ the canonically transformed Hamiltonian contains an effective polaron–polaron interaction. One can show that the interaction potential $v(\boldsymbol{r})$ of (3.9) is short ranged so that we are allowed to neglect $\bar{H}_2$ if the concentration of the charge carriers is not too high.

4. Conductivity

To calculate the symmetric part of the electric conductivity we start from the hopping-current correlation function

$$F_{H\mu\mu}(t) = \langle T\{j_{H\mu}(t)\, j_{H\mu}\}\rangle , \tag{4.1}$$

where T is the time ordering operator and $j_{H\mu}$ denotes the μ-th space component of the hopping-current operator of (3.19). (In the following we shall drop the bars which have been introduced in section 3 to mark the canonically transformed operators.)

Because of the statistical distribution of the impurities $F_{H\mu\mu}(t)$ is isotropic so that we may write

$$F_{H\mu\mu}(t) = F_H(t) = \frac{1}{3} \langle T\{\boldsymbol{j}_H(t)\, \boldsymbol{j}_H\}\rangle = -\frac{1}{3} \sum_{n,n'} \sum_{m,m'} I_{nn'} I_{mm'} \boldsymbol{r}_{nn'} \boldsymbol{r}_{m'm} \times$$

$$\times \langle T\, \{a_n^+(t)\, a_{n'}(t)\, [B_{n'n}(t) - \bar{B}_{n'n}]\, a_m^+ a_{m'}\, [B_{m'm} - \bar{B}_{m'm}]\}\rangle , \tag{4.2}$$

where

$$A(t) = e^{iHt} A\, e^{-iHt} , \qquad \langle A\rangle = \frac{\operatorname{tr}\, (e^{-\beta H} A)}{\operatorname{tr}\, e^{-\beta H}} , \tag{4.3}$$

$$H = H_0 + H_{ep} . \tag{4.4}$$

H_0 and H_{ep} are the free and interaction part of the Hamiltonian as given in (3.4) and (3.13).

From $F_H(t)$ we obtain the conductivity σ_H by introducing the Fourier transform of $F_H(t)$:

$$F_H(\zeta) = \int_0^{-\beta i} dt\, e^{-i\zeta t} F_H(t) , \qquad \zeta = \frac{2\pi}{i\beta} \nu , \qquad \nu = 0, \pm 1, \pm 2, \ldots , \tag{4.5}$$

$$\sigma_H = \frac{1}{2} \Omega^{-1} \lim_{\omega \to 0} \frac{F_H(\omega + i0) - F_H(\omega - i0)}{\omega} , \tag{4.6}$$

where Ω is the total volume.

A first approximation we carry out in (4.2) consists in factorizing the expectation value into its electronic and phonon parts and evaluating the phonon part with respect to $H_{ep} = 0$:

$$\langle T \{a_n^+(t)\, a_{n'}(t)\, [B_{n'n}(t) - \overline{B}_{n'n}]\, a_m^+\, a_{m'}\, [B_{m'm} - \overline{B}_{m'm}]\}\rangle \approx$$
$$\approx \langle T \{a_n^+(t)\, a_{n'}(t)\, a_m^+\, a_{m'}\}\rangle\, [\langle T \{B_{n'n}(t)\, B_{m'm}\}\rangle_0 - \overline{B}_{n'n}\, \overline{B}_{m'm}] \,. \quad (4.7)$$

Now we are left with the calculation of the electronic part of the correlation function in (4.7). In the usual theory of conductivity in metals this is done by expanding the correlation in powers of H_{ep} to obtain the transport cross section which describes the decay of the current correlation due to scattering by phonons.

In our case, however, we are not interested in scattering processes of the carriers but just in the opposite effect: hopping transitions between donor states with different energies are impossible if phonons were absent. Thus, the phonons are playing an "active" or "assisting" role instead of a "scattering" role as in band transitions. This is represented by the fact that the electronic correlation in (4.7) is still multiplied by time-dependent phonon correlations. This enables us to make a further approximation which consists in evaluating even the electronic correlation with respect to $H_{ep} = 0$. We then get

$$F_H(t) = \frac{1}{3} \sum_{m,n} \tilde{I}_{mn}^2\, r_{mn}^2\, G_0(m, t)\, G_0(n, -t)\, \Phi_{mn}(t) \,, \quad (4.8)$$

where $\tilde{I}_{mn}$ is the effective energy overlap defined by (3.14), $G_0(m, t)$ is the free carrier Greens function for $H_{ep} = 0$ the Fourier transform of which reads

$$G_0(m, z) = \frac{1}{z - \eta_m} \,, \quad (4.9)$$

and $\Phi_{mn}(t)$ is the phonon correlation given by

$$\Phi_{mn}^{>}(t) = \langle B_{mn}(t)\, B_{mn}\rangle_0 =$$
$$= \exp\left\{4 \sum_{j=1}^{2} \sum_{q} u_{qj} \sin^2 \frac{1}{2} q\, r_{mn} \frac{\cos \omega_{qj}\left(t + \frac{i}{2}\beta\right)}{\sinh \frac{1}{2} \beta\, \omega_{qj}}\right\} - 1 \quad (4.10)$$

and

$$\Phi_{mn}^{<}(t) = \langle B_{nm}\, B_{mn}(t)\rangle_0 = \Phi_{mn}^{>}(t - i\beta) \,. \quad (4.11)$$

In order to study the effect of the two types of phonons separately we split $\Phi_{mn}(t)$ into three parts

$$\Phi_{mn}(t) = \Phi_{mn}^{(1)}(t)\, \Phi_{mn}^{(2)}(t) + \Phi_{mn}^{(1)}(t) + \Phi_{mn}^{(2)}(t) \,, \quad (4.12)$$

where

$$\Phi_{mn}^{>(j)}(t) = \exp\left\{4 \sum_{q} u_{qj} \sin^2 \frac{1}{2} q\, r_{mn} \frac{\cos \omega_{qj}\left(t + \frac{i}{2}\beta\right)}{\sinh \frac{1}{2} \beta\, \omega_{qj}}\right\} - 1 \quad (4.13)$$

and

$$\Phi_{mn}^{<(j)}(t) = \Phi_{mn}^{>(j)}(t - i\beta) \quad (4.14)$$

describe the j-type phonon correlation only. Equation (4.12) gives rise to three corresponding contributions to the conductivity

$$\sigma_H = \sigma_{12} + \sigma_1 + \sigma_2\,, \tag{4.15}$$

where σ_j describes the j-type phonon-assisted hopping conductivity and σ_{12} is the hopping conductivity assisted by simultaneous exchange of acoustic and polar phonons. In sections 5 and 6 we shall see that either σ_1 or σ_2 is the predominant contribution to σ_H so that we may neglect σ_{12} by assuming that this term is not of great physical significance.

It should be born in mind that the separation in (4.12) only refers to the dynamical or time-dependent influence of the phonons. There is still the static effect of the lattice which is involved in $\tilde{I}_{mn}$, equation (3.14).

By inserting (4.8), (4.12), and (4.13) into (4.5) and (4.6) we finally obtain

$$\sigma_j = \frac{\beta}{24}\,\Omega^{-1} \sum_{m,n} \tilde{I}^2_{mn}\, r^2_{mn} \left[\cosh\frac{1}{2}\beta\,\eta_n \cosh\frac{1}{2}\beta\,\eta_m\right]^{-1} K^{(j)}_{mn}\,(\eta_n - \eta_m)\,, \tag{4.16}$$

where

$$K^{(j)}_{mn}(\omega) = \int_{-\infty}^{+\infty} dt\; e^{i\omega t} \left[\exp\left\{4 \sum_{q} u_{qj} \sin^2\frac{1}{2}\,\boldsymbol{q}\,\boldsymbol{r}_{mn}\; \frac{\cos\omega_{qj}\,t}{\sinh\frac{1}{2}\beta\,\omega_{qj}}\right\} - 1\right]. \tag{4.17}$$

5. The Pure Acoustic Case

If polar phonons are not present we have $u_{q2} = 0$, $\sigma_2 = 0$ and therefore $\sigma_H = \sigma_1$. Let us start with the calculation of the one-phonon exchange contribution $\sigma_{1(1)}$ to σ_1. For this purpose we expand the exponential function in (4.17) up to first order to obtain

$$K^{(1)}_{mn(1)}\,(\eta_n - \eta_m) = 4\,\pi \sum_{q} u_{q1} \frac{\sin^2\frac{1}{2}\,\boldsymbol{q}\,\boldsymbol{r}_{mn}}{\sinh\frac{1}{2}\beta\,\omega_{q1}} \times$$

$$\times\,[\delta\,(\eta_n - \eta_m - \omega_{q1}) + \delta\,(\eta_n - \eta_m + \omega_{q1})]\,. \tag{5.1}$$

Inserting (5.1) into (4.16) for $j = 1$, setting $\sin^2(\boldsymbol{q}\,\boldsymbol{r}_{mn}/2) \approx 1/2$ and assuming a Debye-type spectrum for ω_{q1}, $\omega_{q1} = c\,q$, (c = velocity of sound) we arrive at

$$\sigma_{1(1)} = \frac{g^2}{24\,\pi\,\varrho\,c^5}\,\Omega^{-1} \sum_{m,n} \tilde{I}^2_{mn}\, r^2_{mn}\, \frac{\frac{1}{2}\beta\,(\eta_n - \eta_m)}{\sinh\frac{1}{2}\beta\,(\eta_n - \eta_m)} \times$$

$$\times \left[\cosh\frac{1}{2}\beta\,\eta_n \cosh\frac{1}{2}\beta\,\eta_m\right]^{-1}, \tag{5.2}$$

where $\varrho = M\,N/\Omega$. Since there is no polar coupling we may put $\tilde{I}_{mn} \approx I_{mn}$ in consistency with the first order expansion of (5.1). If the temperature is not too low such that $\beta\,\Delta < 1$ (Δ being the mean spacing of the levels η_n) the expression (5.2) may be simplified further to yield

$$\sigma_{1(1)} = \frac{g^2}{12\,\pi\,\varrho\,c^5}\,\Sigma\,, \tag{5.3}$$

$$\Sigma = \Omega^{-1} \sum_{m,n} I_{mn}^2\, r_{mn}^2\, [1 + \cosh\beta\,\eta_n]^{-1}\,. \tag{5.4}$$

Equations (5.2), (5.3), and (5.4) are identical with the results of Miller and Abrahams [2] and Gosar [3]. To calculate Σ involving the statistical distribution of the impurities Miller and Abrahams have used a network model for the impurities. As far as the temperature dependence is concerned it turns out that

$$\Sigma \sim e^{-\beta A}\,, \tag{5.5}$$

where the activation energy A is a function of compensation K being of order of magnitude of the mean spacing Δ. A has a minimum near $K = 1/2$ and tends logarithmically to inifinity as $K \to 0$ or $K \to 1$.

It is not difficult to evaluate the complete multi-phonon contribution to σ_1. By making use of the same approximation as in the beginning of this section we obtain for $T \ll \theta/2$ (θ being the Debye-temperature)

$$\frac{\sigma_1}{\sigma_{1(1)}} = F\left(\frac{\lambda}{x^2}\right) \exp\left\{-\frac{1}{\pi^2}\frac{\lambda}{x^2}\int_0^x d\xi \cdot \xi \coth\xi\right\}, \tag{5.6}$$

$$F(y) = 2\sum_{l=0}^{\infty} \frac{l!\,y^l}{(2\,l+2)!} = 1 + \frac{y}{12} + \frac{y^2}{180} + \cdots, \tag{5.7}$$

$$x = \frac{\theta}{2\,T}, \qquad \lambda = \frac{g^2(k\,\theta)^2}{2\,\varrho\,c^5}, \tag{5.8}$$

where $\sigma_{1(1)}$ is the one-phonon assisted conductivity of equation (5.3). For very low temperatures, $x \to \infty$ (but still $kT > \Delta$) we have $F(\lambda/x^2) \to 1$ and

$$\frac{\sigma_1}{\sigma_{1(1)}} \to \exp\left(-\frac{\lambda}{2\,\pi^2}\right) < 1\,. \tag{5.9}$$

The numerical evaluation of (5.6) shows that (5.9) is valid already for $T < \theta/4$ if $\lambda \leq 10$. Hence, in this temperature region the temperature dependence of σ_1 is not influenced by multi-phonon processes. The fact that $\sigma_1 < \sigma_{1(1)}$ is due exclusively to the renormalization of the resonance integral $I_{mn} \to \tilde{I}_{mn}$. This result may be considered as a justification for the restrictions to one-phonon exchange processes at low temperatures.

At temperatures $T > \theta/2$ multi-phonon processes will certainly come into play, but it is questionable whether temperatures in this region are of any physical interest at least in the pure acoustic case in semiconductors. The charge carriers will then be thermally excited into the lowest conduction band states which gives rise to intrinsic conduction with a thermally activated temperature dependence of entirely different nature.

6. The Polar Case

In the polar case we have both acoustic and polar phonons so that $u_{q1} \neq 0$ and $u_{q2} \neq 0$ but certainly $u_{q2} \gg u_{q1}$. Nevertheless, we have to take into account both contributions σ_1 and σ_2 as will become clear in the following. Let us start with the polar phonon-assisted part σ_2 to be derived from (4.16) and (4.17) where

$$K^{(2)}_{mn}(\eta_n - \eta_m) \approx K^{(2)}(\eta_n - \eta_m) = \int_{-\infty}^{+\infty} \mathrm{d}t\, \mathrm{e}^{i(\eta_n-\eta_m)t} \times$$

$$\times \left[\exp\left\{2 \sum_{\boldsymbol{q}} u_{\boldsymbol{q}2} \frac{\cos \omega_{\boldsymbol{q}2} t}{\sinh \frac{1}{2} \beta\, \omega_{\boldsymbol{q}2}}\right\} - 1\right]. \tag{6.1}$$

At high temperatures we may evaluate (6.1) by means of the saddle-point method. If we take into account the main saddle-point only we obtain

$$K^{(2)}(\eta_n - \eta_m) =$$

$$= \sqrt{\pi}\, \tau(\beta) \exp\left\{2 \sum_{\boldsymbol{q}} u_{\boldsymbol{q}2} \sinh^{-1} \frac{1}{2}\, \beta\, \omega_{\boldsymbol{q}2}\right\} \exp\left\{-\frac{1}{4}\, \tau^2(\beta)\, (\eta_n - \eta_m)^2\right\}, \tag{6.2}$$

where

$$\tau(\beta) = \left[\sum_{\boldsymbol{q}} u_{\boldsymbol{q}2}\, \omega^2_{\boldsymbol{q}2} \sinh^{-1} \frac{1}{2}\, \beta\, \omega_{\boldsymbol{q}2}\right]^{-1/2} \ll \frac{1}{\omega_0}. \tag{6.3}$$

The inequality in (6.3) is the condition for applicability of the saddle-point method, ω_0 is the mean frequency of the polar vibrations. (6.3) defines a limiting temperature T_1 such that (6.2) is valid for $T \geqq T_1$. From (6.3) we see that T_1 is of the order of $\theta/2$, but may be even lower if $N\, u_{\boldsymbol{q}2} \approx \gamma^2/2\, M\, \omega_0^3 \gg 1$. Since the energy differences $\eta_n - \eta_m$ will always be much smaller than ω_0 we may replace the second exponential function in (6.2) by unity. Inserting (6.2) into (4.17) we finally get

$$\sigma_2 = \frac{\sqrt{\pi}}{12}\, \beta\, \tau(\beta)\, \mathrm{e}^{-\beta\, U(\beta)}\, \Sigma\,, \tag{6.4}$$

where Σ is defined by (5.4) and

$$U(\beta) = \frac{1}{2} \sum_{\boldsymbol{q}} u_{\boldsymbol{q}2}\, \omega_{\boldsymbol{q}2} \frac{\tanh(\beta\, \omega_{\boldsymbol{q}2}/4)}{\beta\, \omega_{\boldsymbol{q}2}/4} \to \frac{1}{2} \sum_{\boldsymbol{q}} u_{\boldsymbol{q}2}\, \omega_{\boldsymbol{q}2} \approx \frac{\gamma^2}{4\, M\, \omega_0^2} \tag{6.5}$$

if $\beta\, \omega_0/4 \ll 1$ or $T \gg \theta/4$. In (6.4) we have included the renormalization of I_{mn} due to the polar coupling whereas the renormalization due to the acoustic coupling has been neglected since $u_{\boldsymbol{q}2} \gg u_{\boldsymbol{q}1}$.

In the region $T > \theta/4$ the energy $U(\beta)$ will be much greater than the activation energy A of Σ (see section 5) such that essentially

$$\sigma_2 \sim \mathrm{e}^{-\beta\, U(\beta)} \tag{6.6}$$

as far as the temperature dependence is concerned.

At temperatures $T < \theta/2$ the condition (6.3) is no longer satisfied so that the saddle-point method becomes a bad approximation. If $\omega_0\, \tau(\beta) \gg 1$ we may evaluate $K^{(2)}(\eta_n - \eta_m)$ by expanding the exponential function. Since the energy differences $\eta_n - \eta_m$ will be much smaller than the optical frequency $\omega_{\boldsymbol{q}2}$ one-phonon processes with polar phonons are impossible. Thus, the lowest non-vanishing contribution comes from the second-order term in the expansion of the exponential function corresponding to a two-phonon process. The result for σ_2 reads

$$\sigma_2 = \frac{\pi}{24} \frac{\beta \gamma^4}{M^2 \omega_0^6 \sinh^2 \frac{1}{2} \beta \omega_0} \exp\left\{-\frac{\gamma^2}{M \omega_0^3} \coth \frac{1}{2} \beta \omega_0\right\} \Sigma . \tag{6.7}$$

From (6.7) we see that the temperature dependence of σ_2 for $T < \theta/2$ is given by

$$\sigma_2 \sim e^{-\beta(\omega_0 + A)} . \tag{6.8}$$

Up to now our treatment of the polar case has been quite analogous to the small polaron theory. Now, at low temperatures $T < \theta/2$, we arrive at an important difference. In the small polaron theory the polar phonon-assisted conductivity is vanishing similar to (6.8). The charge transport is then carried on by polaron band processes. In our case, however, band processes are impossible as has been argued in sections 1 and 3 of this paper. Thus, even at low temperatures the transport processes must be assisted by polar or acoustic phonons. Since the number of polar phonons is decreasing like $\exp(-\beta \omega_0)$ for $T < \theta/2$ we expect that the predominant contribution to σ_H now comes from the acoustic phonon assisted conductivity σ_1 which has been calculated already in section 5. We may apply this result to the polar case if we take into account the renormalization of I_{mn} due to the polar coupling:

$$\left.\begin{aligned} \sigma_1 &= \frac{g^2}{12 \pi \varrho c^5} \exp\left\{-\frac{\gamma^2}{M \omega_0^3} \coth \frac{1}{2} \beta \omega_0\right\} \Sigma , \\ &\sim e^{-\beta A} \quad \text{if} \quad T < \frac{\theta}{2} . \end{aligned}\right\} \tag{6.9}$$

By comparison of (6.7) and (6.9) we see that at temperatures $T < \theta/2$ the main contribution comes from σ_1 whereas at high temperatures $T > \theta/2$ σ_1 is rapidly decreasing since $\coth(\beta \omega_0/2) \to \infty$. Hence, for $T > \theta/2$ the main contribution to σ_H is the polar phonon assisted conductivity σ_2 as given in equation (6.4).

Summarizing our results of the polar case we expect a temperature dependence of σ_H as roughly stretched in Fig. 1. — A temperature dependence of this type has been measured by Owen [6] in a P_2O_5-glass containing V_2O_5 and VO_2.

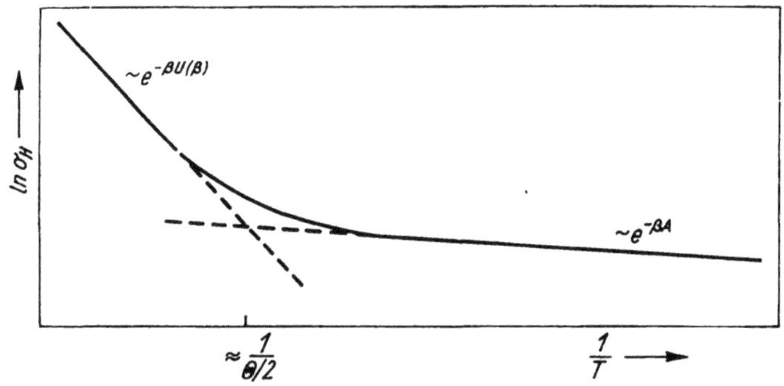

Fig. 1. Expected temperature dependence of the conductivity

Acknowledgements

The author is very much indebted to Professor Sir Nevill Mott for stimulating and encouraging this investigation. He would also like to thank Dr. Eilenberger, Professor Hajdu and Professor Mühlschlegel for many clarifying and helpful discussions.

References

[1] P. W. Anderson, Phys. Rev. **109**, 1492 (1958).
[2] A. Miller and E. Abrahams, Phys. Rev. **120**, 745 (1960).
[3] P. Gosar, phys. stat. sol. **10**, 79 (1965).
[4] T. Holstein, Ann. Phys. (U.S.A.) **8**, **343** (1959).
[5] K. Takeyama, J. Phys. Soc. Japan **23**, 1013 (1967).
[6] A. E. Owen, unpublished.

(Received June 7, 1968)

phys. stat. sol. 28, 635 (1968)

Subject classification: 13 and 18.2

Boris Kidrič Institute of Nuclear Sciences, Beograd

On the Problem of Kinematic Interaction in the Quantum Theory of Magnetism

By

D. I. LALOVIĆ, B. S. TOŠIĆ, and R. B. ŽAKULA

The work considers the problem of the kinematic interaction of elementary excitations in the quantum theory of magnetism. In order to avoid the appearance of so-called non-physical states, a generalization of the Holstein-Primakoff representation is suggested. It is shown that in calculating dynamic characteristics of magnon systems the Holstein-Primakoff representation can be applied consistently only up to terms of the order $2S$, and not up to terms of the order $2S+1$, as it is tacitly assumed. In addition, a Pauli representation of spin operators for spins $S = 1$ and $S = 3/2$ is given.

В работе рассматривается проблема кинематического взаимодействия элементарных возбуждений в квантовой теории магнетизма. Для того, чтобы отклонить вклад от так называемых ,,нефизических состояний" предлагается обобщение известных формул Гольштейна-Примакова. В работе показано, что при вычислениях динамических характеристик магнонной системы, представление Гольштейна-Примакова можно последовательно применять лишь до членов порядка $2S$ по бозонам. Обычно считается, что представление Гольштейна-Примакова применимо до членов порядка $2S+1$. В работе также были выражени спиновые операторы в случае спинов $S = 1$ и $S = 3/2$ через операторы Паули.

1. Introduction

The Hamiltonian used in the quantum theory of magnetism for the description of the magnetic properties of crystals may be expressed in terms of spin operators S_g^α where g is the index of the lattice point[1]), and α is the index of the spin component. The operators S_g^α commute with each other when they correspond to different lattice points, and in one point they satisfy the well-known commutation relations for angular momentum. For such commutation relations the Fourier transformation which performs the transition from configuration to momentum space is not canonical, i.e., it does not preserve the commutation relations. It is therefore necessary to express the operators S_g^α through operators which are transformed canonically and which have definite statistics in momentum space. Such are, for example, Bose and Fermi operators.

In the theory of magnetism the spin operators are usually expressed by means of Bose operators. The best known treatment of this kind is the Holstein-Primakoff representation [1], which gives in lowest approximation the same results as Bloch's theory of spin waves [2]. There is also Dyson's theory [3] of ideal spin waves, which, when translated into the language of second quanti-

[1]) $g \equiv (\boldsymbol{n}, \beta)$ where $\boldsymbol{n}$ and β define positions of an elementary cell and an atom in the cell, respectively.

zation [4], gives the results equivalent to those of the Holstein-Primakoff method [5].

If we consider in the spin Hamiltonian only the second order terms in S_g^+, S_g^- [2]) and we express them in the Bose representation using the Holstein-Primakoff formulae, we shall get beside quadratic terms also terms of higher order in the Bose operators. The appearance of these higher order terms is the exclusive consequence of the specific commutation relations for spin operators S_g^+, S_g^-, i.e., they are of purely kinematic origin, and they are therefore called the Hamiltonian of kinematic interaction of elementary excitations. A similar situation arise in the spin Hamiltonian, when the higher order terms in the operators $S_g^\pm$ are written in Bose representation. When such a term of a given order in $S_g^\pm$ is written in Bose representation, then one gets a term of the same order and terms of higher order in the Bose operators. The former is called the Hamiltonian of dynamic interaction of elementary excitations. The remaining terms we call dynamico-kinematic interaction. For the investigation of interaction effects of elementary excitations (nonlinear or anharmonic effects), which are particularly significant in the vicinity of the transition point, it is very important to find correct expressions for the Hamiltonians of the kinematic and dynamico-kinematic interactions. It is not possible to achieve this by applying the Holstein-Primakoff formulae, since they are approximate and may be used for a description of magnetic phenomena in the crystal only at sufficiently low temperatures, i.e. for low concentrations of elementary excitations when the nonlinear effects are negligible. This limitation in the applicability of the Holstein-Primakoff representation is a consequence of the fact that the spin commutation relations in this representation are satisfied only in the Hilbert space spanned by those boson states for which the boson occupation number does not exceed $2\,S$ (S being the effective spin of an atom of the magnetic lattice). Though the higher boson states in this representation are considered nonphysical, they always give a contribution introducing an uncontrollable error in the physical results. It is obvious, however, that the nearer we are to the transition point the bigger their contribution is, and so the bigger is the error brought about by the use of the Holstein-Primakoff formulae. There is therefore a need for an exact Bose representation of spin operators, which could be applied for the description of magnetic phenomena in crystals in the whole temperature range from the absolute zero up to the transition temperature. The work presented here gives a solution of this problem.

2. The Exact Bose Representation of Spin Operators

The spin operators S_g^α related to one lattice point satisfy the following commutation relations:

$$[S_g^+, S_g^-] = 2\,S_g^z\,, \qquad \{S_g^+, S_g^-\} = 2\,S\,(S+1) - 2\,(S_g^z)^2\,,$$
$$(S_g^+)^{2\,S+1} = (S_g^-)^{2\,S+1} = 0\,, \tag{2.1}$$

where S is the value of the spin, $S_g^\pm = S_g^x \pm i\,S_g^y$, and [] and {} represent the commutator and the anticommutator, respectively. Our aim is to represent the spin operators S_g^α by Bose creation and annihilation operators B_g^+ and B_g in such a way that the commutation relations (2.1) be satisfied in the whole space of boson states. One naturally requires that the operator S_g^z be a function of the operator of boson occupation number $\hat{N}_g = B_g^+\,B_g$ only. Since for a given

[2]) If we choose as axis of quantization the z-axis, then we obtain $S_g^\pm = S_g^x \pm i\,S_g^y$.

spin S the operator $2\,S_g^z$ has $2\,S+1$ eigenvalues, and the operator $\hat{N}_g$ has an infinite number of eigenvalues, it is most convenient to choose this function in such a way that its spectrum be a periodic function with the period $2\,S+1$. On the other hand, this function should be equal to the commutator $[S_g^+, S_g^-]$. Since in the representation in which S_g^z is diagonal the operators S_g^+ and S_g^- are not diagonal, we shall seek the Bose representation of the latter in the form

$$S_g^+ = f(\hat{N}_g)\,B_g\,, \qquad S_g^- = B_g^+\,f(\hat{N}_g)\,. \tag{2.2}$$

From (2.1) and (2.2) it follows

$$2\,S_g^z = (\hat{N}_g+1)\,f^2(\hat{N}_g) - \hat{N}_g\,f^2(\hat{N}_g-1)\,. \tag{2.3}$$

Let us introduce the notation

$$\hat{N}_g\,f^2\,(\hat{N}_g-1) = \Phi(\hat{N}_g)\,. \tag{2.4}$$

Then (2.3) becomes

$$2\,S_g^z = \Phi\,(\hat{N}_g+1) - \Phi(\hat{N}_g)\,. \tag{2.5}$$

In accordance with the above remarks about the spectrum of S_g^z, the spectrum of the expressions on the right-hand side of equation (2.5) must be a periodic function with the period $2\,S+1$, i.e. the following equation must be satisfied:

$$\Phi(N_g) = \Phi\,\big(N_g + \mu\,(2\,S+1)\big)\,; \qquad \mu = 0, 1, 2, \ldots\,, \tag{2.6}$$

where N_g are the eigenvalues of $\hat{N}_g$.

In addition to this property of periodicity, the function $\Phi(N_g)$ must be such that it gives the correct spectrum of the operators $2\,S_g^z$, and also the correct spectrum of the anticommutator

$$\{S_g^+, S_g^-\} = \Phi\,(\hat{N}_g+1) + \Phi(\hat{N}_g)\,. \tag{2.7}$$

From (2.5) and (2.7) we obtain

$$2\,\Phi\,(\hat{N}_g+1) = 2\,S_g^z + \{S_g^+, S_g^-\}\,. \tag{2.8}$$

In order that (2.8) determines uniquely the eigenvalues of the function $\Phi\,(\hat{N}_g+1)$ it is necessary to establish correspondence between the eigenstates of the operators S_g^z and the eigenstates of the operators $\hat{N}_g$. Taking for ground state of the system that one in which the z-projection of the spin in every lattice point has the maximum value S, it is natural to assume that the corresponding state in the boson space is the vacuum, i.e. the state for which $N_g = 0$. The state with the z-projection of the spin being $S-1$ at the lattice point g corresponds to the one boson state with the same index g, and so on, successively, up to the state with projection $-S$, which should be related to the state with $2\,S$ bosons. This correspondence repeats further periodically. It follows from this that the spectrum of the function $\Phi\,(\hat{N}_g+1)$ is defined by the following system of equations:

$$\Phi\,(N_g+1) = 2\,(N_g+1)\,S - N_g\,(N_g+1)\,, \tag{2.9}$$

where $N_g = 0, 1, 2, \ldots, 2\,S$.

The spectrum of the function $\Phi\,(\hat{N}_g+1)$ for $N_g > 2\,S$ is determined by the condition of periodicity.

It is not possible to find the value $\Phi(0)$ from the system of equations (2.9). From (2.5) and (2.7) it is easily seen that $\Phi(0) = 0$, so that instead of the system of equations (2.9) we can take

$$\Phi(N_g) = N_g\,(2\,S + 1 - N_g)\,; \qquad N_g = 0, 1, 2, \ldots, 2\,S + 1\,. \tag{2.9a}$$

The function $\Phi(\hat{N}_g)$, whose spectrum satisfies the conditions (2.6) and (2.9a), is sought in the form

$$\Phi(\hat{N}_g) = \sum_{\nu=0}^{\infty} a_\nu \overset{+}{B}{}_g^{\nu+1} B_g^{\nu+1} = \sum_{\nu=0}^{\infty} a_\nu \hat{N}_g\,(\hat{N}_g - 1) \ldots (\hat{N}_g - \nu)\,. \tag{2.10}$$

Such an expansion of the function $\Phi(\hat{N}_g)$ is most convenient because the spectrum of the operator $\overset{+}{B}{}_g^{\nu} B_g^{\nu}$ is zero in the space of boson states for which $N_g < \nu$.

From (2.10), (2.9a), and (2.6) one can derive the following infinite system of equations:

$$\sum_{\nu=0}^{(2S+1)\mu+N_g-1} a_\nu\,\nu!\binom{(2\,S+1)\,\mu + N_g - 1}{\nu} = \frac{N_g\,(2\,S+1-N_g)}{(2\,S+1)\,\mu + N_g}\,; \tag{2.11}$$

$$N_g = 1, 2, 3, \ldots 2\,S + 1\,; \qquad \mu = 0, 1, 2, \ldots$$

which determines the coefficients a_ν. To determine the first $2\,S + 1$ coefficients a_ν, we put $\mu = 0$ in (2.11) and for N_g we insert the values $1, 2, 3, \ldots, 2\,S + 1$. We obtain the next $2\,S + 1$ coefficients by putting $\mu = 1$, while N_g again takes the values $1, 2, 3, \ldots, 2\,S + 1$, and so on. In this way we can calculate as many coefficients a_ν as we want, and thus find the function $\Phi(\hat{N}_g)$ with the desired accuracy. The knowledge of the function $\Phi(\hat{N}_g)$ enables us to express the spin operators in Bose representation. From (2.2), (2.4), and (2.5) we obtain

$$S_g^+ = \left(\sum_{\nu=0}^{\infty} a_\nu \overset{+}{B}{}_g^{\nu} B^{\nu}\right)^{1/2} B_g\,, \tag{2.12a}$$

$$2\,S_g^z = \sum_{\nu=0}^{\infty} (\nu + 1)\,a_\nu \overset{+}{B}{}_g^{\nu} B_g^{\nu}\,. \tag{2.12b}$$

From (2.12a) and the periodicity of the spectrum of the function $\Phi(N_g)$, it can easily be seen that the following condition is satisfied for all boson states:

$$(S_g^+)^{2S+1} = (S_g^-)^{2S+1} = 0\,.$$

The system of equations (2.11) gives the following values for the first $2\,S + 1$ coefficients a_ν:

$$a_0 = 2\,S\,, \quad a_1 = -1\,, \quad a_\varrho = 0 \quad \text{for} \quad \varrho = 2, 3, 4, \ldots.\,2\,S\,. \tag{2.13}$$

This can be verified by substituting the values (2.13) into the system of equations (2.11) which results when we put $\mu = 0$. Thus, if we limit ourselves to that part of the space of boson states for which $N_g \leqq 2\,S$, the formulae (2.12) become identical with the Holstein-Primakoff formulae. If we want to have an exact Bose representation of the spin operators in a larger space of boson states, i.e. in that space which also involves the states with $N_g > 2\,S$, we have to calculate from (2.11) the remaining coefficients a_ν for $\nu > 2\,S$. Hence, the formulae (2.12) represent a generalization of the Holstein-Primakoff formulae and enable us to find the Hamiltonian of the kinematic and dynamico-kinematic interactions with desired accuracy.

The application of the formulae (2.12a, b) requires a series expansion of the root of the operator function in (2.12a). Up to now in the literature [6] this root has been expanded into a power series in $\hat{N}_g$, but such an expansion is not correct because N_g is not a small parameter. The correct expansion should be done in terms of the operators $\overset{+}{B_g^\nu} B_g^\nu$, because the space of boson states in which they all equal zero will be the larger the larger ν is. As an illustration, let us write the correct expansion of the root appearing in the Holstein-Primakoff formulae:

$$\sqrt{2S - \hat{N}_g} = b_0 + b_1 \overset{+}{B}_g B_g + b_2 \overset{+}{B_g^2} B_g^2 + \cdots + b_{2S} \overset{+}{B_g^{2S}} B_g^{2S} . \qquad (2.14)$$

By putting $N_g = 0, 1, 2, \ldots, 2S$ in the spectrum of this equation we can evaluate the coefficients b_ν. The first several coefficients b_ν are

$$\left.\begin{aligned} b_0 &= \sqrt{2S}\,, \qquad b_1 = \sqrt{2S-1} - \sqrt{2S}\,, \\ b_2 &= \frac{1}{2}\left(\sqrt{2S-2} - \sqrt{2S-1} + \sqrt{2S}\right), \\ b_3 &= \frac{1}{6}\left(\sqrt{2S-3} - 3\sqrt{2S-2} - \sqrt{2S}\right). \end{aligned}\right\} \qquad (2.15)$$

By comparing these coefficients with the corresponding coefficients in [6] we see that the error made in the latter is the smaller the bigger S is.

In conclusion of this section we shall give the Bose representations of the spin operators S_g^+, S_g^- and S_g^z for the spins $S = 1$ and $S = 3/2$, which are exact in that space of boson states for which $N_g \leqq 4S + 1$:

$$\left.\begin{aligned} &S = 1: \\ &S_g^+ = \left(2 - B_g^+ B_g + \frac{1}{4} \overset{+}{B_g^3} B_g^3 - \frac{3}{20} \overset{+}{B_g^4} B_g^4 + \frac{1}{20} \overset{+}{B_g^5} B_g^5\right)^{1/2} B_g\,, \\ &S_g^z = 1 - B_g^+ B_g + \frac{1}{2} \overset{+}{B_g^3} B_g^3 - \frac{3}{8} \overset{+}{B_g^4} B_g^4 + \frac{3}{20} \overset{+}{B_g^5} B_g^5\,; \end{aligned}\right\} \qquad (2.16)$$

$$\left.\begin{aligned} &S = 3/2: \\ &S_g^+ = \left(3 - B_g^+ B_g + \frac{1}{15} \overset{+}{B_g^4} B_g^4 - \frac{2}{45} \overset{+}{B_g^5} B_g^5 + \frac{1}{63} \overset{+}{B_g^6} B_g^6 - \frac{1}{252} \overset{+}{B_g^7} B_g^7\right)^{1/2} B_g\,, \\ &S_g^z = \frac{3}{2} - B_g^+ B_g + \frac{1}{6} \overset{+}{B_g^4} B_g^4 - \frac{2}{15} \overset{+}{B_g^5} B_g^5 + \frac{1}{18} \overset{+}{B_g^6} B_g^6 - \frac{1}{63} \overset{+}{B_g^7} B_g^7\,. \end{aligned}\right\} \qquad (2.17)$$

3. The Pauli Representation of the Spin Operators for the Spins $S = 1$ and $S = 1/2$

For the case of spin $S = 1/2$, the spin operators $S_g^\pm$ and S_g^z can be expressed by means of the Pauli operators P_g in the following way:

$$S_g^+ = P_g\,, \qquad S_g^- = P_g^+\,, \qquad S_g^z = \frac{1}{2} - P_g^+ P_g\,. \qquad (3.1)$$

Since the Pauli operators satisfy the commutation relations

$$[P_g, \overset{+}{P}_{g'}] = (1 - 2\,\overset{+}{P}_g\,P_g)\,\delta_{gg'}\,, \qquad [P_g, P_{g'}] = 0\,, \qquad P_g\,P_g = 0\,, \quad (3.2)$$

where the operator $P_g^+ P_g$ has the eigenvalues 0 and 1, it can easily be seen that the operators S_g^+, S_g^-, and S_g^z given by (3.1) satisfy the commutation relations (2.1) for the spin $S = 1/2$.

There exist exact Bose [7] and Fermi [8] representations for Pauli operators, and they are, respectively

$$P_g = \left(\sum_{\nu=0}^{\infty} \frac{(-2)^\nu}{(\nu+1)!}\,\overset{+}{B}{}_g^\nu\,B_g^\nu\right)^{1/2} B_g\,, \qquad (3.3)$$

$$P_g = F_g\,(f_g + f_g^+)\,. \qquad (3.4)$$

The Fermi operators F and f in (3.4) commute. The formula (3.3) is a special case of (2.12a) for spin $S = 1/2$.

The existence of these explicit and relatively compact representations of Pauli operators suggests that the spin operators for the spin $S > 1/2$ might also be represented by means of Pauli operators. We shall give here such representations for the case of spins $S = 1$ and $S = 3/2$. Those cases are most interesting in practice.

$$\left.\begin{aligned} &S = 1: \\ &S_g^+ = \sqrt{2}\,(P_{1g}\,\overset{+}{P}_{2g} + P_{2g}\,\overset{+}{P}_{3g} + P_{1g}\,P_{2g}\,P_{3g})\,, \\ &S_g^z = 1 - 2\,\hat{L}_{1g} - \hat{L}_{2g} + \hat{L}_{1g}\,\hat{L}_{2g} + \hat{L}_{1g}\,L_{3g} + \hat{L}_{2g}\,\hat{L}_{3g} - 2\,\hat{L}_{1g}\,\hat{L}_{2g}\,\hat{L}_{3g}\,; \end{aligned}\right\} \quad (3.5)$$

$$\left.\begin{aligned} &S = 3/2: \\ &S_g^+ = \sqrt{3}\,P_{1g} + 2\,\overset{+}{P}_{1g}\,P_{2g}\,, \\ &S_g^z = \frac{3}{2} - \hat{L}_{1g} - 2\,\hat{L}_{2g}\,, \end{aligned}\right\} \quad (3.6)$$

where

$$\hat{L}_{ig} = \overset{+}{P}_{ig}\,P_{ig}; \qquad i = 1, 2, 3\,.$$

The Pauli operators P_{ig} in (3.5) and (3.6) commute for various values of i. It can easily be seen that the spin operators given by the expressions (3.5) and (3.6) satisfy the commutation relations (2.1). By combining (3.5) and (3.6) with (3.3) or (3.4) we can readily write the spin Hamiltonians for the spins $S = 1$ and $S = 3/2$ either in Bose or in Fermi representations.

It is clear that this representation of the spin operators by means of several sorts of kinematically independent Bose or Fermi operators means essentially a substitution of a weekly non-ideal magnon gas by an equivalent mixture of weekly non-ideal Bose or Fermi gases. This equivalence is related only to integral physical characteristics, such as magnetization, magnetic susceptibility, internal energy etc.. As to the dispersion law and the life time of elementary excitations such an equivalence need not necessarily exist.

As an illustration for this method Fig. 1 gives the dependence of magnetization on temperature for the case of spin $S = 3/2$, obtained by using the Fermi representation for Pauli operators. We omit the details because the calculation is fully analogous to that of [8]. It is interesting to note that according to this calculation the magnetization should disappear at the temperature $\theta = 4.40149 I$

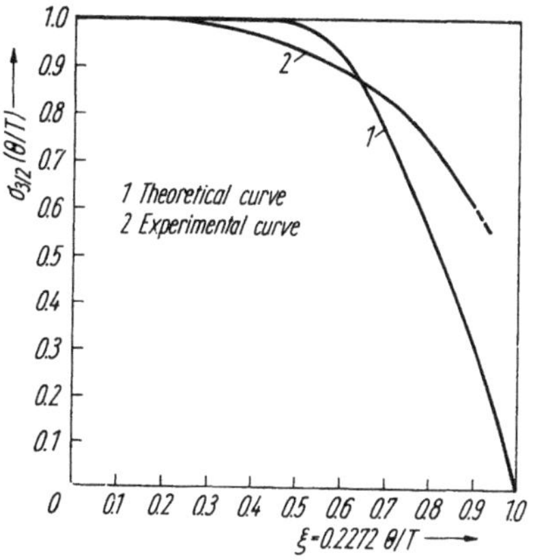

Fig. 1. Theoretical and experimental curves of magnetization as functions of temperature. The theoretical curve is obtained by using formulae (3.6) and (3.4); the experimental curve for iron is taken from [6]

where I is the exchange integral for the nearest neighbours in the simple cubic lattice. According to the theory of molecular field, the disappearance of magnetization is expected at the temperature $\theta = 4.94550\ I$.

4. Conclusion

In Section 2 a method for the exact representation of spin operators by means of Bose operators was described. The expressions obtained by applying this method contain, as one of approximations, the Holstein-Primakoff representation. In order to see more clearly the essence of the new features which this method brings, in comparison with the Holstein-Primakoff method, it is necessary to analyse closely various terms of the spin Hamiltonian, taken with an accuracy up to the fourth order in spin operators:

$$H = H_1 + H_2 + H_3 , \tag{4.1}$$

where

$$H_1 = \left(\mu \mathcal{H} + \frac{1}{2} \sum_{g'} I_{o g'}\right) \sum_{g} (S - S_g^z) , \tag{4.2}$$

$$H_2 = -\frac{1}{2} \sum_{g g'}' I_{g g'} S_g^- S_{g'}^+ , \tag{4.3}$$

$$H_3 = -\frac{1}{2} \sum_{g g'}' I_{g g'} (S - S_g^z)(S - S_{g'}^z) . \tag{4.4}$$

In these equations μ is the magnetic moment of the atom, $\mathcal{H}$ is the external magnetic field, and $I_{gg'}$ is the exchange integral, which enters the theory as an phenomenological parameter.

The most essential consequence of the exact Bose representation of spin operators is the appearance of terms of the kinematic interaction which originate in the Hamiltonian H_1, and of terms of dynamico-kinematic interaction which originate in the Hamiltonian H_3. All these terms cannot appear in the Holstein-Primakoff representation. Moreover, in the Holstein-Primakoff theory there does not exist a concept of dynamico-kinematic interaction for such a Hamiltonian. This deficiency of the Holstein-Primakoff representation is the larger the smaller the spin is, because the first corrections with respect to the Holstein-

Primakoff results are of the order of $\overset{+}{B}{}_g^{2S+1} B_g^{2S+1}$. In addition, the correct series expansion of the root of operator function as discussed in Section 2 shows that the highest order of the terms of the kinematic interaction which can appear in the Holstein-Primakoff representation (in the Hamiltonian H_2) is $\overset{+}{B}{}_g^{2S+1} B_g^{2S+1}$, i.e., the lowest correction is of the order $\overset{+}{B}{}_g^{2S+2} B_g^{2S+2}$.

From the theory presented in this paper one can conclude: The application of the Holstein-Primakoff representation for the Hamiltonian (4.1) is consistent only for terms of the order $2S$, viz., not all terms of the order $2S + 1$ appear in this representation.

Regarding the Pauli representation of spin operators, as described in Section 3, it can be used only for the calculation of integral physical characteristics of magnon systems. It should also be noted that the generalization of this treatment for the case of spins $S > 3/2$ is not a straightforward procedure.

Acknowledgements

The authors are grateful to Dr. F. Herbut for help and valuable discussions. The assistance of Miss V. Nikolić is also gratefully acknowledged.

References

[1] T. HOLSTEIN and H. PRIMAKOFF, Phys. Rev. **58**, 1098 (1940).
[2] F. BLOCH, Z. Phys. **61**, 206 (1930); **74**, 295 (1932).
[3] F. DYSON, Phys. Rev. **102**, 1217 (1956); **102**, 1230 (1956).
[4] S. V. MALEEV, Zh. eksper. teor. Fiz. **33**, 1010 (1957).
[5] T. OGUCHI, Progr. theor. Phys. **25**, 721 (1961).
[6] S. V. TYABLIKOV, The Methods of Quantum Theory in Magnetism, Izd. Nauka, Moscow **1965** (in Russian).
[7] V. M. AGRANOVIČ and B. S. TOŠIĆ, Zh. eksper. teor. Fiz. **53**, 149 (1967).
[8] B. S. TOŠIĆ and R. B. ŽAKULA, phys. stat. sol. **27**, 623 (1968).

(Received April 25, 1968)

phys. stat. sol. 28, 643 (1968)

Subject classification: 6 and 8; 21; 21.1; 21.2; 21.6

Physics Department, University of Allahabad, Allahabad

Thermal Expansion of Cubic Metals in Krebs's Model

By

A. K. SINGH and P. K. SHARMA

A calculation of the temperature dependence of the Grüneisen parameters for copper, silver, gold, aluminium, sodium, and potassium is carried out starting from the experimental values of the pressure dependence of the elastic constants using Krebs's model for the lattice dynamics of metals. Houston's six-term integration procedure is used in evaluating the integrals over the phonon spectrum. The calculated γ–T curves show fairly satisfactory agreement with experiment for copper, but not for other metals.

Die Temperaturabhängigkeit der Grüneisen-Parameter für Kupfer, Silber, Gold, Aluminium, Natrium und Kalium wird aus den experimentellen Werten der Druckabhängigkeit der elastischen Konstanten berechnet, wobei das Krebssche Modell für die Gitterdynamik der Metalle verwendet wird. Zur Berechnung des Integrals über das Phononenspektrum wird Houstons Sechs-Term-Integrationsmethode benutzt. Die berechneten γ–T-Kurven stimmen für Kupfer, aber nicht für andere Metalle befriedigend mit dem Experiment überein.

1. Introduction

Recently Krebs [1] has propounded an elastic force model taking cognizance of electron gas for the study of lattice vibrations in metals. He considered the influence of conduction electrons on ionic motion via the screening of long-range Coulomb interaction between ions. Of the existing electron gas models (Sharma and Joshi [2], Bhatia [3], de Launay [4]), this model is superior in the sense that it satisfies the symmetry requirements of a cubic lattice. This is achieved in it by the inclusion of Umklapp processes as suggested by Lax [5]. Each element of the dynamical matrix is taken to be the sum of two contributions due to ions and electrons, respectively. The ionic term is exactly similar to that in de Launay's model [4] and involves two force constants. The electronic contribution, however, is different, and is calculated from the screened Coulomb interaction between ions embedded in a sea of free electrons. Two parameters enter in the corresponding term of the dynamical matrix: a parameter A depending on the geometry of the lattice, and the screening parameter λ given by the relation

$$\lambda = \beta \left(\frac{r_s}{a_0}\right)^{1/3} k_F ,$$

where r_s is the atomic radius, a_0 the Bohr radius, k_F the Fermi radius, and β a numerical constant. The value of constant β depends upon the description of the screening. When the screening is considered from the Thomas-Fermi stand point, β takes a value of 0.814, while the plasma theory of Bohm and Pines gives it a value of 0.353.

Krebs applied his model to alkali metals (Li, Na, K) to obtain dispersion curves and specific heats using Bohm-Pines β-coefficient, and the agreement with the experimental data was fairly satisfactory. Shukla and Dayal [6] and

Mahesh and Dayal [7] have used this model to the calculation of lattice specific heats of noble metals (Cu, Ag, Au) and alkali metals (Na, K, Rb, Cs), respectively and have obtained very good agreement with experiment by assuming the constant β as an adjustable parameter. The model has also offered a satisfactory explanation of lattice dynamics of aluminium (Shukla and Dayal [8]), nickel (Hautecler and Van Dingenen [9], Shukla and Dayal [10]), chromium and tungsten (Krebs [11]). In a recent study (Singh and Sharma [12]) we found that this model represents fairly well the salient features of the electrical and thermal resistivities of alkali metals. It seems that Krebs's model with an adjustable parameter provides a very satisfactory description of the lattice dynamics of cubic metals.

In view of the success of Krebs's model in study of thermal properties of metals, it was thought worthwhile to compute the thermal expansion on the basis of this model. In this paper we have utilized Krebs's model to calculate the temperature variation of the Grüneisen parameters for some cubic metals from the observed values of the pressure derivatives of elastic constants. The results of calculation are presented for copper, silver, gold, aluminium, sodium, and potassium for which experimental values of the pressure derivatives are available and compared with available experimental data on thermal expansion.

2. Theory

The Grüneisen parameter γ in the theory of thermal expansion of crystals is defined by the relation

$$\gamma = \frac{\alpha V}{\chi C_v}, \tag{1}$$

where α is the volume expansion coefficient, V the volume of the solid, χ the isothermal compressibility and C_v the heat capacity at constant volume. The expansion coefficient α is given by the thermodynamical relation

$$\alpha = -\chi \left(\frac{\partial^2 F}{\partial V\, \partial T}\right), \tag{2}$$

where F is the free energy of crystal and T the absolute temperature. In the quasi-harmonic approximation, the temperature dependent part of the free energy can be written as (Ziman [13])

$$F = k_B T \sum_{q,p} \ln\left[2 \sinh\left(\hbar\, \omega_{q,p}/2\, k_B T\right)\right], \tag{3}$$

where $\omega_{q,p}$ is the angular frequency of lattice wave of wave vector $\boldsymbol{q}$ and polarization p, k_B the Boltzmann constant, and the summation is over all the wave vectors $\boldsymbol{q}$ in the first Brillouin zone and the three polarization branches. Equations (2) and (3) give

$$\frac{\alpha V}{\chi} = \sum_{q,p} \gamma_{q,p}\, E\left(\hbar\, \omega_{q,p}/k_B T\right), \tag{4}$$

where $E(x)$ is the Einstein specific heat function:

$$E(x) = \frac{x^2\, e^x}{(e^x - 1)^2}, \tag{5}$$

and

$$\gamma_{\boldsymbol{q},p} = -\left(\frac{\partial \ln \omega_{\boldsymbol{q},p}}{\partial \ln V}\right)_T . \tag{6}$$

With the help of (1) and (4) we get

$$\gamma = \frac{\sum_{\boldsymbol{q},p} \gamma_{\boldsymbol{q},p}\, E\,(\hbar\, \omega_{\boldsymbol{q},p}/k_{\rm B}T)}{\sum_{\boldsymbol{q},p} E(\hbar\, \omega_{\boldsymbol{q},p}/k_{\rm B}T)} . \tag{7}$$

Grüneisen's theory assumes γ to be a constant independent of temperature. However, (7) tells that γ is a parameter which in general will vary with the temperature. Replacing the summation over $\boldsymbol{q}$ by integration in the usual manner, (7) can be written in the form

$$\gamma = \frac{\sum_p \int \mathrm{d}\Omega \int \mathrm{d}q\, \gamma_p(q)\, E(\hbar\, \omega_p(q)/k_{\rm B}T)\, q^2}{\sum_p \int \mathrm{d}\Omega \int \mathrm{d}q\;\, E\,(\hbar\, \omega_p(q)/k_{\rm B}\, T)\, q^2} . \tag{8}$$

Here Ω is a solid angle in the wave vector space.

3. Numerical Computations

The evaluation of γ from (8) has been made by Houston's method. The integration over q was performed numerically and the integration over Ω was carried out by using a modified Houston's spherical six-term integration procedure as elaborated by Betts et al. [14], the applicability of which has been discussed by various authors (Horton and Schiff [15], Betts [16], Ganesan and Srinivasan [17], Collins [18]). The use of Houston's method to evaluate the integral

$$J = \int I(\theta, \varphi)\, \mathrm{d}\Omega$$

by a six-term approximation leads to the formula

$$J = \frac{4\,\pi}{1081080}\,[117603\; I_{\rm A} + 76544\; I_{\rm B} + \\ + 17496\; I_{\rm C} + 381250\; I_{\rm D} + 311040\; I_{\rm E} + 177147\; I_{\rm F}] . \tag{9}$$

Here the subscripts A, B, C, D, E, and F denote the values of the integrand $I(\theta, \varphi)$ along the directions [100], [110], [111], [210], [211], and [221], respectively. The quantities $\gamma_{\boldsymbol{q},p}$ were expressed in terms of the elastic constants and their pressure derivatives through the solutions of the Krebs's [1] secular determinant for f.c.c. and b.c.c. lattices along the six directions. In the calculation we have taken account of the observed temperature variation of the elastic constants and the lattice parameter. The elastic constants in the high temperature regions were estimated by extrapolation from the existing experimental data. Since data on temperature variation of the pressure derivatives of elastic constants are not available, the room temperature values for them were used in the calculation.

4. Results and Discussion

The calculated values of γ for the six metals copper, silver, gold, aluminium, sodium, and potassium are plotted against reduced temperature (T/θ_0) in Fig. 1 to 6. In these figures we have also shown for comparison the γ-values deduced from the experimental measurements of thermal expansion. The expe-

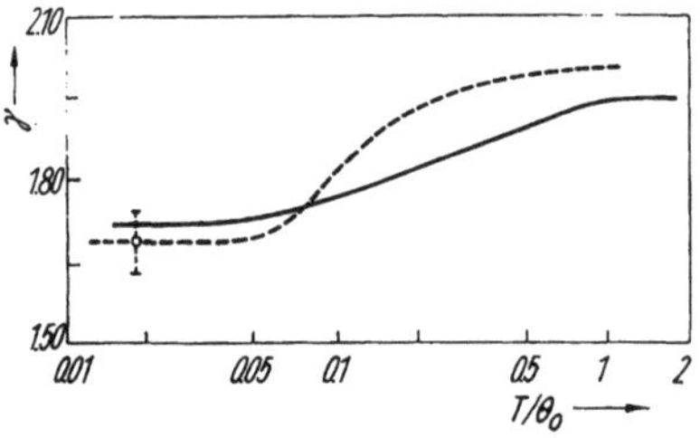

Fig. 1. γ–T curve for copper ($\theta_0 = 331$ °K). Experimental data are marked as dashed curve

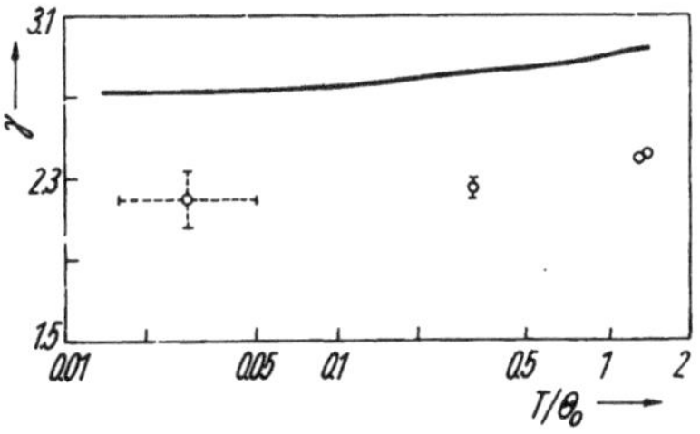

Fig. 2. γ–T curve for silver ($\theta_0 = 217$ °K). Circles are the experimental points

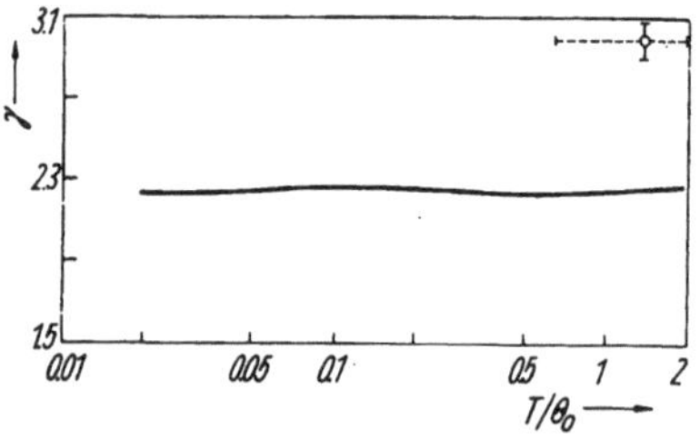

Fig. 3. γ–T curve for gold ($\theta_0 = 156$ °K). Circle is the experimental point

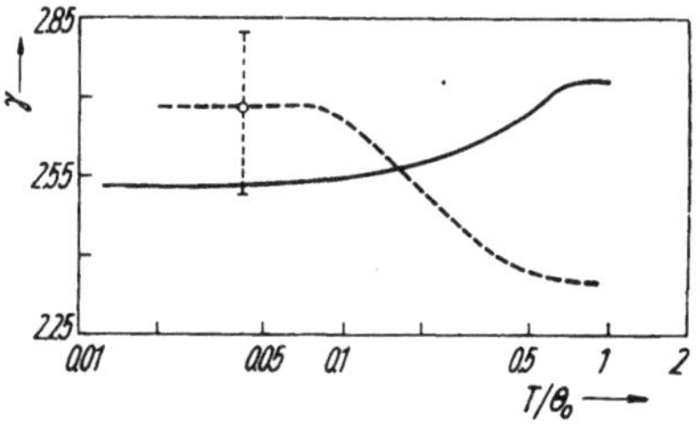

Fig. 4. γ–T curve for aluminium ($\theta_0 = 407$ °K). Experimental data are marked as dashed curve

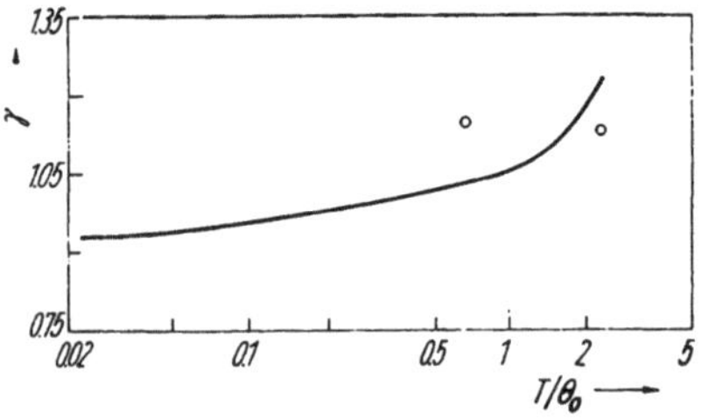

Fig. 5. γ–T curve for sodium ($\theta_0 = 134$ °K). Circles show experimental values

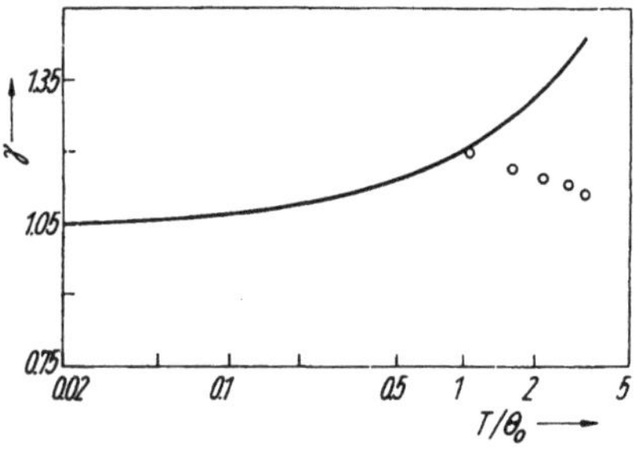

Fig. 6. γ–T curve for potassium ($\theta_0 = 92$ °K). Experimental data are marked as circles

rimental data for thermal expansion for copper are taken from Rubin et al. [19], for silver, gold, aluminium, and sodium from Corruccini and Gniewek [20], and for potassium are due to Monfort and Swenson [21]. The sources of temperature dependence of elastic constants together with the room temperature values of the pressure derivatives used in the calculation are summarized in Table 1. The values of the lattice parameter at different temperature were taken from Pearson [22]. The screening parameter λ was considered as an empirical parameter, the best-fit values of which were taken from the work of Shukla and Dayal [6, 8] and Mahesh and Dayal [7].

A perusal of Fig. 1 to 6 reveals that there is satisfactory agreement between the calculated and experimental values of γ for copper, but not for other metals. In the case of silver the theoretical and experimental γ–T curves are similar, but the theoretical values are higher than the experimental ones. Similar behaviour is evident for gold, though the experimental curve is not precisely known. In view of insufficient experimental data at low temperatures for sodium and potassium, it is difficult to draw definite conclusions for them. The observed

Table 1

Elastic data for cubic metals used in the calculation

Metal	Pressure derivatives of elastic constants $\frac{\partial C_{11}}{\partial P}$	$\frac{\partial C_{12}}{\partial P}$	$\frac{\partial C_{44}}{\partial P}$	Ref.	Temperature variations of elastic constants
Copper	5.940	5.190	2.630	[25]	Overton and Gaffney [29]
Silver	5.117	3.607	3.040	[25]	Neighbours and Alers [30]
Gold	5.717	4.957	1.520	[25]	Neighbours and Alers [30]
Aluminium	7.350	4.110	2.310	[26]	Sutton [31], Kamm and Alers [32]
Sodium	3.901	3.449	1.630	[27]	Quimby and Siegel [33], Diederich and Trivisonno [34]
Potassium	4.305	3.803	1.620	[28]	Marquardt and Trivisonno [35]

discrepancies in aluminium are not surprising and are caused by the assumption of short range interionic interactions in the theory. Recent experiments (Gilat and Nicklow [23]) on inelastic scattering of neutrons from metals suggest that interatomic forces are of fairly long range and noncentral nature.

Daniels and Smith [24] have also measured the pressure derivatives of elastic constants for noble metals. Their values are in approximate agreement with those of Hiki and Granato [25] for copper, but differ considerably in the case of silver and gold. An attempt was made to compute the Grüneisen parameters of silver and gold by using the data of Daniels and Smith. It was found that this considerably improved the agreement between theory and experiment.

The discrepancies between theory and experiment can be attributed to the neglect of temperature variation of pressure derivatives of elastic constants and errors in estimation of elastic constants in the high temperature regions by extrapolation method. More detailed experimental investigations of the dependence of elastic constants on temperature and pressure are desirable to make a decisive comparison possible. As the things stand, it emerges from the present study that Krebs's model provides a reasonable description of Grüneisen parameters of noble and alkali metals.

Acknowledgements

The authors are grateful to the University Grants Commission, New Delhi, and the Department of Atomic Energy, Bombay, for financial assistance, and to the Indian Institute of Technology for the use of IBM 7044 computer.

References

[1] K. Krebs, Phys. Letters (Netherlands), **10**, 12 (1964); Phys. Rev. **138**, A143 (1965).
[2] P. K. Sharma and S. K. Joshi, J. chem. Phys. **39**, 2633 (1963); **40**, 662 (1964).
[3] A. B. Bhatia, Phys. Rev. **97**, 363 (1955).
[4] J. de Launay, J. chem. Phys. **21**, 1975 (1953); Solid State Phys. **2**, 220 (1956).
[5] M. Lax, Proc. Internat. Conf. Lattice Dynamics, Copenhagen 1963, Pergamon Press, 1964 (p. 179).
[6] M. M. Shukla and B. Dayal, J. Phys. Chem. Solids **26**, 1343 (1965).
M. M. Shukla, phys. stat. sol. **8**, 475 (1965).

[7] P. S. MAHESH and B. DAYAL, phys. stat. sol. **9**, 351 (1965).
[8] M. M. SHUKLA and B. DAYAL, phys. stat. sol. **16**, 513 (1966).
[9] S. HAUTECLER and W. VAN DINGENEN, Physica **34**, 257 (1967).
[10] M. M. SHUKLA and B. DAYAL, phys. stat. sol. **19**, 729 (1967).
[11] K. KREBS, unpublished.
[12] A. K. SINGH and P. K. SHARMA, to be published.
[13] J. M. ZIMAN, Electrons and Phonons, Oxford University Press, London 1960 (p. 152).
[14] D .D. BETTS, A. B. BHATIA, and M. WYMAN, Phys. Rev. **104**, 37 (1956).
[15] G. K. HORTON and H. SCHIFF, Proc. Roy. Soc. **A250**, 248 (1959).
[16] D. D. BETTS, Canad. J. Phys. **39**, 233 (1961).
[17] S. GANESAN and R. SRINIVASAN, Canad. J. Phys. **40**, 1153 (1962).
[18] J. G. COLLINS, Phil. Mag. **8**, 323 (1963).
[19] R. RUBIN, H. W. ALTMAN, and H. L. JOHNSTON, J. Amer. Chem. Soc. **76**, 5289 (1954).
[20] R. J. CORRUCCINI and J. J. GNIEWEK, Thermal Expansion of Technical Solids at Low Temperatures, National Bureau of Standards, Washington 1961.
[21] C. E. MONFORT and C. A. SWENSON, J. Phys. Chem. Solids **26**, 291 (1965).
[22] W. B. PEARSON, Hdb. Lattice Spacings and Structures of Metals and Alloys, Pergamon Press, 1958.
[23] G. GILAT and R. M. NICKLOW, Phys. Rev. **143**, 487 (1966).
[24] W. B. DANIELS and C. S. SMITH, Phys. Rev. **111**, 713 (1958).
[25] Y. HIKI and A. V. GRANATO, Phys. Rev. **144**, 411 (1966).
[26] R. E. SCHMUNK and C. S. SMITH, J. Phys. Chem. Solids **9**, 100 (1959).
[27] W. B. DANIELS, Phys. Rev. **119**, 1246 (1960).
[28] P. A. SMITH and C. S. SMITH, J. Phys. Chem. Solids **26**, 279 (1965).
[29] W. C. OVERTON and J. GAFFNEY, Phys. Rev. **98**, 969 (1955).
[30] J. R. NEIGHBOURS and G. A. ALERS, Phys. Rev. **111**, 707 (1958).
[31] P. M. SUTTON, Phys. Rev. **91**, 816 (1953).
[32] G. N. KAMM and G. ALERS, J. appl. Phys. **35**, 327 (1964).
[33] S. L. QUIMBY and S. SIEGEL, Phys. Rev. **54**, 293 (1938).
[34] M. E. DIEDERICH and J. TRIVISONNO, J. Phys. Chem. Solids **27**, 637 (1966).
[35] W. R. MARQUARDT and J. TRIVISONNO, J. Phys. Chem. Solids **26**, 273 (1965).

(Received April 8, 1968)

phys. stat. sol. 28, 649 (1968)

Subject classification: 16; 13.3; 22.4.1

Bereich Halbleiteroptik des Physikalisch-Technischen Instituts der Deutschen Akademie der Wissenschaften zu Berlin und der Sektion Physik der Humboldt-Universität zu Berlin

Surface Photoconductivity of CdS Influenced by Chemisorption and Desorption of Oxygen

By

E.-H. WEBER

Using investigations of the spectral dependence of photoconductivity and of the influence of oxygen on photoconductivity a method is developed to test the properties of space-charge layers at the surface of CdS, at least in some extreme cases. The surface space-charge layer of thick CdS platelets with a photoconductivity greatly sensitive to an oxygen atmosphere is found to be of accumulation type. This makes a re-interpretation necessary of the models so far accepted for the chemisorption and desorption processes. As their rate-limiting steps the direct capture of electrons and holes by neutral and negatively charged oxygen, respectively, is assumed. Theoretical calculations on the basis of this model show qualitative agreement with the experimental results, but desorption seems to be a process more complex than pure photodesorption.

Es wird eine Methode entwickelt, die es gestattet, aus der spektralen Verteilung der Photoleitfähigkeit und aus der Größe des Sauerstoffeinflusses auf die Photoleitfähigkeit den Charakter von Raumladungsrandschichten an der Oberfläche von CdS-artigen Photoleitern zu bestimmen. Es zeigt sich, daß an der Oberfläche dickerer CdS-Plättchen, deren Photoleitfähigkeit empfindlich gegen Änderungen des Sauerstoffdruckes ist, Anreicherungsrandschichten vorhanden sind. Das macht es notwendig, die bisherigen Modellvorstellungen über die Chemisorption und Desorption, die wesentlich auf der Annahme von Verarmungsrandschichten beruhen, so abzuändern, daß der geschwindigkeitsbestimmende Teilschritt der Chemisorption bzw. Desorption direkt der Einfang von Elektronen bzw. Löchern durch neutralen bzw. negativ geladenen Sauerstoff ist. Theoretische Berechnungen der zeitlichen Änderung der Photoleitfähigkeit bei Chemisorption und Desorption und der Abhängigkeit der Photoleitfähigkeit vom Sauerstoffdruck auf der Grundlage eines so abgeänderten Modells ergeben qualitativ Übereinstimmung mit den experimentellen Ergebnissen. Ein quantitativer Vergleich für den Fall der Desorption läßt vermuten, daß die Desorption ein komplizierterer Prozeß als bloße Photodesorption ist.

1. Introduction

As is known from numerous papers [1 to 17], the electrical conductivity of CdS decreases under the influence of an oxygen atmosphere. This decrease depends not only on the electrical surface properties but also on the bulk properties. The phenomenon has been investigated mainly on photoconducting CdS, i.e. crystals isolating in the dark and conducting under appropriate optical excitation. Measurements on single crystals were reported in [1 to 12], on polycrystalline sintered layers in [13 to 15], and on samples with high dark conductivity in [16, 17].

The effect of oxygen on the surface photoconductivity was described by the following model proposed at first by Williams [18] and later on in greater detail by Mark [6, 7] in close proximity to suggestions made by Morrison, Melnick, and Medved [19] in the case of ZnO: Physically adsorbed (physisorbed) oxygen picks up photo-excited electrons from the conduction band and becomes chemisorbed. The photoconductivity then decreases in a region near to the surface and a depletion layer should arise. With growing amount of chemisorbed oxygen the barrier height increases and more and more counteracts an electron transfer into physisorbed oxygen. Simultaneously photodesorption is working against chemisorption. Photodesorption was assumed to occur by photo-generated holes moving in the field of the depletion layer to the surface, there recombining with chemisorbed, negatively charged oxygen ions. After having pumped off the oxygen, this photodesorption process becomes dominant and leads to a photoconductivity increase and to a reduction of the depletion layer.

It is, however, not clear, whether during chemisorption of oxygen a depletion layer is formed at the surface. Previously existing accumulation layers may have been removed as well. In fact, a strong reduction of photoconductivity due to chemisorption of oxygen can be found not only with excitation in the intrinsic absorption region, but also with excitation at wavelengths longer than that corresponding to the absorption edge (Fig. 1 in [10], Table 4 in [2]).

Space-charge layers extending over the whole crystal cannot explain this result in the case of the usually plate-like crystals being 20 to 50 μm thick, since, according to theoretical considerations for photoconductors [20 to 22], the width of the space-charge layer in CdS, even for a trap concentration of only 10^{12} cm^{-3}, should not be larger than about 1 μm. Moreover, to the author's knowledge there is no experimental evidence whether the desorption velocity is essentially determined by the capture of free holes by chemisorbed oxygen, as proposed, or by optical excitation of electrons from chemisorbed oxygen ions.

The aim of this paper is to clarify these questions based on corresponding experiments and on calculations presented in the discussion.

2. Experimental

Investigations were performed on thin monocrystalline CdS platelets grown from the vapour phase by a modified Frerichs method [23]. As to the carrier gas two variants were used: (i) it was composed of nitrogen, hydrogen, and sometimes hydrogen sulfide during the growing and cooling processes, (ii) it consisted only of pure nitrogen during the cooling process.

The crystal thickness ranged from approximately 5 to 100 μm, the breadth from 1 to 4 mm, and the length from 3 to 15 mm. The large faces of the investigated platelets always were $(11\bar{2}0)$ surfaces.

Ohmic contacts to the samples were made with eutectic In–Ga electrodes so that the current flowed parallel to the large illuminated $(11\bar{2}0)$ surfaces. The dark resistivity varied from approximately 10^{10} Ωcm to values less than 10^{13} Ωcm at room temperature. The samples were mounted in a metal chamber which could be evacuated within 2 min to a pressure lower than 10^{-4} Torr (final pressure $\approx 10^{-6}$ Torr) and filled with oxygen within less than 2s. The oxygen was dried by phosphorus pentoxide. The photoconductivity was excited by monochromatic light, using a Zeiss SPM 1 monochromator and a tungsten lamp. The intensity could be reduced by neutral-glass filters (Schott NG 5 or NG 11). The maximum intensity at the surface was about 10^{13} quanta/cm^2 s near the absorption edge.

Before the measurements each crystal was intensely illuminated by light strongly absorbed for at least one day in high vacuum. This procedure causes photodesorption and guarantees a sufficient reproducibility of the results [5].

3. Results

3.1 Non-stationary measurements

Fig. 1a shows the time dependence of photocurrent after admitting (curve (a)) and pumping-off oxygen (curve (b)), pointing out that the influence of oxygen on the photoconductivity is a reversible process. A replot of the photocurrent according to Fig. 1a on a logarithmic time scale (Fig. 1b) manifests this dependence to be linear at intermediate times in the case of chemisorption, in close agreement with Mark [6]. However, also in the case of desorption this dependence turns out to be linear in a certain range (Fig. 1c).

The left part of Fig. 1a and Fig. 1b indicate the existence of a fast and a slow process, especially because curve (a') of Fig. 1b on the left-hand side tends to a "saturation" value less than $I/I_0 = 1$ corresponding to "vacuum". The presence of two processes at intermediate and low illumination intensities becomes clear from Fig. 2 showing the photocurrent after admitting oxygen with successively reducing the excitation intensities from curve (a) to curve (d). In curve (d) an initial, rather fast decrease is followed by a gradual increase of the photocurrent. The total magnitude of the photocurrent decrease in this anomalous range becomes smaller with decreasing photocurrent. At very high excitation after oxygen uptake one finds a photoconductivity change (Fig. 3)

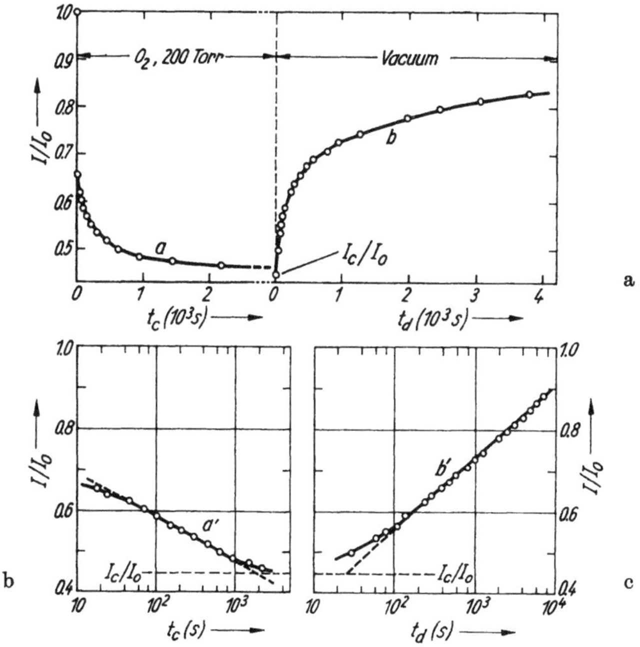

Fig. 1. a) Time dependence of photocurrent after admitting (a) and removal (b) of oxygen. I_0 and I_c denote the steady-state values of the photocurrent in high vacuum and oxygen atmosphere, respectively; t_c time after introduction, t_d time after removal of oxygen. b) Photocurrent during chemisorption (a') and c) photocurrent during desorption (b') versus the logarithm of time. Same measurement as in Fig. 1a

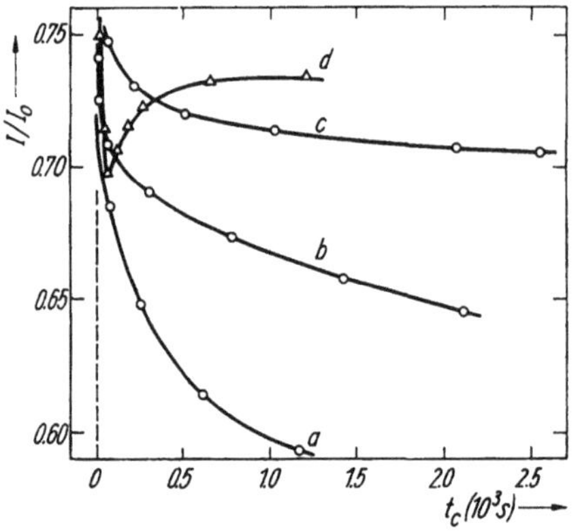

Fig. 2. Time dependence of photocurrent during chemisorption at very low illuminations.
(a) $I_0 = 1.2 \times 10^{-5}$ A, (b) 1.2×10^{-6} A, (c) 1.2×10^{-7} A, (d) 1.2×10^{-8} A; extrinsic excitation; $p_{O_2} = 200$ Torr

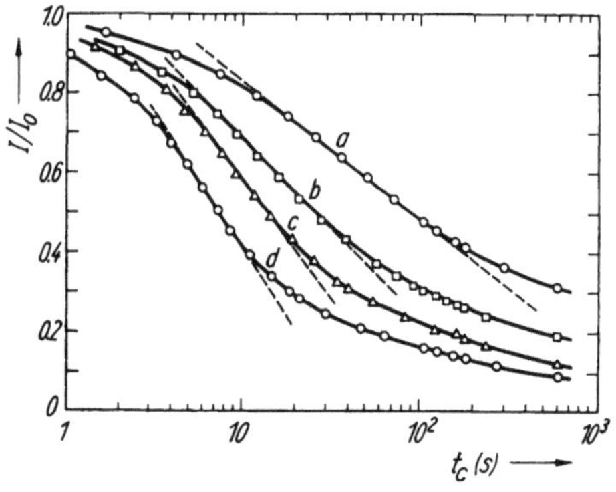

Fig. 3. Time dependence of photocurrent during chemisorption at a high illumination intensity as a function of the oxygen pressure.
(a) 14.5, (b) 70.5, (c) 184.5, (d) 366 Torr

which does not allow a separation into two processes; the curves tend to approach $I/I_0 = 1$ with $t_c \to 0$ without further turning-points. Fig. 3 also shows an increasing velocity and amount of photocurrent change with increasing oxygen pressure (curves (a) to (d)).

Very important for the elucidation of the desorption mechanism is the question how far the desorption velocities differ in the cases of intrinsic or extrinsic excitation. Therefore, a comparison is made of the time dependences of photoconductivity for excitation near the photoconductivity maximum and in the tail. We preferred excitation near the photoconductivity maximum at the absorption edge rather than excitation deep in the intrinsic absorption region to have as well homogeneous excitation conditions as still creation of free holes.[1])

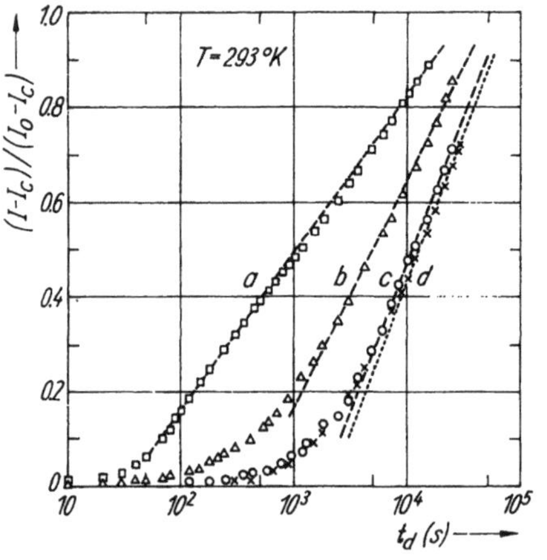

Fig. 4. Normalized photocurrent change (I_0 and I_c stationary photocurrents in vacuum and oxygen atmosphere) versus time during desorption (after evacuating). Curves (a), (b), and (c): excitation near the photoconductivity maximum ($\lambda = 506$ nm), curve (d): excitation in the tail ($\lambda = 575$ nm). Relative illumination intensities (quanta/cm²s): (a) 1, (b) 7×10^{-3}, (c) 2×10^{-4}, (d) 10.
Photocurrents:
(a) $I_0 = 5.78 \times 10^{-7}$ A, $I_c = 1.11 \times 10^{-7}$ A;
(b) $I_0 = 7.78 \times 10^{-9}$ A, $I_c = 1.95 \times 10^{-9}$ A;
(c) $I_0 = 4.88 \times 10^{-10}$ A, $I_c = 1.80 \times 10^{-10}$ A;
(d) $I_0 = 6.25 \times 10^{-10}$ A, $I_c = 2.08 \times 10^{-10}$ A

[1]) The results of Gutsche and Voigt [24] suggest phonon–assisted creation of excitons to be the main absorption mechanism at the absorption edge of CdS. After a short time interval the excitons are thermally decomposed into free electrons and holes which in their turn equilibrate with the centres in the forbidden gap.

The intensities were controlled to get equal photocurrents, i.e. nearly equal electron concentrations in both cases, thereby assuring an equal influence of counteracting chemisorption processes.[2]) The results for one crystal are shown in Fig. 4, curves (c) and (d). To obtain equal photocurrents, the number of exciting quanta must be reduced by a factor of about 10^5 at the photoconductivity maximum ($\lambda = 506$ nm) as compared to the tail ($\lambda = 575$ nm). Nevertheless, curves (c) ($\lambda = 506$ nm) and (d) ($\lambda = 575$ nm) lie very close together, which does not seem to be compatible with the assumption of direct optical excitation as the main desorption mechanism. Curves (b) and (a) of Fig. 4 show the time dependence of the normalized photocurrent change $(I - I_c)/(I_0 - I_c)$ at higher illumination intensities ($\lambda = 506$ nm). It is to notice that, apart from a change in the slope, the curves are shifted against each other on the time scale, but with factors very much lower than the corresponding intensity or photocurrent ratios.

The same results can be obtained with excitation at shorter wavelengths. The velocity of the photocurrent change is primarily determined by the photocurrent level and not by the wavelength.

3.2 Stationary measurements

The investigation of the spectral dependence of photoconductivity of many CdS monocrystals points to two extreme cases: In the first case (case I), starting from the wavelength of the photoconductivity maximum and going on to shorter wavelengths, the photoconductivity, related to an equal excitation level, does not decrease very strongly (Fig. 5a). Under the influence of oxygen one observes a large diminution of the photoconductivity in both the intrinsic and extrinsic absorption regions (Fig. 5a, curve (b)). In the second case (case II) the photoconductivity decreases rather abruptly to shorter wavelengths (Fig. 5b), and no or only an insignificant effect of oxygen on the photoconductivity can be observed (in Fig. 5b both curves are identical). In case II the wavelength

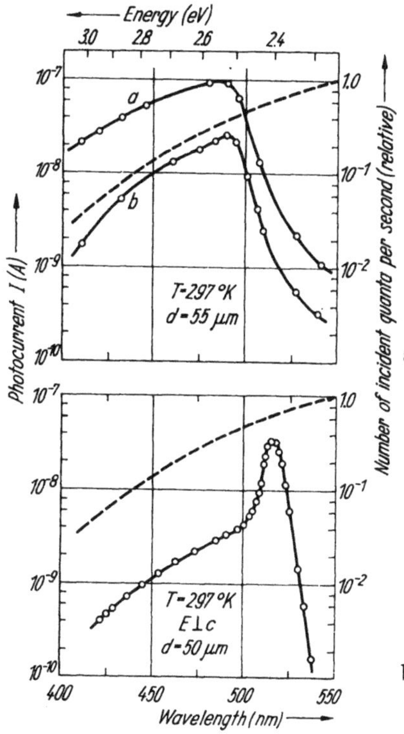

Fig. 5. a) Spectral dependence of photocurrent of a crystal 55 µm thick in high vacuum (a) and for an oxygen pressure of 250 Torr (b). At the wavelength of the photoconductivity maximum the reciprocal absorption coefficient is smaller than the crystal thickness (to an absorption coefficient of $1/(55\ \mu m) = 1.8 \times 10^2\ cm^{-1}$ corresponds a wavelength $\lambda >$ > 515 nm, compared to an observed one of $\lambda \approx 490$ nm; cf. [27]) (case I). b) Spectral dependence of photocurrent of a crystal 50 µm thick unsensitive to an oxygen atmosphere. The wavelength at which the reciprocal absorption coefficient is equal to the crystal thickness is only about 1 nm greater than the wavelength of the photoconductivity maximum (case II)

[2]) This argument becomes clear, if one takes into consideration the result derived in the discussion that at the surface of crystals sensitive to the influence of oxygen, accumulation layers exist.

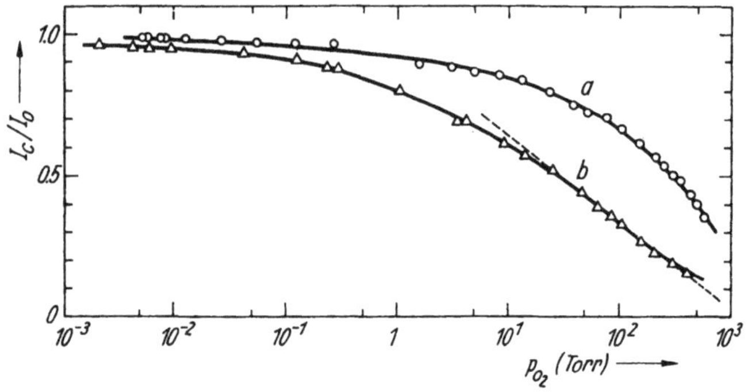

Fig. 6. Photocurrent I_c versus oxygen pressure p_{O_2} for a wide pressure range (I_0 photocurrent in high vacuum)

of the photoconductivity maximum is equal to that wavelength at which the reciprocal absorption coefficient is equal to the crystal thickness. In case I the wavelength of the photoconductivity maximum always is distinctly smaller than to be expected from this estimate. Case I can be particularly frequently observed for crystals grown and cooled in a carrier gas containing hydrogen. Case II most frequently appears, if the carrier gas consists of pure nitrogen during cooling.

If oxygen acts upon crystals belonging to case I at temperatures higher than about 120 °C, the photoelectrical properties of these ctystals irreversibly change, and after this treatment these crystals show a behaviour corresponding to case II.

Most crystals belong to an intermediate type with a photoconductivity maximum at shorter wavelengths than according to the estimate for case II, and with an influence of oxygen on the photoconductivity being more distinct in the intrinsic than in the extrinsic absorption region.

The dependence of the photoconductivity on the oxygen pressure shows a strong variation of the stationary photocurrent within the range from about 1 to 10^3 Torr, while at lower pressures the photocurrent change is very small (Fig. 6). For most of the crystals a nearly linear relation between the photocurrent and the logarithm of oxygen pressure is typical in a range around 100 Torr (Fig. 7). For some crystals at higher pressures a bending off to saturation is indicated (curve (b) in Fig. 6, curve (a) in Fig. 7).

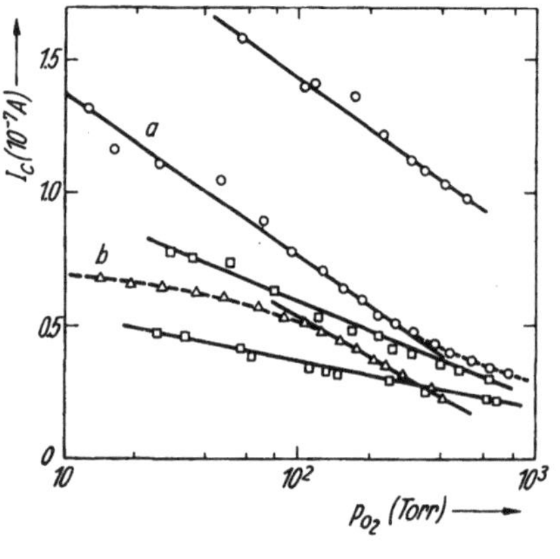

Fig. 7. Dependence of the photocurrent I_c on lg p_{O_2} for higher pressures

4. Discussion

4.1 The properties of the surface space-charge layer as revealed from the spectral dependence of photoconductivity and the influence of oxygen

We confine the discussion to the two extremum cases I and II shown in Fig. 5a and b. Let us begin with case II which is the more common one in the literature.

In case II (Fig. 5b), after reaching a maximum the photoconductivity strongly decreases towards shorter wavelength. This fact was discussed in the past by Fassbender and DeVore. Fassbender [25] assumed bimolecular recombination to occur due to the high excitation rate in the intrinsic absorption region, but this should lead to a slope $\alpha = 1/2$ of the photocurrent-illumination characteristic which was experimentally not observed. DeVore [26] proposed surface recombination to cause low photoconductivity in this region, and really, this concept is a well-established one for Ge and Si, but not for CdS. Photoconductors like CdS are strongly monopolar, i.e., the electron concentration is by many orders of magnitude higher than the hole concentration. Surface recombination under steady-state conditions means that electrons and holes flow to surface recombination centres in equal numbers. Unless the hole capture cross-section of these centres is greater than the electron capture cross-section by many orders of magnitude, a depletion layer must exist at the surface to realize the above-mentioned equal flow of electrons and holes to the surface recombination centres. A depletion layer means that the photoconductor is already desensitized in the space-charge region, i.e. the recombination rate in the space-charge layer is by many orders of magnitude higher than in the bulk. Thereby, and also caused by the low concentration of free carriers in this case, a recombination flow across eventually present surface recombination centres should be only insignificant. The presence of a surface depletion layer already existing under vacuum conditions is supported by the fact that no effect of oxygen on the stationary photoconductivity can be observed.

Therefore, the basic conception in case II would be the following: At the surface a depletion layer exists. The concentration of "effective" recombination centres (concentration of holes in sensitizing centres) in the space-charge region is considerably larger than in the bulk. For higher barrier heights also fast recombination centres must be taken into account. Since in the intrinsic absorption region the optical excitation primarily occurs in the space-charge region, the photoconductivity here is less than at the absorption edge where the absorption coefficient nearly equals to the reciprocal crystal thickness.

The observation that the wavelength of the photoconductivity maximum can be determined from a comparison of the intrinsic absorption coefficient and the reciprocal crystal thickness means that the spectral dependence of photoconductivity is governed by recombination processes, hus confirming the above conception.

In case I (Fig. 5a) the spectral dependence of photoconductivity — at least in the upper part — shows a certain similarity to the spectral dependence of the intrinsic absorption coefficient [27], independent of the crystal thickness. (The falling-off towards shorter wavelength is due to a decreasing number of incident quanta.) Such a behaviour would be expected for the ideal case of a crystal possessing a homogeneous defect distribution, without surface states and with a linear photocurrent-illumination characteristic. But in our case I the effect of surface states cannot be ruled out, since a remarkable effect of oxygen on the photoconductivity exists. An equal diminution of photoconductivity due to oxygen in the intrinsic and extrinsic absorption regions could at most be understood, if the space-charge layers connected with the surface states were comparable to the crystal thickness. But on theoretical arguments [21], using concentrations of 10^{13} to 10^{17} cm^{-3} occupied traps or recombination centres, the width of the space-charge layer is only 10^{-4} to 10^{-6} cm. Since an equal change of photoconductivity in the intrinsic and extrinsic absorption regions was found with crystals up to 100 μm thick, for an explanation only the possibility remains to assume a thin, well photoconducting layer at the surface. Such a layer can originate from surface donors causing an accumulation layer. Then the concentration of "effective" recombination centres at the surface is smaller and the photoconductivity becomes higher than in the bulk. A well photoconducting layer could also arise from a spatially inhomogeneous distribution of defects, but an interpretation in terms of an accumulation layer is supported by observing case I especially frequently on crystals having been in a hydrogen containing carrier gas during the cooling process. Obviously, these crystals are slightly reduced on the surface and a superficial cadmium excess is present. Heating of the crystals to temperatures above 120 °C in an oxygen containing atmosphere leads to an irreversible removal of this excess, thereby producing crystals belonging to case II. In case I, due to the accumulation layer, we have a thin, well photoconducting layer at the surface for both intrinsic and extrinsic excitations. The effect of oxygen consists in a reduction of the barrier height of the accumulation layer, therefore causing a nearly equal decrease of photoconductivity in the whole investigated wavelength range.

These results can be compared with those of Broude and Chikovani [28]. In the main they also found two types of crystals, one with a wavelength of the photoconductivity maximum strongly correlated to the crystal thickness and an other with a nearly constant wavelength independent of thickness.

4.2 Qualitative discussion of the chemisorption and desorption mechanisms

In previous measurements by Mark [7] it was already shown that the velocity of the photocurrent change during oxygen uptake increases with growing illumination intensity, i.e. with growing electron concentration (Table 1 of [7]). This also occurs with growing oxygen pressure (Fig. 3). During desorption the velocity of the photocurrent change increases with growing illumination, indicated by a shift of the curves to shorter times (Fig. 4, curves (c) to (a)). From a comparison of the photocurrent changes during desorption using intrinsic and extrinsic excitation it can be concluded that direct optical excitation of electrons from chemisorbed oxygen ions should not be the principal desorption mechanism. Pure thermal desorption can be excluded on account of the dependence of the desorption velocity on the illumination intensity. So the usually assumed photodesorption caused by hole capture by negatively charged oxygen ions, although not directly proved, seems to be strongly involved in the desorption of oxygen.

Mark [7] made a semi-quantitative calculation of the chemisorption process on the basis of the model quoted in the Introduction. We must choose a somewhat different treatment, because our measurements point to the existence of accumulation layers to which Mark's treatment is not applicable. This requires a new interpretation for the rate-limiting steps of the chemisorption and desorption processes formerly assumed to be the transfer of electrons to the surface across a Schottky barrier formed by the field between the surface ions and the positive space charge below the surface and the transfer of holes to the surface in the field of the Schottky barrier, respectively. In the case of accumulation layers these arguments are not valid, but by slightly modifying them the rate-limiting steps of chemisorption and desorption prove to be the electron capture by neutral oxygen and the hole capture by chemisorbed oxygen ions, respectively. Similar to a proposal made by Sébenne and Balkanski [16] the capture is determined by the respective surface concentrations of the carriers and the concentrations and capture cross-sections of the respective centres. The bulk including the space-charge region must be in equilibrium, i.e., carrier exchange between bands and levels in the gap and all diffusion processes must proceed much faster than the immediate carrier exchange with surface states. Then we have in zeroth approximation the following ansatz:

$$\frac{\mathrm{d}Q_\mathrm{c}}{\mathrm{d}t} = C_\mathrm{n}\, n_\mathrm{s}\, Q_\mathrm{p} - C_\mathrm{p}\, p_\mathrm{s}\, Q_\mathrm{c}\,, \tag{1}$$

where Q_c density (cm^{-2}) of chemisorbed oxygen ions, C_p their hole capture probability, Q_p density of physically adsorbed oxygen molecules, C_n their electron capture probability, n_s and p_s surface concentrations of electrons and holes, respectively. As a result of chemisorption or desorption, i.e. of change of charge Q_ss in surface states, the free carrier concentrations at the surface and, therefore, also the barrier height v_s, the space charge Q_sc, and the excess surface-electron density ΔN change, thus leading to a variation of the photoconductivity in the space-charge region. The photoconductivity change is proportional to a change of the surface conductance $\Delta\sigma = e\,\mu_\mathrm{n}\,\Delta N$, if a change of the electron mobility μ_n can be neglected which seems to be the case for monocrystalline CdS [6].[3])

4.3 Mathematical description of pressure dependence and kinetics of photoconductivity during chemisorption and desorption

In the case of accumulation layers, assuming the simplest ones caused by only one set of completely ionized surface donors, the density of the surface charge Q_ss/e is given by

$$\frac{Q_\mathrm{ss}}{e} = Q_\mathrm{d} - Q_\mathrm{c}\,, \tag{2}$$

where Q_d density of surface donors and Q_c density of chemisorbed oxygen ions as in equation (1). Then equation (1) can be written as (Q_d = const)

$$-\frac{1}{e}\frac{\mathrm{d}Q_\mathrm{ss}}{\mathrm{d}t} = C_\mathrm{n}\, n_\mathrm{s}\, Q_\mathrm{p} - C_\mathrm{p}\, p_\mathrm{s}\left(Q_\mathrm{d} - \frac{Q_\mathrm{ss}}{e}\right). \tag{3}$$

The dependence of the hole concentration p_s on the charge Q_ss in surface states or on the barrier height v_s is not known even for homogeneously excited photoconductors. Restricting to these photoconductors the calculations of [21] can

---

[3]) The stated quantities are exactly defined in [21]; here we use the more customary symbols Q_ss and Q_sc instead of Q_s and Q_r.

be used. Two cases have to be discussed provided that the hole concentration does not greatly disturb the electron concentration:

Case a: The superficial hole concentration varies according to the band bending, i.e. $n_s\, p_s = n_b\, p_b$ (n_b, p_b bulk concentrations).

Case b: The hole concentration is determined by the local lifetime of holes. Since at intermediate excitation rates the sensitizing centres are nearly completely filled with electrons and under the existence of accumulation layers their degree of occupancy is larger than in the bulk, the hole lifetime should be constant according to $\tau_p^{-1} = C_{p1}\, N_{r1} + C_{p2}\, N_{r2}$ and, therefore, $p_s = p_b = \text{const.}$ (N_{r1} and N_{r2} concentrations of fast recombination and sensitizing centres, C_{p1} and C_{p2} their hole capture probabilities, respectively.) In reality, the surface hole concentration will lie between these two limiting cases.

Now we treat the change of excess surface-electron density ΔN, proportional to the photoconductivity change, with change of Q_p (proportional to the oxygen pressure) or time (caused by a temporal change of the chemisorbed oxygen density Q_c).

In the case of a homogeneously excited photoconductor with a linear photocurrent-illumination characteristic the excess surface-electron density ΔN and the surface charge Q_{ss} are given by (cf. equations (3.3) and (3.4) of [21])

$$\Delta N = \mp\, n_b\, L_2 \int_0^{v_s} \frac{(e^v - 1)}{F_2}\, dv\,, \qquad Q_{ss} = -\, Q_{sc} = \mp\, 2\, e\, n_t\, L_2\, F_{2s}\,, \tag{4}$$

where $L_2 = \left(\dfrac{\varepsilon\, \varepsilon_0\, kT}{2\, e^2\, n_t}\right)^{1/2}$ the effective Debye length, $F_2 = (e^{-v} + v - 1)^{1/2}$, and $v = \ln\left(n(z)/n_b\right)$ the band bending as a function of the distance z from the surface. F_{2s} and v_s are the values of F_2 and v at the surface for $z = 0$. In equations (4) and all following formulae the upper sign corresponds to depletion, the lower to accumulation layers. n_t is the concentration of occupied deep electron traps, which is in our approximation equal to the concentration p_{rb} of holes in sensitizing centres assumed to be totally filled with electrons in the dark.[4]) For the following informative numerical calculation we used $p_{rb} = 5 \times 10^{14}\ \text{cm}^{-3}$ and $\varepsilon = 9.8$, i.e. $L_2 \approx 1.2 \times 10^{-5}$ cm.

4.3.1 Dependence of photoconductivity on oxygen pressure

Under steady-state conditions we obtain from equation (3)

$$\frac{C_n\, n_b}{C_p\, p_b}\, Q_p = \left(Q_d - \frac{Q_{ss}}{e}\right) e^{-\beta v_s}\,, \tag{5}$$

where $\beta = 2$ for *case a* and $\beta = 1$ for *case b*.

To get the desired ΔN–Q_p dependence for a given v_s we determine Q_{ss} from equation (4). The left part of equation (5), proportional to Q_p, then can be obtained once having chosen Q_d. Likewise, for the same value of v_s, ΔN is obtained from equation (4). In Fig. 8 $\Delta N/n_b\, L_2$ which is proportional to the relative photoconductivity change, is plotted against $(C_n\, n_b/C_p\, p_b)\, Q_p$.

The curves can be approximated by straight lines over larger ranges of oxygen pressure in agreement with the experimental results (cf. Fig. 7). In a mathematical sense these straight lines are tangents in the turning-points. For

---

[4]) The case of sensitizing centres partially filled with electrons already in thermodynamic equilibrium was discussed by Auth, Krenzke, and Niekisch [22].

Fig. 8. Relative photoconductivity change $\Delta N/n_b L_2$ versus pressure ($\sim (C_n n_b/C_p p_b) Q_p$) for *case a* of a surface hole concentration varying according to the barrier height (curves (1a), (2a), (3a)) and for *case b* of a constant hole concentration (curves (1b), (3b)) with different densities of surface donors Q_d. Curves (1a) and (1b): $Q_d = 0$; (2a): $Q_d = 10^{10}$ cm^{-2}; (3a) and (3b): $Q_d = 2\times10^{10}$ cm^{-2}. The ordinate values of curve (2a) are multipled by 0.5, of curves (3a) and (3b) by 0.1. The arrows mark the values for "high oxygen pressure" and "vacuum" used for the calculations of Fig. 9 (The left-hand arrow for curve (3a) would be at 6×10^4 cm^{-2})

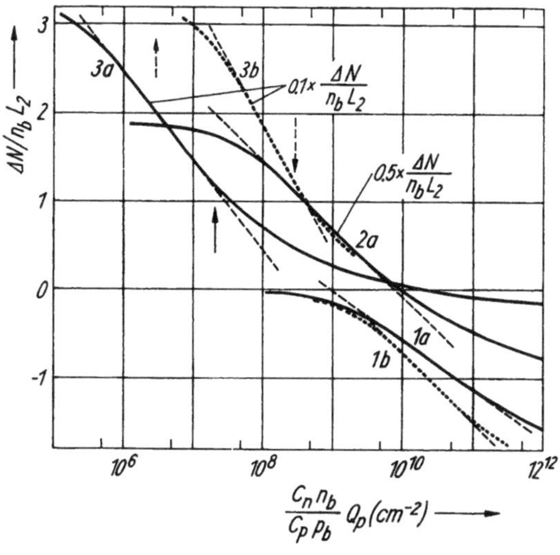

different concentrations of surface donors the curves are shifted against each other with regard to the abscissa. Since such a strong shift was not observed experimentally (cf. Fig. 7), the coverage of the investigated crystals with surface donors should not greatly differ from one another.

4.3.2 *Kinetics of photoconductivity after an abrupt change of oxygen pressure*

By integrating equation (3) one obtains an expression

$$\int_{(F_{2s})_0}^{(F_{2s})} \frac{\pm \,\mathrm{d}F_{2s}}{v_s\left[\dfrac{C_n n_b}{C_p p_b} Q_p - \mathrm{e}^{-\beta v_s}(Q_d \pm 2\, p_{rb} L_2 F_{2s})\right]} = \frac{C_p p_b}{2\, p_{rb} L_2}\, t \qquad (6)$$

which is proportional to the chemisorption or desorption times t_c or t_d, according as the value of Q_p was leveled up or down at $t = 0$. Again $\beta = 2$ for *case a* and $\beta = 1$ for *case b*.

To calculate the photoconductivity decrease during chemisorption, for $(C_n n_b/C_p p_b) Q_p$ a value representative for a "high oxygen pressure" must be selected, as given, for example, by the right-hand arrows in Fig. 8 (according to the experimental results near the ends of the straight line parts of the curves and for an *assumed* value of Q_d not too low to meet case of accumulation layers verified in Section 4.1). For the second part $\mathrm{e}^{-\beta v_s}(Q_d \pm 2\, p_{rb} L_2 F_{2s})$ of the denominator in equation (6) initial values representing "vacuum" conditions given by the left-hand arrows in Fig. 8 are selected. (The corresponding arrow for curve (3a) would be at 6×10^4 cm^{-2}.) From this, with $Q_d = 2\times10^{10}$ cm^{-2}, one determines the initial barrier height $(v_s)_0$ and $(F_{2s})_0$. The right-hand part of equation (6) which is proportional to the chemisorption time can be obtained by starting from $(v_s)_0$ $(t = 0)$ and integrating in the direction of lower values of v_s. For the calculation of the photoconductivity increase during desorption the values and the direction of integration must be exchanged. The corresponding values of ΔN can be obtained from equation (4).

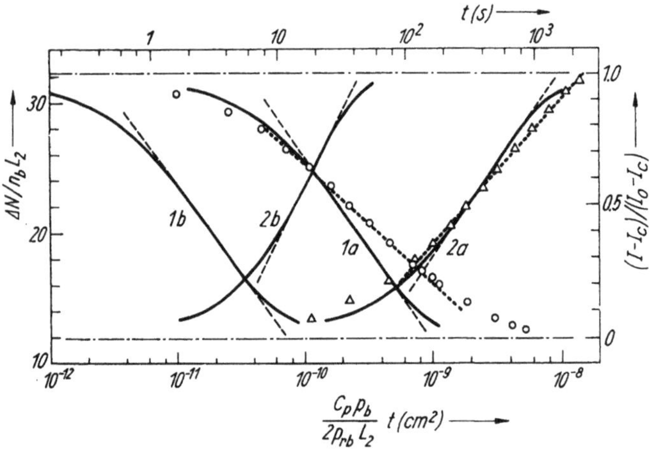

Fig. 9. Calculated photoconductivity change ($\sim \Delta N/n_b\, L_2$) during chemisorption (1a, 1b) and desorption (2a, 2b) for *case a* (1a, 2a) and *case b* (1b, 2b). $Q_d = 2 \times 10^{10}\ cm^{-2}$ corresponding to the curves (3a) and (3b) of Fig. 8. Experimental points: Photocurrent change during chemisorption (same measurement as in Fig. 3, curve (a)) and corresponding desorption normalized to the same ordinate interval as the theoretical curves

Chemisorption: Curves (1a) and (1b) of Fig. 9 show the photoconductivity decrease during chemisorption, calculated for *cases a* and *b*. It is to be seen that the calculated curves satisfactorily describe the experimentally observed behaviour of the photocurrent at high illumination intensities after admitting oxygen (cf. Fig. 3). Curves (1a) and (1b) are shifted against each other mainly due to the different choice of the "pressure" (right-hand arrows) and not on account of the different mechanisms which should manifest themselves primarily in the desorption. The experimental points are taken from curve (a) of Fig. 3. The photocurrent change was normalized to the theoretical change. There are only slight differences in the slope, but since the experimental values of Q_d, Q_p, p_{rb}, etc., are not exactly known, no quantitative adjustment was possible.

On the contrary the kinetics of photocurrent during chemisorption at lower illumination intensities and lower electron concentrations (Fig. 2) must be explained in an other way. In this case it is doubtful whether steady-state conditions exist in the bulk. At low intensities the difference $E_c - E_{fn}$ is so large that the electron traps (traps within the space-charge region or surface traps) cannot come into equilibrium with the conduction band during the initial steps of chemisorption. At first, therefore, only free electrons are extracted from the sub-surface regions and the photocurrent very quickly decays. The following slow diminution should arise from an increasing influence of electron emission from traps into the conduction band. This emission should lead to a rise of photocurrent during the chemisorption process, if the foregoing photocurrent decrease is higher than that corresponding to a final steady-state value of the photocurrent with all centres in equilibrium. Equation (1) cannot be used for a treatment of this behaviour.

Desorption: Curves (2a) and (2b), of Fig. 9 show the "photoconductivity change" during desorption for *cases a* and *b*. Apart from a mutual shift which can be explained, as above, by means of the different "pressures" chosen to have the same ordinate range, the slope of curve (2b) is steeper than that of curve (2a). This occurs on account of the fact that the hole concentration is constant in *case b* and permanently diminishes in *case a*.

As compared to the experimental results (desorption curves in Fig. 9; Fig. 4), the region of constant slope in *case b* is too small with respect to the abscissa, so that *case a* seems to give a better description of the results. It is, however, not to overlook that the agreement between the theoretical calculations and the experimental results is poor with respect to the intensity dependence of desorption. From equation (1) or from Fig. 9 we expect a shift of the photoconductivity curves on the time scale proportional to the hole concentration, i.e. approximately proportional to the illumination intensity or the electron concentration, which was not found experimentally (Fig. 4). The desorption of oxygen from CdS, therefore, seems to be not a simple photodesorption process but a more complicated one.

An order-of-magnitude estimate of the values of C_p and C_n, the capture probabilities of oxygen for holes and electrons in the chemisorbed and neutral state, respectively, is not contradictory to the experience: From a comparison of the theoretical *case a* and the experimental curves (Fig. 9) we find $C_p\, p_b/2\, p_{rb}\, L_2 \approx 10^{-11}$ cm^2 s^{-1}. With the above adopted value $2\, p_{rb}\, L_2 \approx 10^{10}$ cm^{-2} and an assumed $p_b \approx 10^5$ to 10^{10} cm^{-3} we have $C_p \approx 10^{-6}$ to 10^{-11} cm^{3}s^{-1}, i.e. a capture cross-section of 10^{-13} to 10^{-18} cm^2 in agreement with the fact that chemisorbed oxygen forms a negatively charged centre attracting holes. With an estimated $Q_p \approx 10^{12}$cm^{-2} at room temperature and about 1/50 atmospheric pressure, $(C_n\, n_b/C_p\, p_b)\, Q_p \approx 10^7$ cm^{-2} (right-hand arrow in Fig. 8), and an roughly estimated value of $n_b \approx 10^{14}$ cm^{-3} (from photocurrent, mobility, etc.) follows $C_n \approx 10^{-20}$ cm^3 s^{-1}, i.e. a very low capture cross-section ($\approx 10^{-27}$ cm^2) of the neutral adsorbed oxygen for electrons, which makes possible that the hole capture process on the whole is effective in desorption. On account of this low capture cross-section for electrons oxygen does not play a major role in the recombination near the surface but mainly "stores" surface charge.

Acknowledgements

The author would like to thank Dr. E. Gutsche for many helpful discussions, Mr. R. Herbst for the calculation of the theoretical curves, and Dr. H. Lange for a critical reading of the manuscript.

References

[1] R. H. Bube, J. chem. Phys. **21**, 1409 (1953).
[2] R. H. Bube, J. Electrochem. Soc. **113**, 793 (1966).
[3] S. H. Liebson, J. Electrochem. Soc. **101**, 359 (1954); J. chem. Phys. **23**, 1732 (1955).
[4] J. Woods, J. Electronics and Control **5**, 417 (1958).
[5] H. Berger, K. W. Böer, and E.-H. Weber, Z. Phys. **158**, 501 (1960).
[6] P. Mark, J. Phys. Chem. Solids **25**, 911 (1964).
[7] P. Mark, J. Phys. Chem. Solids **26**, 959 (1965).
[8] P. Mark, J. Phys. Chem. Solids **26**, 1767 (1965).
[9] P. Mark, RCA Rev. **26**, 461 (1965).
[10] C. E. Reed and C. G. Scott, Brit. J. appl. Phys. **15**, 1045 (1964).
[11] C. E. Reed and C. G. Scott, Brit. J. appl. Phys. **16**, 471 (1965).
[12] C. E. Bleil and W. A. Albers, Jr., Surface Sci. **2**, 307 (1964).
[13] H. Shear, E. A. Hilton, and R. H. Bube, J. Electrochem. Soc. **112**, 997 (1965).
[14] A. L. Robinson and R. H. Bube, J. Electrochem. Soc. **112**, 1002 (1965).
[15] L. Gombay, J. Lang, and J. Kispéter, Acta Phys. Chem. Szeged (Hung.) **10**, 23 (1964); **12**, 11 (1966).
[16] C. Sébenne and M. Balkanski, Surface Sci. **1**, 22 (1964).
[17] A. Many and A. Katzir, Surface Sci. **6**, 279 (1967).
[18] R. Williams, J. Phys. Chem. Solids **23**, 1057 (1962).

[19] S. R. MORRISON, Adv. Catalysis **7**, 259 (1955).
D. A. MELNICK, J. chem. Phys. **26**, 1136 (1957).
D. B. MEDVED, J. Phys. Chem. Solids **20**, 255 (1961).
[20] S. M. RYWKIN, Photoelektrische Erscheinungen in Halbleitern, Akademie-Verlag, Berlin 1965 (p. 265ff.).
A. ROSE, Concepts in Photoconductivity and Allied Problems, Interscience Publishers, New York/London 1963 (p. 134).
[21] E.-H. WEBER, phys. stat. sol. **17**, 843 (1966).
[22] J. AUTH, G. KRENZKE, and E. A. NIEKISCH, Z. phys. Chem. (Leipzig) **213**, 17 (1960).
[23] K. W. BÖER and K. ZIMMERMANN, Monatsber. Dtsch. Akad. Wiss. Berlin **1**, 336 (1959).
P. FLÖGEL and R. KOEPP, phys. stat. sol. **10**, 765 (1965).
[24] E. GUTSCHE and J. VOIGT, in: II–VI Semiconducting Compounds, Internat. Conf. Providence 1967, Ed. D. G. THOMAS, W. A. Benjamin, Inc., New York/Amsterdam 1967 (p. 337).
[25] J. FASSBENDER, Ann. Phys. (Germany) **5**, 33 (1949).
[26] H. B. DEVORE, Phys. Rev. **102**, 86 (1956).
[27] D. DUTTON, Phys. Rev. **112**, 785 (1958).
[28] V. L. BROUDE and N. N. CHIKOVANI, Ukr. fiz. Zh. **7**, 167 (1962).

(Received May 7, 1968)

phys. stat. sol. 28, 663 (1968)

Subject classification: 6; 22.4.; 22.6

Department of Electrical Engineering, University of Rhode Island, Kingston (a), Air Force Cambridge Research Laboratories, Bedford, Massachusetts (b)

Long-Wavelength Longitudinal Phonons of Multi-Mode Crystals[1])

By

I. F. CHANG (a)[2]), S. S. MITRA (a), J. N. PLENDL (b), and L. C. MANSUR (b)

It is shown that the application of the so-called Drude's rule in obtaining the $\boldsymbol{k} \approx 0$ longitudinal optic mode frequencies (LO's) of multi-resonance damped oscillators is not correct, although it may be used as an approximate method only when a few oscillators, all of appreciable strength, are present. A rigorous procedure suitable for all cases, single or multi-resonance, is shown to be to identify the frequencies at which the minima of the modulus of the complex dielectric constant ($|\varepsilon|$) occur with the LO modes. The frequencies of the maxima of $|\varepsilon|$, on the other hand, approximately represent the $\boldsymbol{k} \approx 0$ transverse optic modes. It is also shown that for a general n-mode crystal it is possible to obtain the LO frequencies individually provided that all the parameters of the Lorentz oscillators needed to represent an observed optical property are known. In the case of a single resonance oscillator the equations reduce to the well-known Lyddane-Sachs-Teller relation. Examples are given where LO's are obtained by the above methods using reflectivity data and compared with whatever available directly measured experimental LO values, e.g., from Raman or neutron scattering measurements.

Es wird gezeigt, daß die Anwendung der sogenannten Drudeschen Regel zur Erlangung der Frequenzen der $\boldsymbol{k} \approx 0$ longitudinalen optischen Moden (LO's) von gedämpften Viel-Resonanzoszillatoren nicht korrekt ist, obgleich sie als Näherungsmethode benutzt werden kann, wenn nur wenige Oszillatoren, alle von genügender Stärke, vorhanden sind. Eine strenge Behandlung, die für Einzel- und Vielresonanz geeignet ist, zeigt, daß die Frequenzen bei denen Minima der Moduln der komplexen Dielektrizitätskonstante ($|\varepsilon|$) auftreten mit den LO-Moden identifiziert werden können. Auf der anderen Seite repräsentieren die Frequenzen der Maxima von $|\varepsilon|$ näherungsweise die $\boldsymbol{k} \approx 0$ transversalen optischen Moden. Es wird ebenfalls gezeigt, daß es für einen allgemeinen n-moden-Kristall möglich ist, die LO-Frequenzen einzeln zu erhalten, vorausgesetzt, daß alle Parameter der Lorentz-Oszillatoren, die zur Darstellung einer beobachteten optischen Eigenschaft benötigt werden, bekannt sind. Im Falle eines Einzelresonanz-Oszillators reduzieren sich die Gleichungen auf die bekannte Lyddane-Sachs-Teller-Beziehung. Beispiele werden angegeben, bei denen die LO's durch die oben angeführten Methoden erhalten werden, wobei Reflexionswerte benutzt werden und mit erhältlichen direkt gemessenen, experimentellen LO Werten, z. B. aus Raman- oder Neutronenstreuungsmessungen, verglichen werden.

1. Introduction

It is well known that the frequencies of the long-wavelength ($\boldsymbol{k} \approx 0$) transverse optic (TO) and the longitudinal optic (LO) modes are not equal in polar crystals. For such crystals the TO modes are usually infrared-active and may be deter-

[1]) Work supported in part by the U.S. Air Force In-House Laboratory Independent Research Fund under Contract No. AF19 (628)—6042.

[2]) Present Address: Components Division, IBM, East Fishkill, New York.

mined directly by absorption measurements on large thin crystals, or reflection measurements on bulk crystals. This is so because the interacting electromagnetic field, being transverse, can couple with transverse phonon fields. This is not true for the longitudinal phonon fields in bulk crystals; thus with light incident normally LO modes are not usually observable. However, there are several direct and indirect methods of obtaining the LO frequencies. In the case of cubic diatomic crystals, LO is usually computed from TO using the Lyddane-Sachs-Teller relation [1]:

$$\frac{\mathrm{LO}}{\mathrm{TO}} = \left(\frac{\varepsilon_0}{\varepsilon_\infty}\right)^{1/2} \tag{1}$$

provided ε_0 and ε_∞, the values of the dielectric constant at very low and high (visible or ultraviolet) frequencies, are available. However, there may be considerable uncertainties as to the values of ε_0 and ε_∞ to be used in many substances. This applies especially to crystals that are not good insulators, and those with more than one reststrahlen peak or with low-energy electronic transitions. LO may also be determined by the so-called Drude's method, which consists in finding the frequency at which the real part of the dielectric constant goes through zero, i.e., where $n = k$, where n and k are, respectively, the refractive index and the extinction coefficient. This method is often used for determining LO from infrared reflection due to a single damped oscillator. For strongly ionic crystals with a vanishingly low value of the damping constant, LO then approximately corresponds to the frequency at which the reflectivity attains a minimum on the high-frequency side of the reststrahlen band.

In some crystals the selection rules permit the $\boldsymbol{k} \approx 0$ LO modes to be active in the Raman scattering. This is the case with the cubic zincblende and the hexagonal wurtzite structures. For such crystals the determination of the LO modes are thus most direct. The LO's of ZnS, ZnSe, CdS, ZnO, ZnTe, and the III–V compounds, for example, have been determined [2 to 9] in this manner. Inelastic neutron scattering studies may also yield dispersion curves for the longitudinal optic mode frequency as a function of the wave vector $\boldsymbol{k}$. A fairly accurate value of the long-wavelength LO can be obtained from the extrapolation of these curves to $\boldsymbol{k} \approx 0$. Brockhouse and associates have obtained LO for several alkali halide crystals by this method [10].

Finally, Berreman [11] has shown that thin films of cubic ionic crystals have sharp, strong infrared absorption and reflection at frequencies characteristic of the long-wavelength LO mode when the radiation is p-polarized and incident obliquely to the surface. The LO modes of LiF [11] and NaF [12], for example, have been determined this way.

For infrared dispersion involving multiple oscillators, although a generalized LST relation,

$$\prod_i \frac{\mathrm{LO}_i}{\mathrm{TO}_i} = \left(\frac{\varepsilon_0}{\varepsilon_\infty}\right)^{1/2}, \tag{2}$$

involving all the long-wavelength longitudinal and transverse optic modes, holds good [13]. Equation (2) does not, however, yield the LO mode frequencies individually. Often Drude's rule has been invoked [14] for the multi-resonance oscillators also, although without any theoretical justification. Furthermore, when some of the oscillators are much weaker than the others the n- and k-curves may not cross at as many points as the expected number of longitudinal optic branches, rendering the application of the Drude's rule impractical. The purpose of this communication is to show that Drude's rule as stated above is not correct for multi-resonance damped oscillators, although approximately true only when a few oscillators, all of appreciable strength, are present. A rigorous procedure suitable for all cases will be shown to be to identify the minima of the modulus of the complex dielectric constant ($|\varepsilon|$) with LO modes. Furthermore, it is possible to obtain the LO frequencies individually as functions of the parameters of Lorentz oscillators needed to represent an observed optical property. The equations reduce to LST relation in the case of a single resonance oscillator. Examples will be given where LO's are obtained by the above methods using reflectivity data and compared with whatever available directly measured experimental LO values, e.g., from Raman or neutron scattering measurements.

2. LO Modes of Single and Multi-Resonance Oscillators

Considering a crystal of single resonance with damping, the equation of motion and the polarization can be written [15] as

$$\ddot{\boldsymbol{W}} + \gamma\, \dot{\boldsymbol{W}} + \omega_0^2\, \boldsymbol{W} = \varrho^{1/2}\, \omega_0\, \boldsymbol{E}\,, \tag{3}$$

$$\boldsymbol{P} = \varrho^{1/2}\, \omega_0\, \boldsymbol{W} + \alpha_\infty\, \boldsymbol{E}\,, \tag{4}$$

where

$$W = u \left(\frac{\mu}{v}\right)^{1/2},$$

u is the phonon displacement amplitude, μ is the reduced mass, v is the unit cell volume. ω_0^2, γ, and $\varrho^{1/2}\,\omega_0$ are the restoring force, damping constant, and effective charge coefficient respectively.[3]) By splitting the phonon amplitude into its solenoidal (transverse) and irrotational (longitudinal) parts and making use of the electrostatic equation $\nabla\, \boldsymbol{D} = \nabla\,(\boldsymbol{E} + 4\,\pi\, \boldsymbol{P}) = 0$, which implies an irrotational $\boldsymbol{E}$-field, one obtains separately the transverse and longitudinal equation shown below:

$$\ddot{W}_\mathrm{t} + \gamma\, \dot{W}_\mathrm{t} + \omega_0^2\, W_\mathrm{t} = 0\,, \tag{5}$$

$$\ddot{W}_\mathrm{l} + \gamma\, \dot{W}_\mathrm{l} + \omega_0^2 \left(1 + \frac{4\,\pi\,\varrho}{\varepsilon_\infty}\right) W_\mathrm{l} = 0\,, \tag{6}$$

where $\varepsilon_\infty = 1 + 4\,\pi\,\alpha_\infty$. If one assumes a sinusoidal solution of the type $W_{\mathrm{t(l)}} = W_{\mathrm{t(l)}}(x)\exp\,(-\,i\,S_{\mathrm{t(l)}}\,t)$, equations (5) and (6) yield two complex frequencies, the transverse and longitudinal mode,

$$S_\mathrm{t} = \pm\left(\omega_0^2 - \frac{\gamma^2}{4}\right)^{1/2} - i\,\frac{\gamma}{2}\,, \tag{7}$$

$$S_\mathrm{l} = \pm\left(\omega_0^2 \left(\frac{\varepsilon_0}{\varepsilon_\infty}\right) - \frac{\gamma^2}{4}\right)^{1/2} - i\,\frac{\gamma}{2}\,, \tag{8}$$

where

$$\varepsilon_0 = \varepsilon_\infty + 4\,\pi\,\varrho\,.$$

Identifying the real parts of equations (7) and (8) as mode frequencies, a modified

---

[3]) Born and Huang have shown that the coefficient of E-term in equation (3) is equal to the coefficient of the displacement in equation (4) and can be related to the strength of the oscillator (see [15]).

LST relation is obtained:

$$\left(\mathrm{Re}(S_l)\right)^2 = \left(\mathrm{RE}(S_t)\right)^2 \left(\frac{\varepsilon_0}{\varepsilon_\infty}\right) + \frac{\gamma^2}{4}\left(\frac{\varepsilon_0}{\varepsilon_\infty} - 1\right) \quad \text{(Re stands for real part)} \tag{9}$$

which reduces to the commonly used LST relation, $\omega_l^2 = \omega_0^2 \left(\frac{\varepsilon_0}{\varepsilon_\infty}\right)$, in the limit of $\gamma \approx 0$.

When a complex crystal has more than one resonance, Huang's theory [15] can be extended. In the general case of n uncoupled damped oscillators, one gets

$$\ddot{\boldsymbol{W}}_{tj} + \gamma_j \dot{\boldsymbol{W}}_{tj} + \omega_j^2 \boldsymbol{W}_j = \sqrt{\varrho_j}\,\omega_j \boldsymbol{E}\,; \qquad j = 1, 2, \ldots, n\,, \tag{10}$$

$$\boldsymbol{P} = \sqrt{\varrho_1}\,\omega_1 \boldsymbol{W}_1 + \sqrt{\varrho_2}\,\omega_2 \boldsymbol{W}_2 + \cdots + \sqrt{\varrho_n}\,\omega_n \boldsymbol{W}_n + \alpha_\infty \boldsymbol{E}\,. \tag{11}$$

These two equations can be separated by the same technique described above into the transverse and longitudinal parts, i.e.,

$$\ddot{W}_{tj} + \gamma \dot{W}_{tj} + \omega_j^2 \dot{W}_{tj} = 0\,; \qquad j = 1, 2, \ldots, n\,, \tag{12}$$

$$\ddot{W}_{lj} + \gamma_j \dot{W}_{lj} + \omega_j^2 W_{lj} = \sqrt{\varrho_j}\,\omega_j \sum_{k=1}^{n} \frac{-4\pi(\varrho_k)^{1/2}\omega_k}{\varepsilon_\infty} W_{lk}; \quad j=1, 2, \ldots, n\,. \tag{13}$$

Equation (12) yields the n transverse mode frequencies as

$$S_{tj} = \pm\left(\omega_j^2 - \frac{\gamma_j^2}{4}\right)^{1/2} - i\,\frac{\gamma_j}{2} \qquad j = 1, 2, \ldots, n\,. \tag{14}$$

However, the longitudinal solutions of equation (13) are interrelated in its secular equation which is given by the symmetrical determinant

$$\begin{vmatrix} -S_l^2 - i\gamma_1 S_l + \omega_1^2\left(1 + \frac{4\pi\varrho_1}{\varepsilon_\infty}\right) & 4\pi(\varrho_1\varrho_2)^{1/2}\omega_1\omega_2 & \cdots & \frac{4\pi(\varrho_1\varrho_n)^{1/2}\omega_1\omega_2}{\varepsilon_\infty} \\ \cdot & -S_l^2 - i\gamma_j S_l + \omega_j^2\left(1 + \frac{4\pi\varrho_j}{\varepsilon_\infty}\right) & \cdots & \frac{4\pi(\varrho_j\varrho_n)^{1/2}\omega_j\omega_n}{\varepsilon_\infty} \\ \cdot & \cdot & \cdots & -S_l^2 - i\gamma_n S_l + \omega_n^2\left(1 + \frac{4\pi\varrho_n}{\varepsilon_\infty}\right) \end{vmatrix} = 0. \tag{15}$$

This equation can be expressed in a different form shown below:

$$\prod_{j=1}^{n} (-S_l^2 - i\gamma_j S_l + \omega_j^2) + \sum_{j=1}^{n} \frac{4\pi\varrho_j\omega_j}{\varepsilon_\infty} \prod_{\substack{k=1\\k\neq j}}^{n} (-S_l^2 - i\gamma_k S_l + \omega_k^2) = 0\,. \tag{16}$$

Using the equation $\varepsilon_0 = \varepsilon_\infty + \sum_{j=1}^{n} 4\pi\varrho_j$ and knowing that the constant term of a polynomial is related to the product of its roots, one obtains a generalized LST relation:

$$\frac{\varepsilon_0}{\varepsilon_\infty} = \prod_j \frac{(S_{lj})(S_{lj})^*}{\omega_j^2} = \prod_j \frac{|S_{lj}|^2}{\omega_j^2}\,. \tag{17}$$

This relation has been obtained by Barker [16] in a different manner and reduces to Cochran's [13] formula $\frac{\varepsilon_0}{\varepsilon_\infty} = \prod_{j=1}^{n} \frac{\omega_{lj}^2}{\omega_j^2}$ in the case of no damping. There are n unknowns in equation (17), thus one cannot obtain the n individual LO frequencies from this relation alone. However, in principle, it is not difficult to solve equation (16) for n LO values provided all the oscillator parameters (which can be obtained by fitting infrared reflectivity data) are known. Some examples are presented in the next section.

Another alternate method that has often been used to determine LO frequencies is the so-called Drude's method, which consists in identifying the frequency at which the real part of the dielectric constant equals to zero. It is a handy method because its result is usually a by-product of the Kramers-Kronig (K–K) analysis [17] or damped oscillator curve fitting, thus saving the trouble of solving equation (16). However, this Drude's rule is not theoretically justified. Secondly it is not applicable for some multi-resonance cases where the oscillator strengths are of substantially different magnitude. Examples of this will be given later.

The dielectric constant is defined as

$$\varepsilon = 1 + 4\pi \frac{P}{E}, \tag{18}$$

where P/E can be obtained by solving equations (10) and (11). Inserting this ratio, equation (18) becomes

$$\varepsilon(S) = \varepsilon_\infty + \sum_{j=1}^{n} \frac{4\pi \varrho_j \omega_j^2}{-S^2 - i\gamma_j S + \omega_j^2}, \tag{19}$$

the usual classical dispersion relation. One notes that the poles of the function $\varepsilon(S)$ occur at

$$S = \pm \left(\omega_j^2 - \frac{\gamma_j^2}{4}\right)^{1/2} - i\frac{\gamma_j}{2}$$

which is the same as equation (14). Therefore, the transverse mode frequencies are those points in the complex S-plane where the dielectric function $\varepsilon(S)$ is infinite or, equivalently,

$$|\varepsilon(S)| = \infty\,. \tag{20}$$

The zeros of $\varepsilon(S)$ are (the same as zero of $\varepsilon(S)/\varepsilon_\infty$) the roots of the following equation:

$$\varepsilon(S) = 0 \quad \text{or} \quad |\varepsilon(S)| = 0\,. \tag{21}$$

Comparing equations (21), (19), and (16), one observes that they are different only by a constant factor ε_∞. By this result, it is justified to state that all the LO's occur at frequencies where the dielectric constant as a function of complex frequency, S, passes through zero. However, equations (20) and (21) are not of much practical use, since one obtains ε experimentally only as a function of real frequency through a K–K analysis or curve fitting procedure. However, one may approximate the complex frequency S by a real ω due to the fact that in real crystals the damping part γ is usually very small. Then the condi-

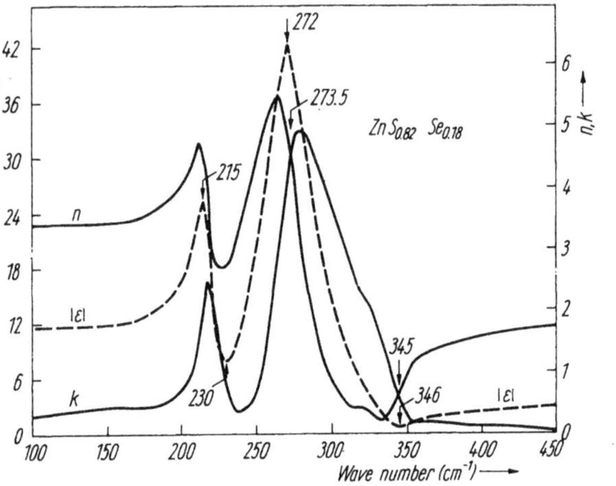

Fig. 1. Optical constants of $ZnS_{0.82}Se_{0.18}$

tions of equations (20) and (21) are relaxed to

$$|\varepsilon(\omega)| = \text{maximum} \qquad \text{(determine TO's)}\,, \tag{22}$$

$$|\varepsilon(\omega)| = \text{minimum} \qquad \text{(determine LO's)}\,. \tag{23}$$

It may be pointed out that the error involved in the approximation is larger for equation (22) than for equation (23). The new relations have one common useful feature, viz., that there always occur maxima and minima wherever there are TO's and LO's, whereas the conventional Drude's rule, on the contrary, is not always realized for every resonance in a multi-resonance system. Fig. 1 shows a typical example where n, k, and $|\varepsilon|$ are plotted for the mixed crystal $ZnS_{0.82}Se_{0.18}$. This crystal has two TO's and two LO's. The TO values given by the maxima of $|\varepsilon|$ are 215 and 272 cm^{-1}. The LO values given by the minima of $|\varepsilon|$ are 230 and 346 cm^{-1}, whereas Drude's rule $(n-k) = 0$ gives only one value, 345.5 cm^{-1}, since there is only one crossing of the n, k curves. These values are to be compared with the experimentally determined LO frequencies by Raman scattering measurement. The latter [18] are 226 and 346 cm^{-1}.

3. Examples

3.1 Mixed crystals

Two sets of mixed crystal systems, $ZnS_{1-x}Se_x$ and $CdS_{1-x}Se_x$, have been studied [18, 19] very thoroughly. The infrared reflectivity data of these two sets of mixed crystal are used first to perform Kramers-Kronig analysis. The rules derived above for determining TO's and LO's are applied on the K–K result, i.e. to identify the TO and LO values at the maximum and minimum points of $|\varepsilon|$. Then the infrared reflectivity data are also fitted by two-resonance classical damped oscillators to yield all the parameters which describe the optical properties of the mixed crystals. TO and LO values are again determined by the maxima–minima rule. The oscillator parameters of these crystals are also fed in equations (14) and (16) to obtain analytical solutions of the TO and LO frequencies. In addition, the Raman data of all these crystals are available [18, 19]. They provide a check on the analytical solution of equation (16) and on the maxima–minima rule performed on the K–K analysis results

Table 1
TO and LO frequencies in ZnS_xSe_{1-x}

x	Mode	Raman measure-ments	Solutions of (14) and (16)	Damped oscillators ω_i	Damped oscillators $\lvert\varepsilon\rvert =$ max	Damped oscillators $\lvert\varepsilon\rvert =$ min	Damped oscillators $\varepsilon' = 0$	Kramers-Kronig analysis $nk\omega =$ max	Kramers-Kronig analysis $\lvert\varepsilon\rvert =$ max	Kramers-Kronig analysis $\lvert\varepsilon\rvert =$ min	Kramers-Kronig analysis $\varepsilon' = 0$
1.0	TO		270.97 $-i$ 4.06	271	270.7			271	269.5		
	LO	351	348.33 $-i$ 4.06			348.5	347.9			350	349.5
0.993	TO							272.5	272		
	LO	350								348	348
0.996	TO							275	273.5		
	LO	349								346.5	347.5
0.816	TO_1		272.95 $-i$ 5.19	273	272.7			274	273		
	LO_1	346	345.14 $-i$ 5.19			345	344.4			346	345.4
	TO_2		217.94 $-i$ 5.23	218	215.5			216.5	215		
	LO_2	226	225.65 $-i$ 5.23			228	XX*)			230	XX
0.597	TO_1		279.87 $-i$ 8.4	280	279			280.5	275		
	LO_1	336	338.07 $-i$ 7.9			338.5	336.5			337	336.5
	TO_2	214	211.95 $-i$ 4.56	212	210.8			213	211		
	LO_2	234	228.83 $-i$ 5.05			230	227.5			236	225.5
0.330	TO_1		287.9 $-i$ 7.92	288	286			287.5	284		
	LO_1	324	324.70 $-i$ 7.29			326	322.6			324	322.5
	TO_2	211	208.97 $-i$ 3.45	209	208.9			210	208		
	LO_1	242	236.3 $-i$ 4.08			236.4	235.8			244.5	236.6
0.072	TO_1		293.92 $-i$ 7.06	294	291			296.5	292		
	LO_1	305.7	304.2 $-i$ 6.7			307	XX			310	XX
	TO_2	207	206.98 $-i$ 2.79	207	206.7			207.5	207.2		
	LO_2	250	247.17 $-i$ 3.14			247.2	247			248.2	248
0.015	TO_1	229**)						298	292		
	LO_1									304	XX
	TO_2	206						208.5	208		
	LO_2	252								250	250.5
0.0	TO	206	205.97 $-i$ 3.6	206	205.5			207.5	207.5		
	LO	253	256.41 $-i$ 3.6			256.5	256.1			254	258

*) n, k do not cross, Drude's rule fails to yield LO's. — **) Local mode.

Table 2

TO and LO frequencies in CdS_xSe_{1-x}

x	Mode *)	Raman**) measurement	Solution of (14) and (16)	Damped oscillators				Kramers-Kronig analysis			
				ω_i	$\lvert\varepsilon\rvert =$ max	$\lvert\varepsilon\rvert =$ min	$\varepsilon' = 0$	$nk\omega =$ max	$\lvert\varepsilon\rvert =$ max	$\lvert\varepsilon\rvert =$ min	$\varepsilon' = 0$
	$TO^{\parallel}$		234.98 $-i\,3.06$	235	234.5			235	234.5		
1.0	$TO^{\perp}$	305	241.99 $-i\,2.7$	242	241.5			239.6	240.1		
	$LO^{\parallel}$		303.12 $-i\,3.06$			303.5	302.7			305	306
	$LO^{\perp}$		305.2 $-i\,2.7$			305.6	305			304.6	305.8
	$TO_1^{\parallel}$		178.9 $-i\,6.09$	179	177			179.6	176.6		
	$TO_1^{\perp}$		182.91 $-i\,5.72$	183	181.3			183	178		
	$LO_1^{\parallel}$	196	194.39 $-i\,6.02$			196	191			198	XX***)
0.77	$LO_1^{\perp}$		195.03 $-i\,5.63$			196.5	191			198.5	XX
	$TO_2^{\parallel}$		241.94 $-i\,5.57$	242	241.5			240.3	239.2		
	$TO_2^{\perp}$		246.94 $-i\,5.31$	247	246			245.5	245		
	$LO_2^{\parallel}$	298	294.3 $-i\,5.63$			294.3	293.2			295	296.7
	$LO_2^{\perp}$		297.37 $-i\,5.35$			298	296.5			302	301.2
	$TO_1^{\parallel}$		173.91 $-i\,5.52$	174	172.2			174	170.5		
	$TO_1^{\perp}$		179.93 $-i\,4.9$	180	178.8			179	176		
	$LO_1^{\parallel}$	200.6	199.10 $-i\,5.75$			200	197			204	196.5
0.54	$LO_1^{\perp}$		201.93 $-i\,5.05$			203	200.5			206.5	200
	$TO_2^{\parallel}$		250.90 $-i\,7.15$	251	249			249.2	251.5		
	$TO_2^{\perp}$		254.93 $-i\,6.18$	255	254.3			252.5	249		
	$LO_2^{\parallel}$	291	290.93 $-i\,6.93$			292	289			294	289
	$LO_2^{\perp}$		291.63 $-i\,6.04$			292.2	292.3			293.6	290.5

*) Superscripts ||, ⊥ refer to the direction relative to the c-axis.
**) Polarization is not determined for Raman measurement.
***) n, k do not cross, Drude's rule fails to yield LO's.

Table 2 (continued)

x	Mode *)	Raman**) measurement	Solution of (14) and (16)	Damped oscillators ω_i	$\|\varepsilon\| =$ max	$\|\varepsilon\| =$ min	$\varepsilon' = 0$	Kramers-Kronig analysis $n k \omega =$ max	$\|\varepsilon\| =$ max	$\|\varepsilon\| =$ min	$\varepsilon' = 0$
0.27	$TO_1^{\parallel}$		170.94 $-i\,4.48$	171	170			168.6	167.6		
	$TO_1^{\perp}$		172.95 $-i\,4.37$	173	172			173	172.8		
	$LO_1^{\parallel}$	206	206.62 $-i\,4.81$			207	205.8			207.5	206.2
	$LO_1^{\perp}$		205.35 $-i\,4.44$			206	204.8			207.7	206.1
	$TO_2^{\parallel}$		257.90 $-i\,7.30$	258	255.5			256	252.6		
	$TO_2^{\perp}$		262.95 $-i\,5.19$	263	261.6			269.5	270		
	$LO_2^{\parallel}$	283	279.62 $-i\,6.96$			281	276.5			287	278.2
	$LO_2^{\perp}$		281.72 $-i\,5.12$			283	280.1			286	280
0.015	$TO_1^{\parallel}$		166.97 $-i\,3.28$	167	166.2			167	166.8		
	$TO_1^{\perp}$		171.99 $-i\,2.12$	172	172			172	171.4		
	$LO_1^{\parallel}$	210.8	211.03 $-i\,3.36$			211.6	210.5			210	211.3
	$LO_1^{\perp}$		210.74 $-i\,2.14$			210.4	210.5			213.2	212.5
	$TO_2^{\parallel}$		265.86 $-i\,8.71$	266	261			267.5	263.8		
	$TO_2^{\perp}$		265.96 $-i\,4.52$	266	263			266	262		
	$LO_2^{\parallel}$	269	267.73 $-i\,8.63$			272	XX			271.4	XX
	$LO_2^{\perp}$		267.15 $-i\,4.5$			270	XX			272	XX

and damped oscillator results. Table 1 presents such results on ZnS_xSe_{1-x} for nine values of x. Similar results on the mixed system CdS_xSe_{1-x} are presented in Table 2. In both tables, column 1 provides all the available Raman data. Column 2 contains the analytical solutions, and curve fitting and K–K analysis result are shown in the last two broad columns.

3.2 α-Al_2O_3 and Cr_2O_3

In this section we provide examples of crystals which have more than one infrared-active lattice mode. In Table 3 the results on α-Al_2O_3 and Cr_2O_3 for the A_{2u} modes [14, 20] are presented. For these crystals the approximate Drude's rule also yields LO values which are not too different from the exact solution.

Table 3

TO and LO frequencies in α-Al_2O_3 and Cr_2O_3

	ω_i	TO solution of (14)	$\lvert\varepsilon\rvert = \max$	LO solution of (16)	$\lvert\varepsilon\rvert = \min$	Drude's rule $\varepsilon' = 0$
α-Al_2O_3 $E \parallel c$	400	$399.98 - i\,4.0$	400			
	583	$582.91 - i\,10.2$	583			
				$512.5 - i\,7.00$	512	512
				$880 - i\,7.19$	880	871
Cr_2O_3 $E \parallel c$	538	$537.92 - i\,9.01$	537			
	613	$612.98 - i\,4.5$	614			
				$602 - i\,5.43$	600.8	602
				$758.6 - i\,8.08$	758.7	759

Table 4

Low-frequency (in cm^{-1}) LO modes in ferroelectric crystals

Crystal	Neutron scattering	Infrared data	$\lvert\varepsilon\rvert = \min$
$SrTiO_3$ (300 °K)		X*)	$LO_1 = 170$
$SrTiO_3$ (85 °K)	$LO_1 = 170$	X*)	$LO = 162$
$KTaO_3$ (300 °K)		$LO_1 = 188$**)	$LO_1 = 181$
		$LO_3 = 423$***)	$LO_3 = 392$
		$LO_4 = 833$***)	$LO_4 = 795$

*) No crossing of n, k curves.
**) Reasonable guess (see ref. [23]).
***) Drude's rule (see ref. [23]).

3.3 Ferroeoectric crystals

Much work has been reported on the ferroelectric crystals. However, as far as the LO frequencies are concerned, some modes have not been determined, simply because Drude's rule fails to yield them. There are complete data of oscillator parameters on $SrTiO_3$ [21] and $KTaO_3$ [22]. Maxima–minima rule is applied on the curve fitting result to give values which cannot be obtained by Drude's rule. The results of these ferroelectric crystals are shown in Table 4, where neutron scattering data [24], whatever available, are also included for comparison.

References

[1] R. H. LYDDANE, R. G. SACHS, and E. TELLER, Phys. Rev. **59**, 673 (1941).
[2] O. BRAFMAN, I. F. CHANG, and S. S. MITRA, Internat. Conf. Localized Excitations in Solids, Plenum Press, New York 1968.
O. BRAFMAN and S. S. MITRA, Phys. Rev., in the press.
[3] S. S. MITRA, J. Phys. Soc. Japan (Suppl.) **21**, 61 (1966).
[4] B. TELL, T. C. DAMEN, and S. P. S. PORTO, Phys. Rev. **144**, 771 (1966).
[5] S. S. MITRA and J. I. BRYANT, Bull. Amer. Phys. Soc. **10**, 333 (1965).
[6] T. C. DAMEN, S. P. S. PORTO, and B. TELL, Phys. Rev. **142**, 570 (1966).
[7] W. TAYLOR, Phys. Letters (Netherlands) **24A**, 556 (1967).
[8] A. MOORADIAN and G. B. WRIGHT, Solid State Commun. **4**, 431 (1966).

[9] O. Brafman, S. S. Mitra, and P. J. Gielisse, J. N. Plendl, and L. C. Mansur, Solid State Commun, in the press.
[10] A. D. B. Woods, W. Cochran, and N. B. Brockhouse, Phys. Rev. **119**, 980 (1960).
A. D. B. Woods, B. N. Brochkouse, R. A. Cowley, and W. Cochran, Phys. Rev. **131**, 1025 (1963).
R. A. Cowley, W. Cochran, B. N. Brockhouse, and A. D. B. Woods, Phys. Rev. **131**, 1030 (1963).
[11] D. W. Berreman, Phys. Rev. **130**, 2193 (1963).
[12] S. S. Mitra, C. Postmus, and J. R. Ferraro, Phys. Rev. Letters **18**, 455 (1967).
I. F. Chang, Ph. D. dissertation, University of Rhode Island, 1968.
[13] W. Cochran, Z. Krist. **112**, 465 (1959).
[14] D. R. Renneke and D. W. Lynch, Phys. Rev. **138**, A530 (1965).
[15] M. Born and K. Huang, Dynamical Theory of Crystal Lattices, Oxford University Press, New York 1954.
[16] A. S. Barker Jr., Phys. Rev. **136**, A1290 (1964).
[17] S. S. Mitra and P. J. Gielisse, Progress in Infrared Spectroscopy, Vol. 2, Plenum Press, New York 1964.
[18] O. Brafman, I. F. Chang, S. S. Mitra, G. Lengyel, and E. Carnall, Phys. Rev. Letters **19**, 1120 (1967).
[19] J. Parrish, C. H. Perry, O. Brafman, I. F. Chang, and S. S. Mitra, Proc. Internat. Conf. Physics of II–VI Semiconductors, W. A. Benjamin, Inc., New York 1967.
[20] A. S. Barker, Phys. Rev. **132**, 1474 (1963).
[21] A. S. Barker, Phys. Rev. **145**, 391 (1966).
[22] R. C. Miller and W. G. Spitzer, Phys. Rev. **129**, 94 (1963).
[23] C. H. Perry, J. H. Fertel, and T. F. McNelly, J. chem. Phys. **47**, 1619 (1967).
[24] R. A. Cowley, Phys. Rev. **134**, A981 (1964).

(Received April 23, 1968)

phys. stat. sol. 28, 675 (1968)

Subject classification: 11; 10.1; 14.1; 21

Atomics International, Division of North American Rockwell Corp., Canoga Park, California (a), Science Center of North American Rockwell Corp., Thousand Oaks, California (b), and University of Utah, Salt Lake City, Utah (c)

Electron Irradiation and Recovery of Tungsten [1)]

By

H. H. NEELY (a), D. W. KEEFER (a), and A. SOSIN (b, c)

Damage rates and recovery of 99.99% pure tungsten have been investigated after electron irradiation near 4 °K by means of electrical resistivity. The damage rate data tend to a value of $T_0 = 38$ to 42 eV. The recovery of the residual electrical resistivity shows three major stages: Stage I from 4.2 to 100 °K, Stage II from 100 to 500 °K, Stage III from 500 to 1000 °K. Stage I is dominated by only one major peak at 70 °K. An analysis based on second-order kinetics yielded a good fit to the data through 95% of the recovery. The energy derived was 0.17 eV. Only ≈ 26% recovery is seen in Stage I and we propose that the suppression is due to interstitial trapping by impurities. Stage III accounts for about 40% recovery of the induced resistivity and has a center temperature of 760 °K. The stage was analysed assuming second-order kinetics; this analysis gave an energy of 1.1 eV.

Messungen der elektrischen Leitfähigkeit wurden benutzt, um Schädigungsraten und Erholungsverhalten von 99,99% reinem Wolfram nach Elektronenbestrahlung bei etwa 4 °K zu untersuchen. Die Meßergebnisse für die Schädigungsraten lassen eine Verlagerungsenergie zwischen 38 und 42 eV vermuten. Die Erholung des elektrischen Restwiderstands weist drei große Bereiche auf: Bereich I von 4,2 bis 100 °K, Bereich II von 100 bis 500 °K, Bereich III von 500 bis 1000 °K. Im Bereich I dominiert eine Stufe bei 70 °K. Die Meßdaten über einen Bereich von 95% der Erholung in dieser Stufe stimmen gut mit einer Reaktion zweiter Ordnung überein. Die Aktivierungsenergie wurde zu 0,17 eV ermittelt. Nur ≈ 26% der bestrahlungsinduzierten Widerstandsänderung heilen in Bereich I aus. Es wird vorgeschlagen, daß der Einfang von Zwischengitteratomen durch Verunreinigungen eine Verkleinerung von Bereich I bewirkt. Etwa 40% der Erholung finden in Bereich III statt, dessen mittlere Temperatur bei 760 °K liegt. Dieser Bereich wurde unter der Annahme einer Kinetik zweiter Ordnung analysiert. Es wurde eine Aktivierungsenergie von 1,1 eV erhalten.

1. Introduction

To date, the only electron irradiation study reported on tungsten is the early work of Lucasson and Walker [1] in which it was found that no increase in resistivity occurred at an electron energy of 1.4 MeV (maximum transfer energy $T_m = 39.67$ eV). They surmised that the threshold displacement energy, T_0, for tungsten was greater than 35 eV. One purpose of the present work was, therefore, to gain preliminary information on the threshold energy for damage production. The main purpose of the work was to investigate the annealing spectrum of point defects produced by electron irradiation. The present results will be compared with observations made previously by other investigators

1) Based on work supported primarily by the Metallurgy Branch, Division of Research, U.S. Atomic Energy Commission, under Contract No. AT (04—3)-701.

using neutron irradiation. Very recently, DiCarlo et al. [2] have presented preliminary data on electron-irradiated tungsten derived mainly from observations on internal friction (see below).

2. Experimental

In this work, all irradiations and subsequent resistivity measurements after anneals were made at 4.2 °K. The material used was 3.0 mil (= 76.2 μm) diameter wire, with a stated purity of 99.99%, produced by Thermionic Products. Since it was anticipated that relatively high annealing temperatures would be required, no solder could be used in mounting the sample. Tungsten potential leads were spot-welded to the sample and the sample was spot-welded to a nickel holder. The sample holder configuration, mounting procedure, and irradiation cryostat have been described by Neely [3]. Prior to irradiation, the cryostat was evacuated and chilled to liquid helium temperature. The sample holder temperature was then raised to room temperature by means of the cryostat heater, and the sample was resistance-annealed for 2 hours at 1700 °C and cooled slowly. The resistivity of the sample at 4.2 °K prior to irradiation was 1.62×10^{-8} Ωcm.

3. Damage Production

The results of a brief study of damage rate are presented in Table 1. Since only a limited number of measurements were made, the data should be considered rather preliminary. The resistivity increase versus electron dose data were linear in each of the measurements.

Table 1

Incident electron energy (MeV)	T_m (eV)	Damage rate $d\varrho/d\Phi$ ($10^{-27}\Omega$ cm/electrons cm^{-2})
1.5	44	1.1
2.0	71	10.8
2.1	77	15.3

The data in this table tend to a value of T_0 = 38 to 42 eV. This estimate can be altered slightly by sample size corrections [4] or by the effect of subthreshold events, as observed in gold and copper by Bauer and Sosin [5]. The only other available data related directly to an evaluation of T_0 are derived from the field ion microscope work of Brandon et al. [6] who predicted a substantially higher value: T_0 = 70 to 85 eV. This estimate of Brandon et al. is clearly too high as demonstrated rather directly in this work and, slightly more indirectly, by the work of Coltman et al. [7]. Coltman et al. observed appreciable damage and a detailed spectrum of post-irradiation annealing in tungsten following irradiations in which recoil displacements produced following thermal neutron capture are responsible for the effects. The average recoil energy, T_r, in tungsten is only 58 eV.

4. Recovery

A typical recovery spectrum is shown in Fig. 1, and a plot of the temperature derivative of these data is shown in Fig. 2. The sample was irradiated to a dose

Fig. 1. Isochronal electrical resistivity recovery of polycrystalline tungsten after 2.1 MeV electron irradiation and a dose of 4.95×10^{17} electrons/cm². The resistivity change was 7.58×10^{-9} Ωcm. The temperature intervals are discussed in the text

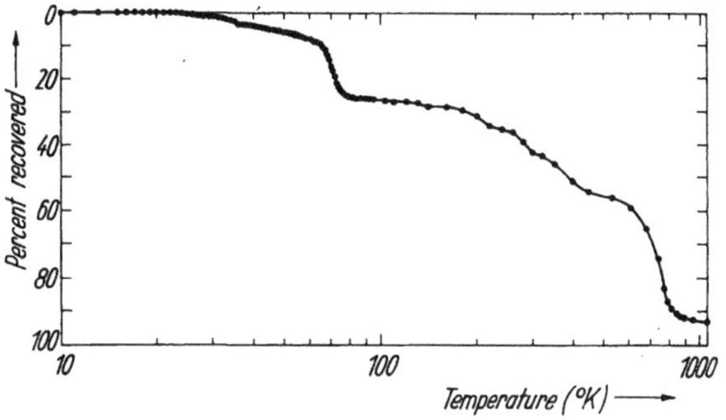

of 4.95×10^{17} electrons/cm², with a resulting resistivity increase of $7.58 \times \times 10^{-9}$ Ωcm. The sample was subsequently annealed at the temperatures indicated in Figs. 1 and 2. The time at each temperature was 5 minutes through 450 °K, and the remaining pulses were for 10 minutes. Below 450 °K, the annealing procedure consisted of raising the entire sample holder with sample to the indicated temperature for the desired period; this procedure is accomplished with ease in our apparatus so that a large number of pulses were made at small, uniformly spaced temperature intervals. Above 450°K, the annealing procedure was altered. The sample holder was heated to 300 °K, then the sample was heated further by passing an electric current through it. The extra difficulty involved in attaining higher temperatures is responsible for the lesser number of data points at higher temperatures, and for the more irregular temperature intervals.

It should be observed that the recovery spectrum shown in Figs. 1 and 2 represents about 100 temperature pulses.

The recovery spectrum is divisible into three distinct stages over the temperature range studied. This division is in agreement with that made by Thompson [8] for data obtained after neutron irradiation. The temperature ranges are as follows: Stage I — from 4.2 to 100 °K; Stage II — from 100 to 500 °K; Stage III — from 500 to 1000 °K.

4.1 Stage I

In Stage I (see particularly Fig. 2), there are several recovery peaks, but the peak centered at about 70 °K clearly dominates the spectrum. This peak constitutes 18% of the recovery. The remaining peaks in Stage I account for only 8%.

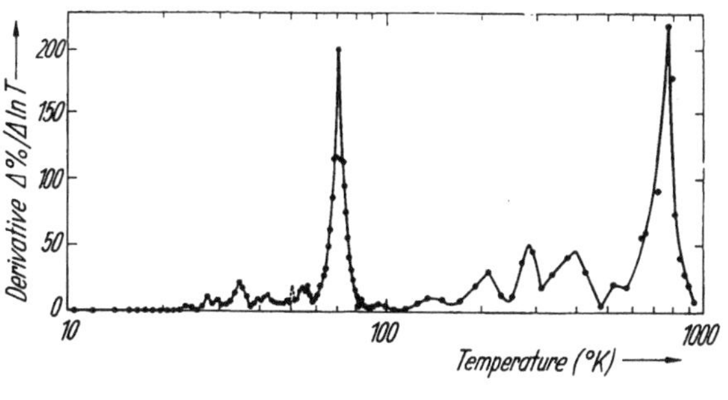

Fig. 2. Derivative versus temperature plot of the irradiation recovery data shown in Fig. 1. Dashed point at 50.5 °K is overnight point

Only the 70° peak is sufficiently large and sufficiently distinct to warrant analysis. In fact, it is recognized that isochronal data, the form of the data presented here, are generally insufficient to produce a detailed analysis of recovery kinetics. However, the rather large number of well-spaced data points available for study in the 70° peak does allow for at least semi-quantitative analysis.

Our initial analysis centered about testing whether the 70° peak should be attributed to the recombination of a particularly close interstitial–vacancy pair or involves more random and extensive diffusion of one of these defects, presumably an interstitial. In the first case, a first-order process would be expected; in the latter case, a process more consistent with a second-order process would be anticipated. Accordingly, the following tests were made.

The chemical rate equation underlying such analysis is taken as

$$\frac{dn}{dt} = -\nu\, n^{\gamma} \exp\left(\frac{-E}{kT}\right), \tag{1}$$

where γ is the order of the reaction. For a first-order process ($\gamma = 1$) in isochronal annealing

$$\ln\left(\frac{n_i}{n_f}\right) = \nu\, \tau \exp\left(\frac{-E}{kT_j}\right), \tag{2}$$

where n_i is the value of the defect concentration at the beginning of the j-th heating pulse at temperature T_j, and n_f is the value at the end of this pulse. (Clearly n_i is also the value at the end of the $(j-1)$-th pulse, as well.) The pulse time is τ. The test of first-order kinetics consists of plotting $\ln\left(\ln\left(n_i/n_f\right)\right)$ vs. $1/T$, and examining for linearity. No such linearity was found.

For a second-order process ($\gamma = 2$) in isochronal annealing

$$n_f^{-1} - n_i^{-1} = \nu\, \tau \exp\left(\frac{-E}{kT_j}\right). \tag{3}$$

Since the actual measurements are those of resistivity, we have further assumed a simple proportionality between defect concentration and the resistivity increase: $\Delta\varrho = \varrho_F\, n$, where ϱ_F is the Frenkel resistivity, the specific resistivity contribution of an interstitial–vacancy couple. Accordingly,

$$\varrho_f^{-1} - \varrho_i^{-1} = \frac{\nu\, \tau}{\varrho_F} \exp\left(\frac{-E}{kT_j}\right) \tag{4}$$

and the test of second-order kinetics is a plot of $\ln\left(\varrho_f^{-1} - \varrho_i^{-1}\right)$ vs. $1/T_j$. Such a plot is presented in Fig. 3.

We regard the second order of Fig. 3 as convincing evidence that the 70° peak is due to a process in which considerable interstitial migration is involved. If it were possible to disregard further factors, we might be further led to claim that the process is indeed a second-order reaction, for the parameters which emerge from this analysis are entirely reasonable, in addition to a good fit.

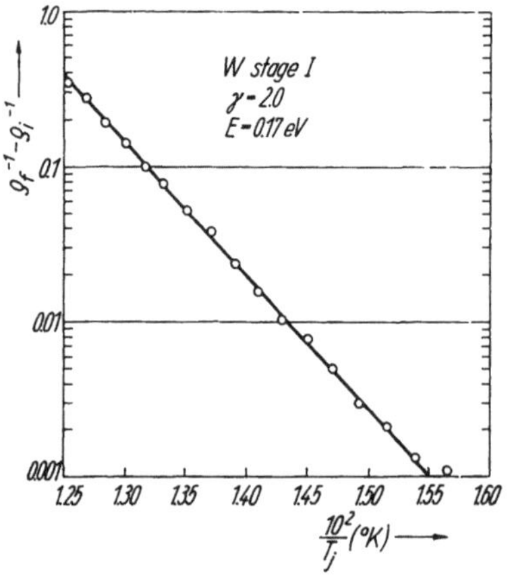

Fig. 3. Determination of activation energy for the 70 °K peak in Stage I

The activation energy deduced is $E = 0.17$ eV; upon somewhat arbitrarily assigning a value of $\varrho_F = 10^{-5}$ Ωcm/at% Frenkel pairs, the frequency factor is $\nu = 2 \times 10^{13}$ s^{-1}.

The purpose of kinetic analysis is to allow one to understand the nature of the underlying processes. Merely to assign the 70° peak to a second-order reaction is of limited value and, in fact, would be made without adequate regard for related considerations. If the 70° peak were a second-order reaction, what atomic process would it represent? The most obvious answer is simple interstitial–vacancy recombination effected through interstitial diffusion. But it is obvious that this model is inadequate: mere interstitial–vacancy recombination should lead to complete recovery. Instead, almost 3/4 of the resistivity increase due to radiation remains beyond the 70° peak. Competitive processes which preserve interstitials (and, of course, vacancies) should be considered, therefore.

We propose that the main features of the 70° peak may be found in an "interstitial-trapping" model. In this model, interstitials, of concentration i, either recombine with vacancies, of concentration v, or are immobilized by traps, of concentration I. The kinetic equations which might describe this model, neglecting considerations of spatial correlation, are

$$\frac{di}{dt} = -K\,i\,v - K\,i\,I\,, \tag{5a}$$

$$\frac{dv}{dt} = -K\,i\,v\,, \tag{5b}$$

$$\frac{d\delta}{dt} = v\,i\,I \tag{5c}$$

with

$$v = i + \delta\,. \tag{5d}$$

In equations (5), i represents the *free* interstitial concentration; the concentration of trapped interstitials is δ. Equations (5a) and (5b) may be combined to give

$$\frac{di}{dv} = 1 + \frac{I}{v} \tag{6}$$

which may be integrated to

$$i = v - I \ln\left(\frac{v_0}{v}\right), \tag{7}$$

where we have taken the initial concentrations of vacancies and interstitials to be v_0. Then (5b) can be written as

$$\frac{dv}{dt} = -\left\{\nu\left[1 - \frac{I}{v}\ln\frac{v_0}{v}\right]\right\} v^2\,. \tag{8}$$

Equation (8) may be interpreted as a second–order reaction with a variable rate constant, indicated in curly brackets. This interpretation is suggestive but should not be over-emphasized since the change in this effective rate is considerable. It is better to recognize that the reaction of (8) starts as second order, then progressively advances toward higher order. Thus it still remains to see why the second-order feature is evidently well preserved in the experiment.

Before continuing, it is worth noting the value of the factors in equation (8). At the end of the reaction, the final vacancy concentration is

$$v_f = I \ln\left(\frac{v_0}{v_f}\right) . \tag{9}$$

Note that the final concentration of vacancies is not zero; this was, of course, built into the model but represents an improvement over the previous, simple second-order description. Assuming that no resistivity change is caused by interstitial trapping, we deduce from experiment that

$$\frac{v_0}{v_f} \approx 1.24 ,$$

so that

$$I \approx 4.5\, v_f \approx 3.6\, v_0 .$$

As noted previously, the resistivity change due to irradiation was 7.6×10^{-9} Ωcm. Taking into account that about 8% of the resistivity recovered below the 70° peak and assuming that the Frenkel resistivity in tungsten is 10 μΩcm/at %, the predicted fractional impurity concentration is about 25×10^{-6}. Recalling that the pre-irradiation value of resistivity at 4.2 °K was 162 μΩcm and arbitrarily taking the average resistivity contribution of impurities in the sample to be 3 μΩcm/at %, the fractional concentration estimated in this manner is 54×10^{-6}. Considering the crudeness of the comparison, the agreement is entirely satisfactory. We are led to believe that the traps are formed by impurities (as well verified in other metals in recovery studies).

In comparing the form of equation (8) with the behavior of the resistivity in the 70° peak, it has been implicitly assumed that no resistivity change is caused by interstitial trapping. There is no difficulty in removing this assumption and we have examined the more general model. However, the resulting form is somewhat complex so that it does not seem profitable to present this added complexity here. The important aspects can be stated in rather physical terms. Since the concentration of trapping centers is a fair amount in excess of the vacancy sink concentration throughout the annealing process of the 70° peak, the interstitial trapping aspects of the annealing are more readily described by a first-order reaction. This serves as an effective "governor" on the tendency espoused in equation (8) for the reaction to progress to orders greater than two. It is this combination of tendencies, weighted by unknown specific resistivity contributions which, we suggest, results in apparent second-order kinetics within the precision of the present data and still preserves most of the defects to more elevated temperatures.

The interstitial-trapping model explored above is, of course, not the only possibility. We have also examined a model in which interstitials self-trap each other. This model seems less likely since, to account for the data reported here, it is necessary to conclude that the ability for an interstitial to self-trap another interstitial is more than one order of magnitude greater than the ability of a vacancy to account for interstitial–vacancy recombination. The greater likelihood of impurity trapping can be justified by noting again the relatively large impurity concentration with respect to point defect concentration. However, there is nothing in the analysis to exclude the possibility that some amount of interstitial self-trapping does occur; rather, it seems to be of less importance than impurity-trapping. It would not be unreasonable to postulate that self-trapping is about equal in importance to interstitial–vancancy recombination.

One more aspect of the 70° peak deserves attention. The peak is characterized by a width which is consistent with a second-order process (as Fig. 3, indeed, implies) and is quite distinct from preceding peaks. This contrasts with the situation in the latter part of Stage I in copper, for example, where a steady transition from recombination of spatially correlated interstitial–vacancy pairs to random recombination is generally observed. In this case (i.e., in copper), the "second-order reaction" is complicated by earlier correlated recovery. The dinstinct character of the recovery observed in tungsten is not understood. Two possibilities occur to us for speculation. First, it is possible that correlation is actually less in tungsten than in copper. A more attractive possibility is that the relatively high impurity concentration is responsible for the present observations, as outlined above in leading to an apparently overall second-order reaction, even in the presence of a fair degree of spatial correlation.

In the thermal neutron experiments of Coltman et al. [7], a rather large peak was observed at 16 °K with a subsidiary one at 29 °K, and a somewhat smaller peak at 62 °K with a subsidiary one at 80 °K; this is in addition to a number of still more minor peaks. They considered, primarily due to its size, that the 16° peak might be due to pair recombination leading to long-range migration at 29° but tended to reject this possibility, based on considerations of the large temperature difference (29 to 16 °K) and the relatively high defect concentrations in their experiment. The present study supports their rejection; neither peak is of any obvious importance following electron irradiation. The situation regarding the 62° and 80° peaks is less clear. Since the defect concentration achieved in the neutron bombardment was about seven times smaller than in the present electron irradiation, it is possible that the 60° peak does represent spatially-correlated pair recovery, and the 82° peak represents random interstitial diffusion. These two peaks would, together, be equivalent to our 70° peak. While this scheme is most attractive, it is also clearly speculative at present.

Recently, DiCarlo et al. [2] have made measurements of internal friction following electron irradiations at $\leqq 20$ °K with 2.5 MeV electrons. They report no evidence for dislocation pinning below 45 °K, although other relaxation effects are observed. Their observations imply that long-range migration occurs above 45 °K, in agreement with our model.

4.2 Stage II

No firm conclusions regarding Stage II can be made based on the present data. It seems likely that release from impurity traps is an important phenomenon and may account for most, if not all, of this stage as it does in other metals. The paucity of data points in any of the apparent peaks in the Stage II region prohibits meaningful speculation concerning the shape of these peaks.

4.3 Stage III

Considerably fewer data are available here in the Stage III region than in the Stage I. Furthermore, some scatter in the data is evident. Nevertheless we have attempted to see if the process can be described by chemical rate theory. Note that recovery *is* essentially completed in Stage III so that we might apply a single chemical rate equation with more confidence in Stage III than in Stage I. The results of such an analysis are presented in Fig. 4, where

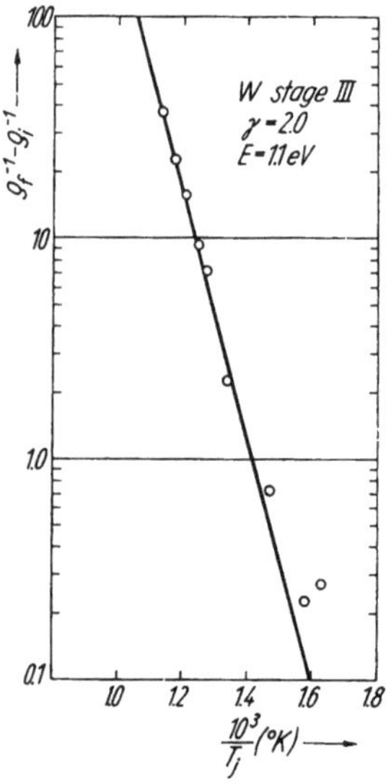

Fig. 4. Determination of activation energy for Stage III

the recovery kinetics were assumed to be second order. Agreement with second order is observed over approximately two decades. The energy derived from the slope in Fig. 4 is 1.1 eV. This is considerably lower than quoted energies of typically 1.7 eV, which have been obtained after neutron irradiation [8, 9] and cold work [10], even though the frequency factor obtained from these data is of the order of $10^{13}\,s^{-1}$. The significance of this discrepancy is not clear at the present time. It is worh mentioning, however, that impurities may play an important role in the Stage III recovery [11].

References

[1] P. G. LUCASSON and R. M. WALKER, Phys. Rev. **127**, 485 (1962).
[2] J. A. DICARLO, C. L. SNEED, JR., and A. N. GOLAND, Bull. Amer. Phys. Soc. **13**, 381 (1968).
[3] H. H. NEELY, North American Aviation SR-MEMO 12404 (1967).
[4] A. SOSIN, Phys. Rev. **126**, 1698 (1962).
[5] W. BAUER and A. SOSIN, J. appl. Phys. **35**, 703 (1964).
[6] D. G. BRANDON, M. WALD, M. J. SOUTHON, and B. RALPH, J. Phys. Soc. Japan **18**, Suppl. II, 324 (1963).
[7] R. R. COLTMAN, C. E. KLABUNDE, and J. K. REDMAN, Phys. Rev. **156**, 715 (1967).
[8] M. W. THOMPSON, Phil. Mag. **5**, 278 (1960).
[9] G. H. KINCHIN and M. W. THOMPSON, J. nuclear Energy **6**, 275 (1958).
[10] H. SCHULTZ, Acta metall. **12**, 649 (1964).
[11] H. H. KUHLMAN and H. SCHULTZ, Acta metall. **14**, 798 (1966).

(Received April 15, 1968)

phys. stat. sol. 28, 683 (1968)

Subject classification: 14.2

Department of Physics, University of Virginia, Charlottesville

Phase-Dependent Superconductive Current Density in a Superconductor – Normal Metal – Superconductor Junction at $T = T_{CN}$[1])

By

R. H. T. YEH and H. MECHETTI[2])

A general equation for the phase-dependent supercurrent density across the weak-link is obtained, from which we find that J_{max} at $T = T_{CN}$ decreases very slowly with d_N when compared with J_{max} at T near T_{CS}.

Une equation générale pour la densité de courrant superconductif que depend de la phase à travers la région N est obtenue. On trouve que J_{max} à $T = T_{CN}$ décroit très lentement avec d_N en comparaison avec J_{max} à T près de T_{CS}.

We consider a SNS sandwich (N layer of thickness $2\,d_N$, and S layer extending to infinity on both sides), and we use subscript S(N) to distinguish the superconducting (normal) component of the sandwich. We assume N is a superconductor with lower transition temperature (T_{CN}) compared with S(T_{CS}). The system is mathematically simple to handle at $T = T_{CN}$ because the linear term in the Ginzburg-Landau equation for the N-region disappears at that temperature. This temperature is also appropriate for investigating the residue phase-transition anomaly in N, e.g., specific heat discontinuity. Recent experiments [1] confirm the remarkable slow-decay behavior of the order parameter Δ in N at this temperature. A related consequence, as we shall demonstrate, is that supercurrent density J_{max} also decays very slowly as a function of d_N.

To compute J_{max} from Ginzburg-Landau's expression, we follow the de Gennes' method [2] who relates $\Delta_S\,|_{\pm d_N}$, $\mathrm{d}\Delta_S/\mathrm{d}x|_{\pm d_N}$ using the even and odd solutions inside the weak-link region. For simplicity, we shall use the Abrikosov's reduced units in terms of the S-layer quantities [3] — i.e., the static small-field penetration depth, $\delta_S(T)$, as the unit of length and $\sqrt{2}$ times the bulk critical field $(\sqrt{2}\,H_{CS}(T))$ as the unit of magnetic field. In order to use the same expression of J for both S- and N-regions, we shall express J in terms of the superconducting electron wave function Ψ. $\Psi_S(\Psi_N)$ is obtained by multiplying $\Delta_S(\Delta_N)$ by a factor $\left[\frac{\chi(\varrho)\,7\,Z\,(3)\,N}{16\,\pi^2\,T_C^2}\right]^{1/2}_{S(N)}$. Notations in the brackets will

---

[1]) Work supported by the Air Force Office of Scientific Research, United States Air Force, under grant contract AFOSR-1138-66. Also supported by the Center for Advanced Studies and by C.N.I.C.T. of Argentina.

[2]) On leave from Universidad Nacional de Tucumán.

be explained below; $Z(3) = 1.202$.

In reduced units Ψ is measured in units of zero-field value of $\Psi_S(T)$. Then, for the one-dimensional case, the expression for J in reduced units is

$$J = -i\,(2\,K_S)^{-1}\left(\Psi^* \frac{d\Psi}{dx} - \Psi \frac{d\Psi^*}{dx}\right), \tag{1}$$

where K_S is the Ginzburg-Landau parameter in the S-region, and the diamagnetic current term $|\Psi|^2 A$, which is generally small, provided T_{CN} is not too small compared with T_{CS}, has been neglected. To evaluate J at the boundary of the weak-link, we have to find an effective boundary condition connecting values of $d\Psi/dx|_\pm$ with $\Psi_\pm$ where the plus and minus signs refers to the right- and left-hand sides of the weak-link region. These effective boundary conditions idealize the weak-link region as a geometrical plane to and is not be confused with the local boundary conditions to be discussed below.

Then the picture of phase locking ($\Psi_+ = |\Psi_+|\, e^{i\Phi_+}$, $\Psi_- = |\Psi_-|\, e^{i\Phi_-}$) immediately gives rise a $\sin(\Phi_+ - \Phi_-)$ factor in J. Since Ginzburg-Landau equations of Ψ relate $d^2\Psi/dx^2$ to lower-order terms, so the most general effective boundary conditions can be written as a linear dependence between $\Psi_\pm$ and $d\Psi/dx|_\pm$. These relations must be satisfied by the even (Ψ_e) and the odd (Ψ_O) solutions in the N-region, from which we can determine the coefficients of the effective boundary conditions, and hence compute J.

To determine Ψ_O and Ψ_e, it is important to use the (local) boundary conditions imposed by the current continuity condition, namely, that Ψ and $d\Psi/dx$ are continuous across the N–S interface. As we have shown elsewhere [3], for T near T_{CS}, the de Gennes' result [2] contains an extra factor ($V_{FS}\, l_S\, N_S/V_{FN}\, l_N\, N_N$) ($V_F$ is the Fermi velocity, l is the mean free path, and N is the density of states). This happens because he uses boundary conditions suitable for proximity effects. These boundary conditions are not consistent with the current-continuity condition. In this connection note that tunneling supercurrent exists quite independently of the proximity effects (e.g., in SIS there exists Josephson current, but no shift in T_C). In particular, each effect requires a different set of boundary condition on Ψ. Using the current-continuity condition we find that

$$J_{max} = \frac{1}{2\,K_S}\,\Psi_e^2\left[\frac{1}{\Psi_O}\frac{d\Psi_O}{dx} - \frac{1}{\Psi_e}\frac{d\Psi_e}{dx}\right]\Bigg|_{d_N}, \tag{2}$$

where Ψ_e and Ψ_O are taken to be real.

As we shall see later, Ψ varies slowly for T close to T_{CN}. So generalized Ginzburg-Landau equations for low temperatures, valid for all free paths, can be used. In reduced units, they are

$$\left.\begin{aligned} &K_S^{-2}\frac{d^2\Psi}{dx^2} + \Psi - \Psi^3 = 0 && \text{in S-region}\,,\\ &\frac{d^2\Psi}{dx^2} - 2\,b_N\,\Psi^3 = 0 \quad (T = T_{CN}) && \text{in N-region}\,,\end{aligned}\right\} \tag{3}$$

where

$$2\,b_N = K_S^2\,\frac{V_{FS}^5}{V_{FN}^5}\left[\frac{\chi(\varrho_S)}{\chi(\varrho_N)}\right]^2, \qquad \chi(\varrho) = \sum_{n=0}^{n=\infty} \frac{0.952}{(2\,n+1)^2\,(2\,n+1+\varrho)},$$

and

$$\varrho = \frac{V_F}{2\pi T_C l}.$$

(The meaning of the symbols has been defined above.)

The even and odd solutions in N, obtained from equation (3), are given respectively by

$$\Psi_e(x) = \frac{\Psi_e}{\cos \Phi_e(x)}, \qquad \Psi_e b_N^{1/2} 2^{1/2} x = \int_0^{\Phi_e} \frac{d\theta}{\left(1 - \frac{1}{2}\sin^2\theta\right)^{1/2}}, \tag{4a}$$

$$\Psi_o(x) = \Psi_o \left(\frac{1 - \cos \Phi_o(x)}{1 + \cos \Phi_o(x)}\right)^{1/2}, \qquad \Psi_o b_N^{1/2} 2 x = \int_0^{\Phi_o} \frac{d\theta}{\left(1 - \frac{1}{2}\sin^2\theta\right)^{1/2}}. \tag{4b}$$

The values of Ψ_e and Ψ_o are determined numerically from

$$K_S\left[\frac{1}{\Psi_e^2} - \frac{1}{\cos^2 \Phi_e(d_N)}\right] = \left[b_N\left(\frac{1}{\cos^4 \Phi_e(d_N)} - 1\right)\right]^{1/2}, \tag{5a}$$

$$K_S\left[\frac{1}{\Psi_o^2} - \frac{1 - \cos \Phi_o(d_N)}{1 + \cos \Phi_o(d_N)}\right] = \left\{b_N\left[\left(\frac{1 - \cos \Phi_o(d_N)}{1 + \cos \Phi_o(d_N)}\right)^2 + 1\right]\right\}^{1/2}, \tag{5b}$$

and the expression for J_{max} is

$$J_{max} = \frac{1}{2K_S}\frac{\sqrt{2}\,\Psi_e^2 b_N^{1/2}}{\cos^2 \Phi_e(d_N)}\left\{\Psi_o\left[\frac{2}{\sin^2 \Phi_o(d_N)} - 1\right]^{1/2} - \Psi_e\left[\frac{1 - \cos^4 \Phi_e(d_N)}{2\cos^2 \Phi_e(d_N)}\right]^{1/2}\right\}. \tag{6}$$

Note that the constant factor 1/2 in front of $\sin^2\theta$ in equation (4) is a simplification which arises from the condition $T = T_{CN}$. From (4) and (5), it is easy to see that, for large d_N, $\Psi_e(d_N)$ decreases monotonically to $(1 + b_N^{1/2}/K_S)^{-1/2}$. This value is quite comparable to $\Psi_e(\infty) = 1$, in contrast with the situation for T near T_{CS} [3], where $\Psi_e/(d_N) \ll 1$. This justifies ou ruse of the generalized Ginzburg-Landau equation. Also, we can see in the expression for J_{max} that the odd (even) term decreases (increases) monotonically to the value $(2 + 2\, b_N^{1/2}/K_S)^{-1/2}$, so $J_{max} \to 0$ for large d_N. For very small d_N, however, J_{max} is limited by the critical current density of the bulk (S) material.

In order to get an idea of how J_{max} varies with d_N we tabulate J_{max} for various d_N, where a typical experimental value of $K_S = 2\, b_N = 1$ is used. It is clear that J_{max} for $T = T_{CN}$ decays much more slowly than J_{max} for T near T_{CS} [3], where $J_{max} \approx 2/\pi \exp(-d_N \pi/2)$. (Here we have used the dirty-limit approximation, and $l_S = l_N$, $V_{FS} = V_{FN}$. This is consistent with $2\, b_N = 1$ above.)

$d_N/\delta_S(T_{CN})$	2	3	4	5	10
$J/J\ (d_N = 2)$ $T = T_{CN}$	1	3.67×10^{-1}	1.43×10^{-1}	9.7×10^{-2}	5.68×10^{-3}
$J/J\ (d_N = 2)$ $T \approx T_{CS}$	1	2.08×10^{-1}	0.432×10^{-1}	0.9×10^{-2}	0.0175×10^{-3}

References

[1] Orsay Group on Superconductivity, Field Dependence of the Superconducting Properties Induces by Proximity Effects: Breakdown Fields (to be published).
[2] P. G. DE GENNES, Rev. mod. Phys. **36**, 225 (1964).
[3] R. H. T. YEH and H. MECHETTI, phys. stat. sol. **25**, K65 (1968).

(Received April 30, 1968)

phys. stat. sol. 28, 687 (1968)

Subject classification: 14.4.1; 13; 13.4; 22.1.1

Institute of Radioengineering and Electronics, Academy of Sciences of the USSR, Moscow

On the Problem of Recombination Autooscillations

By

V. L. Bonch-Bruevich

The conditions are studied under which autooscillations of the concentration of free carriers may take place. A unipolar semiconductor sample is considered containing besides the standard doping impurity some traps of one or of two types. The criteria of the soft self-excitation of the oscillations in question are stated explicitly, together with the conditions of the fluctuational stability of the system. The decisive role of extrinsic illumination is pointed out.

Die Bedingungen werden untersucht, unter denen Oszillationen der Konzentration der freien Ladungsträger auftreten. Es wird eine unipolare Halbleiterprobe betrachtet, die neben der Standartdotierung, Haftstellen eines Typs oder zweier Typen enthält. Die Kriterien für eine weiche Selbsterregung der in Frage stehenden Schwingungen werden explizit angegeben, zusammen mit den Bedingungen für die Stabilität des Systems gegen Fluktuationen. Die entscheidende Rolle der äußeren Belichtung wird herausgestellt.

1. Introduction

If the electron gas is not heated the standard equations describing the recombination kinetics have only damped solutions. It has been shown previously [1] that the situation might be changed by heating the electron gas: if the system is far enough from equilibrium (with respect to the energy) oscillatory (and, in particular, autooscillating) solutions may occur. This phenomenon was called "recombination instability". However, in the particular case of [1] the relevant criteria included the condition that the differential conductivity σ is negative. In principle, spatially homogeneous systems of such a type may exist [2]; but the presence of initial inhomogeneities [3] makes it rather difficult to obtain them: the system may become unstable with respect to the formation of static domains. It was noted in [4] that, if there are two types of charge carriers, the total differential conductivity may still remain positive (thus providing the stability of the homogeneous system), even if one of the "partial" conductivities is negative. However, the unipolar systems might as well be of some experimental interest. It is shown in this paper that in this case one may find autooscillations of the free carrier concentration ("recombination autooscillations") at $\sigma > 0$. To this end one has, besides heating the electron gas, to choose properly the illumination and the doping conditions, that is the intensity and the frequency range of the incident light and the nature of the impurity.[1]) The idea is to regulate the concentration of free carriers (say, electrons) both via field-dependent capture cross-sections and via optical generation from the other impurity.

[1]) It goes without mentioning that the word "impurity" may mean in fact any kind of imperfection producing discrete levels in the forbidden gap.

2. Basic Equations

We consider a macroscopically homogeneous n-type sample containing, besides the standard shallow donors, impurity atoms[2]) of two other types: negatively charged recombination centres (i.e. gold or copper in germanium) and the second impurity to be specified later on (the relevant conditions turn out to be not very hard). In germanium the role of the second impurity might be played, say, by oxygen[3]); a special case is possible, too, when the second impurity, as such is not needed at all, its function being performed by the donors themselves. Let the concentrations of the donors, free electrons, first- and second-type impurities be N_d^*, n, N_1, and N_2 respectively; N_d^* is to be understood as an effective concentration taking account of eventual compensation [1]. The total impurity concentration is $N = N_1 + N_2$. The unheated electron capture coefficients of first- and second-type impurities are c_1 and c_2; in the presence of a constant external electric field E they are changed by the factors $f_1(E)$ and $f_2(E)$. Finally, we denote by n_1 and n_2 the standard expressions describing thermal and optical generation from the first- and second-type impurities. When assuming the recombination to be governed by the electron capture, we have the standard kinetic equations

$$\frac{\mathrm{d}N_i^-}{\mathrm{d}t} = c_i \{n\, f_i(E)\, (N_i - N_i^-) - n_i\, N_i^-\}\,; \qquad i = 1, 2\,. \tag{1}$$

Here N_1^- and N_2^- are the first- and second-type impurity concentrations in the "after the capture" state, respectively. The local electric neutrality condition and the Kirchhoff law are of the form

$$n + N_1^- + N_2^- = N_\mathrm{d}^*\,, \tag{2}$$

$$E = \mathscr{E}\,[L + e\,\mu\, R\, S\, n]^{-1}\,. \tag{3}$$

Here $\mathscr{E}$ is the (constant) e.m.f. of the voltage source, L and S are the sample length and cross-section, μ is the mobility of the charge carriers in the field E, R is the load resistance.

In the particular case with the donors as second impurity we have $N_\mathrm{d} = N_2$, $N = N_1 + N_\mathrm{d}$ and equation (2) takes the form

$$n + N_1 = N_\mathrm{d}^* - N_\mathrm{d}^{*-}\,. \tag{2'}$$

It is convenient to introduce dimensionless quantities measuring the concentrations in terms of N:

$$N_i^- = N\,x_i\,, \quad n = N\,x\,, \quad N_i = N\,y_i\,, \quad n_i = N\,\nu_i\,, \quad c_i\,N = s_i\,, \quad N_\mathrm{d}^* = N\,y\,. \tag{4}$$

Taking into account equation (2) we have by definition

$$y_1 + y_2 = 1\,, \qquad x_i \leqq y_i\,.$$

It will be assumed in the following that $N_1 + N_2 > N_\mathrm{d}^*$, that is

$$y < 1\,. \tag{5}$$

---

[2]) See preceding footnote.

[3]) This remark is of a hypothetical nature since the recombination properties of oxygen in germanium have not yet been studied in detail.

Expressed in the new variables, equations (4), the equations (1) and (2) take the form

$$\frac{dx_i}{dt} = s_i \{x f_i (y_i - x_i) - \nu_i x_i\} , \tag{6}$$

$$x_1 + x_2 + x = y . \tag{7}$$

The problem is now to carry out a qualitative study of this dynamical system.

3. Singular Point

Evidently the singular points of equations (6) correspond to the recombination–generation (r.–g.) equilibrium in the field E. Labelling the equilibrium quantities by a subscript "0" one obtains, using equations (6) and (7)

$$x_{i,0} = \frac{y_i x_0}{x_0 + \nu_i'} ; \quad \nu_i' = \nu_i f_i^{-1}(E_0) , \tag{8}$$

$$x_0^3 + (1 - y + \nu_1' + \nu_2') x_0^2 + [y_1 \nu_2' + y_2 \nu_1' - y (\nu_1' + \nu_2') + \nu_1' \nu_2'] x_0 - \nu_1' \nu_2' y = 0 . \tag{9}$$

The "physical region" corresponds to real and positive values of x_0. In view of equation (5) it is easily seen that the field strength E_0 being given by equation (9) has one and only one solution belonging to this region. However, this does not necessarily mean that there is just one singular point: equation (3) is generally a transcendental equation for E_0. In the following we shall explicitly treat the case of one singular point (it will be seen that the singular point in question must be either a node or a focus).

The possibilities afforded by equation (9) are numerous and the general formula for $x_0(E_0)$ is rather cumbersome. Therefore, we consider just two particular cases where the problem allows considerable simplification.

3.1 Donors and two types of traps — Small but very different generation rates

Let

$$\nu_2' \ll \nu_1' \ll y_i, y; \quad i = 1, 2 \tag{10a}$$

and let

$$y_2 > y \tag{11}$$

(the concentration of the second impurity exceeds that of the donors).

Then one obtains easily

$$x_0 \approx \frac{\nu_2' y}{y_2 - y} . \tag{12a}$$

To satisfy the inequalities (10a) one has to work at not too high temperatures and under the conditions of an optical generation rate from the first-type impurities being strong enough.[4])

[4]) Really, as is well known,

$$n_i = \frac{n^{(0)} (N_i - N_i^{-,(0)})}{N_i^{-,(0)}} + \frac{S_i I_i}{c_i} ; \quad i = 1, 2 ,$$

where the upper index "0" corresponds to thermodynamic equilibrium, S_i is the photoionization cross-section of the i-th impurity, and I_i is the photon density flux of the relevant frequency. If the first- and second-impurity levels are not too close in energy the ratio ν_1'/ν_2' may be varied by varying the fluxes I_1 and I_2.

The differential conductivity is

$$\sigma = e\,N\,\mu_0\,x_0\left[1 + \frac{\mathrm{d}\ln\mu_0\,x_0}{\mathrm{d}\ln E_0}\right] \tag{13}$$

with $\mu_0 \equiv \mu(E_0)$ being the mobility in the field E_0. According to equations (8), (12), and (13) the condition $\sigma > 0$ takes the form

$$1 + \frac{\mathrm{d}\ln(\mu_0\,f_2^{-1})}{\mathrm{d}\ln E_0} > 0\,. \tag{14a}$$

This is certainly the case if, say, the mobility stays constant while $\mathrm{d}f_2/\mathrm{d}E \leqq 0$.

3.2 Donors and traps of one kind — Low temperatures and strong extrinsic illumination

Now let the donors play the role of the second impurity. We consider the low temperature case and let the extrinsic illumination be such that the generation rate from first-type impurities exceeds that from the donors:

$$\nu_2' \ll \nu_1' \ll 1\,, \quad y_1, y_2\,. \tag{10b}$$

Then one obtains from equation (9) neglecting higher powers of the small parameter ν_2'/ν_1',

$$x_0 \approx \left(\frac{y_2\,\nu_1'\,\nu_2'}{1 - y_2}\right)^{1/2}. \tag{12b}$$

The denominator of the right-hand side of equation (12b) is assumed to be of normal order of magnitude (no improbable accidental compensation).

The condition for the differential conductivity to be positive now takes the form

$$1 + \frac{\mathrm{d}\ln[\mu_0(f_1\,f_2)^{-1/2}]}{\mathrm{d}\ln E_0} > 0\,. \tag{14b}$$

4. Phase Plane

Fig. 1 shows the phase plane of equation (6) in the coordinates x_1 and x_2. According to equation (7) the allowed values of these are restricted by the condition

$$x_1 + x_2 = y\,. \tag{7'}$$

Thus the "physical region" lies inside the triangle made of the line given by equation (7′) and the relevant segments of the coordinate axes. Arrows show the directions of the integral curves at the boundaries. It is seen imme-

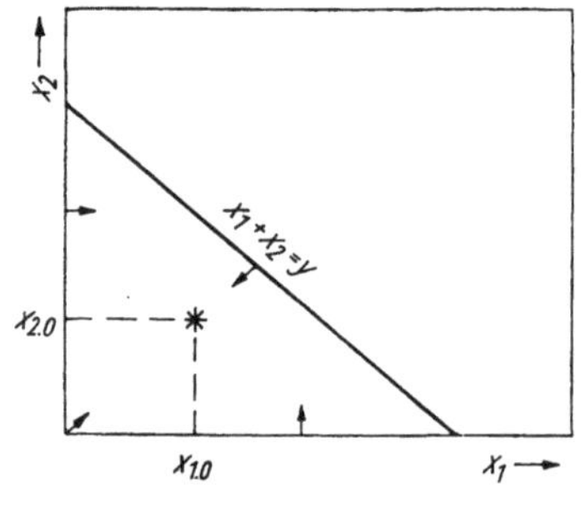

Fig. 1. The phase plane

diately that if the singular point is an unstable node or an unstable focus there certainly is a stable limiting cycle. Thus the self-excitation of the recombination autooscillations must take place in such cases. On the other hand, if the node or the focus in question is stable, the autooscillations either do not appear at all or may be excited under hard regime only. Whichever of these possibilities is realized depends upon the particular form of the functions f_i. The thermodynamic equilibrium case ($f_1 = f_2 = 1$) corresponds to the stable node (with no limiting cycles, of course). However, the case of a stable focus is of interest even if there is no stable limiting cycle. Indeed under such conditions the recombination kinetics would be of the oscillatory type: the approach to r.–g. equilibrium would proceed non-monotonically, leading to the damped-current (or -voltage) oscillations in the circuit.

To study the nature of the singular point one has to linearize equations (6) near it or, which amounts to the same thing, near the point (x_0, E_0). Let

$$\left.\begin{aligned} b_{11} &= s_1 \left\{ y_1 \frac{\nu_1'}{x_0 + \nu_1'} (1 - \varphi_1) + x_0 + \nu_1' \right\} , \\ b_{12} &= s_1 \frac{x_{20}}{x_{10}} y_1 \frac{\nu_1'}{x_0 + \nu_1'} (1 - \varphi_1) , \\ b_{21} &= s_2 \frac{x_{10}}{x_{20}} y_2 \frac{\nu_2'}{x_0 + \nu_2'} (1 - \varphi_2) , \\ b_{22} &= s_2 \left\{ y_2 \frac{\nu_2'}{x_0 + \nu_2'} (1 - \varphi_2) + x_0 + \nu_2' \right\} , \end{aligned}\right\} \tag{15}$$

where

$$\varphi_i = \frac{R}{R + R_0} \frac{\mathrm{d} \ln f_i}{\mathrm{d} \ln E_0} ; \qquad i = 1, 2 \tag{16}$$

and $R_0 = L/e\, \mu_0\, n_0\, S$ is the sample resistance under r.–g. equilibrium conditions.

Then the standard calculation shows that the singular point in question is

a) *a focus* if

$$(b_{11} - b_{22})^2 + 4\, b_{12}\, b_{21} < 0 ; \tag{17}$$

b) *a node* if the strict inequality (17) is not true and, moreover,

$$b_{12}\, b_{21} < b_{11}\, b_{22} . \tag{18}$$

The singular point is unstable if

$$b_{11} + b_{22} < 0 . \tag{19}$$

It is obvious from equations (15) that the last condition may be fulfilled only if one of the quantities $1 - \varphi_i$ ($i = 1, 2$) is negative. To be specific we set

$$\varphi_1 > 1 > \varphi_2 . \tag{20}$$

Note that, according to equation (16), the first of these inequalities is certainly not valid under voltage-controlled conditions ($R \ll R_0$). This was, of course, quite evident from the very beginning.

In the general case equations (15), (17) to (19) are rather cumbersome. The only thing that is quite clear is that the focus may only occur if the generation rate is high enough but not too high: at least, one of the quantities ν_1' and ν_2' must be comparable to y_1 or y_2. It seems, however, that the contrary case of ν_1' and ν_2' being small compared to y_1 and y_2 may be of more immediate experimental interest. In view of the particular models treated in the preceding section we assume that $x_0 \ll y_1, y_2$ and $\nu_1' \gg \nu_2'$. The first condition corresponds, for example, to a compensated sample at not too high temperatures; the second one may be obtained by varying the illumination spectrum. Neglecting small terms of the order, say, ν_i'/y_i, etc., the conditions (18) and (19) may be reduced to the form

$$\frac{s_2\, y_2}{s_1\, y_1} \frac{\nu_2'}{x_0 + \nu_2'} < \frac{\varphi_1 - 1}{1 - \varphi_2} < \frac{y_2}{y_1} \frac{\nu_1'\, \nu_2'}{(x_0 + \nu_2')^2}\,. \tag{21}$$

These inequalities are compatible if

$$\frac{s_2}{s_1} < \frac{\nu_1'}{x_2 + \nu_2'}\,. \tag{22}$$

Relations (21) and (22) define the field strength and the illumination intensity region where the soft self-excitation of the recombination autooscillations takes place. It is seen that the cross-section ratio s_2/s_1 must be small enough, while the concentration of the second-type impurity (x_0 being fixed) must be high enough.

The relations in question are obtained explicitly by substituting some particular values of the equilibrium electron concentration x_0. Using equations (12a, b) one obtains, instead of (21) and (22),

Case a):

$$\frac{s_2}{s_1} \frac{y_2 - y_1}{y_1} < \frac{\varphi_1 - 1}{1 - \varphi_2} < \frac{\nu_1'\, (y_2 - y)^2}{\nu_2'\, y_1\, y_2}\,; \tag{23a}$$

$$\frac{s_2}{s_1} < \frac{\nu_1'\, (y_2 - y)}{\nu_2'\, y_2}\,; \tag{24a}$$

Case b):

$$\frac{s_2}{s_1} \left[\frac{\nu_2'\, y_2\, (1 - y_2)}{\nu_1'\, y_1^2}\right]^{1/2} < \frac{\varphi_1 - 1}{1 - \varphi_2} < \frac{1 - y_2}{y_1}\,, \tag{23b}$$

$$\frac{s_2}{s_1} < \left[\frac{\nu_1'\, (1 - y_2)}{\nu_2'\, y_2}\right]^{1/2}. \tag{24b}$$

Note that equations (23a, b) and (24a, b) do not allow to pass to the limit $y_2 \to 0$ (when there are certainly no oscillations). The point is that the very equations (12a, b) were obtained assuming $y_2 \gg \nu_1', \nu_2'$. The condition (24b) may turn out to be rather a hard one since the cross-section ratio s_2/s_1 is in fact very great in this case. On the other hand, the condition (24a) seems rather easily to be satisfied.

Relations (23a, b) must be compatible with the inequalities (14a, b) providing the positive sign of the differential conductivity.

Neglecting for simplicity the field dependence of the mobility we obtain in line with expressions (14a) and (14b)

$$\varphi_2 < 1 + \frac{R_0}{R}\,, \tag{25a}$$

$$\varphi_2 < \frac{2\,R}{R + R_0} - \varphi_1\,. \tag{25b}$$

These conditions seem to be easily compatible with (23a, b).

5. Concluding Remarks

The physical origin of the autooscillations described above seems to be rather simple. In fact it was qualitatively understood already in [5]. If the fluctuation reduces the free electron concentration below its equilibrium value, then at the same time the concentration of the carriers trapped at the first-type impurities increases. In absence of second-type traps this would lead either to the standard relaxation process or to an instability caused by the capture cross-section field dependence with the field strength being increased during the fluctuation in question. However, the second-type impurity will be as well brought out of equilibrium with the conduction band. Thus the free carrier concentration will start to be re-established by the generation flux from the second impurity. An increase of the free carrier concentration will cause a drop of the field strength in the sample leading to a release of a part of the electrons trapped by the first impurity. The result will be a recombination flux from the band to the second-type impurity, and so on. The process may be an oscillating one if some relations are fulfilled between the parameters of the traps.

The period of the recombination autooscillations (coinciding with that of the electric oscillations in the circuit) is expected to be of the order of the largest of the capture times occuring in equations (6).

Noteworthy is the decisive part of the illumination conditions in establishing the stationary autooscillations. In fact if the differential conductivity is positive and the system is spatially homogeneous, the illumination seems to be the only cause leading to an essentially non-equilibrium situation.

Since the conditions (23a) to (25a) are rather weak, the autooscillations arising in presence of the second-type impurity should not be something exotic, the more so that this impurity may not represent an effective recombination centre. However, it is just this recombination inactivity of the second impurity that may hinder the reliable experimental test of the theory. The properties of the second impurity should in fact be such that it might be rather difficult to study them by conventional methods. In this respect the instability phenomena themselves might provide an interesting tool of studying the impurity.

References

[1] V. L. Bonch-Bruevich and S. G. Kalashnikov, Fiz. tverd. Tela 7, 750 (1965).
[2] V. L. Bonch-Bruevich and Sh. M. Kogan, Fiz. tverd. Tela 7, 23 (1965).
[3] V. L. Bonch-Bruevich, Fiz. tverd. Tela 8, 356 (1966).
[4] M. I. Iglitsyn, E. G. Pel, L. Ya. Pervova, and V. I. Fistul, Fiz. tverd. Tela 8, 3606 (1966).
[5] S. G. Kalashnikov and I. A. Kurova, Fiz. tverd. Tela 5, 3224 (1963).

(Received May 30, 1968)

phys. stat. sol. 28, 695 (1968)

Subject classification: 14.1; 21

Department of Physics, University of Roorkee

On Transverse Electrical Conductivity of Metals

By

R. KISHORE

With the help of the Kubo formula an expression for the transverse electrical conductivity of metals is obtained within the Hartree-Fock approximation using the method of the double-time temperature-dependent Green's function. It is observed that the expression is the same as that obtained by Leribaux with the help of perturbation theory using the Feynman diagram technique if the change in the energy of the electrons due to electron–phonon interaction is neglected.

Mit Hilfe der Formel von Kubo wird in der Hartree-Fock-Näherung ein Ausdruck für die transversale elektrische Leitfähigkeit von Metallen erhalten, wobei die Methode der zweizeitigen, temperaturabhängigen Greenschen Funktion benutzt wird. Es wird gefunden, daß der Ausdruck mit dem von Leribaux mit Hilfe der Störungstheorie unter Benutzung der Feynmanschen Diagrammtechnik erhaltenen übereinstimmt, wenn die Energieänderung der Elektronen durch Elektron–Phonon-Wechselwirkung vernachlässigt wird.

The problem of electrical conductivity is usually treated by two quantum mechanical approaches. These are Boltzmann transport equation [1] and Kubo [2] approaches. Instead of using the Boltzmannn equation or solving in a particular representation the density matrix equation of motion, we shall start from an exact expression (within the linear approximation in the electric field) for the conductivity (the so-called Kubo formula). The advantage of this formalism is its independence of the particular representation. The explicit evaluation of the conductivity from the Kubo formula has been done by many authors [3 to 7] using the thermodynamics Green's function technique. In the conductivity problem however, efforts have been directed mostly towards obtaining the diagonal elements of the doncuctivity tensor. In some cases such as in the evaluation of the Hall coefficient we need transverse (or off diagonal) components of the tensor [7].

In this paper we present the calculation of the transverse electrical conductivity of metals from the Kubo formula within the Hartree-Fock approximation using the method of double-time temperature dependent Green's function. Here the electron transport is assumed to be limited by electron–phonon interaction. It is found that the final expression for the conductivity is the same as that obtained by Leribaux [7] with the help of perturbation theory using diagram technique if the change in energy of the electrons due to electron–phonon interaction is neglected.

The Kubo formula for the electrical conductivity is

$$\sigma_{\mu\nu} = \lim_{\varepsilon \to 0} V \int_0^\infty e^{-\varepsilon t} \, dt \int_0^\beta d\lambda \, \langle J_\nu \, J_\mu \, (t + i\,\lambda) \rangle \,, \tag{1}$$

where $\beta = (k_B T)^{-1}$; V the volume of the system, J_ν is the νth cartesian component of electron current density operator and angular brackets denote the grand canonical ensemble average, namely for any operator

$$\langle O \rangle = \frac{\mathrm{Tr}\, e^{-\beta(H - \mu N)}\, O}{\mathrm{Tr}\, e^{-\beta(H - \mu N)}}\,, \tag{2}$$

where H is the Hamiltonian of the system, N is the number of particles in the system and μ is the chemical potential. Here and in the following we take $\hbar = 1$.

In the second quantized representation, the Hamiltonian of the system of electron interacting with the system of phonon is [8]

$$H = \sum_{\boldsymbol{q}} \omega_{\boldsymbol{q}}\, b_{\boldsymbol{q}}^{+}\, b_{\boldsymbol{q}} + \sum_{l} E_l\, a_l^{+}\, a_l + \sum_{l_1 l_2 \boldsymbol{q}} A_{\boldsymbol{q}}\, a_{l_1}^{+}\, a_{l_2}\, (b_{\boldsymbol{q}} + b_{\boldsymbol{q}}^{+})\,,\; (\boldsymbol{k}_1 - \boldsymbol{k}_2 = \boldsymbol{q})\,, \tag{3}$$

where a_l^{+}, a_l are respectively the creation and annihilation operators of an electron in state l; $b_{\boldsymbol{q}}^{+}$, $b_{\boldsymbol{q}}$ are the same for a phonon with wave-vector $\boldsymbol{q}$, $\omega_{\boldsymbol{q}}$ is the energy of the phonon with wave vector $\boldsymbol{q}$; $l \equiv (n, \boldsymbol{k})$, where n stands both for the band level and the spin quantum number and $\boldsymbol{k}$ is the wave vector of the electron; $E_l = \varepsilon_l - \mu$, where ε_l are the eigenvalues of any suitable chosen one-electron Hamiltonian, $A_{\boldsymbol{q}}$ describes the coupling of the electrons to the phonon field. The second quantized form of current density operator is

$$J_\mu = -\frac{e}{V} \sum_{l l'} (v_\mu)_{l l'}\, a_l^{+}\, a_l\,. \tag{4}$$

Here e is the charge of the electron and v_μ is the μth cartesian component of electron velocity. The diagonal matrix elements are

$$(v_\mu)_l = \frac{\partial \varepsilon_l}{\partial k_\mu}\,. \tag{5}$$

The off diagonal matrix elements may be written in the form [9]

$$(v_\mu)_{l l'} = i\, \omega_{l l'} (\varkappa_\mu)_{l l'} = -\,\omega_{l l'}\, J_\mu^{n n'}(\boldsymbol{k})\, \delta(\boldsymbol{k}, \boldsymbol{k}')\,, \quad l \neq l' \tag{6}$$

with

$$\omega_{l l'} = \varepsilon_l - \varepsilon_{l'}$$

and

$$\tau_\mu^{n n'}(\boldsymbol{k}) = \int_V w_{n k}(\boldsymbol{r})\, \frac{\partial}{\partial R_\mu}\, w_{n' k}(\boldsymbol{r})\, \mathrm{d}^3 r\,, \tag{7}$$

where $w_{n k}$ are the periodic parts of the Bloch eigenfunction of one electron Hamiltonian.

When we combine equation (4) and (1), the expression for the electrical conductivity reduces to

$$\sigma_{\mu\nu} = \lim_{\varepsilon \to 0} \frac{e^2}{V} \int_0^\infty e^{-\varepsilon t}\, \mathrm{d}t \int_0^\beta \mathrm{d}\lambda \sum_{l_1 l_1'} \sum_{l_2 l_2'} (v_\nu)_{l_1 l_1'} (v_\mu)_{l_2 l_2'} \times$$
$$\times \langle a_{l_1}^{+} a_{l_1'}\, a_{l_2}^{+} (t + i\lambda)\, a_{l_2'} (t + i\lambda) \rangle \tag{8}$$

within the Hartree-Fock approximation and neglecting the correlation between creation and annihilation operator, the conductivity is given by [10]

$$\sigma_{\mu\nu} = -\frac{e^2}{V} \lim_{\varepsilon\to 0} \int_0^\infty e^{-\varepsilon t}\, dt \int_0^\beta d\lambda \sum_{l_1 l_1'} \sum_{l_2 l_2'} (v_\nu)_{l_1 l_1'} (v_\mu)_{l_2 l_2'} \times$$
$$\times \int_{-\infty}^{\infty}\int_{-\infty}^{\infty} d\omega_1\, d\omega_2 \frac{e^{\beta\omega_2}}{(e^{\beta\omega_1}+1)(e^{\beta\omega_2}+1)} e^{-i(\omega_1-\omega_2)(t+i\lambda)} \times$$
$$\times \{G_{l_2' l_1}(\omega_1+i\varepsilon) - G_{l_2' l_1}(\omega_1 - i\varepsilon)\}\{G_{l_1' l_2}(\omega_2+i\varepsilon) - G_{l_1' l_2}(\omega_2-i\varepsilon)\}, \tag{9}$$

where $G_{l_1 l_2}(\omega)$ is the Fourier transform of the Green's function

$$G_{l_1 l_2}(t, t') = -i\,\theta(t-t')\,\langle\{a_{l_1}(t), a_{l_2}^+(t')\}\rangle, \tag{10}$$

where $\theta(t)$ is the Heaviside unit step function and curly brackets denote the anticommutator. In the present case when the electron in a crystal is described by the Hamiltonian (3), the expression for the Fourier transform of the Green's function is [8]

$$G_{l_1 l_2}(\omega) = \frac{1}{2\pi} \frac{\delta_{l_1 l_2}}{\omega - E_{l_1} - M_{l_1}(\omega)}, \tag{11}$$

where $M_{l_1}(\omega)$ gives the effect of electron–phonon interaction on the single particle energy.

If we substitute $G(\omega)$ from (11) in (9) and perform the integration over t and λ, we get

$$\sigma_{\mu\nu} = -\frac{i e^2}{\pi^2 V}\left\{\sum_l (v_\nu)_{ll}(v_\mu)_{ll} - \sum_{\substack{l_1 l_2\\ l_1 \neq l_2}} (v_\nu)_{l_1 l_2}(v_\mu)_{l_2 l_1}\right\} \times$$
$$\times \int_{-\infty}^{\infty}\int_{-\infty}^{\infty} d\omega_1\, d\omega_2 \frac{e^{\beta\omega_1} - e^{\beta\omega_2}}{(\omega_1-\omega_2)^2} \frac{1}{(e^{\beta\omega_1}+1)(e^{\beta\omega_2}+1)} \times$$
$$\times \frac{\Gamma_{l_1}(\omega_1)\,\Gamma_{l_2}(\omega_2)}{\{(\omega_1 - E_{l_1}')^2 + \Gamma_{l_1}^2(\omega_1)\}\{(\omega_2 - E_{l_2}')^2 + \Gamma_{l_2}^2(\omega_2)\}}, \tag{12}$$

where $E_l' = E_l + \Delta_l(E_l')$; Δ_l and Γ_l are the real and imaginary parts of the $M_l(\omega)$, it can be shown that for $\nu \neq \mu$, the first term on the right hand side of equation (12) vanishes. After substituting the values of the matrix elements of the components of velocity from equation (6) in equation (12); integrating over ω_1 and ω_2 and assuming $\Gamma_l(\omega)$ a small quantity we get

$$\sigma_{\mu\nu} = -\frac{i e^2}{V} \sum_{l_1 l_2} \frac{\omega_{l_1 l_2}\,\omega_{l_2 l_1}}{(E_{l_1}' - E_{l_2}')^2} \delta(\boldsymbol{k}_1, \boldsymbol{k}_2) \times$$
$$\times [\tau_\nu^{n_2 n_1}(\boldsymbol{k}_1)\,\tau_\mu^{n_1 n_2}(\boldsymbol{k}_1) - \tau_\nu^{n_1 n_2}(\boldsymbol{k}_1)\,\tau_\mu^{n_2 n_1}(\boldsymbol{k}_1)]\, f_{l_1}(1 - f_{l_2}), \tag{13}$$

where

$$f_l = \frac{1}{e^{\beta E_l'} + 1}.$$

The expression (13) gives the transverse electrical conductivity under electron–phonon interaction. If we neglect the effect of electron–phonon interaction i.e. if we take $E_l' \approx E_l = \varepsilon_l - \mu$ we get the same expression as obtained by

Leribaux [7] in Zeroth order in electron–phonon interaction. Leribaux used the perturbation theory and diagram technique. We feel this derivation more satisfactory because it considers the contribution of electron–phonon interaction on the transverse electrical conductivity in a more complete manner. Leribaux study becomes a special case of our investigations.

Acknowledgements

I am grateful to Professor S. K. Joshi for his encouragements and the University Grants Commission, India for financial help.

References

[1] A. SOMMERFELD and H. BETHE, Hdb. Phys., Vol. 24, part 2, Edwards Bros., Ann Arbor **1943** (p. 499).
[2] R. KUBO, J. Phys. Soc. Japan **12**, 570 (1957).
[3] A. A. ABRIKOSOV and L. P. GORKOV, Zh. eksper. teor. Fiz. **35**, 1558 (1958); **36**, 319 (1959) (Engl. transl.: Soviet Phys. — J. exper. theor. Phys. **8**, 1090 (1959); **9**, 220 (1959)).
[4] S. F. EDWARDS, Phil. Mag. **3**, 1020 (1958); **4**, 1171 (1959).
[5] J. S. LANGER, Phys. Rev. **120**, 714 (1960); **124**, 1003 (1961); **127**, 5 (1962).
[6] V. AMBEGAOKAR, Astro-Physics and the Many-Body Problem (Brandies Lectures), W. A. Benjamin, Inc., New York 1962.
[7] H. R. LERIBAUX, Phys. Rev. **150**, 384 (1966).
[8] D. N. ZUBAREV, Soviet Phys.-Uspekhi **3**, 320 (1960).
[9] J. M. LUTTINGER, Phys. Rev. **112**, 739 (1958).
[10] R. KISHORE, phys. stat. sol. **26**, 133 (1968).

(Received May 29, 1968)

phys. stat. sol. 28, 699 (1968)

Subject classification: 13.3; 14.4.1; 22

Institute of Physics of the Rumanian Academy, Bucharest

Stability of Surface-Barriers in Semiconductors

By

A. B. FAZAKAS and P. H. HANDEL[1])

In order to test a recently proposed recurrence method for studying the stability of the semiconductor surface-barrier, we apply it to the currentless state up to second order. We obtain the expected stability at arbitrary values of the surface potential. Furthermore, we investigate the stability in the presence of an applied electrical field at sufficiently small values of the surface potential and restrict ourselves to the first order calculation. The obtained critical field exceeds the breakdown value. Thus, at least for small values of the surface potential ($|V_s| \leqq 0.75\ kT$) stability prevails.

Um eine kürzlich vorgeschlagene zyklische Methode zur Untersuchung der Stabilität von Halbleiteroberflächenbarrieren zu prüfen, wird sie auf den stromlosen Zustand bis zur zweiten Näherung angewandt. Die erwartete Stabilität wird bei beliebigen Werten des Oberflächenpotentials erhalten. Darüber hinaus wird die Stabilität unter angelegtem elektrischen Feld bei genügend kleinen Werten des Oberflächenpotentials untersucht, wobei die Rechnung auf die erste Näherung beschränkt wird. Die errechnete kritische Feldstärke übersteigt den Durchschlagswert. Für kleine Werte des Oberflächenpotentials ($|V_s| \leqq \leqq 0{,}75\ kT$) herrscht also praktisch Stabilität.

1. Introduction

It is usually assumed that the space-charge region below the surface of a semiconductor is stable, independently of the value of the surface potential. There are no doubts about the static stability of the surface-barrier if no electric field is applied. But when a steady current is flowing through the barrier, for example at the contact of two semiconductors, it has been recently shown [1] that some barrier instabilities occur even at values of the current density which are 20 times smaller than the critical pinch-current in the homogeneous material. This new type of instability obtained in [1] arises from the superposition of the strong internal boltzmannian electrical field of the barrier with the relatively weak self magnetic field of the current crossing the barrier. While the pinch-effect is quadratic in the current I, the unstable terms of the above effect are linear in I. The resulting critical current (of approximately 0.5 A/cm^2 in Ge) is lowered by several orders of magnitude if we take into account the inhomogeneities in doping, dislocations, grain boundaries and thermal effects. Thus, without external magnetic field and even at very low applied voltages, the current crossing the surface barrier of a semiconductor will be unstable in some domains. This instability is a local one and is significant only in well confined little regions. It will not change appreciably the current-voltage relation and has to be distinguished from the case of helical instability which appears in the presence of

[1]) Now at the European Institute of Physics Max von Laue—Paul Langevin, Garching bei München, Reak-torgelände.

a longitudinal magnetic field [2 to 4]. Physically it will manifest itself only through a weak turbulence in the plasma of the current carriers. The theory of this turbulence leads to a f^{-1} spectrum of the current fluctuations, and thus to a new interpretation of the flicker noise [5].

The aim of the present paper is to look for the existence of some nonmagnetic instabilities of the surface potential barrier which could also be able to generate turbulent fluctuations and consequently low-frequency noise. In a previous paper Handel [5] developed an approximation procedure for solving the stability equations, i.e. the equations obeyed by the perturbations. The procedure relies on an expansion made with respect to the surface potential which, unfortunately, is not at all a small parameter. So it happened that in [6] where only the first order approximation was explicitly computed, instabilities appeared even in the absence of any current. Now we shall also obtain the second-order approximation in the hope that the unphysical result of the first order calculation will disappear or diminish its extent, i.e. that in the absence of an applied electric field the instability conditions will be more severe, leading finally to the disappearance of the instability in higher orders as required by the convergence condition of the process. Furthermore we shall investigate if the application of a stationary electrical field has a destabilizing effect or not.

2. The Stability Equations

Let us consider a semiconductor plate limited in the z-direction by the planes $z = -L/2$ and $z = L/2$ but unbounded in the x- and y-direction. Along the x-axis an homogeneous stationary electric field E_x is applied. There is a space-charge region below the surface on both sides due to the occupation of the surface states, generating a surface potential V_s of the same value on both sides. The mobilities of the carriers are related to the diffusion coefficients by the Einstein relation. Thus it suffices to introduce only two frictional parameters ν_n and ν_p in order to describe the interaction of the electrons and holes with the phonons of the lattice:

$$\nu_n = \frac{e}{\mu_n} = \frac{kT}{D_n}, \qquad \nu_p = \frac{e}{\mu_p} = \frac{kT}{D_p}.$$

The equations of motion (1), (2), the equations of continuity (3), (4) and the Poisson equation are, respectively

$$\nu_n\, n\, \boldsymbol{v}_n = -e\, n\, \boldsymbol{E} - \nabla(n\, kT)\,, \tag{1}$$

$$\nu_p\, p\, \boldsymbol{v}_p = e\, p\, \boldsymbol{E} - \nabla(p\, kT)\,, \tag{2}$$

$$\frac{\partial n}{\partial t} + \nabla\,(n\, \boldsymbol{v}_n) = -R\,(n\, p - n_i^2)\,, \tag{3}$$

$$\frac{\partial p}{\partial t} + \nabla\,(p\, \boldsymbol{v}_p) = -R\,(n\, p - n_i^2), \tag{4}$$

$$\nabla\, \boldsymbol{E} = \frac{4\,\pi\, e}{\varkappa}\,(p - n - N_A + N_D)\,. \tag{5}$$

Here n, p, $\boldsymbol{v}_n$, $\boldsymbol{v}_p$ are the concentrations and the drift velocities of electrons and holes. R is a recombination coefficient, $\varkappa$ is the dielectric constant of the lattice and N_A and N_D are the concentrations of the ionized acceptors and donors.

The unperturbed (laminar) solution is

$$\boldsymbol{E}^0 = \boldsymbol{E}_x^0 + \boldsymbol{E}_z^0;\; E_z^0 = -\frac{\mathrm{d}V^0}{\mathrm{d}z} = -\frac{kT}{e}\frac{\mathrm{d}\ln n^0}{\mathrm{d}z} = \frac{kT}{e}\frac{\mathrm{d}\ln p^0}{\mathrm{d}z},$$

$$\nu_\mathrm{p}\,\boldsymbol{v}_\mathrm{p}^0 = -\nu_\mathrm{n}\,\boldsymbol{v}_\mathrm{n}^0 = e\,\boldsymbol{E}_x^0 = \text{const}\,. \tag{6}$$

Denoting with a prime the little perturbations superposed on the stationary solution, we get

$$\nu_\mathrm{n}\,(n\,\boldsymbol{v}_\mathrm{n})' = -e\,n^0\,\boldsymbol{E}' - e\,n'\,\boldsymbol{E}^0 - kT\,\nabla\,n'\,, \tag{7}$$

$$\nu_\mathrm{p}\,(p\,\boldsymbol{v}_\mathrm{p})' = e\,p^0\,\boldsymbol{E}' + e\,p'\,\boldsymbol{E}^0 - kT\,\nabla\,p'\,, \tag{8}$$

$$\frac{\partial n'}{\partial t} + \nabla\,(n\,\boldsymbol{v}_\mathrm{n})' = -R\,(n^0\,p' + p^0\,n')\,, \tag{9}$$

$$\frac{\partial p'}{\partial t} + \nabla\,(p\,\boldsymbol{v}_\mathrm{p})' = -R\,(n^0\,p' + p^0\,n')\,, \tag{10}$$

$$\nabla\,\boldsymbol{E}' = \frac{4\,\pi\,e}{\varkappa}\,(p' - n')\,. \tag{11}$$

Here the occupation of the acceptor and donor centres is assumed not to be affected by the perturbation, corresponding to the choice of the recombination term in (3) and (4). In the homogeneous case ($V = 0$) this effect was taken into account by Konstantinov and Perel [7].

Introducing (7) and (8) in (9) and (10) yields the system of equations

$$\frac{\nu_\mathrm{n}}{T}\frac{\partial n'}{\partial t} - \beta\,\mathrm{e}^v\,(p' - n') - n'\,[\alpha\,(\mathrm{e}^{-v} - 1) - \beta\,(\mathrm{e}^v - 1)] + \frac{\mathrm{d}v}{\mathrm{d}z}\frac{\partial n'}{\partial z} - \varepsilon_x\,\frac{\partial n'}{\partial x} -$$
$$- \nabla^2\,n' + \beta\,\mathrm{e}^v\,\frac{\mathrm{d}v}{\mathrm{d}z}\frac{\partial f}{\partial z} = -\varrho_\mathrm{n}\,(\beta\,\mathrm{e}^v\,p' + \alpha\,\mathrm{e}^{-v}\,n')\,, \tag{12}$$

$$\frac{\nu_\mathrm{n}}{T}\frac{\partial p'}{\partial t} + \alpha\,\mathrm{e}^{-v}\,(p' - n') + p'\,[\alpha\,(\mathrm{e}^{-v} - 1) - \beta\,(\mathrm{e}^v - 1)] - \frac{\mathrm{d}v}{\mathrm{d}z}\frac{\partial p'}{\partial x} + \varepsilon_x\,\frac{\partial p'}{\partial x} -$$
$$- \nabla^2\,p' - \alpha\,\mathrm{e}^{-v}\,\frac{\mathrm{d}v}{\mathrm{d}z}\frac{\partial f}{\partial z} = -\,\varrho_\mathrm{p}\,(\beta\,\mathrm{e}^v\,p' + \alpha\,\mathrm{e}^{-v}\,n')\,, \tag{13}$$

$$-\nabla^2\,f = p' - n'\,. \tag{14}$$

Here we introduced the notations

$$\frac{4\,\pi\,e^2}{\varkappa\,kT}\,p_\infty \equiv \alpha;\quad \frac{4\,\pi\,e^2}{\varkappa\,kT}\,n_\infty \equiv \beta;\quad \frac{e\,E_x^0}{kT} \equiv \varepsilon_x;\quad \frac{e\,E_z^0}{kT} \equiv \varepsilon_z = -\frac{\mathrm{d}v}{\mathrm{d}z}; \tag{15}$$

$$\frac{\varkappa}{4\,\pi\,e}\,\boldsymbol{E}' \equiv -\nabla\,f;\; \varrho_\mathrm{p} \equiv \frac{\varkappa\,R}{4\,\pi\,e^2}\,\nu_\mathrm{p};\quad \varrho_\mathrm{n} \equiv \frac{\varkappa\,R}{4\,\pi\,e^2}\,\nu_\mathrm{n}\,, \tag{16}$$

where p_∞ and n_∞ are the bulk concentrations of the carriers.

If the occupation of the surface is not changed by the perturbations, the following boundary conditions are valid at $z = \mp L/2$:

$$n'\,\frac{\mathrm{d}v}{\mathrm{d}z} - \frac{\partial n'}{\partial z} + \beta\,\mathrm{e}^v\,\frac{\partial f}{\partial z} = \mp\,\sigma_\mathrm{n}\,(\beta\,\mathrm{e}^v\,p' + \alpha\,\mathrm{e}^{-v}\,n')\,, \tag{17}$$

$$-p'\,\frac{\mathrm{d}v}{\mathrm{d}z} - \frac{\partial p'}{\partial z} - \alpha\,\mathrm{e}^{-v}\,\frac{\partial f}{\partial z} = \mp\,\sigma_\mathrm{p}\,(\beta\,\mathrm{e}^v\,p' + \alpha\,\mathrm{e}^{-v}\,n')\,. \tag{18}$$

Here

$$\sigma_\mathrm{n} \equiv \frac{\varkappa\,\Gamma}{4\,\pi\,e^2}\,\nu_\mathrm{n};\quad \sigma_\mathrm{p} \equiv \frac{\varkappa\,\Gamma}{4\,\pi\,e^2}\,\nu_\mathrm{p}\,, \tag{19}$$

and Γ is a coefficient of superficial recombination.

3. A Recurrence Method for Solving the Stability Equations

In order to solve the system (12) to (14) with the boundary conditions (17), (18) it seems necessary to obtain first a solution for the static potential v in the surface barrier. From Poisson's equation

$$\frac{d^2v}{dz^2} = -\alpha\,(e^{-v} - 1) + \beta\,(e^{v} - 1) \equiv a'(v) \tag{20}$$

it follows that

$$\frac{1}{2}\left(\frac{dv}{dz}\right)^2 = a(v)\,. \tag{21}$$

Hereby is

$$a(v) \equiv \alpha\,(e^{-v} + v - 1) + \beta\,(e^{v} - v - 1) \quad \text{and} \quad a' = \frac{da}{dv}\,. \tag{22}$$

The potential v is not an elementary function of z. To avoid this difficulty without solving equation (21) we pass from z to v taking v as a new independent variable in (12) to (14) and (17), (18). Denoting

$$\alpha\, n'\, e^{-v} \equiv \psi;\ \beta\, p'\, e^{v} \equiv \varphi;\ \alpha\,\beta\, f \equiv \chi\,, \tag{23}$$

and using (20), (21) we obtain

$$\left[\frac{\nu_n}{T}\frac{\partial}{\partial t} + k^2 + \beta\, e^{v} + \alpha\, e^{-v}\varrho_n - i\,\varepsilon_x k_x - (a' + 2\,a)\frac{\partial}{\partial v} - 2\,a\,\frac{\partial^2}{\partial v^2}\right]\psi -$$

$$- \alpha\, e^{-v}(1 - \varrho_n)\,\varphi = -2\,a\,\frac{\partial\chi}{\partial v}\,, \tag{24}$$

$$-\beta\, e^{v}(1 - \varrho_p)\,\psi + \left[\frac{\nu_p}{T}\frac{\partial}{\partial t} + k^2 + \alpha\, e^{-v} + \beta\, e^{v}\varrho_p + i\,\varepsilon_x k_x -\right.$$

$$\left. - (a' - 2\,a)\frac{\partial}{\partial v} - 2\,a\,\frac{\partial^2}{\partial v^2}\right]\varphi = -2\,a\,\frac{\partial\chi}{\partial v}\,, \tag{25}$$

$$\left(k^2 - a'\frac{\partial}{\partial v} - 2\,a\,\frac{\partial^2}{\partial v^2}\right)\chi = \alpha\, e^{-v}\varphi - \beta\, e^{v}\psi \tag{25'}$$

with the boundary conditions

φ, ψ, χ finite at $v = 0$, and

$$\frac{\partial\,(\varphi + \psi)}{\partial v} = -\frac{\alpha\, e^{-v}\sigma_n + \beta\, e^{v}\sigma_p}{\sqrt{2\,a}}\,(\varphi + \psi)\ \text{sign}\ v \quad \text{at} \quad v = v_s\,. \tag{26}$$

If the thickness of the semiconductor plate exceeds the Debye-length, $L \gg L_D$, the value of v at $z = 0$ will approach zero. In any case v is varying monotonically from a certain small value $v(0) \approx 0$ to $v(\pm L/2) = v_s$. Note that in the case $v_s = 0$ the change of variable from z to v is no longer permitted, whereas the use of

$$u \equiv \frac{v}{v_s} \tag{27}$$

remains valid because $u \to \exp\left[\sqrt{\alpha + \beta}\,(|z| - L/2)\right]$ as $v_s \to 0$.

Seeking for solutions of the form

$$\psi = e^{st}\, u^r \sum_{l=0}^{\infty} b_l\, v^l\,, \quad \varphi = e^{st}\, u^r \sum_{l=0}^{\infty} c_l\, v^l\,, \quad \chi = e^{st}\, u^r \sum_{l=0}^{\infty} a_l\, v^l \tag{28}$$

with the condition

$$\operatorname{Re} r \geqq 0 , \tag{29}$$

which follows from the finiteness-condition in (26), we get an infinite chain of algebraic equations for the coefficients a_0, b_0, c_0, a_1, b_1, c_1, ... developing the coefficients of (24) and (25) in powers of v:

$$a = \frac{\alpha + \beta}{2!} v^2 + \frac{\beta - \alpha}{3!} v^3 + \frac{\alpha + \beta}{4!} v^4 + \cdots ,$$

$$a' = (\alpha + \beta) v + \frac{\beta - \alpha}{2!} v^2 + \frac{\alpha + \beta}{3!} v^3 + \cdots .$$

Eliminating the a_l-coefficients from each system we write them down restricting ourselves to the case $\varrho_n = \varrho_p = 0$ of negligible bulk recombination because each instability obtained in these conditions will surely persist for sufficiently small recombination rates.

$$\left.\begin{aligned} &\left[\frac{\nu_n}{T} s + k^2 + \beta - i \varepsilon_x k_x - (\alpha + \beta) r^2\right] b_0 - \alpha c_0 = 0 , \\ &-\beta b_0 + \left[\frac{\nu_p}{T} s + k^2 + \alpha + i \varepsilon_x k_x - (\alpha + \beta) r^2\right] c_0 = 0 , \end{aligned}\right\} \tag{30}$$

$$\left.\begin{aligned} &\left[\frac{\nu_n}{T} s + k^2 + \beta - i \varepsilon_x k_x - (\alpha + \beta) (r + 1)^2\right] b_1 - \alpha c_1 = \\ &= \left[\frac{5\alpha + 7\beta}{6} r + \frac{\beta - \alpha}{3} r^2 - \beta\left(1 - \frac{(\alpha + \beta) r}{k^2 - (\alpha + \beta) r^2}\right)\right] b_0 - \alpha\left[1 + \frac{(\alpha + \beta) r}{k^2 - (\alpha + \beta) r^2}\right] c_0 , \\ &-\beta b_1 \left[\frac{\nu_p}{T} s + k^2 + \alpha + i \varepsilon_x k_x - (\alpha + \beta) (r + 1)^2\right] c_1 = \beta\left[1 + \frac{(\alpha + \beta) r}{k^2 - (\alpha + \beta) r^2}\right] b_0 + \\ &\quad + \left[-\frac{7\alpha + 5\beta}{6} r + \frac{\beta - \alpha}{3} r^2 + \alpha\left(1 - \frac{(\alpha + \beta) r}{k^2 - (\alpha + \beta) r^2}\right)\right] c_0 , \end{aligned}\right\} \tag{31}$$

$$\left.\begin{aligned} &\left[\frac{\nu_n}{T} s + k^2 + \beta - i \varepsilon_x k_x - (\alpha + \beta) (r + n)^2\right] b_n - \alpha c_n = \\ &= G_n(b_0, c_0, b_1, c_1, \ldots, b_{n-1}, c_{n-1}, r, k, \alpha, \beta) , \\ &-\beta b_n + \left[\frac{\nu_p}{T} + k^2 + \alpha + i \varepsilon_x k_x - (\alpha + \beta) (r + n)^2\right] c_n = \\ &= H_n(b_0, c_0, b_1, c_1, \ldots, b_{n-1}, c_{n-1}, r, k, \alpha, \beta) . \end{aligned}\right\} \tag{32}$$

Here G_n and H_n are rational functions of their arguments. We obtained them explicitly. They contain sums over the coefficients up to the order $n - 1$. A solution can exist only if the determinant of the zeroth-order system (30) vanishes because b_0, c_0, and a_0 can not be simultaneously zero in (28). Thus there emerge two possible values of the exponential growth rate s:

$$\frac{\nu}{T} s_{1,2} = (\alpha + \beta)\left(r^2 - \frac{1}{2}\right) - k^2 \pm \sqrt{\left(\frac{\alpha + \beta}{2}\right)^2 - \varepsilon_x^2 k_x^2 - i \varepsilon_x k_x (\beta - \alpha)} . \tag{33}$$

For the sake of simplicity we considered $\nu_n = \nu_p = \nu$ here (equal mobilities). Instability will occur if the condition

$$\operatorname{Re} s > 0 \tag{34}$$

is satisfied for values of r and k satisfying the conditions (26), (29) as well as a supplementary physical limitation:

$$|r| < 1\,, \quad |k| < \sqrt{\alpha + \beta}\,. \tag{35}$$

Indeed solutions with $|r| > 1$ or $|k| > \sqrt{\alpha + \beta}$ correspond to the very small eddies, smaller than the Debye-length $1/\sqrt{\alpha + \beta}$. Such short-wave perturbations are strongly damped. Taking into account that ε_x is usually also small (1.16×10^3 cm^{-1} for $E_x = 30$ V/cm) compared with $\sqrt{\alpha + \beta}$ (1.45×10^4 cm^{-1} in intrinsic Ge and larger in the extrinsic one), we can use a power series expansion of the root in (33):

$$\frac{\nu}{T} s_1 = (\alpha + \beta)\, r^2 - k^2 - 2\, \varepsilon_x^2\, k_x^2 \frac{\alpha^2 + \beta^2}{(\alpha + \beta)^3} - 2\, \varepsilon_x^3\, k_x^3 \frac{(\alpha - \beta)^3}{(\alpha + \beta)^5} + \\ + i\, \varepsilon_x\, k_x \frac{\alpha - \beta}{\alpha + \beta} + 2\, i\, \varepsilon_x^3\, k_x^3 \frac{\alpha - \beta}{(\alpha + \beta)^3} + \cdots. \tag{36}$$

We have restricted ourselves to the positive sign of the root (corresponding cf. [6] to the so called acoustical or neutral solution). The optical or electrical solution is obtained by taking the minus sign of the root and is always stable according to (35).

No instability will occur if $\operatorname{Re} r = 0$; $\operatorname{Re} r < 0$ is also forbidden by (29). From (34), (36) and (35) the necessary condition

$$(\alpha + \beta) \operatorname{Re} r^2 - k^2 > 0 \tag{37}$$

follows. For a semiconductor plate unlimited in the x and y directions k can be arbitrarily small. Thus (37) and (29) yield the condition

$$|\arg r| < \frac{\pi}{4}\,; \quad |r| < 1\,. \tag{37'}$$

The boundary condition (26) which determines the allowed r-values yields with (28) and $\sigma_n \approx \sigma_p = \sigma$

$$r \sum_{l=0}^{\infty} (b_l + c_l)\, v_s^l = - \sum_{l=0}^{\infty} (b_l + c_l)\, v_s^l \left[l + \frac{\alpha\, e^{-v_s} + \beta\, e^{v_s}}{\sqrt{2\, a\, v_s}}\, \sigma\, |v_s| \right]. \tag{38}$$

Developing also the expression in the last bracket on the right hand side in powers of v_s the condition which determines r will be (to the second order in v_s)

$$r + \sigma \sqrt{\alpha + \beta} \left[1 + \frac{5}{6} \frac{\beta - \alpha}{\beta + \alpha}\, v_s + \frac{11}{24}\, v_s^2 - \frac{(\alpha - \beta)^2}{8\, (\alpha + \beta)^2}\, v_s^2 \right] + \\ + \left[r + 1 + \sigma \sqrt{\alpha + \beta} \left(1 + \frac{5}{6} \frac{\beta - \alpha}{\beta + \alpha}\, v_s \right) \right] v_s \frac{b_1 + c_1}{b_0 + c_0} + \\ + \left(r + 2 + \sigma \sqrt{\alpha + \beta} \right) v_s^2 \frac{b_2 + c_2}{b_0 + c_0} = 0\,. \tag{39}$$

To calculate $(b_1 + c_1)/(b_0 + c_0)$ it is necessary to solve equation (30) and (31) but for $(b_2 + c_2)/(b_0 + c_0)$ the solution of (32) with $n = 2$ is also needed. These fractions will be rational functions of r so that equation (39) leads to an algebraic equation in r:

$$\Phi(r) = 0\,, \tag{40}$$

where $\Phi(r)$ is a polynomial in r. If there is a root of (40) which satisfies the instability condition (37), the configuration is unstable; otherwise it is stable.

4. The First and Second-Order Approximation in the Absence of Any Applied Field

In equations (30) to (32) the term $i\,\varepsilon_x\,k_x$ introduces much trouble and so we shall first discuss the case $\varepsilon_x = 0$, in which no instability should occur, in order to verify the convergence of the v_s-expansions. If $\varepsilon_x = 0$ equation (40) has real coefficients. If for $r = 1$ the sign of $\Phi(r)$ is different from that for $r = 0$, there will be at least a root of $\Phi(r)$ in this interval. That would mean instability because the condition (37′) would be then satisfied. Thus a sufficient condition for instability is

$$\operatorname{sign}\Phi(0) = -\operatorname{sign}\Phi(1)\,. \tag{41}$$

a) To first order in v_s the calculation gives

$$\Phi^{\mathrm{I}}_{(0)} = \xi\,(1 - 5\,\eta) \quad \text{and} \quad \Phi^{\mathrm{I}}_{(1)} = 1 + \xi - \eta\,(6\,\xi + 2) \tag{42}$$

with

$$\xi \equiv \sigma\sqrt{\alpha+\beta}\,, \quad \eta \equiv \frac{v_s}{6}\,\frac{\alpha-\beta}{\alpha+\beta}\,. \tag{43}$$

Thus instability is predicted by the first order calculation in a small region of the $\xi\,\eta$-semiplane ($\xi > 0$) between the curves

$$\eta = \frac{1}{5}\,, \quad \eta = \frac{(\xi+1)}{(6\,\xi+2)}\,.$$

We conclude that to first order: 1. No instability occurs if an accumulation layer is present at the surface. 2. If there is an inversion- or depletion-layer at the surface, instability will occur under the condition $\eta > 1/5$, i.e. with usual notations if

$$\frac{p_\infty - n_\infty}{p_\infty + n_\infty}\,V_s > \frac{kT}{e}\,. \tag{44}$$

This is identical with the result obtained in [6]. Usually if the semiconductor is not quite intrinsic, the factor in front of V_s approaches unity and the condition (44) needs $|v_s| \geqq 1.2$.

b) To second order in v_s we obtain

$$\Phi^{\mathrm{II}}_{(0)} = \xi\left(1 + \frac{17}{24}\,v_s^2 - 5\,\eta - \frac{27}{2}\,\eta^2\right) + \frac{v_s^2}{2} - 18\,\eta^2\,, \tag{45}$$

where we considered k^2 limited by the condition (37) and

$$\Phi^{\mathrm{II}}_{(1)} = \xi\left(1 + \frac{v_s^2}{2} - 6\,\eta\right) + 1 + \frac{v_s^2}{8} - 2\,\eta - \frac{3}{2}\,\eta^2\,. \tag{46}$$

From the definition (43) of η we see that the whole discussion is restricted to the strip

$$|\eta| < |v_s|/6$$

of the $\xi\,\eta$-semiplane with $\xi > 0$. It is easy to see that within this strip of thickness determined by v_s the inequalities

$$\Phi^{II}_{(0)} > 0\,, \quad \Phi^{II}_{(1)} > 0 \tag{47}$$

are always satisfied. Thus we conclude that in the second order approximation no instabilities are predicted in the absence of an applied electric field. We believe that this result is valid in all higher orders too.

5. On the Validity of the Expansion Method

The fact that the unphysical result obtained in the first approximation (section 4 a), is corrected by the second, makes the expansion method subject to some doubts. However, it is easy to show that at high orders ($n \gg 1$) of our expansion method the expansions (28) with coefficients obtained from (30) to (32) converge like expansions of exponential functions, i.e. for arbitrarily large values of the expansion parameter v. We state that for high order terms the behaviour of the coefficients in (28) is given by

$$\psi(v) \sim \sum_l \left[b^{(1)} + (-1)^l\, b^{(2)}\right] \frac{v^l}{2^l\, l!} = b^{(1)}\, e^{v/2} + b^{(2)}\, e^{-v/2}\,, \tag{48}$$

$$\varphi(v) \sim \sum_l \left[c^{(1)} + (-1)^l\, c^{(2)}\right] \frac{v^l}{z^l\, l!} = c^{(1)}\, e^{v/2} + c^{(2)}\, e^{-v/2}\,, \tag{49}$$

$$\chi(v) \sim \sum_l \left[b^{(1)} - (-1)^l\, c^{(2)}\right] \frac{v^l}{2^l\, l!} = b^{(1)}\, e^{v/2} - c^{(2)}\, e^{-v/2}\,. \tag{50}$$

Here $b^{(1)}$, $b^{(2)}$, $c^{(1)}$ and $c^{(2)}$ are arbitrary constants to be determined by solving equations (31), (32) ... consecutively. Let us assume that the expansions (48) to (50) are valid to the order $l = n - 1$; ($n \gg 1$). Then they are valid for $l = n$ too. Indeed, introducing the expansion coefficients from (48) to (50) in the explicite form of the right hand side of (32) as well as in the right hand side of a similar relation for a_n obtained from (25′), we get

$$\begin{aligned}
&\beta \sum_{j=0}^{n} \frac{b^{(1)} + (-1)^n\, b^{(2)}}{j!\,(n-j)!\,2^{n-j}} - (\alpha+\beta)\, n\,(n-1)\, b_n + \sum_{j=0}^{n} \frac{[\alpha(-1)^j - \beta]\,[b^{(1)} - (-1)^{n-j}\, b^{(2)}]}{j!\,(n-j)!\,2^{n-j+1}} - \\
&- \sum_{j=0}^{n} \frac{[\alpha(-1)^j + \beta]\,[b^{(1)} - (-1)^{n-j}\, b^{(2)}]}{j!\,(n-j)!\,2^{n-j}} + \frac{\alpha+\beta}{(n-2)!\,2^n}\,[b^{(1)} + (-1)^n\, b^{(2)}] - \\
&- \sum_{j=0}^{n} \frac{[\alpha\,(-1)^j + \beta]\,[b^{(1)} + (-1)^{n-j}\, b^{(2)}]}{j!\,(n-j)!\,2^{n-j+1}} - \alpha \sum_{j=0}^{n} \frac{(-1)^j\, c^{(1)} + (-1)^n\, c^{(2)}}{j!\,(n-j)!\,2^{n-j}} = \\
&= - \sum_{j=0}^{n} \frac{[\alpha(-1)^j + \beta]\,[b^{(1)} + (-1)^{n-j}\, c^{(2)}]}{j!\,(n-j)!\,2^{n-j}}\,,
\end{aligned} \tag{51}$$

and two similar equations for c_n and a_n. Here we put $r = 0$, for only the values $r = 0$ and $r = 1$ are needed in computing equation (45) and (46) to order n. In (51) we neglected terms which are at least n times smaller than each of the written ones. Performing the summations and again neglecting terms which are at least of order n^{-1}, we see that all summations cancel mutually. Finally, we obtain from (51)

$$b_n = \frac{b^{(1)} + (-1)^n\, b^{(2)}}{2^n\, n!} \tag{52}$$

and from the other two equations

$$c_n = \frac{c^{(1)} + (-1)^n c^{(2)}}{2^n\, n!} \quad \text{and} \quad a_n = \frac{b^{(1)} - (-1)^n c^{(2)}}{2^n\, n!}. \tag{53}$$

Thus, our statement (48) to (50) is extensible from $n - 1$ to n and, therefore, by induction, to any large value of l. An analogous demonstration holds for $r = 1$.

All expansions made in the boundary condition (38) were also expansions of exponential functions. Therefore, $\Phi(0)$ and $\Phi(1)$ (section 4) are power series of v_s which converge in higher orders like expansions of exponential functions.

The convergence of the exponential series is particularly slow at large negative values of the argument which correspond to inversion layers. To get the correct relative sign of $\Phi(0)$ and $\Phi(1)$, the error in their expansions has to be less than 100%. The corresponding error is easily calculated by introducing the asymptotic exponentials (48), (49) in equation (26) and affecting all exponentials by the error $\delta\,(n, -|v_s|)$ which is the rest of the exponential expansion. This yields the necessary order of approximation for each v_s.

Performing the calculation, we see that the first order is sufficient if $|v_s| \leqq 0.75$. The second order holds up to $|v_s| = 1.2$ and the third up to $|v_s| = 1.5$. The instabilities obtained in the first order for $|v_s| \geqq 1.2$ are beyond the domain of validity of the first order approximation. Thus they are unphysical.

6. Influence of the Electric Field

If an electric field ε_x is applied in the x-direction, a current will flow along the surface. To first order in v_s and second order in $\varepsilon_x k_x$[2]) a calculation yields the following form for equation (40) at $\xi = 0$[3]):

$$\begin{aligned} 2\,(\alpha + \beta)^2\,(1+\eta)\,r^3 + (1-5\,\eta)\,(\alpha+\beta)^2\,r^2 = \Bigg\{ 4\,\varepsilon_x^2 k_x^2 \left[3\,\eta - (1+\eta)\,\frac{\alpha^2+\beta^2}{(\alpha+\beta)^2} \right] + \\ + 2\,i\,\varepsilon_x\,k_x \left[(1+\eta)\,(\alpha-\beta) + \frac{\alpha^2+\beta^2}{\alpha+\beta}\,v_s \right] + 12\,i\,\eta\,\varepsilon_x^2 k_x^2 \left(\frac{\alpha-\beta}{\alpha+\beta} \right)^2 \Bigg\}\,r^2 + \\ + \left\{ -2\,(1-5\,\eta)\,\varepsilon_x^2 k_x^2\,\frac{\alpha^2+\beta^2}{(\alpha+\beta)^2} + i\,\varepsilon_x\,k_x \left[(1-2\,\eta)\,(\alpha-\beta) - (\alpha+\beta)\,\frac{v_s}{2} \right] \right\} r + \\ + 6\,\eta\,\varepsilon_x^2 k_x^2\,\frac{\alpha^2+\beta^2}{(\alpha+\beta)^2} + 2\,i\,\varepsilon_x\,k_x\,\frac{\alpha\beta}{\alpha+\beta}\,v_s + \frac{6\,\eta\,\varepsilon_x^2 k_x^2\,r}{k^2 - (\alpha+\beta)\,r^2} \times \\ \times \left[\frac{\alpha^2+\beta^2}{\alpha+\beta} + 2\,r\,(\alpha+\beta) + 2\,i\,r\,(\alpha^2+\beta^2)\,\frac{v_s}{\varepsilon_x\,k_x} + 2\,i\,r\,\frac{(\alpha-\beta)^2}{\alpha+\beta} \right]. \end{aligned} \tag{54}$$

To have confidence in the validity of the first-order equation, we used for a numerical investigation of the stability problem a relatively small $v_s = -0.122$. We took the values of α and β for n-germanium: $\alpha = 1.05 \times 10^3\ \text{cm}^{-2}$, $\beta = 1.05 \times 10^9\ \text{cm}^{-2}$. Then $\eta = 0.02$, corresponding to a weak depletion layer, and

[2]) The development parameter used for ε_x is small up to $\varepsilon_x\,k_x \approx 10^8$ for usual α and β values.

[3]) Usual values of ξ at germanium surfaces are very small. With $\Gamma\,(p_\infty + n_\infty) = 100$ cm/s, $p_\infty + n_\infty = 5 \times 10^{13}\ \text{cm}^{-3}$ (intrinsic Ge), $\nu = 5.62 \times 10^{-16}$ erg s/cm², $\varkappa = 16$, $kT = 4 \times 10^{-14}$ erg we obtained $\xi = 9.34 \times 10^{-5} \approx 10^{-4}$. An instability obtained with $\xi = 0$ will be still valid for small $\xi \neq 0$ values.

for $k = 10^2$ cm^{-2} we are led to the following equation:

$$2.29\times10^{18}\,r^3 + (1.01\times10^{18} + 3.804\,x^2 + 4.66\times10^9\,i\,x - 0.235\,i\,x^2)\,r^2 +$$
$$+ (1.78\,x^2 + 0.9352\times10^9\,i\,x)\,r - 0.119\,x^2 + 3.54\times10^6\,i\,x - \frac{x^2\,r}{1-1.06\times10^5\,r^2}\times$$
$$\times\left[1.25\times10^4 + 2.55\times10^4\,r - 3.24\times10^{12}\frac{i\,r}{x} + 2.45\times10^4\,i\,r\right] = 0;\; x\equiv \varepsilon_x\,k_x\,. \tag{55}$$

An "unstable" r-value satisfying the requirements (37') first appears for $x = -1.32\times10^8$. This corresponds to an electric field of about 30 kV/cm. Thus the critical electric field exceeds the breakdown value by several orders of magnitude.

We can conclude that, even in the presence of a strong electrical field, no instability occurs, at least for small values of the surface potential parameter v_s. For $0.02 < \eta < 0.2$ e.g. at values of $|v_s|$ above 0.12, the threshold value of x (the critical electric field obtained from a study of equation (54)) diminishes up to zero. But at values of η comparable to 0.125 ($|v_s| \sim 0.75$) we know from section 5 that the second order v_s-approximation is already needed in order to obtain physically meaningful reliable results. To calculate the second v_s-order analogue of equation (54) is much more difficult but it would give the correct answer to the stability problem for η-values up to 0.2 ($|v_s| \approx 1.2$).

Thus we conclude that at least for small values of v_s ($|v_s| \leqq 0.75$) stability prevails for all physically realizable applied fields. An investigation of the case of greater values of v_s requires the computation of several following orders of approximation which is unfortunately a formidable calculatory task.

7. Conclusions

Although the surface barrier of a semiconductor is an inhomogeneous physical system, the recurrence method used in this work permits the study of its stability in the presence of a stationary current, reducing it to the solution of linear algebraic systems and of algebraic equations.

To test the method, in section 4 we applied it to the currentless (equilibrium) state which must be stable a priori. The first-order approximation gave stability except for strong inversion or depletion layers with surface potential $|V_s| \geqq \geqq 1.2\,kT$. The second-order result was stability in each case corresponding to our expectations. Owing to the fact that only exponential functions of v_s were subjected to series expansions, it was possible to show in section 5 that the method is convergent for each value of the expansion parameter v_s. This would mean that for an arbitrary given v_s there is a finite order of approximation which gives reliable results, e.g. the value of the root r or growth rate s do not change significantly further if we pass to higher orders. If $|v_s| \leqq 0.75$ the first order is sufficient. The second order holds up to $|v_s| = 1.2$ and the third up to $|v_s| = 1.5$.

In Section 6 we showed that at least up to $|v_s| \leqq 0.75$ no instability occurs independently of the existence of the electrical field. The absence of instabilities at $v_s = 0$ was already proved in [6]. At higher values of $|v_s|$ the information obtained from equation (54), which predicted the lowering of the critical field, is not valid and the second order v_s-approximation must be taken into account.

Acknowledgements

We are grateful to Dr. V. Sergiescu as well as to Prof. V. Novacu and Prof. P. Fulde for helpful suggestions.

References

[1] P. H. Handel, Z. Naturf. **21a**, 573 (1966).
[2] M. Glicksmann, Plasma Effects in Solids, Dunod, Paris 1965 (p. 149).
[3] B. Ancker-Johnson, Phys. Rev. **135A**, 1423 (1964).
[4] C. E. Hurwitz and A. L. McWhorter, Phys. Rev. **134A**, 1033 (1964).
[5] P. H. Handel, Z. Naturf. **21a**, 579 (1966).
[6] P. H. Handel, Z. Naturf. **21a**, 561 (1966).
[7] O. V. Konstantinov and V. I. Perel, Fiz. tverd. Tela **6**, 3364 (1964).

(Received May 29, 1968)

phys. stat. sol. 28, 711 (1968)

Subject classification: 14.3; 14.4.1; 22.2.3

Microwave Department, Royal Institute of Technology, Stockholm

Conductivity of Intrinsic InSb in High Electric Fields

By

P. WEISSGLAS and K. BLÖTEKJAER

A theory is developed for the carrier transport properties and impact ionization rate in an intrinsic polar semiconductor with high carrier density. Polar scattering prevails in weak fields but it is shown that intercarrier interactions become important at higher field strengths. The carrier concentration is found to vary approximately exponentially with applied field. Numerical results for data representative of InSb are presented.

Es wird eine Theorie entwickelt für die Ladungstransporteigenschaften und die Ionisierungsrate in eigenleitenden polaren Halbleitern mit hoher Trägerdichte. Bei schwachen Feldern überwiegt polare Streuung, es wird jedoch gezeigt, daß bei höheren Feldstärken Wechselwirkung zwischen den Ladungsträgern bedeutsam wird. Es wird gefunden, daß die Trägerkonzentration näherungsweise exponentiell mit der angelegten Feldstärke variiert. Numerische Ergebnisse, die für InSb repräsentativ sind, werden angegeben.

1. Introduction

It has been established by Ehrenreich [1] that the low field mobility of InSb at and well below room temperature is controlled by optical phonon scattering.[1]) It is also known that nonohmic effects set in at comparatively low electric fields. There has been a considerable interest [2 to 5] in high field phenomena in InSb and a number of experiments extending into the avalanche breakdown region have been carried out. It appears that the mobility variation before avalanching occurs is well explained by Stratton's theory [6] which includes only optical phonon scattering and is based on the assumption that the electron velocity distribution is represented by a displaced drifting maxwellian. As soon as avalanching occurs the situation becomes more complicated [2]. The purpose of this paper is to present the results of calculations using a particular model for the avalanche. The starting point is Stratton's assumption that electron–electron collisions are sufficiently frequent to maintain a maxwellian velocity distribution. If this is true no single electron can gain an exceptionally large amount of energy from the field and cause impact ionization but the energy picked up from the field is rapidly shared among the electrons. The increased mean energy of the electrons manifests itself in an increased electron temperature and in an intrinsic material this leads to a higher electron density. If the material is initially extrinsic it will become quasiintrinsic at some value of the applied field even when sufficiently efficient cooling is supplied to maintain the lattice at a constant temperature. The time required for the establishment of an equilibrium electron density at any particular field should be of the order of the mean carrier lifetime or about 10 ns in room temperature InSb. As a result of this mechanism there is no particular value of the applied field in an

[1]) Recent work by Tsidilkovski [8] has thrown some doubt on this assertion.

intrinsic material where the avalanche occurs but instead a continuous exponential increase in electron density takes place. Ehrenreich has shown that at low fields electron-impurity (hole) scattering is negligable up to very high carrier concentrations. However, when the lattice temperature is kept constant, and only the electron temperature is increased the polar mobility increases. It will also appear from below that the electron and hole densities in an intrinsic sample increases rapidly with increasing electron temperature. It is therefore likely that electron–hole scattering will turn out to be important when a strong electric field is applied.

The calculations to be presented include polar optical phonon scattering as well as electron–hole scattering. In order to calculate the latter a theory giving the electron density as a function of the applied electric field is also needed. When the sample is in thermal equilibrium the intrinsic carrier concentration can be calculated from a thermodynamic argument and the result is independent of the particular ionization-recombination process which maintains this carrier concentration. When the equilibrium is perturbed this approach fails and it is necessary to go into the details of the ionization processes to calculate the carrier concentration. It has been established by Beattie and Landsberg [7] that above about 200 °K the dominant recombination process is Auger recombination directly across the bandgap. Due to detailed balance this also means that in thermal equilibrium the inverse process i.e. electron or hole impact ionization is the dominant carrier producing mechanism. At high applied electric fields the dominance of these processes should be further enhanced. The rate of change of carrier concentration n is then given by

$$\frac{\mathrm{d}n}{\mathrm{d}t} = \alpha(T_\mathrm{v}, T_\mathrm{c}, v_\mathrm{v}, v_\mathrm{c})\, n - \beta(T_\mathrm{v}, T_\mathrm{c}, v_\mathrm{v}, v_\mathrm{c})\, n^3 , \tag{1}$$

where we have explicitly shown a dependence of the ionization coefficient α and the recombination coefficient β on the electron (c) and hole (v) temperatures T and drift velocities v. It will be seen however that both α and β are independent of n. In a steady state we then have

$$n^2 = \frac{\alpha}{\beta} \tag{2}$$

and our main problem will be to calculate α and β as functions of v and T. The following section contains an outline of such a theory for a parabolic and isotropic semiconductor in which Auger recombination predominates. Then follows a section where the relevant formulas for calculating T and v as function of the applied field and n are collected. The final section contains numerical results for InSb-data and a discussion of these.

2. Nonequilibrium Auger Recombination

Following Beattie and Landsberg we illustrate the relevant Auger ionization and recombination processes in Fig. 1. It has been shown that when the electron to hole mobility ratio is small the processes c) and d) contribute negligibly and they will therefore be neglected henceforth. The net recombination probability $P(\mathrm{a, b})$ due to effects a) and b) is given by

$$P(\mathrm{a, b}) = \int [\theta(1)\, \theta(1')\, \theta(2)\, \theta(2') - \varrho(1)\, \varrho(1')\, \varrho(2)\, \varrho(2')]\, T_{\mathrm{if}}\, \mathrm{d}S , \tag{3}$$

where the integration is over all possible one-electron states and the arguments

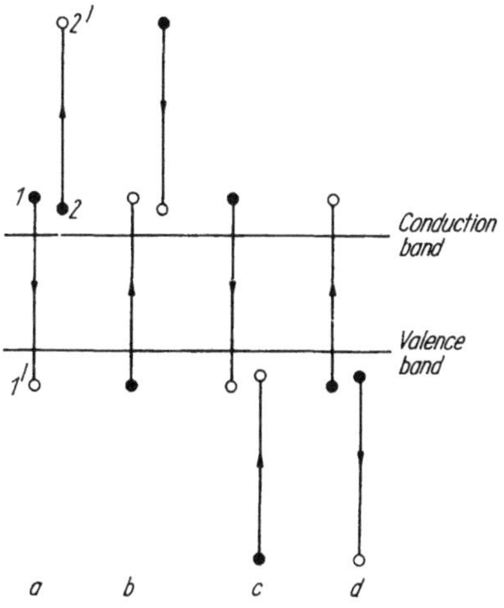

Fig. 1. Auger recombination and ionization mechanisms

1, 1′ etc. refer to the states of Fig. 1. T_{if} is the transition probability and θ is the appropriate probability for electron occupation or vacancy. Also

$$\varrho(i) = 1 - \theta(i) \tag{4}$$

is a corresponding probability for the reverse effect. In thermal equilibrium energy conservation implies that $P(\text{a, b})$ vanishes. In a nonequilibrium steady state, however, the vanishing of $P(\text{a, b})$ is equivalent to our equation 2 and in fact determines the Fermi levels or the carrier density. As we will only consider nondegenerate statistics it turns out simpler to work directly with carrier density than with Fermi levels. The transition probability T_{if} has been computed by Beattie and Landsberg and therefore we only need expressions for the occupation probabilities $\theta(i)$ to define our problem. Assuming that electron–electron and hole–hole collision are sufficiently rapid to ensure a displaced maxwellian velocity distribution we have

$$\left.\begin{aligned}
\theta(1) &\approx n\left(\frac{m_c\, k_B\, T_c}{2\pi\hbar^2}\right)^{-3/2} \exp\left[-\frac{\hbar^2\,(\boldsymbol{k}_1 - \boldsymbol{k}_c)^2}{2\, m_c\, k_B\, T_c}\right),\\
\theta(1') &\approx n\left(\frac{m_v\, k_B T_v}{2\pi\hbar^2}\right)^{-3/2} \exp\left[-\frac{\hbar^2\,(\boldsymbol{k}_{1'} - \boldsymbol{k}_v)^2}{2\, m_v\, k_B\, T_v}\right],\\
\theta(2) &\approx n\left(\frac{m_c\, k_B\, T_c}{2\pi\hbar^2}\right)^{-3/2} \exp\left[-\frac{\hbar^2\,(\boldsymbol{k}_2 - \boldsymbol{k}_c)^2}{2\, m_c\, k_B\, T_c}\right],\\
\theta(2') &\approx 1 - n\left(\frac{m_c\, k_B\, T_c}{2\pi\hbar^2}\right)^{-3/2} \exp\left[-\frac{\hbar^2(\boldsymbol{k}_{2'} - \boldsymbol{k}_c)^2}{2\, m_c\, k_B\, T_c}\right] \approx 1\,,
\end{aligned}\right\} \tag{5}$$

where $\boldsymbol{k}_i$ is the wave vector for an electron in state i and k_B Boltzmann's constant. m_c and m_v are the electron and hole masses respectively and $\boldsymbol{k}_c$ and $\boldsymbol{k}_v$ are related to the electron and hole drift velocities by

$$\hbar\, \boldsymbol{k}_c = m_c\, \boldsymbol{v}_c\,, \qquad \hbar\, \boldsymbol{k}_v = m_v\, \boldsymbol{v}_v\,. \tag{6}$$

The integration element $\mathrm{d}S$ in equation 3 is now given by

$$\mathrm{d}S = \mathrm{d}^3\boldsymbol{k}_1\, \mathrm{d}^3\boldsymbol{k}_{1'}\, \mathrm{d}^3\boldsymbol{k}_{2'}\,. \tag{7}$$

Four of the integrals in equation 3 drop out because of conservation of energy and momentum but an exact treatment involves carrying out the remaining five integrals. Such a calculation is difficult even in the equilibrium case and appears impossible in the present situation. Beattie and Landsberg has suggested an approximate procedure based on the fact that the integrand of equation 3, consists of rapidly varying exponential factors contained in the θ:s and ϱ:s and more slowly varying factors in T_{if}. For each of the two terms of the integrand in equation 3 the dominant variation with temperatures and drift velocities is therefore obtained if the product of the occupation probabilities θ or ϱ is replaced by its maximum value subject to energy and momentum conservation. Beattie and Landsberg also manage to find a nonexponential temperature dependent factor by a simple but nonrigorous procedure. In the equilibrium case they were able to carry out an alternative more rigorous calculation to verify their results. In the present case although such a calculation may be possible it will be extremely complicated and will not be attempted here. Instead we will rely on their nonrigorous method which is believed to give the correct temperature variation apart from factors of the form $A + B(T_c/T_v)$ with the constants A and B of the same order of magnitude. Our problem then is to maximize the products $\theta(1)\,\theta(1')\,\theta(2)\,\theta(2')$ and $\varrho(1')\,\varrho(1')\,\varrho(2)\,\varrho(2')$ subject to the conditions

$$\left.\begin{aligned} \boldsymbol{k}_1 + \boldsymbol{k}_2 &= \boldsymbol{k}_{1'} + \boldsymbol{k}_{2'}\,, \\ \frac{\hbar^2 k_1^2}{2\,m_c} + \frac{\hbar^2 k_2^2}{2\,m_c} + \frac{\hbar^2 k_{1'}^2}{2\,m_v} + \frac{\hbar\, k_g^2}{2\,m_c} &= \frac{\hbar^2 k_{2'}^2}{2\,m_c} \end{aligned}\right\} \tag{8}$$

expressing conservation of momentum and energy. Here k_g is related to the energy gap E_g by

$$\frac{\hbar^2 k_g^2}{2\,m_c} = E_g\,. \tag{9}$$

It can be shown that these products obtain their maximum values both in the equilibrium and nonequilibrium cases when all the $\boldsymbol{k}$-vectors are parallel. Still following Beattie and Landsberg we write

$$\boldsymbol{k}_1 = a\,\boldsymbol{k}_{1'}\,, \qquad \boldsymbol{k}_2 = b\,\boldsymbol{k}_{1'}\,, \tag{10}$$

and minimize

$$Y = \frac{\hbar^2 (\boldsymbol{k}_1 - \boldsymbol{k}_c)^2}{2\,m_c\,k_B\,T_c} + \frac{\hbar^2 (\boldsymbol{k}_2 - \boldsymbol{k}_c)^2}{2\,m_c\,k_B\,T_c} + \frac{\hbar^2 (\boldsymbol{k}_{1'} - \boldsymbol{k}_v)^2}{2\,m_v\,k_B\,T_v} \tag{11a}$$

and

$$Z = \frac{\hbar^2 (\boldsymbol{k}_{2'} - \boldsymbol{k}_c)^2}{2\,m_c\,k_B\,T_c} \tag{11b}$$

subject to the conditions given by equation (8). This is a straightforward variational problem and it is easily seen that both Y and Z are minimum when $a = b$. It is not possible to determine analytically the common value of these parameters but they will be determined by a numerical analysis. Once a and b are known all k_i:s are easily eliminated from (11a) and (11b) by using (8) and (10). By copying the arguments of Beattie and Landsberg we also find that integration over any particular k_i in equation (3) results in a factor $T_c^{-1/2}$ or $T_v^{-1/2}$ whatever is appropriate for that i-value. We then find that $P(\mathrm{a, b}) = 0$ is equivalent to

$$n^3\,T_c^{-1}\,T_v^{-1/2} \exp\,(-\,Y_{\min}) \approx \text{const}\; n\,T_c^{1/2}\,T_v \exp\,(-\,Z_{\min}) \tag{12}$$

or

$$n^2 = \text{const}\; T_c^{3/2}\,T_v^{3/2} \exp\,(Y_{\min} - Z_{\min})\,. \tag{13}$$

In the absence of an applied electric field we have

$$\left.\begin{aligned} a = b &= m_c/m_v\,, \quad T_c = T_v = T\,, \\ Y_{\min} &= \frac{\hbar^2 k_g^2}{2\,m_c\,k_B\,T}\,\frac{m_c}{m_c + m_v}\,, \\ Z_{\min} &= \frac{\hbar^2 k_g^2}{2\,m_c\,k_B\,T}\,\frac{2\,m_c + m_v}{m_c + m_v} \end{aligned}\right\} \tag{14}$$

which implies an intrinsic carrier concentration of

$$n = \text{const}\ T^{3/2} \exp(-E_g/2\,k_B\,T)\,. \tag{15}$$

According to the discussion above the const in (15) should actually be weakly dependent on temperatures and drift velocities but this dependence will be neglected and the value of the constant can then be obtained from the conventional formula for intrinsic carrier concentration

$$n = \frac{2}{\hbar^3}\left(\frac{k_B\,T}{2\,\pi}\right)^{3/2} (m_c\,m_v)^{3/4} \exp(-E_g/2\,k_B\,T)\,. \tag{16}$$

In the numerical calculations the nonparabolicity of the conduction band was taken into account by using a temperature dependent effective mass although this approximation is certainly hard to motivate in the present context. For consistency also the m_c in equation (16) was allowed to vary with T_c. A straight line fit to the effective mass values of Glicksman [2] was used giving

$$m_c = m\,(0.0145 + 1.13\times 10^{-5}\,T_c) \tag{17}$$

with T_c in °K.

3. Calculation of Carrier Drift Velocities and Temperatures

This section only contains a compilation of known results and will therefore be kept very brief. The momentum balance for each species of carrier requires that the momentum gained from the field equals the sum of the momentum $P_{\alpha p}$ lost to the lattice through the phonon interaction and the momentum $P_{\alpha\alpha'}$ lost to the opposite kind of carrier. The subscript α is equal to c for electrons and v for holes. We thus get

$$-\,n\,e\,E = P_{cv} + P_{cp}\,, \quad n\,e\,E = P_{vc} + P_{vp}\,, \quad P_{cv} = -\,P_{vc}\,. \tag{18}$$

Similarly for the rate of change of energy

$$n\,e\,v_c\,E = W_{cv} + W_{cp}\,, \quad n\,e\,v_v\,E = W_{vc} + W_{vp}\,, \quad W_{cv} = -\,W_{vc}\,. \tag{19}$$

Following Stratton [6] we have

$$\left.\begin{aligned}
P_{cp} &= -\frac{4\,n\,e\,v_c}{3\sqrt{\pi}\sqrt{k_B\,\theta/m_c}}\,F_{0c}\left(\sinh\frac{\theta}{2\,T_0}\right)^{-1}\left(\frac{\theta}{2\,T_c}\right)^{3/2}\times\\
&\quad\times\left[K_0\left(\frac{\theta}{2\,T_c}\right)\sinh\left(\frac{\theta}{2\,T_0}-\frac{\theta}{2\,T_c}\right)+K_1\left(\frac{\theta}{2\,T_c}\right)\cosh\left(\frac{\theta}{2\,T_0}-\frac{\theta}{2\,T_c}\right)\right],\\
P_{vp} &= \frac{4\,n\,e\,v_v}{3\sqrt{\pi}\sqrt{k_B\,\theta/m_v}}\,F_{0v}\left(\sinh\frac{\theta}{2\,T_0}\right)^{-1}\left(\frac{\theta}{2\,T_v}\right)^{3/2}\times\\
&\quad\times\left[K_0\left(\frac{\theta}{2\,T_v}\right)\sinh\left(\frac{\theta}{2\,T_0}-\frac{\theta}{2\,T_v}\right)+K_1\left(\frac{\theta}{2\,T_v}\right)\cosh\left(\frac{\theta}{2\,T_0}-\frac{\theta}{2\,T_v}\right)\right],\\
W_{cp} &= \frac{2}{\sqrt{\pi}}\,n\,e\,F_{0c}\sqrt{\frac{k_B\,\theta}{m_c}}\left(\sinh\frac{\theta}{2\,T_0}\right)^{-1}\left(\frac{\theta}{2\,T_c}\right)^{1/2}\times\\
&\quad\times K_0\left(\frac{\theta}{2\,T_c}\right)\sinh\left(\frac{\theta}{2\,T_0}-\frac{\theta}{2\,T_c}\right),\\
W_{vp} &= \frac{2}{\sqrt{\pi}}\,n\,e\,F_{0v}\sqrt{\frac{k_B\,\theta}{m_v}}\left(\sinh\frac{\theta}{2\,T_0}\right)^{-1}\left(\frac{\theta}{2\,T_v}\right)^{1/2}\times\\
&\quad\times K_0\left(\frac{\theta}{2\,T_v}\right)\sinh\left(\frac{\theta}{2\,T_0}-\frac{\theta}{2\,T_v}\right),
\end{aligned}\right\} \tag{20}$$

where θ is the Debye temperature, T_0 the lattice temperature and F_0 is given by

$$F_0 = \frac{m\, e\, k_B\, \theta}{4\, \pi\, \hbar^2} \left(\frac{1}{\varepsilon_\infty} - \frac{1}{\varepsilon_s} \right). \tag{21}$$

ε_∞ and ε_s are the high frequency and static values of the dielectric constant. The momentum exchange between the electrons and holes is essentially given by the Brooks-Herring formula but in the present case we have also taken the finite hole mass into account. The result is

$$P_{cv} = -\frac{e^4\, \Lambda\, n^2}{6\, (2\, \pi)^{3/2}\, \varepsilon_s^2}\, \frac{m_c + m_v}{m_c\, m_v} \left(\frac{k_B\, T_c}{m_c} + \frac{k_B\, T_v}{m_v} \right)^{-3/2} (v_c + v_v)\,, \tag{22}$$

where

$$\Lambda = \ln\, (\xi + 1) - \frac{\xi}{1 + \xi}\,,$$

$$\xi = \frac{4\, m_c\, m_v}{(m_c + m_v)^2}\, \frac{m_c\, k_B\, T_v + m_v\, k_B\, T_c}{n\, (T_c + T_v)}\, \frac{\varepsilon_s\, k_B\, T_c\, T_v}{\hbar^2\, e^2}\,. \tag{23}$$

Analogously we get for the energy exchange

$$W_{cv} = \frac{e^4\, \Lambda}{2\, (2\, \pi)^{3/2}\, \varepsilon_s^2}\, \frac{n^2}{m_c\, m_v} \left(\frac{k_B\, T_c}{m_c} + \frac{k_B\, T_v}{m_v} \right)^{-3/2} k_B\, (T_c - T_v)\,. \tag{24}$$

Equations (13), (18), and (19) now define a set of coupled equations for the variables n, T_c, T_v, v_c, and v_v as functions of the electric field E. A numerical solution has been obtained for data representative of InSb.

4. Results and Discussion

Table 1 summarizes the data for InSb which have been used in the numerical calculations. The results for a lattice temperature of 293 °K are given in Fig. 2 and 3.

The carrier density curve (Fig. 3) shows an exponential rise similar to a plot vs. lattice temperature. It should be pointed out that this curve is irrelevant for calculated carrier densities lower than those corresponding to the net donor

Table 1
Numerical data for InSb

ε_∞	15.7
ε_s	18.7
m_c (77 °K)	0.015 m_0
m_c (300 °K)	0.018 m_0
m_v	0.4 m_0
E_g	0.26 eV
n_i (300 °K)	1.6×10^{16} cm^{-3}
θ	260 °K

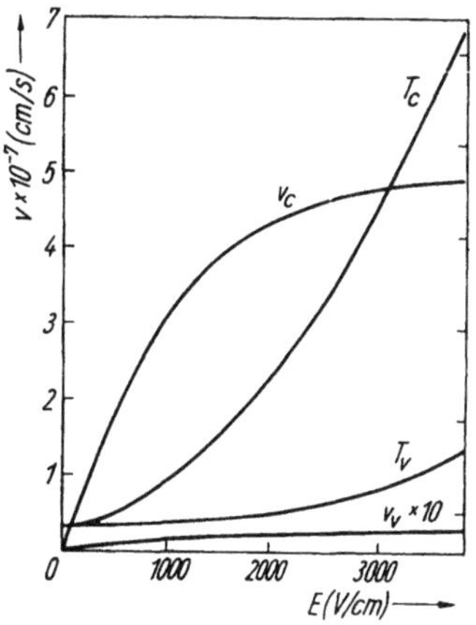

Fig. 2. Carrier temperatures and velocities vs. electric field. $T_L = 290$ °K

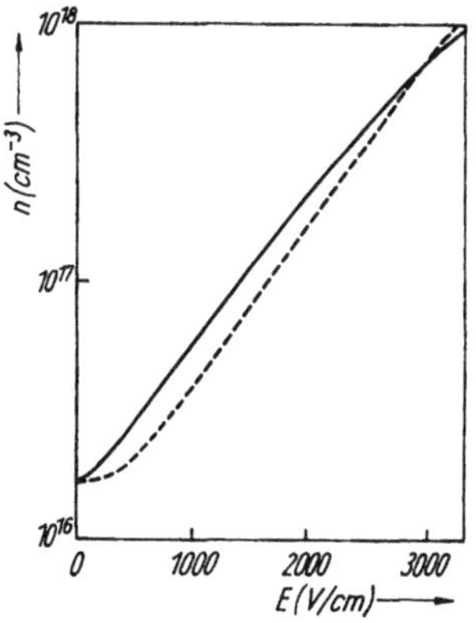

Fig. 3. Carrier concentration vs. electric field (solid line). Intrinsic carrier concentration at sample temperature equal to electron temperature shown for comparison (broken line). $T_L = 290$ °K

or acceptor density. For comparison we have also in the same figure given the carrier concentration resulting from the calculated electron temperature inserted into equation (15).

At moderate electric fields the temperature and velocity curves in Fig. 2 agree with those given by Stratton's theory but at higher fields there is no occurrence of dielectric breakdown as would be expected if polar scattering were dominant. We can therefore conclude that intercarrier scattering predominates. It should be pointed out that there is considerable uncertainity in the data which determine the dielectric breakdown field for InSb and a different choice might have resulted in a upward concave portion of the v_c vs. E curve.

An attempt was also made to perform calculations for a lattice temperature of 77 °K but numerical difficulties limited the results in this case to electric fields lower than about 400 V/cm. Fig. 4 and 5 give the results in this case.

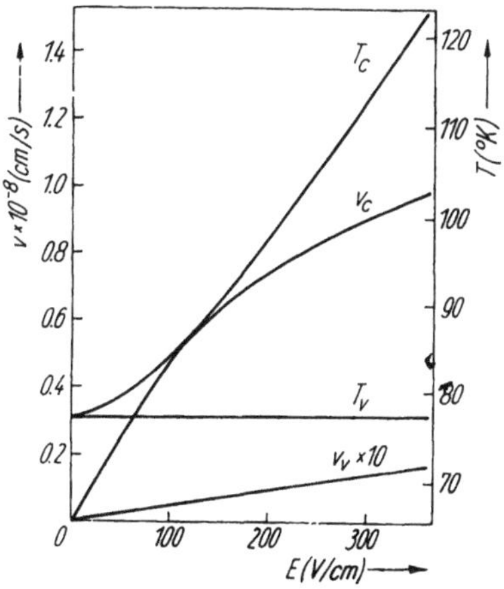

Fig. 4. Carrier temperatures and velocities vs. electric field. $T_L = 77$ °K

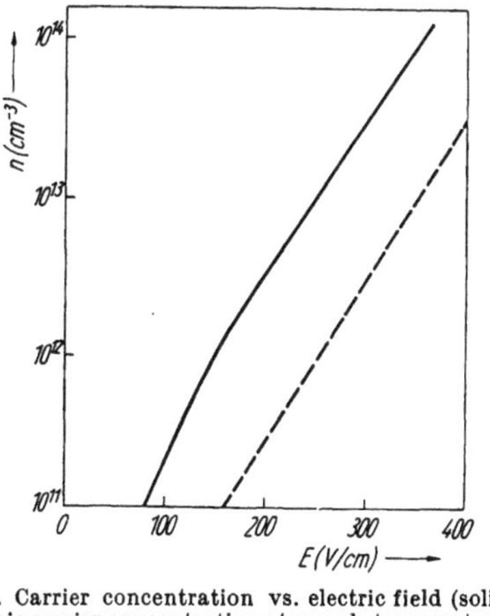

Fig. 5. Carrier concentration vs. electric field (solid line). Intrinsic carrier concentration at sample temperature equal to electron temperature shown for comparison (broken line). $T_L = 77$ °K

In Fig. 4 and 5 the electron density is sufficiently low to make intercarrier scattering almost negligible.

It should be pointed out that at the lower carrier concentrations shown in Fig. 5 neither the assumption of a displaced maxwellian velocity distribution nor of a dominant Auger process is valid. However, if we consider an extrinsic sample with an electron concentration of e.g. 10^{14} cm^{-3} Stratton's theory ought to be applicable. When the quasi-intrinsic carrier concentration due to the applied field reaches the value of the extrinsic concentration, we should have an Auger recombination rate comparable to that of other processes. For n-type material it is also at such field-strengths that the field-induced increase in concentration begins to have an effect on the conductivity. Defining a breakdown field E_B through the relation

$$n\,(E_B) = n_{ex}\,, \tag{25}$$

where n_{ex} is the extrinsic carrier concentration, we find from Fig. 5 $E_B \approx \approx 300$ V/cm for $n_{ex} \approx 10^{14}$ cm^{-3}. This value of E_B is in fair agreement with experimental results. It should also be noted that equation (15) with $T = T_c$ is a reasonable approximation to the carrier density at not too low fields and room temperature but unsatisfactory at liquid nitrogen temperature.

References

[1] H. EHRENREICH, J. Phys. Chem. Solids **2**, 131 (1957).
[2] M. GLICKSMAN and W. A. HICINBOTHEM, Phys. Rev. **129**, 1572 (1963).
[3] H. MORISAKI and Y. INUISKI, J. Phys. Soc. Japan **20**, 1814 (1965).
[4] M. C. STEELE and S. TOSHIMA, Japan J. appl. Phys. **2**, 381 (1963).
[5] B. ANCKER-JOHNSON, J. Phys. Soc. Japan **21**, 694 (1966).
[6] R. STRATTON, Proc. Roy. Soc. **246**, 406 (1958).
[7] A. R. BEATTIE and P. T. LANDSBERG, Proc. Roy Soc. **A 249**, 16 (1959).
[8] I. M. TSIDILKOVSKI, phys. stat. sol. **8**, 253 (1965).

(Received April 30, 1968)

phys. stat. sol. 28, 719 (1968)

Subject classification: 16; 22

Faculty of Physics, University of Bucharest

Role of Gradual Capture in the Photoelectromagnetic and Photoconductive Effects

By

V. DOLOCAN

A theoretical study is made of the role of the gradual capture in the photoelectromagnetic and photoconductive effects. The results obtained show that the photoelectromagnetic lifetime is not affected, while the photoconductive lifetime is affected by gradual trapping. The gradual recombination affects both the photoelectromagnetic and photoconductive lifetime.

Die Bedeutung des stufenweisen Einfangs für den photoelektromagnetischen Effekt und die Photoleitung wird theoretisch untersucht. Die erhaltenen Ergebnisse zeigen, daß die photoelektromagnetische Lebensdauer nicht verändert wird, während die Photoleitungslebensdauer durch stufenweisen Ladungsträgereinfang beeinflußt wird. Die stufenweise Rekombination beeinflußt sowohl die photoelektromagnetische als auch die Photoleitungslebensdauer.

1. Introduction

Zitter [1] has studied the role of trapping in the photoelectromagnetic (PEM) and photoconductive (PC) effects and his theoretical results are in good agreement with the experimental results. In the presence of trapping the photoconductive lifetime is different from the photoelectromagnetic lifetime. The photoelectromagnetic lifetime is independent of trapping [2] and is determined by the minority carrier lifetime. The photoconductive lifetime, generally, is determined by the majority carrier lifetime and depends on doping and trap concentration, on temperature, and on the electron and hole effective masses [3].

In his theoretical derivation, Zitter has assumed that the trap concentration is constant in the whole semiconductor bulk. But, according to the impurity diffusion law in semiconductors it is possible that the concentration is spatially dependent.

In this paper we have studied the role of the gradual capture in the photoelectromagnetic and photoconductive effects.

2. General Equations and Gradual Capture

Let us consider a semiconductor (Fig. 1) illuminated in y-direction and located in a magnetic field B in z-direction and in an electric field E_x in x-direction. It is assumed that a) the light is strongly absorbed at the illuminated ($y = 0$) surface; b) the nonequilibrium carrier concentrations n and p created by illumination are much smaller than the equilibrium carrier concentrations n_0 and p_0; c) boundary effects and the effect of surface states can be neglected; d) the steady-state conditions are fulfilled.

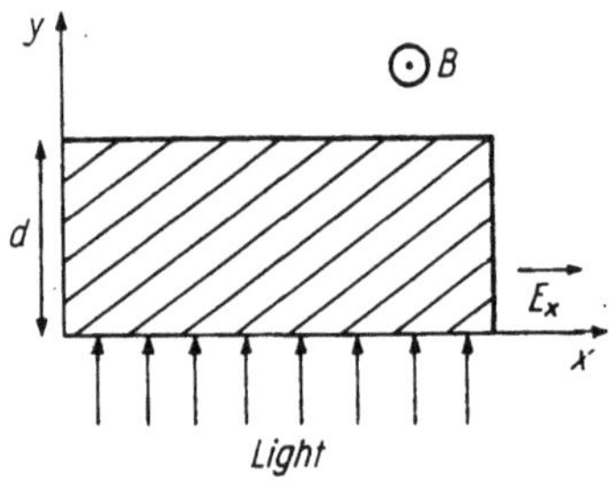

Fig. 1. Illuminated sample in an electric and magnetic field

If $\boldsymbol{J}_n, \boldsymbol{J}_p$ are the currents of electrons and holes, respectively, they have the form

$$\boldsymbol{J}_n = e\,D\,\nabla n - \mu\,\boldsymbol{J}_n \times \boldsymbol{B} + e\,\mu\,(n_0 + n)\,\boldsymbol{E}\,, \tag{1}$$

$$b\,\boldsymbol{J}_p = -\,e\,D\,\nabla p + \mu\,\boldsymbol{J}_p \times \boldsymbol{B} + e\,\mu\,(p_0 + p)\,\boldsymbol{E}\,, \tag{2}$$

where D is the electron diffusion constant, μ is the electron mobility, b is the ratio of electron to hole mobilities.

The continuity equations are

$$\nabla \boldsymbol{J}_n = \frac{e\,n}{\tau_n} = -\,\nabla \boldsymbol{J}_p = \frac{e\,p}{\tau_p}\,, \tag{3}$$

where τ_n and τ_p are electron and hole lifetimes, respectively.

By denoting with n_t the concentration of excess electrons in traps, the charge neutrality condition is

$$p = n + n_t\,. \tag{4}$$

In addition, $J_{ny} + J_{py} = 0$.

If boundary effects are ignored, for a plate-shaped sample a two-dimensional model can be used with the functions $\boldsymbol{J}_n, \boldsymbol{J}_p, \boldsymbol{E}, n$, and p varying only in the y-direction.

By writing equations (1) to (3) in component form and eliminating J_{nx}, J_{px} and E_y from the obtained equations system, one obtains the differential equation for J_{ny}, which in the case $n \ll n_0$, $p \ll p_0$ and using equation (4) has the form

$$\begin{aligned} J''_{ny}\,[n_0\,\tau_p + p_0\,\tau_n]\,b\,D + J'_{ny}\,[n_0\,\tau'_p + p_0\,\tau'_n]\,b\,D - \\ - J_{ny}\,[b\,p_0\,(1 + \mu^2\,B^2) + n_0\,(b^2 + \mu^2\,B^2)] = \\ = -\,e\,\mu^2\,B\,n_0\,p_0\,(1 + b)\,E_x\,. \end{aligned} \tag{5}$$

The normal component of the electron and hole densities at the illuminated and non-illuminated surfaces is assumed to obey the boundary conditions

$$\left.\begin{aligned} J_{ny} &= -\,J_{py} = -\,e\,g + e\,s_n\,n = -\,e\,g + e\,s_p\,p\,; && y = 0\,,\\ J_{ny} &= -\,J_{py} = -\,e\,s_n\,n = -\,e\,s_p\,p\,; && y = d\,, \end{aligned}\right\} \tag{6}$$

where g is the generation rate of electron–hole pairs by illumination and s_n, s_p are the electron and hole surface recombination velocities; d is the dimension of the sample in the y-direction.

In order to find the solution of equation (5) with the boundary conditions (6) it is necessary to specify the spatial dependence of the concentration of the capture centres. By capture centres we understand recombination and trapping centres. If, according to some authors [5, 6], we assume that the trapping centres are the centres for which the capture cross-section for one carrier type is negligible we find easily [3] that $\beta = n_0 \tau_p + p_0 \tau_n$ is independent on the trap concentration and is reversely proportional to the concentration of the recombination centres.

Even without making this assumption it was shown that β is independent on trapping [2] and therefore reversely proportional to the concentration of the capture centres. In the following sections we will write the expressions of the photoelectromagnetic and photoconductive lifetimes in the case if the concentration of the capture centres changes exponentially in the y-direction (that includes an exponential variation of β in this direction).

3. PEM Short-Circuit Current

In this case E_x vanishes and the short-circuit current per unit sample width, i_{PEM}, is given by

$$i_{\text{PEM}} = -\mu B \left(1 + \frac{1}{b}\right) \int_0^d J_{ny} \, dy \,. \tag{7}$$

Using J_{ny} obtained from equations (5) and (6), it results

$$i_{\text{PEM}} = \frac{2 e g}{\alpha A} \mu B \left(1 + \frac{1}{b}\right) \eta_1, \tag{8}$$

where

$$A = [\gamma_0 I_1(\gamma_0) - s_n \tau_{n0} \gamma_0^2 I_0(\gamma_0)] \left[K_1(\gamma_1) - s_n \tau_{n1} \frac{\alpha}{2} \gamma_1 K_0(\gamma_1)\right] - \\ - \left[I_1(\gamma_1) + s_n \tau_{n1} \frac{\alpha}{2} I_0(\gamma_1)\right] \left[\gamma_0 K_1(\gamma_0) + s_n \tau_{n0} \frac{\alpha}{2} \gamma_0^2 K_0(\gamma_0)\right], \tag{9}$$

$$\eta_1 = \frac{1}{\gamma_1} - [I_0(\gamma_0) K_1(\gamma_1) + K_0(\gamma_0) I_1(\gamma_1)] + \\ + s_n \tau_{n1} \frac{\alpha}{2} \gamma_1 [K_0(\gamma_1) I_0(\gamma_0) - I_0(\gamma_1) K_0(\gamma_0)] \,,$$

$$\alpha = \frac{1}{d} \ln \left(\frac{\beta_0}{\beta_1}\right),$$

and

$$\gamma = \frac{2}{\alpha} \left[\frac{p_0 (1 + \mu^2 B^2) + n_0 \left(b + \frac{\mu^2 B^2}{b}\right)}{D}\right]^{1/2} \beta^{-1/2} \,. \tag{10}$$

$\gamma_0, \beta_0, \tau_{n0}$ and $\gamma_1, \beta_1, \tau_{n1}$ are the values of γ, β, τ_n at $y = 0$ and $y = d$ surfaces, respectively. $K_j(\gamma)$ and $I_j(\gamma)$ ($j = 0, 1$) are the modified Bessel functions. For $\beta_0 \gg \beta_1$ (that involving $\gamma_1 \gg \gamma_0$) and negligible surface recombination velocities, equation (8) is simplified and we obtain

$$i_{\text{PEM}} = e g \mu B \left(1 + \frac{1}{b}\right) L_G \,, \tag{11}$$

where

$$L_G = L_D \frac{K_1(\gamma_0)}{K_0(\gamma_0)} \tag{12}$$

is the "gradual effective ambipolar diffusion length" and

$$L_{\mathrm{D}} = \left[\frac{D\,(n_0\,\tau_{\mathrm{p}0} + p_0\,\tau_{\mathrm{n}0})}{p_0\,(1 + \mu^2 B^2) + n_0\left(b + \dfrac{\mu^2 B^2}{b}\right)}\right]^{1/2} \tag{13}$$

is the "ungradual effective ambipolar diffusion length" [1].

The gradual photoelectromagnetic lifetime, $\tau_{\mathrm{PEM}}^{\mathrm{G}}$, is given by the expression

$$\tau_{\mathrm{PEM}}^{\mathrm{G}} = \tau_{\mathrm{PEM}}\left[\frac{K_1(\gamma_0)}{K_0(\gamma_0)}\right]^2 \tag{14}$$

where τ_{PEM} is the ungradual photoelectromagnetic lifetime [1].

Since β is independent of trapping [2] it follows that γ (equation (10)) is independent of trapping and, therefore, the photoelectromagnetic lifetime is independent of trapping, irrespective of the existence of the gradual or ungradual trapping. This conclusion shows that the photoelectromagnetic effect is determined by an ambipolar diffusion process and the trapping of a certain carrier type makes a further displacement of the pair impossible. The photoelectromagnetic lifetime depends on the gradual recombination, as it is observed from relation (14). For $\beta_0 = \beta_1$ it results that $L_{\mathrm{G}} = L_{\mathrm{D}}$ and $\tau_{\mathrm{PEM}}^{\mathrm{G}} = \tau_{\mathrm{PEM}}$.

4. PC Short-Circuit Current

In this case B vanishes and the short-circuit current per unit sample width, i_{PC}, is given by

$$i_{\mathrm{PC}} = e\,\mu\,E_x \int_0^d \left(n + \frac{p}{b}\right) \mathrm{d}y\,. \tag{15}$$

Using equation (5) and the expression for $J_{\mathrm{n}y}$, the integral from equation (15) may be easily resolved in two cases: a) the trapping is negligible; b) the concentration of the recombination centres is spatially independent while the trap concentration depends on the spatial coordinates (by assuming that the trapping centres are the centres for which the capture cross-section of one carrier type is negligible).

In the first case, for large values of γ, we obtain

$$i_{\mathrm{PC}} = \mu\,E_x\left(\tau_{\mathrm{n}0} + \frac{\tau_{\mathrm{p}0}}{b}\right)\frac{\eta_2}{A}\,\gamma_0^2\,, \tag{16}$$

where

$$\eta_2 = \frac{1}{\gamma_0}\,[I_1(\gamma_0)\,K_1(\gamma_1) - K_1(\gamma_0)\,I_0(\gamma_1)] + \frac{1}{\gamma_1}\,s_{\mathrm{n}}\,\tau_{\mathrm{n}1}\,\frac{\alpha}{2} - \\ - \frac{\gamma_1}{\gamma_0}\,s_{\mathrm{n}}\,\tau_{\mathrm{n}1}\,\frac{\alpha}{2}\,[K_0(\gamma_1)\,I_1(\gamma_0) + K_1(\gamma_0)\,I_0(\gamma_1)]\,. \tag{17}$$

A and γ are given by relations (9) and (10), respectively, for $B = 0$. $\tau_{\mathrm{p}0}$ is the value of τ_{p} at $y = 0$ surface.

For $\beta_0 \gg \beta_1$ and when the surface recombination velocities are negligible, results

$$i_{\mathrm{PC}} = \mu\,E_x\,e\,g\,\tau_{\mathrm{PC}}^{\mathrm{G}}\,, \tag{18}$$

where

$$\tau_{\mathrm{PC}}^{\mathrm{G}} = \tau_{\mathrm{PC}}\,\frac{I_0(\gamma_1)}{I_1(\gamma_1)} \tag{19}$$

is the gradual photoconductive lifetime. τ_{PC} is the ungradual photoconductive lifetime [1].

In the second case according to the fact that in this case β is independent of the trap concentration, it results that β is independent of the spatial coordinates, and therefore equation (5) reduces to Zitter's equation [1]. The dependence of the lifetime on the trap concentration may be linear in a large temperature range [6, 7]. For an exponential variation of the trap concentration in the y-direction, taking into account the above simplifying conditions, τ_{PC}^{G} becomes

$$\tau_{PC}^{G} = \frac{1}{2}\left(\tau_{n0} + \frac{\tau_{p0}}{b}\right)\left(1 + \frac{1}{\alpha_1 L_D + 1}\right), \tag{20}$$

where

$$\alpha_1 = \frac{1}{d}\ln\frac{\tau_{PC_0}}{2\,\tau_{PC_1} - \tau_{PC_0}}.$$

τ_{PC_0} and τ_{PC_1} are the values of τ_{PC} at the $y = 0$ and $y = d$ surfaces, respectively. For a variation of the trap concentration, N_t, of the form $N_t \sim y^n$ ($n = 1, 2, 3, \ldots$), in the same simplifying conditions, we obtain

$$\tau_{PC}^{G} = \tau_{n0} + \frac{\tau_{p0}}{b} + n!\left(\frac{L_D}{d}\right)^n\left(\Delta\tau_n + \frac{\Delta\tau_p}{b}\right), \tag{21}$$

where $\Delta\tau_n = \tau_{n1} - \tau_{n0}$, $\Delta\tau_p = \tau_{p1} - \tau_{p0}$.

L_D from (20) and (21) is given by (13) for $B = 0$. The equations (19), (20), and (21) show that the photoconductive lifetime is affected by both gradual recombination and gradual trapping.

References

[1] R. N. Zitter, Phys. Rev. **112**, 852 (1958).
[2] S. G. Kalashnikov and K. P. Thiessen, Fiz. tverd. Tela **2**, 2743 (1960).
[3] C. Constantinescu and V. Dolocan, to be published in Rev. Roumaine Phys. (1968).
[4] S. W. Kurnick and R. N. Zitter, J. appl. Phys. **27**, 278 (1956).
[5] S. M. Ryvkin, Fotoelektricheskie yavleniya v poluprovodnikakh, Fizmatgiz, Moskva 1963.
[6] J. Shirafuji, Japan. J. appl. Phys. **5**, 469 (1966).
[7] C. Constantinescu and V. Dolocan, to be published in Fiz. Tekh. Poluprov. (1968).

(Received May 28, 1968)

phys. stat. sol. 28, 725 (1968)

Subject classification: 9; 10.2; 22.5.1

Institute of Physical Chemistry, Bulgarian Academy of Sciences, Sofia

Temperature Dependence of the Diffusion of Photoexcited Holes in AgBr

By

M. GEORGIEV, E. ZAGORSKA, and J. MALINOWSKI

To Professor R. KAISCHEW, on the occasion of his sixtieth birthday

A detection technique based on the bleaching of a thin silver deposit is used to investigate the diffusion of photoinduced holes between −35 and +35 °C. The apparent diffusion coefficient is found to follow an Arrhenius-type temperature dependence exhibiting an activation energy of 0.53 eV. Following previous suggestions, it is again concluded that the holes become engaged in multiple trapping while migrating within the crystal lattice. Some trapping models are discussed.

Eine Nachweistechnik, die auf dem Ausbleichen einer dünnen Silberschicht beruht, wird zur Untersuchung der Diffusion von photoinduzierten Defektelektronen zwischen − 35 und + 35 °C benutzt. Der scheinbare Diffusionskoeffizient zeigt eine Arrhenius-Abhängigkeit von der absoluten Temperatur mit einer Aktivierungsenergie von 0,53 eV. Früheren Annahmen folgend, wird wieder angenommen, daß die Defektelektronen an Mehrfachtrapping während der Wanderung im Kristallgitter beteiligt sind. Einige Haftstellenmodelle werden diskutiert.

1. Introduction

Studies of the diffusion of photoexcited holes in single crystals of AgBr were recently carried out at room temperature by means of development techniques [1, 2]. The apparent diffusion coefficient found from these experiments was about 3×10^{-7} cm^2 s^{-1}. A hole drift mobility of about 1.2×10^{-5} cm^2 V^{-1} s^{-1} was calculated from the above value by means of the Einstein relation. However, the latter estimate proved to be sharply inconsistent, by several orders of magnitude, with the Hall mobility data yielding about 2 cm^2 V^{-1} s^{-1} at room temperature. It was suggested that the observed discrepancy between diffusion and Hall data at 300 °K might indicate that the holes were engaged in multiple trapping while moving through the crystal.

Prior to the above mentioned work, investigations carried out on crystals having a stoichiometric excess of bromine [3, 4] showed that diffusion was well correlated with the drift of holes in relatively weak external electric fields. The temperature dependence of the hole drift mobility was investigated between 250 °C and room temperature [4], and, as result, the discrepancy between diffusion and Hall data could be followed up within a rather wide temperature range. This discrepancy was found to deepen as the temperature was lowered below 150 °C. Various types of trapping models were considered [4, 5] to account for the observed deviations.

The present paper describes an investigation of the temperature dependence of the hole diffusion coefficient between − 35 and +35 °C. This investigation was a natural extension of earlier works [1, 2] at room temperature. In view of the lack of reliable information for most of the above region, it was considered desirable to investigate the trapping process at moderated temperatures.

2. Principle of the Method

The method employed [1, 2] may be described briefly as follows: The cathetus side of a wedge-shaped sample was illuminated at a fixed temperature through narrow slits, each of these giving the crystal a different exposure (Fig. 1). The light which was completely absorbed within some microns from the surface of incidence, produced a steep concentration gradient resulting in a diffusion flow of electronic defects into the interior. Holes emerging at the hypothenuse side were recorded there by the bleaching of a monoatomic silver deposit. Slit-like bleached images of different lengths revealed by development indicated the depths penetrated by the diffusing holes. The experimentally established dependence of the penetration depths, corresponding to the upper terminals of the images, on the duration of the exposure was then used to determine the apparent diffusion coefficient D of the holes.

As shown earlier [1], the dependence of the penetration depths x_c on the time of exposure t might be represented approximately by the simple equation

$$x_c(t) = (\lambda^2 + 6\,D\,t)^{1/2} - \lambda \tag{1}$$

provided that the penetration depths are "sufficiently small". Here λ is a parameter which depends on the boundary condition at the illuminated surface and on the density of the silver deposit on the hypotenuse side. The term "small" means that x_c obeys an inequality of the type

$$x_c \leqq \pi\,(D\,t/m)^{1/2} \tag{2}$$

over the whole range of exposure investigated. Here m is a positive number greater than unity, and as shown in [1] the error introduced on using approximation (1) decreases very quickly as m is increased. The apparent diffusion coefficient D is derived from the best fit of equation (1) to the experimental plot $x_c = x_c(t)$ for each temperature separately. Once D is derived, one can test the validity of inequality (2) and evaluate m. In the present case, for different temperatures m was found to range between 1.6 and 5.7 showing that the calculation error due to the approximation was between 25 and 0.3 per cent.

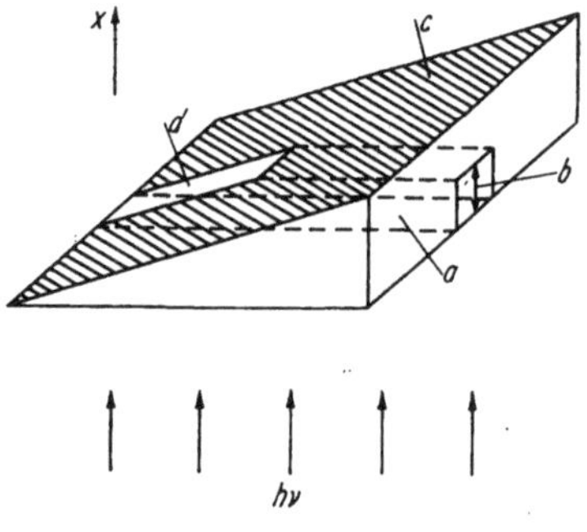

Fig. 1. Schematic plot of the wedge-shaped sample: a – illuminated slit, b – depth penetrating the diffusing holes, c – silver deposit used as an indicator, d – bleached image of the slit

3. Experimental

The preparation and further treatment of the samples has already been described in detail [1]. In order to minimize the effect of the surface evolution of photolytic bromine on the measuring accuracy during exposure, each sample was immersed in a bromine acceptor. Alcohol was found to be a good agent for this purpose, chiefly because of its relatively low freezing point. The depths of penetration corresponding to the upper terminals of the bleached images, observed after development, were determined under a microscope.

During exposure, the temperature of the sample was kept constant by placing the whole arrangement in a simple cryostat cooled by liquid nitrogen [6]. Temperature control was secured by using a copper–constantan thermocouple mounted on the brass plate supporting the sample. The rate of heating was not higher than 20 °C per hour and the temperature was maintained constant within ± 2 °C.

4. Results

Experimental results at five temperatures are presented in Fig. 2 where the hole penetration depths x_c in two samples are plotted against the square root of the time of exposure t. The solid curves there represent equation (1) at five pairs of values of D and λ which were obtained by means of a computer analysis of the experimental data. It will be seen from equation (1) that D is proportional to the square of the slope of the linear part of the experimental curve at larger values of t. The curves presented in Fig. 2, therefore, clearly reveal a decrease of the apparent diffusion coefficient as the temperature is lowered.

The values of the hole diffusion coefficient obtained presently are next shown in Fig. 3 in logarithmic plot against the reciprocal of the absolute temperature. These values are seen to fit well a linear dependence on T^{-1}. The slope corresponds to an activation energy for hole diffusion of 0.525 eV. The solid line in Fig. 3 is plotted according to the equation

$$D = \exp(5.44 - 0.525\ \text{eV}/kT)\ \text{cm}^2\ \text{s}^{-1}\,. \tag{3}$$

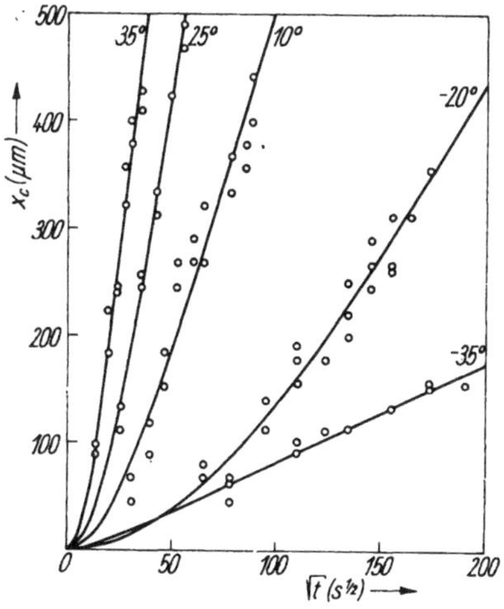

Fig. 2. Hole penetration depths vs. square root of the exposure time at five different temperatures

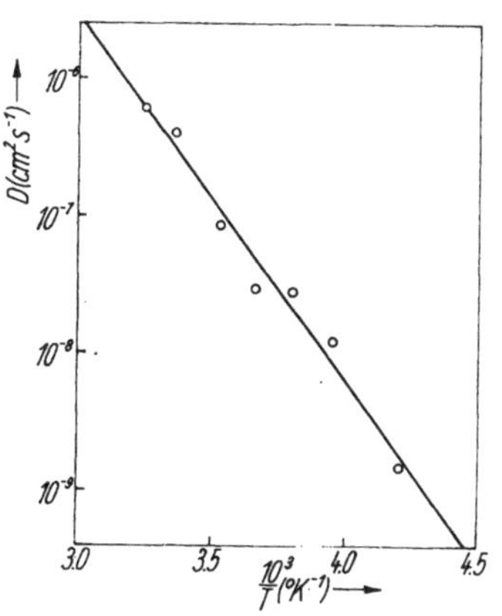

Fig. 3. Temperature dependence of the apparent diffusion coefficient of photoexcited holes in AgBr

5. Discussion

As mentioned earlier, a hole drift mobility μ_d could be related to the apparent diffusion coefficient D by means of the Einstein relation. Following previous suggestions [1], the deviations of μ_d from the microscopic mobility μ_m of the holes can be formally explained by introducing a correction term δ of the form

$$1 + \delta = (kT / e) (\mu_m / D) = \mu_m / \mu_d , \tag{4}$$

where e, k, and T have their usual meanings. Deviations between μ_d and μ_m occurring in some temperature region always indicate that the hole migration in this region differs from the motion of free carriers. Such a difference might arise as a result of the interaction of holes with some other defects of the crystal lattice.

The temperature dependence of δ in the range of the present investigation can easily be found from equations (3) and (4). At least for a part of this range, Hanson's data [7] for the Hall mobility of the holes yield [4, 5]

$$\mu_H = \exp(-2.01 + 0.07 \text{ eV}/kT) \quad \text{cm}^2/\text{Vs} . \tag{5}$$

Now, assuming identity between microscopic and Hall mobility one obtains

$$\delta = 6.2 \times 10^{-5} \exp(0.56 \text{ eV}/kT) . \tag{6}$$

Here, in deriving the expression (6) the following approximation has also been used:

$$T = 1.235 \times 10^3 \exp(-0.0366 \text{ eV}/kT) \text{ deg.}$$

At the same time, the temperature dependence of the hole drift mobility within the range between 150 °C and room temperature as found by Eisenmann and Jaenicke [4]

$$\mu_d = 1 \times 10^8 \exp(-0.71 \text{ eV}/kT) \quad \text{cm}^2/\text{Vs} \tag{7}$$

leads to

$$\delta = 1.34 \times 10^{-9} \exp(0.78 \text{ eV}/kT) . \tag{8}$$

The difference of 0.22 eV in the activation energies ascribed to equations (8) and (6) might be related to differences in the properties of AgBr in the two neighbouring temperature regions studied.

The possible nature of the interaction leading to $\delta \neq 0$ has already been subjected to a careful consideration [1, 4, 5]. It was shown that space charge effects resulting from long range electrostatic interactions of the migrating holes with mobile ionic defects are relatively less important. At the same time recent investigations of the diffusion of electrons and holes at room temperature [8] have indicated that these species migrate independently of each other. As a result, it was concluded that multiple trapping schemes alone can reliably account for the reduced hole drift mobility at moderated temperatures [1, 4, 5]. In this respect, various trapping models were discussed [4, 5].

It has been estimated much earlier [9] that trapping of electrons in AgBr crystals should be followed by an appropriate ionic migration aimed at compensating the excess negative charge. According to a widely accepted view, the photographic products in these crystals are formed as a result of alternative processes of trapping and compensation.

Similar trapping schemes have also been proposed for holes [4, 5, 10]. In one of these [5], the additional charge produced by trapping a hole h$^{\cdot}$ at a bulk site $J_h^\times$ is assumed to be neutralized by a silver ion vacancy V'_{Ag} which combines with the trapped hole $J_h^\cdot$ to form a relatively stable complex $(V_{Ag}\,J_h)^\times$. Under the influence of thermal excitation, however, this complex is sooner or later disintegrated to be formed again at another site. Recently, the latter model obtained further experimental support. It was shown unambiguously [6] that vacancy-type compensation actually predominates in the impurity region of temperature.

Now, in accordance with the latter remarks, one may attempt to analyse the deviations between diffusion and Hall data in terms of the above-mentioned model:

$$\left.\begin{aligned} h^\cdot + J_h^\times &\rightleftharpoons J_h^\cdot\,; & K_t^{-1}, \\ V'_{Ag} + J_h^\cdot &\rightleftharpoons (V_{Ag}\,J_h)^\times\,; & K_p^{-1}\,. \end{aligned}\right\} \qquad (9)$$

Here cross, dash, and dot stand for a net charge of 0, $-e$, and $+e$ respectively. According to statistical theories [11], the equilibrium constants K_t and K_p are given by

$$\left.\begin{aligned} K_t &= \tfrac{1}{2}\, g_v\, (2\,\pi\, m_h\, kT/h^2)^{3/2} \exp\,(-\,E_t/kT) = \\ &= 6.1\times 10^{-2}\, (m_h/m)^{3/2} \exp\left(-\,(E_t + 0.055)\ \mathrm{eV}/kT\right), \\ K_p &= g^{-1} \exp\,(-\,H_p/kT) \qquad \text{(mole fraction)}\,. \end{aligned}\right\} \qquad (10)$$

Here E_t is the trapping energy and H_p is the enthalpy for pairing of a trapped hole with a silver ion vacancy; m_h is the hole effective mass, $m_h = 0.1\,m$ [5]; $g_v = 12$ is the number of valleys in the valence band of AgBr [5]; m and h are the usual atomic constants; g is the number of orientations of the hole–vacancy complex. Assuming that holes in AgBr are trapped by substitutional metal cations, $J_h \equiv Me_{Ag}$, one obtains $g = 12$.

It will be seen that in the above model the correction term δ appears in the form

$$\delta = \left([J_h^\cdot] + [(V_{Ag}\,J_h)^\times]\right)/[h^\cdot] = ([J_h^\times].\,K_t^{-1})\,(1 + [V'_{Ag}].\,K_p^{-1})\,, \qquad (11)$$

where square brackets denote mole fractions. The latter expression has been derived by applying the law of mass action to equation (9). Equation (11) is to be compared next with the experimental equation (6). Assuming a strong hole–vacancy association, the second term inside the brackets in (11) may become much larger than unity. At the same time, the mole fraction of cationic vacancies in the impurity region of temperature is constant, being determined entirely by the divalent metal impurity content. If finally one assumes a low excitation level, when the density of traps exceeds largely the density of holes trapped, it follows

$$[J_h^\times]\,[V'_{Ag}] = 1.2\times 10^{-7}\, g^{-1}\,, \qquad E_t + H_p = 0.51\ \mathrm{eV}\,. \qquad (12)$$

Recent measurements of the ionic conductivity of AgBr [12] have revealed enthalpies of 0.16 eV for the pairing of cationic vacancies with divalent substitutional cations. On the other hand, drift mobility measurements made under pulse regime, disturbing the second equilibrium in equations (9), have pointed to the presence of hole traps as high as 0.40 eV above the top of the valence band [6]. It is to be noted that the latter values are not too different from those to be expected from [12].

Recent spectrochemical analysis of the crystals used here revealed the presence of Fe ions and other heavy metal impurities of mole fractions not exceeding

7×10^{-8}. Bearing in mind that for such relatively pure crystals the range of temperatures between −35 and +35 °C already belongs to the impurity region of ionic conductivity [12], one comes to the conclusion that the mole fraction of cationic vacancies in the present samples was probably about 10^{-7}. But now, from equation (12) a tremendously large fraction of hole traps as high as 1.2 g^{-1} follows. Such a large trap density is not consistent with the impurity content of the crystals and is likely to indicate that certain self-trapping events could occur in the AgBr crystals even around room temperature. However, any self-trapping scheme has to be dealt with cautiously, inasmuch as this type of trapping is most likely to occur at lower temperatures, as found for alkali halides [13]. Nevertheless, Höhne and Stasiw [14] seem to have proved recently by ESR measurements the occurrence of self-trapping in the silver sublattice leading to the formation of $Ag^{\cdot}_{Ag}$.

6. Conclusion

Following recent suggestions [1, 4 to 6, 8, 10], it seems reliable to conclude that near room temperature the photoholes in AgBr become effectively trapped at some sites in the bulk and are then neutralized by silver ion vacancies. The identity of the trapping centres has not yet been reliably established, but whatever their nature, it seems very likely that a rather stable complex is formed between a trapped hole and a silver ion vacancy. It is natural to identify these complexes, the formation of which is revealed by the discrepancy between diffusion and Hall data, as primary products of photodecomposition in the AgBr crystals. The latter products are unstable, having a lifetime as much as some tenths of a second, but might aggregate during the later stages of photolysis giving rise to the occurrence of a "bromine image" in the crystals. Perhaps it is worth noting finally that the above-mentioned scheme of hole trapping and compensation, if further confirmed, would eventually lead to a symmetrical extension of the primary Gurney-Mott scheme in which electrons and holes would become independently and separately trapped and then neutralized correspondingly by the two mobile constituents of ionic disorder in AgBr.

References

[1] M. Georgiev and J. Malinowski, J. Phys. Chem. Solids **28**, 931 (1967).
[2] M. Georgiev and J. Malinowski, C.R. Acad. Bulg. Sci. **20**, 45 (1967).
[3] G. W. Luckey, Disc. Faraday Soc. **28**, 113 (1959).
[4] E. Eisenmann and W. Jaenicke, Z. phys. Chem. (N.F.) **49**, 1 (1966).
[5] F. A. Kröger, J. Phys. Chem. Solids **27**, 1967 (1966).
[6] A. Buroff and J. Malinowski, phys. stat. sol. **26**, 267 (1968).
[7] R. C. Hanson, J. phys. Chem. **66**, 2376 (1962).
[8] M. Georgiev and J. Malinowski, Proc. Internat. Congr. Photographic Science, Tokyo 1967.
[9] R. W. Gurney and N. F. Mott, Proc. Roy. Soc. **164**, 151 (1938).
[10] W. F. Berg, Phot. Sci. Eng. **11**, 242 (1967).
[11] O. Stasiw, Elektronen und Ionenprozesse in Ionenkristallen, Springer-Verlag, 1959.
[12] P. Müller, phys. stat. sol. **9**, 193 (1965).
[13] W. Känzig., Proc. Internat. Conf. Semicond. Phys., Prague 1960.
[14] M. Höhne and M. Stasiw, phys. stat. sol. **25**, K55 (1968).

(Received June 4, 1968)

phys. stat. sol. 28, 731 (1968)

Subject classification: 1.2 and 6; 14.4.2; 22

Institute of Physics and Astronomy,
Academy of Sciences of the Estonian SSR, Tartu

Displacive Vibronic Phase Transitions in Narrow-Gap Semiconductors

By

N. KRISTOFFEL and P. KONSIN

Displacive phase transitions are considered in small-gap semiconductors due to the electron–phonon interaction of two bands. The Helmholtz free energy of the system is obtained on the basis of the electron spectrum renormalized by the interaction. The temperature dependent low-symmetry lattice distortion, renormalized frequencies of the active "soft" optical branch in both phases, the Curie temperature, etc., are obtained. By means of the Green's function method combinative vibronic frequencies of the system are found. The dielectric susceptibility which is essentially determined by active vibrations is calculated and the general picture of similar (ferroelectric) phase transitions is discussed.

Рассмотрены фазовые переходы типа смещения в полупроводниках с узкой щелью в результате электрон-фононного взаимодействия двух зон. Свободная энергия системы получена на основе перенормированного взаимодействием электронного спектра. Получены зависящее от температуры низкосимметричное искажение решетки, перенормированные частоты активной „мягкой" оптической ветви в обоих фазах, температура Кюри и т.д. Методом функций Грина найдены сложные электрон-фононные частоты системы. Вычислена диэлектрическая проницаемость, существенно определяемая активными колебаниями и обсуждена общая картина подобных фазовых переходов (сегнетоэлектрических).

Recent investigations have shown that the electron–phonon (vibronic) interaction for degenerate or close electron states of a crystal (the Jahn-Teller and pseudo-Jahn-Teller effects respectively) may give rise to displacive phase transitions [1 to 10]. This problem is particularly essential for explaining the nature of ferroelectricity. Using a simple model of two extremely narrow electron bands mixed by cut-off vibrations some qualitative aspects of such kind of transitions were considered in [6]. The present paper gives a more thorough treatment for a more realistic model.

1. General Picture of Phase Transition

Assume a crystal with two electron bands separated by the energy gap Δ, where the lower ($\sigma = 1$) is fully occupied at $T = 0$ and the upper ($\sigma = 2$) is empty. These bands are mixed by suitable lattice vibrations, i.e. vibronic interaction takes place. The Hamiltonian of the system can be written as

$$\mathcal{H} = \sum_{\sigma,\boldsymbol{k}} \varepsilon_\sigma(\boldsymbol{k})\, a^+_{\sigma\boldsymbol{k}}\, a_{\sigma\boldsymbol{k}} + \frac{1}{2}\sum_{\boldsymbol{q},j} (p^2_{\boldsymbol{q}j} + \omega^2_{\boldsymbol{q}j} Q^2_{\boldsymbol{q}j}) +$$

$$+ \hbar \sum_{\sigma,\boldsymbol{k}} \sum_{\sigma',\boldsymbol{k}'} \sum_{\boldsymbol{q},j} C^j_{\sigma\sigma'}\begin{pmatrix} \boldsymbol{q} \\ \boldsymbol{k}, \boldsymbol{k}' \end{pmatrix} a^+_{\sigma\boldsymbol{k}}\, a_{\sigma'\boldsymbol{k}'}\, Q_{\boldsymbol{q}j}\,. \tag{1}$$

Here $Q_{\boldsymbol{q}j}$, $p_{\boldsymbol{q}j}$ are the normal coordinates of vibrations and their conjugated momenta, respectively, a^+ and a are the creation and annihilation operators of electrons and j is the index of a vibrational branch. The last term in equation (1) describes the usual linear electron–phonon interaction. The calculation of the terms $C^j_{\sigma\sigma'}\begin{pmatrix} \boldsymbol{q} \\ \boldsymbol{k}, \boldsymbol{k}' \end{pmatrix}$ (also presence of such bands, etc.) is a special problem for concrete crystal models and in the following consideration only some probable general properties of C are used.

First we compute the electron spectrum $\bar{\varepsilon}_\sigma(\boldsymbol{k})$ of the Hamiltonian $\mathcal{H}$ considering the $Q_{\boldsymbol{q}j}$ as c-numbers. The application of double-time Green functions method or the usual variational procedure of the perturbation theory gives the renormalized one-electron spectrum in the form

$$\bar{\varepsilon}_{1,2}(\boldsymbol{k}) = \frac{\varepsilon_1(\boldsymbol{k}) + \varepsilon_2(\boldsymbol{k})}{2} \mp$$

$$\mp \frac{1}{2}\left\{[\varepsilon_1(\boldsymbol{k}) - \varepsilon_2(\boldsymbol{k})]^2 + 4\,\hbar^2 \sum_{\boldsymbol{q}jj'} C^j_{12}\begin{pmatrix} \boldsymbol{q} \\ \boldsymbol{k}, \boldsymbol{k}-\boldsymbol{q} \end{pmatrix} C^{j'}_{12}\begin{pmatrix} \boldsymbol{q} \\ \boldsymbol{k}, \boldsymbol{k}-\boldsymbol{q} \end{pmatrix} Q_{\boldsymbol{q}j}\, Q_{\boldsymbol{q}j'}\right\}^{1/2}. \tag{2}$$

We have assumed that $\varepsilon_\sigma(\boldsymbol{k} - \boldsymbol{q}) \approx \varepsilon_\sigma(\boldsymbol{k})$, which restricts our consideration to a sufficiently small range of $\boldsymbol{q}$ around $\boldsymbol{q} = 0$ and that the vibronic interaction constants for intraband processes are equal to zero.[1]) We shall only examine the case where N electrons of the system (N is of the order of the number of unit cells) are in interaction with one vibrational branch. We also suppose that [11]

$$C_{12}\begin{pmatrix} \boldsymbol{q} \\ \boldsymbol{k}, \boldsymbol{k}-\boldsymbol{q} \end{pmatrix} = C_{12}(\boldsymbol{q}) \tag{3}$$

and introduce the designations

$$V(\boldsymbol{q}) = \hbar\sqrt{M\,N}\; C_{12}(\boldsymbol{q})\,, \qquad y_{\boldsymbol{q}} = \frac{Q_{\boldsymbol{q}}}{\sqrt{M}}\,, \tag{4}$$

where M is the appropriate ion-mass factor.

Now we write the Helmholtz free energy for a crystal by adding to the electron part $F_{\text{e}} = \sum_\sigma F_\sigma$ the phonon part $F_{\text{p}} = \frac{1}{2}\sum_{\boldsymbol{q}} M\,\omega^2_{\boldsymbol{q}}\, y^2_{\boldsymbol{q}}$ of the active vibrational branch (the contribution of other branches and the phonon entropy terms are of no interest here and are neglected). Then

$$F = \sum_\sigma \left\{ n_\sigma\,\mu - k_{\text{B}}T \int \ln\left(1 + \mathrm{e}^{\frac{\mu - \bar{\varepsilon}_\sigma}{k_{\text{B}}T}}\right) g(\bar{\varepsilon}_\sigma)\, \mathrm{d}\bar{\varepsilon}_\sigma \right\} + F_{\text{p}} \tag{5}$$

---

[1]) This is in accordance with what is supposed below (the diminishing of $C(\boldsymbol{q})$ from the maximum value at $\boldsymbol{q} = 0$). The influence of intraband processes is discussed in Section 3.

and

$$n_\sigma = k_B T \frac{\partial}{\partial \mu} \int \ln\left(1 + e^{\frac{\mu - \bar{\varepsilon}_\sigma}{k_B T}}\right) g(\bar{\varepsilon}_\sigma) \, d\bar{\varepsilon}_\sigma , \tag{6}$$

where $g(\bar{\varepsilon}_\sigma)$ are band density functions. The chemical potential is determined by the condition

$$n_1 + n_2 = N . \tag{7}$$

From now on we assume retaining the qualitative aspect of the problem, that the electrons in band 2 and the holes in band 1 have no great difference in effective masses (for narrow bands it is certainly not a bad approximation) so that an average for the undisturbed gap

$$\bar{\Delta} = \varepsilon_2(\boldsymbol{k}) - \varepsilon_1(\boldsymbol{k}) \tag{8}$$

may be introduced. Then $\mu = \Delta/2$ and the necessary part of the free energy, which depends on $y_{\boldsymbol{q}}$, reads

$$F = -N \, k_B T \ln \left\{\left[1 + \exp\left(-\frac{1}{k_B T}\left[\left(\frac{\bar{\Delta}}{2}\right)^2 + \frac{1}{N}\sum_{\boldsymbol{q}} V^2(\boldsymbol{q}) \, y_{\boldsymbol{q}}^2\right]^{1/2}\right)\right] \times \right.$$
$$\left. \times \left[1 + \exp\left(\frac{1}{k_B T}\left[\left(\frac{\bar{\Delta}}{2}\right)^2 + \frac{1}{N}\sum_{\boldsymbol{q}} V^2(\boldsymbol{q}) \, y_{\boldsymbol{q}}^2\right]^{1/2}\right)\right]\right\} + \frac{1}{2}\sum_{\boldsymbol{q}} M \, \omega_{\boldsymbol{q}}^2 \, y_{\boldsymbol{q}}^2 . \tag{9}$$

The electron–phonon interaction results in a lattice distortion when the equilibrium values $y_{\boldsymbol{q}0}$, determined by equation

$$\left.\frac{\partial F}{\partial y_{\boldsymbol{q}}}\right|_{y_{\boldsymbol{q}0}} = 0 \tag{10}$$

are real and not equal to zero. In equation (9) $y_{\boldsymbol{q}}^2$ stands under the root and we may say that the system possesses an essential anharmonicity caused by the vibronic interaction. That allows the determination of the low-symmetry distortion ($y_{\boldsymbol{q}}$ is generally not a totally symmetrical vibration) without using the own vibrational anharmonicity [8] for the stabilization of the lattice. In our case equation (10) gives

$$\left\{\left(\frac{\bar{\Delta}}{2}\right)^2 - \left[\frac{[f_2(y_{\boldsymbol{q}0}) - f_1(y_{\boldsymbol{q}0})] \, V^2(\boldsymbol{q})}{M \, \omega_{\boldsymbol{q}}^2}\right]^2 + \frac{1}{N}\sum_{\boldsymbol{q}} V^2(\boldsymbol{q}) \, y_{\boldsymbol{q}0}^2\right\} y_{\boldsymbol{q}0}^2 = 0 , \tag{11}$$

where

$$\left.\begin{aligned} f_1(y_{\boldsymbol{q}0}) &= \left\{\exp\left[-\frac{1}{k_B T}\left(\left(\frac{\bar{\Delta}}{2}\right)^2 + \frac{1}{N}\sum_{\boldsymbol{q}} V^2(\boldsymbol{q}) \, y_{\boldsymbol{q}0}^2\right)^{1/2}\right] + 1\right\}^{-1}, \\ f_2(y_{\boldsymbol{q}0}) &= \left\{\exp\left[\frac{1}{k_B T}\left(\left(\frac{\bar{\Delta}}{2}\right)^2 + \frac{1}{N}\sum_{\boldsymbol{q}} V^2(\boldsymbol{q}) \, y_{\boldsymbol{q}0}^2\right)^{1/2}\right] + 1\right\}^{-1} . \end{aligned}\right\} \tag{12}$$

The solution of equation (11) $y_{\boldsymbol{q}0} \neq 0$ must have the properties $y_{\boldsymbol{q}'0} \neq 0$ and $y_{\boldsymbol{q}+\boldsymbol{q}',0} = 0$. Only one of these solutions is interesting where y_{00}, which corresponds to the cut-off optical vibration, is not equal to zero and for the others it is valid $y_{\boldsymbol{q}0} = 0$. In this case the crystal is distorted by means of a mutual displacement of sub-lattices [1]. If the crystal is at least partially ionic and the active cut-off vibration[2]) y_0 is odd there may rise a state of spontaneous

[2]) If this vibration is degenerate (at least one of the mixing bands is a degenerate one) a number of various low-symmetry configurations is possible (see, e.g. [4, 12, 13]).

polarization (ferroelectric). The other possible solutions of equation (11) correspond to a fluctuating distortion of the lattice from cell to cell and to the absence of a macroscopic dipole moment.

The formal solution of (11) can be written as

$$y_{00}^2 = N\left\{\frac{[f_1(y_{00}) - f_2(y_{00})]^2\ V^2(0)}{(M\ \omega_0^2)^2} - \frac{\bar{\Delta}^2}{4\ V^2(0)}\right\}. \tag{13}$$

From expression (13) y_{00} can easily be obtained for $T = 0$. Then, if the necessary condition

$$\frac{2\ V^2(0)}{M\ \omega_0^2\ \bar{\Delta}} > 1 \tag{14}$$

for the existence of the lattice distortion is fulfilled, it is

$$y_{00}^2 = N\left[\frac{V^2(0)}{(M\ \omega_0^2)^2} - \frac{\bar{\Delta}^2}{4\ V^2(0)}\right]. \tag{15}$$

It can be seen that with rising temperature y_{00} diminishes and the Curie temperature $T = T_c$ exists where $y_{00} = 0$. From equations (13) and (12) we obtain

$$k_B T_c = \frac{\bar{\Delta}}{2}\left[\ln\frac{2\ V^2(0) + M\ \omega_0^2\ \bar{\Delta}}{2\ V^2(0) - M\ \omega_0^2\ \bar{\Delta}}\right]^{-1}. \tag{16}$$

From these equations it also follows that the amount of lattice distortion is not dependent on T and is equal to (15) if $\bar{\Delta} \gg k_B T$ and, consequently, broad-gap systems don't show such phase transitions.

For $T \neq 0$, equation (13) must be solved numerically. The results are that the distortion is maximum at $T = 0$ and tends to zero at $T = T_c$. This is illustrated in Fig. 1 by $y_{00}^2(T)$ for some values of the parameters. The fact that f_1 and f_2 in equation (13) are themselves dependent on $y_{00}(T)$ takes into account the variation of the band gap $\bar{\varepsilon}_2 - \bar{\varepsilon}_1$ with temperature. At $T = 0$ the gap is maximum and equals to $\frac{2\ V^2(0)}{M\ \omega_0^2}$ and diminishes to $\bar{\Delta}$ at $T = T_c$.

In our preceding paper [6] f_1 and f_2 were determined by the gap obtained by means of y_{00} which is found separately for the bands (at $T = 0$). Recently, an interesting paper of Adler and Brooks [10] appeared, which considers in some respect analogous features. These authors consider F as a function of independent variables n and ε (the occupation numbers and deformation, respectively); the bands are considered as independent phases. An analogous procedure (also that carried out in [6]) for our case would give results not far from those obtained here.

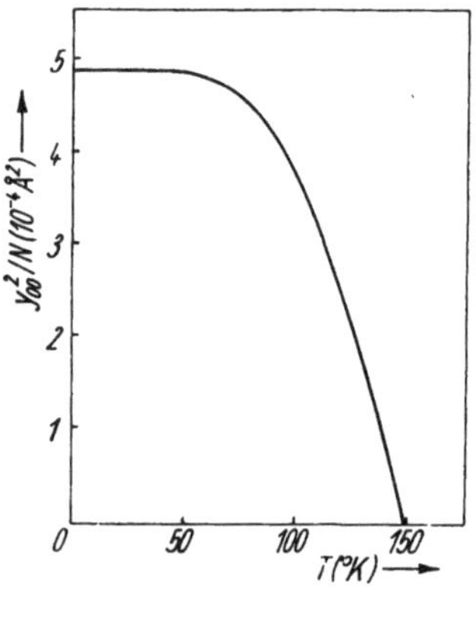

Fig. 1. Crystal structure distortion y_{00}^2 as function of temperature at chosen parameters $\bar{\Delta} = 0.05$ eV, $M\omega_0^2 = 30$ eV/Å², $V(0) = 1$ eV/Å

According to equation (11) the free energy has two equivalent (orientationally degenerated) minima. That appears to be possible for crystals having a domain structure [1].

The active vibration y_0 represents a "soft" mode which occurs in the Ginzburg-Anderson-Cochran theory of ferroelectricity [12, 14, 15]. From equation (14) it also follows that the instability of the lattice occurs earlier for the lower optical branch.

The vibronic interaction causes also an essential temperature dependent renormalization of the frequencies in the active vibrational branch. In an expansion of equation (9) in a power series with respect to $y_{\boldsymbol{q}}$ around $y_{\boldsymbol{q}0}$, the quadratic terms give for the high symmetry phase

$$\Omega_{\mathrm{hs}}^2(\boldsymbol{q}) = \omega_{\boldsymbol{q}}^2 + \frac{2\,[f_2(0) - f_1(0)]\,V^2(\boldsymbol{q})}{M\,\bar{\Delta}} \tag{17}$$

and for the low symmetry phase at $T < T_\mathrm{c}$

$$\begin{aligned} \Omega_{\mathrm{ls}}^2(\boldsymbol{q}) &= \omega_{\boldsymbol{q}}^2 - \left[\frac{M\,\omega_0^3}{[f_1(y_{00}) - f_2(y_{00})]\,V^3(0)}\right]^2 \times \\ &\times \left\{V^2(\boldsymbol{q})\left(\frac{\bar{\Delta}}{2}\right)^2 - \left[V^4(0) - V^2(0)\,V^2(\boldsymbol{q}) - \frac{2\,V^6(0)}{M\,\omega_0^2\,k_\mathrm{B}T}\,f_1^2(y_{00}) \times \right.\right. \\ &\times \left.\left. \exp\left[-\frac{V^2(0)\,[f_1(y_{00}) - f_2(y_{00})]}{M\,\omega_0^2\,k_\mathrm{B}T}\right]\right]\left[\frac{[f_1(y_{00}) - f_2(y_{00})]^2\,V^2(0)}{(M\,\omega_0^2)^2} - \frac{\bar{\Delta}^2}{4\,V^2(0)}\right]\right\}. \end{aligned} \tag{18}$$

From equation (17) it follows that with decreasing temperature towards T_c the frequencies of the active branch begin to decrease from its high-temperature limits $\omega_{\boldsymbol{q}}$. This is really observed (see e.g. [16]). At the Curie point it is $\Omega_{\mathrm{hs}}(0) = \Omega_{\mathrm{ls}}(0) = 0$ and $\Omega_{\mathrm{hs}}(\boldsymbol{q}) = \Omega_{\mathrm{ls}}(\boldsymbol{q})$, i.e. we have a cut-off phonon instability while a second kind phase transition takes place. The frequencies with $\boldsymbol{q} \neq 0$ reach the minimum values at the transition point but do not turn to zero supposing $V(0) > V(\boldsymbol{q})$. The decrease of V with the increase of $\boldsymbol{q}$ is expected to take place in our case analogously to [17]. The dependence of Ω in both phases on temperature is illustrated in Fig. 2 for the active vibration y_0.

The calculation of $V(\boldsymbol{q})$ itself remains to be an interesting problem. That is necessary for calculating the current[3]) in such systems, too, because the renormalized masses for dressed carriers depend on it according to equation (2).

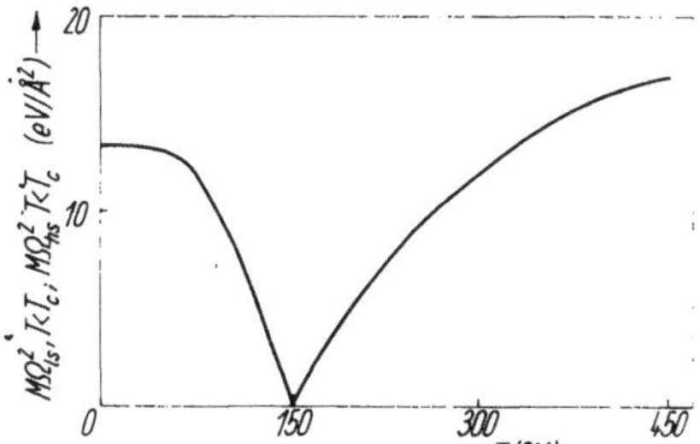

Fig. 2. Temperature dependence of renormalized frequencies in both phases for active vibration y_0. The values of parameters are the same as in Fig. 1. (The second term on the ordinate should read $M\Omega_{\mathrm{hs}}^2$, $T > T_\mathrm{c}$)

---

[3]) Semiconductor properties have been found for a number of ferroelectrics [18 to 20]. It is also remarkable that an exoelectron emission peak is observed at the phase transition point [21].

2. Vibronic Frequencies

Now we calculate the renormalization of phonon frequencies of the active branch j on the basis of equation (1) introducing the retarded phonon Green's function

$$G(\boldsymbol{q}\, j, \boldsymbol{f}\, j; t) = \langle\langle Q_{\boldsymbol{q} j}(t), Q_{\boldsymbol{f} j}(0)\rangle\rangle_r \,. \tag{19}$$

Forming the corresponding chain of equations and carrying out the interpolation decoupling [22] of functions of type $\langle\langle a^+_{\sigma \boldsymbol{k}}\, a_{\sigma' \boldsymbol{k}'}\, Q_{\boldsymbol{q} j}, Q_{\boldsymbol{f} j}\rangle\rangle$ we get

$$G(\boldsymbol{q}\, j, \boldsymbol{f}\, j; \omega) = \frac{\delta_{\boldsymbol{q}\boldsymbol{f}}}{2\pi\, P_{\boldsymbol{q} j}(\omega)}\,, \tag{20}$$

$$\begin{aligned} P_{\boldsymbol{q} j}(\omega) = {} & [b_{\boldsymbol{q} j}(\omega)]^{-1} - \hbar \sum_{\boldsymbol{k}} \left\{ \left[C^j_{11}\begin{pmatrix} & \boldsymbol{q} & \\ \boldsymbol{k}, & \boldsymbol{k} - & \boldsymbol{q}\end{pmatrix}\right]^2 \times \right. \\ & \times \frac{f_1(\boldsymbol{k}) - f_1(\boldsymbol{k}-\boldsymbol{q})}{\hbar\omega - [\varepsilon_1(\boldsymbol{k}-\boldsymbol{q}) - \varepsilon_1(\boldsymbol{k})]} + \left[C^j_{22}\begin{pmatrix} & \boldsymbol{q} & \\ \boldsymbol{k}, & \boldsymbol{k} - & \boldsymbol{q}\end{pmatrix}\right]^2 \frac{f_2(\boldsymbol{k}) - f_2(\boldsymbol{k}-\boldsymbol{q})}{\hbar\omega - [\varepsilon_2(\boldsymbol{k}-\boldsymbol{q}) - \varepsilon_2(\boldsymbol{k})]} + \\ & \left. + \left[C^j_{12}\begin{pmatrix} & \boldsymbol{q} & \\ \boldsymbol{k}, & \boldsymbol{k} - & \boldsymbol{q}\end{pmatrix}\right]^2 \left[\frac{f_1(\boldsymbol{k}) - f_2(\boldsymbol{k}-\boldsymbol{q})}{\hbar\omega - [\varepsilon_2(\boldsymbol{k}-\boldsymbol{q}) - \varepsilon_1(\boldsymbol{k})]} - \frac{f_1(\boldsymbol{k}-\boldsymbol{q}) - f_2(\boldsymbol{k})}{\hbar\omega - [\varepsilon_1(\boldsymbol{k}-\boldsymbol{q}) - \varepsilon_2(\boldsymbol{k})]}\right]\right\}, \end{aligned} \tag{21}$$

where $b_{\boldsymbol{q} j}(\omega) = \hbar\,[\omega^2 - \omega^2_{\boldsymbol{q} j}]^{-1}$ and $f_\sigma(\boldsymbol{k})$ are the occupation numbers of $\sigma\, \boldsymbol{k}$ states. The phonon spectrum altered by the interaction is determined by equation

$$P_{\boldsymbol{q} j}(\omega) = 0\,. \tag{22}$$

From equation (21) it follows that the renormalization of frequencies of the cut-off vibration at $\omega \neq 0$ is caused only by terms connected with interband transitions of electrons. Taking into account equations (3) and (8) and neglecting intraband processes equation (22) can be rewritten as follows:

$$\omega^2 - \omega^2_{\boldsymbol{q}} - \hbar^2 \sum_{\boldsymbol{k}} C^2_{12}(\boldsymbol{q}) \frac{2\,\bar{\Delta}}{\hbar^2\,\omega^2 - \bar{\Delta}^2} \left(f_1(0) - f_2(0)\right) = 0\,, \tag{23}$$

where $f_1(0)$ and $f_2(0)$ are determined by equation (12).

In (23) $\sum_{\boldsymbol{k}}$ may cancel the multiplier $\sim N^{-1}$ in $C^2_{12}(\boldsymbol{q})$, if the number of electrons determining the effect is of the order N. Then the shift of frequencies is finite. This condition may be weakened in the resonance case when $\hbar\,\omega \sim \bar{\Delta}$.

At $\hbar\,\omega \neq \bar{\Delta}$ the solution of equation (23) has the form

$$\omega^2 = (\Omega^{\pm}_{\boldsymbol{q}})^2 = \frac{\hbar^2\,\omega^2_{\boldsymbol{q}} + \bar{\Delta}^2}{2\,\hbar^2} \pm \left\{\left(\frac{\hbar^2\,\omega^2_{\boldsymbol{q}} - \bar{\Delta}^2}{2\,\hbar^2}\right)^2 + \frac{2\,V^2(\boldsymbol{q})}{M\,\hbar^2}\,\bar{\Delta}\,[f_1(0) - f_2(0)]\right\}^{1/2}. \tag{24}$$

Hence, we have two mixed vibronic energy bands.

We expand expression (24) in a series at $\bar{\Delta} > \hbar\,\omega_{\boldsymbol{q}}$. Then

$$\left.\begin{aligned} (\Omega^-_{\boldsymbol{q}})^2 &= \omega^2_{\boldsymbol{q}} + \frac{2\,V^2(\boldsymbol{q})}{M\,\bar{\Delta}}\,[f_2(0) - f_1(0)] - \cdots, \\ (\Omega^+_{\boldsymbol{q}})^2 &= \left(\frac{\bar{\Delta}}{\hbar}\right)^2 + \frac{2\,V^2(\boldsymbol{q})}{M\,\bar{\Delta}}\,[f_1(0) - f_2(0)] - \cdots. \end{aligned}\right\} \tag{25}$$

The first two terms in $(\Omega_q^-)^2$ correspond to phonon frequencies in the high-symmetry phase in accordance with (17). Ω_q^+ are the "electron" frequencies altered by electron–phonon interaction and they increase with temperature (the lowering one). At $\bar{\Delta} < \hbar\,\omega_q$ we obtain from (24)

$$\left.\begin{aligned} (\Omega_q^-)^2 &= \left(\frac{\bar{\Delta}}{\hbar}\right)^2 + \frac{2\,V^2(q)\,\bar{\Delta}}{M\,\hbar^2\,\omega_q^2}\,[f_2(0) - f_1(0)] - \cdots, \\ (\Omega_q^+)^2 &= \omega_q^2 + \frac{2\,V^2(q)\,\bar{\Delta}}{M\,\hbar^2\,\omega_q^2}\,[f_1(0) - f_2(0)] - \cdots, \end{aligned}\right\} \tag{26}$$

i.e. the electron frequencies decrease and the phonon frequencies increase with temperature. In the case $\bar{\Delta} = \hbar\,\omega_q$ the electron frequencies fall into the phonon spectrum.

If we take the activities of long-wave vibrations into account formulae (14) and (16) follow from (23) as before.

3. Dielectric Susceptibility

Now we discuss the dielectric properties of our model.

Taking the first term of the electric moment of the crystal and using the method developed in [23] it is possible to show that the dielectric susceptibility χ is proportional to the Fourier components of the retarded phonon Green's function

$$\chi_{\alpha\beta}(\omega) = -\frac{2\,\pi}{\hbar\,v}\sum_{j j'} \mathfrak{M}_\alpha\begin{pmatrix}0\\ j\end{pmatrix}\mathfrak{M}_\beta\begin{pmatrix}0\\ j'\end{pmatrix} G(0\,j, 0\,j'; \omega + i\,\delta)\,, \tag{27}$$

where v is the volume of the crystal. Let the crystal have a cubic symmetry in the high-symmetry phase. Then the part of $\chi_{\alpha\beta}(\omega)$ renormalized by the vibronic interaction is

$$\chi(\omega) = -\frac{1}{3\,\hbar\,\tau}\cdot\frac{\overline{\mathfrak{M}}^2\begin{pmatrix}0\\ j\end{pmatrix}}{P_{0\,j}(\omega) - i\,\gamma_0(\omega)}\,, \tag{28}$$

τ is the volume of a unit cell and

$$\overline{\mathfrak{M}}\begin{pmatrix}0\\ j\end{pmatrix} = \sum_s \frac{e_s}{\sqrt{M_s}}\,q\begin{pmatrix}s, & 0\\ & j\end{pmatrix}, \qquad \overline{\mathfrak{M}}\begin{pmatrix}0\\ j\end{pmatrix} = \frac{1}{\sqrt{N}}\,\mathfrak{M}\begin{pmatrix}0\\ j\end{pmatrix}, \tag{29}$$

where e_s and M_s are the charge and the mass of the s-th ion of the cell, respectively, $q\begin{pmatrix}s, & 0\\ & j\end{pmatrix}$ is the absolute value of the polarization vector. The damping factor of phonons is determined by the expression

$$\begin{aligned} \gamma_0(\omega) = \pi\,\hbar \sum_k \left[C_{12}\begin{pmatrix}0\\ k, k\end{pmatrix}\right]^2 [f_1(k) - f_2(k)] \times \\ \times\, 2\,[\varepsilon_2(k) - \varepsilon_1(k)]\,\delta\,(\hbar^2\,\omega^2 - [\varepsilon_2(k) - \varepsilon_1(k)]^2)\,. \end{aligned} \tag{30}$$

From equation (30) one can conclude that $\gamma_0(0) = 0$. To the occurrence of the polarization catastrophe corresponds

$$\chi(0) = \infty\,, \tag{31}$$

where $\chi(0)$ is the static dielectric susceptibility. From (28) it follows that the behaviour of the static dielectric susceptibility is determined by the frequency

$$\Omega^2_{\text{h s}}(0) = \omega_0^2 + \frac{2\,V^2(0)}{M\,\bar{\Delta}}\,[f_2(0) - f_1(0)] + \frac{1}{k_{\text{B}}T\;M}\left[V^2_{11}(0)\;f_1^2(0)\;\mathrm{e}^{-\frac{\bar{\Delta}}{2\,k_{\text{B}}T}} + \right.$$

$$\left. + \;V^2_{22}(0)\;f_2^2(0)\;\mathrm{e}^{\frac{\bar{\Delta}}{2\,k_{\text{B}}T}}\right] \tag{32}$$

which coincides with the frequency given by (17) if intraband transitions are neglected. These transitions give a contribution to the stabilization of the lattice.

From equation (28) we get the complex dielectric susceptibility

$$\chi(\omega) = \chi_1(\omega) - i\,\chi_2(\omega)\,, \tag{33}$$

where

$$\chi_1(\omega) = -\frac{1}{3\,\tau\,\hbar}\,\overline{\mathfrak{M}^2}\binom{0}{j}\frac{P_{0j}(\omega)}{[P_{0j}(\omega)]^2 + [\gamma_0(\omega)]^2}\,, \tag{34}$$

$$\chi_2(\omega) = \frac{1}{3\,\tau\,\hbar}\,\overline{\mathfrak{M}^2}\binom{0}{j}\frac{\gamma_0(\omega)}{[P_{0j}(\omega)]^2 + [\gamma_0(\omega)]^2}\,. \tag{35}$$

Since

$$P_{0j}(\omega) = \frac{\hbar}{\hbar^2\,\omega^2 - \bar{\Delta}^2}\,[(\Omega_0^+)^2 - \omega^2]\,[(\Omega_0^-)^2 - \omega^2]\,,$$

then for $\omega^2 = (\Omega_0^-)^2$ and $\omega^2 = (\Omega_0^+)^2$ χ_1 turns to zero if $\omega \neq 0$. Consequently, the vibrations which are active in the phase transition determine the dielectric properties of the crystal to a considerable degree. Formulae (34) and (35) represent the possibility of examining experimentally the frequencies of "soft" modes on the basis of the dielectric susceptibility.

If our system is situated in external field $\boldsymbol{E}$ it is easy to show that the corresponding perturbation term in F which depends on y_0 has the form

$$-\alpha\,\boldsymbol{y}_0\,\boldsymbol{E}\,, \tag{36}$$

where α is the ionic polarizability. The perturbation makes the orientationally degenerate minima of $F(y_0)$ non-equivalent. The corresponding existence of stable and metastable states of the crystal may be the basis for explaining the hysteresis [24]. In the case of sufficiently large field one of the minima vanishes altogether, determining the "absolute" limit of the hysteresis.

Generally speaking, it is important for the present problem to take also the own anharmonicity of active vibrations into consideration (see in more detail [8]). Then a first kind phase transition may be obtained instead of the second one, which is close to the latter. Besides, the anharmonicity will also contribute to the stabilization of the lattice.

Acknowledgements

The authors express their gratitude to G. S. Zavt for helpful discussions and to E. Jaigma for technical assistance.

References

[1] N. Kristoffel, Trudy IFA Akad. Nauk Est. SSR № 27, 85 (1964); Fiz. tverd. Tela **6**, 3266 (1964).
[2] J. Birman, Phys. Rev. **125**, 1959 (1962); **127**, 1093 (1962).
[3] J. Birman, Phys. Rev. Letters **17**, 1216 (1966).
[4] I. B. Bersuker, Phys. Letters (Netherlands) **20**, 589 (1966).
[5] I. B. Bersuker and B. G. Vekhter, Fiz. tverd. Tela **9**, 2652 (1967).
[6] N. Kristoffel and P. Konsin, phys. stat. sol. **21**, K39 (1967).
[7] K. Sinha and A. Sinha, Indian J. pure appl. Phys. **2**, 91 (1964).
[8] G. Shukla and K. Sinha, J. Phys. Chem. Solids **27**, 1837 (1966).
[9] J. Labbe and J. Friedel, J. Phys. Radium **27**, 153 and 303 (1966).
[10] D. Adler and H. Brooks, Phys. Rev. **155**, 826 (1967).
[11] R. Knox, Theory of Excitons, Izd. "Mir", Moscow 1966.
[12] W. Cochran, Phys. Rev. Letters **3**, 412 (1959).
[13] N. Kristoffel and G. Zavt, Fiz. tverd. Tela **9**, 1582 (1967).
[14] V. L. Ginzburg, Uspekhi fiz. Nauk **38**, 490 (1949).
[15] P. W. Anderson, in: Proc. All-Union Conf. Physics of Dielectrics, Academy of Sciences of the USSR, Moscow 1958 (p. 290).
[16] G. Shirane, B. Fraser, V. Minkiewicz, I. Leake, and A. Linz, Phys. Rev. Letters **18**, 234 (1967).
[17] A. S. Davydov, phys. stat. sol. **20**, 143 (1967).
[18] H. Frederikse, W. Thurber, and W. Hosler, Phys. Rev. **134**, A442 (1964).
[19] R. Nitsche, H. Roetschi, and P. Wild, Appl. Phys. Letters **4**, 210 (1964).
[20] G. Pawley, W. Cochran, R. Cowley, and G. Dolling, Phys. Rev. Letters **17**, 753 (1966).
[21] V. S. Kortov and R. I. Mints, Fiz. tverd. Tela **9**, 1828 (1967).
[22] D. N. Zubarev, Uspekhi fiz. Nauk **71**, 71 (1960).
[23] R. Kubo, J. Phys. Soc. Japan **12**, 570 (1957).
[24] L. D. Landau and E. M. Lifshits, Electrodynamics of Continuous Media, Gosud. Izd. Tekh.-Teor. Literat., Moscow 1959.

(Received May 2, 1968)

phys. stat. sol. 28, 741 (1968)

Subject classification: 13

Institut für Physik am Max-Planck-Institut für Metallforschung, Stuttgart

Die Unterscheidung zwischen Leitungs- und Rumpfelektronen bei der feldtheoretischen Behandlung des Vielelektronen-Problems

Von

P. BRAND

Für die Greensche Funktion der Elektronen eines Festkörpers wird eine reelle Spektralfunktion aufgestellt, welche die Bedeutung einer verallgemeinerten Zustandsdichte besitzt. Die Beiträge der Leitungs- und Rumpfelektronen eines Metalls zu dieser Spektralfunktion lassen sich voneinander trennen, ohne daß dabei die Austauschkräfte zwischen Leitungs- und Rumpfelektronen vernachlässigt werden. Der Zusammenhang mit der Pseudopotential-Methode wird aufgezeigt; die Pseudopotential-Methode wird kritisiert.

A real spectral function of the electron Green's function for a solid is developed. This spectral function has the meaning of a generalized density of states. For a metal it is possible to distinguish between contributions to the spectral functions, caused either by conduction or by core electrons. Exchange between conduction and core-electrons is fully accounted for. Finally the pseudopotential method and its connection with the present formalism is considered.

1. Einleitung

Die elektronischen Eigenschaften eines Festkörpers werden in der Regel ausschließlich von den energetisch höheren Zuständen der bindenden Elektronen (bei Metallen von den Leitungselektronen) geprägt; die energetisch tiefsten Elektronenzustände (bei Metallen die Rumpfzustände) unterscheiden sich im allgemeinen nur wenig von den entsprechenden Zuständen des zugehörigen freien Atoms.

Da auch alle interessanten Vielelektronen-Effekte in Metallen nur die Leitungselektronen betreffen, sollte eine rationelle Rechenmethode soweit wie möglich auf die Behandlung der Leitungselektronen beschränkt sein. Dieser Forderung steht bei einer feldtheoretischen Methode jedoch die Schwierigkeit im Weg, daß die einzuführenden Feldoperatoren, wie auch die Vertauschungsrelationen, immer für das Feld aller Elektronen gelten.

In der vorliegenden Arbeit soll gezeigt werden, wie man auch bei einer feldtheoretischen Behandlung der wechselwirkenden Elektronen eines Metalls zwischen Leitungs- und Rumpfzuständen unterscheiden kann.

Am einfachsten ist dies möglich in der Spektralfunktion der Elektronen-Green-Funktion. Dazu wird zunächst eine reelle Spektralfunktion für die Elektronen eines inhomogenen Körpers, z. B. eines Kristalls, eingeführt, wobei jedoch Periodizität nicht vorausgesetzt zu werden braucht.

Die Eigenschaften dieser Spektralfunktion, die von Ort und Impuls abhängt, werden untersucht. Es wird insbesondere gezeigt, wie man zwischen lokalisierten und nichtlokalisierten Zuständen unterscheiden kann, und welchen Einschränkungen das Spektrum der Leitungselektronen durch das Spektrum der Rumpfelektronen infolge der Vertauschungsrelation unterworfen ist.

Liegt das Spektrum der Rumpfelektronen schon fest (z. B. näherungsweise bestimmt aus den Wellenfunktionen des freien Atoms), so kann man einen einfachen Ansatz für das Teilspektrum der Leitungselektronen machen. Dieser Ansatz enthält schon alle von den Austauschkräften allein herrührenden Effekte.

Die vorliegende Methode liefert den exakten theoretischen Hintergrund der sogenannten Pseudopotential-Methode, die in ihrer derzeitigen Form nicht nur begriffliche, sondern auch schwerwiegende formale Mängel aufweist, wie im letzten Abschnitt gezeigt wird.

2. Spektralfunktionen für Kristallelektronen

Der Hamilton-Operator der Elektronen eines Kristalls lautet in der Formulierung mit Feldoperatoren

$$H = -\frac{\hbar^2}{2m}\int d\tau_{\boldsymbol{x}}\, \Psi^+(x)\, \Delta\Psi(x) + \\ + \int d\tau_{\boldsymbol{x}}\, V(\boldsymbol{x})\, \Psi^+(x)\, \Psi(x) - e\int d\tau_{\boldsymbol{x}} \int d\tau_{\boldsymbol{y}}\, \Psi^+(x)\, \Psi^+(y)\, \Psi(y)\, \Psi(x)\,. \tag{2.1}$$

Dabei sind $\Psi^+(x)$ und $\Psi(x)$ die Erzeugungs- bzw. Vernichtungsoperatoren für ein Elektron am Ort $\boldsymbol{x}$ zur Zeit x_0.

Der erste Term in (2.1) stellt die kinetische Energie, der zweite die potentielle Energie der Elektronen im Potential $V(\boldsymbol{x})$ der Kerne dar. Im dritten Beitrag zum Hamilton-Operator wird die Elektron–Elektron-Wechselwirkung berücksichtigt, wobei die Formulierung als direkte Teilchen–Teilchen-Wechselwirkung gewählt ist. Alles Folgende gilt jedoch ohne Einschränkung auch für die indirekte, durch ein elektromagnetisches Feld übertragene Wechselwirkung der Elektronen untereinander.

Die kausale thermodynamische Greensche Funktion für die Feldoperatoren der Elektronen [1] ist definiert als

$$G(x, x') = -i\langle T\{\Psi(x)\, \Psi^+(x')\}\rangle \equiv -i\, \frac{\sum\limits_n (n|\, T\,\{\Psi(x)\, \Psi^+(x')\}\; e^{-\beta(H-\mu N)}|n)}{\sum\limits_l (l|\; e^{-\beta(H-\mu N)}\,|l)}\,. \tag{2.2}$$

T ist das Zeitordnungssymbol mit Vorzeichenumkehr bei Vertauschung von Fermi-Operatoren; $|n)$ bzw. $|l)$ sind die stationären Eigenzustände des ganzen Systems mit festen Energiewerten E_n bzw. E_l und mit bestimmten Teilchenzahlen N_n bzw. N_l; $\beta = 1/kT$.

Die Greensche Funktion (2.2) ist der thermodynamisch-quantenmechanische Mittelwert des zeitgeordneten Operatorprodukts $\Psi(x)\, \Psi^+(x')$ für ein großkanonisches Ensemble. Wegen des Zeitordnungssymbols zerfällt $G(x, x')$ in die beiden analytischen Stücke

$$G(x, x') = \begin{cases} G^>(x, x') = -i\,\langle\Psi(x)\, \Psi^+(x')\rangle & \text{für } x_0 > x_0', \\ G^<(x, x') = +i\,\langle\Psi^+(x')\, \Psi(x)\rangle & \text{für } x_0 < x_0', \end{cases} \tag{2.3}$$

so daß $G(x, x')$ bei $x_0' = x_0$ einen Sprung aufweist, der wegen der Vertauschungsrelation gerade gleich der dreidimensionalen δ-Funktion ist:

$$G(\boldsymbol{x}, x_0; \boldsymbol{x}', x_0^+) - G(\boldsymbol{x}, x_0; \boldsymbol{x}', x_0^-) = G^<(\boldsymbol{x}, x_0; \boldsymbol{x}', x_0) - G^>(\boldsymbol{x}, x_0; \boldsymbol{x}', x_0) = \\ = i\,\langle\{\Psi(\boldsymbol{x}), \Psi^+(\boldsymbol{x}')\}_+\rangle = i\,\delta\,(\boldsymbol{x} - \boldsymbol{x}')\,. \tag{2.4}$$

Um zu den Spektralfunktionen der Greenschen Funktion $G(x, x')$ zu gelangen, bildet man zunächst den Ausdruck

$$A(x, x') = -i\,[G^{<}(x, x') - G^{>}(x, x')] = \langle\{\Psi(x), \Psi^{+}(x')\}_{+}\rangle\,, \qquad (2.5)$$

der wegen des thermischen Gleichgewichts nur von $t = x_0 - x_0'$ abhängt.

Die Spektralfunktion in der Ortsdarstellung ist die Fourier-Transformierte von $A(\boldsymbol{x}, \boldsymbol{x}'; t)$ nach der Zeit t:

$$A(\boldsymbol{x}, \boldsymbol{x}'; \omega) = \int\limits_{-\infty}^{+\infty} dt\, e^{i\omega t} \langle\{\Psi(\boldsymbol{x}, t), \Psi^{+}(\boldsymbol{x}', 0)\}_{+}\rangle$$

$$= \int\limits_{-\infty}^{+\infty} dt\, e^{i\omega t} \left\langle \left\{ e^{i\frac{Ht}{\hbar}} \Psi(\boldsymbol{x})\, e^{-i\frac{Ht}{\hbar}}, \Psi^{+}(\boldsymbol{x}') \right\}_{+} \right\rangle . \qquad (2.6)$$

Unter Beachtung von (2.2) folgt aus (2.6)

$$A(\boldsymbol{x}, \boldsymbol{x}'; \omega) = 2\pi\hbar \sum_{m,n} e^{-\beta(E_n - \mu N_n)} \times$$

$$\times \frac{(n|\Psi(\boldsymbol{x})|m)(m|\Psi^{+}(\boldsymbol{x}')|n)\,\delta(\hbar\omega - (E_m - E_n)) + (n|\Psi^{+}(\boldsymbol{x}')|m)(m|\Psi(\boldsymbol{x})|n)\,\delta(\hbar\omega - (E_n - E_m))}{\sum_l (l|\, e^{-\beta(H - \mu N)}\,|l)} . \qquad (2.7)$$

Aus (2.7) kann man eine Spektralfunktion in der Impulsdarstellung gewinnen durch die Fourier-Transformation

$$A(\boldsymbol{k}, \boldsymbol{k}'; \omega) = \int d\tau_{\boldsymbol{x}} \int d\tau_{\boldsymbol{x}'}\, e^{-i(\boldsymbol{k}+\boldsymbol{k}')\boldsymbol{x}}\, e^{+i(\boldsymbol{k}-\boldsymbol{k}')\boldsymbol{x}'} A(\boldsymbol{x}, \boldsymbol{x}'; \omega) =$$

$$= \frac{2\pi\hbar}{\sum_l (l|\, e^{-\beta(H-\mu N)}\,|l)} \sum_{m,n} e^{-\beta(E_n - \mu N_n)} \{(n|\Psi(\boldsymbol{k} + \boldsymbol{k}')\, m)\,(m|\Psi^{+}(\boldsymbol{k} - \boldsymbol{k}'|n) \times$$

$$\times\, \delta\left(\hbar\omega - (E_m - E_n)\right) + (n|\Psi^{+}(\boldsymbol{k} - \boldsymbol{k}')|m)\,(m|\Psi(\boldsymbol{k} + \boldsymbol{k}')|n)\,\delta\left(\hbar\omega - (E_n - E_m)\right)\} \qquad (2.8)$$

mit der Umkehrbeziehung

$$A(\boldsymbol{x}, \boldsymbol{x}'; \omega) = \frac{1}{(2\pi)^6} \int d\tau_{\boldsymbol{k}} \int d\tau_{\boldsymbol{k}'}\, e^{i\boldsymbol{k}\boldsymbol{x}} e^{-i\boldsymbol{k}'\boldsymbol{x}'} A\left(\frac{\boldsymbol{k}+\boldsymbol{k}'}{2}, \frac{\boldsymbol{k}-\boldsymbol{k}'}{2}; \omega\right) . \qquad (2.8\,a)$$

Führt man in (2.8a) die Substitutionen $\boldsymbol{g} = (\boldsymbol{k} + \boldsymbol{k}')/2$, $\boldsymbol{g}' = (\boldsymbol{k} - \boldsymbol{k}')/2$ durch, so kann man die Umkehrformel auch wie folgt schreiben:

$$A(\boldsymbol{x}, \boldsymbol{x}'; \omega) = \frac{1}{(2\pi)^6} \int d\tau_{\boldsymbol{g}} \int d\tau_{\boldsymbol{g}'}\, e^{i(\boldsymbol{g}+\boldsymbol{g}')\boldsymbol{x}}\, e^{-i(\boldsymbol{g}-\boldsymbol{g}')\boldsymbol{x}'} A(\boldsymbol{g}, \boldsymbol{g}'; \omega) =$$

$$= \frac{1}{(2\pi)^3} \int d\tau_{\boldsymbol{g}}\, e^{i\boldsymbol{g}(\boldsymbol{x}-\boldsymbol{x}')} \cdot \frac{1}{(2\pi)^3} \int d\tau_{\boldsymbol{g}'}\, e^{i\boldsymbol{g}'(\boldsymbol{x}+\boldsymbol{x}')} A(\boldsymbol{g}, \boldsymbol{g}'; \omega) . \qquad (2.8\,b)$$

Neben den Spektralfunktionen in der Orts- und Impulsdarstellung läßt sich auch eine gemischte Spektralfunktion definieren:

$$A(\boldsymbol{k}, \boldsymbol{x} + \boldsymbol{x}'; \omega) = \frac{1}{(2\pi)^3} \int d\tau_{\boldsymbol{k}'}\, e^{i\boldsymbol{k}'(\boldsymbol{x}+\boldsymbol{x}')} A(\boldsymbol{k}, \boldsymbol{k}'; \omega) . \qquad (2.9)$$

Wegen (2.8b) gilt die Beziehung

$$A(\boldsymbol{x}, \boldsymbol{x}'; \omega) = \frac{1}{(2\pi)^3} \int d\tau_{\boldsymbol{k}}\, e^{i\boldsymbol{k}(\boldsymbol{x}-\boldsymbol{x}')} A(\boldsymbol{k}, \boldsymbol{x} + \boldsymbol{x}'; \omega) . \qquad (2.10)$$

Die Eigenschaften dieser gemischten Spektralfunktion sind, verglichen mit den beiden anderen Spektraldarstellungen, besonders einfach.

3. Mathematische Eigenschaften der Spektralfunktionen

Für $A(\boldsymbol{x}, \boldsymbol{x}'; \omega)$ und $A(\boldsymbol{k}, \boldsymbol{k}'; \omega)$ ergeben sich aus (2.7) bzw. (2.8) unmittelbar folgende Eigenschaften:

$$A^*(\boldsymbol{x}, \boldsymbol{x}', \omega) = A(\boldsymbol{x}', \boldsymbol{x}; \omega) , \tag{3.1}$$

$$A^*(\boldsymbol{k}, \boldsymbol{k}'; \omega) = A(\boldsymbol{k}, -\boldsymbol{k}'; \omega) . \tag{3.2}$$

Daraus folgt für $A\ (\boldsymbol{k}, \boldsymbol{x} + \boldsymbol{x}'; \omega)$ wegen (2.9) oder (2.10)

$$A^*\ (\boldsymbol{k}, \boldsymbol{x} + \boldsymbol{x}'; \omega) = A\ (\boldsymbol{k}, \boldsymbol{x} + \boldsymbol{x}'; \omega) . \tag{3.3}$$

Die gemischte Spektralfunktion ist also, im Gegensatz zu den beiden Spektralfunktionen in der Orts- bzw. Impulsdarstellung, reell.

Eine weitere Eigenschaft der Spektralfunktionen ergibt sich aus der Vertauschungsrelation

$$\{\Psi(\boldsymbol{x}), \Psi^+(\boldsymbol{x}')\}_+ = \delta\ (\boldsymbol{x} - \boldsymbol{x}') , \tag{3.4}$$

die man wegen (2.5) in der Form

$$A\ (\boldsymbol{x}, \boldsymbol{x}'; t = 0) = \delta\ (\boldsymbol{x} - \boldsymbol{x}') \tag{3.4a}$$

schreiben kann.

Mit Hilfe der Umkehrformel zu (2.6) folgt aus (3.4a) für $A\ (\boldsymbol{x}, \boldsymbol{x}'; \omega)$ die Bedingung

$$\frac{1}{2\pi} \int_{-\infty}^{+\infty} d\omega\ A\ (\boldsymbol{x}, \boldsymbol{x}'; \omega) = A\ (\boldsymbol{x}, \boldsymbol{x}'; t = 0) = \delta\ (\boldsymbol{x} - \boldsymbol{x}') . \tag{3.5}$$

Die entsprechenden Formeln für die beiden anderen Spektralfunktionen erhält man aus (3.5) über die Beziehungen (2.8b) und (2.10). Sie lauten

$$\frac{1}{2\pi} \int_{-\infty}^{+\infty} d\omega\ A(\boldsymbol{k}, \boldsymbol{k}'; \omega) = \delta(\boldsymbol{k}') \tag{3.6}$$

und

$$\frac{1}{2\pi} \int_{-\infty}^{+\infty} d\omega\ A\ (\boldsymbol{k}, \boldsymbol{x} + \boldsymbol{x}'; \omega) = 1 . \tag{3.7}$$

Vernachlässigt man die Wechselwirkung der Elektronen so weit, daß man mit stationären Einteilchen-Zuständen $\psi_i(\boldsymbol{x})$ rechnen kann, so lautet die Spektralfunktion (2.7) einfach

$$A(\boldsymbol{x}, \boldsymbol{x}'; \omega) = 2\pi\hbar \sum_{i=1}^{\infty} \psi_i^*(\boldsymbol{x}')\ \psi_i(\boldsymbol{x})\ \delta\ (\hbar\,\omega - \varepsilon_i) . \tag{3.8}$$

Die Bedingung (3.5) führt damit auf

$$\frac{1}{2\pi} \int_{-\infty}^{+\infty} d\omega\ A(\boldsymbol{x}, \boldsymbol{x}'; \omega) = \sum_{i=1}^{\infty} \psi_i^*(\boldsymbol{x}')\ \psi_i(\boldsymbol{x}) = \delta\ (\boldsymbol{x} - \boldsymbol{x}') , \tag{3.9}$$

d. h. auf die Vollständigkeitsrelation für das normierte Orthogonalsystem der Einteilchen-Wellenfunktionen $\psi_i(\boldsymbol{x})$. Das bedeutet, daß die Vollständigkeit eines Systems von Einteilchen-Wellenfunktionen notwendig und hinreichend ist für die Gültigkeit der Vertauschungsrelation (3.4).

Aus der Normierung der Wellenfunktionen, $(\psi_i, \psi_i) = 1$, folgt für die gemischte Spektralfunktion unter Beachtung von (2.10)

$$\frac{1}{(2\pi)^3}\int d\tau_{\boldsymbol{k}} \int d\tau_{\boldsymbol{x}}\, A(\boldsymbol{k}, 2\,\boldsymbol{x}; \omega_i) = \int d\tau_{\boldsymbol{x}}\, A(\boldsymbol{x}, \boldsymbol{x}; \omega_i) =$$
$$= 2\pi\hbar\,(\psi_i, \psi_i)\,\delta\,(\hbar\,\omega - \varepsilon_i) = 2\pi\hbar\,\delta\,(\hbar\,\omega - \varepsilon_i)\,. \tag{3.10}$$

4. Physikalische Eigenschaften der Spektralfunktionen

Läßt man im Hamilton-Operator das äußere Potential $V(\boldsymbol{x})$ weg und führt statt dessen neben dem Feld $\Psi(x)$ der Elektronen auch ein Feld $\Phi(x)$ für die Kerne ein, so erhält man einen Hamilton-Operator, der invariant ist gegenüber beliebigen Translationen und Rotationen. Auch die Einteilchen-Greenschen- bzw. Spektralfunktionen sind dann translations- und rotationsinvariant und hängen nur von $|\boldsymbol{x} - \boldsymbol{x}'|$ ab. (Das heißt jedoch keineswegs, daß der durch den Hamilton-Operator beschriebene physikalische Körper Translations- oder Rotationssymmetrie aufweisen muß.)

Für eine solche homogene Spektralfunktion in der Ortsdarstellung tritt an die Stelle von (2.8b) die Beziehung

$$A(\boldsymbol{x}, \boldsymbol{x}'; \omega) = \frac{1}{(2\pi)^3}\int d\tau_{\boldsymbol{k}}\, e^{i\boldsymbol{k}(\boldsymbol{x}-\boldsymbol{x}')}\, A(\boldsymbol{k}, \omega)\,, \tag{4.1}$$

wobei die Spektralfunktion

$$A(\boldsymbol{k}, \omega) = 2\,\pi\,\hbar \sum_{m,n} e^{-\beta(E_n - \mu N_n)} \times$$
$$\times \frac{|(m|\,\Psi^+(\boldsymbol{k})\,|n)|^2\,\delta\,(\hbar\,\omega - (E_m - E_n)) + |(m|\,\Psi(\boldsymbol{k})\,|n)|^2\,\delta\,(\hbar\,\omega - (E_n - E_m))}{\sum_l e^{-\beta(E_l - \mu N_l)}} \tag{4.2}$$

offensichtlich reell und positiv ist:

$$A^*(\boldsymbol{k}, \omega) = A(\boldsymbol{k}, \omega)\,, \tag{4.3}$$

$$A(\boldsymbol{k}, \omega) \geqq 0\,. \tag{4.4}$$

Weiter genügt $A(\boldsymbol{k}, \omega)$ der Bedingung

$$\frac{1}{2\pi}\int_{-\infty}^{+\infty} d\omega\, A(\boldsymbol{k}, \omega) = 1\,. \tag{4.5}$$

Wie bekannt ist [2], läßt sich $A(\boldsymbol{k}, \omega)$, der Eigenschaften (4.3) bis (4.5) wegen, physikalisch interpretieren als die Wahrscheinlichkeit (pro Energieintervall), mit der zum Impuls $\hbar\,\boldsymbol{k}$ die Energie $\hbar\,\omega$ gehört. Anders ausgedrückt: $A(\boldsymbol{k}, \omega)$ stellt eine verallgemeinerte Zustandsdichte dar, die nicht nur, wie die übliche Zustandsfunktion $N(E)$ für stationäre Einteilchen-Zustände, von der Energie, sondern auch noch vom Impuls abhängt. Es besteht der Zusammenhang

$$n(\omega) = \frac{N(\omega)}{V} = \frac{1}{(2\pi)^3}\int d\tau_{\boldsymbol{k}}\, A(\boldsymbol{k}, \omega)\,. \tag{4.6}$$

Im Falle des Hamilton-Operators (2.1) kommt eine $A(\boldsymbol{k}, \omega)$ entsprechende Bedeutung nur der gemischten Spektralfunktion $A\,(\boldsymbol{k}, \boldsymbol{x} + \boldsymbol{x}'; \omega)$ zu, wobei nun aber zusätzlich noch eine Ortsabhängigkeit auftritt.

Den Eigenschaften (4.3) und (4.5) von $A(\boldsymbol{k}, \omega)$ entsprechen bei $A\,(\boldsymbol{k}, \boldsymbol{x} + \boldsymbol{x}'; \omega)$ genau die Eigenschaften (3.3) und (3.5).

Wie im Anhang gezeigt wird, besitzt außerdem die gemischte Spektralfunktion die Gestalt

$$A\,(\boldsymbol{k}, \boldsymbol{x}+\boldsymbol{x}'\,;\,\omega) = \frac{2\,\pi\,\hbar}{\sum_l \mathrm{e}^{-\beta(E_l-\mu N_l)}} \sum_{m,n} \mathrm{e}^{-\beta(E_n-\mu N_n)} \times$$
$$\times \left\{ \int \mathrm{d}\tau_{\boldsymbol{y}}\, \psi_{nm}(\boldsymbol{k}, \boldsymbol{y})\, \psi^*_{nm}\left(\boldsymbol{k}, \frac{\boldsymbol{x}+\boldsymbol{x}'}{2} - \boldsymbol{y}\right) \delta\left(\hbar\,\omega - (E_m - E_n)\right) + \right.$$
$$\left. + \int \mathrm{d}\tau_{\boldsymbol{y}}\, \psi_{mn}(\boldsymbol{k}, \boldsymbol{y})\, \psi^*_{mn}\left(\boldsymbol{k}, \frac{\boldsymbol{x}+\boldsymbol{x}'}{2} - \boldsymbol{y}\right) \delta\left(\hbar\,\omega - (E_n - E_m)\right) \right\} \tag{4.7}$$

mit

$$\left.\begin{aligned} \psi_{mn}(\boldsymbol{k}, \boldsymbol{z}) &\equiv (m|\Psi(\boldsymbol{k}, \boldsymbol{z})|n) = \mathrm{e}^{-i\boldsymbol{k}\boldsymbol{z}}\,(m|\Psi(\boldsymbol{z})|n)\,, \\ \psi^*_{mn}(\boldsymbol{k}, \boldsymbol{z}) &\equiv (n|\Psi^+(\boldsymbol{k}, \boldsymbol{z})|m) = \mathrm{e}^{i\boldsymbol{k}\boldsymbol{z}}\,(n|\Psi^+(\boldsymbol{z})|m)\,. \end{aligned}\right\} \tag{4.8}$$

Aus der Darstellung (4.7) läßt sich die physikalische Bedeutung der gemischten Spektralfunktion leicht ablesen, wenn man die Integrale noch etwas umformt. Das erste Integral etwa kann man auch in der Form

$$\int \mathrm{d}\tau_{\boldsymbol{y}}\, \mathrm{e}^{-i\boldsymbol{k}\cdot 2\boldsymbol{y}}\, \psi_{nm}\left(\frac{\boldsymbol{x}+\boldsymbol{x}'}{2} + \boldsymbol{y}\right) \psi^*_{nm}\left(\frac{\boldsymbol{x}+\boldsymbol{x}'}{2} - \boldsymbol{y}\right) \tag{4.9}$$

schreiben. Aus der Darstellung (4.9) geht hervor, daß $A\,(\boldsymbol{k}, \boldsymbol{x}+\boldsymbol{x}'\,;\,\omega)$ die Wahrscheinlichkeit (pro Energieintervall) angibt, mit der zum Impuls $\hbar\,\boldsymbol{k}$ und zum Schwerpunkt $(\boldsymbol{x}+\boldsymbol{x}')/2$ irgend zweier Elektronen die Energie $\hbar\,\omega$ gehört. Dies soll anhand von Fig. 1 veranschaulicht werden:

Das Integral (4.9) stellt eine Überlagerung von ebenen Wellen $\exp\,(-i\,\boldsymbol{k}\cdot 2\,\boldsymbol{y})$ mit dem Impuls $\hbar\,\boldsymbol{k}$ dar, und zwar für alle möglichen Abstände $2\,y$ zweier betrachteter Elektronen. Die Amplitude der einzelnen ebenen Wellen ist gegeben durch das Produkt der Wahrscheinlichkeits-Amplituden für die Erzeugung eines Elektrons am Ort $\left((\boldsymbol{x}+\boldsymbol{x}')/2\right) - \boldsymbol{y}$ und die Vernichtung eines zweiten Elektrons am Ort $\left((\boldsymbol{x}+\boldsymbol{x}')/2\right) + \boldsymbol{y}$ im Abstand $2\,y$ vom ersten Elektron.

Da bei der Integration zu jeder Amplitude mit $+\boldsymbol{y}$ auch die entsprechende Amplitude mit $-\boldsymbol{y}$ vorkommt, wird aus der komplexen Streuamplitude eine reelle Aufenthaltswahrscheinlichkeit für den Schwerpunkt zweier Elektronen (Anhang). Die Orte der einzelnen Elektronen bleiben völlig unbestimmt, so daß kein Verstoß gegen die Orts–Impuls-Unschärferelation vorliegt.

Für einen idealen Kristall wird $A(\boldsymbol{k}, \boldsymbol{r}\,;\,\omega)$ eine periodische Funktion des Schwerpunkts $\boldsymbol{R} = (\boldsymbol{x}+\boldsymbol{x}')/2$. In diesem Fall wird aus (2.9) eine Fourier-Reihe

$$A(\boldsymbol{k}, \boldsymbol{r}\,;\,\omega) = \sum_{\nu} \mathrm{e}^{i\boldsymbol{K}_\nu \boldsymbol{r}}\, A(\boldsymbol{k}, \boldsymbol{K}_\nu;\,\omega)\,, \tag{4.10}$$

wobei $\boldsymbol{K}_\nu$ ein reziproker Gittervektor ist.

Die durch (4.10) für einen idealen Kristall definierte Spektralfunktion $A(\boldsymbol{k}, \boldsymbol{K}_\nu;\,\omega)$ besitzt ebenfalls die Bedeutung einer Wahrscheinlichkeit.

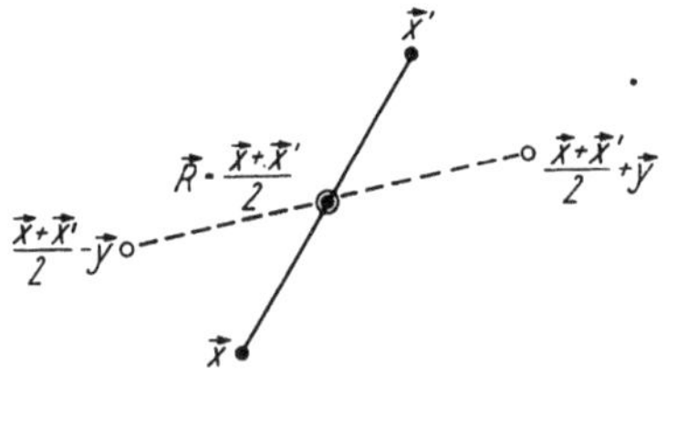

Fig. 1. Integrationsgebiet zu (4.9): Die hellen Kreise stellen die Orte dar, von denen die Beiträge zum Integral (4.9) herrühren. Die dunklen Kreise sind die bei der Integration festen Orte $\boldsymbol{x}$ und $\boldsymbol{x}'$, festgelegt durch die Greensche Funktion $G(x, x')$

In einem Kristall mit Inversionssymmetrie gilt

$$A(\boldsymbol{k}, -\boldsymbol{r}; \omega) = A(\boldsymbol{k}, \boldsymbol{r}; \omega) . \tag{4.11}$$

Daraus folgt mit (2.9)

$$A(\boldsymbol{k}, \boldsymbol{K}_\nu; \omega) = A(\boldsymbol{k}, -\boldsymbol{K}_\nu; \omega) \quad \text{für beliebiges } \nu , \tag{4.12}$$

oder, wegen (2.8),

$$A^*(\boldsymbol{k}, \boldsymbol{K}_\nu; \omega) = A(\boldsymbol{k}, \boldsymbol{K}_\nu; \omega) . \tag{4.13}$$

Aus (2.8) ergibt sich damit für $A(\boldsymbol{k}, \boldsymbol{K}_\nu; \omega)$ die Gestalt

$$A(\boldsymbol{k}, \boldsymbol{K}_\nu; \omega) = 2\pi\hbar \sum_{m,n} \mathrm{e}^{-\beta(E_n - \mu N_n)} \times$$
$$\times \frac{|(m|\,\Psi^+(\boldsymbol{k} \pm \boldsymbol{K}_\nu)\,|n)|^2\,\delta(\hbar\omega - (E_m - E_n)) + |(m|\,\Psi(\boldsymbol{k} \pm \boldsymbol{K}_\nu)\,|n)|^2\,\delta(\hbar\omega - (E_n - E_m))}{\sum_l \mathrm{e}^{-\beta(E_l - \mu N_l)}} . \tag{4.14}$$

Die Spektralfunktion $A(\boldsymbol{k}, \boldsymbol{K}_\nu; \omega)$ stellt die Wahrscheinlichkeit dar, mit der zum Impuls $\hbar\boldsymbol{k}$ die Energie $\hbar\omega$ gehört. Die Variable $\boldsymbol{K}_\nu$ gibt dabei nur an, daß $A(\boldsymbol{k}, \boldsymbol{K}_\nu; \omega)$ in $\boldsymbol{k}$ periodisch ist mit der Periode $\boldsymbol{K}_\nu$.

In der Sprache des Bändermodells ausgedrückt heißt dies, daß $A(\boldsymbol{k}, \boldsymbol{K}_\nu; \omega)$ das Energiespektrum im „ausgedehnten Wellenzahlraum" wiedergibt, so daß $\hbar\boldsymbol{k}$ nicht ein Quasi-Impuls, sondern der physikalische Impuls ist. Es bleibt zu beachten, daß das Energiespektrum bei Wechselwirkung verbreitert ist, d. h., daß es sich nicht mehr um $E(\boldsymbol{k})$-Linien, sondern um Bänder handelt (Fig. 2).

Eine weitere wichtige physikalische Eigenschaft der Spektralfunktionen beantwortet die Frage nach sogenannten gebundenen Zuständen. Darunter möchte ich alle lokalisierten Zustände verstehen (z. B. die Zustände der lokalisierten Rumpfelektronen), wobei jedoch die Teilchen nicht um einen festen Ort herum lokalisiert sein müssen, sondern an anderen Teilchen lokalisiert sein können (Lokalisierung im körperfesten, nicht im raumfesten Koordinatensystem). Dies leuchtet ein, wenn man z. B. an die Bewegung der Ionen denkt: Die Rumpfelektronen sind an den *bewegten* Kernen lokalisiert.

Bei einer streng feldtheoretischen Behandlung auch der Kerne entfällt die Möglichkeit einer raumfesten Lokalisierbarkeit ohnehin ganz, denn durch den dann zu verwendenden Hamilton-Operator wird kein Raumpunkt besonders ausgezeichnet.

Aus allen expliziten Darstellungen der Spektralfunktionen, von (2.7) bis (4.14), kann man herauslesen, daß $A(\lambda, \lambda'; \omega)$ für $\omega < 0$ im allgemeinen verschwindet; denn $\hbar\omega = E_2 - E_1$ ist jeweils die Energiedifferenz zweier Eigenzustände des Gesamtsystems mit $N_2 = N_1 + 1$. Da dem im Zustand 2 gegenüber dem Zustand 1 zusätzlich vorhandenen Elektron kinetische Energie zukommt, gilt im allgemeinen $E_2 > E_1$, d. h. $\hbar\omega > 0$.

Nur für den Fall, daß das zusätzliche Elektron eine solche Wechselwirkung mit Teilen des Systems aufweist, daß die potentielle Energie die kinetische dem Betrage nach übertrifft, ist $E_2 < E_1$, d. h. $\hbar\omega < 0$

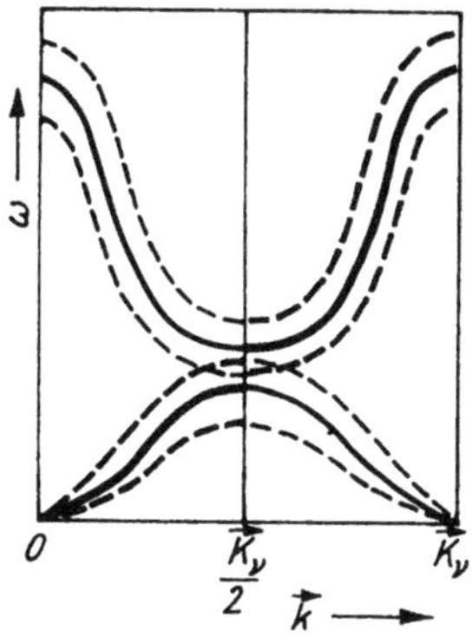

Fig. 2. Verbreiterte Energiebänder im Energie-Impuls- bzw. Energie Wellenzahl-Raum (schematisch)

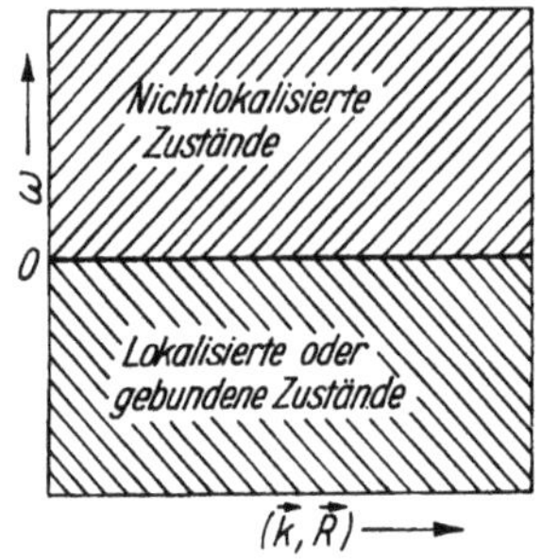

Fig. 3. Symbolische (zweidimensionale) Darstellung der Lage der lokalisierten Zustände ($\boldsymbol{k}$, $\boldsymbol{R}$) im verallgemeinerten (siebendimensionalen Energie-Impuls-Schwerpunkts-) Zustandsraum

möglich. Tritt für irgendein gerade betrachtetes Teilchen dieser Fall ein, so befindet sich das Teilchen in einem gebundenen Zustand.

Die Spektralfunktion $A(\boldsymbol{k}, \boldsymbol{x} + \boldsymbol{x}'; \omega)$, für ein ideales Gitter auch $A(\boldsymbol{k}, \boldsymbol{K}_\nu; \omega)$, gibt also für $\omega > 0$ die verallgemeinerte Zustandsdichte nichtlokalisierter Zustände, für $\omega < 0$ die Zustandsdichte lokalisierter Zustände wieder (Fig. 3).

5. Aufspaltung in Rumpf- und Leitungselektronen

Betrachtet man speziell ein Metall, so bietet sich nun die Möglichkeit, zwischen Leitungs- und Rumpfelektronen zu unterscheiden: Die Rumpfelektronen befinden sich in lokalisierten, die Leitungselektronen in nichtlokalisierten Zuständen (vorläufige Definition).

Da sich die Zustände der inneren Elektronen beim Übergang vom freien Atom zum Kristall nur wenig ändern, liegt es nahe, für die Rumpfzustände der Elektronen eines Metalls etwa die ersten N Wellenfunktionen des freien Atoms zu benützen. Damit läßt sich ein Teil des Elektronenspektrums, das der Rumpfelektronen, näherungsweise festlegen, so daß nur noch das Spektrum der Leitungselektronen zu berechnen ist.

Bestünde zwischen den Elektronen keine Wechselwirkung, so gäbe es dabei keinerlei Schwierigkeiten; in Wirklichkeit existieren aber zwei formal verschiedene Ursachen einer Wechselwirkung zwischen Rumpf- und Leitungselektronen, nämlich erstens die Coulomb-Kräfte und zweitens die wegen der Vertauschungsrelation wirkenden Austauschkräfte.

Die Coulomb-Kräfte kann man in guter Näherung dadurch berücksichtigen, daß man im Potential $V(\boldsymbol{x})$ in (2.1) neben dem Potential der Kerne auch das der Rumpfelektronen mitnimmt. Dadurch wird zwar vernachlässigt, daß sich nicht nur die Zustände der Leitungselektronen unter dem Einfluß der Rumpfelektronen, sondern auch die der Rumpfelektronen verändern können. Da aber der zuletzt aufgeführte Effekt, wie bereits betont, klein ist, ist diese Vernachlässigung gerechtfertigt.

Was wirklich Schwierigkeiten macht, ist die Berücksichtigung der Austauschkräfte zwischen Leitungs- und Rumpfelektronen. Hier zeigt sich nun der Vorteil der feldtheoretischen Methode der Spektralfunktion. Die Beeinflussung der Leitungselektronen durch die Rumpfelektronen infolge der Fermi-Statistik läßt sich nämlich sehr einfach angeben:

Die Vertauschungsrelation erfordert, daß (3.7) erfüllt ist. Hat man nun schon das Teilspektrum $A_R(\boldsymbol{k}, \boldsymbol{r}; \omega)$ der Rumpfelektronen, so kommen für die Leitungselektronen nur noch solche Spektren $A_L(\boldsymbol{k}, \boldsymbol{r}; \omega)$ in Frage, für die gilt

$$\frac{1}{2\pi}\int_{-\infty}^{+\infty} d\omega\, A_L(\boldsymbol{k}, \boldsymbol{r}, \omega) = 1 - \frac{1}{2\pi}\int_{-\infty}^{+\infty} d\omega\, A_R(\boldsymbol{k}, \boldsymbol{r}; \omega)\,. \tag{5.1}$$

Das Spektrum der Leitungselektronen muß also so beschaffen sein, daß es das schon vorhandene Gewicht der Rumpfelektronen für jeden Zustand $(\boldsymbol{k}, \boldsymbol{r})$ vollends auf 1 ergänzt.

Mit (3.8) und wegen (4.7) bis (4.9) lautet die Bedingung (5.1)

$$\frac{1}{2\pi}\int_{-\infty}^{+\infty} d\omega\, A_L(\boldsymbol{k}, \boldsymbol{r}; \omega) = 1 - \alpha(\boldsymbol{k}, \boldsymbol{r}) =$$

$$= 1 - \int d\tau_{\boldsymbol{y}}\, e^{-2i\boldsymbol{k}\boldsymbol{y}} \sum_{i=1}^{N} \sum_{\nu} \psi_i^*\left(\frac{\boldsymbol{x}+\boldsymbol{x}'}{2} - \boldsymbol{y} - \boldsymbol{R}_\nu\right)\psi_i\left(\frac{\boldsymbol{x}+\boldsymbol{x}'}{2} + \boldsymbol{y} + \boldsymbol{R}_\nu\right). \tag{5.2}$$

Diese Bedingung ist erfüllt, wenn man für $A_L(\boldsymbol{k}, \boldsymbol{r}; \omega)$ den Ansatz macht

$$A_L(\boldsymbol{k}, \boldsymbol{r}; \omega) = \frac{1 - \alpha(\boldsymbol{k}, \boldsymbol{r})}{\sqrt{\pi}\, \Gamma(\boldsymbol{k}, \boldsymbol{r})}\, 2\pi\hbar \cdot e^{-\frac{(\hbar\omega - E(\boldsymbol{k}, \boldsymbol{r}))^2}{\Gamma(\boldsymbol{k}, \boldsymbol{r})}}\,. \tag{5.3}$$

$E(\boldsymbol{k}, \boldsymbol{r})$ gibt die Stelle des Maximums der Zustandsdichte für den Zustand $(\boldsymbol{k}, \boldsymbol{r})$ an, $\Gamma(\boldsymbol{k}, \boldsymbol{r})$ die Halbwertsbreite des Maximums.

Wie aus Fig. 4 zu ersehen ist, können die durch den Ansatz (5.3) beschriebenen Zustände der Leitungselektronen je nach den Werten von $E(\boldsymbol{k}, \boldsymbol{r})$ und $\Gamma(\boldsymbol{k}, \boldsymbol{r})$ teilweise auch noch gebundene Zustände mit $\hbar\omega < 0$ sein, ergänzend zur vorläufigen Definition.

Dies besagt, daß die Leitungselektronen noch einen gewissen Bindungsgrad aufweisen können, wenn man als Bindungsgrad $B(\boldsymbol{k}, \boldsymbol{r})$ eines Zustands $(\boldsymbol{k}, \boldsymbol{r})$ den Ausdruck

$$B(\boldsymbol{k}, \boldsymbol{r}) = \frac{1}{2\pi}\int_{-\infty}^{0} d\omega\, A(\boldsymbol{k}, \boldsymbol{r}; \omega) \tag{5.4}$$

einführt. Als Bindungsgrad pro Atom kann man definieren:

$$B = \frac{1}{2\pi}\int_{-\infty}^{0} d\omega_{\text{El.-Zelle}} \int d\tau_{\boldsymbol{r}} \frac{1}{(2\pi)^3}\int d\tau_{\boldsymbol{k}}\; A(\boldsymbol{k}, \boldsymbol{r}; \omega)\,. \tag{5.5}$$

B gibt die Zahl der pro Atom lokalisierten Elektronen an.

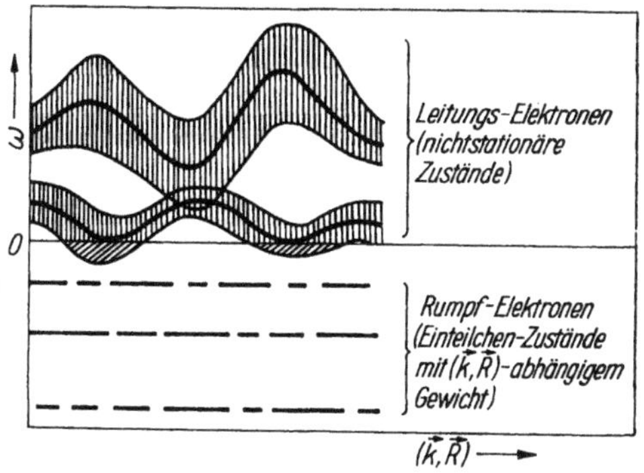

Fig. 4. Schematische Übersicht über die verallgemeinerte Zustandsdichte für die Elektronen eines Metalls. In den schraffierten Gebieten liegen lokalisierte Zustände der Leitungselektronen

Erreicht Ausdruck (5.5) mit $A_L(\boldsymbol{k}, \boldsymbol{r}, \omega)$ für ein Atom einen Wert $\geqq 1$, so ist an dem betreffenden Atom ein Leitungselektron zusätzlich gebunden (etwa an einer positiv geladenen Störstelle).

6. Zusammenhang mit der Pseudopotential-Methode

Mit der Abtrennung der Leitungselektronen von den Rumpfelektronen befaßt sich, im Rahmen der Einteilchentheorie, auch die sogenannte Pseudopotential-Methode [3]. Bei diesem Verfahren werden die Zustände der Leitungselektronen durch Wellenfunktionen $\psi_L(\boldsymbol{x})$ approximiert, die jedoch nicht aus der Schrödinger-Gleichung

$$H\,\psi_L(\boldsymbol{x}) = E_L\,\psi_L(\boldsymbol{x}) \tag{6.1}$$

bestimmt werden, sondern für die der Ansatz (Schmidtsches Orthogonalisierungsverfahren) [4]

$$\psi_L(\boldsymbol{x}) = \varphi_L(\boldsymbol{x}) - \sum_R (\psi_R, \varphi_L)\,\psi_R(\boldsymbol{x}) \tag{6.2}$$

gemacht wird, wobei die $\psi_R(\boldsymbol{x})$ die Rumpfzustände sind und die $\varphi_L(\boldsymbol{x})$ aus der nichtlinearen Differentialgleichung

$$H'\,\varphi_L(\boldsymbol{x}) = \{H + V_0(\boldsymbol{x})\}\,\varphi_L(\boldsymbol{x}) = \left\{-\frac{h^2}{2m} + V_I(\boldsymbol{x}) + V_0(\boldsymbol{x})\right\}\varphi_L(\boldsymbol{x}) = E_L\,\varphi_L(\boldsymbol{x}) \tag{6.3}$$

mit

$$V_0(\boldsymbol{x}) = \sum_R (E_L - E_R)\,|\psi_R)\,(\psi_R| \tag{6.4}$$

zu berechnen sind.

In der Sprache der vorliegenden Arbeit ausgedrückt, besteht das Ziel der Pseudopotential-Methode darin, den Teil von $A_L(\boldsymbol{k}, \boldsymbol{r}; \omega)$ bzw. von $A_L(\boldsymbol{k}, \boldsymbol{K}_\nu; \omega)$ zu berechnen, der zu $\omega < 0$ gehört, d. h. den nichtlokalisierten Teil der Zustände der Leitungselektronen. Die Wellenfunktionen, aus denen dieser Anteil aufgebaut werden kann, müssen verhältnismäßig glatt sein. Dies ist nur möglich, wenn das sogenannte Pseudopotential $V_P = V_I + V_0$ sehr klein ist, was dadurch zustande kommen soll, daß das positive Orthogonalitätspotential in Kernnähe das dort stark negative Ionenpotential mehr oder weniger vollständig kompensiert.

Gegen diese Methode sind verschiedene Einwände zu erheben, von denen der folgende am schwersten wiegt: Nur wenn die $\varphi_L(\boldsymbol{x})$ zusammen mit den $\psi_R(\boldsymbol{x})$ denselben Hilbert-Raum H darstellen, den die $\psi_n(\boldsymbol{x}) = \psi_L(\boldsymbol{x}) \cup \psi_R(\boldsymbol{x})$ aufspannen, kann die lineare Transformation (6.2) die richtigen $\psi_L(\boldsymbol{x})$ liefern. Die $\varphi_n(\boldsymbol{x}) = \varphi_L(\boldsymbol{x}) \cup \psi_R(\boldsymbol{x})$ stellen genau dann den Hilbert-Raum H dar, wenn sie umgekehrt durch eine lineare Transformation aus den $\psi_n(\boldsymbol{x})$ hervorgehen. Da die Umkehrtransformation zu (6.2) wegen des Auftretens von Projektionsoperatoren in (6.2) nicht eindeutig existiert, bleibt nur übrig, zu untersuchen, ob die durch Gleichung (6.3) implizierte Transformation linear ist.

Wäre die Transformation $\psi_n(\boldsymbol{x}) \rightarrow \varphi_n(\boldsymbol{x})$ linear, so müßte (6.1) in die Gleichung

$$H'\,\varphi_n(\boldsymbol{x}) = \sum_k E'_{nk}\,\varphi_k(\boldsymbol{x}) \tag{6.5}$$

übergehen, oder, im Spezialfall einer unitären Transformation, in

$$H' \varphi_n(\boldsymbol{x}) = E'_n \varphi_n(\boldsymbol{x}) \quad \text{mit} \quad E'_n = E_n . \tag{6.6}$$

Die Beziehung (6.3) stimmt zwar mit (6.6) überein, ist jedoch nicht unitär.

Aus diesem Widerspruch folgt, daß die durch (6.3) implizierte Transformation $\psi_n(\boldsymbol{x}) \to \varphi_n(\boldsymbol{x})$ nicht linear sein kann, und daß die durch (6.3) bestimmten $\varphi_L(\boldsymbol{x})$ nicht dem Hilbert-Raum H zugehören. Das mit Hilfe der Pseudopotential-Methode gewonnene Funktionssystem $\psi_n(\boldsymbol{x})$ ist also nicht vollständig.

Nach (3.9) stellt aber die Nichtvollständigkeit eine Verletzung der Vertauschungsrelation dar. Das bedeutet, daß zumindest an der Stelle, an der die Teilräume zusammenstoßen, die Vertauschungsrelationen nicht erfüllt sein können. Physikalisch ausgedrückt: Die Austauschkräfte zwischen den Rumpf- und den Leitungselektronen können mit der Pseudopotentialmethode nicht richtig beschrieben werden.

Ein anderer Einwand, der eng mit dem ersten zusammenhängt, betrifft die Vieldeutigkeit des Pseudopotentials, eine Folge der Nichtlinearität der Differentialgleichung (6.3). Durch Minimalisieren des Pseudopotentials oder durch andere Nebenbedingungen kann man zwar die Vieldeutigkeit beseitigen, verliert aber unter Umständen gebundene Lösungen.

Der dritte Einwand bezieht sich auf den Begriff der Lokalisierung von Einteilchen-Zuständen. Man kann nämlich im Einteilchen-Bild den Begriff des Bindungsgrades (5.4) gar nicht definieren, so daß nicht klar ist, was die geglätteten nichtnormierten Wellenfunktionen eigentlich darstellen sollen.

Da alle diese Schwierigkeiten bei der feldtheoretischen Behandlung mit Wechselwirkung nicht auftauchen (der Bindungsgrad des nichtstationären Vielteilchen-Zustandes $(\boldsymbol{k}, \boldsymbol{r})$ kann genau definiert werden), liegt die Vermutung nahe, daß sich der physikalische Inhalt der Pseudopotential-Methode im Rahmen einer Einteilchen-Näherung gar nicht präzise formulieren läßt, sondern erst in einer Vielteilchen-Theorie verständlich wird.

Herrn P. Lorch und besonders Herrn H. Glock danke ich für ausführliche Diskussionen zum Thema Pseudopotential.

Anhang

Durch Fourier-Transformation der Matrixelemente $\psi_{mn}(\boldsymbol{x})$, (vgl. (4.8)) wird

$$\psi_{mn}(\boldsymbol{k} + \boldsymbol{k}') = \int d\tau_{\boldsymbol{x}}\, e^{-i(\boldsymbol{k}+\boldsymbol{k}')\boldsymbol{x}}\, \psi_{mn}(\boldsymbol{x}) \tag{A 1}$$

definiert, durch unvollständige Rücktransformation außerdem

$$\psi_{mn}(\boldsymbol{k}, \boldsymbol{x}) = \frac{1}{(2\pi)^3} \int d\tau_{\boldsymbol{k}'}\, e^{i\boldsymbol{k}'\boldsymbol{x}}\, \psi_{mn}(\boldsymbol{k} + \boldsymbol{k}') =$$
$$= \frac{1}{(2\pi)^3} \int d\tau_{\boldsymbol{k}'} \int d\tau_{\boldsymbol{y}}\, e^{i\boldsymbol{k}'(\boldsymbol{x}-\boldsymbol{y})}\, e^{-i\boldsymbol{k}\boldsymbol{y}}\, \psi_{mn}(\boldsymbol{y}) = e^{-i\boldsymbol{k}\boldsymbol{x}}\, \psi_{mn}(\boldsymbol{x}) . \tag{A 2}$$

Die Umkehrung zu (A 2) lautet

$$\psi_{mn}(\boldsymbol{k}, \boldsymbol{k}') = \int d\tau_{\boldsymbol{x}}\, e^{-i\boldsymbol{k}'\boldsymbol{x}}\, \psi_{mn}(\boldsymbol{k}, \boldsymbol{x}) = \int d\tau_{\boldsymbol{k}}\, e^{-i(\boldsymbol{k}+\boldsymbol{k}')}\, \psi_{mn}(\boldsymbol{x}) = \psi_{mn}(\boldsymbol{k} + \boldsymbol{k}') . \tag{A 3}$$

Aus der Funktion

$$A_{mn}(\boldsymbol{k}, \boldsymbol{x}) = \frac{1}{(2\pi)^3} \int d\tau_{\boldsymbol{k}'}\, e^{i\boldsymbol{k}'\boldsymbol{x}}\, \psi^*_{mn}(\boldsymbol{k} - \boldsymbol{k}')\, \psi_{mn}(\boldsymbol{k} + \boldsymbol{k}') \tag{A 4}$$

kann man wegen (2.8), (2.9) und (4.8) die gemischte Spektralfunktion wie folgt aufbauen:

$$A(\boldsymbol{k}, \boldsymbol{x}; \omega) = \frac{2\pi\hbar}{\sum_l e^{-\beta(E_l - \mu N_l)}} \sum_{m,n} e^{-\beta(E_n - \mu N_n)} \times$$
$$\times \{A_{nm}(\boldsymbol{k}, \boldsymbol{x})\, \delta(\hbar\omega - (E_m - E_n)) + A_{mn}(\boldsymbol{k}, \boldsymbol{x})\, \delta(\hbar\omega - (E_n - E_m))\}. \quad \text{(A 5)}$$

Unter Beachtung von (A 1) und (A 2) läßt sich (A 4) umformen:

$$A_{mn}(\boldsymbol{k}, \boldsymbol{x}) = \frac{1}{(2\pi)^3} \int d\tau_{\boldsymbol{k}'}\, e^{i\boldsymbol{k}'\boldsymbol{x}} \int d\tau_{\boldsymbol{y}}\, e^{i(\boldsymbol{k}-\boldsymbol{k}')\boldsymbol{y}}\, \psi^*_{mn}(\boldsymbol{y}) \int d\tau_{\boldsymbol{z}}\, e^{-i(\boldsymbol{k}+\boldsymbol{k}')\boldsymbol{z}}\, \psi_{mn}(\boldsymbol{z}) =$$
$$= \int d\tau_{\boldsymbol{y}}\, e^{i\boldsymbol{k}\boldsymbol{y}}\, \psi^*_{mn}(\boldsymbol{y}) \cdot e^{-i\boldsymbol{k}(\boldsymbol{x}-\boldsymbol{y})}\, \psi_{mn}(\boldsymbol{x}-\boldsymbol{y}) = \int d\tau_{\boldsymbol{y}}\, \psi^*_{mn}(\boldsymbol{k}, \boldsymbol{y})\, \psi_{mn}(\boldsymbol{k}, \boldsymbol{x}-\boldsymbol{y}). \quad \text{(A 6)}$$

Setzt man (A 6) in (A 5) ein, so erhält man die Darstellung (4.7).

Die Spektralfunktion $A(\boldsymbol{x}, \boldsymbol{x}'; \omega)$ kann als kontinuierliche, unendliche, hermitische Matrix aufgefaßt werden:

$$\mathsf{A}(\omega) = \begin{pmatrix} A(\boldsymbol{x}, \boldsymbol{x}; \omega) & A(\boldsymbol{x}, \boldsymbol{x}'; \omega) \\ A(\boldsymbol{x}', \boldsymbol{x}; \omega) & A(\boldsymbol{x}', \boldsymbol{x}'; \omega) \end{pmatrix} = \begin{pmatrix} A(\boldsymbol{x}, \boldsymbol{x}; \omega) & A(\boldsymbol{x}, \boldsymbol{x}'; \omega) \\ A^*(\boldsymbol{x}, \boldsymbol{x}'; \omega) & A(\boldsymbol{x}', \boldsymbol{x}'; \omega) \end{pmatrix}, \quad \text{(A 7)}$$

ebenso $A\left(\frac{\boldsymbol{k}+\boldsymbol{k}'}{2}, \frac{\boldsymbol{k}-\boldsymbol{k}'}{2}; \omega\right)$ als

$$\mathsf{B}(\omega) = \begin{pmatrix} A(\boldsymbol{g}, \boldsymbol{g}; \omega) & A(\boldsymbol{g}, \boldsymbol{g}'; \omega) \\ A(\boldsymbol{g}', \boldsymbol{g}; \omega) & A(\boldsymbol{g}', \boldsymbol{g}'; \omega) \end{pmatrix} = \begin{pmatrix} A(\boldsymbol{g}, \boldsymbol{g}; \omega) & A(\boldsymbol{g}, \boldsymbol{g}'; \omega) \\ A^*(\boldsymbol{g}, \boldsymbol{g}'; \omega) & A(\boldsymbol{g}', \boldsymbol{g}'; \omega) \end{pmatrix}. \quad \text{(A 8)}$$

Die Matrix (A 7) geht aus der Matrix (A 8) gemäß (2.8b) durch eine unitäre Transformation hervor:

$$\mathsf{A}(\omega) = \mathsf{U}^{-1}\mathsf{B}(\omega)\,\mathsf{U}, \quad A(\boldsymbol{x}, \boldsymbol{x}'; \omega) = \frac{1}{(2\pi)^6} \int d\tau_{\boldsymbol{k}} \int d\tau_{\boldsymbol{k}'}\, e^{i\boldsymbol{k}\boldsymbol{x}} A\left(\frac{\boldsymbol{k}+\boldsymbol{k}'}{2}, \frac{\boldsymbol{k}-\boldsymbol{k}'}{2}, \omega\right) e^{-i\boldsymbol{k}'\boldsymbol{x}'}, \quad \text{(A 9)}$$

wobei die Unitaritätsbedingung

$$\frac{1}{(2\pi)^3} \int d\tau_{\boldsymbol{k}}\, e^{i\boldsymbol{k}(\boldsymbol{x}-\boldsymbol{x}')} = \delta(\boldsymbol{x} - \boldsymbol{x}') \quad \text{(A 10)}$$

besteht.

Zu jeder hermitischen Matrix gibt es eine unitäre Hauptachsentransformation U_{H}, die die hermitische Matrix zu einer reellen Diagonalmatrix macht:

$$A(\boldsymbol{x}, \boldsymbol{x}'; \omega) = \int d\tau_{\boldsymbol{k}}\, U_{\mathrm{H}}^{-1}(\boldsymbol{k}, \boldsymbol{x}) \cdot D(\boldsymbol{k}; \omega) \cdot U_{\mathrm{H}}(\boldsymbol{k}, \boldsymbol{x}'), \quad \text{(A 11)}$$

als Grenzfall der Beziehung

$$A_{kl}(\omega) = \sum_i U_{ik}^{-1}\, D_i(\omega)\, U_{il} \qquad \text{oder} \qquad \mathsf{A}(\omega) = \mathsf{U}_{\mathrm{H}}^{-1}\, \mathsf{D}(\omega)\, \mathsf{U}_{\mathrm{H}}. \quad \text{(A 11a)}$$

Die Transformation (A 11) ist nur für freie Teilchen eine Fourier-Transformation; nur in diesem Fall besitzt die Variable $\hbar\boldsymbol{k}$ die physikalische Bedeutung eines Impulses. Im vorliegenden Fall wäre die Diagonalisierung nur durch Blochfunktionen bzw. noch allgemeinere Funktionen zu erzielen; die Variable $h\boldsymbol{k}$ ist also nicht mehr der Impuls.

Man kann jedoch eine Art Hauptachsentransformation auch mit Hilfe der Fourier-Transformation durchführen, wenn man wie in (2.10) vorgeht. Diese Quasi-Hauptachsentransformation besitzt die Gestalt

$$A(\boldsymbol{x}, \boldsymbol{x}'; \omega) = \int d\tau_{\boldsymbol{k}}\, U_{\mathrm{QH}}^{-1}(\boldsymbol{k}, \boldsymbol{x}) \cdot D_{\mathrm{Q}}\left(\boldsymbol{k}, \frac{\boldsymbol{x}+\boldsymbol{x}'}{2}; \omega\right) \cdot U_{\mathrm{QH}}(\boldsymbol{k}, \boldsymbol{x}') \quad \text{(A 12)}$$

mit

$$D_{\mathrm{Q}}\left(\boldsymbol{k}, \frac{\boldsymbol{x}+\boldsymbol{x}'}{2}; \omega\right) = A\left(\boldsymbol{k}, \frac{\boldsymbol{x}+\boldsymbol{x}'}{2}; \omega\right).$$

Die Matrix $D_Q\left(\boldsymbol{k}, \frac{\boldsymbol{x}+\boldsymbol{x}'}{2}; \omega\right)$ ist wie $D(\boldsymbol{k}; \omega)$ reell. Betrachtet man $(\boldsymbol{x}+\boldsymbol{x}')/2$ als einen Parameter, so ist $D_Q\left(\boldsymbol{k}, \frac{\boldsymbol{x}+\boldsymbol{x}'}{2}; \omega\right)$ diagonal, was in dem Wegfallen eines Integrals seinen Ausdruck findet. (A 12) ist jedoch keine echte Hauptachsendarstellung von $A(\boldsymbol{x}, \boldsymbol{x}'; \omega)$, denn zu jedem Matrixelement $A(\boldsymbol{x}, \boldsymbol{x}'; \omega)$ gehört im allgemeinen eine andere Quasi-Diagonalmatrix; nur zu zwei Elementen mit demselben Schwerpunkt gehört dieselbe Diagonalmatrix.

Der Vorteil der Darstellung (A 12) besteht darin, daß man $\hbar\, \boldsymbol{k}$ weiterhin als Impuls interpretieren kann, daß aber dennoch alle Matrixelemente reell sind und, genau wie die Diagonalelemente in (A 7) und (A 8), als Wahrscheinlichkeiten interpretiert werden können dafür, daß zu einem Impuls $\hbar\, \boldsymbol{k}$ und zu dem Schwerpunkt $(\boldsymbol{x}+\boldsymbol{x}')/2$ die Energie $\hbar\, \omega$ gehört.

Literatur

[1] L. P. Kadanoff und G. Baym, Quantum-Statistical Mechanics, Benjamin, New York 1962.
[2] P. Brand, Z. Phys., im Druck.
[3] W. A. Harrison, Pseudopotentials in the Theory of Metals, Benjamin, New York 1966.
[4] W. I. Smirnow, Lehrgang der höheren Mathematik, Teil V, VEB Deutscher Verlag der Wissenschaften, Berlin 1962.

(Received June 5, 1968)

phys. stat. sol. 28, 755 (1968)

Subject classification: 18.2; 10.1; 21.1

Institut für Physik am Max-Planck-Institut für Metallforschung, Stuttgart

Der Beitrag der Versetzungen zur Anisotropie des magnetischen Einmündungsgesetzes in Nickel

Von

A. HOLZ und H. KRONMÜLLER

Das magnetische Einmündungsgesetz bei Anwesenheit von Stufenversetzungen wird in Abhängigkeit von der Magnetfeldrichtung für Nickel als numerisches Beispiel berechnet. Die Suszeptibilität läßt sich in einen streufeldfreien Anteil χ_0 und einen vom Streufeld hervorgerufenen negativen Beitrag χ_s zerlegen. Numerische Rechnungen wurden für die durch Burgersvektor, Gleitebenennormalen und Versetzungslinien bestimmten Ebenen ausgeführt. Im Bereich bis 2000 Oe wird in allen drei Ebenen eine ausgeprägte Anisotropie gefunden, die im wesentlichen durch χ_s bestimmt ist. Mit großem Magnetfeld H nimmt χ_s proportional $1/H^4$ ab, während χ_0 nur mit $1/H^3$ abfällt. Im Feldbereich $H \geqq 4\,\pi\, I_s$ ist deshalb die Anisotropie von χ vornehmlich durch χ_0 bestimmt.

The magnetic saturation curve in the presence of edge-dislocations is determined as a function of the direction of the magnetic field using nickel as a numerical example. The susceptibility can be represented by two parts χ_0 and χ_s, where χ_0 is positive and independent of the stray field whereas χ_s is negative and takes into account the homogenizing effect of the stray field. Numerical calculations are done for the planes determined by the Burgers vector, the normal of the slip plane, and the dislocation line. At fields smaller than 2000 Oe a strong anisotropy is found in all three planes. It is essentially determined by χ_s. In large magnetic fields χ_s decreases as $1/H^4$ whereas χ_0 decreases only with $1/H^3$. For fields $H \geqq 4\,\pi\, I_s$ the anisotropy of χ is therefore essentially determined by χ_0.

1. Einleitung

Der Bereich der Einmündung in die ferromagnetische Sättigung darf heute wohl als der am besten untersuchte Teil der Magnetisierungskurve der Ferromagnetika angesehen werden. Dieser Sachverhalt beruht im wesentlichen auf zwei Tatsachen: 1. Im Einmündungsgebiet erfolgt die Magnetisierungszunahme allein durch reversible Drehprozesse, 2. Die Auslenkungen der spontanen Magnetisierung aus der Richtung des angelegten Feldes dürfen als klein betrachtet werden. Auf Grund dieser beiden Sachverhalte kann man den Einmündungsbereich mit Hilfe der linearisierten mikromagnetischen Grundgleichungen behandeln [1 bis 4]. Die Untersuchung des Einmündungsbereiches besitzt den besonderen Vorteil, daß nicht nur fehlerfreie Kristalle, sondern auch Realkristalle, die innere Spannungen in Form von Versetzungen enthalten, einer theoretischen Behandlung zugänglich sind. Der letztgenannte Problemkreis war in den letzten Jahren Gegenstand zahlreicher experimenteller und theoretischer Arbeiten [3 bis 7]. Unter anderem wurde eine Theorie entwickelt, die es erlaubt, für beliebige Versetzungsverteilungen die reversible Suszeptibilität im Einmündungsbereich zu berechnen [3].

Während in früheren Arbeiten [4 bis 7] die Bestimmung der Amplitude der inneren Spannungen in plastisch verformten Einkristallen aus Messungen der

Suszeptibilität χ im Vordergrund stand, soll in dieser Arbeit der theoretische Zusammenhang zwischen der *Orientierung* der Versetzungen und der *Anisotropie* des Einmündungsgesetzes behandelt werden. Derartige Untersuchungen gewinnen im Zusammenhang mit den aus elektronenmikroskopischen Durchstrahlungsaufnahmen bestimmten Versetzungsanordnungen besonderes Interesse, da man hofft, aus Messungen der Anisotropie des Einmündungsgesetzes die Anisotropie der inneren Spannungen bestimmen zu können, die dann mit elektronenmikroskopischen Ergebnissen verglichen werden kann. Auf diese Weise sollte es möglich sein zu entscheiden, welche Versetzungen für die bei der plastischen Verformung auftretenden inneren Spannungen vorwiegend verantwortlich sind. Daß man die magnetischen Eigenschaften zur Untersuchung anisotroper innerer Spannungen heranziehen kann, haben neuere Untersuchungen von Bhattacharyya et al. [8], Geiss und Silcox [9] sowie Gessinger et al. [10] gezeigt.

Im allgemeinen ist die von anisotropen Versetzungsverteilungen hervorgerufene Anisotropie magnetischer Eigenschaften derjenigen Anisotropie überlagert, die von der auch in idealen Kristallen vorhandenen Kristallenergie herrührt. Auch äußere Spannungen können auf Grund der magnetoelastischen Kopplung einen Beitrag zur Anisotropie der magnetischen Eigenschaften liefern. Dies bedeutet, daß bei der Untersuchung des Einmündungsgesetzes der Beitrag des idealen Kristalls zur Anisotropie bekannt sein muß. Für eine langgestreckte zylinderförmige Probe, deren Stabachse mit den drei kubischen $\langle 100 \rangle$-Richtungen die Richtungskosinus $\gamma_1, \gamma_2, \gamma_3$ bildet und die einer homogenen Zugspannung σ parallel zur Stabachse ausgesetzt wird, lautet das Einmündungsgesetz [11] bei Vernachlässigung der zweiten Kristallenergiekonstanten K_2

$$\chi H^3 = \frac{[2 K_1 + 3 \sigma (\lambda_{100} - \lambda_{111})]^2}{I_s} \left[\sum_{i=1}^{3} \gamma_i^6 - \left(\sum_{i=1}^{3} \gamma_i^4 \right)^2 \right] \tag{1.1}$$

(K_1 = Kristallenergiekonstante, λ_{100} und λ_{111} magnetostriktive Konstanten, I_s = spontane Magnetisierung, H = angelegte Gleichfeldstärke). Aus (1.1) geht unter anderem hervor, daß bei magnetostriktiver Isotropie ($\lambda_{100} = \lambda_{111}$) die äußeren Spannungen keinen Einfluß mehr auf die Anisotropie der Suszeptibilität im Einmündungsgebiet ausüben. Zur Diskussion der von Versetzungen hervorgerufenen Anisotropie kann (1.1) jedoch nicht herangezogen werden, da der Volummittelwert innerer Spannungen verschwindet und in (1.1) im Falle inhomogener Spannungen in erster Näherung deren Volummittelwert einzusetzen ist. Da innere Spannungen im allgemeinen stark inhomogen sind, ist die Orientierung der spontanen Magnetisierung auch durch die Austauschenergie und die Streufeldenergie bestimmt.

Eine Behandlung dieses allgemeineren Problems ist mit Hilfe der für den Fall magnetostriktiver Anisotropie von Seeger und Kronmüller [3, 4] abgeleiteten mikromagnetischen Grundgleichungen möglich. Diese Theorie wird den folgenden Betrachtungen zugrundegelegt werden. Im einzelnen werden wir das Einmündungsgesetz für geradlinige Stufenversetzungen, die in der (111)-Ebene liegen, berechnen. Vereinfachend wird dabei angenommen werden, daß keine Wechselwirkung zwischen den Versetzungen besteht. Der von einer Überlagerung der Spannungsfelder benachbarter Versetzungen herrührende Einfluß auf das Einmündungsgesetz wurde in früheren Arbeiten ausführlich diskutiert [4, 7, 12].

2. Zur Messung der Anisotropie des Einmündungsgesetzes

Die Anisotropie des Einmündungsgesetzes läßt sich sowohl mit einer statischen als auch mit einer dynamischen Methode bestimmen. Die in beiden Fällen gemessene physikalische Größe bei vorgegebenem Magnetfeld $\boldsymbol{H}$ ist die reversible Suszeptibilität $\boldsymbol{\chi}$. Diese stellt den Zusammenhang zwischen einer infinitisimalen Änderung $\mathrm{d}\boldsymbol{H}$ des angelegten Magnetfeldes und der dabei auftretenden Änderung $\mathrm{d}\boldsymbol{I}$ der Magnetisierung her. Es gilt

$$\mathrm{d}\boldsymbol{I} = \boldsymbol{\chi} \cdot \mathrm{d}\boldsymbol{H} . \tag{2.1}$$

Im allgemeinen bedeutet $\boldsymbol{\chi}$ einen Tensor zweiter Stufe. Im Einmündungsbereich, d. h. bei Feldstärken $H > |K_1|/I_s$, besitzt $\boldsymbol{\chi}$ eine besonders einfache Form, wenn wir uns auf ein kartesisches Koordinatensystem beziehen, dessen y-Achse parallel zur angelegten Feldstärke liegt. In diesem sogenannten Brownschen Koordinatensystem lautet der Suszeptibilitätstensor

$$\boldsymbol{\chi} = \begin{pmatrix} 0 & 0 & 0 \\ 0 & \chi_{yy} & 0 \\ 0 & 0 & 0 \end{pmatrix} . \tag{2.2}$$

Die Komponenten χ_{xx} und χ_{zz} sind Null, da diese sich als lineare Funktion der Komponenten des Tensors der inneren Spannungen darstellen lassen [3] und deren Mittelwert nach dem Satz von Albenga [13] Null ist.

Die Anisotropie der Suszeptibilitätskomponente χ_{yy}, die wir im folgenden als Einmündungssuszeptibilität χ bezeichnen wollen, erhält man bei der statischen Methode, indem man bei konstantem äußeren Felde die Orientierung des Kristalls bezüglich der Feldrichtung verändert und jeweils die Suszeptibilität gemäß Gleichung (2.1) bestimmt. Im Falle der dynamischen Methode wird bei festgehaltener Feldrichtung (Kristallorientierung) die Orientierung des Kristalls (Feldes) periodisch verändert (z. B. Rotation flacher zylinderförmiger Kristallscheiben um ihre Zylinderachse [14]). Da sich hierbei die Komponente der pauschalen Magnetisierung parallel zum angelegten Feld verändert, wird in einer Spule, deren Windungsfläche senkrecht zum Felde steht, eine Spannung induziert. Nach dem Induktionsgesetz ergibt sich für die in der Spule induzierte Spannung

$$U_{\mathrm{ind}} = 4\,\pi\, n\, F_{\mathrm{eff}} \frac{\mathrm{d}I(\varphi, H)}{\mathrm{d}t} , \tag{2.3}$$

wo n die Windungszahl und F_{eff} die effektive Windungsfläche der Spule bedeutet. Wird die Kristallscheibe mit der Kreisfrequenz ω um eine Achse senkrecht zu $\boldsymbol{H}$ rotiert, so besteht zwischen der Winkeländerung $\mathrm{d}\varphi$ und dem Zeitintervall $\mathrm{d}t$ der Zusammenhang $\mathrm{d}\varphi = \omega\,\mathrm{d}t$, so daß wir für Gleichung (2.3) auch

$$U_{\mathrm{ind}}(\varphi, H) = 4\,\pi\, n\, F_{\mathrm{eff}}\, \omega \frac{\mathrm{d}I(\varphi, H)}{\mathrm{d}\varphi} \tag{2.4}$$

schreiben können, oder

$$\mathrm{d}I(\varphi, H) = \frac{1}{4\,\pi\, n\, \omega\, F_{\mathrm{eff}}} U_{\mathrm{ind}}(\varphi, H)\,\mathrm{d}\varphi . \tag{2.5}$$

Aus Gleichung (2.5) erhalten wir für die Magnetisierungsänderung bei einer Drehung von $\varphi = 0$ bis $\varphi = \varphi$

$$\Delta I(\varphi, H) = I(\varphi, H) - I(0, H) = \frac{1}{4\,\pi\, n\, \omega\, F_{\mathrm{eff}}} \int\limits_0^{\varphi} U_{\mathrm{ind}}(\varphi, H)\,\mathrm{d}\varphi . \tag{2.6}$$

Die Magnetisierungsänderung ΔI läßt sich in der Form

$$\Delta I(\varphi, H) = \int_{\infty}^{H} (\chi(\varphi, H) - \chi(0, H))\,\mathrm{d}H \tag{2.7}$$

schreiben. Durch Gleichsetzen von (2.7) und (2.6) erhalten wir durch Differentiation nach H und φ

$$\frac{\mathrm{d}\Delta I(\varphi, H)}{\mathrm{d}H} = \frac{1}{4\,\pi\,n\,F_{\mathrm{eff}}\,\omega} \int_{0}^{\varphi} \frac{\mathrm{d}U_{\mathrm{ind}}(\varphi, H)}{\mathrm{d}H}\,\mathrm{d}\varphi = \chi(\varphi, H) - \chi(0, H)\,, \tag{2.8}$$

$$\frac{\mathrm{d}^2\Delta I(\varphi, H)}{\mathrm{d}\varphi\,\mathrm{d}H} = \frac{1}{4\,\pi\,n\,F_{\mathrm{eff}}\,\omega} \frac{\mathrm{d}U_{\mathrm{ind}}(\varphi, H)}{\mathrm{d}H} = \frac{\mathrm{d}\chi(\varphi, H)}{\mathrm{d}H}\,. \tag{2.9}$$

Gleichung (2.8) stellt den Zusammenhang zwischen der induzierten Spannung und der statisch gemessenen Suszeptibilität her, während Gleichung (2.9) die Feldabhängigkeit der induzierten Spannung in Zusammenhang zur Winkelabhängigkeit der Suszeptibilität bringt. Im allgemeinen wird U_{ind} in Form einer Fourierreihe

$$U_{\mathrm{ind}} = \sum_{m} A_m(H) \cos m\,\varphi \tag{2.10}$$

bestimmt, wo die Fourierkoeffizienten $A_m(H)$ von der angelegten Feldstärke abhängig sind. Wird (2.10) in (2.8) eingesetzt, so erhält man folgende Beziehung zwischen den experimentell zu bestimmenden Koeffizienten $\mathrm{d}A_m(H)/\mathrm{d}H$ und der Suszeptibilität

$$\frac{\mathrm{d}\Delta I(\varphi, H)}{\mathrm{d}H} = \chi(\varphi, H) - \chi(0, H) = \frac{1}{4\,\pi\,n\,\omega\,F_{\mathrm{eff}}} \sum_{m} \frac{1}{m} \frac{\mathrm{d}A_m(H)}{\mathrm{d}H} \sin m\,\varphi\,. \tag{2.11}$$

3. Berechnung der Einmündungssuszeptibilität

3.1 Theoretische Grundlagen

Im Bereich der Einmündung sind die Elektronenspins einer idealen einkristallinen Probe parallel ausgerichtet und liegen nahezu in Richtung des angelegten Magnetfeldes. In Realkristallen treten Abweichungen von dieser homogenen Verteilung der Elektronenspins in der Umgebung von Versetzungen auf, deren Eigenspannungen auf Grund der magnetoelastischen Kopplung auf die Elektronenspins ein Drehmoment ausüben und daher eine mehr oder weniger inhomogene Magnetisierungsverteilung erzeugen. Eine Berechnung dieser inhomogenen Magnetisierungsverteilung ist mit Hilfe der mikromagnetischen Theorie von Brown [1, 2], insbesondere aber mit der Einmündungstheorie, wie sie von Seeger und Kronmüller [3, 4] für magnetostriktive Anisotropie und beliebige Versetzungsverteilungen entwickelt wurde, möglich.

Führt man im Kristall ein sogenanntes Brownsches Koordinatensystem ein, dessen y-Achse mit der Richtung des äußeren Feldes $\boldsymbol{H}$ zusammenfällt, dann lautet die zum Feld $\boldsymbol{H}$ parallele Magnetisierungskomponente I_y näherungsweise

$$I_y = I_s\,\overline{\gamma}_2 = I_s\,(1 - \overline{\gamma_1^2} - \overline{\gamma_3^2})^{1/2} \approx I_s\left(1 - \frac{1}{2}\,(\overline{\gamma_1^2} + \overline{\gamma_3^2})\right), \tag{3.1}$$

wobei I_s die Sättigungsmagnetisierung bedeutet und γ_1 und γ_3 die Richtungskosinus der spontanen Magnetisierung bezüglich der x- und z-Achse des Brownschen Koordinatensystems sind. Quergestrichene Größen sind Mittelwerte über das Probenvolumen. Für die magnetische Suszeptibilität gilt dann

$$\chi = \frac{\mathrm{d}I_y}{\mathrm{d}H} = -\frac{I_s}{2}\frac{\mathrm{d}}{\mathrm{d}H}\,(\overline{\gamma_1^2} + \overline{\gamma_3^2})\,. \tag{3.2}$$

Die Berechnung der quadratischen Mittelwerte $\overline{\gamma_i^2}$ erfolgt mit Hilfe des Parsevaalschen Theorems [15]. Dieses lautet in der für uns zweckmäßigen Form

$$\overline{\gamma_i^2} = \frac{1}{V} \int \gamma_i^2(\boldsymbol{r})\, d\tau_{\boldsymbol{r}} = \frac{1}{V} \int |\tilde{\gamma}_i(\boldsymbol{k})|^2\, d\tau_{\boldsymbol{k}}, \tag{3.3}$$

wo V das Probenvolumen und $\tilde{\gamma}_i(\boldsymbol{k})$ die Fouriertransformierte von $\gamma_i(\boldsymbol{r})$ darstellt. Die Größen $\tilde{\gamma}_i$ findet man durch Fouriertransformation der mikromagnetischen Grundgleichungen. Seeger und Kronmüller [3] erhielten folgende Ergebnisse:

$$\tilde{\gamma}_i = \frac{-\tilde{g}_i(k)}{k^2 + \varkappa_H^2} + \frac{h^2 k_i (\tilde{g}_1 k_1 + \tilde{g}_3 k_3)}{(k^2 + \varkappa_H^2)^2 \{k^4 + (k^2 + h^2) k^2 - h^2 k_2^2\}}, \tag{3.4}$$

$$|\tilde{\gamma}_1|^2 + |\tilde{\gamma}_3|^2 = \frac{|\tilde{g}_1|^2 + |\tilde{g}_3|^2}{(k^2 + \varkappa_H^2)^2} - \left\{2 h^2 - \frac{h^4 (k_1^2 + k_3^2)}{k^4 + k^2 (\varkappa_H^2 + h^2) - h^2 k_2^2} \times \right.$$
$$\left. \times \frac{|\tilde{g}_1|^2 k_1^2 + |\tilde{g}_3|^2 k_3^2 + 2 k_1 k_3 (\mathrm{Re}\, \tilde{g}_1 \,\mathrm{Re}\, \tilde{g}_3 + \mathrm{Im}\, \tilde{g}_1 \,\mathrm{Im}\, \tilde{g}_3)}{(k^2 + \varkappa_H^2)^2 \{k^4 + k^2 (\varkappa_H^2 + h^2) - h^2 k_2^2\}}\right\}. \tag{3.5}$$

Dabei gilt (A Austauschkonstante):

$$h^2 = \frac{2\pi I_s^2}{A}, \quad \varkappa_H^2 = \frac{H I_s}{2A}, \tag{3.6a}$$

$$\left.\begin{aligned} g_i(\boldsymbol{r}) &= \frac{1}{2A} \frac{\partial}{\partial \gamma_i} (\Phi_K + \Phi_M) \Big|_{\gamma_{1,3} = 0}^{\gamma_2 = 1}, \\ \tilde{g}_i(\boldsymbol{k}) &= (2\pi)^{-3/2} \int g_i(\boldsymbol{r}) \exp(-i\, \boldsymbol{k} \cdot \boldsymbol{r})\, d\tau_{\boldsymbol{r}}. \end{aligned}\right\} \tag{3.6b}$$

Φ_K bedeutet die Kristallanisotropieenergiedichte und Φ_M die Energiedichte der magnetoelastischen Kopplungsenergie. Im folgenden werden wir uns auf Φ_M beschränken, da Φ_K im Einmündungsgebiet vernachlässigt werden kann. In kubischen Kristallen lautet Φ_M, wenn die Richtungskosinus γ_i und der Spannungstensor σ' auf die kubischen Achsen bezogen werden,

$$\Phi_M = -\frac{3}{2} \lambda_{100} (\sigma'_{xx} \gamma_1'^2 + \sigma_{yy} \gamma_2'^2 + \sigma_{zz} \gamma_3'^2) -$$
$$- 3 \lambda_{111} (\sigma'_{xy} \gamma'_1 \gamma'_2 + \sigma_{xz} \gamma'_1 \gamma'_3 + \sigma'_{yz} \gamma'_2 \gamma'_3), \tag{3.7}$$

wobei λ_{100} und λ_{111} die Magnetostriktionskonstanten und σ'_{ij} die Komponenten des Spannungstensors bedeuten. Um zu einer möglichst einfachen Formulierung des Problems zu kommen, führen wir noch ein drittes Koordinatensystem (x'', y'', z'') ein, dessen x''-Achse parallel zum Burgersvektor verläuft. Die y''-Achse steht senkrecht auf der Gleitebene und die z''-Achse soll parallel zur Versetzungslinie verlaufen. In diesem Koordinatensystem besitzt der Spannungstensor σ''_{ij} geradliniger Versetzungen seine einfachste Form. Die Transformation der Richtungskosinus zwischen den verschiedenen Koordinatensystemen erfolgt durch die Matrizen (β_{ij}), (β'_{ij}) und (β''_{ij}) in folgender Weise

$$\gamma'_i = \beta_{ik} \gamma_k, \quad \gamma''_i = \beta'_{ik} \gamma'_k, \quad \gamma''_i = \beta''_{ik} \gamma_k, \tag{3.8}$$

wobei entsprechend der Einsteinschen Summenkonvention über gleichlautende Indizes von 1 bis 3 summiert wird. Berechnet man mit Hilfe von Gleichung (3.8) die Transformation von Φ_M, dann erhält man in quadratischer Näherung in den

Größen γ_i und σ''_{ij}

$$\Phi_M = -\frac{1}{2}\beta'_{il}\beta'_{km} C_{lm}\{(\beta_{l2}\beta_{m1}+\beta_{l1}\beta_{m2})\gamma_1\sigma''_{ik}+(\beta_{l2}\beta_{m3}+\beta_{l3}\beta_{m2})\gamma_3\sigma''_{ik}\} \tag{3.9}$$

mit

$$C_{ii} = 3\lambda_{100} \quad \text{und} \quad C_{ij} = 3\lambda_{111} \quad \text{für} \quad i \neq j\,.$$

Berechnet man mittels dieser Darstellung von Φ_M den Ausdruck (3.6) und setzt (3.6) in (3.5) ein, so ergibt sich für die Suszeptibilität, wenn man die Integration von Gleichung (3.5) in der früher beschriebenen Weise ausführt [3, 4]

$$\chi = \frac{B_1^2 N}{16\, I_s H^3}\pi\left(2\ln\frac{1}{2}\varkappa_H R - 0{,}34566\right)\int\limits_0^{2\pi}\left\{\left(B^1_{ik}\tilde{\zeta}''_{ik}(\varphi)\right)^2 + \left(B^3_{ik}\tilde{\zeta}''_{ik}(\varphi)\right)^2\right\}d\varphi\, +$$

$$+\frac{2\pi I_s}{H}\int\limits_0^{2\pi}[\tilde{\zeta}''_{ik}(\varphi)\,A^1_{ik}\cos\varphi + \tilde{\zeta}''_{ik}(\varphi)\,A^3_{ik}\sin\varphi]^2\, J(\Theta,\psi,\varphi)\,d\varphi\,. \tag{3.10}$$

Hierbei gilt

$$\left.\begin{aligned}
B^\nu_{ik} &= \beta'_{il}\beta'_{km}C_{lm}(\beta_{l\nu}\beta_{m2}+\beta_{l2}\beta_{m\nu})\,,\\
A^\nu_{ik} &= B^1_{ik}\beta''_{1\nu} + B^3_{ik}\beta''_{3\nu}\,,\\
B_1 &= \frac{b\,G}{2\pi(1-\nu)}\,,\\
\zeta''_{11} &= 3\sin\varphi - \sin 3\varphi\,, \qquad \zeta''_{33} = 4\nu\sin\varphi\,,\\
\zeta''_{22} &= \sin\varphi + \sin 3\varphi\,, \qquad \zeta''_{12} = \cos\varphi - \cos 3\varphi\,.
\end{aligned}\right\} \tag{3.11}$$

G ist der Schubmodul, ν die Poissonsche Konstante, N die Versetzungsdichte pro cm². $J(\Theta, \psi, \varphi)$ wird im Anhang berechnet. Θ und ψ sind in Fig. 1 definiert.

Bevor wir auf die Integration der obigen Ausdrücke eingehen, soll eine für die numerische Rechnung geeignete Symmetriebetrachtung angestellt werden.

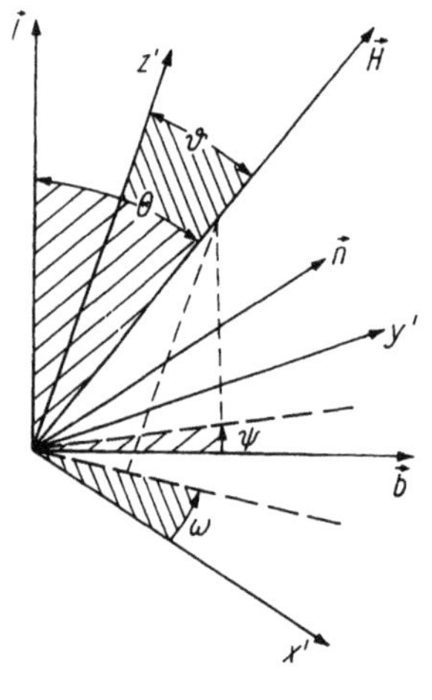

Fig. 1. Zur Veranschaulichung der Polarwinkel Θ, ψ im Koordinatensystem der Versetzung und der Polarwinkel ϑ, ω im kubischen Koordinatensystem

3.2 Symmetriebetrachtung

Die Einmündungssuszeptibilität ist im allgemeinen eine Funktion der Orientierung der Versetzungslinie und der Richtung des angelegten äußeren Feldes. Da χ invariant gegenüber einer Inversion von $\boldsymbol{H}$ oder $\boldsymbol{l}$ ist und bei Stufenversetzungen $\boldsymbol{l}$ und $\boldsymbol{n}$, d. h., Linienvektor und Normalenvektor der Gleitebene durch den Burgersvektor bestimmt sind, läßt sich für χ folgende Abhängigkeit angeben:

$$\chi = \chi(\boldsymbol{b}, \boldsymbol{H}) = \chi(-\boldsymbol{b}, -\boldsymbol{H}) = \chi(\boldsymbol{b}, -\boldsymbol{H}) . \tag{3.12}$$

Bezeichnet man mit T_i eine kubische Symmetrieoperation, dann gilt

$$\chi (T_i \boldsymbol{b}, T_i \boldsymbol{H}) = \chi(\boldsymbol{b}, \boldsymbol{H}) \tag{3.13a}$$

oder

$$\chi (T_i \boldsymbol{b}, \boldsymbol{H}) = \chi (\boldsymbol{b}, T_i^{-1} \boldsymbol{H}) . \tag{3.13b}$$

Aus Gleichung (3.13) geht hervor, daß, wenn $\chi(\boldsymbol{b}, \boldsymbol{H})$ für eine Versetzung mit dem Burgersvektor $\boldsymbol{b}$ als Funktion von $\boldsymbol{H}$ bekannt ist, $\chi(\boldsymbol{b}, \boldsymbol{H})$ auch für alle anderen Versetzungen angegeben werden kann, die durch eine kubische Transformation auseinander hervorgehen. Diese Eigenschaft läßt sich jedoch nur für den vom Streufeld unabhängigen ersten Teil von (3.10) verwerten. Wenn wir das im ersten Term von (3.10) auftretende Integral mit $f(\boldsymbol{b}, \boldsymbol{H})$ abkürzen, so ergibt sich hierfür

$$\begin{aligned} f(\boldsymbol{b}, \boldsymbol{H}) = \int_0^{2\pi} \Big\{[B_{ik}^1 \tilde{\zeta}_{ik}''(\varphi)]^2 + [B_{ik}^3 \tilde{\zeta}_{ik}''(\varphi)]^2\Big\} \, d\varphi = \\ = 4 \, [(B_{12}^1)^2 + (B_{12}^3)^2] + 2 \, [B_{11}^1 \, B_{22}^1 + B_{11}^3 \, B_{22}^3] + \\ + 12 \, \nu \, [B_{11}^1 \, B_{33}^3 + B_{11}^3 \, B_{33}^3] + 4 \, \nu \, [B_{22}^1 \, B_{33}^1 + B_{22}^3 \, B_{33}^3] + \\ + 5 \, [(B_{11}^1)^2 + (B_{11}^3)^2] + [(B_{22}^1)^2 + (B_{22}^3)^2] + 8 \, \nu^2 \, [(B_{33}^1)^2 + (B_{33}^3)^2] . \end{aligned} \tag{3.14}$$

Durch Einsetzen von (3.11) in (3.14) ergibt sich, daß die Richtungskosinus des äußeren Feldes in (3.14) in Potenzen 0,2 und 4 auftreten können.

Zur Darstellung von Gleichung (3.14) genügen daher die Kugelfunktionen Y_4^m, Y_2^m und Y_0^0. Allgemein gilt somit

$$f(\boldsymbol{b}, \boldsymbol{H}) = \sum_{l=0}^{2} \sum_{m=-2l}^{2l} b_{lm} \, Y_{2l}^m(\vartheta, \omega) \tag{3.15}$$

(ϑ und ω siehe Fig. 1). Insgesamt treten 15 Koeffizienten b_{lm} auf, die man in der Weise bestimmt, daß man $f(\boldsymbol{b}, \boldsymbol{H})$ für 15 verschiedene Richtungen von $\boldsymbol{H}$ berechnet und das so sich ergebende Gleichungssystem nach den b_{lm} auflöst. Die Koeffizienten b_{lm} lassen sich folgendermaßen darstellen:

$$b_{lm} = 9 \, b_{lm}^{(1)} \lambda_{111}^2 + 9 \, b_{lm}^{(2)} \lambda_{100} \lambda_{111} + 9 \, b_{lm}^{(3)} \lambda_{100}^2 . \tag{3.16}$$

In Tabelle 1 sind die Konstanlen $b_{lm}^{(i)}$ für Ni angegeben. Die zugehörigen Kugelfunktionen $Y_{2l}^{\pm m}$ sind durch $\frac{1}{\sqrt{M}} \begin{Bmatrix} \cos m\omega \\ \sin m\omega \end{Bmatrix} P_{2l}^{\pm m} (\cos \vartheta)$ definiert, wo P_{2l}^m die zugeordneten Legendrefunktionen sind und $M = \int_0^{\pi} \int_0^{2\pi} [\cos m\omega \, P_{2l}^m (\cos \vartheta)]^2 \times \sin \vartheta \, d\vartheta \, d\omega$ gilt. Der numerischen Rechnung wurden die Materialkonstanten

Tabelle 1

Entwicklungskoeffizienten des streufeldfreien Beitrags der Einmündungssuszeptibilität für eine Versetzung mit dem Burgersvektor [$\bar{1}$10]

	$b^{(1)}_{l\,m}$	$b^{(2)}_{l\,m}$	$b^{(3)}_{l\,m}$
Y_0	2,835906297	12,47183756	0
Y^0_2	0,301974138	0,03621431222	0
X^{+1}_2	0	−0,6492967809	1,86392864
Y^{-1}_2	0	2,309309091	1,86392864
Y^{+2}_2	0	0	0
Y^{-2}_2	0	−2,634734899	2,18941664
Y^0_4	−1,170380331	2,807509517	0
Y^{+1}_4	0	−0,353432401	−0,45095177
Y^{-1}_4	0	1,257028644	−0,45095177
Y^{+2}_4	0	0	0
Y^{-2}_4	0	4,056439744	−2,52812058
Y^{+3}_4	0	−0,935094244	−2,13043972
Y^{-3}_4	0	−3,325785202	−2,13043972
Y^{+4}_4	−0,75322418	3,252376436	0
Y^{-4}_4	0	0	−1,50644838

von Nickel bei Zimmertemperatur zugrunde gelegt. Diese lauten: $I_s = 480$ G, $K_1 = -4{,}8 \times 10^4$ erg/cm³, $A = 0{,}86 \times 10^{-6}$ erg/cm, $\lambda_{100} = -53{,}5 \times 10^{-6}$, $\lambda_{111} = -23 \times 10^{-6}$, $B_1 = 0{,}769$ dyn/cm. Sämtliche Berechnungen wurden für eine Versetzungsdichte von 10^8 cm^{-2} durchgeführt, wobei Interferenzeffekte zwischen den Spannungsfeldern der Versetzungen nicht berücksichtigt wurden (siehe hierzu [4, 12]).

Wenden wir uns nun dem streufeldabhängigen Anteil der Suszeptibilität zu. In diesem Falle läßt sich für den in der Klammer des zweiten Termes von Gleichung (3.10) stehenden Ausdruck $g(\boldsymbol{b}, \boldsymbol{H}, \varphi)$ eine ähnliche Entwicklung wie für die Funktion $f(\boldsymbol{b}, \boldsymbol{H})$ angeben, und zwar gilt

$$g(\boldsymbol{b}, \boldsymbol{H}, \varphi) = (\tilde{\zeta}''_{ik} A^{(1)}_{ik} \cos\varphi + \tilde{\zeta}''_{ik} A^{(3)}_{ik} \sin\varphi)^2 =$$

$$= a^0 + \sum_{p=1}^{4} (a^{p,c} \cos 2\,p\,\varphi + a^{p,s} \sin 2\,p\,\varphi) \qquad (3.17)$$

mit

$$a^{p,c,s} = \sum_{l=0} \sum_{m=2l} a^{p,c,s}_{lm}\, Y^m_{2l}(\vartheta, \omega)\,. \qquad (3.18)$$

Die Koeffizienten $a^{p,c,s}_{lm}$ werden nach demselben Verfahren berechnet wie die Koeffizienten $b_{l\,m}$. Nach Ausführung der φ-Integration ergibt sich

$$\chi = \chi_0 + \chi_s\,,$$

wobei der vom Streufeld unabhängige Anteil

$$\chi_0 = \frac{I_s B_1^2 N}{64\, A^2 H \varkappa_H^4} \left\{\pi \left(2 \ln \frac{1}{2} R_0 \varkappa_H - 0{,}345668\right) \sum_{l=0}^{2} \sum_{m=-2l}^{2l} b_{l\,m}\, Y^m_{2l}(\vartheta, \omega)\right\} \quad (3.19)$$

und der streufeldabhängige Anteil

$$\chi_s = \frac{\pi I_s^2 B_1^2 N}{32 A^2 H^2 \varkappa_H^4} \sum_{l=0}^{3} \sum_{m=-2l}^{2l} \left\{ Y_{2l}^m(\vartheta, \omega) \left[a_{lm}^0 J_0(\Theta, \psi) + \right.\right.$$

$$\left.\left. + \sum_{p=1}^{4} (a_{lm}^{p,c} \cos 2 p \psi + a_{lm}^{p,s} \sin 2 p \psi) J_p(\Theta, \psi) \right] \right\} \tag{3.20}$$

beträgt. Die Winkelpaare (Θ, ψ) und (ϑ, ω) sind in Fig. 1 erklärt und bedeuten die Polarwinkel des äußeren Feldes H bezüglich des Versetzungsdreibeins und der kubischen Achsen des Kristalls. Explizite Ausdrücke für die Integrale $J_p(\Theta, \psi)$ werden im Anhang angegeben. Zur numerischen Berechnung der Suszeptibilität mit Hilfe der obigen Formel sei bemerkt, daß auf einer elektronischen Großrechenanlage vergleichbar der TR4 von Telefunken die Zeit für eine χ-Berechnung ungefähr 5 s erfordert. Die Ursache hierfür liegt darin, daß $J_p(\Theta, \psi)$ teilweise durch numerische Integration bestimmt werden muß.

4. Diskussion

Die Berechnung der Anisotropie des Einmündungsgesetzes erfolgte für geradlinige Stufenversetzungen in Nickel, wobei die Versetzungen als Einzelversetzungen behandelt wurden. Um einen repräsentativen Eindruck von der Anisotropie der Suszeptibilität bezüglich der Magnetfeldorientierung zu vermitteln, wurde die Rechnung für die drei Ebenen durchgeführt, die von Burgersvektor, Linienvektor und Normalenvektor einer Stufenversetzung aufgespannt werden. Der numerischen Rechnung wurden dabei die oben angegebenen Materialkonstanten zugrunde gelegt. Die Versetzungsdichte wurde $N = 10^8\ \text{cm}^{-2}$ gesetzt. Ein Vergleich zwischen der berechneten Suszeptibilität χ und den entsprechend dem in Abschnitt 2 behandelten Meßverfahren bestimmten Koeffizienten A_m sollte Rückschlüsse auf die Lage der Versetzungslinien zulassen.

In den Fig. 2, 3 und 4 ist die Suszeptibilität χ für ein äußeres Feld von 1000 Oe in einem Polardiagramm für die erwähnten von den Versetzungen ausgezeichneten Ebenen $(\boldsymbol{b}, \boldsymbol{l})$, $(\boldsymbol{l}, \boldsymbol{n})$, $(\boldsymbol{b}, \boldsymbol{n})$ darsgestellt. Angegeben ist die Abhängigkeit der Suszeptibilität von den Polarwinkeln Θ und ψ. Außerdem ist in den Fig. 2, 3 jeweils die Suszeptibilität χ_0, wie man sie bei Vernachlässigung des Streufeldtermes erhalten würde, mit eingezeichnet. Die Differenz zwischen den beiden eingezeichneten Kurven stellt den in Gleichung (3.20) definierten Anteil des Streufeldes zur Gesamtsuszeptibilität dar. Wie aus den Abbildungen hervorgeht, bedingt die Mitnahme des Streufeldes eine starke Verringerung der Suszeptibilität. Die physikalische Begründung hierfür ist, daß die Streufeldenergie genauso wie die Austauschenergie bestrebt ist, die von der magnetoelastischen Kopplungsenergie hervorgerufenen starken Inhomogenitäten auszugleichen, um allzu große magnetische Raumladungen zu verhindern. Zum Verlauf der Anisotropiekurven sei bemerkt, daß bei magnetostriktiver Isotropie, d. h. $\lambda_{100} = \lambda_{111}$, wie in einer vorhergehenden Arbeit gezeigt wurde [16], die Suszeptibilität bezüglich den Achsen $\boldsymbol{b}$, $\boldsymbol{l}$ und $\boldsymbol{n}$ spiegelsymmetrisch sein würde. Sämtliche Kurven sind punktsymmetrisch zum Ursprung, da die Gesamtenthalpie invariant gegen eine Inversion der Magnetfeldrichtung ist. In Fig. 5 ist die Suszeptibilität für alle drei Ebenen gemeinsam aufgetragen. Das Verhältnis des Betrags der Suszeptibilität in ihren Extremalpunkten ist, wie aus Fig. 5 hervorgeht, von der Größenordnung 1 : 10 und müßte daher meßtechnisch gut

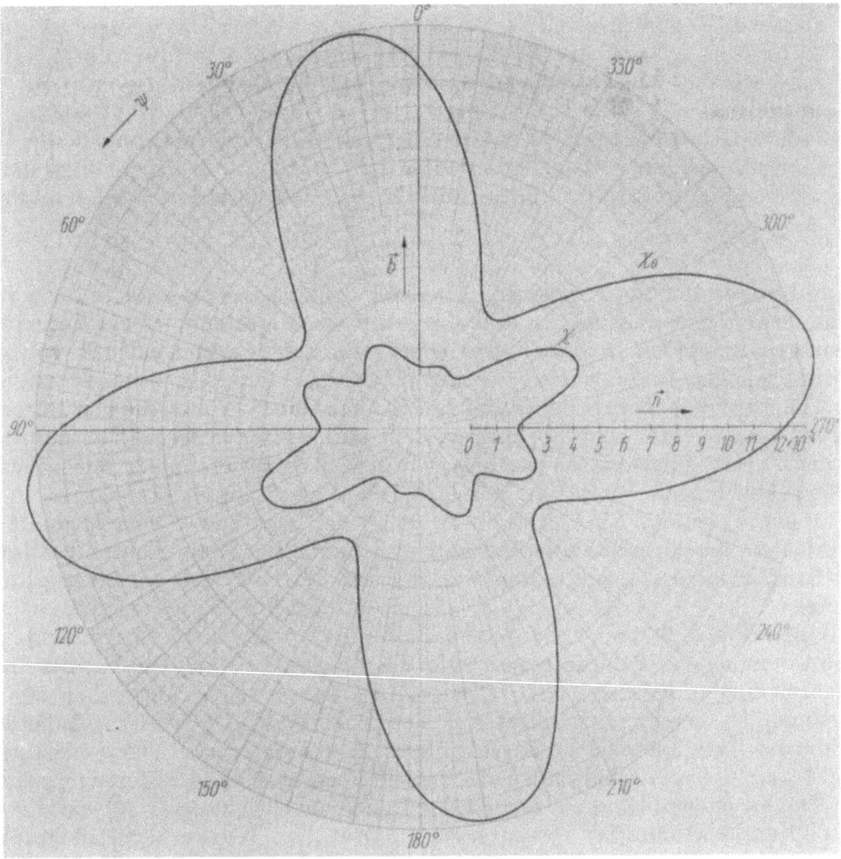

Fig. 2. Abhängigkeit der Suszeptibilitäten χ_0 und χ von der Magnetfeldrichtung in der von Burgersvektor $\boldsymbol{b}$ und Normalenvektor $\boldsymbol{n}$ aufgespannten Ebene ((112)-Ebene). χ ist in Einheiten [G/Oe] für $H = 1000$ Oe und eine Versetzungsdichte von $N = 10^8\ \text{cm}^{-2}$ angegeben

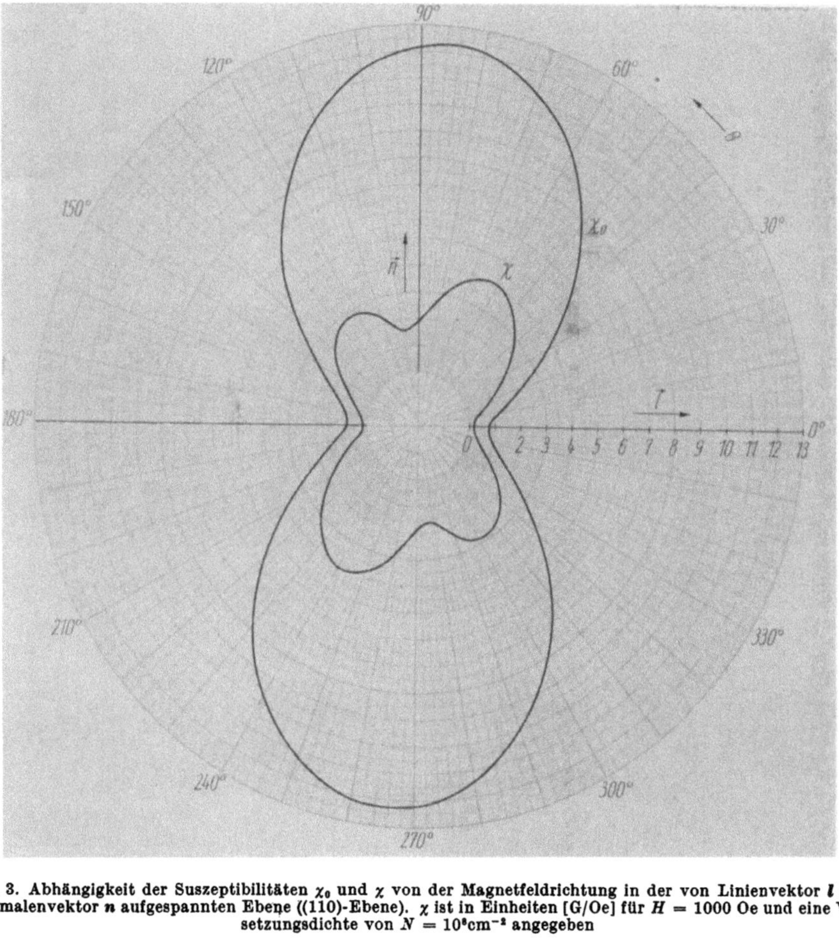

Fig. 3. Abhängigkeit der Suszeptibilitäten χ_0 und χ von der Magnetfeldrichtung in der von Linienvektor $\boldsymbol{l}$ und Normalenvektor $\boldsymbol{n}$ aufgespannten Ebene ((110)-Ebene). χ ist in Einheiten [G/Oe] für $H = 1000$ Oe und eine Versetzungsdichte von $N = 10^8 \mathrm{cm}^{-2}$ angegeben

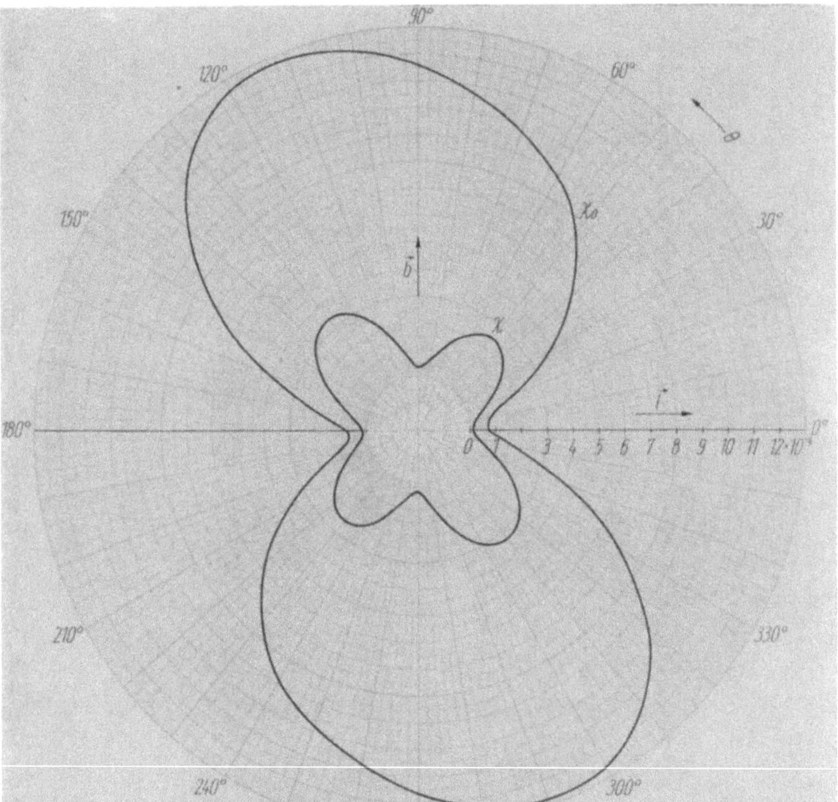

Fig. 4. Abhängigkeit der Suszeptibilitäten χ_0 und χ von der Magnetfeldrichtung in der von Linienvektor $\boldsymbol{l}$ und Burgersvektor $\boldsymbol{b}$ aufgespannten Ebene ((111)-Ebene). χ ist in Einheiten [G/Oe] für $H = 1000$ Oe und eine Versetzungsdichte von $N = 10^8$ cm^{-2} angegeben

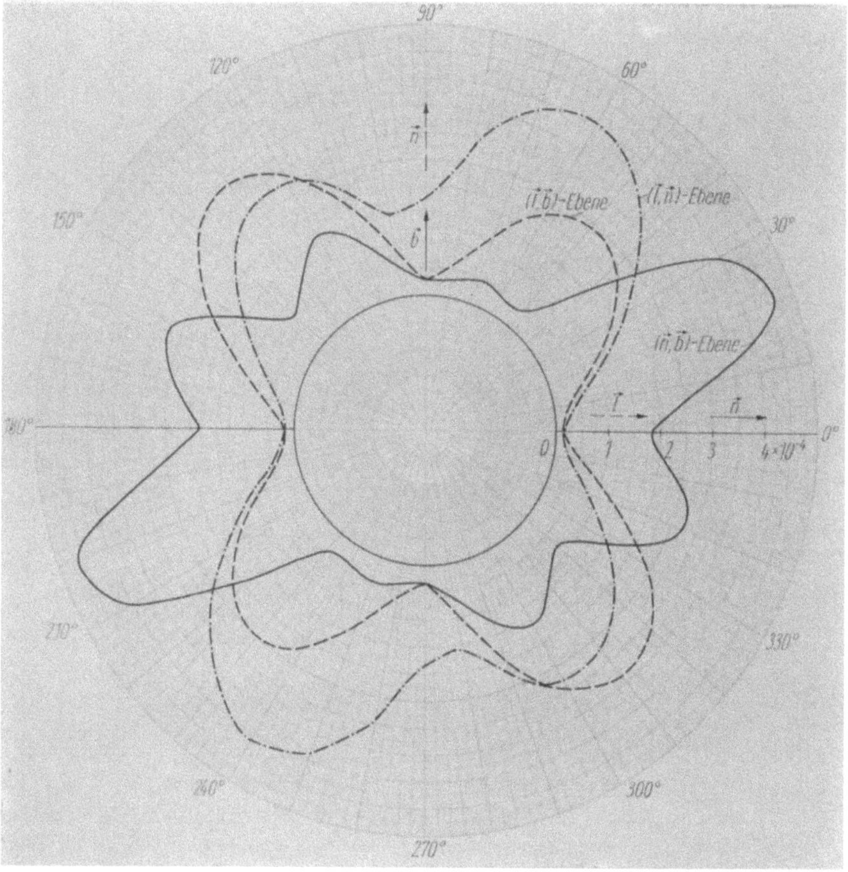

Fig. 5. Die Abhängigkeit der Gesamtsuszeptibilität χ von der Magnetfeldrichtung in den drei ausgezeichneten Ebenen der Stufenversetzung. ($H = 1000$ Oe; $N = 10^8$ cm^{-2})

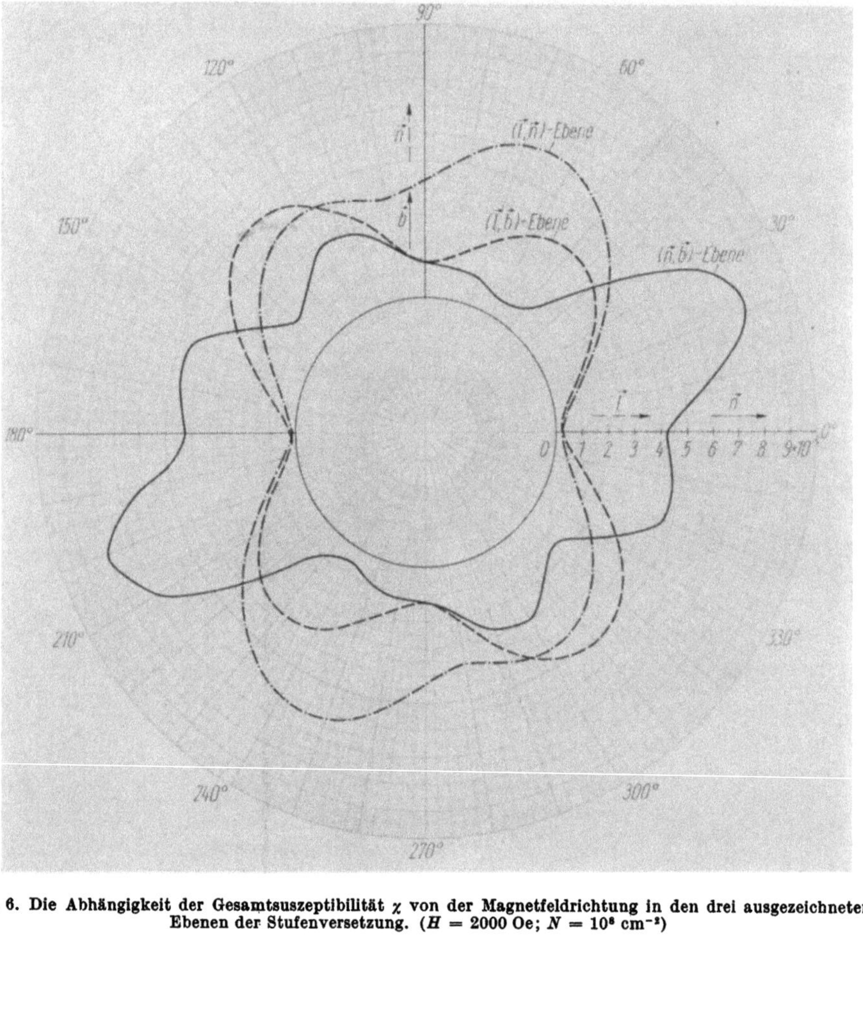

Fig. 6. Die Abhängigkeit der Gesamtsuszeptibilität χ von der Magnetfeldrichtung in den drei ausgezeichneten Ebenen der Stufenversetzung. (H = 2000 Oe; $N = 10^8$ cm^{-2})

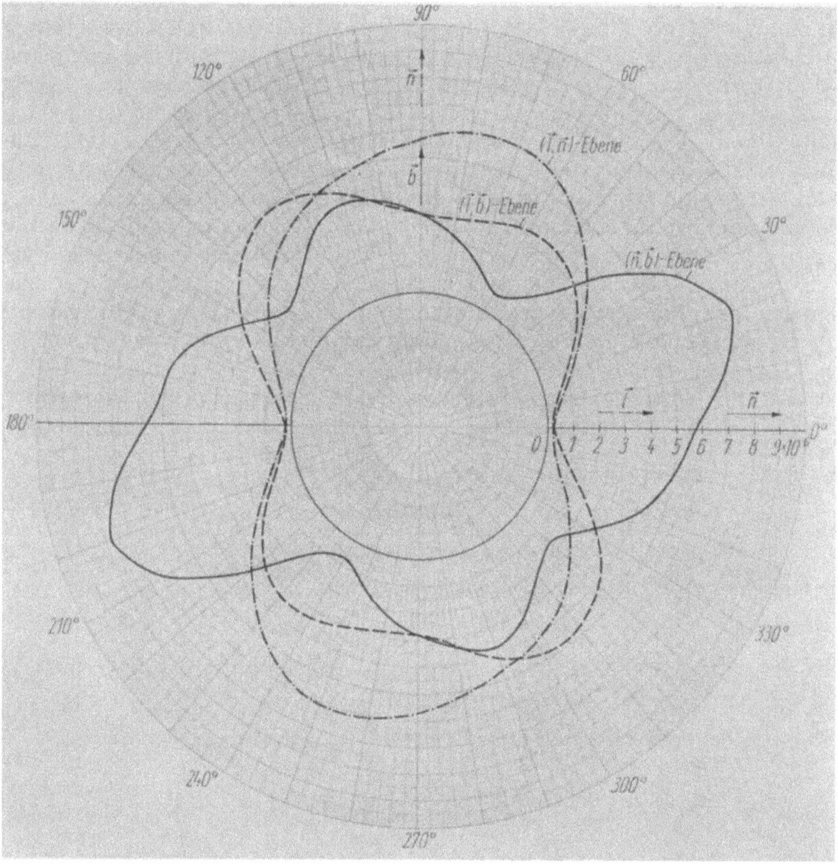

Fig. 7. Die Abhängigkeit der Gesamtsuszeptibilität χ von der Magnetfeldrichtung in den drei ausgezeichneten Ebenen der Stufenversetzung. ($H = 5000$ Oe; $N = 10^8$ cm^{-2})

zu erfassen sein. In Fig. 6 und 7 ist die Suszeptibilität für ein Magnetfeld von 2 und 5 kOe aufgetragen. Diese Abbildungen lassen erkennen, daß die Anisotropie der Suszeptibilität stark feldabhängig ist. Mit zunehmendem Magnetfeld wird die Anisotropie der Suszeptibilität mehr und mehr durch den streufeldfreien Anteil χ_0 bestimmt. Die Ursache hierfür ist darin zu suchen, daß mit wachsender Feldstärke χ_0 wie $1/H^3$ und χ_s nach (3.20) wie $1/H^4$ abnimmt. Die Anisotropie von χ_0 allein ist feldstärkeunabhängig.

Die in Fig. 3 und 4 dargestellte Abhängigkeit der Suszeptibilität von dem Polarwinkel Θ (vgl. Fig. 1) läßt sich im wesentlichen durch Glieder $\cos 2\,\Theta$, $\cos 4\,\Theta$, $\sin 2\,\Theta$ und $\sin 4\,\Theta$ darstellen. Dasselbe gilt auch für die ψ-Abhängigkeit von χ in der $(\boldsymbol{n}, \boldsymbol{b})$-Ebene, die in Fig. 2 wiedergegeben ist. Aus der Phasenlage der Maxima von χ können aus experimentellen Ergebnissen, die gemäß dem in Abschnitt 2 angegebenen Verfahren auszuwerten sind, Rückschlüsse auf die Richtung der Versetzungslinien gezogen werden. Die Amplitude der Anisotropiekurven erlaubt Aussagen über die Versetzungsdichte und die Versetzungsanordnung zu machen. Letzterer Problemkreis wurde in früheren Arbeiten schon für eine spezielle Kristallorientierung experimentell und theoretisch untersucht [4, 7]. Die jetzt vorliegende Arbeit ermöglicht es, neuere Messungen der Anisotropie des Einmündungsgesetzes [7] zur Untersuchung von Versetzungsstrukturen in plastisch verformten Metallen heranzuziehen.

Die Autoren möchten Herrn Professor Dr. A. Seeger für sein Interesse und seine zahlreichen Anregungen zu der vorliegenden Arbeit herzlich danken.

Anhang

Berechnung der Integrale $J_p(\Theta, \psi)$

Zur Berechnung der Integrale $J_p(\Theta, \psi)$ geht man von

$$J(\Theta, \psi, \varphi) = -\frac{I_s}{A} H \varkappa_H^6 \int_0^\infty \frac{dk_\varrho\,(J_0(k_\varrho R_0) - 1)^2}{k_\varrho\,\alpha^2} [(k_\varrho^2 + \varkappa_H^2)^{-3} - (k_\varrho^2 + \varkappa_H^2 + \alpha^2)^{-3}]$$

aus, wobei R_0 der halbe mittlere Abstand zwischen den Versetzungen ist und

$$\alpha^2 = h^2 \left(1 - \frac{1}{2} \sin^2 \Theta - \frac{1}{2} \sin^2 \Theta \cos 2\,(\varphi - \psi)\right)$$

bedeutet. Integriert man über k_ϱ, dann ergibt sich nach Seeger und Kronmüller [3, 4]

$$J(\Theta, \psi, \varphi) = -\frac{I_s}{A} H \varkappa_H^6 \left\{\alpha^{-2} \varkappa_H^{-6} \left(\ln \frac{1}{2} R_0 \varkappa_H + z\right) - \right.$$

$$\left. - \alpha^{-2} (\varkappa_H^2 + \alpha^2)^{-3} \left(\ln \frac{1}{2} R_0 \sqrt{\varkappa_H^2 + \alpha^2} + z\right)\right\}, \quad z = -0{,}172784$$

und damit erhält man für die Integrale $J_p(\Theta, \psi)$

$$J_p(\Theta, \psi) = \int_0^{2\pi} J(\Theta, \psi, \varphi) \cos 2\,p\,(\varphi - \psi)\, d\varphi =$$

$$= -\left(\ln \frac{1}{2} R_0 \varkappa_H + z + \frac{1}{4}\right) (2\,\Phi_0^p + 3\,\Phi_1^p + \Phi_2^p) + \frac{1}{2} (\Phi_0^p + \Phi_1^p + \Phi_2^p) + L^p\,;$$

hierbei gilt

$$\Phi_0^p = \frac{2\,\pi\,(a^2 - b^2 - a)^p}{b^p \sqrt{a^2 - b^2}}\,,$$

$$\Phi_0^p = \Phi_2^p\,(a^2 - b^2)^{-1/2}\left(p + \frac{a}{\sqrt{a^2 - b^2}}\right),$$

$$\Phi_2^p = 0{,}5\,\Phi_0^p\,(a^2 - b^2)^{-3/2}\left(3\,p\,a + \frac{3\,a^2}{\sqrt{a^2 - b^2}} + (p^2 - 1)\sqrt{a^2 - b^2}\right),$$

$$L^p = 2\int\limits_0^{\pi} \frac{\ln\,(a + b\cos\varphi)\cos p\,\varphi}{(a - 1 + b\cos\varphi)\,(a + b\cos\varphi)}\,\mathrm{d}\varphi\,,$$

$$a = 1 + \frac{4\,\pi\,I_s}{H}\left(1 - \frac{1}{2}\sin^2\Theta\right),$$

$$b = -\frac{2\,\pi\,I_s}{H}\sin^2\Theta\,.$$

Das Integral L^p kann nur durch numerische Integration bestimmt werden.

Literatur

[1] W. F. Brown, Jr., Phys. Rev. **58**, 736 (1940).
[2] W. F. Brown, Jr., Phys. Rev. **60**, 132 (1941).
[3] A. Seeger und H. Kronmüller, J. Phys. Chem. Solids **12**, 298 (1960).
[4] H. Kronmüller und A. Seeger, J. Phys. Chem. Solids **18**, 93 (1961).
[5] T. H. Huzimura, Sci. Rep. Res. Inst. Tôhoku Univ. **8**, 71 (1956).
[6] H. Dietrich und E. Kneller, Z. Metallk. **47**, 672 (1956).
[7] H. Kronmüller, Z. Phys. **154**, 574 (1959).
[8] A. Bhattacharyya, S. R. Pollack und R. Maddin, J. appl. Phys. **37**, 4443 (1966).
[9] R. H. Geiss und J. Silcox, J. appl. Phys. **39**, 982 (1968).
[10] H. Gessinger, E. Köster und H. Kronmüller, J. appl. Phys. **39**, 986 (1968).
[11] H. Bauer, Diplomarbeit, Universität Stuttgart (1967).
[12] H. Kronmüller, Canad. J. Phys. **45**, 631 (1967).
[13] S. Albenga, Atti Accad. Sci. Torino. Cl. Sci. fis. mat. natur. **54**, 864 (1917/18).
[14] H. Kronmüller, Z. angew. Phys. **23**, 130 (1967).
[15] H. Schneidler, Integralgleichungen, Akademische Verlagsgesellschaft, Leipzig 1950 (S. 17).
[16] H. Kronmüller, J. appl. Phys. **38**, 1314 (1967).

(Received June 5, 1968)

phys. stat. sol. 28, 773 (1968)

Subject classification: 10.1; 14.1; 21.6

Institute of Metallurgy, Czechoslovak Academy of Sciences, Brno

Thermal Quasi-Equilibrium of Vacancies, Divacancies, and Impurities in Platinum

By

J. Polák

The change in the electrical resistivity of platinum quenched in air and in water is measured. The transient phenomenon at the very beginning of the isothermal annealing at a temperature of about 250 °C is observed. The possible interactions of point defects in quenched platinum are analysed. Various interactions are quantitatively compared with the measured data and the probable interpretation of the results is discussed.

Es wird die Änderung des spezifischen elektrischen Widerstandes von in Wasser und in Luft abgeschrecktem Platin gemessen. Im Anfangsbereich der isothermen Ausheilkurve wird um 250 °C ein Übergang beobachtet. Die möglichen Wechselwirkungen zwischen Punktdefekten im abgeschreckten Platin werden analysiert und mit experimentellen Angaben verglichen. Wahrscheinliche Interpretationen der Ergebnisse werden diskutiert.

1. Introduction

The concept of the thermal quasi-equilibrium between single and divacancies was used for the explanation of the migration of the vacant lattice sites after quenching of various metals [1 to 3]. The achievement of this equilibrium was studied only scarcely. De Jong and Koehler [1, 4] measured the electric resistivity of quenched gold during the annealing near room temperature and observed the transient process at the very beginning of the annealing. They have discussed this phenomenon in terms of vacancies and divacancies coming into thermal quasi-equilibrium. The initial decrease of the resistivity was also observed by Benci and Gasparini [5] in the Au_3–Cu alloy.

In this paper we shall present data showing the similar phenomenon in platinum quenched in air and also in water. We shall also derive the relations for obtaining thermal quasi-equilibrium of vacancies, divacancies, and impurities both during isothermal annealing and during quenching from high temperatures. The experimental data will be discussed in terms of these interactions.

2. Experimental Procedure and Results

The material and experimental procedure used in this work were similar to those in the earlier one [3]. The specimens made from 99.999% pure platinum wires (supplied by Johnson & Matthey, London) were heated in air by passing a direct current through them and quenched from about 1550 °C either by switching off the current or by rapid immersing into distilled water held at room temperature. The cooling curves for both kinds of quenching are shown in Fig. 1.

In Fig. 2 the isothermal annealing curves after quenching in air and in water are shown. At the beginning of every annealing curve which is S-shaped altogether [3], there is a transient phenomenon. This transient phenomenon is

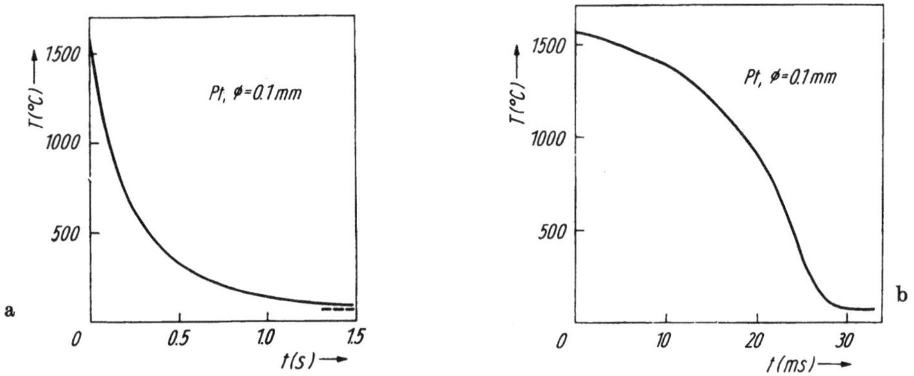

Fig. 1. Cooling curves for 99.999% Pt wires of 0.1 mm diameter during quenching from T_q = 1560 °C a) in air, b) in distilled water

characterized by a more rapid decrease of the resistivity with time (dotted line), which is superimposed by the usual S-shape of the curve (full line). This initial decrease of the resistivity is nearly exponential and amounts only to about 0.3% of the total quenched-in resistivity. However, it is well reproducible and appears both after quenching in water and in air.

The rate of annealing after the termination of the transient phenomenon is greater in specimens quenched in water than in specimens quenched in air as can be seen by comparing Figs. 2a and b. Most probably this can be explained by an increased density of sinks in the specimens quenched in water as a result of small plastic deformation during quenching [6], or by the appearance of secondary defects which act as sinks for vacancies. However, in both cases the transient phenomenon is of the same magnitude and has the same characteristic time.

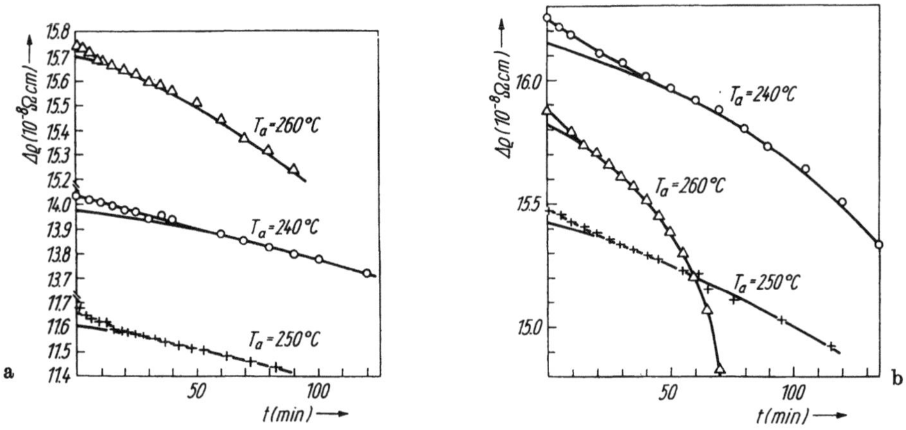

Fig. 2. Change of electric resistivity during isothermal annealing after quenching from temperatures around 1560 °C a) in air, b) in distilled water

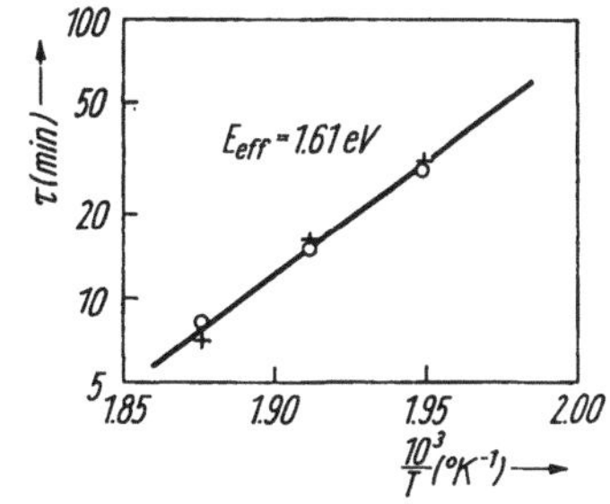

Fig. 3. Dependence of characteristic time τ of transient process on reciprocal annealing temperature (+ air quenching, ○ water quenching)

We have plotted the characteristic time τ of the transient process (i.e. the time necessary to reach $1/e$ of the initial value of the excess resistivity) versus the reciprocal absolute temperature in Fig. 3. From the slope of the straight line which was fitted to the experimental points the effective activation energy of this process was determined to $E_{eff} = (1.61 \pm 0.10)$ eV.

3. Interaction of Vacancies, Divacancies, and Impurities

It was found experimentally that the transient phenomenon does not depend on the kind of quenching and takes place always in the same temperature interval independently of the number of sinks in the specimen. These facts support the hypothesis that the observed phenomenon is caused by the interaction of point defects, one of which being able to migrate in the mentioned temperature interval. Point defects present in the specimen after quenching from high temperatures are single and double vacancies and possibly their complexes with impurities.

The mutual interaction of these defects during quenching was studied by de Jong and Koehler [1], Koehler et al. [4], and Doyama [7, 8]. They have shown the existence of critical temperatures. After reaching the critical temperature during quenching the interaction of defects becomes so slow that the thermal quasi-equilibrium cannot be attained.

In order to investigate the achievement of the quasi-equilibrium between various kinds of defects during isothermal annealing it would be necessary to solve the system of coupled differential equations describing the process (see, e.g. [4, 8]).

Such system of equations can be solved only numerically. A better insight into the whole process can be obtained by treating the individual reactions separately and finding the characteristic quantities which depend on the intrinsic parameters of the defects taking part in the reaction.

3.1 Single vacancies and divacancies

The achievement of the quasi-equilibrium between single and divacancies was treated by de Jong and Koehler [1] and by Koehler et al. [4]. We shall present only the result for a more general case using elementary functions.

We shall assume that the total concentration of vacant lattice sites, $c_{V,t}$ does not change during annealing. The initial concentration of divacancies will be denoted by $c_{2V,i}$. By solving the appropriate differential equation it is possible to obtain the time dependence of the concentration of divacancies in the form

$$c_{2V} = c_{2V,e} \frac{1 - \frac{H_{2V}}{K_{2V}} \exp\left(-\frac{t}{\tau_{2V}}\right)}{1 - H_{2V} \exp\left(-\frac{t}{\tau_{2V}}\right)}, \tag{1}$$

where

$$c_{2V,e} = \frac{1 + 2\,p\,c_{V,t} - \sqrt{1 + 4\,p\,c_{V,t}}}{4\,p} \tag{2}$$

is the concentration of divacancies in quasi-equilibrium with single vacancies at the temperature T. The quantity p characterizes the binding of a divacancy:

$$p = 12 \exp\left(\frac{S^B_{2V}}{k}\right) \exp\left(\frac{E^B_{2V}}{kT}\right). \tag{3}$$

S^B_{2V} and E^B_{2V} are the binding entropy and energy of a divacancy, respectively.

The characteristic time τ_{2V} is given by the relation

$$\tau^{-1}_{2V} = 14\,\nu_1 \exp\left(\frac{S^M_V - S^B_{2V}}{k}\right) \exp\left(-\frac{E^M_V + E^B_{2V}}{kT}\right) \sqrt{1 + 4\,p\,c_{V,t}}\,. \tag{4}$$

ν_1 is the frequency of vibration of the atoms near the single vacancy, S^M_V and E^M_V are the migration entropy and energy of a single vacancy, respectively.

The quantities K_{2V} and H_{2V} are

$$K_{2V} = \frac{4\,c^2_{2V,e}}{c^2_{V,t}}, \tag{5}$$

$$H_{2V} = \frac{4\,c_{2V,e}\,(c_{2V,e} - c_{2V,i})}{c^2_{V,t} - 4\,c_{2V,e}\,c_{2V,i}}. \tag{6}$$

If the initial concentration of divacancies is zero then it follows $H_{2V} = K_{2V}$ and the relation (1) becomes simpler

$$c_{2V} = c_{2V,e} \frac{1 - \exp\left(-\frac{t}{\tau_{2V}}\right)}{1 - K_{2V} \exp\left(-\frac{t}{\tau_{2V}}\right)}. \tag{1a}$$

$K_{2V} \ll 1$ holds for a small concentration of vacant sites or for weak binding and the second term in the denominator of expression (1a) can be neglected. In this case the equilibrium value $c_{2V,e}$ can be exponentially approached with the characteristic time τ_{2V}. τ_{2V} is the time during which $(1 - 1/e)$ of the equilibrium concentration of divacancies is formed.

Assuming that $\varrho_d > 0$ is the difference between the resistivity of two single vacancies and a divacancy then the drop in resistivity corresponds to the increment of the concentration of divacancies. The achievement of quasi-equilibrium can thus be followed experimentally by measuring the resistivity. Taking the zero level of the resistivity at $t \to \infty$ then at the beginning of the annealing the resistivity is increased by $\Delta\varrho_0 = \varrho_d c_{2V,e}$, which diminishes according to the relation

$$\frac{\Delta\varrho}{\Delta\varrho_0} = 1 - \frac{c_{2V}}{c_{2V,e}} = \frac{(1 - K_{2V}) \exp\left(-\frac{t}{\tau_{2V}}\right)}{1 - K_{2V} \exp\left(-\frac{t}{\tau_{2V}}\right)}. \tag{7}$$

This decrease is exponential for small K_{2V}.

3.2 Single vacancies and impurities

Considering the interaction of vacancies and impurities only, we shall assume that the total concentration of vacant lattice sites $c_{V,t}$ remains constant. The initial concentration of vacancy–impurity pairs will be denoted by $c_{VI,i}$, the total concentration of impurities by $c_{I,t}$. By solving the appropriate differential equation the approach to quasi-equilibrium can be written in a form analogous to divacancies:

$$c_{VI} = c_{VI,e} \frac{1 - \frac{H_{VI}}{K_{VI}} \exp\left(-\frac{t}{\tau_{VI}}\right)}{1 - H_{VI} \exp\left(-\frac{t}{\tau_{VI}}\right)}, \tag{8}$$

where

$$c_{VI,e} = \frac{1}{2}\left[c_{V,t} + c_{I,t} + \frac{1}{q} - \sqrt{\left(c_{V,t} + c_{I,t} + \frac{1}{q}\right)^2 - 4\, c_{V,t}\, c_{I,t}}\right] \tag{9}$$

is the concentration of vacancy–impurity pairs in quasi-equilibrium. The quantity q characterizes the binding of the pair:

$$q = 12 \exp\left(\frac{S^B_{VI}}{k}\right) \exp\left(\frac{E^B_{VI}}{kT}\right). \tag{10}$$

S^B_{VI} and E^B_{VI} are the binding entropy and energy of a vacancy–impurity pair, respectively.

The characteristic time τ_{VI} is given by the relation

$$\tau^{-1}_{VI} = 84\, \nu_1 \exp\left(\frac{S^M_V}{k}\right) \exp\left(-\frac{E^M_V}{kT}\right) \sqrt{\left(c_{V,t} + c_{I,t} + \frac{1}{q}\right)^2 - 4\, c_{V,t}\, c_{I,t}}\,. \tag{11}$$

The quantities K_{VI} and H_{VI} are

$$K_{VI} = \frac{c^2_{VI,e}}{c_{V,t}\, c_{I,t}}, \tag{12}$$

$$H_{VI} = \frac{c_{VI,e}\,(c_{VI,e} - c_{VI,i})}{c_{V,t}\, c_{I,t} - c_{VI,e}\, c_{VI,i}}. \tag{13}$$

In the case $c_{VI,i} = 0$ the relation (8) may be simplified.

If the resistivity of a vacancy–impurity pair is smaller than the sum of the resistivitiy of the isolated defects, the achievement of quasi-equilibrium will be accompanied by a resistivity drop according to a relation analogous to (7) with the characteristic time τ_{VI}.

3.3 Divacancies and impurities

The interaction of divacancies and impurities can be followed similarly as in Section 3.2. For the sake of simplicity one complex type will be considered that will be denoted by „D". The total concentration of divacancies $c_{2\,\mathrm{V,t}}$ is assumed to be constant and the initial concentration of divacancy–impurity complexes is put equal to zero.

The concentration of divacancy–impurity complexes c_D reaches the quasi-equilibrium value according to the relation

$$c_\mathrm{D} = c_\mathrm{D,e} \frac{1 - \exp\left(-\frac{t}{\tau_\mathrm{D}}\right)}{1 - K_\mathrm{D} \exp\left(-\frac{t}{\tau_\mathrm{D}}\right)}, \tag{14}$$

where

$$c_\mathrm{D,e} = \frac{1}{2}\left[c_\mathrm{I,t} + c_{2\,\mathrm{V,t}} + \frac{1}{r} - \sqrt{\left(c_\mathrm{I,t} + c_{2\,\mathrm{V,t}} + \frac{1}{r}\right)^2 - 4\, c_\mathrm{I,t}\, c_{2\,\mathrm{V,t}}}\right] \tag{15}$$

is the concentration of divacancy–impurity complexes in quasi-equilibrium. r is given by the relation

$$r = 18 \exp\left(\frac{S_\mathrm{D}^\mathrm{B} - S_{2\,\mathrm{V}}^\mathrm{B}}{k}\right) \exp\left(\frac{E_\mathrm{D}^\mathrm{B} - E_{2\,\mathrm{V}}^\mathrm{B}}{kT}\right). \tag{16}$$

S_D^B and E_D^B are the binding entropy and energy of a divacancy–impurity complex, respectively.

The characteristic time is given by the relation

$$\tau_\mathrm{D}^{-1} = 180\, \nu_2 \exp\left(\frac{S_{2\,\mathrm{V}}^\mathrm{M}}{k}\right) \exp\left(-\frac{E_{2\,\mathrm{V}}^\mathrm{M}}{kT}\right) \sqrt{\left(c_\mathrm{I,t} + c_{2\,\mathrm{V,t}} + \frac{1}{r}\right)^2 - 4\, c_\mathrm{I,t}\, c_{2\,\mathrm{V,t}}}\,, \tag{17}$$

where ν_2 is the vibration frequency of the atom near the divacancy, $S_{2\,\mathrm{V}}^\mathrm{M}$ and $E_{2\,\mathrm{V}}^\mathrm{M}$ are the migration entropy and energy of a divacancy, respectively.

The quantity K_D is given by

$$K_\mathrm{D} = \frac{c_\mathrm{D,e}^2}{c_\mathrm{I,t}\, c_{2\,\mathrm{V,t}}}. \tag{18}$$

If the resistivity of the divacancy–impurity complex is smaller than the sum of the resistivities of each of the divacancies and of the impurities separately, the achievement of this equilibrium can also be accompanied by a drop in the resistivity. The characteristic time for this process will be τ_D.

3.4 Critical temperatures and critical cooling rates

The thermal quasi-equilibrium between the individual defects is arising already during quenching. At the so-called critical cooling rate v^* there exists the critical temperature T^* below which the movement of defects is so slow that the number of complexes after quenching is equal to the number of complexes in quasi-equilibrium at this critical temperature T^*. The expressions for critical cooling rates at various interactions of defects was derived by Doyama [8]. It is important to point out the simple relation between the critical cooling rate v^* of the process and the characteristic time τ. It holds

$$v_{\rm j}^* = \left(\frac{{\rm d}T}{{\rm d}t}\right)_{T=T_{\rm j}} = -\frac{kT_{\rm j}^{*2}}{E_{\rm j}^{\rm B}}\frac{1}{\tau_{\rm j}}, \tag{19}$$

where the index "j" means either 2 V, VI, or D.

Knowing the critical temperature T^* for the measured quenching curve, the initial concentration of divacancies immediately after quenching is

$$c_{2\,{\rm V,i}} = \frac{1 + 2\,c_{\rm V,t}\,p(T_{2\rm V}^*) - \sqrt{1 + 4\,c_{\rm V,t}\,p(T_{2\rm V}^*)}}{4\,p(T_{2\rm V}^*)}. \tag{20}$$

Analogous relations hold for other kinds of interactions.

4. Discussion

In the previous section we have analysed the possible reactions which can cause the observed drop of electric resistivity at the beginning of the annealing curve. In the following we want to compare quantitatively the measured quantities with those of various models and discuss the interpretation of the observed phenomenon.

The quantities that we shall compare are the characteristic time τ, critical quenching rate v^*, and effective activation energy $E_{\rm eff}$. It would be also possible to compare the magnitude of the excess resistivity and its temperature dependence; however, the precision of its determination is not good enough.

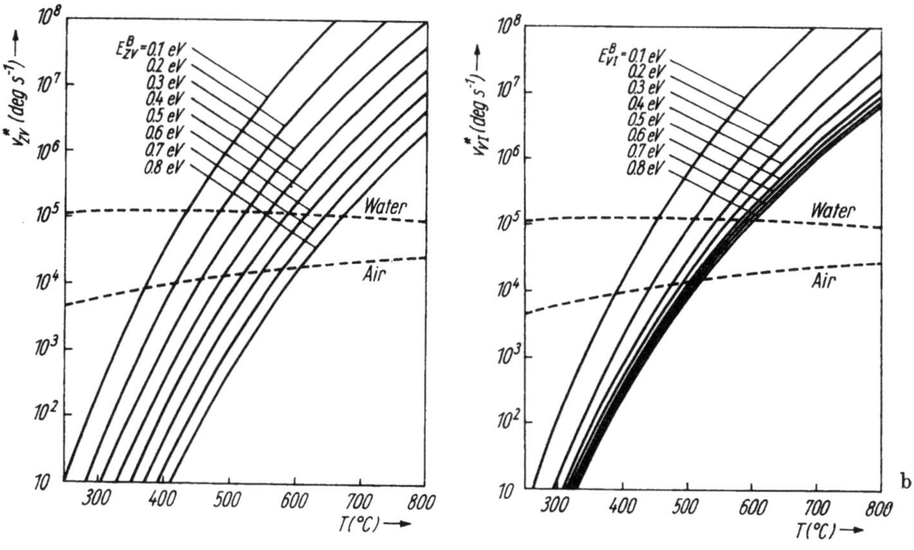

Fig. 4. Dependence of critical cooling rate v^* on temperature for various values of binding energy $E^{\rm B}$ a) for divacancies, b) for vacancy–impurity pairs. (Read index 2 V instead of ZV)

Let us first consider the characteristic time τ of the transient process. The maximum value of the quantity τ_D is $\tau_{D,\max} = \left[180 \nu_2 c_{I,t} \exp\left(\frac{S^M_{2V}}{k}\right) \exp\left(-\frac{E^M_{2V}}{kT}\right)\right]^{-1}$. Assuminng the values $E^M_{2V} = 1.11$, $\nu_2 \exp\left(\frac{S^M_{2V}}{k}\right) = 5 \times 10^{12}\ s^{-1}$ [3], and $c_{I,t} = 1 \times 10^{-5}$ evaluated from the residual resistivity of our specimens, we obtain for the temperature $T_a = 260$ °C $\tau_D \approx 3$ s. This value is considerably lower than the value $\tau = 450$ s observed experimentally. Therefore, this reaction can be excluded from further considerations.

Now we have to decide between the formation of divacancies and the formation of vacancy–impurity pairs. We shall take the following numerical values of the parameters for calculating the characteristic quantities [3]: $c_{V,t} = 8 \times 10^{-4}$, $c_{I,t} = 1 \times 10^{-5}$, $\nu_1 \exp\left(\frac{S^M_V}{k}\right) = 5 \times 10^{12}\ s^{-1}$, $S^B_{2V} = S^B_{VI} = k$, $E^M_V = 1.45$ eV. The most important parameter is the migration energy of single vacancies E^M_V. The influence of the magnitude of E^M_V on the calculated quantities will be discussed later.

In Figs. 4a and b there is a plot of the critical cooling rate versus temperature for various values of the binding energy. The points of intersection of these curves with the experimentally determined dependence of the cooling rate on the temperature yield the critical temperatures. These are shown in Table 1 in

Table 1

Critical temperatures T^*_{2V} and T^*_{VI} for quenching in air and in water

E^B (eV)	Air		Water	
	T^*_{2V} (°C)	T^*_{VI} (°C)	T^*_{2V} (°C)	T^*_{VI} (°C)
0.1	371	390	448	458
0.2	418	442	484	513
0.3	457	475	524	547
0.4	489	492	558	567
0.5	520	503	586	581
0.6	550	511	618	589
0.7	580	519	646	596
0.8	610	522	673	602

dependence on the binding energy E^B. The critical temperatures for both reactions and for both kinds of quenching are considerably higher than the annealing temperatures. From this point of view both reactions are equivalent. A change of the value E^M_V by 0.05 eV would result in a change of about 20 °C, in both cases in the same direction. Therefore, the above mentioned conclusion is not influenced.

Fig. 5 gives τ for both processes as function of the binding energy E^B at $T = 260$ °C supposing either $E^M_V = 1.45$ eV or $E^M_V = 1.55$ eV. The dotted line designates the experimental data. It is apparent that the agreement with the experiment can be obtained only if the value of the migration energy E^M_V is as high as 1.5 eV.

The temperature dependence of the characteristic time can be describe in terms of the effective activation energy.

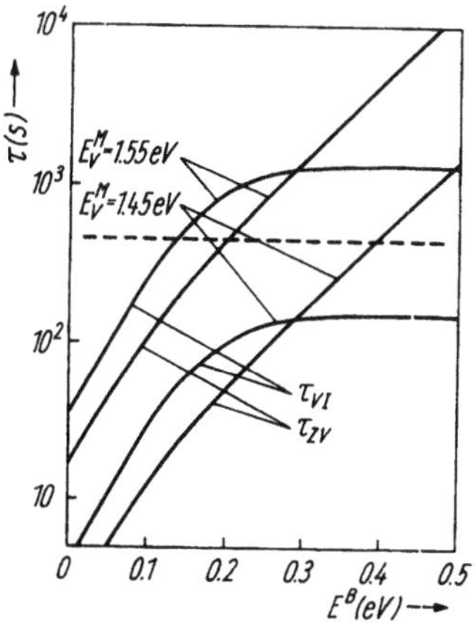

Fig. 5. Characteristic times $\tau_{2\mathrm{V}}$ and τ_{VI} versus binding energies $E^{\mathrm{B}}_{2\mathrm{V}}$ and $E^{\mathrm{B}}_{\mathrm{VI}}$, respectively, at $T_{\mathrm{a}} = 260$ °C. (Read index 2V instead of ZV)

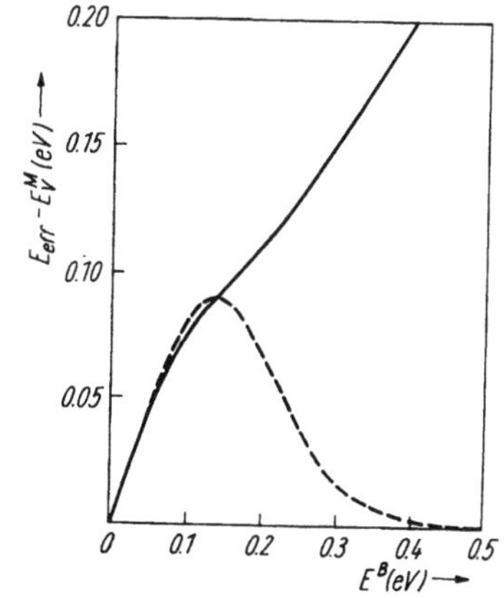

Fig. 6. Difference between effective activation energy of the process, $E^{\mathrm{M}}_{\mathrm{eff}}$, and migration energy of single vacancies, E^{M}, versus binding energy for divacancies (full line) and for vacancy–impurity pairs (dotted line) at $T_{\mathrm{a}} = 260$ °C

The effective activation energy of this process is

$$E_{\mathrm{eff},\,2\mathrm{V}} = E^{\mathrm{M}}_{\mathrm{V}} + E^{\mathrm{B}}_{2\mathrm{V}} \frac{1 + 2\,p\,c_{\mathrm{V,t}}}{1 + 4\,p\,c_{\mathrm{V,t}}} \tag{21}$$

for the case of divacancies and

$$E_{\mathrm{eff},\,\mathrm{VI}} = E^{\mathrm{M}}_{\mathrm{V}} + E^{\mathrm{B}}_{\mathrm{VI}} \frac{c_{\mathrm{V,t}} + c_{\mathrm{I,t}} + \dfrac{1}{q}}{q\left[\left(c_{\mathrm{V,t}} + c_{\mathrm{I,t}} + \dfrac{1}{q}\right)^2 - 4\,c_{\mathrm{V,t}}\,c_{\mathrm{I,t}}\right]} \tag{22}$$

for the formation of the vacancy–impurities pairs. The quantities $E_{\mathrm{eff},\,2\mathrm{V}} - E^{\mathrm{M}}_{\mathrm{V}}$ and $E_{\mathrm{eff},\,\mathrm{VI}} - E^{\mathrm{M}}_{\mathrm{V}}$ are plotted in Fig. 6 in dependence on the binding energy of a divacancy or of a vacancy–impurity pair.

The agreement between the calculated effective energies and the measured activation energy can be obtained only for higher values of the migration energy $E^{\mathrm{M}}_{\mathrm{V}} \approx 1.5$ eV mainly in the case of vacancy–impurity pairs.

From the previous discussion it is obvious that, if we do not know exactly the value of $E^{\mathrm{M}}_{\mathrm{V}}$, it is not possible from the present experimental data to decide unambiguously which interaction causes the observed phenomenon. For this purpose it would be necessary to perform similar experiments on a number of specimens of different purities.

Considering the model of divacancies, the best fit with regard to both the time constant and the effective activation energy can be obtained by choosing the following energy values: $E^{\mathrm{M}}_{\mathrm{V}} = 1.5$ eV and $E^{\mathrm{B}}_{2\mathrm{V}} = 0.3$ eV.

Taking into account the possible errors in experimental data and the uncertainty of the frequency and entropy factors, the present estimate of these quantities is not in contradiction with the latest values found from annealing experiments [3, 6, 10 to 12]. We conclude therefore that the explanation by the quasi-equilibrium between single and divacancies seems to be most probable.

5. Conclusion

Analysing the possible reactions between point defects in a metal quenched from high temperatures it was shown that the observed transient phenomenon in platinum can be explained in terms of these interactions. By comparing the calculated and measured values, the interaction between divacancies and impurities was excluded, and it was deduced that this phenomenon is most probably due to the achievement of thermal quasi-equilibrium between single and divacancies.

References

[1] M. DE JONG and J. S. KOEHLER, Phys. Rev. **129**, 40 (1963); **129**, 49 (1963).
[2] M. DOYAMA and J. S. KOEHLER, Phys. Rev. **134**, 1522 (1964).
[3] J. POLÁK, phys. stat. sol. **21**, 581 (1967).
[4] J. S. KOEHLER, M. DE JONG, and F. SEITZ, J. Phys. Soc. Japan **18**, Suppl. III, 1 (1963).
[5] S. BENCI and G. GASPARINI, Phys. Letters (Netherlands) **24A**, 37 (1967).
[6] J. J. JACKSON, in: Lattice Defects in Quenched Metals, Ed. R. M. J. COTTERILL, M. DOYAMA, J. J. JACKSON, and M. MESHII, Academic Press, New York 1965 (pp. 467 and 479).
[7] M. DOYAMA, in: Lattice Defects . . ., l.c. (pp. 167 and 185).
[8] M. DOYAMA, Phys. Rev. **148**, 681 (1966).
[9] A. C. DAMASK and G. J. DIENES, Point Defects in Metals, Gordon & Breach, New York/London 1963.
[10] J. POLÁK, Phys. Letters (Netherlands) **24A**, 649 (1967).
[11] V. S. KOPAN and M. SKOROKHOD, in: Isledovanie nesoverstvenstv kristallicheskovo stroeniya, Naukovaya dumka, Kiev 1965 (p. 99).
[12] D. SCHUMACHER, A. SEEGER, and O. HÄRLIN, phys. stat. sol. **25**, 359 (1968).

(Received May 22, 1968)

phys. stat. sol. 28, 783 (1968)

Subject classification: 14.3; 13.1; 22

Institute of Physics, Academy of Sciences of the Azerbaidzhan SSR, Baku and Azerbaidzhan S. M. Kirov State University, Baku

The Transverse Magnetic Resistance in Semiconductors with Non-Standard Energy Band

By

B. M. Askerov and F. M. Gashimzade

The transverse electrical conductivity σ_{xx} in a strong magnetic field is calculated for semiconductors with spherical but nonparabolic energy band. The calculation is carried out for the case of electron scattering by acoustical phonons. In the quantum limit non-degenerate and degenerate semiconductors are considered. It is shown that in the formula for the transverse electrical conductivity in the non-degenerate case the effective mass at the bottom of the parabola shifted by the magnetic field and in the degenerate case the effective mass at the Fermi surface are included. The temperature dependence is the same as for a standard energy band. Therefore an assumed non-parabolicity does not change the dependence of σ_{xx} on the electrical field. The effect of the non-parabolicity is reduced to the above renormalization of the effective mass.

Вычислена поперечная электропроводность σ_{xx} в квантующем магнитном поле для полупроводников со сферической, но непараболической зоной. Расчет проведен для случая рассеяния электронов на акустических фононах. В квантовом пределе рассмотрены невырожденные и вырожденные полупроводники. Показано, что в невырожденном случае в формулу для поперечной электропроводности входит эффективная масса на дне смещенной магнитным полем параболы, а в вырожденном случае эффективная масса на поверхности ферми. Температурная зависимость такая же как и для стандартной зоны, поэтому учет нестандартности не изменяет зависимость σ_{xx} от электрического поля в сильном электрическом поле. Влияние нестандартности сводится к вышеуказанной перенормировке эффективной массы.

1. Introduction

The theory of the transverse magnetic resistance developed by Adams and Holstein [2] was generalized to the case of an isotropic but non-standard energy band of the type n-InSb [1]. It was shown that the non-parabolicity of the energy band appeared in the expression for electric conductivity in a transverse magnetic field only through the energy conservation law in the scattering probability. The formula obtained was used for the analysis of the Shubnikov-de Haas oscillations and magnetophonon resonance in n-InSb.

However, in [1] only the conditions for the Shubnikov-de Haas oscillation and magnetophonon resonance were derived. For that, one had to know the general form of the electric conductivity tensor.

In the present work the transverse electric conductivity is calculated for semiconductors with spherical but non-parabolic energy bands being in a strong magnetic field. The calculation is carried out for the case of electron scattering on acoustical phonons. In the quantum limit non-degenerate and degenerate semiconductors are considered. The temperature and field dependence of the electric conductivity are found for semiconductors obeying Kane's dispersion law.

2. Transverse Magnetoresistance for Scattering on Acoustical Phonons

On the basis of the formula for the transverse electric conductivity [1] it is

$$\sigma_{xx} = e^2 \sum_{\alpha\alpha'\sigma} \left(-\frac{\partial f_\alpha}{\partial \varepsilon_\alpha}\right) \frac{(x_\alpha - x_{\alpha'})^2}{2} W_{\alpha\alpha'}\,, \tag{1}$$

where $f_\alpha = f(\varepsilon_\alpha)$ is the Fermi distribution function for equilibrium, $\alpha = n$, k_y, k_z is a set of quantum numbers, $x_\alpha = -R^2 k_y$ is the centre of the magnetic Landau oscillator, $R = \left(\frac{\hbar c}{e H}\right)^{1/2}$ is the magnetic length, and $\sigma = \pm\frac{1}{2}$ the spin quantum number.

In the case of scattering on acoustical phonons the transition probability is of the form

$$W_{\alpha\alpha'} = W(q)\,|I_{nn'}|^2 \{N_q\, \delta_{k_y',k_y+q_y}\, \delta_{k_z',k_z+q_z}\, \delta(\varepsilon_{\alpha'} - \varepsilon_\alpha - \hbar\omega_q) + (N_q + 1)\, \delta_{k_y',k_y-q_y}\, \delta_{k_z',k_z-q_z}\, \delta(\varepsilon_{\alpha'} - \varepsilon_\alpha + \hbar\omega_q)\}\,. \tag{2}$$

Here it is

$$W(q) = \frac{\pi E_0^2 q^2}{\varrho\, \omega_q}, \tag{3}$$

where ϱ is the density, E_0 the deformation potential constant, $\omega_q = s\,q$ the frequency of an acoustical phonon, $N_q = \left[\exp\left(\frac{\hbar\omega_q}{k_0 T}\right) - 1\right]^{-1}$ the Planck function,

$$I_{nn'} = \int \varphi_{n'}(x - x_{\alpha'})\, e^{i q_x x}\, \varphi_n(x - x_\alpha)\, dx\,. \tag{4}$$

Inserting (2) into (1) and summarizing over k_y', k_z', k_y we find

$$\sigma_{xx} = \frac{e^2 R^2}{2(2\pi)^5} \sum_{nn'\sigma} \int dk_z \int d\boldsymbol{q}\, W(q)\,(2N_q + 1)\,|I_{nn'}|^2\, q_y^2 \left(-\frac{\partial f_\alpha}{\partial \varepsilon_\alpha}\right) \times \delta(\varepsilon_{n'\sigma,k_z+q_z} - \varepsilon_{n\sigma,k_z} - \hbar\omega_q)\,. \tag{5}$$

We integrate over dq_z by means of a δ-function. We have to calculate an integral of the type

$$\int F(q)\,\delta(\varepsilon_{n'\sigma,k_z+q_z} - \varepsilon_{n\sigma,k_z} - \hbar\omega_q)\,dq_z = \frac{F(q_x, q_y, q_{z1})}{\frac{\partial\varepsilon}{\partial k_z^2}\,2(k_z + q_{z1})} + \frac{F(q_x, q_y, q_{z2})}{\frac{\partial\varepsilon}{\partial k_z^2}\,2(k_z + q_{z2})}, \tag{6}$$

where

$$q_{z1,2} = -k_z \pm \sqrt{k_z^2 + \frac{2m_0^*}{\hbar^2}[B(\varepsilon + \hbar s q_\perp) - B(\varepsilon) - (n' - n)\hbar\Omega]}\,. \tag{7}$$

In the derivation of equation (7) we used the dispersion law in the form

$$k_z(\varepsilon) = \frac{\sqrt{2m_0^*}}{\hbar} \sqrt{B(\varepsilon) - \left(n + \frac{1}{2}\right)\hbar\Omega - \sigma g^* \mu_B H} \tag{8}$$

and took into account that $q_z \ll q_\perp$, i.e. $\hbar\omega_q \approx \hbar s q_\perp$. The following designations are introduced in equation (8): m_0^* is the effective mass at the bottom of the conduction band, $\mu_B = \frac{e\hbar}{2 m_0 c}$ is Bohr's magneton, $\Omega = \frac{eH}{m_0^* c}$, g^* is the factor of spin splitting, $B(\varepsilon)$ is an arbitrary function of energy.

As $|I_{nn'}|^2$ depends only on $q_\perp^2$ [3], we introduce a polar coordinate system in the plane (q_x, q_y) and integrate over the polar angle. Then we shall take into account that

$$\frac{\partial \varepsilon}{\partial k_z^2} = \frac{\hbar^2}{2 m_0^*} \left(\frac{\partial B}{\partial \varepsilon}\right)^{-1} \tag{9}$$

and pass from the integral over dk_z to the integral over $d\varepsilon$. Then it is

$$\sigma_{xx} = \frac{e^2 R^2 m_0^*}{4 (2\pi)^4 \hbar^2} \sum_{nn'\sigma} \int \left(\frac{\partial B}{\partial \varepsilon}\right)^2 \left(-\frac{\partial f}{\partial \varepsilon}\right) d\varepsilon \int dq_\perp \, q_\perp^3 \, |I_{nn'}(q_\perp^2)|^2 \times$$

$$\times \frac{[W(q)(2N_q+1)]_{q_z=q_{z1}} + [W(q)(2N_q+1)]_{q_z=q_{z2}}}{\sqrt{B(\varepsilon + \hbar s q_\perp) - \left(n' + \frac{1}{2}\right)\hbar\Omega - \sigma g^* \mu_B H}\ \sqrt{B(\varepsilon) - \left(n+\frac{1}{2}\right)\hbar\Omega - \sigma g^* \mu_B H}}. \tag{10}$$

The consideration of inelasticity is necessary only for eliminating the integral divergence over the energy in the quantum limit ($n = n' = 0$). As in quantum limit the phonons having impulses $q_\perp \sim R^{-1}$ interact with electrons $\left(\text{the transition probability is proportional to } \exp\left(-\frac{q_\perp^2 R^2}{2}\right)\right)$ we can substitute into equation (10) $B(\varepsilon + \hbar s q_\perp) \approx B(\varepsilon + \hbar s R^{-1})$. Then this expression will be of the form

$$\sigma_{xx} = \frac{e^2 R^2 m_0^*}{4 (2\pi)^4 \hbar^2} \sum_{nn'\sigma} \int \left(-\frac{\partial f}{\partial \varepsilon}\right) \left(\frac{\partial B}{\partial \varepsilon}\right)^2 \frac{[I_{nn'}(q_{z1}) + I_{nn'}(q_{z2})]}{\sqrt{B(\varepsilon + \hbar s R^{-1}) - \left(n' + \frac{1}{2}\right)\hbar\Omega - \sigma g^* \mu_B H}} \times$$

$$\times \frac{d\varepsilon}{\sqrt{B(\varepsilon) - \left(n + \frac{1}{2}\right)\hbar\Omega - \sigma g^* \mu_B H}}, \tag{11}$$

where

$$I_{nn'}(q_{z1,2}) = \int q_\perp^3 \, |I_{nn'}(q_\perp^2)|^2 \, [W(q)(2N_q+1)]_{q_z = q_{z1,2}} \, dq_\perp \,. \tag{12}$$

In the case of electron scattering on acoustical phonons we have for the exponent of Planck's function

$$\frac{\hbar \omega_q}{k_0 T} = \frac{\hbar s q}{k_0 T} \approx \frac{\hbar s R^{-1}}{k_0 T} = \left(\frac{\hbar \Omega}{k_0 T}\right)^{1/2} \left(\frac{m_0^* s^2}{k_0 T}\right)^{1/2} \ll 1 \tag{13}$$

as $\frac{m_0^* s^2}{k_0 T} \ll 1$. The inequality (13) takes place even in the quantum limit when $\frac{\hbar \Omega}{k_0 T} > 1$. Therefore, one can set

$$2 N_q + 1 \approx \frac{2 k_0 T}{\hbar s q}. \tag{14}$$

From equations (12), (13), and (14) it follows that

$$I_{nn'}(q_{z1}) = I_{nn'}(q_{z2}) = \frac{8 \pi E_0^2 k_0 T}{\hbar s^2 \varrho R^4} (n + n' + 1). \tag{15}$$

Inserting (15) into (11) for σ_{xx} one obtains

$$\sigma_{xx} = \frac{2\, e^2\, m_0^*\, E_0^2\, k_0 T}{(2\,\pi)^3\, \hbar^3\, s^2\, \varrho\, R^2} \sum_{n n' \sigma} \int \left(-\frac{\partial f}{\partial \varepsilon}\right) \left(\frac{\partial B}{\partial \varepsilon}\right)^2 \times$$

$$\times \frac{(n + n' + 1)\, \mathrm{d}\varepsilon}{\sqrt{B\,(\varepsilon + \hbar\, s\, R^{-1}) - \left(n' + \frac{1}{2}\right) \hbar\, \Omega - \sigma\, g^*\, \mu_{\mathrm{B}}\, H}\ \sqrt{B(\varepsilon) - \left(n + \frac{1}{2}\right) \hbar\, \Omega - \sigma\, g^*\, \mu_{\mathrm{B}}\, H}}\,. \tag{16}$$

For a non-degenerate semiconductor it is $\frac{\partial f}{\partial \varepsilon} = -\frac{1}{k_0 T}\, \mathrm{e}^{\frac{\zeta - \varepsilon}{k_0 T}}$. In this case the integral over ε in formula (16) can be approximately calculated. As result one finds

$$\sigma_{xx} = \frac{2\, e^2\, E_0^2\, m_0^*}{(2\,\pi)^3\, \hbar^3\, s^2\, \varrho\, R^2}\, \mathrm{e}^{\frac{\zeta}{k_0 T}} \sum_{n n' \sigma} \mathrm{e}^{-\frac{\varepsilon_1}{k_0 T}} \left(\frac{\partial B}{\partial \varepsilon_1}\right) (n + n' + 1) \ln \left[\frac{(k_z + k_z')^2}{2\,(n - n')\, R^{-2} + \delta}\right], \tag{17}$$

where ε_1 is the root of the equation $k_z(\varepsilon) = 0$, i.e.

$$\left.\begin{aligned} B(\varepsilon) - \left(n + \frac{1}{2}\right) \hbar\, \Omega - \sigma\, g^*\, \mu_{\mathrm{B}}\, H = 0\,, \\ \delta = \frac{2\, m_0^*\, s}{\hbar\, R} \left(\frac{\partial B}{\partial \varepsilon}\right)_{\varepsilon = \varepsilon_1 + k_0 T}. \end{aligned}\right\} \tag{18}$$

In the derivation of equation (17) we have expanded $B\,(\varepsilon + \hbar\, s\, R^{-1}) \approx B(\varepsilon) + \hbar\, s\, R^{-1} \left(\frac{\partial B}{\partial \varepsilon}\right)$ and in the correction due to inelasticity we set $\varepsilon = \varepsilon_1 + k_0 T$, as the integral involves $\exp\left(-\frac{\varepsilon}{k_0 T}\right)$ and split from the integral a logarithmic function, assuming in it $\varepsilon = \varepsilon_1 + k_0 T$.

The chemical potential ζ appearing in equation (17) can be determined from the normalization condition for the electron concentration

$$N = \frac{2}{(2\,\pi\, R)^2} \sum_{n \sigma} \left(-\frac{\partial f}{\partial \varepsilon}\right) k_z(\varepsilon)\, \mathrm{d}\varepsilon\,. \tag{19}$$

Inserting $k_z(\varepsilon)$ from equation (8) into (19) and integrating approximately one finds

$$\exp\left(\frac{\zeta}{k_0 T}\right) = \frac{(2\,\pi\, R)^2\, \hbar\, N}{2 \sqrt{2\, m_0^*}} \left[\sum_{n \sigma} \mathrm{e}^{-\frac{\varepsilon_1}{k_0 T}} \sqrt{B\,(\varepsilon_1 + k_0 T) - \left(n + \frac{1}{2}\right) \hbar\, \Omega - \sigma\, g^*\, \mu_{\mathrm{B}}\, H}\right]^{-1}. \tag{20}$$

3. Non-Degenerate Semiconductor in the Quantum Limit

We consider now the quantum limit ($n = n' = 0$). As in the quantum limit it is $\varepsilon_1 \approx \hbar\, \Omega \gg k_0 T$, we have $B\,(\varepsilon_1 + k_0 T) \approx B(\varepsilon_1) + k_0 T \left(\frac{\partial B}{\partial \varepsilon_1}\right)$. Then from equation (20) we obtain

$$\exp\left(\frac{\zeta}{k_0 T}\right) = \frac{(2\,\pi\, R)^2\, \hbar\, N}{2 \sqrt{2\, m_0^*\, k_0 T}} \left[\sum_{\sigma} \mathrm{e}^{-\frac{\varepsilon_1}{k_0 T}} \left(\frac{\partial B}{\partial \varepsilon_1}\right)^{1/2}\right]^{-1}. \tag{21}$$

For a standard energy band it is $B(\varepsilon) = \varepsilon$ and $g^* = 2$; from equation (18) one finds $\varepsilon_1 = \frac{\hbar\,\Omega}{2} - 2\,\sigma\,\mu_B\,H$. Under the condition $\frac{\hbar\,\Omega}{2} \gg k_0 T \gg \mu_B\,H$ from equation (21) we get the well-known result which differs by the factor $\sqrt{\pi}/2$ (see formula (7) in [4]).

Further we shall use Kane's dispersion law in the two-band approximation [5, 6]:

$$B(\varepsilon) = \varepsilon\left(1 + \frac{\varepsilon}{\varepsilon_g}\right). \tag{22}$$

From equations (22) and (18) we determine ε_1:

$$\varepsilon_1 = -\frac{\varepsilon_g}{2}\left[1 - \sqrt{1 + \frac{2\,\hbar\,\Omega}{\varepsilon_g} + \frac{4\,\sigma\,g^*\,\mu_B\,H}{\varepsilon_g}}\right]. \tag{23}$$

Then it is

$$\frac{\partial B}{\partial \varepsilon_1} = 1 + \frac{2\,\varepsilon_1}{\varepsilon_g} = \sqrt{1 + \frac{2\,\hbar\,\Omega}{\varepsilon_g} + \frac{4\,\sigma\,g^*\,\mu_B\,H}{\varepsilon_g}}\,. \tag{24}$$

If the spin splitting of subbands is large, i.e. $\frac{\sigma\,g^*\,\mu_B\,H}{k_0 T} \gg 1$, then the role of the upper subband can be neglected and it is

$$\sigma_{xx} = \frac{e^2\,E_0^2\,N}{4\,\pi\,\hbar^2\,s^2\,\varrho}\left(\frac{m_0^*}{2\,k_0 T}\right)^{1/2}\left(1 + \frac{2\,\hbar\,\Omega}{\varepsilon_g} - \frac{2\,|g^*|\,\mu_B\,H}{\varepsilon_g}\right)^{1/4}\ln\left(\frac{2\,k_0 T}{\hbar\,\Omega}\,\frac{2\,k_0 T}{m_0^*\,s^2}\right). \tag{25}$$

If the spin splitting is small in comparison with $\hbar\,\Omega$, then both of the subbands give an equal contribution and in equation (25) the factor of non-parabolicity is of the form $2\left(1 + \frac{2\,\hbar\,\Omega}{\varepsilon_g}\right)^{1/4}$. It is seen that in both cases the non-standard character of the energy band appears only through the effective mass renormalization, that is, it occurs in σ_{xx} the effective mass at the bottom of the parabola shifted by the magnetic field.

For semiconductors with a very narrow forbidden band as in solid solutions of the type $Cd_xHg_{1-x}Te$ the unit in the factor of non-parabolicity can be neglected $\left(\frac{2\,\hbar\,\Omega}{\varepsilon_g} \gg 1\right)$ and σ_{xx} has a dependence on the magnetic field proportional to $H^{1/4}$.

Note, as in the quantum limit the temperature dependence of the transverse conductivity is the same as that for a standard energy band, that taking into account of the non-standard character of the energy band does not change the non-linear character in strong magnetic and electric fields [7].

4. Degenerate Semiconductor in the Quantum Limit

Now we consider a degenerate semiconductor in the quantum limit. From equation (16) we have

$$\sigma_{xx} = \frac{2\,e^2\,m_0^*\,E_0^2\,k_0 T}{(2\,\pi)^3\,\hbar^3\,s^2\,\varrho\,R^2}\sum_\sigma\left(\frac{\partial B}{\partial \zeta}\right)^2\left[B(\zeta) - \frac{\hbar\,\Omega}{2} - \sigma\,g^*\,\mu_B\,H\right]^{-1}. \tag{26}$$

In the two-band approximation ζ is defined by the electron concentration according to equations (19), (8), and (22) in the following way:

$$N = \frac{2}{(2\pi R)^2} \frac{\sqrt{2 m_0^*}}{\hbar} \sum_\sigma \sqrt{\zeta\left(1+\frac{\zeta}{\varepsilon_g}\right) - \frac{\hbar\Omega}{2} - \sigma g^* \mu_B H}\,. \tag{27}$$

If the spin splitting of energy band is greater than the Fermi boundary energy, that is $\frac{1}{2} |g^*| \mu_B H > \zeta$, then in equations (26) and (27) we can confine ourselves only to the lower subband. From (27) we get

$$\zeta = -\frac{\varepsilon_g}{2}\left[1 - \sqrt{1 + \frac{2\hbar\Omega}{\varepsilon_g} - \frac{2|g^*|\mu_B H}{\varepsilon_g} + \frac{N^2 \hbar^2 (2\pi R)^4}{2 m_0^* \varepsilon_g}}\right]. \tag{28}$$

Then finally we have

$$\sigma_{xx} = \frac{e^2 E_0^2 k_0 T m_0^{*2}}{8\pi^7 \hbar^5 s^2 \varrho R^6 N^2}\left[1 + \frac{2\hbar\Omega}{\varepsilon_g} - \frac{2|g^*|\mu_B H}{\varepsilon_g} + \frac{N^2\hbar^2(2\pi R)^4}{2 m_0^* \varepsilon_g}\right]. \tag{29}$$

We see that taking into account the non-parabolicity results in replacing the constant effective mass in the model of a standard energy band by the effective mass on the Fermi surface, which essentially depends on electron concentration.

5. Conclusions

1. For a non-degenerate semiconductor with non-parabolic energy band in the expression for σ_{xx} the effective mass appears at the bottom of the parabola shifted by a magnetic field, instead of the constant effective mass, i.e. it appears the additional factor

$$\left(1 + \frac{2\hbar\Omega}{\varepsilon_g} - \frac{2|g^*|\mu_B H}{\varepsilon_g}\right)^{1/4}.$$

2. For a non-degenerate semiconductor with a very narrow forbidden band, as in solid solutions of the type $Cd_xHg_{1-x}Te$, σ_{xx} is dependent on a magnetic field proportional to $H^{1/4}$.

3. For a degenerate semiconductor in the expression for σ_{xx} the constant effective mass is replaced by the effective mass on the Fermi surface, which essentially depends on the electron concentration.

4. In all cases the temperature dependence of σ_{xx} is the same as that for a standard energy band.

It can be shown that just the same results are obtained in the cases of electron scattering on optical phonons, piezoacoustical waves at high temperatures, and on impurity ions.

References

[1] B. M. Askerov and F. M. Gashimzade, phys. stat. sol. **18**, 667 (1966).
[2] E. Adams and T. Holstein, J. Phys. Chem. Solids **10**, 254 (1959).
[3] V. L. Gurevich and Yu. A. Firsov, Zh. eksper. teor. Fiz. **47**, 734 (1964).
[4] A. I. Anselm and B. M. Askerov, Fiz. tverd. Tela **9**, 31 (1967).
[5] E. O. Kane, J. Phys. Chem. Solids **1**, 249 (1957).
[6] R. Bowers and Y. Yafet, Phys. Rev. **115**, 1165 (1959).
[7] R. F. Kazarinov and V. G. Skobov, Zh. eksper. teor. Fiz. **42**, 1047 (1962).

(Received May 8, 1968)

phys. stat. sol. 28, 789 (1968)

Subject classification: 6; 22.5.2

Physics Department of Surrey University, London

Calculation of the Point Lattice Potential of a Bounded Crystal

By

O. G. VENDIK[1])

The eigenfunctions of the electrostatic potential in a rectangular box were used for the calculation of the potential of a point lattice of an ionic crystal. The potential within the bounded crystal and in its surface region was found. The final result was obtained in the form of an expansion in spherical harmonics referred to coordinates whose origin could be at any lattice point of the crystal. The method was used to explain an experiment [8] dealing with electron diffraction from the surface layers of $BaTiO_3$.

Электростатический потенциал точечной решётки ионного кристалла рассчитывается путём применения собственных функции прямоугольного объёма. Потенциал определяется как внутри ограниченного кристалла, так и в области, близкой к его поверхности. Окончательный результат получен в форме разложения по сферическим гармоникам, начало координат которых может быть отнесено к любому центру кристаллической решётки. Предложенный метод был использован для объяснения результатов эксперемента [8] по диффракции электронов от поверхностных слоёв $BaTiO_3$.

1. Introduction

The aim of this paper is to present a method for the calculation of the potential of an ionic crystal with a definite form using a point lattice approximation. The method is based on the expansion of the lattice potential in the eigenfunctions of a rectangular box with perfectly conducting walls. The crystal considered is assumed to be situated within this box far from each wall. In the limit the size of the box can be supposed to be infinite, but the crystal remaining definite in form and size.

The calculation of the electric potential distribution will be made both within the inner region of the crystal and on its bounding faces. The final result of the calculation will be obtained in the form of the expansion in spherical harmonics, and therefore can be easily used for investigation of the perturbation of the ion electron wave functions which is produced by the lattice field. The coordinates of the spherical harmonics can be referred to any lattice point. It is obvious that the expansion coefficients are changed as the position of the chosen lattice point is moved from the inside of the crystal to its boundary.

This method generalizes to a certain extent the known method of the calculation of the lattice potential by summation of the potentials created by ions which are situated on several coordinate spheres in an infinite crystal [1, 2], and the method of summation of the potentials created by ions which are situated on the parallel layers in a semi-infinite crystal [3, 4].

---

[1]) Permanent address: Special Physics Department, V. I. Ulyanov (Lenin) Electrical Engineering Institute, Leningrad.

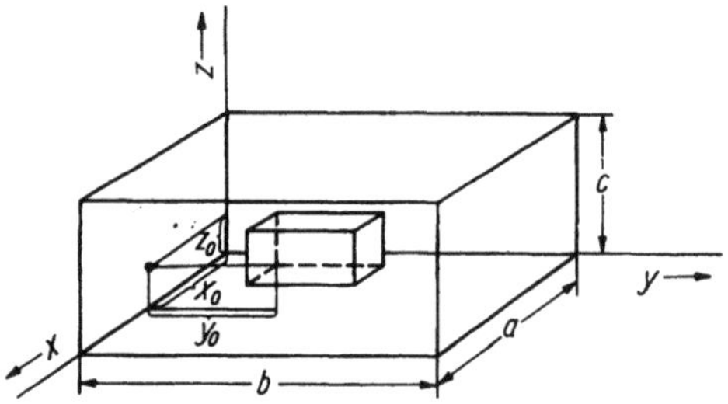

Fig. 1. The position of the crystal in a box with perfectly conducting walls

Let us locate the finite crystal in the rectangular box (Fig. 1). Solving the Poisson equation for zero boundary conditions we find the potential inside the box:

$$\Phi(x, y, z) = \frac{e_0}{\varepsilon_0} \sum_{\varrho, \tau, \sigma} \frac{1}{k^2_{\varrho\tau\sigma}} \Phi_{\varrho\tau\sigma}(x, y, z) \sum_{\lambda, \mu, \nu} Z_{\lambda\mu\nu} \Phi_{\varrho\tau\sigma}(x_{\lambda,\mu,\nu}, y_{\lambda,\mu,\nu}, z_{\lambda,\mu,\nu}), \quad (1)$$

where e_0 is the electronic charge, ε_0 the dielectric constant of the vacuum, $Z_{\lambda\mu\nu}$ the normalized value of the charge of the ion number λ, μ, ν having the coordinates $x_{\lambda,\mu,\nu}$, $y_{\lambda,\mu,\nu}$, and $z_{\lambda,\mu,\nu}$;

$$\Phi_{\varrho\tau\sigma}(x, y, z) = 2\sqrt{\frac{2}{a\,b\,c}} \sin\frac{\varrho\pi x}{a} \sin\frac{\tau\pi y}{b} \sin\frac{\sigma\pi z}{c} \quad (2)$$

are the eigenfunctions of the box with dimensions a, b, c;

$$k^2_{\varrho\tau\sigma} = \left(\frac{\varrho\pi}{a}\right)^2 + \left(\frac{\tau\pi}{b}\right)^2 + \left(\frac{\sigma\pi}{c}\right)^2. \quad (3)$$

For simplicity, we shall consider a sodium chloride crystal with rectangular faces perpendicular to the axes x, y, z. The coordinates and the charge of the ion with numbers λ, μ, ν are given by

$$\left.\begin{aligned} &x_{\lambda\mu\nu} = x_0 + \lambda d, \quad y_{\lambda\mu\nu} = y_0 + \mu d, \quad z_{\lambda\mu\nu} = z_0 + \nu d; \\ &Z_{\lambda\mu\nu} = (-1)^{\lambda+\mu+\nu}. \end{aligned}\right\} \quad (4)$$

Substituting (2) and (4) into (1) and summing over λ, μ, ν we obtain

$$\Phi(x, y, z) =$$

$$= \frac{8\,e_0}{a\,b\,c\,\varepsilon_0} \sum_{\varrho\tau\sigma} \left\{ \frac{1}{k^2_{\varrho\tau\sigma}} \frac{\sin\left[\frac{\varrho\pi}{a} x_0 + \frac{L-1}{2}\left(\frac{\varrho\pi}{a} + \pi\right)\right] \sin\left[\frac{L}{2}\left(\frac{\varrho\pi d}{a} + \pi\right)\right]}{\sin\left[\frac{1}{2}\left(\frac{\varrho\pi}{a} + \pi\right)\right]} \times \cdots \times \cdots \right\}, \quad (5)$$

where L, M, N are the numbers of the layers in the crystal along the axes x, y, z, respectively. Dots in the expressoin (5) mean analogous fractions for y and z.

The expression (5) will now be analysed and simplified for two cases which correspond to the location of a particular lattice point within the crystal or near its bounding face.

2. The Potential within the Crystal

The most important fact which makes it possible to simplify the expression (5) is that the ratio of the trigonometric functions is very small, and alternates in sign for the majority of the values of ϱ, τ, σ, except those values of ϱ, τ, σ which cause the sine in the denominators to be zero.

These latter values can be determined in the following way:

$$\varrho_i = \frac{a}{d}(2\,i-1)\,, \qquad \tau_j = \frac{b}{d}(2\,j-1)\,, \qquad \sigma_k = \frac{c}{d}(2\,k-1)\,.$$

Now we can make summation in (5) in two stages: over the integers i, j, k which determine the reference points of the summation, and over the numbers ϱ', τ', σ' which determine the summation near the reference points and later will be used as variables, for integration. Thus we have

$$\varrho = \varrho_i + \varrho'\,, \qquad \tau = \tau_j + \tau'\,, \qquad \sigma = \sigma_k + \sigma'\,.$$

Rejecting integers of π, we may replace sines by their arguments in the denominators of the fractions in (5).

Now we introduce the following notation:

$$\left.\begin{aligned} \xi &= x - (x_0 + \lambda\, d) \qquad |\xi| \leqq d/2\,,\\ \eta &= y - (y_0 + \mu\, d) \qquad |\eta| \leqq d/2\,,\\ \zeta &= z - (z_0 + \nu\, d) \qquad |\zeta| \leqq d/2\,, \end{aligned}\right\} \tag{6}$$

where λ, μ, ν are integers which determine the position of the lattice point.

After the trigonometrical transformation, the numerator in the term of (5) which is connected with x, may be written in the following way:

$$\begin{aligned} \frac{1}{4}\Big\{\sin\Big[\Big(\frac{\varrho\,\pi\,d}{a}+\pi\Big)\Big(L-\lambda-\frac{1}{2}\Big)-\frac{\varrho\,\pi}{a}\,\xi\Big] + \sin\Big[\Big(\frac{\varrho\,\pi\,d}{a}+\pi\Big)\Big(\lambda+\frac{1}{2}\Big)+\frac{\varrho\,\pi}{a}\,\xi\Big] - \\ - \sin\Big[\Big(\frac{\varrho\,\pi\,d}{a}+\pi\Big)\Big(L+\lambda-\frac{1}{2}\Big)+\frac{\varrho\,\pi}{a}\,(2\,x_0+\xi)\Big] + \\ + \sin\Big[\Big(\frac{\varrho\,\pi\,d}{a}+\pi\Big)\Big(\lambda-\frac{1}{2}\Big)+\frac{\varrho\,\pi}{a}\,(2\,x_0+\xi)\Big]\Big\}\,. \end{aligned} \tag{7}$$

Since $x_0 \gg d$ the summation over ϱ' in (5) for the last two terms may be replaced by an integration. We can then use the integral

$$\frac{1}{\pi}\int\limits_{-\infty}^{\infty}\frac{\sin q\,x}{x}\,\mathrm{d}x = \operatorname{sign} q\,; \qquad q \neq 0\,.$$

Because these terms have different signs, they cancel out when substituted into (5).

Now we further transform the first two terms of (7):

$$\begin{aligned} \frac{1}{4}\Big\{\sin\Big[\Big(\frac{\varrho\,\pi\,d}{a}+\pi\Big)\Big(\lambda+\frac{1}{2}\Big)\Big] + \sin\Big[\Big(\frac{\varrho\,\pi\,d}{a}+\pi\Big)\Big(L-\lambda-\frac{1}{2}\Big)\Big]\Big\}\cos\frac{\varrho\,\pi}{a}\,\xi\, + \\ + \frac{1}{4}\Big\{\cos\Big[\Big(\frac{\varrho\,\pi\,d}{a}+\pi\Big)\Big(\lambda+\frac{1}{2}\Big)\Big] - \cos\Big[\Big(\frac{\varrho\,\pi\,d}{a}+\pi\Big)\Big(L-\lambda-\frac{1}{2}\Big)\Big]\Big\}\sin\frac{\varrho\,\pi}{a}\,\xi\,. \end{aligned} \tag{8}$$

Because the lattice point considered is at sufficient distance from any face of the crystal we can write

$$\left.\begin{array}{ll} \lambda \gg 1\,, & L-\lambda \gg 1\,; \\ \mu \gg 1\,, & M-\mu \gg 1\,; \\ \nu \gg 1\,, & N-\nu \gg 1\,. \end{array}\right\} \tag{9}$$

It is easy to see that the conditions (9) are followed by the fact that, after summation in (5), the second line in (8) is zero. The term $1/k^2_{\varrho\tau\sigma}$ is a slowly varying function of ϱ, τ, σ, therefore we may use the average theorem and take out this term from the sum or from the integral. The summation over ϱ', τ', σ' now can be easily replaced by integration and finally we obtain the following expression for the potential around the ion number λ, μ, ν:

$$\Phi_{\lambda,\mu,\nu}(\xi,\eta,\zeta) = \frac{8\,e\,(-1)^{\lambda+\mu+\nu}}{\varepsilon_0\,\pi^2\,d} \sum_{i,j,k=1}^{\infty} \frac{1}{(2\,i-1)^2+(2\,j-1)^2+(2\,k-1)^2} \times$$
$$\times \cos\,(2\,i-1)\,\frac{\pi\,\xi}{d}\,\cos\,(2\,j-1)\,\frac{\pi\,\eta}{d}\,\cos\,(2\,k-1)\,\frac{\pi\,\zeta}{d}\,. \tag{10}$$

The potential determined by (10) is the result of the summation of the contribution to the total potential of all the ions of the lattice. One of these ions is the ion number λ, μ, ν. Therefore, when $\xi, \eta, \zeta \to 0$, expression (10) becomes a divergent series. We must eliminate the potential of this ion and consider the potential distribution around and at the lattice point number λ, μ, ν which is produced by all other ions of the lattice excluding the ion at this particular point. It is reasonable for this purpose to replace the Cartesian coordinates ξ, η, ζ by the spherical coordinates r, θ, φ with the centre at the lattice point number λ, μ, ν.

Now we surround the point considered by a sphere of a radius $R_0 < d$, and subtract from the potential on the sphere calculated by means of (10) the potential produced on the sphere by the ion located at its centre. After this operation no charges remain inside the sphere and the potential distribution within the sphere can be found by solving the Laplace equation for the given potential distribution on the bounding surface.

The solution of the Laplace equation may be written in the following way:

$$\Phi(r,\theta,\varphi) = \sum_{l=0}^{\infty} \sum_{m=-l}^{l} A_{l,m}\, r^l\, Y^*_{l,m}(\theta,\varphi)\,. \tag{11}$$

The expansion coefficients $A_{l,m}$ have to be found by equating (10) and (11) at $r = R_0$. For this, we must expand cosines in (10) in spherical harmonics [5] and subtract the potential of the point charge in the centre of the sphere. We then obtain

$$A_{l,m} = (-1)^{\lambda+\mu+\nu} \left\{ \frac{32\,e}{\varepsilon_0\,\pi\,d}\, N_{l,m} \sum_{i,j,k=1}^{\infty} \frac{i^l\, J_l\left(k_{ijk}\,\dfrac{\pi\,R_0}{d}\right)}{R_0^l\, k_{ijk}^2} \times \right.$$
$$\left. \times\, P_l^m\,(\cos\theta_{ijk})\,\cos m\,\varphi_{ij} - \delta_{0,l}\,\frac{e_0}{\sqrt{4\,\pi}\,\varepsilon_0\,R_0} \right\}, \tag{12}$$

where $\delta_{0,l}$ is Kronecker's delta, $J_l(\ )$ Bessel spherical function,

$$\left.\begin{aligned} \theta_{ijk} &= \arccos \frac{2k-1}{\sqrt{(2i-1)^2+(2j-1)^2+(2k-1)^2}}, \\ \varphi_{ij} &= \operatorname{arctg} \frac{2j-1}{2i-1}, \quad k_{ijk}^2 = (2i-1)^2+(2j-1)^2+(2k-1)^2, \\ N_{l,m} &= \sqrt{\frac{(2l+1)(l-m)!}{4\pi(l+m)!}}. \end{aligned}\right\} \tag{13}$$

The term with $\delta_{0,l}$ in (12) appeared as a result of subtracting the potential of the point charge in the centre of the sphere. Taking into account this form of (12) one can see that series (11) is no longer divergent.

From the expansion of the cosines in spherical harmonics we find that the numbers l, m are even. However, it can be shown, taking into account the symmetry of a cubic crystal, that in fact the coefficients $A_{l,m}$ are not equal to zero only when numbers l, m are divisible by four.

3. The Lattice Potential near the Boundary of the Crystal

Now we assume that one of the six inequalities (9) is not satisfied. Let ν be of the order of unity. This means that we consider the ion located near the bounding face perpendicular to the z-axis but far from the edges of this face.

Now for the z-direction we must replace the letters ξ, ϱ, a, λ in (8) by ζ, σ, c, ν. If we take into account (9) we can see that the result of the summation in (5) for the first line of (8) in the first approximation is independent of ν. The result of the summation of the terms in the second line of (8) is no longer zero and requires a special investigation. Taking into account the component of (8) which does not vanish near the crystal face, and making summation in (5) we find in analogy with (10) that the extra component of the lattice potential associated with the surface region can be written as

$$\begin{aligned} \Phi'_{\lambda\mu\nu}(\xi,\eta,\zeta) &= \frac{4\,e_0\,(-1)^{\lambda+\mu+\nu}}{\varepsilon_0\,\pi^2\,d} \sum_{i,j,\,k=1}^{\infty} I_{ijk} \times \\ &\times \cos\,(2i-1)\frac{\pi\,\xi}{d}\cos\,(2j-1)\frac{\pi\,\eta}{d}\sin\,(2k-1)\frac{\pi\,\zeta}{d}, \end{aligned} \tag{14}$$

where

$$I_{i,j,k} = \frac{d}{c} \sum_{\sigma''=-\frac{c}{d}}^{\frac{c}{d}} \frac{\cos\left[\frac{\pi\,d\,\sigma''}{2\,c}(2\nu+1)\right]}{\left\{(2i-1)^2+(2j-1)^2+\left(2k-1+\frac{d}{c}\sigma''\right)^2\right\}\sin\frac{\pi\,d\,\sigma''}{2\,c}}; \tag{15}$$

$$\sigma'' \neq 0.$$

The term $\sigma''=0$ is excluded from the sum (15) because in the case $\sigma''=0$ the difference of the cosines in the second line of (8) gives zero after summation in (5).

The Taylor series expansion in powers of σ'' of the expression in the curly brackets gives

$$I_{i,j,k} = \frac{-(2k-1)}{[(2i-1)^2+(2j-1)^2+(2k-1)^2]^2} \sum_{\sigma''=-\frac{c}{d}}^{\frac{c}{d}} \frac{\frac{2d}{c}\sigma'' \cos\left[\frac{\pi d \sigma''}{2c}(2\nu+1)\right]}{\sin\frac{\pi d \sigma''}{2c}} \frac{d}{c}. \tag{16}$$

The summation can be replaced by integration, and one obtains

$$I_{i,j,k} = -\frac{16}{\pi^2} \frac{2k-1}{[(2i-1)^2+(2j-1)^2+(2k-1)^2]^2} \frac{\sin\frac{\pi}{2}(2\nu+1)}{2\nu+1} \times$$
$$\times \left[1 + \left(\frac{\pi^2}{24} - \frac{1}{3(2\nu+1)^2}\right) + \cdots\right]. \tag{17}$$

Using the expansion of the trigonometric function of (14) in terms of spherical harmonics we obtain the following expression for $A'_{l,m}$ which corresponds to the near-face region component of the lattice potential:

$$A'_{l,m} = (-1)^{\lambda+\mu+1} \frac{16 e_0}{\varepsilon_0 \pi d} N_{l,m} \frac{\frac{16}{\pi^2}}{2\nu+1} \sum_{i,j,k=1}^{\infty} \frac{i^{l+1} J_l\left(k_{ijk}\frac{\pi R_0}{d}\right)}{R_0^l k_{ijk}^4} \times$$
$$\times (2k-1) P_l^m(\cos\theta_{ijk}) \cos m\varphi_{ij}, \tag{18}$$

where the notation is similar to (13), the numbers l are odd, and m divisible by 4.

The most important properties of the coefficients $A'_{l,m}$ are that they do not change their sign with changing the number of layers ν, and decrease the further the point considered gets from the crystal face.

4. Discussion

The final results of the present calculation of the lattice potential are given by the expressions (11), (12), and (18).

Putting $r = 0$ in (11) one can obtain

$$\Phi(r, \theta, \varphi)\,|_{r=0} = A_{0,0}. \tag{19}$$

This can be used for the calculation of the Madelung constant. The calculation of this for NaCl [6] coincides with the known value for any choice of the radius R_0 within the limits $0 < R_0 < d$.

Using the formulae usually applied for deriving the selection rules [7] one finds

$$\left.\frac{\partial\Phi}{\partial z}\right|_{r=0} = -\sqrt{\frac{3}{4\pi}}\, A_{1,0}, \tag{20}$$

$$\left.\frac{\partial^2\Phi}{\partial z^2}\right|_{r=0} = -\sqrt{\frac{5}{\pi}}\, A_{2,0}. \tag{21}$$

Here we neglect the value $A_{2,0}$, but for a crystal not of cubic symmetry it has to be taken into account for the calculation of the dynamic properties of the lattice.

Fig. 2. The normalized value of the field calculated by (20) and (22) and the normalized value of the ion polarization in the surface region of certain alkali halide crystals
—○—○— the field
—△—△— NaF
—▽—▽— NaCl
—□—□— NaBr
—×—×— NaI
(NaF, NaCl, NaBr, NaI: polarization)

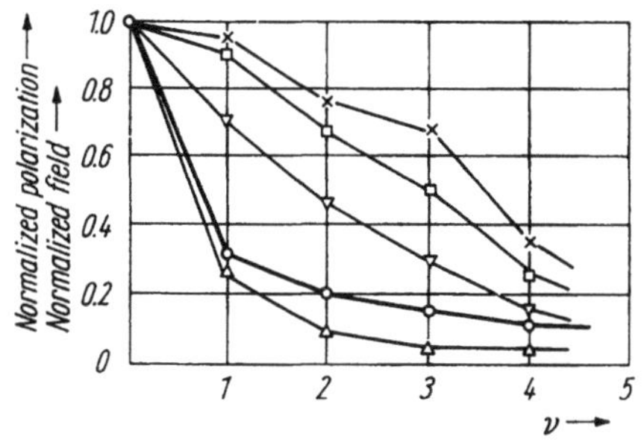

The expression (20) gives the value of the electric field which emerges near the face of the crystal and causes the change of the inter-ionic distance and the ion polarization.

The calculation of this field can be considered only as a first approximation because it is not a self-consistent solution which should take into account the changed position of the ions and their polarization.

We consider the following example in order to estimate the accuracy of this field calculation in its present form. From the expressions (20) and (18) we find that the electric field at the lattice points decreases inversely proportional to the value of $2\nu + 1$. This field produces the polarization of the ions. The polarization of the ions in the surface region has been calculated [4] for the alkali halide crystal by minimizing the total energy of the system of the movable and polarizable ions. Fig. 2 shows the normalized values of the electric field from the present calculation and the normalized values of the ion polarizations which are proportional to the field calculated more precisely [4]. One can conclude that the present calculation gives the values of the field in the surface region which can be used as the first approximation, and can then be made more accurate by the self-consistent iteration procedure.

The electric field near the surface has been calculated for a barium titanate crystal [9] and the result was used to explain of the experimental fact [8] that the boundary layers of a barium titanate crystal have tetragonal structure and are stable above the Curie point. The calculation [9] has given qualitative agreement with the result of the electron diffraction measurements [8].

This method can be applied to the problem of the properties of ferroelectrics near a metal electrode. For this purpose the distance between the box wall and the crystal face z_0 has to be zero and all other calculations made according to the presented scheme.

Acknowledgements

I should like to thank Professor A. I. Gubanov for the initiating this problem and Professor W. Cochran for helpful discussions.

References

[1] I. V. Abarenkov, Proc. Leningrad University **16**, 5 (1958).
[2] D. S. McClure, J. chem. Phys. **36**, 2757 (1962).
[3] B. J. Alder, J. R. Vaisnys, and G. Jura, J. Phys. Chem. Solids **11**, 182 (1959).
[4] G. C. Beuson, P. I. Freeman, and E. Dempsey, J. chem. Phys. **39**, 302 (1963).
[5] J. D. Jackson, Classical Electrodynamics, Wiley, 1962.

[6] O. G. VENDIK and L. A. ROSENBERG, Proc. Leningrad Electrical Engineering Institute **63**, 123 (1968).
[7] H. A. BETHE and E. E. SALPETER, Quantum Mechanics of One- and Two-Electron Atoms, Springer-Verlag, 1957.
[8] M. ANLIKER, N. R. BRUGGER, and W. KANZIG, Helv. phys. Acta **27**, 29 (1954).
[9] L. A. ROSENBERG, Diploma Thesis, Leningrad Electrical Engineering Institute, 1967.

(Received April 8, 1968)

phys. stat. sol. 28, 797 (1968)

Subject classification: 16; 14.3.1; 22

Institute of Semiconductors, Academy of Sciences of the Ukrainian SSR, Kiev

On the Origin of High Photovoltages in Semiconductors

By

I. P. ZHADKO and V. A. ROMANOV

A new possible mechanism is proposed for explaining the high photovoltage arising in semiconductor films assuming an anisotropic conductivity of the photosensitive crystallites of which the films consist. In this case the high photovoltage is the sum of the elementary transverse Dember photo-e.m.f.'s. This model may explain practically all peculiarities of the high-photovoltage effect.

Предложен новый возможный механизм возникновения высоковольтного фотонапряжения в пленках различных полупроводников. В основе этого механизма лежит предположение об анизотропии электропроводности фотоактивных ячееккристаллитов, из которых состоит пленка. Высокое фотонапряжение возникает в этом случае благодаря суммированию элементарных поперечных фотоэдс Дембера. С помощью этой модели можно обьяснить практически все особенности эффекта высоковольтного фотонапряжения.

1. Introduction

Investigations of the high photovoltage effect (HPVE) where the photovoltage much exceeds the band gap were carried out on films of monoatomic semiconductors [1 to 3] and on films of many semiconducting materials of the types $A^{II}B^{VI}$ and $A^{III}B^{V}$ [4 to 15]. The results of these investigations are intricate, contradictory, and can be reproduced only qualitatively. In consequence, about the origin of the HPVE were set up only different hypotheses. A point of view of the most disseminated hypotheses is the following: The HPVE is a result of the addition of elementary photovoltages. They arise on micro p–n junctions or due to the diffusion of photocarriers (Dember effect) in homogeneous cells of special geometrical form (see, e.g., [16]).

It should be noted that none of the hypotheses connected the formation of the HPVE with a conductivity anisotropy of photosensitive cell-crystallites. In our opinion, however, with this supposition one can explain practically all the HPVE peculiarities.

Indeed, the Dember field arising due to the diffusion of photocarriers in anisotropic semiconductors in the common case is directed under some angle with the concentration gradient. The transverse component of this field is analogous to the photomagnetic one in an isotropic semiconductor. The magnitude of this transverse component may be of the same order as the longitudinal one for an appreciable anisotropy and suitable orientation of the crystal [18].

2. Basic Peculiarities of the HPVE in Semiconductors

We suppose that the film consists of photosensitive anisotropic cells having the same crystallographic orientation. These cells are separated by layers which hinder the motion of free carriers between cells. The HPVE is in this case the result of the addition of elementary transverse Dember e.m.f.

The conditions of the same orientation of the cell-crystallites and their conductivity anisotropy may seem hard to be fulfilled for the films of all semiconductors on which the HPVE was observed, but it is not so, namely by the following reasons:

a) The HPVE is observed, as a rule, on films with well-developed texture [5, 6, 9, 19, 20], that is on such films which have the same orientation of crystallites.

b) Some semiconductors (Te, Se, CdSe, etc), on films of which the HPVE was observed, have the natural anisotropy of conductivity.

c) In films prepared by vacuum evaporation of different materials on substrates, there may exist very high mechanical tensions up to 10^9 dyn/cm^2 and more [20 to 25]. They can cause the considerable artificial conductivity anisotropy. (The piezoresistances of many semiconductors are strongly anisotropic; this is connected with their complicate band structure [26, 27].)

Thus, there is a high probability that the HPVE is caused by the anisotropy of conductivity of the crystallites of which the films consist. We shall analyse the basic peculiarities of the HPVE proceeding from this assumption.

1. The distinctive feature of the transverse Dember effect like the photomagnetic effect is the change of its sign under change of the illumination from the front surface of a specimen to the back one [18]. The HPVE on films has exactly the same feature [1, 9, 10, 11, 14, 17].

2. It is difficult to explain the dependence of the sign of HPVE on wavelength of the exciting light. In our opinion this property may be explained by the conception on the anomalous transverse Dember effect [28]. If the surface recombination on the illuminated side of the film is greater than on an unilluminated one, the diffusion flow of photocarriers in the cells changes its sign when the strong absorption changes to weak one.

3. There is a strong dependence of the HPVE on the technology of film preparation, namely on the deposition angle, substrate and emitter temperatures, atmosphere of residual gas in the vacuum chamber, and evaporation intensity [2, 4, 6, 11, 14, 15]. However, just these technological parameters determine the value and character of mechanical tensions [20 to 23, 25] and, therefore, the conductivity anisotropy of the cells.

One of the main causes of such tensions is the difference between the expansion coefficients of the substrate and semiconducting material. Owing to this difference the mechanical tensions and, hence, the HPVE must strongly depend on the temperature. Results of the papers [1] to [4] and [11] confirm this conclusion.

It is known that the HPVE strongly depends on treatments in the atmosphere of different gases. Some authors consider the HPVE to be of surface origin (see, e.g., [2]). One must be very careful in estimating such conclusions because the mechanical tensions in films can strongly depend on such treatments in accord with [22], [23], and [25].

4. The most important peculiarity of the HPVE is its dependence on the angle of deposition (see, e.g., [7]). The magnitude of the HPVE, as a rule, has a maximum value for oblique deposition angle and becomes very small for normal deposition direction. Usually this peculiarity is considered as the most convincing proof that the HPVE is caused by the distinctive needle-shaped

structure either of p–n junctions or Dember cells in the films. However, it is difficult by the help of such models to explain the next fact: the photovoltage on films deposited normally and illuminated under oblique angle is considerably less than the photovoltage on films deposited obliquely and illuminated normally (see, e.g., [11]).

According to our model such dependence of the HPVE on the deposition angle can be caused by the following circumstances:

a) The mechanical tensions in films can show a strong anisotropy and depend on the deposition angle like in SiO films [23, 25]. Namely, with increasing deposition angle beginning with normal incidence the tensions increase by several times; they reach a maximum value between 30 and 50° and then decrease.

As the value of anisotropy is proportional to the mechanical tensions, the dependence of the HPVE on deposition angle must be similar to the angular dependence of the tension in films.

b) The transverse Dember effect depends also on the orientation of the illuminated surface relatively to the main crystallographic axes [18]. This orientation in films is set by the deposition angle (see, e.g., [5, 9]). For example, for germanium films the planes of primary growth are (110) or (111) depending on substrate temperature [19]. On the other hand, the directions [111] and [110] in germanium are those of the maximum values of piezoresistance [27]. Therefore, for normal deposition direction the HPVE must be absent in germanium films because the direction of photocarrier diffusion in this case coincides with that of one of the main components of the electron or hole conductivity tensor. With high probability the same situation may take place in Si films; such a situation is not impossible in films of other semiconductors.

5. Some authors [3, 29] observed the change in the spectral distribution of the HPVE for a turning of the polarization plane of the incident light. In our opinion, this effect may be connected with the artificial or natural structure anisotropy of the semiconductor lattice. This causes the dependence of the absorption coefficient on turning the polarization plane and, consequently, the change in the spectral distribution of the photo-e.m.f.

6. The HPVE caused by the conductivity anisotropy must decrease at additional illumination of the opposite side of the film because in this case the diffusion flows of photocarriers partly compensate one another. Such behaviour was observed experimentally [7].

3. Conclusion

We have examined a model attributing the HPVE to a simple addition of elementary transverse Dember e.m.f. in homogeneous photosensitive cells. However, a model of the film consisting of an aligned array of many p–n junctions may be analysed, too. In contrast to a system of p–n junctions consisting of an isotropic semiconductor when the photo-e.m.f. of the neighbouring p–n junctions is mutually annihilated, in a many-elemental p–n–p. . .–n system consisting of an anisotropic semiconductor the HPVE must be observed like the odd photomagnetic e.m.f. in a many-layer isotropic p–n–p. . .–n system [30].

Acknowledgements

The authors are indebted to Z. S. Gribnikov and E. I. Rashba for discussions.

References

[1] H. Kallmann, B. Kramer, E. Haidemenakis, W. J. McAler, H. Barkemeyer, and P. I. Pollak, J. Elektrochem. Soc. **108**, 247 (1961).
[2] P. P. Kokorev and K. Lubits, Fiz. tverd. Tela **6**, 71 (1964).
[3] E. I. Adirovich, V. M. Rubinov, and Yu. M. Yuabov, Dokl. Akad. Nauk SSSR **157**, 76 (1964).
[4] B. Goldstein and L. Pensak, J. appl. Phys. **30**, 155 (1959).
[5] S. A. Semiletov, Fiz. tverd. Tela **4**, 1241 (1962).
[6] F. T. Novik, Fiz. tverd. Tela **45**, 3142 (1963).
[7] E. I. Adirovich, V. M. Rubinov, and Yu. M. Yuabov, Izv. Akad. Nauk UzSSR, Ser. fiz.-mat., No. 6, 63 (1964).
[8] V. A. Ignatyuk and F. T. Novik, Fiz. tverd. Tela **8**, 3661 (1966).
[9] R. Ya. Berlaga, M. A. Rumsh, and L. P. Strakhov, Radiotekhnika i Elektronika **2**, 287 (1957).
[10] V. M. Lyubin and G. A. Fedorova, Fiz. tverd. Tela **4**, 2026 (1962).
[11] I. A. Karpovich and M. V. Shilova, Fiz. tverd. Tela **5**, 3560 (1963).
[12] A. A. Smilga and Yu. K. Vishchakas, Lit. Fiz. Sbornik, **5**, 154 (1965).
[13] M. D. Uspenskii, N. G. Ivanova, and I. E. Malkis, Fiz. Tekh. Poluprov. **1**, 1268 (1967).
[14] M. I. Korsunskii and V. M. Fridman, Dokl. Akad. Nauk SSSR **168**, 318 (1966).
[15] L. S. Palatnik and V. K. Sorokin, Fiz. tverd. Tela **8**, 2795 (1966).
[16] E. I. Adirovich, V. M. Rubinov, and Yu. M. Yuabov, Dokl. Akad. Nauk SSSR **168**, 1037 (1966).
[17] E. I. Adirovich, V. M. Rubinov, and Yu. M. Yuabov, Dokl. Akad. Nauk SSSR, **174**, 545 (1967).
[18] I. P. Zhadko, E. I. Rashba, V. A. Romanov, I. M. Stachira, and K. D. Tovstyuk, Fiz. tverd. Tela **7**, 1777 (1965).
W. van Roosbroeck and W. G. Pfann, J. appl. Phys. **33**, 2304 (1962).
I. K. Kikoin and S. D. Lasarev, Zh. eksper. teor. Fiz. **47**, 780 (1964).
[19] J. E. Davey, J. appl. Phys. **32**, 877 (1961).
[20] R. E. Tun, Physics of Thin Films, Vol. I, Mir, Moscow 1967 (p. 225).
[21] J. P. Priest, Rev. sci. Instrum. **32**, 1349 (1961).
[22] M. A. Novice, Brit. J. appl. Phys. **13**, 561 (1962).
[23] J. Priest, H. L. Caswell, and Y. Budo, J. appl. Phys. **34**, 347 (1963).
[24] J. H. van der Merwe, Phil. Mag. **7**, 1433 (1962).
[25] A. E. Hill and G. R. Hoffman, Brit. J. appl. Phys. **18**, 13 (1967).
[26] O. N. Tufte and D. Long, Solid State Electronics **6**, 323 (1963).
[27] C. S. Smith, Phys. Rev. **94**, 42 (1954).
[28] W. Gärtner, Phys. Rev. **105**, 823 (1957).
M. Azizov, Izv. Akad. Nauk Uz. SSR, Ser. fiz.-mat. No. 4, 62 (1962).
[29] G. Cheroff, R. C. Enck, and S. P. Keller, Phys. Rev. **116**, 1091 (1959).
[30] Yu. I. Ravich, Photomagnetic Effect in Semiconductors and Its Applications, Sovetskoe radio, Moscow 1967 (pp. 72 and 88).
W. Dunstan, Proc. Phys. Soc. **77**, 459 (1961).
I. K. Kikoin and I. N. Nikolaev, Zh. eksper. teor. Fiz. **41**, 1692 (1961).

(Received April 16, 1968)

phys. stat. sol. 28, 801 (1968)

Subject classification: 3.2; 13.1; 21.1; 21.6

A. F. Ioffe Physico-Technical Institute, Academy of Sciences of the USSR, Leningrad

The Theory of Cohesion of Noble Metals (II)

By

V. K. Nikulin and M. B. Trzhaskovskaya

On the basis of the results obtained in previous papers a calculation is made of the cohesive energy, Wigner-Seitz sphere radius, and compressibility of Cu and Ag using the nonlocal model potential. An isotropic band effective mass is also estimated for these elements.

Auf Grund von Ergebnissen aus früheren Arbeiten wird eine Berechnung der Bindungsenergie, des Radius der Wigner-Seitz-Kugel und der Kompressibilität von Cu und Ag durchgeführt, wobei ein nichtlokales Modellpotential benutzt wird. Für diese Elemente wird auch eine isotrope effektive Masse berechnet.

1. Introduction

In earlier papers [1, 2] an expression for the ground-state energy of a metal has been obtained, using the many-particle theory and the pseudopotential method. The calculated values of the cohesion energy, lattice constant, and compressibility of Cu are in good agreement with experimental data. The calculations have been performed using a local model pseudopotential determined [3] from spectroscopic terms. However, in an attempt to make the same calculations for other noble metals we obtained unsatisfactory results.

The purpose of this paper is to make calculations by use of a non-local model potential and to discuss the basic assumptions made in [1, 2].

The expression for the ground-state energy E of a metal in approximation of linear response of the electron system to the perturbation produced by the ions has been obtained [1, 2] in the following form[1]):

$$E = E_0 - \frac{3}{20}\frac{N}{r_S} + \frac{1}{2} N \overline{V}_M + N U_0 + \frac{1}{2}\sum_{mn} [U_{mn} + W_{mn}] . \tag{1.1}$$

Here E_0 is an energy of the electron gas in the Hartree-Fock (H–F) approximation. The second term represents the potential interaction energy of uniformly distributed electrons in the field of point ions (N is the number of unit cells, r_S is the radius of the cell with volume Ω_0). $\overline{V}_M$ is the mean value of the deviation of the model potential from the potential of the point ion. According to the expression (3.7) of [2], $\overline{V}_M$ can be represented adequately as follows:

$$\overline{V}_M = \frac{1}{\Omega_0} [V_M(q) - \mathcal{V}(q)]_{q=0} . \tag{1.2}$$

$V_M(q)$ and $\mathcal{V}(q)$ are the Fourier transforms of the model potential and of the

[1]) Atomic units are used.

potential of a point charge, respectively;

$$V_M(q) = -\,\mathcal{V}(q)\,\varrho(q) = \int e^{-i\boldsymbol{q}\boldsymbol{r}}\, V_M(r)\, \mathrm{d}\boldsymbol{r}\,, \tag{1.3}$$

where $\varrho(q)$ is the form factor of the bare ion model potential. U_0 is the self-energy of an ion:

$$U_0 = \frac{1}{(2\,\pi)^2} \int \frac{\varrho^2(q)}{q^2} \left(\frac{1}{\varepsilon(q)} - 1\right) \mathrm{d}\boldsymbol{q}\,. \tag{1.4}$$

In this expression, $\varepsilon(q)$ is the dielectric susceptibility of the electron gas. U_{mn} represents the interaction between the screened ions with charge $Z = 1$:

$$U_{mn} = \frac{2}{(2\,\pi)^2} \int \left[1 + \varrho^2(q) \left(\frac{1}{\varepsilon(q)} - 1\right)\right] \frac{\exp\,[i\,\boldsymbol{q}\,(\boldsymbol{R}_m - \boldsymbol{R}_n)}{q^2}\, \mathrm{d}\boldsymbol{q}\,. \tag{1.5}$$

$\boldsymbol{R}_m$ is the position of the m-th ion; W_{mn} is the overlapping energy of the ion cores. In its calculation we followed Gombás (formulae (4.7) and (4.8) of [2]).

2. The Model Potential

The interaction of a conduction electron with the ion core is represented [4] by the operator

$$V_M(\boldsymbol{r}) = \sum_l V_l(r)\, \hat{P}_l\,, \tag{2.1}$$

$$V_l(r) = -\,A_l\, \theta\,(r_l - r) - \frac{Z}{r}\, \theta\,(r - r_l)\,. \tag{2.2}$$

$\hat{P}_l$ is the projection operator on the subspace of spherical harmonics, $\theta(r)$ is the Heaviside function, A_l and r_l are parameters. Besides being different for different l, A_l appear [4] to be dependent on the energy of the free-atomic terms. For arbitrary r_l, A_l must be extended by interpolation to energies in the band. However, one can [3] choose r_l so that A_l is equal for different terms with the same l. With this choice of r_l the model potential is practically independent of the energy in the band. In such a way the parameters A_0 and r_0 for noble metals were determined for $l = 0$ [3]. It was also suggested that all V_l were equal to V_0, so that we have a local potential independent of energy. This approximation is actually rather good for Cu, but not for Ag and Au. Using tables of A_l for $l = 1$ [4] the parameters A_1 and r_1 may be determined as suggested in [3]. The parameters for $l = 2$ cannot be determined in such a way since the terms correspond to the highly excited states of a valence electron. However, we will regard strictly only V_0 and V_1 putting $V_l = V_1$ for $l \geqq 2$. This approximation is rather good for Cu and Ag but not for Au as may be established by extrapolating A_2 using tables for $l = 2$ [4]. The parameters for Cu and Ag are given in Table 1.

Table 1

Parameters for model potential (The footnote [1]) in [2] is wrong. The r_0-value for Cu determined in [3] was correct)

	A_0 (at. un.)	r_0 (at. un.)	A_1 (at. un.)	r_1 (at. un.)
Cu	0.465	3.47	0.509	3.24
Ag	0.425	4.05	0.448	3.60

With the approximation $V_l = V_1$ for $l \geqq 2$ we have from (2.1)

$$V_{\mathrm{M}}(\boldsymbol{r}) = V_1(r) + [V_0(r) - V_1(r)]\, \hat{P}_0 \,. \tag{2.3}$$

The second term on the right-hand side represents the non-local part of the model potential. In the plane-wave representation the matrix element of the model potential between states with wave vectors $\boldsymbol{k}$ and $\boldsymbol{k} + \boldsymbol{q}$ can be written as follows:

$$\langle \boldsymbol{k} + \boldsymbol{q} \,|V_{\mathrm{M}}|\, \boldsymbol{k}\rangle = V_1(q) + V_{\mathrm{M}}^{n}\,(|\boldsymbol{k} + \boldsymbol{q}|, k)\,. \tag{2.4}$$

It depends not only on q (because of the non-locality of the potential) but on k and $|\boldsymbol{k} + \boldsymbol{q}|$. We can, after some computation, write the non-local part as $(Z = 1)$

$$V_{\mathrm{M}}^{n}(k, k) = -\frac{2\pi}{k^2}\Big[A_0\, r_0\, \big(1 - j_0(2\,k\,r_0)\big) - A_1\, r_1\, \big(1 - j_0(2\,k\,r_1)\big) - - \ln\frac{r_0}{r_1} + \mathrm{Ci}\,(2\,k\,r_0) - \mathrm{Ci}\,(2\,k\,r_1)\Big]\,, \tag{2.5}$$

$$\begin{aligned} V_{\mathrm{M}}^{n}\,(|\boldsymbol{k} + \boldsymbol{q}|, k) = -\frac{2\pi}{|\boldsymbol{k} + \boldsymbol{q}|\, k}\,\{&A_0\, r_0\big(j_0\,[(|\boldsymbol{k}+\boldsymbol{q}| - k)\, r_0] - j_0\,[(|\boldsymbol{k}+\boldsymbol{q}| + k)\, r_0]\big) - \\ &- A_1\, r_1\, \big(j_0\,[(|\boldsymbol{k} + \boldsymbol{q}| - k)\, r_1] - j_0\,[(|\boldsymbol{k} + \boldsymbol{q}| + k)\, r_1]\big) - \\ &- \mathrm{Ci}\,[(|\boldsymbol{k} + \boldsymbol{q}| - k)\, r_0] + \mathrm{Ci}\,[(|\boldsymbol{k} + \boldsymbol{q}| - k)\, r_1] + \\ &+ \mathrm{Ci}\,[(|\boldsymbol{k} + \boldsymbol{q}| + k)\, r_0] - \mathrm{Ci}\,[(|\boldsymbol{k} + \boldsymbol{q}| + k)\, r_1]\}\,, \end{aligned} \tag{2.6}$$

where

$$j_0(x) = \frac{\sin x}{x}\,, \qquad \mathrm{Ci}\,(x) = \int\limits_{\infty}^{x} \frac{\cos t}{t}\, \mathrm{d}t\,.$$

The calculation of the self-consistent screening of this non-local potential is rather difficult. It should be noted that this potential is more clumsy than that quoted in [5] on account of $r_0 \neq r_1$. Therefore we will consider the screening of the model potential treating it as a local potential. This approximation has proved [6] capable of giving accurate results over a wide range of q, except $q \approx k_{\mathrm{F}}$ (k_{F} is the wave number at the Fermi level). We shall put [5] now in (2.5) and (2.6) $k = k_{\mathrm{F}}$, for $q \leqq 2\,k_{\mathrm{F}}$ we shall put $|\boldsymbol{k} + \boldsymbol{q}| = k_{\mathrm{F}}$, for $q > 2\,k_{\mathrm{F}}$ we shall put $|\boldsymbol{k} + \boldsymbol{q}| = q - k_{\mathrm{F}}$.

In the approximation of the local screening the results of [1, 2] remain completely valid.

3. The Results of the Calculation for Cu and Ag

We now proceed to an evaluation of the ground-state energy. On taking into account (2.1), (2.4), and expression (4.2) of [2] $\overline{V}_{\mathrm{M}}$ may be written in the form[2])

$$\overline{V}_{\mathrm{M}} = -\frac{r_1^2}{r_{\mathrm{S}}^3}\left(A_1\, r_1 - \frac{3}{2}\right) + \frac{1}{\Omega_0}\, V_{\mathrm{M}}^{n}(k_{\mathrm{F}}, k_{\mathrm{F}})\,. \tag{3.1}$$

The form factor of the model potential can be written according to (1.3) and

[2]) In the expression (4.2) of [2] a minus sign has been omitted, but in the calculations it has been taken into account.

(2.4) as the sum of two parts:

$$\varrho(q) = \varrho_1(q) + \varrho^n(k_F, k_F, q) , \tag{3.2}$$

where ϱ_1 is given by expression (4.3) of [2] and ϱ^n may be obtained directly from (2.5) or (2.6).

Performing the numerical calculations of (1.4) and (1.5) we have used the Hubbard approximation with screened exchange for the static dielectric constant

$$\varepsilon(q) = g(q) + \big(1 - g(q)\big)\, \varepsilon^{\mathrm{RPA}}(q) , \tag{3.3}$$

where $\varepsilon^{\mathrm{RPA}}$ is the dielectric constant in the random-phase approximation. In contrast to [1, 2] we have used here $g(q)$ as suggested by Kleinman [7]:

$$g(q) = \frac{1}{4}\left(\frac{q^2}{q^2 + k_F^2 + \beta^2} + \frac{q^2}{k_F^2 + \beta^2}\right); \quad \beta^2 = \frac{4}{\pi}\, k_F . \tag{3.4}$$

We also note [7] that (3.3) is ε_{el} but not ε_{test} being used in [1, 2].

In calculating W we employed data obtained by a modified Hartree-Fock-Slater (H-F-S) method [8]; the calculations were performed with the help of an electronic computer. The magnitude of W obtained for Cu slightly differs

Table 2

Total repulsive energy per one ion

	Cu					Ag				
r_S (at. un.)	2.5	2.54	2.58	2.62	2.66	2.76	2.82	2.88	2.94	3.00
W (at.un.)	0.057	0.049	0.043	0.037	0.033	0.038	0.030	0.024	0.019	0.015

Table 3

Values of the self-energy U_0 and of the interaction energy of screened ions, U_{0m}, in atomic units ($U_{00} = U_0$, m is the number of the coordination sphere)

m	0	1	2	3	4	5	6
Cu ($r_S = 2.52$)	−0.23971	−0.02106	−0.00230	+0.00043	+0.00006	+0.00012	+0.00008
Ag ($r_S = 2.78$)	−0.21300	−0.01778	−0.00209	+0.00041	+0.00011	+0.00005	+0.00003

Table 4

Values of ground-state energy ε, cohesion energy ε_C, r_S, and compressibility of Cu and Ag (experimental values from [9])

	ε (at. un.)	ε_C (at. un.)	r_S (at. un.)	$\varkappa \times 10^{12}$ (cm²/dyn)	m^*/m	
Cu	−0.416	0.132	2.51	0.55	1.05	theoret.
	−0.413	0.129	2.66	0.72		exper.
Ag	−0.389	0.110	2.77	0.72	0.96	theoret.
	−0.388	0.109	3.00	1.00		exper.

from that given in [2]. The total repulsion energy of ions connected with the overlap of d-shells is given in Table 2.

The values of the self-energy U_0 and of the interaction energies with its neighbours in six coordination spheres for a f.c.c. lattice are given in Table 3.

The calculations of $\varepsilon = E/N$, r_S, and compressibility $\varkappa$ were performed analogously as in [2] and the final results are listed in Table 4.

An isotropic-band effective mass m^*, as given by first-order perturbation theory, may be calculated [5] by use of the non-local part of the model potential

$$\frac{\hbar k}{m^*} = \frac{\hbar k}{m} + \frac{1}{\hbar \Omega_0} \frac{\mathrm{d} V_M^n(k, k)}{\mathrm{d}k} . \tag{3.5}$$

In our case of an energy independent model potential we have[1]), according to (2.5) and (3.5),

$$\frac{m}{m^*} = 1 + \frac{3}{k_F^4 r_S^3} \Big[A_0 r_0 \Big(1 - \frac{3}{2} j_0 (2 k_F r_0)\Big) - A_1 r_1 \Big(1 - \frac{3}{2} j_0 (2 k_F r_1)\Big) - \ln \frac{r_0}{r_1} + + \mathrm{Ci} (2 k_F r_0) - \mathrm{Ci} (2 k_F r_1) + \frac{1}{2} (A_0 r_0 - 1) \cos 2 k_F r_0 - \frac{1}{2} (A_1 r_1 - 1) \cos 2 k_F r_1 \Big] . \tag{3.6}$$

The evaluated values of m^*/m are listed in Table 4. However, they are not used in calculating the dielectric constant.

4. Conclusion

We can see from Table 4 that the theoretical results are in good agreement with the observed values. The somewhat larger relative error of the compressibility is caused by its strong dependence on the curvature of $W(r_S)$. For r_S-values close to the experimental ones very good results may be obtained [2].

We will now discuss the basic assumptions made in [1, 2]. They have been considered in part in [1, 2].

1. E_0 has been taken in H-F approximation, as the correlation hole contributes to V_M (the correlation hole in the free atom and in the metal are supposed to be the same). This assumption is satisfactory for noble metals having large and even overlapping ion cores. According to our preliminary calculations for alkali metals good results may be obtained when the correlation energy E_{corr} would be taken into account in the form $(1 - \alpha) E_{corr}$ where $\alpha = V_i/\Omega_0$ and V_i is the volume of the ion core.

2. In calculating of the screened form factor for the model potential a depletion hole [10] has not been taken into account. We believe that this approximation is quite suitable as far as A_l is practically independent of the energy. Also in the approximation of local screening it is more consistent not to take into account this correction at all since in this case only the correct long-wavelength limit [10] of the screened form factor may be obtained.

3. Values of the overlapping energy W depend strongly on the choice of the functions of the d-electrons. We believe [2] that the employment of those obtained [8] by the H-F-S method for the free atom is more justified than of those obtained by the H-F method for the free ion. The former have more correct asymptotic behaviour at large distance inasmuch as the corresponding energy levels are closer to the experimental values. From this point of view the use of the semi-empirical [11, 4] wave functions calculated by adjusting

the one-electron energy levels to the corresponding experimental values would be even more justified. This would result[3]) in an increase of W, i.e. in improvement of r_s and of compressibility.

Acknowledgements

The authors wish to thank A. I. Gubanov and I. V. Abarenkov for valuable discussions.

References

[1] V. K. NIKULIN, phys. stat. sol. **16**, K125 (1966).
[2] A. I. GUBANOV and V. K. NIKULIN, phys. stat. sol. **17**, 815 (1966).
[3] V. K. NIKULIN, Fiz. tverd. Tela **7**, 2708 (1965); Soviet Phys. — Solid State **7**, 2189 (1966).
[4] I. V. ABARENKOV and V. HEINE, Phil. Mag. **12**, 529 (1965).
[5] A. O. E. ANIMALU and V. HEINE, Phil. Mag. **12**, 1249 (1965).
[6] A. O. E. ANIMALU, Phil. Mag. **11**, 379 (1965).
[7] L. KLEINMAN, Phys. Rev. **160**, 585 (1967).
[8] F. HERMAN and S. SKILLMAN, Atomic Structure Calculations, Engelwood Cliffs, New Jersey 1963.
[9] G. LEIBFRIED, Gittertheorie der mechanischen und thermischen Eigenschaften der Kristalle, in: Hdb. Phys. Bd. 7, Teil 1, Springer-Verlag 1955.
[10] R. W. SHAW JR. and W. A. HARRISON, Phys. Rev. **163**, 604 (1967).
[11] M. I. PETRASHEN and I. V. ABARENKOV, Vestnik LGU **5**, 141 (1954).

(Received May 9, 1968)

[3]) We are indebted to I. V. Abarenkov for this comment.

phys. stat. sol. 28, 807 (1968)

Subject classification: 13.2; 13.1; 22; 22.5.2

Donetsk Physico-Technical Institute and Institute of Semiconductors, Ukrainian Academy of Sciences, Kiev

Exciton Theory in Crystals with Complicated Band Structure with Electron–Phonon Interaction

By

T. A. KUDYKINA, K. B. TOLPYGO, and S. M. ZUBKOVA

The theory of the exciton is developed on the basis of the many-electron model. It is supposed that the excited electron can pass from any occupied state of one atom to a free state of this atom or of another one. Two limiting cases are considered: the large-radius exciton in many-valley semiconductors with the value of the many-valley splitting of its level being estimated, and the small-radius exciton in alkali halide crystals. In the last case the energy of the exciton, composed of Känzig's autolocalized hole and an electron, moving over ten nearest neighbours, is calculated.

Теория экситонов строится на основе многоэлектронного рассмотрения кристалла и предположения, что в результате возбуждения электрон может перейти с любого занятого состояния на любое свободное этого или другого атома. Рассмотрены предельные случаи экситонов большого радиуса в многодолинных полупроводниках и оценена величина многодолинного расщепления его уровня, а также случай экситона малого радиуса в щелочно-галоидных кристаллах. В последнем случае рассчитана знергия экситона, образованного автолокализованной по Кенцигу дыркой и электроном, движущимся по 10-ти ближайшим соседям.

1. Introduction

For the overwhelming majority of semiconductors the existence of complicated bands is a rule (bands with some equivalent minima and with non-isotropic masses or being degenerated in the minimum) and simple parabolic ones are rather an exception.

For complicated bands the exciton theory is constructed according to Dresselhaus [1] on the base of the equation for the smoothed function:

$$\left[\mathcal{E}_1(-i\,\nabla_1) + \mathcal{E}_2(-i\,\nabla_2) - \frac{e^2}{\varepsilon_\infty\, r_{12}}\right] \Psi(r_1, r_2) = E\,\Psi(r_1, r_2) \tag{1}$$

which is obtained if the wave vectors $\boldsymbol{k}_1$ and $\boldsymbol{k}_2$ are replaced by the differentiation operators over their coordinates $-i\,\nabla_1$ and $-i\,\nabla_2$ in the expressions for the energy bands of electrons and holes, $\mathcal{E}_1(\boldsymbol{k}_1)$ and $\mathcal{E}_2(\boldsymbol{k}_2)$. Additionally, there are introduced electron–phonon interaction terms in equation (1), both for electrons and holes and, besides, in long wave approximation. It should be noted, however, that equation (1) requires itself a foundation and as it will be seen later this foundation is not always possible. Besides, the long wave approximation can hardly be considered as sufficient because the neutral exciton interaction with optical phonons trends towards zero for $\lambda \to \infty$; the deformation potential of type γ div $\boldsymbol{u}$, where $\boldsymbol{u}$ is the displacement vector, is increasing when the wavelength decreases already due to its definition.

Below, the method of theoretical exciton consideration is given in the case of complicated band structure and for taking into account electron–phonon interaction. Estimations are presented for the energies of the lowest states.

2. General Scheme of Exciton Consideration

Suppose that the semiconductor band structure and band wave functions are known. Obviously, valence band functions may be approximately considered as formed by isolated atom (ion) functions suitably combined to obtain valence bonds [2].

Let ψ_s^l be the wave function of normal sites having a complete set of electrons and φ_{si}^l be the function with an electron removed from the i-th state. Later on, for shortening the sort of the site and the released state number will be denoted by $h = 1, 2, \ldots, n$. Similarly, the function of the $\binom{l}{s}$-th atom with an excess electron in the state j is denoted by $\xi_{sj}^l \to \xi_g^l$ and the function of the excited $\binom{l}{s}$-th atom by $\chi_{sij}^l \to \chi_{gh}^l$; the latter being received with the help of replacing the i-th line by the j-th one in the Fock determinant describing the state ψ_s^l.

It is evident that the exact exciton consideration is possible only on the base of the multi-electron crystal function because the introduction of the high frequency dielectric constant ε_∞ is connected with electron interaction and correlation effects. Now we define the auxiliary function of all crystal electrons $\Psi_{hg}^{ll'}$ as the state where φ_h^l is instead of ψ_s^l in the $\binom{l}{s}$-th site and $\xi_g^{l'}$ is in the $\binom{l'}{s'}$-th one. In the case, when $l = l'$ and $s = s'$, we suppose that $\psi_s^l \to \chi_{gh}^l$ in the $\binom{l}{s}$-th site.

Suppose that all functions ψ_s^l, φ_h^l, and ξ_g^l may slightly vary because of electron and hole fields, and their exact form is defined by the minimum condition of the Hamiltonian diagonal matrix element $H_{hghg}^{ll'll'}$.

Considering this change as a polarization of all sites, using dipol moments $\boldsymbol{P}_s^l$ and earlier results [3] we have

$$H_{hghg}^{ll'll'} = \sum_{s''l''} \frac{(\boldsymbol{P}_{s''}^{l''})^2}{2\,\alpha_{s''}} - \frac{1}{2} \sum_{s''l''} \boldsymbol{P}_{s''}^{l''} \cdot \boldsymbol{E}(\boldsymbol{R}_{s''}^{l''}) - e \sum_{s''l''} \frac{\boldsymbol{P}_{s''}^{l''} \cdot \boldsymbol{R}_{s''s}^{l''l}}{|\boldsymbol{R}_{s''s}^{l''l}|^3} +$$

$$+ e \sum_{s''l''} \frac{\boldsymbol{P}_{s''}^{l''} \cdot \boldsymbol{R}_{s''s'}^{l''l'}}{|\boldsymbol{R}_{s''s'}^{l''l'}|^3} - \frac{e^2}{|\boldsymbol{R}_{ss'}^{ll'}|} + e\,V_{\mathrm{m}}(s) - e\,V_{\mathrm{m}}(s') + I_{hg}^{ll'} + \Delta E_{si} + \Delta E_{s'j} + \text{const}\,. \quad (2)$$

Here, $\boldsymbol{E}(\boldsymbol{R}_{s''}^{l''})$ is the field in the $\binom{l''}{s''}$-th site which is formed by all other dipols; $\alpha_{s''}$ is the polarizability of the $\binom{l''}{s''}$-th site. The third and fourth terms are hole and electron interaction energies with all dipoles $\boldsymbol{P}_{s''}^{l''}$, respectively, $I_{hg}^{ll'}$ is the exchange energy change of all crystal electrons due to electron transitions $\binom{l}{s\,i} \to \binom{l'}{s'\,j}$ minus the deviation of the Coulomb integral between these states from $-\dfrac{e^2}{|\boldsymbol{R}_{ss'}^{ll'}|}$ [4]; ΔE_{si} is the energy necessary for releasing an electron from the $\binom{l}{s}$-th site; $\Delta E_{s'j}$ is the energy necessary for connecting an electron with the $\binom{l'}{s'}$-th site, $V_m(s)$ is the Madelung potential depending only on the site sort s; const is the ideal crystal energy. In the case of $s = s'$ and $l = l'$ dipoles do not appear and it is $H^{l\,l\,l\,l}_{h\,g\,h\,g} = \Delta_{sij} +$ const, where Δ_{sij} is the s-th site excitation energy.

Minimization of equation (2) over all $\boldsymbol{P}^{l''}_{s''}$ allows to calculate all of them with the dependence on the site $\binom{l}{s}$ and $\binom{l'}{s'}$ position and to find both the corrections to the electron and hole Coulomb energy and the interaction energy sum of electrons and holes with dipoles induced by them. These latter quantities as well as the Madelung potentials may be summed with Δ_{si} and $\Delta_{s'j}$. Shortly,

$$H^{l\,l'\,l\,l'}_{h\,g\,h\,g} = -\frac{e^2}{\left|\boldsymbol{R}^{l\,l'}_{s\,s'}\right|\,\varepsilon_\infty\left(\boldsymbol{R}^{l\,l'}_{s\,s'}\right)} + \Delta E_{si} + \Delta E_{s'j} + \text{const}\,. \tag{3}$$

The resulting change in the Coulomb and exchange energy can be described by means of an effective dielectric constant $\varepsilon_\infty(r_{12})$ depending on the distance between charges [5, 6]. The nuclei are shifted from their equilibrium positions by quantities $\boldsymbol{u}^l_s$; to the expression (2) should be added the energy term for the ideal lattice with shifted nuclei and it is necessary to replace the shell dipoles $\boldsymbol{P}^{l''}_{s''}$ interacting with the electron and hole by the complete dipoles $\boldsymbol{\Pi}^{l''}_{s''} = \boldsymbol{P}^{l''}_{s''} + e_{s''}\cdot\boldsymbol{u}^{l''}_{s''}$, where $e_{s''}$ is the charge of the $\binom{l''}{s''}$-th site. In this case, as shown in [3] it is convenient to introduce lattice normal coordinates $q_{\alpha\varkappa}$ and to subdivide the electron dipoles $\boldsymbol{P}^{l''}_{s''}$ into parts $\widetilde{\boldsymbol{P}}^{l''}_{s''}$ determined only by nuclei shifts and into parts $\boldsymbol{R}^{l''}_{s''}$ determined by the electron and hole field. For the latter the minimization procedure mentioned above is carried out and as result the summands should be added to the right-hand part of formula (3):

$$\frac{N\,e^2}{2\,a}\sum_{\alpha\varkappa}\Omega^2_{\alpha\varkappa}\,q_{\alpha\varkappa}\,q_{\alpha-\varkappa} + \frac{2\,i\,e^2}{a}\sum_{\alpha\varkappa}Q_{\alpha\varkappa}\,q_{\alpha-\varkappa}\,, \tag{4}$$

where $\Omega_{\alpha\varkappa}$ are the dimensionless eigenfrequencies of the lattice, a is the lattice constant.

It is

$$Q_{\alpha\varkappa} = \sum_{s''}\left(\boldsymbol{F}^{\varkappa}_{s''}\cdot\boldsymbol{p}^{\alpha-\varkappa}_{s''} + \boldsymbol{G}^{\varkappa}_{s''}\cdot\boldsymbol{P}^{\alpha-\varkappa}_{s''}\right); \tag{5}$$

$$(\boldsymbol{p}^{l''}_{s''} = e_{s''}\,\boldsymbol{u}^{l''}_{s''})$$

are generalized forces acting on the coordinates $q_{\alpha\varkappa}$ from the site of the electron and hole; $\boldsymbol{p}^{\alpha-\varkappa}_{s''}$ and $\boldsymbol{P}^{\alpha-\varkappa}_{s''}$ are orthonormalized amplitudes of the eigenoscillations and $\boldsymbol{F}^{\varkappa}_{s''}$ and $\boldsymbol{G}^{\varkappa}_{s''}$ are Fourier components of electron and hole fields acting on the dipoles $\boldsymbol{p}_{s''}$ and $\boldsymbol{P}_{s''}$ [3].

After defining in such a way the basis function system $\Psi^{ll'}_{hg}$ we study the crystal state represented as expansion taking into account the translational symmetry, being reduced to

$$\Psi_{\boldsymbol{K}} = \frac{1}{\sqrt{N}}\sum_{hgll'} e^{i\boldsymbol{K}\cdot\boldsymbol{R}^l_s}\,C_{hg}(\boldsymbol{R}^{ll'}_{ss'})\,\Psi^{ll'}_{hg}\,, \tag{6}$$

where $\boldsymbol{K}$ is a the exciton wave vector.

Thus, the intrinsic motion is separated from the translation one and the coefficients C_{hg} will be determined from the variational principle:

$$\min \bar{H} = \min \sum_{\substack{hg\bar{h}\bar{g} \\ LL'\bar{L}'}} e^{i\boldsymbol{K}\cdot\boldsymbol{R}^{L0}_{s\bar{s}}} C^{*}_{\bar{h}\bar{g}}\left(\boldsymbol{R}^{0\bar{L}'}_{\bar{s}\bar{s}'}\right) H^{0\bar{L}'LL'}_{\bar{h}\bar{g}hg} C_{hg}(\boldsymbol{R}^{LL'}_{ss'}) \,, \tag{7}$$

thereby

$$\sum_{\bar{h}\bar{g}\bar{L}'} \left|C_{\bar{h}\bar{g}}\left(\boldsymbol{R}^{0\bar{L}'}_{\bar{s}\bar{s}'}\right)\right|^2 = 1 \,, \tag{8}$$

where it is denoted $l - \bar{l} = L$, $l' - \bar{l} = L'$, $\bar{l}' - \bar{l} = \bar{L}'$, and the summation over $\bar{l}$ is performed taken into account the translation symmetry.

3. Small Radius Excitons

For small radius excitons, Dexter's excitation model will be the limiting case [7], when only $C_{hg}(0)$ and some coefficients $C_{hg}(\boldsymbol{R})$ for the nearest neighbours with $R^{ll'}_{ss'} \approx a$ will be nonzero. Dyckman [8] has investigated an exciton model in KCl according to which a hole is located in any state of the 3p -shell of the Cl^- ion and an electron is moving with equal probability over the 6 nearest K^+ ions. Here, the electron–phonon interaction calculation over all oscillation branches by means of lattice dynamics theory [9] has given the correct value of the exciton absorption band width.

On the basis of the results of [10, 11] and Känzig's work on the hole autolocalization [12], Tolpygo set forward another small radius exciton model [13] according to which the hole is moving over the two nearest anion sites placed in the direction [110], and the electron over ten cations nearest to them and also over these anions. Such a hole delocalization must lead to a considerable exciton energy reduction.

The calculation was carried out for the case when the atom nuclei are not shifted and the electron can move only over cations at $\boldsymbol{K} = 0$. In this case, there are 6 possible hole motion states and 10 states for the electron motion over the 3s-shells of 10 cations. This gives $6 \times 10 = 60$ functions $\Psi^{ll'}_{hg}$. For the exciton state where the phototransition under the action of the field directed along the Z-axis (orthogonal to the plane of the two anions along which the hole is moving) is allowed there are 7 basic functions $\Psi^{ll'}_{hg}$. The matrix elements $H^{0\bar{L}'LL'}_{\bar{h}\bar{g}hg}$ were determined for valence band functions from [11] and the electron band ones from [14].

The lowest exciton energy in NaCl relative to the bottom of the conduction band [11, 14] is proved to be -1.85 eV. The same calculation has given for Dyckman's model [8] the value -1.48 eV. Thus the model suggested in [13] is in fact energetically more advantageous and, therefore, it must be closer to the true case due to the variational principle, while the model [7, 8] must be rejected.

4. Large Radius Excitons

For large radius excitons it is advisable to perform the functional minimization $\left(\text{by expansion of } C_{\bar{h}\bar{g}}\left(\boldsymbol{R}^{\bar{l}\bar{l}'}_{\bar{s}\bar{s}'}\right)\right)$ over band state coefficients. Here we receive equations which generalize the equation of the effective mass method.

There was assumed that the matrix elements for one hole transition $H^{0L'LL'}_{\bar{h}g\,hg} \to H^{0L}_{\bar{h}h}$ slightly depend on the electron position and its state, and the same picture is found for the electron transition $H^{0\bar{L}'0L'}_{h\bar{g}hg} \to H^{\bar{L}'L'}_{\bar{g}g}$. Simultaneous electron and hole transitions are neglected as well as the difference in longitudinal and transverse exciton energies [15, 16]. It is supposed, further, that the solution of the system of equations describing hole and electron band states is known.

$$\left.\begin{aligned} &\sum_{hL} H^{0L}_{\bar{h}h}\, e^{i\varkappa\cdot R^{L0}_{s\bar{s}}}\, b^{\alpha}_{h}(\varkappa) = E_{\alpha}(\varkappa)\, b^{\alpha}_{\bar{h}}(\varkappa)\,; \\ &\sum_{gL'} H^{\bar{L}'L'}_{\bar{g}g}\, e^{i\varkappa\cdot R^{L'\bar{L}'}_{s'\bar{s}'}}\, b^{\beta}_{g}(\varkappa) = E_{\beta}(\varkappa)\, b^{\beta}_{\bar{g}}(\varkappa), \end{aligned}\right\} \tag{9}$$

where α, β are the electron and hole band numbers. It is assumed that the diagonal elements $H^{0\bar{L}'0\bar{L}'}_{\bar{h}\bar{g}\bar{h}\bar{g}} = H^{0\bar{L}'}_{\bar{h}\bar{g}}$ depend only on the distance between the electron and hole and (additively) on the hole formation energy in the h-th state and that of the electron in the g-th state. Then the solution of the secular equation corresponding to the minimum of equation(7) under the additional condition (8) is found in the form

$$C_{hg}(R^{ll'}_{ss'}) = \sum_{\varkappa} b_{hg}(\varkappa)\, e^{i\varkappa\cdot R^{ll'}_{ss'}} = \sum_{\varkappa}\sum_{\alpha\beta} D_{\alpha\beta}(\varkappa)\, b^{\alpha}_{h}(K+\varkappa)\, b^{\beta}_{g}(-\varkappa)\, e^{i\varkappa\cdot R^{ll'}_{ss'}}\,. \tag{10}$$

Substitution of equation (10) into the secular equation above (which is not necessary to write down) with the account of equation (9) (after summation over $\bar{L}'$) multiplication by $\overset{*}{b}{}^{\alpha'}_{\bar{h}}(K+\varkappa')\cdot \overset{*}{b}{}^{\beta'}_{\bar{g}}(-\varkappa')$ and summation over $\bar{h}\,\bar{g}$ and taking into account orthogonality coefficients $b^{\cdot}_{\cdot}$ for $\alpha \neq \alpha'$ and $\beta \neq \beta'$ gives

$$D_{\alpha'\beta'}(\varkappa')\,[E_{\alpha'}(K+\varkappa') + E_{\beta'}(-\varkappa')] + \frac{1}{N}\sum_{\alpha\beta\varkappa} H^{\varkappa'\varkappa}_{\alpha'\beta'\alpha\beta}\, D_{\alpha\beta}(\varkappa) = E\, D_{\alpha'\beta'}(\varkappa')\,, \tag{11}$$

where

$$H^{\varkappa'\varkappa}_{\alpha'\beta'\alpha\beta} = \sum_{\bar{h}\bar{g}\bar{L}'} \overset{*}{b}{}^{\alpha'}_{\bar{h}}(K+\varkappa')\, \overset{*}{b}{}^{\beta'}_{\bar{g}}(-\varkappa')\, H^{0\bar{L}'}_{\bar{h}\bar{g}}\, e^{i(\varkappa-\varkappa')\cdot R^{0\bar{L}'}_{\bar{s}\bar{s}'}}\, b^{\alpha}_{\bar{h}}(K+\varkappa)\, b^{\beta}_{\bar{g}}(-\varkappa)\,. \tag{12}$$

We can interpret equation (14) as obtained as the result of a Fourier transformation for a two -quasiparticle movement with the dispersion laws $E_{\alpha}(\varkappa)$ and $E_{\beta}(\varkappa)$, and the interaction operator $\hat{V}(r_1, r_2)$, the Fourier coefficient matrix elements of which between the band states α, β and α', β' give $H^{\varkappa'\varkappa}_{\alpha'\beta'\alpha\beta}$. In the common case, however, this operator is not diagonal what corresponds to non-local electron and hole interaction because the matrix element $H^{\varkappa'\varkappa}_{\alpha'\beta'\alpha\beta}$ (12) depends not only on the difference $\varkappa - \varkappa'$ but on $\varkappa$ and $\varkappa'$ separately.

Let us consider now sufficiently large radius excitons. Then, at fixed K the biggest magnitudes will have those coefficients $D_{\alpha\beta}(\varkappa')$ which correspond to the neighbourhood $\varkappa \approx \varkappa_0$ of the energy minimum $E_{\alpha}(K+\varkappa_0) + E_{\beta}(-\varkappa_0)$ and particularly for those bands α, β where this minimum is the least (these bands will be called actual). If in the direct neighbourhood of the point $K + \varkappa_0$ and $-\varkappa_0$ overlapping or sticking of the electron or hole bands does not take place, i.e. the minimum $\varkappa_0$ is "sufficiently isolated" then in (12) we may approximately change $\varkappa$ by $\varkappa_0$ and $\varkappa'$ by $\varkappa_0$ in all factors except the exponential one. It is obvious that at the actual band formation from φ^{l}_{h} or $\xi^{l'}_{h}$ states according to (9) the main contribution will be given by those states h (or g) for which the hole (or excessive electron) formation energy on the atom $\binom{l}{s}$ $\left(\text{or } \binom{l'}{s'}\right)$ is the least. Generally, such states are degenerated (for example p-bands), but different groups of states differ in energies strongly enough. Then within the degenerated group of states for $R_{ss'}^{ll'} \gg a$ one can neglect the $H_{\bar{h}\bar{g}}^{0\bar{L}'}$ dependence. on the $\bar{h}, \bar{g}$ indexes and perform the summation over them in equation (12). Due to the orthonormality of the b coefficients we have

$$H_{\alpha'\beta'\alpha\beta}^{\varkappa'\varkappa} \to \delta_{\alpha\alpha'}\,\delta_{\beta\beta'} \sum_{\bar{L}'} e^{i(\varkappa-\varkappa')\cdot R_{ss'}^{ll'}}\, H_{\text{act}}^{0\bar{L}'} \,. \tag{13}$$

Sign "act" in $H_{\text{act}}^{0\bar{L}'}$ means that $H_{\bar{h}\bar{g}}^{0\bar{L}'}$ is taken just for the actual indexes $\bar{h}\,\bar{g}$. Then (13) seems to depend only on the difference $\varkappa - \varkappa'$ and must be retained only for actual bands α, β. Then, performing the reverse Fourier transformation of equation (11) we obtained the equation of the effective mass method

$$[E_\alpha(-i\,\nabla_1) + E_\beta(-i\,\nabla_2) + V_{12}(r_{12})]\,\Psi(r_1, r_2) = E\,\Psi(r_1, r_2)\,, \tag{14}$$

where V_{12} is given by the right-hand part of equation (11) under change of the discrete variable $\boldsymbol{R}_{ss'}^{ll'}$ by the continuous one.

However, all that construction is not valid if near the point $\boldsymbol{K} + \varkappa_0$, $-\varkappa_0$ besides actual modes there are other modes of hole and electron bands being close in energy. For example, in the point $\boldsymbol{K} = 0$ for many semiconductors a degeneration of the hole bands takes place. As a result $b_h^\alpha(\varkappa)$ depends essentially on the direction $\varkappa$ even at $\varkappa \to 0$ and the change procedure of setting $\varkappa$ and $\varkappa'$ by $\varkappa_0$ in equation (12) becomes incorrect.

There was considered a more simple problem concerning many-valley splitting of the exciton spectrum. For example, there is one actual electron band with the minimum at $\boldsymbol{K} = 0$ and one hole band with $p > 1$ equivalent minima in the point $\varkappa_r^0$, $r = 1, 2, \dots p$, located symmetrically relative to the origin of coordinates. Then omitting band indexes α, β we rewrite equation (11) in a more simple form:

$$[E_{\text{h}}(\boldsymbol{K} + \varkappa') + E_{\text{e}}(-\varkappa')]\,D(\varkappa') + \frac{1}{N}\sum_{\varkappa} H^{\varkappa'\varkappa}\,D(\varkappa) = E\,D(\varkappa')\,. \tag{15}$$

The lowest states will be characterized by a set of band functions with $\varkappa$ close to the extremum points $\varkappa = \varkappa_r$ determined from equation

$$\nabla_\varkappa E_{\text{h}}(\boldsymbol{K} + \varkappa) - \nabla_\varkappa E_{\text{e}}(-\varkappa) = 0 \tag{16}$$

(they can be found from a plot).

For the vector $\boldsymbol{K}$ of general position the minimum $\varkappa_r(\boldsymbol{K})$ is a single one and for $\boldsymbol{K}$ lying in one of the symmetry directions (or symmetry plane) there are some minima $p' \leqq p$ (for $\boldsymbol{K} = 0$ it will be $\varkappa_r^0$).

Expanding $E_{\text{h}} + E_{\text{e}}$ in a series with respect to $\varkappa - \varkappa_r(\boldsymbol{K})$ up to square terms inclusively, we get p' equation systems of the effective mass method in the neighbourhood of each minimum and p' equivalent solutions turning into each other under symmetry transformations and corresponding to equal energy E. For $\varkappa'$ being close to some minimum $\varkappa_r(\boldsymbol{K})$, $r = 1, 2, \dots, p'$, non-diagonal elements $H^{\varkappa'\varkappa}$ will be large, when $\varkappa$ is located near the same minimum, and they will be small when $\varkappa$ is close to any other minimum $\varkappa_t(\boldsymbol{K})$, $t \neq r$. (In the case when $\varkappa$ is located between the minima $\varkappa_r$ and $\varkappa_t$ the coefficients $D(\varkappa)$ appear to be small.) Thus, in zero approximation relatively to $H^{\varkappa_r \varkappa_t}$ the system (15) splits into p' independent systems, each of them being equivalent to the equation of the effective mass method with one extremum $\varkappa_r(K)$. Supposing all of them to be solved we shall get p' systems of the $D_r^0(\varkappa)$ coefficients and p' degenerated (in zero approximation) exciton functions. For $\varkappa' \approx \varkappa_r$ we obtain in first approximation

$$-\frac{\hbar}{2}\sum_{xy}\frac{1}{\mu_{xy}^{(r)}}(\varkappa_x'-\varkappa_{rx})(\varkappa_y'-\varkappa_{ry})\,D^{(1)}(\varkappa')+\frac{1}{N}\sum_{\varkappa\approx\varkappa'\approx\varkappa_r}\sum H_0^{\varkappa'\varkappa}\,D^{(1)}(\varkappa')-$$

$$-\mathscr{E}_0\,D^{(1)}(\varkappa')=\mathscr{E}'\sum_{t=1}C_t\,D_t^0(\varkappa')-\frac{1}{N}\sum_{t=r}'\sum_{\varkappa\approx\varkappa_t}H_{(1)}^{\varkappa'\varkappa}\,C_t\,D_t^0(\varkappa)\,. \tag{17}$$

The orthogonality condition of the right-hand part of this equation with respect to the solution of the corresponding homogeneous equation gives

$$\mathscr{E}'\,C_r\sum_{\varkappa'\approx\varkappa_r}|D_r^0(\varkappa')|^2-\frac{1}{N}\sum_{t\neq r}'\sum_{\varkappa\approx\varkappa_r}\sum_{\varkappa'\approx\varkappa_t}C_t\,\overset{*}{D_r^0}(\varkappa')\,H_{(1)}^{\varkappa'\varkappa}\,D_t^0(\varkappa)=0\,. \tag{18}$$

We consider $\sum_{\varkappa'}|D_r^0(\varkappa')|^2=1$; then the corrections to the energy $\mathscr{E}'$ are obtained from the secular equation having only non-diagonal elements:

$$W_{rt}=\frac{1}{N}\sum_{\varkappa'\approx\varkappa_r}\sum_{\varkappa\approx\varkappa_t}\overset{*}{D_r^0}(\varkappa')\,H_{(1)}^{\varkappa'\varkappa}\,D_t^0(\varkappa)\,. \tag{19}$$

It is possible to introduce smoothed functions of the effective mass method $\varphi_r(\boldsymbol{r})$, the Fourier expansion coefficients of which are $D_r^0(\varkappa)$. Then, assuming, the radii of all $\varphi_r(\boldsymbol{r})$ states to be many times greater than the lattice constant, we have approximately

$$W_{rt}=\Delta\sum_{ghL'}b_h^*(\boldsymbol{K}+\varkappa_r)\,b_h(\boldsymbol{K}+\varkappa_t)\,b_g^*(-\varkappa_r)\,b_g(-\varkappa_t)\times$$

$$\times\,\varphi_r^*(\boldsymbol{R}_{ss'}^{0L'})\,\varphi_t(\boldsymbol{R}_{ss'}^{0L'})\,H_{hg}^{0L'}\,e^{-i(\varkappa_r-\varkappa_t)\cdot\boldsymbol{R}_{ss'}^{0L'}}\,, \tag{20}$$

where Δ is the elementary cell volume.

If $\varkappa_r-\varkappa_t$ has the order of the reverse lattice constant $1/a$, the main contribution in the sum over L' comes from terms with $R_{ss'}^{0L'}\approx a$ and the argument $\boldsymbol{r}$ in $\varphi_r^*(\boldsymbol{r})$ and $\varphi_t(\boldsymbol{r})$ may be substituted by zero.

Supposing $H_{hg}^{0L'}=-\dfrac{e^2}{\varepsilon_\infty R_{ss'}^{0L'}}$, the summation over L' can be carried out according to the Ewald method and the sum over h and g from the product of four coefficients is considered having the order of unity. That gives a rough estimation of the order of the value of coefficient $W_{rt}\approx\dfrac{\Delta}{r_{\rm ex}^3}\dfrac{e^2}{\varepsilon_\infty a}\approx\mathscr{E}_{\rm ex}\left(\dfrac{a}{r_{\rm ex}}\right)^2$, where $r_{\rm ex}$ is the average radius of the exciton and $\mathscr{E}_{\rm ex}$ is its binding energy.

It was performed the calculation of the value W_{rt} for $K=0$ using the b-coefficients corresponding to the band structure of NaCl under different assumptions on $\varkappa_r$ minima arrangements. It was assumed that the hole may be located in one of the 3p anion states and the electron is in the state of the type 3s-Na^+ and 4s-Cl^-. The results are given in [2].

In the particular case of 6 equivalent extrema situated on the coordinate axes the symmetry considerations give the singlet (all C_t are the same), the doublet: $C_1=C_2=-C_3=-C_4$, $C_5=C_6=0$, $C_1=C_2=-C_5=-C_6$,

$C_3 = C_4 = 0$; and triplet: $C_1 = -C_2$, $C_3 = -C_4$, $C_5 = -C_6$ (the others are zero).

The correction values of the first order in the case of $\varkappa_r = 0.6\,\pi/a$ are for $r_{ex} = 3\,a$: $E^{(')}_{singl} = 0.0391$, $E^{(')}_{doubl} = -0.0467$, $E^{(')}_{trip} = 0.0181$; for $r_{ex} = 10\,a$: $E^{(')}_{singl} = 0.0007$, $E^{(')}_{doubl} = -0.0011$, $E^{(')}_{trip} = 0.0005$; and in the case $\varkappa_r = 0.82\,\pi/a$ for $r_{ex} = 3\,a$: $E^{(')}_{singl} = -0.1250$, $E^{(')}_{doubl} = -0.2600$, $E^{(')}_{trip} = -0.2150$; for $r_{ex} = 10\,a$: $E^{(')}_{sing} = -0.0034$, $E^{(')}_{doubl} = -0.0070$, $E^{(')}_{trip} = -0.0058$ (in eV).

Thus, the exciton spectrum alterations due to many-valley splitting play a substantial role in distinguishing this spectrum from the hydrogen-like one, and even exceeds the shift due to electron–phonon interaction [17].

References

[1] G. Dresselhaus, J. Phys. Chem. Solids **1**, 14 (1956).
[2] S. M. Zubkova and K. B. Tolpygo, Ukr. fiz. Zh. **12**, 1489 (1967); **12**, 1839 (1967).
[3] K. B. Tolpygo, Ukr. fiz. Zh. **2**, 242 (1957).
[4] S. A. Moskalenko and K. B. Tolpygo, Zh. eksper. teor. Fiz. **36**, 149 (1959).
[5] K. B. Tolpygo, Uspekhi fiz. Nauk **74**, 269 (1961).
T. I. Kucher, Zh. eksper. teor. Fiz. **30**, 724 (1956).
[6] K. B. Tolpygo and T. I. Kucher, Fiz. Tekh. Poluprov. **1**, 77 (1967).
T. I. Kucher, Fiz. Tekh. Poluprov. **1**, 1406 (1967).
[7] D. L. Dexter, Phys. Rev. **108**, 3 (1957).
[8] I. M. Dyckman, Proc. Inst. Phys. Acad. Sci. Ukr. SSR **5**, 48 (1954).
[9] K. B. Tolpygo, Proc. Kiev State Univ., Math., **5**, 99 (1951).
[10] L. P. Howland, Phys. Rev. **109**, 1927 (1958).
[11] T. I. Kucher and K. B. Tolpygo, Fiz. tverd. Tela **2**, 2301 (1960).
[12] W. Känzig, Phys. Rev. **90**, 1890 (1955).
[13] K. B. Tolpygo, Physical Properties of Alkali-Halide Crystals, Proc. 2nd All-Union Conf., Riga 1962 (p. 15).
[14] Z. Ya. Evseev, Fiz. tverd. Tela **5**, 2345 (1963).
[15] W. R. Heller and A. Marcus, Phys. Rev. **108**, 3 (1957).
[16] E. I. Rashba, Zh. eksper. teor. Fiz. **36**, 1703 (1959).
[17] I. G. Zaslavskaya, Dissertation, Inst. Semicond., Kiev 1965.

(Received May 22, 1968)

phys. stat. sol. 28, 815 (1968)

Subject classification: 14.4.1; 22.1.1

Institute of Semiconductors, Academy of Sciences of the Ukrainian SSR, Kiev

Injection of Current Carriers in Anisotropic Semiconductor Plates and the Magnetodiode Effect

By

Z. S. Gribnikov, G. I. Lomova, and V. A. Romanov

It is shown that the current–voltage characteristics in of plate- or filament-shaped long diodes are essentially different for the isotropic and anisotropic cases. For the effective lifetime of excess carriers in anisotropic case is inversely proportional to the external electric field, the direct current through such a diode increases slower than $I \sim U^2$, that is "ohmically" $I \sim U$. As a result of such difference the characteristics of the isotropic plate-shaped diodes strongly depend on pressure and magnetic field (the effect of a small magnetic field may be considered to be equivalent to the effect of pressure). This indicates that significant tensodiode and magnetodiode effects take place. A satisfactory agreement of the experimental data with the theory has been obtained.

Показано, что прямые характеристики длинных диодов в форме пластин (или нитей) существенно различны для изотропного и анизотропного случаев. В силу того, что во втором случае эффективное время жизни неравновесных носителей уменьшается с ростом электрического поля обратно пропорционально величине его напряженности, ток через такие образцы нарастает замедленно, не по параболическому закону $I \sim U^2$, а ,,омически" $I \sim U$. Из-за этого различия характеристики изотропных диодов-пластин сильно зависят от давления и от магнитного поля (действие которого при малой величине поля эквиваленто), то-есть имеют место значительные тензодиодный и магнитодиодный эффекты. Получено хорошее согласие теоретических и экспериментальных результатов.

1. Introduction

1. Size effects obliged to the redistribution of current carriers in a specimen take place in anisotropic semiconductors [1]. In semiconductor plates ($-d \leqq y \leqq d$), which are anisotropic in the $x\,y$ plane (x is the current direction), and for high electric fields E_x, the electron–hole pairs are concentrated near one of the surfaces; the volume becomes completely monopolar and in the case of an intrinsic semiconductor fully depleted. A similar effect takes place in isotropic plates exposed to a magnetic field. The redistribution of current carriers and phenomena connected with that were studied in [2, 3]. The case of a macroscopically anisotropic intrinsic semiconductor was investigated in the papers [4], the redistribution of photocarriers in anisotropic semiconductors in [5].

Besides magnetic fields, natural and artificial anisotropy (created by mechanical tension), the redistribution of current carriers may be due to the anisotropy caused by strong electric fields and the many-valley spectrum for one type of carriers [6, 7].

2. In the papers mentioned above the bipolarity of the semiconductor was either intrinsic or was created by homogeneous illumination along the x-direction. However, an analogous picture must be found in the case of injection of non-equilibrium carriers distributed inhomogeneously along the x-direction. The sweeping of the injected carriers to one of thc surfaces due to the magnetic field H_z is known as the Suhl effect [8]. It is evident that phenomena like the Suhl effect have to take place without magnetic field in anisotropic semiconductors.

At low injection levels the forward conductivity is weakly influenced by redistribution of the injected carriers but for high injection levels the influence becomes strong. In such case the forward characteristics of diodes in form of plates with one injecting contact and the other being an ohmic one essentially differ for isotropic and anisotropic materials. The direct currents of such diodes must be very sensitive to magnetic fields (magnetodiode effect) and to the uniaxial deformation (tensodiode effect).

3. The magnetodiode effect was predicted by Stafeev [9] who had paid attention to the opportunity of changing of the ratio l/L (l is the diode length, L is the bipolar diffusion length) and the potential drop on the diode being dependent very sharply-exponentially upon this ratio (for $l/L \gg 1$). The experimental and theoretical investigation of the magnetodiode effect [10] (on diodes of nearly intrinsic germanium) led to the conclusion that the magnetic field increases the ratio l/L. This results from the decrease of L due to the decrease of the mobility and the increase of the effective length of the diode owing to the distortion of current lines. The effect of current carrier redistribution caused by this distortion has been observed by measuring the variation of the infrared absorption due to free carriers. The authors of [10] consider the state of surface to have immaterial influence on the magnetodiode effect.[1])

4. It is known, however, that the direct characteristic of a 'long' diode is not always very strongly sensitive to l/L. The analysis of paper [11] shows that this characteristic may be roughly divided into three parts (Fig. 1) 1. 'ohmic', 2. 'parabolic', 3. region of injection breakdown. The ohmic region corresponds to small injection levels and is terminating in the point with the coordinates

$$\varphi_1 \sim \left(\frac{l}{L}\right)^2, \quad j_1 \sim N\frac{l}{\tau}, \tag{1}$$

where τ is the effective lifetime, N is the concentration of donors (to be definite,

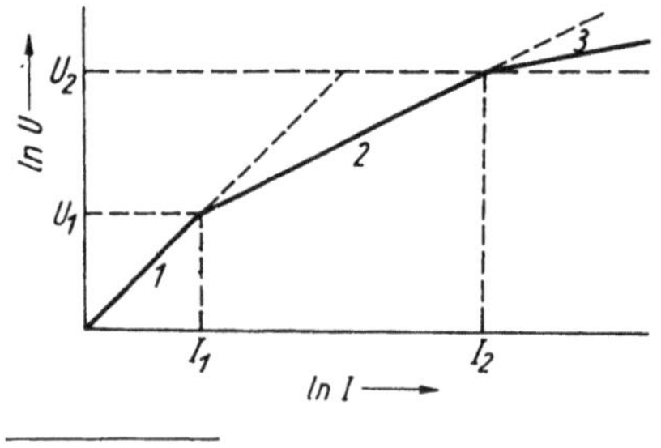

Fig. 1. Schematic direct volt–ampere characteristic of "long" diode

[1]) In a recent article [16] these authors discovered the strong influence of the surface state on the magnetodiode effect. They interpreted this influence to be caused by the change of the ratio l/L due to the change of the surface recombination contribution for different H_z.

we consider n-type Ge with hole injecting contact), $\varphi = \frac{e\,U}{kT}$ (U is the voltage). The parabolic region of the voltage–current characteristic is described by the formula of Rashba and Tolpygo [12, 13]

$$j = \frac{9}{8} \frac{N\,D_n\,D_p\,\tau}{l^3} \varphi^2 , \tag{2}$$

where $D_{n,p} = \frac{kT}{e} \mu_{n,p}$ are the diffusion coefficients of electrons and holes, respectively. This region is terminating in the point

$$\varphi_2 \sim e^{l/2L} \quad \text{or} \quad e^{l/L} , \quad j_2 \sim \frac{N\,L}{\tau} (e^{l/L} \quad \text{or} \quad e^{2l/L}) . \tag{3}$$

In formula (3) the first quantities correspond to contacts of type n–n$^+$, and the second ones to an ohmic contact ($S_c = \infty$). In the first cases the voltage in the region of injection breakdown is saturated beginning from φ_2. In the second one the voltage increases slowly (logarithmically) with increasing current. It is to be noted here that the injection breakdown takes place only, if for the current is valid $j \gtrsim j_2$ and the current through the injecting contact is carried only by holes. For limited emissivity of the contact (see [14]) and high values of j_2 the third region may be absent and the volt–ampere characteristic remained parabolic. Practically the injection breakdown region takes place for moderate values of l/L and weakly doped semiconductors (small N). It is in this region where the magnetodiode mechanism proposed in [9] is realized due to the exponential dependence of φ_2 on l/L. In sufficiently heavily doped semiconductors for great l/L, when equation (2) is valid over a wide current range, the mechanism mentioned above does not take place because according to equation (2) the small changes of l and $D_{n,p}$ result in a small change of j.

5. But even in this case as it will be shown below the magnetodiode effect may be sizable. This may be related to the decrease of the non-equilibrium carrier lifetimes in the specimen in form of a plate thanks to the transverse current carrier redistribution. This phenomenon is analogous to that investigated in [1 to 5] for a homogeneous distribution of carriers along the x-direction.

In specimens with limited dimensions in the $y\,z$ plane the lifetime τ in equation (2) does not coincide with the volume one τ_0 even for the isotropic case. The calculation of τ was carried out [15] for an isotropic plate $-d \leqq y \leqq d$ with the same surface recombination velocities S on both surfaces and it was obtained

$$\tau^{-1} = \tau_0^{-1} + \left[\frac{d}{S} + \tau_0 \left(\frac{d}{L} \operatorname{cth} \frac{d}{L} - 1\right)\right]^{-1} . \tag{4}$$

For the anisotropic case the field E_y[2]) exists in the plate, the value of it being equal to

$$|E_y| = a\,|E_x| , \tag{5}$$

where a is the parameter of anisotropy.

---

[2]) In this case the talk is about the effective field resulting in the transverse drift of electrons and holes (see below formula (17)). Its value may essentially differ from the "true" electric field, that is equal to the sum of the Hall and Dember fields. In the special case of intrinsic semiconductor with equal mobilities of electrons and holes this effective field does not equal zero although the "true" field is absent.

This field sweeps the carriers injected over all the width $2\,d$ to the surface. The average drift time of the injected carriers is

$$\tau_1 \approx \frac{d}{D_{yy}\,|E_y|} \approx \frac{d}{D_{yy} a\,|E_x|}\,, \tag{6}$$

where D_{yy} is the coefficient of bipolar diffusion in 'y-direction'; $\boldsymbol{E} = -\nabla\varphi$.

Carriers being swept on the surface, are concentrated in a layer of the thickness

$$d_{\mathrm{E}} \approx \frac{1}{a\,|E_x|}\,. \tag{7}$$

Their lifetime in this layer is according to equation (4)

$$\tau_2 \approx \frac{1}{a\,S\,|E_x|}\,. \tag{8}$$

So the effective lifetime of carriers in a specimen is equal to

$$\tau = \tau_1 + \tau_2 = \frac{1}{D_{yy}\,a\,|E_x|}\left(d + \frac{D_{yy}}{S}\right). \tag{9}$$

From equations (9) and (2) and considering $|E_x| = \varphi/l$ one obtains

$$j \sim \frac{D_{\mathrm{n}}\,D_{\mathrm{p}}}{D_{yy}}\,\frac{N\,\varphi}{a\,l^2}\left(d + \frac{D_{yy}}{S}\right). \tag{10}$$

As it will be shown below formula (10) obtained for qualitative reasons is very close to the precise one.

It follows from equation (10) that in an anisotropic plate the current increases proportionally to φ but not to φ^2 on account of $\tau \sim \frac{1}{|E_x|}$. So the current through the specimen can be substantionally decreased by a transformation of the plate material from isotropic to anisotropic one; it can be done by the use of uniaxial deformation or magnetic field ($a \sim H_z$). In consequence of this, strong magnetodiode and tensodiode effects take place.

2. Theory

In this section the distribution of carriers and of the field in plate-type diodes is calculated.

1. The study is restricted to the case of small anisotropy:

$$a \ll 1\,, \tag{11}$$

because otherwise a two-dimensional problem is to be solved. In regard to magnetodiodes, equation (11) corresponds to the case of small Hall angles. For isotropic material and $H_z = 0$ the use of a quasi-one-dimensional approximation [15] is always possible because the diffusion length L, determining the carrier distribution in y-direction, is much smaller than the increased diffusion length $D\,E_x\,\tau$ for the fields used; this length is determining the carrier distribution in x-direction. In anisotropic specimens or $H_z \neq 0$ the distribution in y-direction for sufficiently high E_x is determined by two lengths: shortened by about $\frac{1}{a\,E_x}$ and increased by about $a\,D\,E_x\,\tau$; that is why the quasi-one-dimensionality is valid only if condition (11) is satisfied.

2. Further we shall neglect effects that are proportional to a^2 (like the magnetoresistance), because their contribution is much smaller than that of the change of τ due to carrier redistribution. Up to terms of second order in a, the x-components of the carrier flows may be written as in the isotropic case [12]:

$$j_{nx} = - D_{nxx} (p + N) E_x , \quad j_{px} = D_{pxx} \, p \, E_x , \tag{12}$$

where p is the hole concentration, D_{nxx} and D_{pxx} are components of the diffusion coefficient tensors for electrons and holes, respectively. Further assumptions follow which are likely well justified in specimens with $l \gg 2\,d$:

$$j_y = j_{py} - j_{ny} = 0 . \tag{13}$$

Then

$$\frac{\partial j_x}{\partial x} = \frac{\partial j_{px}}{\partial x} - \frac{\partial j_{nx}}{\partial x} = 0 \quad \text{and} \quad E_x = E_x(x) . \tag{14}$$

From equations (12), (13), and (14) one obtains

$$\frac{\partial j_{px}}{\partial x} = - \frac{N \, D_{xx}}{2} \frac{\mathrm{d}E_x}{\mathrm{d}x} \equiv - G(x) , \tag{15}$$

where $D_{xx} = \dfrac{2 \, D_{pxx} \, D_{nxx}}{D_{pxx} + D_{nxx}}$ is the bipolar diffusion coefficient. Taking into account equation (15), the continuity equation for the hole flow is written

$$\frac{\partial j_{py}}{\partial y} = G(x) - \frac{p - p_0}{\tau_0} \tag{16}$$

$\left(\text{it is considered that } \dfrac{\partial j_{pz}}{\partial z} = 0\right)$. Equation (16) is coinciding with the continuity equation for the case of homogeneous generation of carriers along the y-direction with the intensity $G(x)$. The flow j_{py} with respect to equation (13) is equal to

$$j_{py} = - D_{yy}(p) \left(\frac{\partial p}{\partial y} + a(p) \, p \, E_x \right) , \tag{17}$$

where

$$D_{yy}(p) = \frac{D_{pyy} \, D_{nyy} \, (N + 2\,p)}{(D_{pyy} + D_{nyy}) \, p + D_{nyy} \, N} ,$$

$$a(p) = \frac{p + N}{2\,p + N} (a_n - a_p) ,$$

$$a_n = \frac{D_{nyx}}{D_{nyy}} , \quad a_p = \frac{D_{pyx}}{D_{pyy}} .$$

In the isotropic case it is $D_{n,pxx} = D_{n,pyy}$, $a_{n,p} = \pm r_{n,p} H_z$. In the anisotropic case the coefficients $a_{n,p}$ differ from zero for $H_z = 0$, if the x- and y-directions do not coincide with the main directions of mobility ellipses in the $x\,y$ plane.

We shall solve the system of equations (16) and (17) only for the case when the coefficients D_{yy} and a are constant. This is correct only in the extreme case of very high and low injection levels, when they are equal to $D_{yy}(0) = D_{pyy}$, $D_{yy}(\infty) = \dfrac{2 \, D_{nyy} \, D_{pyy}}{D_{nyy} + D_{pyy}}$, respectively, and

$$a(0) = a_n - a_p , \quad a(\infty) = \frac{a_n - a_p}{2} . \tag{18}$$

Further we shall neglect volume and surface generation in comparison with the injection ($p_0 \approx 0$) that in extrinsic semiconductor may be done. If on the surface

$y = \pm d$ takes place:

$$j_{py}(\pm d) = \pm S^{\pm} (p(\pm d) - p_0) \tag{19}$$

we find

$$p(x, y) = G \tau_0 \left(1 + \frac{\delta_1 e^{\beta_1 y} + \delta_2 e^{\beta_2 y}}{\delta}\right), \tag{20}$$

$$\delta = (\beta_1 + \sigma^-)(\beta_2 - \sigma^+) e^{(\beta_1 - \beta_2) d} - (\beta_2 + \sigma^-)(\beta_1 - \sigma^+) e^{-(\beta_1 - \beta_2) d},$$
$$\delta_1 = (\beta_1 + \sigma^-)(2\gamma + \sigma^+) e^{-\beta_2 d} - (2\gamma - \sigma^-)(\beta_1 - \sigma^+) e^{\beta_2 d},$$
$$\delta_2 = (2\gamma - \sigma^-)(\beta_2 - \sigma^+) e^{\beta_1 d} - (2\gamma + \sigma^+)(\beta_2 + \sigma^-) e^{-\beta_1 d},$$
$$\sigma^{\pm} \equiv \frac{1}{d^{\pm}} = \frac{S^{\pm}}{D_{yy}}, \quad \gamma = \frac{a E_x}{2}, \quad \frac{1}{L_y^2} \equiv \alpha^2 = \frac{1}{D_{yy} \tau_0}, \quad \beta_{1,2} = -\gamma \pm \sqrt{\gamma^2 + \alpha^2}.$$

From equation (12) the total current through a specimen is

$$I = e\,[(D_{pxx} + D_{nxx}) P + 2 d \mathsf{N} D_{nxx}] E_x, \tag{21}$$

where

$$P = \int_{-d}^{d} p(x, y)\, dy = 2 d G \tau_0 \Psi(E_x); \qquad \Psi(E_x) = 1 + \frac{\delta_1}{\delta} \frac{\sinh \beta_1 d}{\beta_1 d} + \frac{\delta_2}{\delta} \frac{\sinh \beta_2 d}{\beta_2 d}. \tag{22}$$

The volt–ampere characteristic in parametric form is given by

$$l = D_{pxx} \tau_0 \int_0^{E(l)} \frac{E_x \Psi(E_x)\, dE_x}{E_i - E_x}, \tag{23}$$

$$\varphi = D_{pxx} \tau_0 \int_0^{E(l)} \frac{E_x^2 \Psi(E_x)\, dE_x}{E_i - E_x}, \tag{24}$$

where

$$E_i = \frac{I}{2 d \mathsf{N} D_{nxx}}.$$

3. Because of complexity of the function $\Psi(E_x)$ we shall examine special cases. First of all, the asymptotic behaviour for large values of φ will be considered. If φ is so large as the strong inequalities are fulfilled:

$$a E(l) L_y \gg 1, \frac{L_y}{d}, \frac{d}{L_y}, \frac{L_y}{d^-}, \frac{d^-}{L_y}, \tag{25}$$

in the integrals (23) and (24) $\Psi(E_x)$ with sufficient accuracy may be replaced by the approximate expression

$$\Psi(E_x) \approx \frac{\alpha^2 (d + d^-)}{a E_x}. \tag{26}$$

In this case from equation (23) one obtains

$$E(l) = E_i (1 - e^{-l/l_0}), \tag{27}$$

where[3])

$$l_0 = \frac{d + d^-}{a \left(\frac{D_{yy}}{D_{pxx}}\right)}. \tag{28}$$

[3]) The length l_0 has the meaning of the maximum drift length. Its existence shows that in this situation the drift length of injected carriers does not increase infinitely with increasing field as in the isotropic case, but is limited due to the decrease of the effective lifetime when the field increases.

From equation (24) with respect to (27) and (28) the 'ohmic' volt–ampere charcteristic is

$$i = \frac{I}{2\,d} = \frac{e\,N\,D_{n\,x\,x}\,\varphi}{l - l_0 + l_0\,e^{-l/l_0}} = \sigma_0\,\varphi\,f\left(\frac{l_0}{l}\right), \tag{29}$$

where

$$\sigma_0 = \frac{N\,D_{n\,x\,x}}{l} \quad \text{and} \quad f(\chi) = \frac{1}{1 - \chi + \chi \exp\left(-\frac{1}{\chi}\right)}. \tag{29'}$$

For $l \gg l_0$ the diode conductivity coincides with that of the specimen without injecting contact and with length $l - l_0$. For $l \ll l_0$ from equation (29) follows

$$i \approx 2\,\frac{D_{n\,x\,x}\,D_{p\,x\,x}}{D_{y\,y}}\,\frac{N\,\varphi}{a\,l^2}\,(d + d^-) \tag{30}$$

which differs from equation (10) by the factor 2. In calculating the volt–ampere characteristic the values $a(\infty)$ and $D(\infty)$ have to be substituted into equations (29) and (30). The function $f(\chi)$ in equation (29) is well approximated for not too very small values of χ by the expression

$$f(\chi) \approx 2\left(\chi + \frac{1}{3}\right) \tag{29''}$$

(the first and second term of the asymptotic expansion). For example, for $\chi = 0.5$ the error $\Delta f/f$ amounts to 6%. Thus the dependence $i(l_0)$ is nearly linear if the length l_0 is not considerably smaller than l. If l is replaced by x formula (27) gives the distribution $E_x(x)$; this formula remains correct for the region of high injection as well as far from the injecting contact where the injection level may be low. For the last case the values $a(0)$ and $D_{yy}(0)$ have to be used. This asymptotic formula is suitable also for the intermediate region, where in the layer near one surface of the specimen the injection level is high while in the depth it is small. For this case equation (28) may be rewritten:

$$l_0 = \frac{1}{D_{p\,x\,x}}\left[\frac{d}{a(0)\,D_{y\,y}(0)} + \frac{d^-(\infty)}{a(\infty)\,D_{i\,y}(\infty)}\right]. \tag{31}$$

4. It is of interest to compare the asymptotic characteristic (29) with the characteristic for $a = 0$ (from equations (23) and (24) with $\gamma \to 0$). For $E(l) = \frac{3\,\varphi}{2\,l} \gg \frac{2\,l}{D_{p\,x\,x}\,\tau}$ it is given by expression (2) with $\tau = \tau_0\,\Psi(0)$, $D_{n,p} = D_{n,p\,x\,x}$. The ratio of currents for the same biases φ is

$$\varkappa = \frac{i(a)}{i(0)} = \frac{16}{9}\,\frac{1}{\varphi}\,\frac{l\,l_0}{D_{p\,x\,x}\,\tau_0\,\Psi(0)} = \frac{16}{9}\,\frac{1}{\varphi}\,\frac{l\,l_0}{L_p^2\,\Psi(0)}, \tag{32}$$

where

$$\Psi(0) = 1 - \frac{\sinh \alpha\,d}{\alpha\,d}\,\frac{\alpha\,(\sigma^+ + \sigma^-)\cosh \alpha\,d + 2\,\sigma^+\,\sigma^-\sinh \alpha\,d}{(\alpha^2 + \sigma^+\,\sigma^-)\sinh 2\,\alpha\,d + \alpha\,(\sigma^+ + \sigma^-)\cosh 2\,\alpha\,d}.$$

Formula (32) may be used, if

$$\varphi \gg \frac{l^2}{L_p^2\,\Psi(0)}, \frac{l\,l_0}{a\,L_p^2}, \frac{l}{a\,d}, \frac{l}{a\,d^-} \tag{33}$$

and $l_0 > l$. The first of the strong inequalities (33) is bound with the use of (2), the others are ensued from equation (25). In thick specimens ($\alpha\,d \gg 1$) and in specimens where the surface recombination velocity is small ($\sigma^+, \sigma^- \ll \alpha^2\,d$)

it is $\Psi(0) \approx 1$, and we find from the inequalities (33)

$$\varkappa < 1 . \tag{34}$$

Thus, magnetodiode and tensodiode effects cause the decrease of the current through the diode for any values of φ. The same situation takes place if the relation

$$\sigma^- \gtrapprox \sigma^+ \tag{35}$$

is fulfilled. The other situation corresponds to

$$\sigma^+ \gg \sigma^- , \quad 2\,\alpha^2\,d\,; \quad 2\,\alpha\,d \lessapprox 1 . \tag{36}$$

Under these conditions over the range $\varphi < \varphi_0$ the magneto- and tensodiode effects alter their signs. This is associated with carriers being swept from the surface with high recombination velocity ($y = +\,d$). The voltage φ_0 is easy to calculate for the case $\alpha\,d \ll 1$:

$$\varphi_0 = \frac{8}{9}\,l\,\frac{(1 + 2\,\sigma^-\,d)\,(1 + \sigma^-\,d)}{a\,\sigma^-\,d} . \tag{37}$$

To sum up, in initially isotropic samples under condition (35) when increasing the magnetic field H_z having suitable sign or under compression, the current at first increases for every φ, then increases for small φ and decreases for large φ, the value of electric field at which sign change of the effect takes place, decreases. At least the current decreases at every φ becoming less than the initial current for $H_z = 0$.

3. Experimental

The aim of the experimental part of this work was to verify the following conclusions of the theory:

1. The existence of strong magnetodiode and tensodiode effects in long plate-type diodes.
2. The strong dependence of this effects on surface recombination velocity.
3. The 'ohmicness' of asymptotic volt–ampere characteristics for high values of φ.
4. The existence of the region with inverse sign of the effect for $\sigma^+ \gg \sigma^-$.

Our investigation is restricted to only the magnetodiode effect because it is experimentally more simple. The tensodiode effect is quite analogous to the magnetodiode one and was verified by control experiments.

Measurements were carried out on six specimens of n-Ge doped with Sb having $\varrho = 5.7\ \Omega$cm at 300 °K. The diffusion length was found to be 0.8 mm. The typical size of the specimens was: $l \approx 1.2$ cm, $2\,d \approx 0.1$ cm, $d_z \approx 0.5$ cm. The p–n junctions have been created by fusing of In into a small hollow on one of the butt-ends of the specimen; for the "ohmic" contact tin was soldered on the other butt-end. The electric field was applied in form of pulses (pulse duration varies from 1.0 to 0.1 ms decreasing as the field increases; repetition frequency was 10 Hz). A special water cooling was used to avoid sample heating. The current amplitude was measured with an oscilloscope after the current had became stationary.

For each of the specimens two groups of volt–ampere characteristics were obtained: the first (a) corresponds to small and equal surface recombination rates on both sides (specimens were etched and exposed to air for 24 hours),

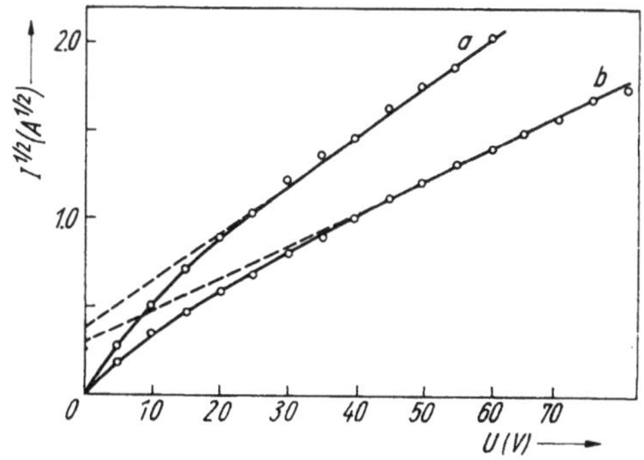

Fig. 2. Direct volt–ampere characteristics for a Ge specimen with p–n junction. The size is $12 \times 0.94 \times 5.7$ mm^3; $H_z = 0$. (a) $S^- \approx S^+ \approx$ $\approx 10^2$ cm/s; (b) $S^- \approx 10^2$ cm/s; $S^+ \approx 10^4$ cm/s

the second (b) to large surface recombination rate on one side (this side was roughly grinded). The magnetic field during the measurements was changed from $+6500$ to -6500 Oe.

As it is seen in Fig. 2 for one of the specimens the volt–ampere characteristic for high biases and $H_z = 0$ is always strictly quadratic.

In Fig. 3 the two groups of volt–ampere characteristics mentioned above are shown for the same specimen. It is seen that for $H_z = 0$ the asymptotic behaviour of characteristics for high values of U is substantially changed and is converted from a parabolic into an 'ohmic' one according to the theory. The asymptotic conductivity $\sigma(H_z)$ decreases with increasing H_z. $\sigma(H_z)$ is shown in

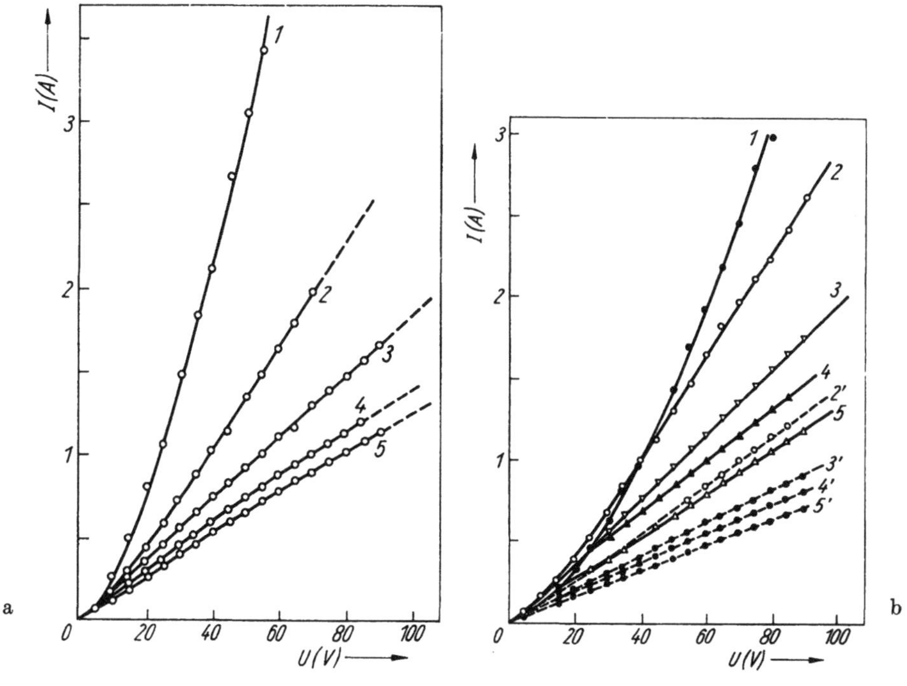

Fig. 3. Direct volt–ampere characteristics for the same sample as in Fig. 2 at different values of magnetic field. a) $S^- \approx S^+ \approx 10^2$ cm/s; b) $S^- \approx 10^2$ cm/s, $S^+ \approx 10^4$ cm/s. (1) $H_z = 0$, (2) and (2') $H_z = 1500$ Oe, (3) and (3') $H_z = 3000$ Oe, (4) and (4') $H_z = 4500$ Oe, (5) and (5') $H_z = 6500$ Oe

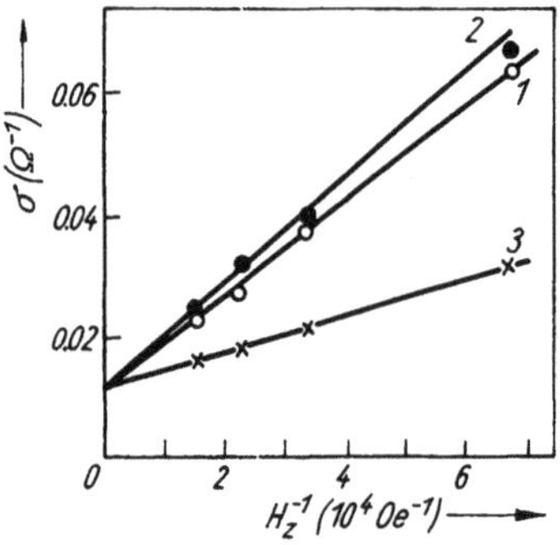

Fig. 4. Conductivity for the same sample as in Fig. 2 at large voltage vs. $1/H_z$. Curve 1: $S^- \approx S^+ \approx 10^2$ cm/s; curves 2 and 3: $S^- \approx 10^2$ cm/s, $S^+ \approx 10^4$ cm/s. Curve 2: The carriers are deflected to the surface with small recombination rate S^-. Curve 3: The carriers are deflected to the surface with high recombination rate S^+

Fig. 4 to be linear dependent on $1/H_z$ in agreement with equation (29) because of $l_0 \sim \frac{1}{a} \sim \frac{1}{H_z}$. The characteristics of group a) in Fig. 3, corresponding to the same value of H_z and various signs, approximately coincide. Therefore, in Fig. 3a only characteristics for one sign of H_z are shown.

The characteristics of group b) corresponding to various signs of H_z are very different. The half-group of characteristics corresponding to the deflection of carriers to the surface with high surface recombination velocity (dashed lines), qualitatively resemble the characteristics of group a). (In this specimen the parameters $\Psi(0)$ and $l_0(H_z)$ are changed equally; therefore, both groups of characteristics are approximately similar.) The half-group of characteristics corresponding to the deflection to surface with small surface recombination velocity (solid lines) for $H_z \neq 0$ approximately coincides with that of group a) because the lengths $l_0(H_z)$ did not change. So long as $\Psi(0)$ in case b) decreased strongly in comparison with case a) and the characteristics for $H_z = 0$ are placed considerably lower than in case a), the curves of this half-group on its initial parts are placed higher than the characteristics for $H_z = 0$. Thus the magnetodiode effect changes sign depending on H_z and U. Because of $2\alpha\, d \approx \approx 1$ the magnetodiode effect of inverse sign is small, but it takes place in case b) for all specimens (for one of the specimens this effect was considerably greater than in Fig. 3b and for other specimens by vague causes).

Thus, the experiments confirmed all qualitative results of the theory. It is worthwhile to note in general that the small initial anisotropy of the specimens resulting from accidental causes (like the mounting of the specimen, soldering of the contacts, etc.) may strongly affect their behaviour at large direct biases.

The tensodiode effect may be used for performing tensodevices, with the tensosensitivity being not smaller than in [17].

Acknowledgements

The authors are indebted to E. I. Rashba for reviewing the manuscript and remarks and to G. A. Abramchuk for assistance.

References

[1] E. I. RASHBA, Fiz. tverd. Tela **6**, 2347 (1964).
[2] H. WELKER, Z. Naturf. **6a**, 184 (1951).
E. WEISSHAAR and H. WELKER, Z. Naturf. **8a**, 681 (1953).

O. Madelung, I. Teword, and H. Welker, Z. Naturf. **10a**, 476 (1955).
E. Weisshaar, Z. Naturf. **10a**, 486 (1955).
[3] G. E. Pikus, Zh. tekh. Fiz. **26**, 22 (1956); **26**, 36 (1956).
V. P. Zhuze, G. E. Pikus, and O. V. Sorokin, Fiz. tverd. Tela **1**, 1420 (1959).
[4] I. I. Boiko, I. P. Zhadko, E. I. Rashba, and V. A. Romanov, Fiz. tverd. Tela **7**, 2239 (1965).
E. I. Rashba, V. A. Romanov, I. I. Boiko, and I. P. Zhadko, phys. stat. sol. **16**, 43 (1966).
[5] V. A. Romanov, I. P. Zhadko, and I. I. Boiko, phys. stat. sol. **17**, 389 (1966).
[6] Z. S. Gribnikov and E. I. Rashba, Fiz. tverd. Tela **9**, 967 (1967).
[7] E. I. Rashba, Zh. eksper. teor. Fiz. **48**, 1427 (1965).
Z. S. Gribnikov, V. A. Kochelap, and E. I. Rashba, Zh. eksper. teor. Fiz. **51**, 266 (1966); Fiz. tverd. Tela **8**, 2479 (1966).
[8] H. Suhl and W. Shockley, Phys. Rev. **75**, 1617 (1949); **76**, 180 (1949).
W. Shockley, Electrons and Holes in Semiconductors, Van Nostrand, New York 1950.
[9] V. I. Stafeev, Fiz. tverd. Tela **1**, 841 (1959).
[10] E. I. Karakushan and V. I. Stafeev, Fiz. tverd. Tela **3**, 677 (1961); **3**, 2031 (1961).
I. E. Vorobev, E. I. Karakushan, and V. I. Stafeev, Fiz. tverd. Tela **5**, 982 (1963).
[11] Z. S. Gribnikov, Fiz. tverd. Tela **7**, 251 (1965).
[12] E. I. Rashba and K. B. Tolpygo, Zh. tekh. Fiz. **26**, 1419 (1956).
[13] M. A. Lampert and A. Rose, Phys. Rev. **121**, 26 (1961).
[14] Z. S. Gribnikov, Radiotekhnika i Elektronika **9**, 163 (1964); **9**, 851 (1964).
[15] R. Hirota, S. Tosima, and M. A. Lampert, J. Phys. Soc. Japan **18**, 535 (1963).
[16] E. I. Karakushan and V. I. Stafeev, Electronic Processes on Surface and in Monocrystal Layers of Semiconductors, Izd. Nauka Novosibirsk 1967 (p. 49) (in Russian).
[17] I. P. Zhadko, V. A. Romanov, E. I. Rashba, and I. I. Boiko, Fiz. Tekh. Poluprov. **1**, 1174 (1967).

(Received April 16, 1968)

phys. stat. sol. 28, 827 (1968)

Subject classification: 14.4.1; 22

A. F. Ioffe Physico-Technical Institute, Academy of Sciences of the USSR, Leningrad

Behaviour of the High-Field Domains below the Voltage of the Nucleation Threshold

By

M. E. LEVINSTEIN and M. S. SHUR

On the basis of the phenomenological theory it is shown that the high-field domain in a Gunn diode does not annihilate if the applied voltage becomes lower than the threshold for domain nucleation. The dependence of the threshold for domain annihilation on the diode parameters is obtained. It is found that the domain annihilation amplitude is equal to half of the domain amplitude when the applied voltage corresponds to that at the beginning of the region of negative differential resistivity in the diode. The results obtained are important if an additional microwave voltage is applied, for example if the diode operates in a resonator.

Сформировавшийся домен сильного поля в диоде Ганна не исчезает, если напряжение на образце становится меньше, чем пороговое напряжение возникновения домена. Этот эффект исследован теоретически в рамках феменологической теории. Найдена зависимость порогового напряжения рассасывания домена от параметров образца. Показано, что амплитуда домена при пороге рассасывания равна половине его амплитуды при напряжении, соответствуюцем началу участка отрицательного дифференциального сопротивления в материале. Отмечается, что учет полученных результатов существенно необходим при наличии дополнительного С.В.Ч. смещения, например, при работе диода в резонаторе.

In the paper [1] it was experimentally shown that the high-field domains formed in Gunn diodes continuied to exist even if the voltage was decreased to less than the nucleation voltage of the domain U_n. In connection with this, in paper [1] the existence of a domain annihilation voltage U_a was noted which in those experiments had the value $U_a \approx U_n/2$.

If the voltage is lowered to U while the domain is propagating along the sample and if $U_a < U < U_n$, then it reaches the anode without annihilation. If $U < U_a$ the domain disappears in Maxwell's time.

Later D. E. McCumber and A. G. Chynoweth showed by computer calculations that this effect could be explained within the framework of their theory [2]. More detailed computer calculations were made by J. A. Copeland [3]. Using the model of McCumber and Chynoweth he also calculated (together with other domain parameters) the dependence of the annihilation field $E_a = U_a/L$ on the length of the sample L for the values of the equilibrium carrier concentration $n_{01} = 10^{14}$ cm^{-3}, $n_{02} = 10^{15}$ cm^{-3}.

The qualitative dependence of E_a on the parameter $n_0 L$ was also discussed in the paper of W. Heinle [4].

However it is necessary to emphasize that the quantitative results of the numerical calculations [2, 3] are principally determined by the dependence of the mean drift velocity $V(E)$ on the field. — But it is well known that the curves

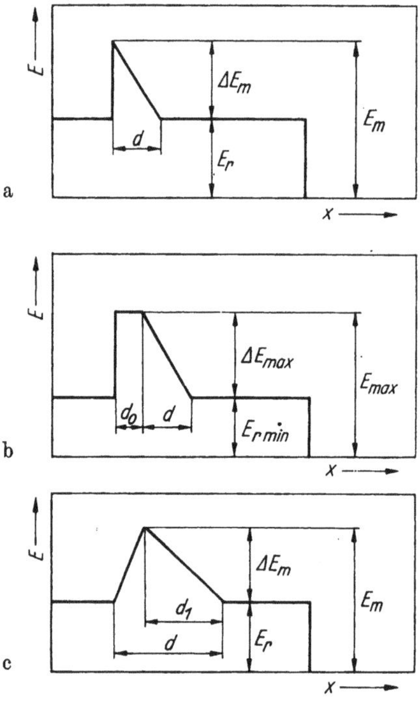

Fig. 1. The form of the high-field domain. a) unsaturated rectangular domain, b) saturated rectangular domain, c) triangular domain

$V(E)$ obtained experimentally contradict one another. For example the values of the important parameter, that is the mean negative differential mobility μ_3, which were determined in the papers [5 to 8] differ by more than a factor of 10. The theoretical curves for $V(E)$ [2, 9, 10] differ also one from another. So, at the present time, there is no reason to hope for the construction of an exact quantitative theory of the Gunn diode. The detailed theoretical analysis corrobarates this consideration [11].

That is why it is possible to insist that the quantitative results of the calculations [2, 3] are model-dependent though they give a correct qualitative (and even semi-quantitative) description of the domain behaviour.

In this work we consider a simple analytical theory which also gives a semi-quantitative description of the domain behaviour. The qualitative conclusions of our theory agree completely with the results of the papers [2 to 4]. The main advantage of our theory (besides the simplicity of analytical results) is that the results may be used directly for an analytical calculation of the operating regime for the Gunn diode when there is an additional ultra-high frequency bias. For example, the results of this work together with the methodes proposed by W. Heinle [12] permit an analytical description of the operation of the Gunn diode in a resonator.

First of all we consider a domain which has the form of a rectangular triangle (Fig. 1a) if the applied voltage is less than the saturation voltage U_a. If the voltage is higher than the saturation voltage U_s then the domain is assumed to have the form of a rectangular trapezoid (Fig. 1b).

The depletion layer is assumed to be fully depleted. These assumptions are valid for any carrier concentration for the case when diffusion can be neglected [13]. It follows from the work of [14] that even if the diffusion is taken into account then these assumptions are also valid for GaAs diodes when the equilibrium concentration of carriers $n_0 < 10^{15}$ cm^{-3}.

The equations describing the stable unsaturated domain with a fully depleted depletion layer (Fig. 2a) have the form

$$U = E_r L + \Delta E_m \frac{d}{2}, \tag{1}$$

where U is the applied voltage, L is the length of the sample. The rest of the symbols are clear from Fig. 2a.

$$d = \frac{\varepsilon_0}{4 \pi e n_0} \Delta E_{\mathrm{m}} = 2 b \Delta E_{\mathrm{m}} , \tag{2}$$

where ε_0 is the static dielectric constant and e is the electron charge.

The equation (2) follows immediately from the Poisson equation if the depletion layer is assumed to be fully depleted.

$$E_{\mathrm{r}} = E_{\mathrm{t}} - k \Delta E_{\mathrm{m}} , \tag{3}$$

where

$$k = \frac{E_{\mathrm{t}} - E_{\mathrm{r\,min}}}{\Delta E_{\max}} .$$

Equation (3) connects the field outside the domain E_{r} with the amplitude of the domain field ΔE_{m}.

It would be exact if the "dynamic curve" [13], which determines the amplitude of the domain field as a function of the mean drift velocity of the carriers outside the domain, would be a straight line.

Using a "square law" [13] we have made numerical calculations for the curves for $U(E)$ of Mc.Cumber and Chynoweth [2], Butcher and Fawcett [9], Convell and Vassel [10] and for the N-shaped approximation in order to estimate the error due to the substitution of the "dynamic curve" by a straight line.

The latter curve, which is to be used for the numerical estimates was obtained by such considerations. We have taken the mobility of the carriers in a weak field $\mu_1 = 6000\ \mathrm{cm^2\,V^{-1}\,s^{-1}}$ and the value of the threshold field $E_{\mathrm{t}} = 3\ \mathrm{kV\,cm^{-1}}$.[1]) The amplitude of the field in the saturated domain was taken as $\Delta E_{\max} = 75\ \mathrm{kV\,cm^{-1}}$ [15]. (The experimental values for this parameter are within the limits of 40 kV cm^{-1} [16] and 75 kV cm^{-1}.) We have supposed (as well as in paper [2]) that the mobility in strong field was $\mu_2 = 100\ \mathrm{cm^2\,V^{-1}\,s^{-1}}$.

We have supposed also that the negative differential mobility was constant and we have determined it using the "square law" for the saturated domain. (The value found is $\mu_3 = 530\ \mathrm{cm^2\,V^{-1}\,s^{-1}}$.)

This curve for $V(E)$ is shown in Fig. 2. Our approximation to the "dynamic curve" is shown by the solid line, the "dynamical curve" by the dashed line. It can be seen from the figure that the error is about 30%. (The error for the theoretical curves for $U(E)$ enumerated above has the same order of magnitude.)

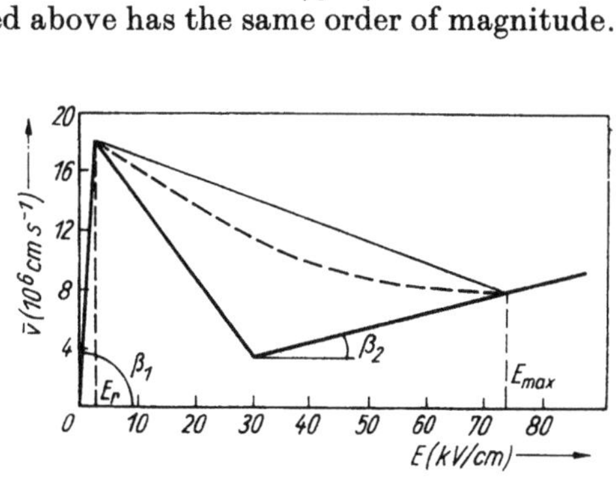

Fig. 2. The dependence of the mean drift velocity of the carriers on the field $V(E)$. (Read E_{t} instead of E_{r})

---

[1]) If $n_0 L > 10^{12}\ \mathrm{cm^{-2}}$ the voltage of the nucleation threshold U_{n} is equal to $E_{\mathrm{t}} L$. If $n_0 L < 10^{12}\ \mathrm{cm^{-2}}$ U_{n} increases weakly with decreasing $n_0 L$ and if $n_0 L$ is as low as possible ($5 \times 10^{11}\ \mathrm{cm^{-2}}$) then because of Kramer's criterion, U_{n} is more than $E_{\mathrm{t}} L$ by 25% [4].

From the discussion given above it follows that such an error is completely acceptable for the present state of the problem.

If a more exact calculation is desired it is necessary only to substitute the dynamic curve by the broken line. The only change in our results would be the substitution of the factor K by K_{eff}, where K_{eff} is the corresponding factor for the 1th segment of the broken line and etc.. The "second" approximation is shown in Fig. 2 by the dotted line. It can be seen that the error of this approximation is not larger than 5%. We checked it also by numerical calculations for the other $U(E)$ curves enumerated above.

All the other approximations are rather general. They are discussed, tor example, in paper [17].

The solution of equations (1) to (3) determines the dependence on the applied voltage of the amplitude of the domain ΔE_m, the width of the domain d and the field outside the domain E_r. The solution for ΔE_m has the form

$$\Delta E_m = a \pm \sqrt{a^2 + \frac{U - E_t L}{b}}\,, \tag{4}$$

where

$$a = \frac{K L}{2 b} = \frac{E_t - E_{r\,min}}{\Delta E_{max}} \cdot \frac{4 \pi e n_0 L}{\varepsilon_0}\,. \tag{4a}$$

The solutions for $d_{1,2}$ and $E_{r1,2}$ are determined from equations (2), (3).

It can be seen from equation (4) that there are non-zero solutions for the domain amplitude and width in the region of the applied voltage $U < E_t$. The domain becomes unstable only if

$$a^2 + \frac{U - E_t L}{b} < 0\,. \tag{5}$$

We have found from (5) that the difference between the voltage $U_t = E_t L$ and the annihilation threshold U_a is equal to

$$\Delta U_{ta} = U_t - U_a = \frac{K^2 L^2}{4 b} = \left(\frac{E_t - E_{r\,min}}{\Delta E_{max}}\right)^2 \cdot \frac{2 \pi e n_0 L^2}{\varepsilon_0}\,. \tag{6}$$

The estimates for the typical parameters of GaAs diodes ($E_t = 3$ kV/cm; $E_{max} = 75$ kV/cm; $\mu_1 = 6 \times 10^3$ cm^2 V^{-1} s^{-1}; $\mu_2 = 100$ cm^2 V^{-1} s^{-1}; $\varepsilon_0 = 12.5$;

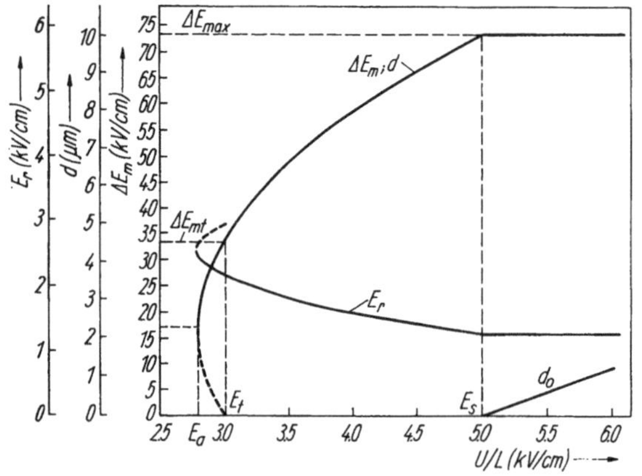

Fig. 3. The dependence of the domain parameters on the applied field

$n_0 = 5 \times 10^{14}$ cm^{-3}; $L = 10^{-2}$ cm) give:

$$\Delta U_{ta} = 2.04 \text{ V}; \quad \Delta E_{ta} = \Delta U_{ta}/L = 204 \text{ V/cm.}^{2)}$$

The dependence of the domain parameters on the applied voltage is shown on Fig. 3. The scale of the Fig. 3 is chosen so that the curves for the width of the domain d (which is proportional to ΔE_m) and the domain amplitude ΔE_m coincide. The branches of the solution which are shown by the dotted lines in Fig. 3 would correspond to the solutions describing the domain where the field is less than that outside provided that $U > U_t$. The possible existence of such domains was shown theoretically [18, 19] but they were not observed experimentally. In Fig. 3, the dependence of the width of the flat top of the domain d_0 (Fig. 1b) on the applied voltage is also shown. Such a top appears if the applied voltage is higher than the saturation voltage U_s

$$d_0 = \frac{U - U_S}{\Delta E_{max}} \,.$$

It can be seen from the Fig. 3 that in the region between the voltage U_t and the annihilation threshold U_a, the domain amplitude is reduced by half (this result does not depend upon the diode parameters).

Now let us consider the problem of the value for $n_0 L$ in the limit for which expressions (4), (4a) and (6) are valid. Because the domain amplitude cannot exceed the saturation amplitude ΔE_{max}, we have the following inequality for the domain amplitude at the moment of annihilation:

$$a \leqq \Delta E_{max} \,. \tag{7}$$

Expression (6) together with (4a) can be rewritten as

$$\frac{\Delta U_{ta}}{L} = \frac{k\, a}{2} = \frac{E_t - E_{r\,min}}{\Delta E_{max}} \frac{a}{2} \tag{8}$$

Including (7) we obtain

$$\frac{\Delta U_{ta}}{L} \leqq \frac{E_t - E_{r\,min}}{2} \tag{9}$$

and, correspondingly

$$\frac{U_a}{L} = E_a \geqq \frac{E_{r\,min} + E_t}{2} \,. \tag{10}$$

The limiting case with the equality in (7) to (10) corresponds to the case where the domain not only nucleates but even annihilates like the saturated one. Using (6) and taking (9) into account we found that it is realized if

$$n_0 L \geqq (n_0 L)_{lim} = \frac{\Delta E_{max}^2 \varepsilon_0}{4 \pi e (E_t - E_{r\,min})} = \frac{\Delta E_{max} \varepsilon_0}{4 \pi e K} \,. \tag{11}$$

Using the expressions (1) and (11) it is easy to check that if $n_0 L$ is equal to $(n_0 L)_{lim}$ then the threshold for domain annihilation is equal to the saturation field E_s

$$E_a = E_s = E_{r\,min} + \frac{b}{L} \Delta E_{max}^2 \,. \tag{12}$$

---

[2]) Unfortunately the data given in the papers [1, 2] are not sufficient to allow quantitative comparison of our results with these experiments. However, qualitative comparison using typical values for the missing parameters gives reasonable agreement.

(For the parameters chosen $(n_0\, l)_{\text{lim}}$ is equal to 2.17×10^{13} cm^{-2}.)

If $n_0\, L > (n_0\, L)_{\text{lim}}$ our estimates are not valid. In this case the domain annihilation being saturated, and its' annihilation threshold is equal to the minimal field for which (for a given magnitude of $n_0\, L$) the saturated domain can exist — to the saturation field (12). If $n_0\, L \to \infty$ E_{a} tends to $E_{\text{r min}}$.

Thus our results are in accord with the qualitative description of Gunn diodes proposed in [4] and they permit us to find its parameters via the diode characteristics.

Another limiting value for $\Delta U_{\text{ta}}/L$ can be obtained by using a criterion which is analogous to Kramer's criterion [20]. Accordingly, if $n_0\, L \leqq (n_0\, L)_{\text{min}} = 5 \times 10^{11}$ cm^{-2} the domain is not nucleated. Correspondingly

$$\left(\frac{\Delta U_{\text{ta}}}{L}\right)_{\text{min}} = \left(\frac{E_{\text{t}} - E_{\text{r min}}}{\Delta E_{\text{max}}}\right)^2 \frac{2\,\pi\, e (n_0\, L)_{\text{min}}}{\varepsilon_0} \tag{13}$$

which gives for the parameters chosen, $(\Delta U_{\text{ta}}/L)_{\text{lim}} = 20.4$ V/cm.

The dependence of the domain amplitude on applied voltage for different values of the parameter $n_0\, L$ is shown in Fig. 4a. The corresponding dependence of the field E_{r} outside the domain is shown in Fig. 4b. The diode current density is proportional to E_{r}:

$$I = e\, n_0\, \mu_1\, E_{\text{r}} \tag{14}$$

(where μ_1 is the mobility in weak fields) and therefore the curves for the dependence of the current on applied voltage, for different values of $n_0\, L$ with an appropriate scale, coincide with the curves shown in Fig. 4b.

The dependence of the threshold for domain annihilation on the parameter $n_0\, L$ which is obtained from equations (6) and (12) with the parameters chosen as in the curve $V(E)$, is shown in Fig. 5.

Let us now estimate the corrections to our results for the case when $n_0 > 10^{15}$ cm^{-3} and the simple approximation to the domain form by the rectangular triangle is not valid [14]. In this case we approximate the domain form by a non-rectangular triangle (Fig. 1c). Certainly such an approximation is not self-consistent because if the diffusion is not taken into account the domain has a form of rectangular triangle. If the diffusion (even small) is taken into account then the triangle approximation is not valid because the diffusion current would

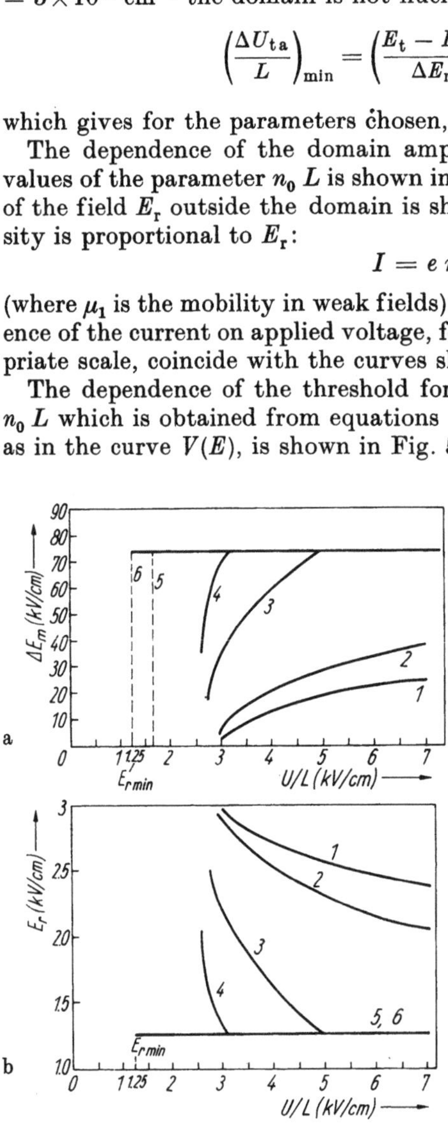

Fig. 4. a) the dependence of the domain amplitude ΔE_{m} on the applied voltage for different parameters $n_0\, L$. b) the analogous dependence for the field outside the domain E_{r}

1) $n_0\, L = 5 \times 10^{11}$ cm^{-2}
2) $n_0\, L = 10^{12}$ cm^{-2}
3) $n_0\, L = 5 \times 10^{12}$ cm^{-2}
4) $n_0\, L = 10^{13}$ cm^{-2}
5) $n_0\, L = 5 \times 10^{13}$ cm^{-2}
6) $n_0\, L \to \infty$

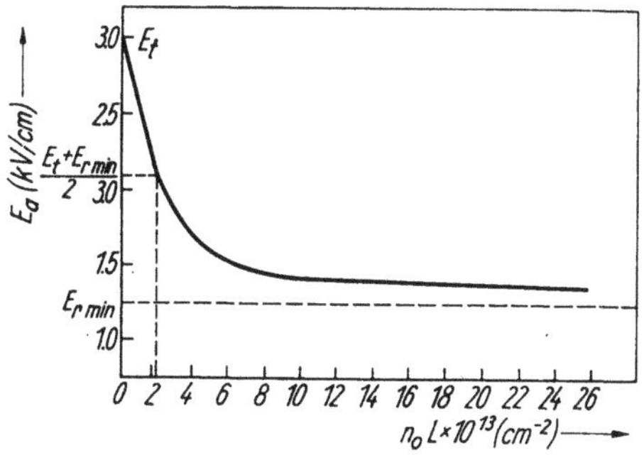

Fig. 5. The dependence of the threshold of the domain annihilation E_a on the parameter $n_0 L$

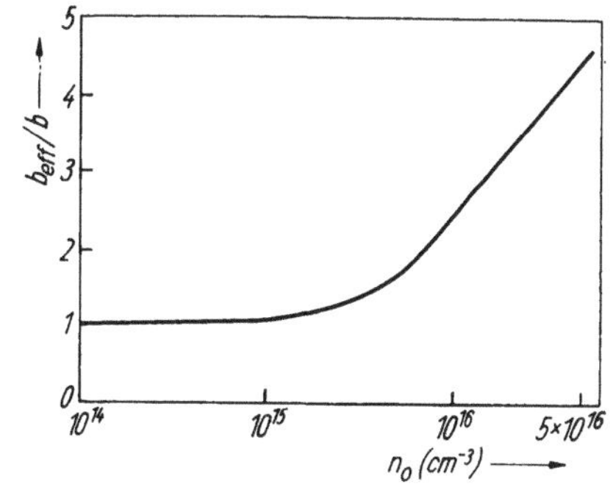

Fig. 6. The dependence b_{eff}/b on the parameter n_0. The scale for n_0 is logarithmic

be infinite at the corners of the domain triangle. But as shown by the experimental data, such a domain form can be used for the rough estimate.

In this case it follows from Poisson's equation for the depletion layer

$$d_1 = \frac{\varepsilon_0}{4\pi e (n_0 - n_{min})} \Delta E_m , \tag{15}$$

where n_{min} is the carrier concentration in the depletion layer. Expression (15) can be formally rewritten:

$$d = \left(\frac{d}{d_1}\right) d_1 = \frac{d}{d_1} \cdot \frac{n_0}{n_0 - n_{min}} \cdot \frac{\varepsilon_0}{4\pi e n_0} \Delta E_m \tag{16}$$

or

$$d = 2\, b_{eff}\, \Delta E_m , \tag{16a}$$

where

$$b_{eff} = \frac{d}{d_1} \frac{n_0}{n_0 - n_{min}} b . \tag{16b}$$

As equations (1) and (3) are still valid for this case, it follows from expression (16a) that for the case of an arbitrary domain it is possible to use all the results obtained for the rectangular domain if we substitute b by b_{eff} or, equivalently, n_0 by $n_0 \cdot b/b_{eff}$.

The values d/d_1 and $n_0/(n_0 - n_{min})$ for various equilibrium carrier concentrations can be calculated using the results of the work [14].

The curve b_{eff}/b $(n_0 L)$ calculated from the data [14] is shown in Fig. 6.

The results obtained can be used for the calculation of the regime of the Gunn diode if there is additional ultra-high frequency (uhf) bias, for the analysis of the problem of the phase-synchronization, for the consideration of the operation of the Gunn diode in the uhf circuits and, possibly for the description of the analogous devices suggested by Sandbank [21].

Acknowledgements

The authors thank A. I. Gubanov and V. I. Perel for valuable discussions of the results.

References

[1] J. S. HEEKS, A. D. WOODE, and C. P. SANDBANK, Proc. IEEE **3**, 554 (1965); Radio and Electron. Eng. **30**, 377 (1965).
[2] D. E. MCCUMBER and A. G. CHYNOWETH, IEEE Trans. Electron Devices **13**, 4 (1966).
[3] J. A. COPELAND, IEEE Trans. Electron Devices **13**, 189 (1966); J. appl. Phys. **37**, 3602 (1966).
[4] W. HEINLE, Electronics Letters **2**, 417 (1966).
[5] J. B. GUNN and B. J. ELLIOTT, Phys. Lett. **22**, 369 (1966).
[6] H. W. THIM, Electronics Letters **2**, 403 (1966).
[7] I. C. RUCH and G. S. KINO, Appl. Phys. Letters **10**, 400 (1967).
[8] C. HAMAGUCHI, A. KONO, and Y. INUISHI, Phys. Letters (Netherlands) **24A**, 500 (1967).
[9] P. N. BUTCHER and W. FAWCETT, Phys. Letters (Netherlands) **21**, 489 (1966).
[10] E. M. CONWELL and M. O. VASSEL, Appl. Phys. Letters **9**, 411 (1966).
[11] L. STENFLO, Proc. IEEE **54**, 1970 (1966); **55**, 1088 (1967).
[12] W. HEINLE, Electronics Letters **3**, 52 (1967).
[13] P. N. BUTCHER, W. FAWCETT, and C. HILSUM, Brit. J. appl. Phys. **17**, 841 (1966).
[14] J. W. ALLEN, W. SHOCKLEY, and G. L. PEARSON, J. appl. Phys. **37**, 3191 (1966).
[15] J. S. HEEKS, IEEE Trans. Electron Devices **13**, 68 (1966).
[16] M. C. COHEN, S. KNIGHT, and J. P. ELWARD, Appl. Phys. Letters **8**, 269 (1966).
[17] V. L. BONCH-BRUEVICH, Proc. Internat. Conf. Kyoto 1966 (p. 509).
[18] A. F. VOLKOV, Phys. Letters (Netherlands) **20**, 598 (1966).
[19] B. W. KNIGHT and G. A. PETERSON, Phys. Rev. **155**, 393 (1967).
[20] H. KROEMER, Proc. IEEE **52**, 1736 (1964).
[21] C. P. SANDBANK, Solid State Electronics **10**, 369 (1967).

(Received April 15, 1968)

phys. stat. sol. 28, 835 (1968)

Subject classification: 18.2 and 18.4; 5; 22.8

Institute of Semiconductors, Academy of Sciences of the USSR, Leningrad

Effect of Lattice Distortions on the Magnetic Behaviour of Perovskite-Type Manganites

By

V. A. Bokov, N. A. Grigoryan, M. F. Bryzhina, and V. V. Tikhonov

Some series of solid solutions with perovskite-type structure are synthesized and investigated, viz. the binary systems Nd_{1-x}, Me_xMnO_3 where Me = Ca^{2+}, Sr^{2+}, Ba^{2+}, and the ternary systems $(Bi_{1-x}La_x)_{0.5}Ca_{0.5}MnO_3$ and $Bi_{0.5}(Ca_{1-x}Me_x)_{0.5}MnO_3$ where Me = = Sr^{2+}, Pb^{2+}, Ba^{2+}. Distortions from the cubic perovskite structure are detected at room and liquid nitrogen temperature. Magnetic measurements are made in the temperature interval 77 to 800 °K. For all the compounds, a correlation between lattice distortions and magnetic behaviour is observed; 1. cubic and slightly distorted compounds exhibits a pure ferromagnetism; 2. distortions from the cubic structure give reduced spontaneous moments and in some cases antiferromagnetism developes. The ferromagnetic nature of the Mn^{3+}-O^{2-}-Mn^{3+} and Mn^{3+}-O^{2-}-Mn^{4+} interactions which are found in the non-distorted compounds is interpreted according to the Goodenough-Kanamori rules assuming that the quasi-static hypothesis is satisfied. The lattice distortions observed are thought to give rise to a static distortion in the Mn^{3+} ion environment and an ordering of covalent bonds. As a result, some part of exchange couplings becomes antiferromagnetic. The influence of the decreased subtended angle in the Mn-O^{2-}-Mn linkage is discussed. This angle decreases from 180° when the lattice becomes distorted and this fact disturbs the orbital orthogonality. This results in an increase of the antiferromagnetic t_{2g}-t_{2g} interaction via σ-bonding and in the occurence of the other superexchange antiferromagnetic interactions which couple both the e_g and t_{2g} orbitals and the t_{2g} and t_{2g} orbitals via π-bonding. The contribution of double exchange to the ferromagnetism of mixed manganites is believed to be not important.

Синтезированы и исследованы системы твердых растворов со структурой типа перовскита: двойные — Nd_{1-x} Me_x MnO_3, где Me = Ca^{2+}, Sr^{2+}, Ba^{2+}, и тройные — $(Bi_{1-x}La_x)_{0,5}Ca_{0,5}MnO_3$ и $Bi_{0,5}(Ca_{1-x}Me_x)_{0,5}MnO_3$, где Me = Sr^{2+}, Pb^{2+}, Ba^{2+}. Рентгенографически определены искажения структуры перовскита при комнатной температуре и при температуре жидкого азота. Магнитные измерения проводились в интервале температур от 77 до 800 °К. Обнаружена корреляция между искажением решетки и магнитными свойствами этих соединений: 1. кубические и мало искаженные составы обладают чистым ферромагнетизмом; 2. искажение решетки перовскита приводит к уменьшению спонтанного магнитного момента, а иногда и к появлению антиферромагнетизма. Ферромагнитный характер взаимодействий Mn^{3+}–O^{2-}–Mn^{3+} и Mn^{3+}–O^{2-}–Mn^{4+} у неискаженных составов интерпретирован на основе правил Гуденафа-Канамори в предположении о справедливости квазистатической гипотезы. Считается, что искажение решетки приводит к статическому искажению окружения иона Mn^{3+} и к упорядочению ковалентных связей, вследствие чего некоторая часть обменных связей становится антиферромагнитной. Рассматривается влияние изменения угла в цепочке Mn–O^{2-}–Mn: искажение решетки уменьшает этот угол и в результате из-за нарушения ортогональности возникают сверхобменные антиферромагнитные взаимодействия между e_g–t_{2g} и t_{2g}–t_{2g} орбитами через σ-связь и усиливается антиферромагнитное взаимодействие между t_{2g}–t_{2g} орбитами через π-связь. Что касается двойного обмена, то его роль в возникновении ферромагнетизма в смешанных манганитах считается несущественной.

1. Introduction

In compounds ABO_3 with perovskite-type structure, the magnetic cation–anion–cation exchange interactions are dependent on the electron configuration at the paramagnetic B cations, on the distances between the ions, and on the subtended angle of the cation–anion–cation linkage (γ). When this structure is cubic, all the cation–anion distances are equivalent and the angle γ is 180°. Deviations in the perovskite structure from cubic symmetry may cause both a reduction of the angle γ and a non-equivalence of the cation–anion distances.

During the study of compounds where the octahedral cations are trivalent manganese ions, it was observed by Havinga [1] that the lattice distortions have an effect on the magnetic properties of perovskites. It was shown that the strength of the magnetic exchange interaction between trivalent manganese ions is dependent on the magnitude of γ.

In perovskites of type $LaMn_{1-x}Me_xO_{3+x}$ (Me = Ga^{3+}, Co^{III}, Ni^{III}), Goodenough et al. [2] have interpreted the correlation between the magnetic properties and lattice distortions in terms of the electron configuration of Mn^{3+} ions. A distortion to orthorhombic symmetry is here considered to result from an ordering of e_g-orbitals caused by the cooperative Jahn-Teller effect [3]. This gives rise to anisotropic Mn^{3+}–O^{2-}–Mn^{3+} exchange interaction. (In the compound $LaMnO_3$, the interaction in the (001) plane is ferromagnetic but antiferromagnetic along the [001] direction [5].) This static cooperative distortion may be destroyed at high temperatures or by dilution of the cation B sublattice, if only the diluting ions are not Jahn-Teller ones [6]. Then under certain conditions, the interaction may become isotropic.

Experimental data on systems containing Mn^{3+} and Mn^{4+} [4 to 10] permit to suggest that the Mn^{3+}–O^{2-}–Mn^{4+} interaction also must be dependent on the lattice distortion. In systems of solid solutions of type $La_{1-x}Me_xMnO_3$ (Me = Ca^{2+}, Sr^{2+}, Ba^{2+}) which have been studied by Jonker and van Santen [4, 5], ferromagnetic compounds with cubic symmetry were observed for middle Mn^{4+} concentrations. When $x \geqq 0.5$, a decrease of the temperature is followed by development of a considerable distortion from the cubic structure, and in these cases, the ferromagnetic order was not found [7, 8]. However, Zener [11] explained the ferromagnetism in these systems by a mechanism of "double exchange", coupling Mn^{3+} and Mn^{4+} ions [12, 13]. The Zener mechanism is basing on the fact that in the $La_{1-x}Sr_xMnO_3$ system the ferromagnetic compositions have a rather high electric conductivity [14].

Our first work [15] on the $Bi_{1-x}Ca_xMnO_3$ system has shown that in this system a purely ferromagnetic order cannot be observed. For compositions ranging from $0.2 \leqq x \leqq 0.85$, a structural phase transition was shown to take place from the orthorhombic modification to the monoclinic one, and at this transition an extrapolated Curie temperature decreases and in some cases becomes negative (Fig. 13 in [15]). This structure transition has been explained by an electron ordering on manganese ions which must exclude the double exchange. But later, a more detailed treatment of this problem confirmed the supposition that the distortions in the $Bi_{1-x}Ca_xMnO_3$ system might have other origin. We have tried to examine the contribution of A ions to the lattice distortion by measuring ternary systems of type $(Bi_{1-x}La_x)_{0,5}Ca_{0.5}MnO_3$ and $Bi_{0.5}(Ca_{1-x}Me_x)_{0.5}MnO_3$ ($Me = Pb^{2+}$, Sr^{2+}, Ba^{2+}). We have also investigated binary systems $Nd_{1-x}Me_xMnO_3$ ($Me = Ca^{2+}$, Sr^{2+}, Ba^{2+}) to get a further information about magnetic and crystallographic behaviour of mixed manganites. The results of our investigations and proper discussion of them will be given below.

2. X-Ray Diffraction Measurements

Crystallographic properties in all the investigated ceramic systems were measured by X-ray diffraction technique. An apparent symmetry of the perovskite lattice and its distortion ratio were estimated from powder patterns recorded by the conventional diffractometer URS-50I (FeK_α radiation, Mn filter). The "point-by-point" measurement technique and the analysis of results were the same as in [15]. The linear parameters were measured with a precision of $\pm$ 0.01 Å and the angular parameters with a precision of $\pm$ 2 to 3′ (it is the low limiting case). In all the cases where the reflections were not splitted, we were obliged to describe the symmetry as pseudocubic (PC). Note that the rhombohedral perovskite structure is everywhere designated as "R", the orthorhombic and monoclinic ones of type $a < c/\sqrt{2} < b$ as "O" and "M" and of type $c/\sqrt{2} < a < b$ as "O′" and "M′", respectively (in analogy with [2]).

We were interested in a temperature displacement of the structure transitions which accompanies the change in composition in the series of solid solutions. We confine ourselves to measurements at room (T_{rt}) and liquid-nitrogen temperature (T_{ln}), and the temperature of corresponding structure transition was estimated from magnetic measurements. For a quantitative estimate of the lattice deviation from cubic symmetry, we have introduced a so-called average distortion ratio which is given by the formula

$$D = \frac{1}{6} \sum_{i=1,2,3} \left[\frac{|\Delta a_i|}{\tilde{a}\sqrt{2}} + \frac{|\Delta \alpha_i|}{90°} \right],$$

where $\tilde{a}$ is a cubic root of the volume of a single perovskite unit cell ($\tilde{a} = \sqrt[3]{V/4}$, V is the volume of the unit cell with $O^{(I)}$ or $M^{(I)}$ symmetry), $\Delta a_i/\sqrt{2}$ and $\Delta\alpha_i$ are increments of linear and angular parameters of the unit cell, respectively (the linear one is reduced to the parameter of a single unit cell). This quantity (D) was found only for liquid-nitrogen temperature. Thus, we got the possibility to make an obvious quantitative comparison of the lattice distortions abstracting from their symmetry. Note that $D = 0$ means lack of distortion.

2.1 Ternary systems $Bi_{0.5}(Ca_{1-x}Me_x)_{0.5}MnO_3$ ($Me = Sr^{2+}$, Pb^{2+}, Ba^{2+}) and $(Bi_{1-x}La_x)_{0.5}Ca_{0.5}MnO_3$

As room temperature measurements of the $Bi_{0.5}(Ca_{1-x}Sr_x)_{0.5}MnO_3$ system have shown, for samples with $x = 0.5, 0.7, 0.85$, and 1 the M′-distortion increases suggesting the increase of the $M' \rightleftharpoons O$ transition temperature which was found in the sample $Bi_{0.5}Ca_{0.5}MnO_3$ [15]. In the systems $Bi_{0.5}(Ca_{1-x}Ba_x)_{0.5}MnO_3$ and $Bi_{0.5}(Ca_{1-x}Pb_x)_{0.5}MnO_3$, the Ba^{2+} and Pb^{2+} substitutions lead to some decrease of the lattice distortions, the effect being more pronounced in the case of Ba^{2+} substitution.

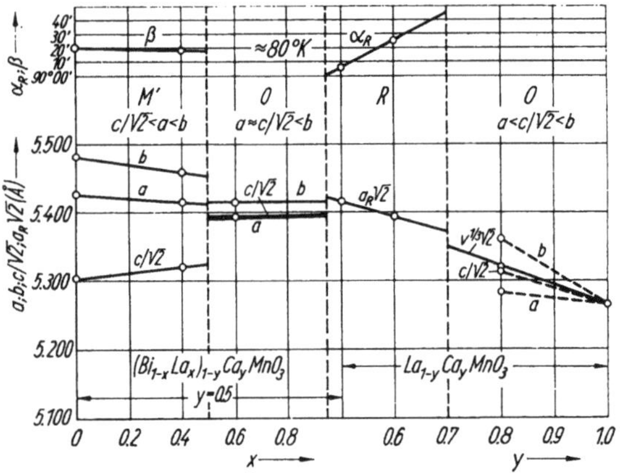

Fig. 1. The cell parameters at liquidnitrogen temperature (at about 80 °K) for systems of solid solutions $(Bi_{1-x}La_x)_{1-y}Ca_yMnO_3$ ($y = 0.5$) and $La_{1-y}Ca_yMnO_3$ ($0.5 \leq y \leq 1$)

The $(Bi_{1-x}La_x)_{0.5}Ca_{0.5}MnO_3$ system was studied in more detail. Cell parameters were measured both at T_{rt} and at T_{ln} (the results for T_{ln} are given in Fig. 1). It is shown that the La^{3+} substitution also gives rise to some sharp decrease of the $M' \rightleftharpoons O$ transition temperatures (see also Fig. 10).

From the fact that in all the ternary systems a substitution for A ions markedly displaces the $M' \rightleftharpoons O$ transition temperature, it obviously follows that this transition depends on the degree of close-packing of large cations rather than on electron configuration of manganese ions which remains constant during all the substitutions.

2.2 $La_{1-y}Ca_yMnO_3$ system

In this system, the ferromagnetic sample with $y = 0.4$ has no distortion both at T_{rt} and T_{ln} ($a = 3.849$ and 3.842 Å, respectively), but non-ferromagnetic samples with $y = 0.6$ and 0.8 (see Fig. 1) show at T_{ln} an R- and O-distortion, respectively. Our T_{rt}-data are in agreement with the data of [5, 7], but for T_{ln} it was found a more reasonable symmetry than in [7].

2.3 Binary systems $Nd_{1-x}Me_xMnO_3$ ($Me = Ca^{2+}, Sr^{2+}, Ba^{2+}$)

In these systems, the cell parameters were measured at T_{rt} and T_{ln} (the T_{ln}-data are given in Figs. 2 to 4, respectively). Opposite to Bertaut et al. [16], who defined the $NdMnO_3$ structure as orthorhombic, we have found for our sample a monoclinic perovskite structure with the parameters $a = 5.762$, $b = 5.410$, $c = 7.544$ Å, $\beta = 89°46'$ (M'). This discrepancy may be explained by the fact that the Mn^{3+} content of our samples was more close to the stoichiometric one.

In systems on the basis of $NdMnO_3$, the Jahn-Teller Mn^{3+} ions are substituted by Mn^{4+} ions. Therefore, the Jahn-Teller M'-distortion decreases rapidly with increasing x and gives rise to the pseudocubic phase. But simultaneously, Nd^{3+} ions are substituted by divalent ions that have larger radii than the former. If we consider the ideal perovskite structure as cubic close-packing of oxygen ions together with large A cations, then the closer the radius of the A ion to that of the oxygen ion, the smaller is the lattice distortion. Therefore, the larger the radius of substituting ion, the sooner the pseudocubic phase occurs in relation to composition and temperature. Thus, in systems with Sr^{2+} and Ba^{2+}

substitution, as distinguished from the system with Ca^{2+} substitution, there are rather large low-temperature pseudocubic regions where the lattice distortion is either negligible or doesn't exist at all (Figs. 2 to 4). This property is clearly illustrated by plots of the average distortion ratio versus composition (cf. *D*-curves in Figs. 5 to 7).

These compositions which are rich of Mn^{4+} have revealed another property: the larger the radius of the divalent ion, the sooner the non-consistence is developed between the too large radius of the A ion and the sufficiently small radius of the Mn^{4+} ion. It is expressed by the fact that although the perovskite lattice is formed in systems with Ca^{2+} and Sr^{2+} substitution when $x \geqq 0.4$, but it has an appreciable lattice distortion and in the system with Ba^{2+} substitution the compositions with $x > 0.4$ don't exist at all.

Note that in the $Nd_{1-x}Ca_xMnO_3$ system, compositions with $x = 0.4$, 0.5, 0.6, and 0.7 undergo a structure transition from the pseudocubic phase to the monoclinic one (M′) which is between T_{rt} and T_{ln}. In the $Nd_{1-x}Sr_xMnO_3$ system,

Fig. 2. Cell parameters at about 80 °K for the solid solution system $Nd_{1-x}Ca_xMnO_3$

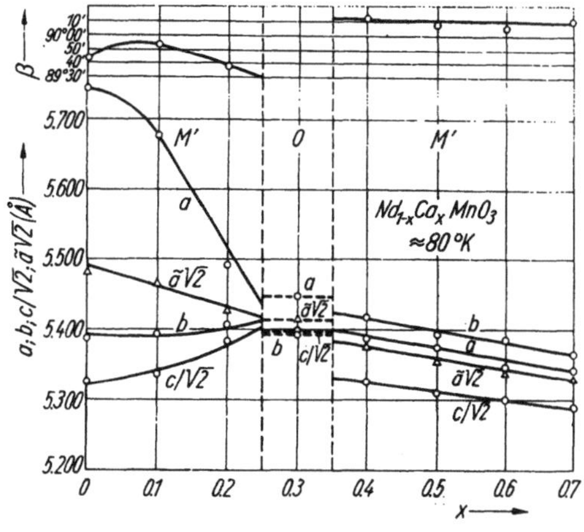

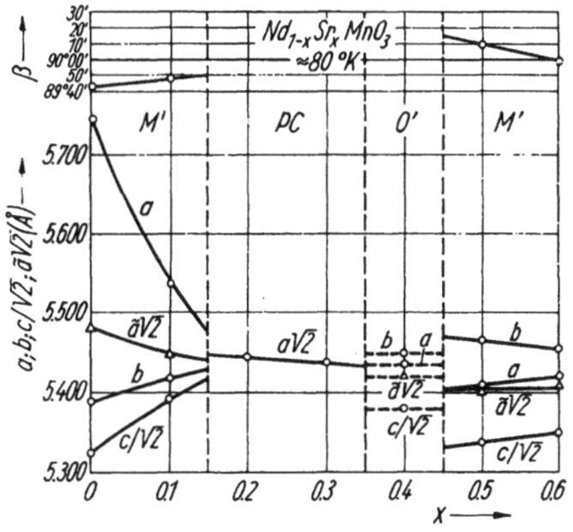

Fig. 3. Cell parameters at about 80 °K for the solid solution system $Nd_{1-x}Sr_xMnO_3$

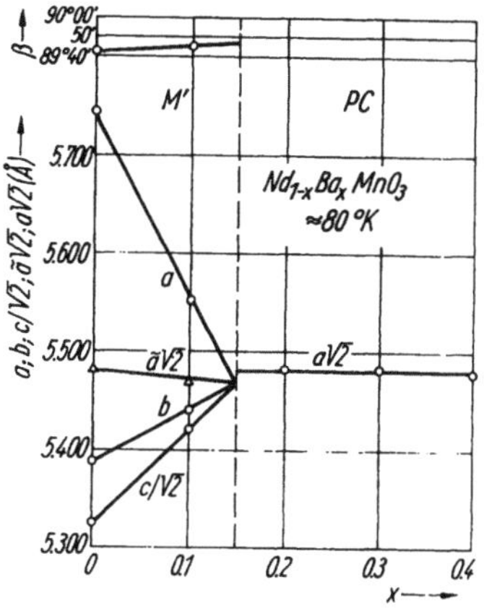

Fig. 4. Cell parameters at about 80 °K for the solid solution system $Nd_{1-x}Ba_xMnO_3$

the composition with $x = 0.5$ undergoes the O′ ⇌ M′ structure transition also between T_{rt} and T_{ln} (compositions with $x = 0.4$ and 0.6 preserve their symmetry unchanged).

In all the orthorhombic and monoclinic diffraction patterns, superstructure reflections were present, suggestive of the corresponding reflections of $GdFeO_3$ [17]. Therefore, we can believe that our orthorhombic and ever monoclinic manganites are "isostructural" with $GdFeO_3$ what gives us some reason to consider that octahedral rotation takes place, giving rise to a decrease of the angle γ.

3. Magnetic Measurements

The magnetization of samples (σ) was measured in the temperature interval from 77 to 800 °K with the aid of pendelum magnetometer using a maximum magnetic field of 13 kOe. The spontaneous magnetization (σ_s) was determined by extrapolating the $\sigma(H)$ isothermes to $H = 0$, and the paramagnetic Curie temperatures (θ_p) were obtained by extrapolating the linear high-temperature plots of the reciprocal susceptibility versus temperature. The temperature of magnetic ordering (T_M) was obtained as a point of interception of the tangent to the steep part of the $\sigma(T)$ dependence with the temperature axis.

3.1 Binary systems

For $NdMnO_3$, the extrapolation of the $1/\chi(T)$ dependence gives $\theta_p \approx 30$ °K, unlike the data of Vickery and Klann [18] where $\theta_p \approx 112$ °K. Figs. 5 to 7 show the extrapolated Curie temperature, the magnetic-ordering temperature, and the spontaneous magnetic moment at 77 °K for the $Nd_{1-x}Ca_xMnO_3$, $Nd_{1-x}Sr_xMnO_3$, and $Nd_{1-x}Ba_xMnO_3$ systems, respectively. In all the systems, θ_p increases rapidly with the increase of Mn^{4+} concentration. If we try to estimate the character of a net exchange interaction from the magnitude of extrapolated Curie temperature, it might be considered that at high temperatures the ferromagnetic exchange interaction is predominating.

In all the systems studied, $1/\chi(T)$ shows a linear high-temperature behaviour following the Curie-Weiss law (with $\theta_p > 0$). However, for some perovskites a low temperature plot $1/\chi(T)$ deviates from the linear law. In the

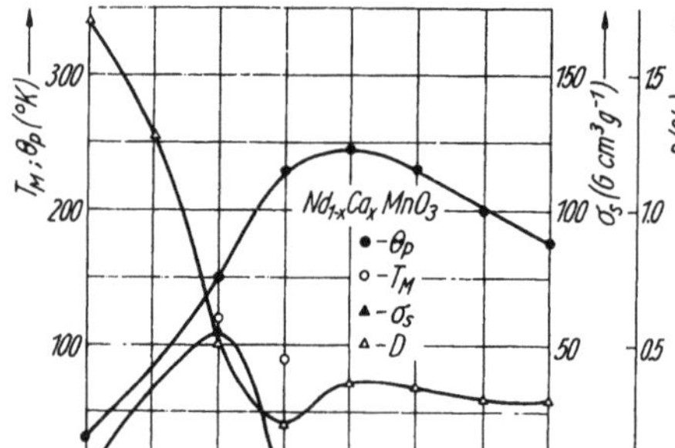

Fig. 5. Composition dependences of the paramagnetic Curie temperature (θ_p) of the spontaneous magnetic moment at 77 °K (σ_s), and of the average distortion ratio at about 80 °K (D) for the $Nd_{1-x}Ca_xMnO_3$ system, and the temperature of magnetic ordering (T_M) for two compositions with $x = 0.2$ and 0.3

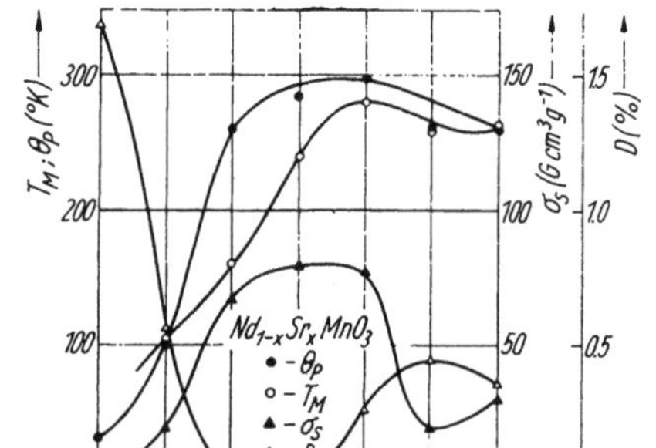

Fig. 6. Composition dependences of the paramagnetic Curie temperature (θ_p), of the magnetic-ordering temperature (T_M), of the spontaneous magnetic moment at 77 °K (σ_s), and of the average distortion ratio at about 80 °K (D) for the $Nd_{1-x}Sr_xMnO_3$ system

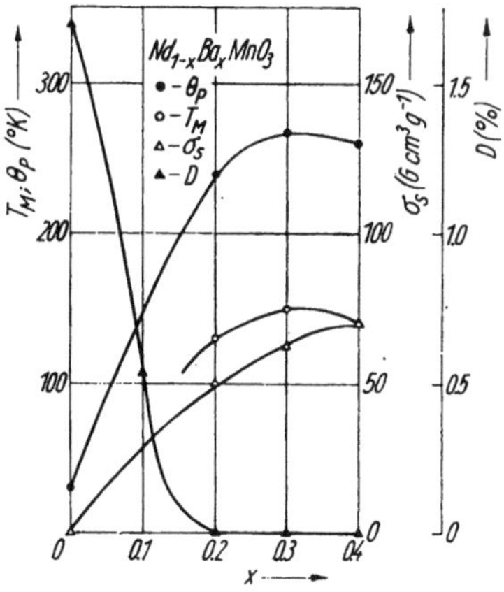

Fig. 7. Temperature dependences of the paramagnetic Curie temperature (θ_p), of the magnetic-ordering temperature (T_M), of the spontaneous magnetic moment at 77 °K (σ_s), and of the average distortion ratio at about 80 °K (D) for the $Nd_{1-x}Ba_xMnO_3$ system

$Nd_{1-x}Ca_xMnO_3$ system, the compositions with $x = 0.5$, 0.6, and 0.7 were found to cause anomalies in $1/\chi(T)$ in the form of sharp slope changes (Fig. 8). Below these anomalies, θ_p becomes negative. As shown by X-ray diffraction investigations, these sharp changes in the slope are accompanied by a structure transition to the M′-phase.

In the $Nd_{1-x}Sr_xMnO_3$ system with compositions in the range $0.2 \leqq x \leqq 0.4$, the decrease of temperature gives rise to a true ferromagnetic ordering (Fig. 6). The T_M-values are rather high and due to that fact $\sigma_s(T)$ can be extrapolated to 0 °K. The extrapolated values of σ_s are essentially in accord with theoretical ones, calculated for a pure spin contribution of Mn^{3+} and Mn^{4+} ions. Although θ_p and T_M practically show no decrease at $x > 0.4$, σ_s is much lower than the theoretical value. Moreover, while the sample with $x = 0.5$ is cooled below the magnetic-ordering temperature down to 180 °K, σ_s at first increases rapidly but drops sharply during further cooling (Fig. 11). This sharp drop of $\sigma_s(T)$ begins at the O′ ⇌ M′ structure-transition temperature.

In the $Nd_{1-x}Ba_xMnO_3$ system, the temperatures of magnetic ordering are not high (Fig. 7). To judge from a rapid increase of σ_s with decreasing temperature, it can be assumed that for compositions with $0.2 \leqq x \leqq 0.4$ the values of σ_s at 0 °K will be close to the theoretical ones.

3.2 Ternary systems

In the $Bi_{1-x}Ca_xMnO_3$ system, middle compositions exhibit the O ⇌ M′ structure transition at temperatures close to room temperature [15]. At the transition temperature, the temperature dependence of $1/\chi$ shows sharp changes in the slope, and at temperatures below the transition point, for the M′-phase $\chi(T)$ goes through a maximum (see Figs. 9 to 11 in [15]). These maxima were assumed to correspond to the temperatures of antiferromagnetic ordering. Consequently, anomalies of the specific heat must be found at these tempera-

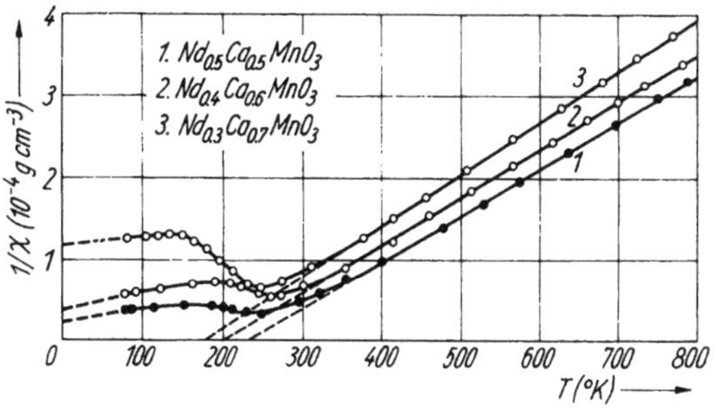

Fig. 8. Temperature dependence of the reciprocal magnetic susceptibility ($1/\chi$) for three compositions of the system $Nd_{1-x}Ca_xMnO_3$. $x = 0.5$, 0.6, and 0.7

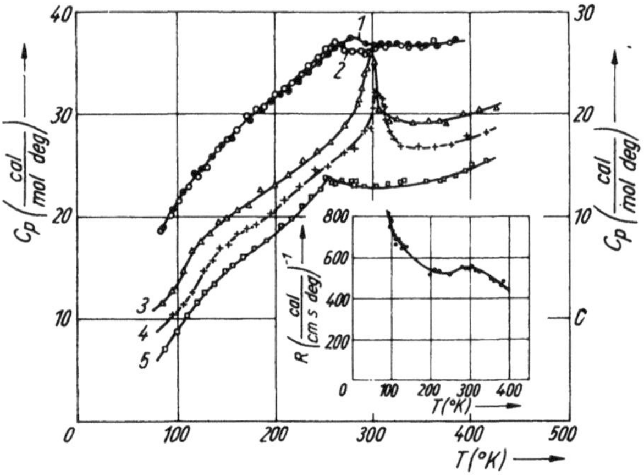

Fig. 9. Temperature dependence of the specific heat (C_p) for two compositions of the system $(Bi_{1-x}La_x)_{0.5}Ca_{0.5}MnO_3$ ($x = 0.1$ and 0.2 – curves 1 and 2, right scale) and for three compositions of the system $Bi_{1-x}Ca_xMnO_3$ ($x = 0.4$, 0.5, 0.7 – curves 3 to 5, left scale).
Notation; Curves 3 and 5 are displaced by 2 cal mol^{-1} deg^{-1} up and down in relation to curve 4, respectively. In the low right corner of the figure the temperature dependence of the heat resistivity (R) is shown for the $Bi_{0.6}Ca_{0.4}MnO_3$ composition. – The specific heat is measured with the aid of an adiabatic calorimeter [26] in the temperature interval from 80 to 420 °K

tures. Therefore, we have studied the thermal properties of some samples of the $Bi_{1-x}Ca_xMnO_3$ and $(Bi_{1-x}La_x)_{0.5}Ca_{0.5}MnO_3$ systems (samples with $x = 0.2$ and 0.4). The results are given in Fig. 9. The figure shows that 1. at the temperature of the $O \rightleftharpoons M'$ structure transition, for the compositions $Bi_{0.5}Ca_{0.5}MnO_3$ and $Bi_{0.6}Ca_{0.4}MnO_3$, a sharply defined maximum of specific heat is found; 2. the specific heat of the M′-phase has an anomaly associated with the antiferromagnetic transition; 3. the temperature of the specific heat maximum corresponding to the structure transition drops rapidly with introduction of La^{3+} ions (cf. curves in Fig. 9); 4. for the composition $Bi_{0.6}Ca_{0.4}MnO_3$, the heat resistance rises at the structure transition temperature.

In Fig. 10, the composition dependence of paramagnetic Curie temperature, of spontaneous magnetic moment, and of $O \rightleftharpoons M'$ transition temperature for the $(Bi_{1-x}La_x)_{0.5}Ca_{0.5};MnO_3$ and $La_{1-y}Ca_yMnO_3$ systems are given. In the $(Bi_{1-x}La_x)_{0.5}Ca_{0.5}MnO_3$ system, the temperature of the sharp change in $1/\chi(T)$ slope, corresponding to the $O \rightleftharpoons M'$ structure transition, decreases with increasing La^{3+} concentration. The temperature of antiferromagnetic ordering also decreases. At $x \geqq 0.3$, a spontaneous magnetic moment occurs.

In the $Bi_{0.5}(Ca_{1-x}Ba_x)_{0.5}MnO_3$ system, 5% of Ba^{2+} ions with respect to the total A ion amount is sufficient to drop the temperature of the sharp change in the $1/\chi(T)$ slope below liquid-nitrogen temperature. This composition shows a

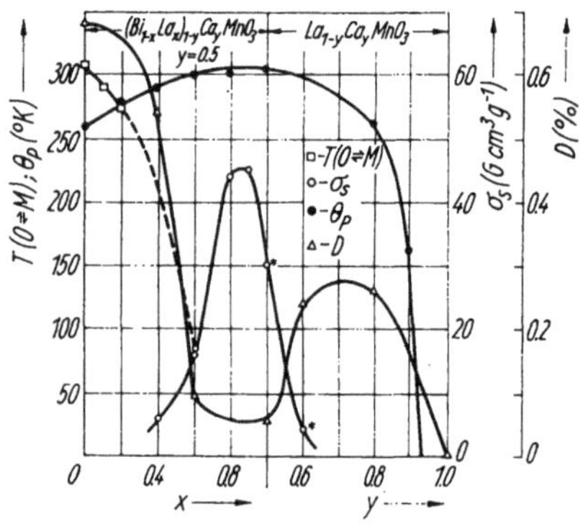

Fig. 10. Composition dependences of the paramagnetic Curie temperature (θ_p) of the spontaneous magnetic moment at 77 °K (σ_s), of the structure transition temperature (T) ($O \rightleftharpoons M'$), and of the average distortion ratio at about 80 °K (D) for the systems $(Bi_{1-x}La_x)_{1-y}Ca_yMnO_3$ ($y = 0.5$) and $La_{1-y}Ca_yMnO_3$ ($0.5 \leqq y \leqq 1$).
Notation: * the spontaneous magnetic moments for $La_{0.5}Ca_{0.5}MnO_3$ and $La_{0.4}Ca_{0.6}MnO_3$ which are quoted from [5]

marked deviation from the Curie-Weiss law in the temperature range below about 500 °K and the magnetic ordering is not observed. In the $Bi_{0.5}(Ca_{1-x}Pb_x)_{0.5}MnO_3$ system, the temperature of the sharp change in the $1\chi/(T)$ slope is lowered also with introduction of Pb^{2+} but here this effect slightly depends on the concentration. The temperature of antiferromagnetic ordering is probably below liquid nitrogen one. Unlike these cases, in the $Bi_{0.5}(Ca_{1-x}Sr_x)_{0.5}MnO_3$ system the Sr^{2+} ion substitution for Ca^{2+} ions leads to an increase in the temperature of the $1/\chi(T)$ anomaly, and hence of the structure transition temperature. Simultaneously the antiferromagnetic ordering temperature is lowered.

4. Discussion

After a review of the results both of our magnetic and crystallographic studies and of the works [5] to [8] and [15], the following conclusions may be deduced:

1. *In mixed manganites with perovskite-type structure, the electron configuration of Mn ions has a minor importance in the stabilization of either point symmetry (cubic, rhombohedral, orthorhombic, or monoclinic)*, except for the cases when a high Mn^{3+} concentration gives rise to the cooperative Jahn-Teller distortion [19]. A good evidence to this statement is given by changes of the $O \rightleftharpoons M'$ transition temperature in ternary systems with fixed manganese concentration and varying A ions (see, for example, Fig. 10). How strongly the lattice symmetry changes with A ion substitution can be seen from all the known binary systems of mixed manganites if comparing various compositions with fixed x ($La_{1-x}Me_xMnO_3$ where Me = Ca^{2+}, Sr^{2+}, Ba^{2+} [5], $Nd_{1-x}Me_xMnO_3$ where Me = Ca^{2+}, Sr^{2+}, Ba^{2+} — see Figs. 2 to 4 — and $Bi_{1-x}Ca_xMnO_3$ [15]).

The above lattice distortions appear to be caused primarily by some spatial effects connected with the radii of A ions, but without doubt a proper contribution to the distortion of the perovskite structure and to the stabilization of a given lattice symmetry comes from all physical parameters of A and B cations. A similar conclusion was done by Gilleo [20] on the basis of investigations of the perovskites $A^{3+}B^{3+}O_3$ (where A^{3+} is a rare earth ion), which crystallize in the orthorhombic structure whether or not the B^{3+} ion has a full or partly-filled 3d shell.

2. *Magnetic behaviour of mixed manganites is affected by lattice distortions of the perovskite structure.* This conclusion is suggested by the comparison of the magnetic with crystallographic properties of known systems of solid solutions containing Mn^{3+} and Mn^{4+} ions. In all the systems, a purely ferromagnetism is obtained only in those cases when the crystal lattice is either undistorted at all or has a small distortion. For the system $La_{1-x}Me_xMnO_3$ (Me = Ca^{2+}, Sr^{2+}, Ba^{2+}) this phenomenon was observed by some authors [5, 7]. Similarly, in our $Nd_{1-x}Me_xMnO_3$, (Me = Sr^{2+}, Ba^{2+}) and $(Bi_{1-x}La_x)_{0.5}Ca_{0.5}MnO_3$ systems the ferromagnetism occurs only in that composition range where the average distortion ratio D is close to zero (Figs. 6, 7, and 10). Also the spontaneous magnetic moment is reduced below the theoretical value for those compositions which have a distorted lattice at low temperatures although their extrapolated Curie temperatures are sufficiently high.

The lattice distortion, that arises at temperatures above the extrapolated Curie temperature, influences the temperature dependence of the magnetic susceptibility in that way that some deviation from the Curie-Weiss law occurs. For example, in some cases, the structure transition to the monoclinic phase

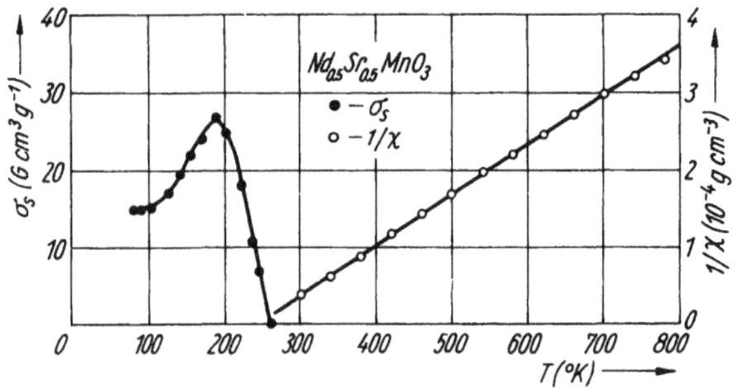

Fig. 11. Temperature dependences of the reciprocal magnetic susceptibility ($1/\chi$) and of the spontaneous magnetic moment (σ_s) for the composition $Nd_{0.5}Sr_{0.5}MnO_3$

causes the extrapolated Curie temperature to become negative and gives rise to antiferromagnetic order. This takes place in the systems $Bi_{1-x}Ca_xMnO_3$ (Figs. 10 and 11 in [15]), $Nd_{1-x}Ca_xMnO_3$ (Fig. 8), and in all our ternary stystems $Bi_{0.5}(Ca_{1-x}Me_x)_{0.5}MnO_3$ (Me = Sr^{2+}, Pb^{2+}, Ba^{2+}).

On the other hand, if structure transitions occur at temperatures below the point of ferromagnetic ordering, they appreciably decrease the magnitude of spontaneous magnetic moment as compared with the theoretical value (see the composition $Nd_{0.5}Sr_{0.5}MnO_3$ in Fig. 11).

3. *In mixed manganites, there exists a correlation both of the magnetic properties and of the conductivity with lattice distortion: but a direct correlation between conductivity and magnetic properties doesn't exist. That was taken for this correlation so far only reflects the dependence of both these properties on lattice distortion. This means that there are no reasons to attribute the existence of ferromagnetism to the double exchange.*

As it was mentioned above, the existence of ferromagnetism in the systems $La_{1-x}Me_xMnO_3$ (Me = Ca^{2+}, Sr^{2+}, Ba^{2+}) was previously interpreted as the result of Zener double exchange [11]. This interpretation seemed reasonable since ferromagnetic compositions have an elevated conductivity. Therefore, the unexpected lack of ferromagnetism in some distorted manganites ($La_{1-x}Ca_xMnO_3$ at $x \geqq 0.5$ [6] and $Bi_{1-x}Ca_xMnO_3$ [15]) was attributed to the electron ordering on manganese ions which quenches the ferromagnetic double exchange and simultaneously explains the transition into the more distorted perovskite phase.

If actually this structure transition was produced by the above mentioned electron ordering, it should be succeeded by sufficiently sharp changes in the

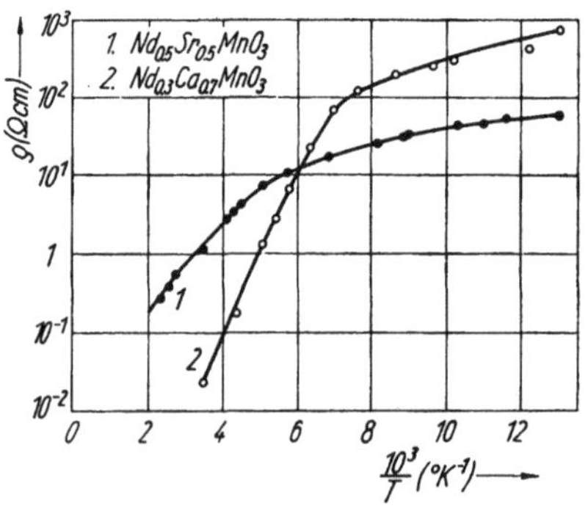

Fig. 12. Logarithm of specific resistivity (log ϱ) versus reciprocal temperature ($1/T$) for the compositions $Nd_{0.5}Sr_{0.5}MnO_3$ and $Nd_{0.3}Ca_{0.7}MnO_3$

magnitude of conductivity. For the samples of composition $Nd_{0.7}Ca_{0.3}MnO_3$ and $Nd_{0.5}Sr_{0.5}MnO_3$ (Fig. 12), the structure transition into the M′-phase gives only a sharp change in activation energy but doesn't influence the magnitude of conductivity. The observed change in activation energy may be accounted for by the change in the width of the energy gap between the valence and conduction bands resulting from the lattice deformation, and by the fact that there is an admixtured band between the above bands which mainly contributes to the low-temperature conductivity.

4. *The magnetic behaviour of mixed manganites can be explained on the basis of the Goodenough-Kanamori rules [21, 22], the lattice distortion being included into consideration.* These rules state that the exchange interaction between paramagnetic cations are antiferromagnetic ones if the cation orbitals which are directed towards a diamagnetic oxygen anion are half filled and of the same symmetry. The case of symmetry e_g gives the maximum orbital overlapping, therefore the exchange interactions are the most strong ones. But the exchange interactions between paramagnetic cations are ferromagnetic when one of the two cation orbitals, directed towards the opposite sides of intervening anion, is empty and the other is half filled. As it was shown by Anderson [21], these rules result from the superexchange theory according to which a superexchange interaction can occur only in the case if the two orbitals overlap themselves (an exchange integral is proportional to S^4 [23] where S is an overlapping integral). If the orbitals are orthogonal, the electrons can have a direct exchange integral and that can produce a ferromagnetic interaction.

We shall consider the Mn^{3+}–O^{2-}–Mn^{3+} interaction following Goodenough et al. [2] who have explained the existence of ferromagnetism in the systems $LaMn_{1-x}Me_xO_{3+x}$ (Me = Co^{III}, Ni^{III}) using the occupancy of Mn^{3+} ion orbitals only. The ferromagnetic samples were here considered to have no static local Jahn-Teller distortions. If that is the fact, the Mn^{3+} ion can be in a degenerate state. Therefore, its e_g-orbitals are such as if they were half filled. However, the orbital degeneracy is partly removed by dynamic correlation of Mn^{3+} ions (the quasi-static hypothesis) and the exchange interaction becomes ferromagnetic. If besides Mn^{3+} ions non Jahn-Teller ions appear in the lattice (particularly, Mn^{4+} ions), the dynamic correlation of these ions vanishes [2]. The Mn^{3+} ions interact in that case if their e_g-orbitals were half filled. In agreement with the Goodenough-Kanamori rules, the Mn^{3+} ions will interact with the Mn^{4+} ions ferromagnetically. In other words, the Mn^{3+}–O^{2-}–Mn^{3+} and Mn^{3+}–O^{2-}–Mn^{4+} interactions are both ferromagnetic with no static or cooperative Jahn-Teller distortion.

From this point of view, it can be understood that at high temperatures when the lattice is either cubic or has a small distortion, ferromagnetic exchange interactions are dominant in all systems of solid solutions. If with decreasing temperature lattice distortions occur, this can result in a static distortion of the Mn^{3+} environment because of which both the ground-state degeneracy of this ion will be removed and an ordering of covalent bonds will occur. This means that some fraction of exchange coupling becomes antiferromagnetic. If we consider the fact that the paramagnetic Curie temperature is subsequently defined by the concentration of ferro- and antiferromagnetic coupling, then the change of this concentration must be followed by a deviation of the magnetic susceptibility from the Curie-Weiss law $\chi = C/(T - \theta)$ and by a reduction of the spontaneous magnetic moment.

5. *The magnetic properties of mixed manganites in their connection to the distortion of the perovskite structure can be reasonably explained on the basis of the "hard-octahedra" model.* In the distorted mixed manganites the subtended angle γ is probably different from 180° as it takes place in orthoferrites [17, 24, 25]. The decrease of γ gives rise to some change in the exchange interactions. This assumtion can be indirectly confirmed by the fact that at high concentrations of Mn^{4+} ions, especially when the cell parameters are not large, the exchange interactions become more sensitive to the lattice distortion, for example in the $Nd_{1-x}Ca_xMnO_3$ system (see Fig. 2).

We cannot determine the angle γ experimentally because our compounds are not synthesized in monocrystalline form. But it might be assumed from superstructure reflections that in the presence of a distortion the oxygen octahedra are rotated. An estimation of γ from the cell parameters, as it was done by some authors [3, 17, 24], is not possible since the choice of a critical distance between the manganese ions is hindered by the presence of Mn ions with different valency. Therefore, the influence of γ on the exchange interactions can be considered only qualitatively.

When the distance between Mn ions decreases, there can take place a strengthening of the antiferromagnetic superexchange interaction between half-filled orbitals t_{2g} via $p\pi$–t_{2g} bonding. In the 180° linkage, the anion orbitals $p\sigma$ are orthogonal to the cation orbitals t_{2g} and the anion orbitals $p\pi$ are orthogonal to the cation orbitals e_g. The reduction of the angle γ can disturb the orthogonality both of $p\sigma$– to t_{2g}–orbitals and of $p\pi$- to e_g-orbitals. Consequently, independently of the removal of the degeneracy of the Mn^{3+} ion ground state, deformations and rotations of the octahedra can in general result in the occurrence of superexchange interaction between t_{2g}-orbitals via σ-bonding and between t_{2g}- and e_g-orbitals both via σ-bonding and via π-bonding, and that will compete with the ferromagnetic interaction. From this point of view, it will be easily understood why the observed extrapolated Curie temperature decreases with increasing distortion.

6. It is not possible at present to make a definite choice between the model based on the Goodenough-Kanamori rules and the "hard-octahedra" model. We believe that the static removal of degeneracy and the octahedra rotation simultaneously result from the lattice distortion and that the experimental data reflect the summary effect. It may be assumed that, if the unit cell is larger (Mn^{3+} concentration is high), the major contribution is due to the removal of the ground-state degeneracy of Mn^{3+} ions; and if the unit cell is small (Mn^{4+} concentration is high), the major contribution is due to the change in the interionic distances and in the subtended angle.

Acknowledgement

The authors would like to thank Prof. G. A. Smolensky for his interest and encouragement during all phases of this work and for stimulating discussions.

References

[1] E. E. Havinga, Phil. Res. Rep. **21**, 432 (1966).
[2] J. B. Goodenough, A. Wold, R. J. Arnott, and N. Menjuk, Phys. Rev. **124**, 373 (1961).
[3] J. Kanamori, J. appl. Phys. **31**, 14S (1960).
[4] G. H. Jonker and J. H. van Santen, Physica **16**, 337 (1950).

[5] G. H. Jonker, Physica **22**, 707 (1956).
[6] J. B. Goodenough, Phys. Rev. **100**, 564 (1955).
[7] E. O. Wollan and W. C. Kochler, Phys. Rev. **100**, 545 (1955).
[8] H. L. Yakel, Acta cryst. **8**, 394 (1955).
[9] M. G. Harwood, Proc. Phys. Soc. **B68**, 586 (1955).
[10] H. Watanabe, J. Phys. Soc. Japan **16**, 433 (1961).
[11] C. Zener, Phys. Rev. **82**, 403 (1951).
[12] P. W. Anderson and H. Hasegawa, Phys. Rev. **100**, 675 (1955).
[13] P. G. de Gennes, Phys. Rev. **118**, 141 (1960).
[14] J. H. van Santen and G. H. Jonker, Physica **16**, 599 (1950).
[15] V. A. Bokov, N. A. Grigoryan, and M. F. Bryzhina, phys. stat. sol. **20**, 745 (1967).
[16] F. Bertaut, and F. Forrat, J. Phys. Radium **17**, 129 (1956).
[17] S. Geller, J. chem. Phys. **24**, 1236 (1956).
[18] R. C. Vickery and A. Klann, J. chem. Phys. **37**, 1161 (1957).
[19] H. A. Jahn and E. Teller, Proc. Phys. Soc. **A161**, 220 (1937).
[20] M. A. Gilleo, Acta cryst. **10**, 161 (1957).
[21] P. W. Anderson, Magnetism, Ed. G. T. Rado and H. Suhl, Vol. I, Academic Press, New York 1965.
[22] J. B. Goodenough, Magnetism and the Chemical Bond, Wiley, New York/London 1963.
[23] J. Kondo, Progr. theor. Phys. (Kyoto) **22**, 41 (1951).
[24] P. Coppens and M. Eibschütz, Acta cryst. **19**, 531 (1965).
[25] D. Treves, M. Eibschütz, and P. Coppens, Phys. Letters (Netherlands) **18**, 216 (1965).
[26] E. D. Devyatkova and V. V. Tikhonov, Fiz. tverd. Tela **9**, 742 (1967).

(Received May 13, 1968)

Erratum

To the paper by B. A. Strukov, M. Amin, and V. A. Kopchik

Comparative Investigation of the Specific Heat of KH_2PO_4 (KDP) and KD_2PO_4 (DKDP) Single Crystals

phys. stat. sol. **27**, 741 (1968)

the authors request the following rectification:

On pages 744 and 745 ΔQ in the fourth column of Tables 1 and 2 must be multiplied by the factor 10^{-1}.

See Short Note by H. BERGER et al.

a b

Fig. 2. Crystallite size in dependence on film thickness a) film thickness 0.16 μm, b) film thickness 5.0 μm

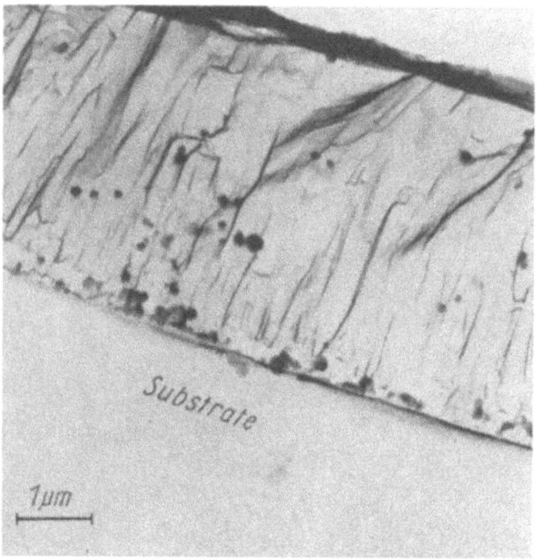

Fig. 3. Electron micrograph of a replica from the cross section of the CdS film shown in Fig. 2b

phys. stat. sol. 28, K97 (1968)
Subject classification: 14.3.1; 1.4; 22.4.1

Bereich Halbleiteroptik
des Physikalisch-Technischen Instituts
der Deutschen Akademie der Wissenschaften zu Berlin (a)
and
Bereich Halbleiteroptik der Sektion Physik
der Humboldt-Universität zu Berlin (b)

Thickness Dependence of Conductivity due to the Polycrystalline Structure in Evaporated CdS Thin Films

By
H. BERGER, W. KAHLE (a), and G. JÄNICHE (b)

In investigations of conductivity of semiconducting thin films it is generally supposed that the concentration of electrons is independent of film thickness and that the mobility of electrons does not depend on film thickness if the mean free path of the electrons is smaller than film thickness. Experimental results, e.g. on thin films of germanium (1), lead telluride (2), tellurium (3), and cadmium sulphide (4), in which a dependence of carrier concentration and mobility on film thickness was observed for thicknesses up to 10 μm, show that this assumption is not always satisfied for thicker films, too. In the mentioned papers the investigations were performed on several specimens of different thickness obtained from different evaporations. Such specimens are commonly not identical in real structure.

Therefore we have measured the conductivity in dependence on growing film thickness during evaporation. The thickness was determined by optical interference fringes. The glass substrate was held at a temperature of 160 $^{\circ}$C.

The current as a function of film thickness had the characteristic behaviour shown in Fig. 1. After a curved part follows a straight line for film thicknesses

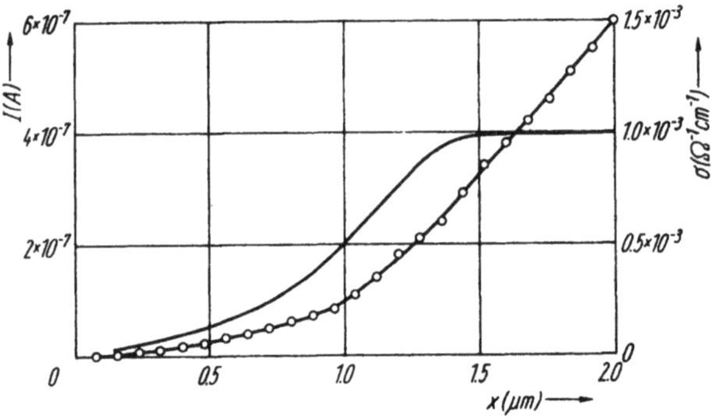

Fig. 1. Increase of current with growing film thickness (lower curve) and conductivity in dependence on film thickness obtained from the current curve by graphical differentation (upper curve), $\sigma(x) = \frac{1}{Eb}\frac{dI}{dx}$

greater than about 1 μm; the latter behaviour is expected for a conductivity independent of film thickness. In the case of a thickness dependent conductivity one gets $\sigma(x)$ from the measured current

$$I(x) = Eb\int_0^x \sigma(\xi)\,d\xi$$

by differentating with respect to film thickness x (E electric field and b gap length). Fig. 1 also sh ws the film thickness dependence of conductivity obtained by graphical differentation. From this plot it is visible that the conductivity has a gradient within the specimen.

The observed dependence of conductivity on film thickness can have the following reasons:

1. By interaction between substrate and the growing CdS film a depletion layer is forming at the interface substrate-CdS film.

2. The stoichiometric composition of the film changes during the deposition process. The excess of cadmium generally assumed to be the reason for a high dark conductivity, increases with growing film thickness.

3. By a decrease of the number of crystallite boundaries with film thickness the number of potential barriers, which can influence the current in polycrystalline films of semiconductors (5), decreases too.

In the present state of investigations it is not possible to exclude one of the above reasons or to make responsible only one of them for the observed gradient of conductivity.

As the first two reasons are leading to a gradient in carrier concentration their influence can be tested by Hall effect investigation. Such investigations are realized at present. First results received from CdS and CdSe films are revealing a connection of the conductivity with crystallite size.

To clarify this connection we have investigated films of different thickness deposited under identical evaporation conditions, by electron microscopy using replica technique. The typical behaviour, a growth in crystallite size with increasing film thickness, shows Fig. 2.

A diminution of the number of crystallite boundaries with growing film thickness can be realized by the growth mechanism of these films. It is known that CdS films of greater thicknesses made by normal deposition onto amorphous substrates are showing a texture in which the hexagonal axis is oriented normal to the substrate. Electron diffraction investigations performed by Addiss (6) and Shallcross (7) on very thin CdS films ($d < 1000$ Å), however, show that the crystallites are still randomly oriented.

This fact and the observed increase of crystallite size with film thickness are pointing out that the texture of CdS films originates from growth orientation. Crystallites grown with their basic plane on the substrate, show a strong areal growth in consequence of the very fast growth speed normal to the c-axis and they hinder crystallites with an inclined c-axis in further growth. By this growth selection the number of further growing crystallites decreases with increasing film thickness. The electron micrograph of a replica of a cross section through a film (Fig. 3), showing that the number of crystallite boundaries decreases with growing film thickness, points to such a growth mechanism, too.

From this results that the diminution of the number of crystallite boundaries can be made responsible for the observed increase in conductivity.

References

(1) J.E. DAVEY, Appl. Phys. Letters 8, 164 (1966).

(2) A. GOSWAMI and S.S. KOLI, Grundprobleme der Physik dünner Schichten, Proc. Internat. Symp., Göttingen 1966 (p. 646).

(3) V.D. DEOKAR and A. GOSWAMI, Grundprobleme ..., l.c. (p. 635).

(4) K.V. SHALIMOVA, A.F. ANDRUSHKO, V.A. DMITRIEV, and L.P. PAVLOV, Izv. vuzov (Moskva) 3, 134 (1964).

(5) H. BERGER, phys. stat. sol. 1, 740 (1961).

(6) R.R. ADDISS, Trans. 10th Nat. Vac. Symp., Macmillan Co., New York 1963 (p. 354).

(7) F.V. SHALLCROSS, Trans. AIME 236, 309 (1966).

(Received June 18, 1968)

phys. stat. sol. 28, K101 (1968)
Subject classification: 14.3; 22.1.3

Institut für angewandte Physik, Universität Karlsruhe

Halleffektsmessungen an Tellur in Magnetfeldern bis 150 kG

Von
W. ZIMMERMANN und F. STÖCKMANN

In einer systematischen Untersuchung wurden die Hallkoeffizienten und alle Koeffizienten des Tensors der magnetischen Widerstandsänderung an Te-Einkristallen mit einer Dotierungskonzentration von 10^{14} bis 10^{15} cm^{-3} gemessen. Das Magnetfeld betrug bis zu B = 150 kG (= $1,5 \times 10^{-3}$ $Vs\,cm^{-2}$), die Temperaturen T lagen zwischen 150 und 350 ^{o}K. Hier werden als Teilergebnis die B- und T-Abhängigkeiten der Hallkoeffizienten mitgeteilt. Eine ausführliche Darstellung der Gesamtergebnisse und ihrer Diskussion folgt später.

Die Proben wurden aus einem großen Einkristall nach den üblichen Verfahren (1) herauspräpariert und behandelt. Ihre Größe war etwa $0,5 \times 0,1 \times 0,1$ cm^3. Sie wurden vor dem Kontaktieren 320 h in Argon bei 380 ^{o}C getempert. Nach dem Anbringen der Hallsonden wurden die Kristalle in UHU-PLUS eingebettet, um mechanische Beanspruchungen möglichst gering zu halten.

Das Magnetfeld wurde durch Entladung einer Kondensatorbatterie über eine mechanisch sehr stabile Spule (2) erzeugt und durch die in einer Probespule induzierte Spannung gemessen. Die Zeitdauer für den Aufbau des Magnetfeldes betrug etwa 70×10^{-6} s. Die Temperatur konnte durch Anblasen des Kristalls mit erwärmter oder frisch verdampfter kalter Luft eingestellt werden. Magnetfeld und Hallspannung wurden an zwei Eingängen eines Oszillographen gemessen, wodurch eine genaue Zuordnung beider gewährleistet war.

In Fig. 1 sind die Hallkoeffizienten $R_1 = R_2$ (Magnetfeld senkrecht zur c-Achse) und R_3 (Magnetfeld parallel zur c-Achse) in Abhängigkeit von der Temperatur mit B als Parameter aufgetragen. Bei tiefen Temperaturen sind R_1 und R_3 nahezu gleich und für $B > 5 \times 10^{-4}$ $Vs\,cm^{-2}$ vom Feld unabhängig. Mit steigender Temperatur werden im Übergangsbereich zur Eigenleitung die Hallkoeffizienten kleiner und wechseln schließlich das Vorzeichen. Dieser Nulldurchgang ist von B abhängig, er erfolgt für größere Felder erst bei höheren Temperaturen. In Fig. 2 sind die Tem-

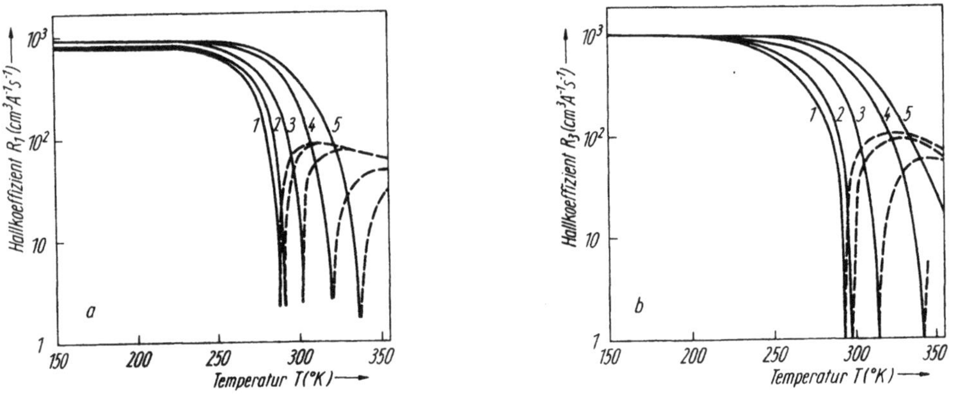

Fig. 1. Temperaturabhängigkeit der Hallkoeffizienten R_1 (a) und R_3 (b); ausgezogen: R > 0, gestrichelt: R < 0.

1 : B = 10×10^{-5} $Vscm^{-2}$,

2 : B = 20×10^{-5} $Vscm^{-2}$,

3 : B = 50×10^{-5} $Vscm^{-2}$,

4 : B = 100×10^{-5} $Vscm^{-2}$,

5 : B = 150×10^{-5} $Vscm^{-2}$; 10^{-5} $Vscm^{-2}$ = 10^3 G.

Es läßt sich beim bisherigen Stand der Messungen noch nicht mit Sicherheit entscheiden, ob die bei niedrigen Temperaturen gemessene B-Abhängigkeit von R_1 reell ist

peraturen T_o der Nulldurchgänge in Abhängigkeit vom Feld für beide Hallkoeffizienten aufgetragen. Sie steigen linear mit B und liegen bei verschwindendem Feld für R_1 bei T_o = 284 ^{o}K und für R_3 bei 286 ^{o}K.

Eine Vorzeichenumkehr des Hallkoeffizienten als Funktion des Magnetfeldes ist bei p-leitenden Materialien verschiedentlich beobachtet worden, z. B. bei p-Ge (3), p-InSb (4), p-Si (5). Sie läßt sich darauf zurückführen, daß beim Vorhandensein vergleichbarer Mengen mehrerer Trägersorten (Elektronen und Löcher) in kleinen Feldern diejenigen mit der höheren Beweglichkeit - i. a. Elektronen - das Vorzeichen der Hallspannung bestimmen, während für höhere Felder die weniger beweglichen Träger immer stärker an ihrem Zustandekommen beteiligt werden. Bei hohen Magnetfeldern sollte der Halleffekt von den einzelnen Beweglichkeiten ganz unabhängig sein und nur noch von der Gesamtkonzentration der beweglichen Ladungen

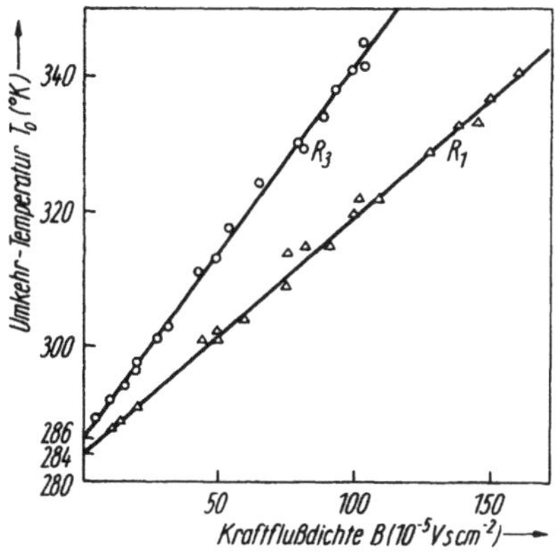

Fig. 2. Temperatur T_o des Nulldurchgangs von R_1 (△) und R_3 (o) in Abhängigkeit vom Magnetfeld

bestimmt werden (6):

$$R = 1/e\,(p - n) \qquad \text{für großes B.} \quad (1)$$

Die vorliegenden Meßergebnisse sind mit dieser Behauptung in voller Übereinstimmung. Bei tiefen Temperaturen sind die Hallkoeffizienten von T unabhängig, die vorhandenen Akzeptoren also vollständig ionisiert. Darum bleibt als Folge der Neutralitätsbedingung die Differenz (p - n) auch beim Übergang zur Eigenleitung, der den Vorzeichenwechsel des Halleffekts verursacht, konstant. Wie man gemäß Gleichung (1) erwartet, können also mit genügend hohen Magnetfeldern auch bei höheren Temperaturen die Abweichungen der Hallkoeffizienten von den konstanten Werten klein gehalten werden. Daher verschiebt sich auch der Vorzeichenwechsel der Hallkoeffizienten mit wachsendem Feld zu höheren Temperaturen. Die beiden Hallkoeffizienten sind bei tiefen Temperaturen und hohen Magnetfeldern ($B > 5\text{x}10^{-4}$ $Vscm^{-2}$) innerhalb der Meßgenauigkeit (besser als 10%) gleich: R_1 = = 930 $cm^3A^{-1}s^{-1}$, R_3 = 970 $cm^3A^{-1}s^{-1}$. Die daraus berechnete Konzentration p = = $(R\,e)^{-1} = 6,6\text{x}10^{15}$ cm^{-3} ist höher als die vom Hersteller angegebene Störstellenkonzentration von 10^{14} bis 10^{15} cm^{-3}. Das ist vielleicht darauf zurückzuführen, daß nicht alle bei der Präparation erzeugten akzeptorartigen Gitterfehler durch die nachfolgende Wärmebehandlung ausgeheilt worden sind.

In der starken Feldabhängigkeit der Hallkoeffizienten tritt der unterschiedliche Einfluß des Magnetfeldes auf die Transportgrößen der verschiedenen Ladungsträger deutlich zutage. Das zeigt sich besonders im Übergangsbereich zwischen Stör- und Eigenleitung. Hier ist der Zusammenhang zwischen Hallspannung und Magnetfeld auch für niedrige Magnetfelder nicht linear.

Herrn Dr. P. Grosse (Köln) danken wir für die Überlassung des Einkristalls.

Literatur

(1) P. GROSSE, Z. Phys. 193, 318 (1966).

(2) L. HOFFMANN und V. SCHEUING, Bericht CERN-63-36 (1963).

(3) R.K. WILLARDSON, T.C. HARMAN und A.C. BEER, Phys. Rev. 96, 1512 (1954).

(4) KH.I. AMIRKHANOV, R.I. BASHIROV und Z.A. ISMAILOV, Soviet Phys. - Solid State 5, 2071 (1964).

(5) B. ORAZGULYEV, Soviet Phys. - Solid State 8, 1569 (1966).

(6) C. HERRING und E. VOGT, Phys. Rev. 101, 944 (1956).

(Received June 14, 1968)

phys. stat. sol. 28, K105 (1968)
Subject classification: 18.2; 13

Laboratory of Theoretical Physics,
Joint Institute for Nuclear Research, Moscow

On the Use of Bose Operators for Ferromagnets at General Temperatures

By
E. PRAVECZKI

One of the problems in the theory of ferromagnets is that the spin operators are neither Bose operators nor Fermi operators, thus the methods developed in terms of quantum field theory cannot be applied directly.

Dyson (1) and Maleev (2) have shown a way how to replace the spin operators by Bose operators. Here we will apply the Bose operators for the Heisenberg model at general temperatures.

On the basis of Dyson's and Maleev's work we can express the relative magnetization as

$$\sigma = 1 - \frac{\mathrm{Sp}\,(a_f^+ a_f e^{-\beta \tilde{\mathcal{H}}} P)}{\mathrm{Sp}\,(S e^{-\beta \tilde{\mathcal{H}}} P)} \quad , \qquad (1)$$

where a_f, a_f^+ are the Bose operators; $\tilde{\mathcal{H}}$ is the Maleev transform of the Hamiltonian of the Heisenberg model, and P is the metric operator which excludes from physical quantities any non-physical contributions.

The relative magnetization will be determined with the use of the Green's function method. Let us define the Green's function as

$$G_\varkappa (t, x) = \langle\langle a_\varkappa (t) \; ; \; P a_\varkappa^+ [x] \rangle\rangle ,$$

where $a_\varkappa (t)$ and $a_\varkappa^+ [x]$ are the Fourier transforms of $a_f(t)$ and $\exp(x a_f^+ a_f) a_f^+$, respectively. The differential equation for $G_\varkappa (t, x)$ is

$$\frac{d}{dt} G_\varkappa (t, x) = \varrho (x) \delta (t) - i \varepsilon_\varkappa G_\varkappa (t, x) + i (SN)^{-1} \Sigma_{\mu\nu} (\varepsilon_{\varkappa+\mu-\nu} - \varepsilon_{\mu-\nu}) G_{\varkappa\mu\nu} (t, x), \qquad (2)$$

where $\varepsilon_{\varkappa}$ is the energy of an ideal spin wave, $\varepsilon_{\varkappa} = \mu\varkappa + 2S\omega_{\varkappa}$; ϱ (x) denotes the average $\left\langle \left[a_{\varkappa}, a^{+}_{\varkappa} [x] \right] \right\rangle$; and $G_{\varkappa\mu\nu}$ (t, x) is the higher order Green's function

$$G_{\varkappa\mu\nu}(t, x) = \left\langle\!\left\langle a^{+}_{\mu}(t)\, a_{\nu}(t)\, a_{\mu-\nu+\varkappa}(t) \, ; \, P a^{+}_{\varkappa}[x] \right\rangle\!\right\rangle .$$

The higher order Green's function can be reduced approximately to the former on the basis of Wick's theorem. Really, we get

$$G_{\varkappa\mu\nu}(t, x) \approx \left[\delta_{\mu\nu} \left\langle a^{+}_{\mu} a_{\nu} \right\rangle + \delta_{\mu,\varkappa+\mu-\nu} \left\langle a^{+}_{\mu} a_{\mu-\nu+\varkappa} \right\rangle \right] G_{\varkappa}(t, x) .$$

Using this approximation, we have, instead of equation (2)

$$\frac{d}{dt} G_{\varkappa}(t, x) = \varrho(x)\, \delta(t) - i\, \bar{\varepsilon}_{\varkappa} G_{\varkappa}(t, x) , \qquad (3)$$

where $\bar{\varepsilon}_{\varkappa}$ denotes the renormalized energy of the ideal spin wave, $\bar{\varepsilon}_{\varkappa} = \mu\varkappa + 2S\sigma\omega_{\varkappa}$. It is easy to solve equation (3). We write down only the result as

$$\left\langle a^{+}_{f}\, e^{x a^{+}_{f} a_{f}}\, P a_{f} \right\rangle = \varrho(x)\, \bar{n} , \qquad (4)$$

where we use the notation

$$\bar{n} = N^{-1} \sum_{\varkappa} \left[e^{\beta \bar{\varepsilon}_{\varkappa}} - 1 \right]^{-1} .$$

Considering the quantity ϱ (x), we can write, instead of equation (3)

$$\left[1 - \bar{n}(e^{x} - 1) \right] \left\langle a^{+}_{f} a_{f}\, e^{x a^{+}_{f} a_{f}} \right\rangle - \bar{n} e^{x} \left\langle e^{x a^{+}_{f} a_{f}} Q_{f} \right\rangle = 0 , \qquad (5)$$

where $Q_f = \left[a_f, P a^{+}_{f} \right]$. The metric operator contained by Q_f can be eliminated from (5) differentiating with respect to x. We get

$$\left\{ \frac{\partial^2}{\partial x^2} - \left[2S + 1 - \frac{2S\bar{n}e^{x}}{1 - \bar{n}(e^{x} - 1)} \right] \frac{\partial}{\partial x} + \frac{2S\bar{n}}{1 - \bar{n}(e^{x} - 1)} \right\} \left\langle e^{x a^{+}_{f} a_{f}} \right\rangle . \qquad (6)$$

The differential equation is analogous to that derived by Callen (3). Using the solution of the equation, we get for the relative magnetization defined by equation (1)

$$\sigma = 1 - \frac{2S+1}{S} \, \frac{\bar{n}^{2S+1}}{(1+\bar{n})^{2S+1} - \bar{n}^{2S+1}} .$$

Thereby the results achieved by Tyablikov (4), Tahir-Kheli and ter Haar (5) and the author (6) are reproduced. We can conclude that Bose operators can be used in the theory of ferromagnets even at high temperatures if the kinematical interactions of ideal spin waves are taken into account.

References

(1) F. J. DYSON, Phys. Rev. 102, 1217, 1230 (1956).

(2) S. V. MALEEV, Zh. eksper. teor. Fiz. 33, 1010 (1957).

(3) H. B. CALLEN, Phys. Rev. 130, 890 (1963).

(4) S. V. TYABLIKOV, Dokl. Akad. Nauk SSSR 148, 153 (1961).

(5) R. A. TAHIR-KHELI and D. TER HAAR, Phys. Rev. 127, 88 (1962).

(6) E. PRAVECZKI, phys. stat. sol. 5, 481 (1964).

(Received June 17, 1968)

phys. stat. sol. 28, K109 (1968)
Subject classification: 14.4.1 and 20.1; 13.1; 22.1.1

Istituto di Fisica dell' Università di Messina, Messina

Electroreflectance at Oblique Incidence in Ge[1)]

By

S. BALLARÒ, A. BALZAROTTI[2)], and V. GRASSO

Normal incidence electroreflectance experiments in semiconductors have recently provided detailed information on the optical transitions beyond the fundamental edge where the samples are strongly absorbent (1). At normal incidence the optical electric field is perpendicular to the modulating field and this configuration prevents studying the effects with parallel fields. Other methods of field application have been developed (2), but they can be successfully applied only to materials of high resistivity. In this note we propose a simple way to look at polarization effects, which is particularly suitable for conducting materials. If a beam of polarized light, reflected from the sample, makes an oblique angle with respect to the normal at the surface, the radiation field has a component in the direction of the external field when the electric vector of light lies in the plane of incidence. In principle one can expect that for a critical point occurring at $k \neq 0$ the selection rules for dipole transitions depend on the polarization of light even if the crystal has cubic symmetry. In the case of the critical points of Ge in [111] direction, the $\Lambda_3 \rightarrow \Lambda_1$ transitions are allowed (or forbidden) when the polarization vector of light is perpendicular (or parallel) to the [111] direction. The electroreflectance signal also depends on the direction of the field with respect to the direction of the negative effective masses of all the equivalent interband edges involved in the transition (3).

The sensitivity of the method was tested by measuring the modulated electroreflectance of Ge in the region 2 to 3.6 eV using an incidence angle φ of 45°. The modulation and detection technique is similar to that described by others (4). Typical curves of $\Delta R/R$ are shown in Fig. 1 for a (100) surface.

The measured structure is similar to that reported by other authors (5). The

---

1) Partially supported by the Consiglio Nazionale delle Ricerche through the G.N.S.M.

2) Present address: Istituto di Fisica dell' Università, L'Aquila, Italy.

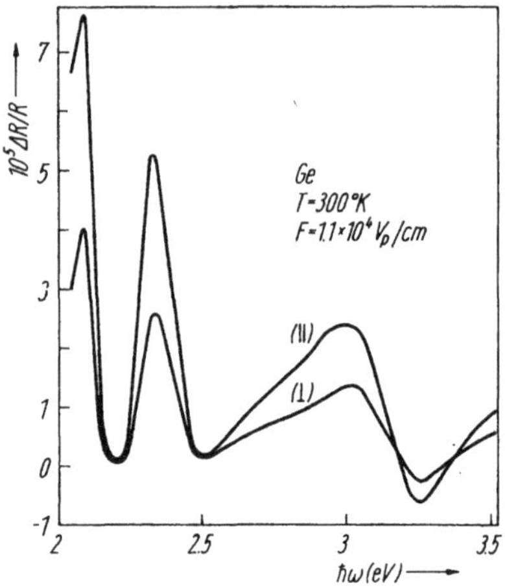

Fig. 1

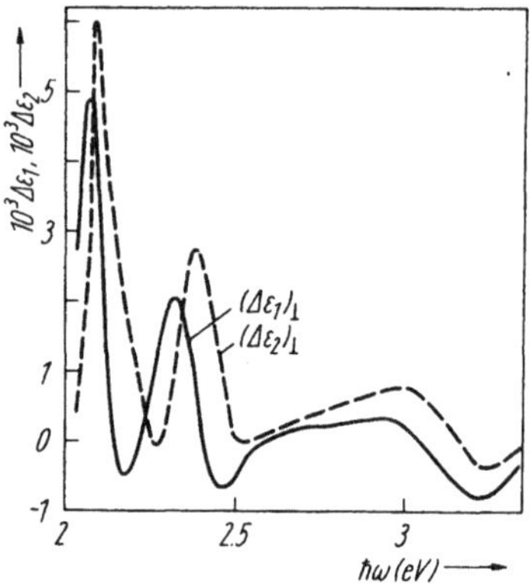

Fig. 2

Fig. 1. Relative change of reflectivity vs. photon energy of Ge with light vector having (∥) or not (⊥) a component in the direction of the external field (applied in [100] direction)

Fig. 2. $\Delta\varepsilon_1$ and $\Delta\varepsilon_2$ obtained by the Kramers-Kronig analysis of the data of Fig. 1

two main peaks refer to the spin-orbit split (0.23 eV) $\Lambda_3 \rightarrow \Lambda_1$ transition while the structure between 2.9 and 3.5 eV was tentatively associated (6) to the overlapping of the transitions $\Delta_5 \rightarrow \Delta_1$ (M_1 and M_2) and $\Gamma'_{25} \rightarrow \Gamma_{15}$ (M_0). For the case of perpendicular polarization the $\Delta\varepsilon_1$ and $\Delta\varepsilon_2$ spectra were obtained by the Kramers-Kronig analysis (Fig. 2) and they agree with those calculated from normal incidence electroreflectance results (5). Similar results are obtained with light polarized parallel to the electric field.

The lowering of $(\Delta R/R)_\perp$ curve with respect to $(\Delta R/R)_\parallel$ can be understood by application of Fresnel's formulas for $R_\perp$ and $R_\parallel$ (see, for example, (7)). Total differentiation of $R_\perp$ gives

$$\left(\frac{\Delta R}{R}\right)_\perp = \alpha_\perp (n, k, \varphi) \, \Delta n + \beta_\perp (n, k, \varphi) \, \Delta k \, .$$

A similar equation can be written for $(\Delta R/R)_\parallel$. The coefficients α and β for the two cases were calculated as a function of φ using the experimental values

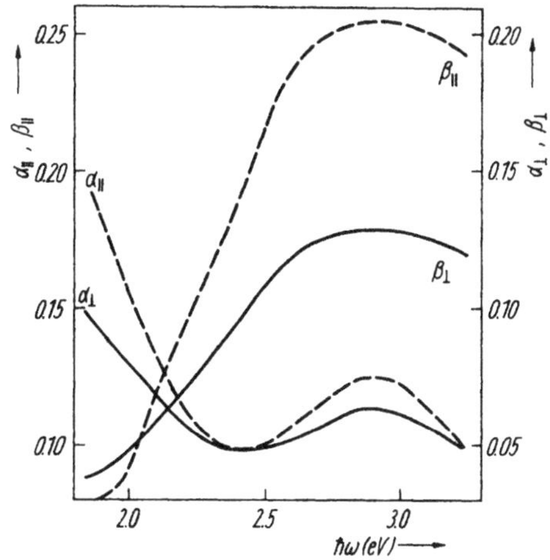

Fig. 3. Fractional coefficients α and β for an angle of incidence of 45° in the two cases of polarization

of n and k of Marple and Ehrenreich (8). These coefficients (for both polarization directions) intersect at 2.12 eV irrespective of φ, so that they weigh equally in the region of the main structure. In Fig. 3 we show the case of $\varphi = 45^\circ$. One can see that $\alpha_\parallel / \alpha_\perp = \beta_\parallel / \beta_\perp \approx 2$ and this accounts, within the accuracy ($\pm 1^\circ$) of the angle of incidence, for the lowering of $(\Delta R/R)_\perp$ curve with respect to $(\Delta R/R)_\parallel$ shown in Fig. 1.

In conclusion the structure near 2 eV in Ge looks the same for both polarization modes when the field is along a fourfold symmetry axis. The transition at 3.4 eV in Si is also insensitive to the polarization of light when the field is in [001] direction (5). This behaviour should be expected if the electric field is applied in a direction symmetric with respect to the eight equivalent hyperbolic saddle-points in k-space (9). On the other hand the effects on the higher-energy transitions are probably complicated by the overlapping of edges. Polarization effects should be larger with the field applied along the [110] axis (6) or in crystals of lower symmetry.

References

(1) B.O. SERAPHIN and R.B. HESS, Phys. Rev. Letters 14, 138 (1965).
K.L. SHAKLEE, F.H. POLLAK, and M. CARDONA, Phys. Rev. Letters 15, 883 (1965).

(2) V. REHN and D.S. KYSER, Phys. Rev. Letters 18, 848 (1967).

(3) N. BOTTKA and U. ROESSLER, Solid State Commun. 5, 939 (1967).

(4) B.O. SERAPHIN, Phys. Rev. 140, A1716 (1965).

(5) M. CARDONA, K.L. SHAKLEE, and F.H. POLLAK, Phys. Rev. 154, 696 (1967).

(6) D. BRUST, Phys. Rev. 134, A1337 (1964).

A.K. GHOSH, Phys. Rev. 165, 888 (1968).

(7) F. STERN, Solid State 15, 320 (1963).

(8) D.T.F. MARPLE and H. EHRENREICH, Phys. Rev. Letters 8, 87 (1962).

(9) F. AYMERICH and F. BASSANI, to be published.

(Received June 18, 1968)

phys. stat. sol. 28, K113 (1968)
Subject classification: 8; 22

Institute of Physics, Academy of Sciences
of the Azerbaidzhan SSR, Baku

Thermal Properties of Tin Sulphide Single Crystals

By
YA. S. FEIZIEV, YA. N. NASIROV, and O. N. NOVRUZOV

This paper is concerned with the investigation of the thermal diffusivity ($\varkappa$) and heat capacity (C_p) of a single-crystal SnS specimen by the light-pulse method (1 to 3), and from these parameters the thermal conductivity (λ) has been calculated.

As a standard we used melting quartz for checking the accuracy of $\varkappa$ and C_p measurements, whose thermal characteristics are well known.

The absolute accuracy of $\varkappa$ and C_p determination was 7%.

Measurements of $\varkappa$ and C_p on tin sulphide were carried out in an inert gas atmosphere.

The dependence of $\varkappa$ on temperature is presented in Fig. 1a. The thermal diffusivity is related to the thermal conductivity, specific heat, and density by

$$\varkappa = \frac{\lambda}{C_p \varrho} \quad .$$

Using the values of $\varkappa$ (experimental curves) and the heat capacity and density data for SnS the heat conductivity was calculated. The variation of density with temperature was not taken into account in the calculation. The heat capacity of SnS has been measured in the range from 200 to 900 $^{\circ}$K.

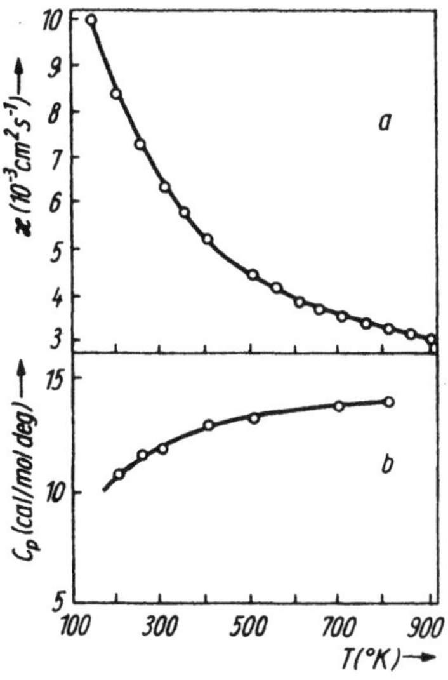

Fig. 1. a) Dependence of thermal diffusivity ($\varkappa$) on temperature T ($^{\circ}$K).
b) Dependence of heat capacity (C_p) on temperature T ($^{\circ}$K)

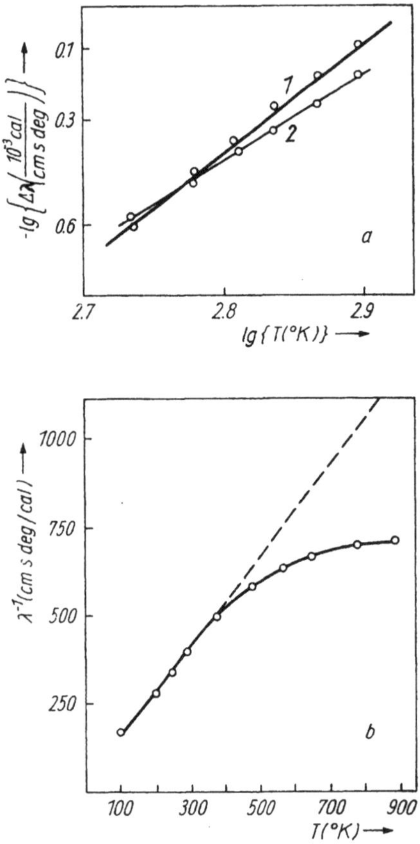

Fig. 2. a) Plot of the complementary heat conductivity (lg $\Delta\lambda$) vs. temperature (lg T); curve 1 - theoretical, curve 2 - experimental. b) Dependence of the thermal resistance ($1/\lambda$) on temperature T (°K)

The temperature dependence of C_p is presented in Fig. 1b. The presence of a large number of structural defects in the crystal under study is also substantiated by the consideration of the temperature dependence of thermal resistance (Fig. 2b). Below 400 °K this dependence holds well, which indicates that it is determined only by phonon scattering. Above 400 °K a complementary heat conductivity mechanism is observed. This complementary heat conductivity will be apparently that of photon heat transfer, which according to Genzel (5) is given by

$$\Delta\lambda = \frac{16}{3}\,\frac{\sigma_o}{\alpha}\,T^3.$$

Here, n is the refraction index (n = 3.6), α is the absorption coefficient (α = = 29 cm^{-1}), and σ_o = 1.38×10^{-12} cal/cm^{2}deg^4 is the Stefan-Boltzmann constant. It should be noted that α for this crystal has been determined in our laboratory and the value of n has been taken from the paper of other investigators (6).

In Fig. 2a theoretical (curve 1) and experimental (curve 2) values of the complementary heat conductivity are shown as a function of temperature in a double-logarithmic plot. From this figure it is seen that the complementary heat conductivity is proportional to T^3 and $T^{2.7}$, respectively.

The comparison of the experimentally observed complementary heat conductivities with those calculated from the Genzel formula shows that a heat transfer mechanism due to radiation appears in SnS above 400 °K (Table 1).

Table 1

T (^{o}K)	lg T	$\Delta\lambda_{calc} \times 10^3$ $\left(\frac{cal}{cm\,s\,deg}\right)$	$-lg(\Delta\lambda_{calc})$	$\Delta\lambda_{exp} \times 10^3$ $\left(\frac{cal}{cm\,s\,deg}\right)$	$-lg(\Delta\lambda_{exp})$
550	2.74	0.25	0.60	0.27	0.57
600	2.78	0.43	0.48	0.35	0.46
650	2.81	0.41	0.39	0.40	0.40
700	2.84	0.53	0.28	0.45	0.35
750	2.87	0.65	0.19	0.50	0.30
800	2.92	0.75	0.10	0.64	0.19

In conclusion the authors express their thanks to Prof. V.P. Zhuze and Prof. M.I. Aliev for their help and discussion of the results of this work.

References

(1) W.I. PARKER, R.I. JENKINS, C.P. BUTTLER, and G.Z. ABBOTT, J. appl. Phys. 32, 1679 (1961).

(2) W.I. PARKER and R.I. JENKINS, Advaced Energy Conversion 2, 87 (1962).

(3) O.N. NOVRUZOV, Izv. Akad. Nauk Azerbaidzhan SSR, Ser. fiz. 2, 91 (1966).

(4) E.D. DEVYATKOVA, A.B. PETROV, Y.A. SMIRNOV, and B.Y. MOIGESS, Fiz. tverd. Tela 2, 738 (1960).

(5) L. GENZEL, Z. Phys. 135, 177 (1953).

(6) W. ALBERS, C. HAAS, I.H. VINK, and I.D. WASSEHER, J. appl. Phys., Suppl. 32, 220 (1961).

(Received May 28, 1968)

phys. stat. sol. 28, K117 (1968)
Subject classification: 17; 21

Jena, Carl-Zeiss-Straße 1

On the Photostimulation of the Exoelectron Emission in the Infrared Spectral Region

By
P. GÖRLICH and M. LUDWIG

Although several authors (1) have tried to solve experimentally the problem of photostimulation of the exoelectron emission (PEE), no satisfactory clarification could so far be achieved, especially with regard to the threshold wavelength of the PEE as compared with the normal photoemission. The stimulation may be extended as far into the near infrared region, as was in particular shown by Bohun. We had the shift of the long-wavelength threshold of the PEE checked by Gubner (2), who was able to confirm the main results achieved so far.

During the last years two- and multiple-photon effects (e.g. in absorption, photoemission, photoconduction, luminescence) could experimentally be shown to exist, so that it was not unreasonable to assume multiple-photon effects also for the PEE. In our investigations we used the light of a pulsed $CaWO_4$:Nd laser (1060 nm; pulse energy 0.1 Ws) for photostimulation. The samples were pure aluminium wafers which were excited by scraping with a ≈ 5 mm broad edge of fracture of a ruby crystal and brought into the measuring position under an open methane point flow counter of the Fintelmann-Hartung type (3) 10 s after the end of excitation.

The yields of this stimulation (in electrons per quantum) were compared with the yields at stimulations with a ≈ 20 mm broad spectral region centred at 530 nm (radiation power $12x10^{-6}$ W). Measurements were in each case carried out over a period of 15 s. For three series of measurements the following yield ratios η (1060) : : η (530) were obtained for 53 different samples:

1 : 11 000, 1 : 20 000, 1 : 25 000.

After stimulation with 1060 nm an average of 144 electrons was emitted, the spread amounting to $\sigma = 37$. Each sample was first investigated while still unexcited. By having a laser pulse acting on the unexcited sample 1.6 electrons were released on an average by this pulse, with the spread amounting to $\sigma = 1.2$. The null

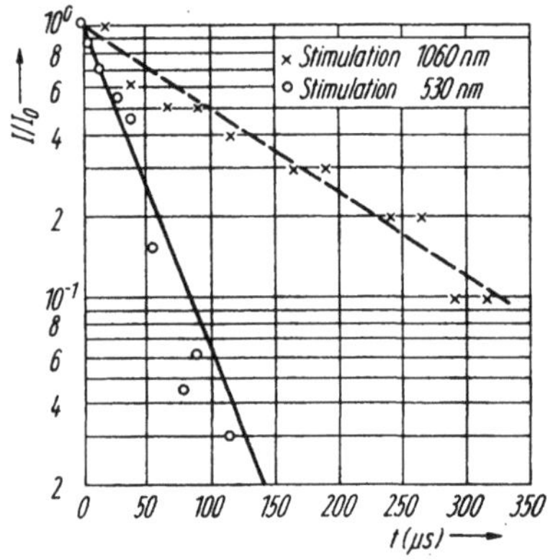

Fig. 1

effect of the counter was less than 1 pulse per minute.

No correlation between the spikes of the laser pulse and the electron emission was revealed by the oscillographic investigation of the counting pulses. This observation militates against the assumption that the emission is a result of the heating - either thermostimulated exoelectron emission or even thermionic emission - for, according to Lichtman and Ready (4) the temperature of the surface follows directly the spikes of the laser pulse. Another result of the oscillographic investigations was the observation that electrons were strongly emitted even after the stimulation had ceased. A similar behaviour could be observed for stimulations in the visible range with the integral light of an incandescent lamp. Fig. 1 shows the decay behaviour of these processes.

Worthy of note is the difference between the decay constants. Let the exoelectron pulse density at the end of the light pulse be I_o and the pulse density at the time t I_t.

Then $I_t / I_o = 1/e$

for t = 141 μs at stimulation with 1060 nm and

t = 34 μs at stimulation with 530 nm.

This fact suggests that the emission centres for both cases have different energetic levels and that possibly tunnel effects may play some role. For the filling of these centres one has - in the case of a stimulation with 1060 nm - to take into consideration an internal multiple-photon process. For clarifying the situation more investigations are being carried out.

References

(1) A. BOHUN, Czech. J. Phys. 5, 224 (1955).

A. BOHUN, phys. stat. sol. 3, 779 (1963).

J. KRAMER, Acta phys. Austriaca 10, 327 (1957).

B. SUJAK, Acta phys. Austriaca 10, 460 (1957).

(2) E. GUBNER, unpublished measurements.

(3) D. FINTELMANN and I. HARTUNG, Exper. Tech. Phys. 12, 36 (1964).

(4) D. LICHTMAN and F. READY, Phys. Rev. Letters 10, 342 (1963).

(Received June 28, 1968)

phys. stat. sol. 28, K121 (1968)
Subject classification: 5 and 8; 22.4.3

Institute of Physics, Polish Academy of Sciences, Warsaw (a),
and Institute of Precision Mechanics, Warsaw (b)

X-Ray Measurements of Lattice Dilatation in HgTe

By

L. ŚNIADOWER (a), M. PSODA (b), and R.R. GAŁĄZKA (a)

Recent magnetooptical measurements of mercury telluride exibit a strong temperature dependence of the basic band parameters. Investigations of the inter-band magnetoreflection at several temperatures between 1.5 and 77 °K (1) show that E_o (Γ_6 - Γ_8 energy separation for an inverted band structure (2), widely accepted for HgTe (3)) considerably increases with decreasing temperature. At the same time studies of the magnetoplasma reflection at 8, 85, and 296 °K (4), (5) have given evidence that the effective mass of electrons taken at the Fermi level strongly increases with decreasing temperature.

On the other side, results of pressure transport experiments are known (3), indicating a decrease of $|E_o|$ with increasing hydrostatic pressure. These facts attracted attention to the problem of the lattice dilatation in HgTe which is important for an interpretation of results of temperature and pressure experiments.

Two inconsistent sources of information concerning this problem are known. Measurements of Novikova and Abrikosov (6) made by means of a quartz dilatometer have shown an anomalous thermal expansion (decrease of temperature results in volume increase) below 60 °K; measurements performed by Alper and Saunders (7) by the same technique have not confirmed these results, they have given different values of the linear thermal expansion coefficient in the temperature region 77 to 300 °K.

In this note direct measurements of the lattice constant of mercury telluride between 77 and 300 °K are presented. A monocrystalline sample of HgTe - crystallizing in the zinc-blende structure - was placed in a low temperature camera of an X-ray diffractometer (diffractometer Rigaku-Denki Geiger-flex with radiation CuK_β). The dimensions of HgTe samples were 12x15x2 mm^3. The samples were optically polished before the experiment. A reflex from the (660) plane was observed

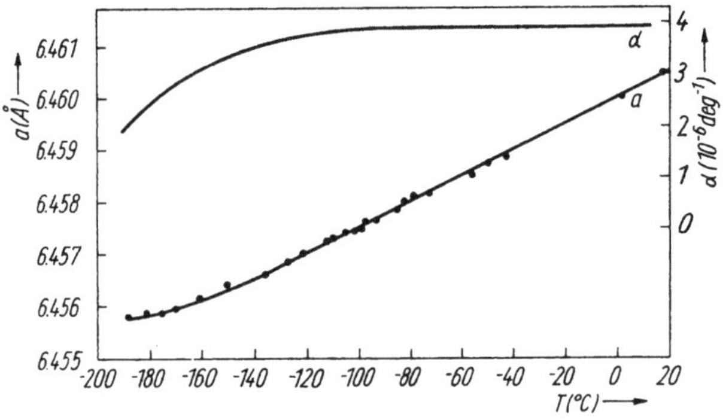

Fig. 1. The lattice constant a and thermal expansion coefficient α of mercury telluride as a function of temperature

at the angle $2\vartheta = 132^{\circ}$.

The temperature dependence of the lattice constant a is shown in Fig. 1. In the region from room temperature to 140 $^{\circ}$K the dependence is linear, while below 140 $^{\circ}$K a non-linearity can be observed. The α curve represents the thermal expansion coefficient calculated from experimental data for a. In the plateau region α is equal to 3.9×10^{-6} deg^{-1}.

In the investigated temperature region no anomalous thermal expansion was found. For temperatures between 300 to 120 $^{\circ}$K our values of α agree with those of Novikova and Abrikosov, obtained by means of a different experimental technique (quartz dilatometer). For temperatures lower than 120 $^{\circ}$K, a strong decrease of α and a transition through zero has not been found.

The authors are very grateful to Prof. Dr. L. Sosnowski for his constant interest in the present investigations. They are indebted to Dr. Dr. J. Lampe and J. Raułuszkiewicz for their valuable advice.

References

(1) C.R. PIDGEON and S.H. GROVES, Internat. Conf. II-VI-Compounds, Providence 1967, New York 1967 (p. 1080).

(2) S.H. GROVES and W. PAUL, Proc. 7th Internat. Conf. Phys. Semicond., Paris 1964 (p. 41).

(3) R. PIOTRZKOWSKI, S. POROWSKI, Z. DZIUBA, J. GINTER, W. GIRIAT,

and L. SOSNOWSKI, phys. stat. sol. 8, K135 (1965).

(4) L. ŚNIADOWER and R. KOWALCZYK, phys. stat. sol. 25, 233 (1968).

(5) L. ŚNIADOWER, phys. stat. sol. 28, K125 (1968).

(6) S.I. NOVIKOVA and N.KH. ABRIKOSOV, Fiz. tverd. Tela 5, 2138 (1963).

(7) T. ALPER and G.A. SAUNDERS, J. Phys. Chem. Solids 28, 1637 (1967).

(Received July 1, 1968)

phys. stat. sol. 28, K125 (1968)
Subject classification: 18 and 20.1; 13.1; 22.4.3

Institute of Physics, Polish Academy of Sciences, Warsaw

Temperature Studies of Magnetoplasma Reflection in Mercury Telluride

By
L. ŚNIADOWER

In the previous paper (1) a detailed analysis of magnetoplasma reflection in strongly degenerate semiconductors was presented. Theoretical curves were given to show the effect of the effective mass m^* taken at the Fermi level, the relaxation time τ, and the dielectric constant ε on the shape of the plasma edge. A method was given for the simultaneous determination of m^*, τ, ε, and the concentration of free carriers N, on the basis of experimental variations of reflectivity in the plasma edge region. This method was applied to the analysis of results obtained for n-type HgTe at 296 and 85 $^{\circ}$K.

In this paper similar measurements performed at 8 $^{\circ}$K are described. Magnetoplasma reflection measurements in the infrared region (15 to 25 nm) were made in the standard optical set-up (1). The samples were placed in a metallic helium cryostat with a KRS-5 window (Fig. 1). It was possible to move an internal part of the cryostat with respect to an external one in vacuum and at low temperature conditions. Thus, it was possible to perform measurements on different samples as well as on the reference mirror without any disturbances in the optical set-up. The temperature of the samples was measured by an Au + Fe - chromel thermocouple (Johnson, Matthey). Allen-Bradley resistors were used as helium level indicators.

Plasma and magnetoplasma reflection measurements were performed on three HgTe samples with approximately equal current carrier concentration (1.3 to 1.5x $\times 10^{18}$ cm^{-3}). After being polished the samples were etched in aqua regia.

Points in Figs. 2 and 3 represent results obtained for one of the samples for magnetic field values H = 0 and H = 22.4 kG respectively. Results of measurements performed at 300 and 85 $^{\circ}$K are also shown. We observe that a decrease of temperature shifts the whole plasma edge towards lower photon energies, reducing, at the same time, the distance between minima ($H \neq 0$) and sharpening the extrema. Sub-

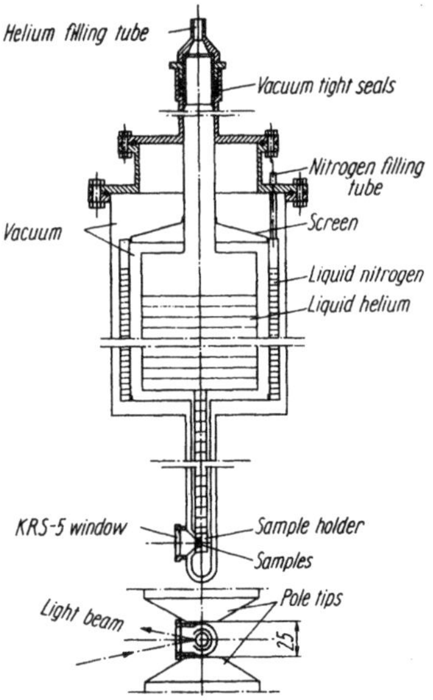

Fig. 1. Cross-section of the helium cryostat for magnetoreflection measurements in the infrared region

stantial part of the effect takes place between 300 and 85 $^{\circ}$K, while a transition to 8 $^{\circ}$K influences the curves to a much smaller degree.

It follows from a simplified theory of the plasma and magnetoplasma reflection (2 to 4) that for H = 0 the reflection coefficient has a minimum for a frequency $\omega_{min} = \omega_p(\varepsilon/\varepsilon - 1)^{1/2}$ where $\omega_p = (4\pi q^2 N/\varepsilon m^*)^{1/2}$ is the plasma frequency. For $H \neq 0$ the difference between the split minima is approximately equal to the cyclotron frequency $\omega_c = qH/m^*c$. Knowing from Hall constant measurement the concentration N, one can determine m^* and ε. Such a qualitative analysis of the results shown in Figs. 2 and 3 indicates that a decrease of temperature is followed by an increase of the effective mass and a decrease of the dielectric constant (in heavily doped samples of HgTe the electron concentration N is not dependent on temperature).

More detailed quantitative conclusions concerning changes of the above mentioned parameters as well as the relaxation time τ can be found on the basis of the theory presented in the previous paper (1). The reflectivity coefficient of the light beam normal to a reflecting surface is given by the formula

$$R = \frac{2\left[\mathrm{Re}\bar{N}^2 + \sqrt{(\mathrm{Re}\bar{N}^2)^2 + (\mathrm{Im}\bar{N}^2)^2}\right]\left[\sqrt{\frac{1}{2}\left(\mathrm{Re}\bar{N}^2 + \sqrt{(\mathrm{Re}\bar{N}^2)^2 + (\mathrm{Im}\bar{N}^2)^2}\right)} - 1\right]^2 + (\mathrm{Im}\bar{N}^2)^2}{2\left[\mathrm{Re}\bar{N}^2 + \sqrt{(\mathrm{Re}\bar{N}^2)^2 + (\mathrm{Im}\bar{N}^2)^2}\right]\left[\sqrt{\frac{1}{2}\left(\mathrm{Re}\bar{N}^2 + \sqrt{(\mathrm{Re}\bar{N}^2)^2 + (\mathrm{Im}\bar{N}^2)^2}\right)} + 1\right] + (\mathrm{Im}\bar{N}^2)^2} , \qquad (1)$$

where

$$\mathrm{Re}\bar{N}^2 = -\frac{1}{D}\left[(b + \varepsilon)(x^2 - y^2) - 2axy\right] + b + \varepsilon \qquad (2)$$

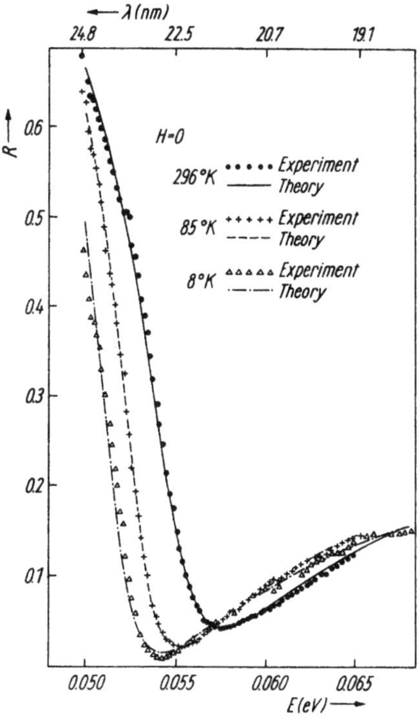

Fig. 2

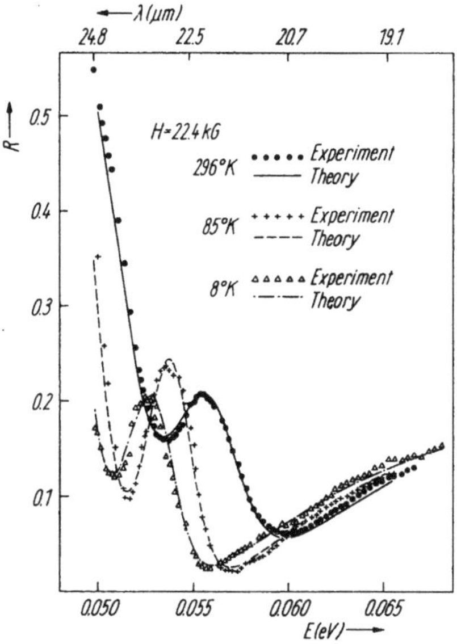

Fig. 3

Fig. 2. Influence of temperature on the plasma edge in HgTe for H = 0. Points represent experimental results, curves are found theoretically for parameter values given in Table 1 (H = 0)

Fig. 3. Influence of temperature on the magnetoplasma reflection for H = 22.4 kG. Points represent experimental results, curves are found theoretically for parameter values given in Table 1 (H ≠ 0)

$$\mathrm{Im}\bar{N}^2 = -\frac{1}{D}\left[a(x^2 - y^2) + 2xy(b + \varepsilon)\right] - a \quad , \tag{2}$$

and

$$\begin{aligned} a &= \frac{2\pi q^2 N\tau}{m^*\omega}\left[\frac{1}{1+(\omega+\omega_c)^2\tau^2} + \frac{1}{1+(\omega-\omega_c)^2\tau^2}\right] , \\ b &= \frac{2\pi q^2 N\tau}{m^*\omega}\left[\frac{(\omega+\omega_c)\tau}{1+(\omega+\omega_c)^2\tau^2} + \frac{(\omega-\omega_c)\tau}{1+(\omega-\omega_c)^2\tau^2}\right] , \end{aligned} \tag{3}$$

$$x = \frac{2\pi q^2 N\tau}{m^* \omega} \left[\frac{(\omega + \omega_c)\tau}{1 + (\omega + \omega_c)^2 \tau^2} - \frac{(\omega - \omega_c)\tau}{1 + (\omega - \omega_c)^2 \tau^2} \right] ,$$

$$y = \frac{2\pi q^2 N\tau}{m^* \omega} \left[\frac{1}{1 + (\omega + \omega_c)^2 \tau^2} - \frac{1}{1 + (\omega - \omega_c)^2 \tau^2} \right] , \tag{3}$$

$$D = (b + \varepsilon)^2 + a^2 .$$

$\overline{N}$ is the modulus of the propagation vector expressed in terms of the refraction index n and the extinction coefficient k: $\overline{N} = n - ik$.

To determine values of the parameters m^*, ε , τ , N from these measurements, the least square method was applied to approximate the experimental curve by a theoretical one. The form of the function (1 to 3) describing the reflectivity in the plasma edge region allows us to calculate all four parameters simultaneously for $H \neq 0$. For $H = 0$, i.e. $\omega_c = 0$, it can be seen from relation (3) that the parameters N and m^* appear only as a ratio N/m^* which prevents us from finding any one of them separately. Table 1 gives values of the parameters obtained in this way. The fit for H = 0 was made with the value of m^* taken from the fit for $H \neq 0$. Values of the electron concentrations N_H found from Hall constant measurements at 296 and 85 $^{\circ}$K are given for comparison.

The curves shown in Figs. 2 and 3 are just the theoretical curves corresponding to the fitted values of the parameters.

As can be found from Table 1 a change of temperature has a significant influence on the effective mass, the dielectric constant, and the relaxation time. The strong increase of m^* connected with a decrease of temperature is in qualitative agreement with the temperature variation of the $\Gamma_6 - \Gamma_8$ energy separation determined by interband magnetoreflection measurements (5). Taking into account a normal dilatation of HgTe (6) and knowing the influence of the lattice dilatation on the energy separation $\Gamma_6 - \Gamma_8$ from pressure experiments (8), one must look for an explanation of the strong temperature dependence of the conduction band structure in the presence of a significant electron-phonon interaction. The strength of this

Table 1

$H \neq 0$					
T (^{o}K)	ε	$\frac{m^*}{m_o}$	τ(s)	N (cm^{-3})	N_H (cm^{-3})
296	14.0 ± 0.2	0.046 ± 0.001	$(1.55 \pm 0.05) x 10^{-13}$	$(1.39 \pm 0.02) x 10^{18}$	$1.42 x 10^{18}$
85	12.8 ± 0.2	0.054 ± 0.001	$(2.64 \pm 0.1) x 10^{-13}$	$(1.40 \pm 0.02) x 10^{18}$	$1.41 x 10^{18}$
8	12.6 ± 0.2	0.057 ± 0.002	$(2.32 \pm 0.1) x 10^{-13}$	$(1.40 \pm 0.02) x 10^{18}$	-
$H = 0$					
T (^{o}K)	ε	$\frac{m^*}{m_o}$	τ(s)	N (cm^{-3})	N_H (cm^{-3})
296	14.1 ± 0.1	-	$(1.54 \pm 0.05) x 10^{-13}$	$(1.39 \pm 0.02) x 10^{18}$	$1.42 x 10^{18}$
85	12.9 ± 0.1	-	$(2.30 \pm 0.1) x 10^{-13}$	$(1.40 \pm 0.02) x 10^{18}$	$1.41 x 10^{18}$
8	12.7 ± 0.1	-	$(2.38 \pm 0.1) x 10^{-13}$	$(1.40 \pm 0.02) x 10^{18}$	-

interaction in HgTe seems to undermine the assumption (7) that a curvature of the band is not influenced by temperature alterations of the energy gap resulting from electron-phonon interaction.

The author is greatly indebted to Prof. Dr. L. Sosnowski for valuable discussions. He is also very grateful to Dr. J. Raułuszkiewicz for his valuable advice.

References

(1) L. ŚNIADOWER and R. KOWALCZYK, phys. stat. sol. 25, 233 (1968).

(2) W.G. SPITZER and H.Y. FAN, Phys. Rev. 106, 882 (1957).

(3) B. LAX and G.B. WRIGHT, Phys. Rev. Letters 4, 16 (1960).

(4) L. ŚNIADOWER, J. RAUŁUSZKIEWICZ, and R.R. GAŁĄZKA, phys. stat. sol. 6, 549 (1964).

(5) C.R. PIDGEON and S.H. GROVES, Internat. Conf. II-VI-Compounds, Providence 1967, New York 1967 (p. 1080).

(6) L. ŚNIADOWER, M. PSODA, and R.R. GAŁĄZKA, phys. stat. sol. 28, K121 (1968).

(7) H. EHRENREICH, J. Phys. Chem. Solids 2, 131 (1957).

(8) R. PIOTRZKOWSKI, S. POROWSKI, Z. DZIUBA, J. GINTER, W. GIRIAT, and L. SOSNOWSKI, phys. stat. sol. 8, K135 (1965).

(Received July 1, 1968)

phys. stat. sol. 28, K131 (1968)
Subject classification: 14.4.1 and 20.1; 13.1; 13.2; 22.6

Institute of Applied Physics,
Academy of Sciences of the Moldavian SSR, Kishinev

Electroreflectance Spectrum of Cuprous Oxide Crystals

By
S. N. SHESTATSKII and V. V. SOBOLEV

Cu_2O crystals have an important place in investigations of semiconductors (1, 2). In the region of the long-wavelength fundamental absorption edge of this compound four exciton series are known due to the doublet structure of the highest valence band and the lowest conduction band at the centre of the Brillouin zone. According to (1, 2) at T = 77 $^{\circ}$K the most intensive exciton absorption lines occur at λ = = 6125 Å (yellow series) and at λ = 5496 Å (green series) and the most intensive exciton reflection lines of quantum number n = 1 of the blue-green and blue series are placed at λ = 4796 Å and λ = 4569 Å, respectively. By extrapolation of the two latter exciton series to $n = \infty$ their dissociation limits equal to the band gaps were approximately determined: λ = 4712 Å (E = 2.631 eV) for the blue-green and λ = = 4480 Å (E = 2.765 eV) for the blue series.

During the last years the electroreflectance method was effectively applied for study of interband transitions due to its high resolution and sensitivity (3, 4). In the present work an attempt was made to study the electroreflectance of Cu_2O. The measurements were performed at room temperature in the wavelength region 600 to 400 nm with the aid of a registration apparatus using a monochromator with dispersion of 29 Å/mm at λ = 271 nm. The block-scheme of the experimental set-up is the same as that of Cardona et al. (4). The electric field arised at the contact of the Cu_2O crystal (1 mm thick and 3x15 mm^2 area) with an electrolyte (0.1 M KCl water solution). The modulating voltage of 4 $V_{eff.}$ superimposed on the negative dc bias of 1.5 V was applied between the sample and the Pt electrode. The gold film deposited on the back surface of the sample and covered by epoxy resin was used as ohmic contact.

The Cu_2O electroreflectance spectrum obtained is represented in Fig. 1. The spectrum consists of two narrow and strong negative peaks: the peak I at λ = 457 nm

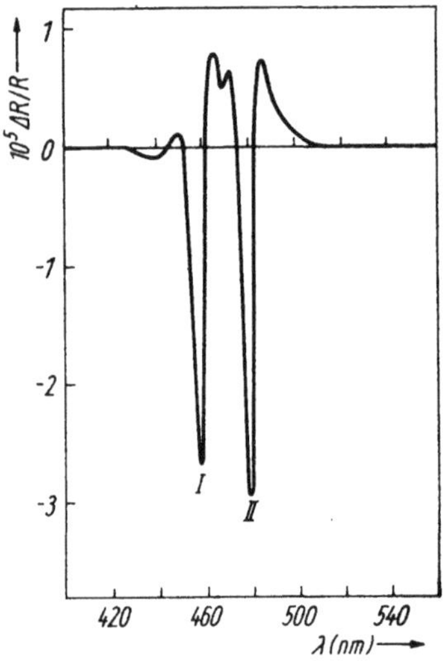

Fig. 1

with two positive oscillations at each side at λ = 450 nm and λ = 465 nm and the peak II at λ = 479 nm with the positive oscillations at λ = 472 nm and λ = 486 nm. The line shape of the spectrum agrees qualitatively with theoretical predictions (5).

The room temperature reflectivity spectrum of Cu_2O in the region of the blue-green and blue exciton series is unknown. However, from the temperature dependence of the position of the yellow and green series in the interval T = = 77 to 293 ^oK (6, 7) and of the blue-green and blue series in the interval T = 4 to 169 ^oK (8, 9) one can assume that the spectrum in the range under investigations should shift to longer wavelengths by 80 to 100 Å when passing from 77 to 293 ^oK. Thus at room temperature the most intensive exciton lines with n = 1 and the respective interband gaps would be located approximately at λ = 489 nm (exciton) and λ = 479 nm (interband) for the blue-green and at λ = 466 nm (exciton) and λ = 457 nm (interband) for the blue series.

As is seen from Fig. 1 the positions of peaks I and II agree well with the above mentioned positions of interband transitions. Therefore it is natural to suppose that the peaks at λ = 457 nm (E = 2.71 eV) and λ = 479 nm (E = 2.59 eV) are caused by direct interband transitions from the upper doublet valence band to the second conduction band.

In the range of the yellow and green exciton series no structure was observed. This is possibly caused by the fact that the absorption coefficient in this range is by some orders smaller than that for the blue and blue-green series (6 to 8).

For a more convincing interpretation of the electroreflectance spectrum of Cu_2O investigations at different temperatures and electric fields are nessesary; they are carried out at present.

References

(1) E.F. GROSS, Uspekhi fiz. Nauk 63, 575 (1957).

(2) E. F. GROSS, Uspekhi fiz. Nauk 86, 433 (1962).

(3) B.O. SERAPHIN, J. appl. Phys. 37, 721 (1966).

(4) M. CARDONA, L. KERRY, K. L. SHAKLEE, and F.H. POLLAK, Phys. Rev. 154, 696 (1967).

(5) D. ASPNES, Phys. Rev. 147, 554 (1966).

(6) E. F. GROSS and CH. GUAN-IN, Fiz. tverd. Tela 4, 261 (1962).

(7) V.V. SOBOLEV, Dokl. Akad. Nauk SSSR 165, 803 (1965).

(8) S. NIKITINE, Proc. Internat. Conf. Phys. Semicond., Prague 1960.

(9) J. PASTRNJAK, Proc. Internat. Conf. Phys. Semicond., Prague 1960.

(Received July 1, 1968)

phys. stat. sol. 28, K135 (1968)
Subject classification: 20.1; 13.2; 22.5; 22.5.2; 22.5.3

School of Physics, University of Warwick, Coventry

The Variation of Exciton Peak Position with Lattice Parameter in Metal Iodide Crystals

By
M.R. TUBBS

Experiments have shown that the photon energies of interband reflectivity peaks in the intrinsic spectra of group IV-VI-semiconductors are inversely proportional to the lattice parameter (1, 2). In this note it is pointed out that a relation of this kind also exists for intrinsic exciton peaks in many partially ionic metal iodide crystals. In these materials the photon energies of exciton peaks are inversely proportional to the square of the interionic distance.

The exciton spectra of ten metal di- and tri-iodides with layer structures have recently been described in some detail (3). The results of this study are summarized in Fig. 1 which shows how the photon energies of exciton peaks vary with the inverse square of the interionic distance. The lines marked X_1 and X_2 in Fig. 1 represent exciton peaks belonging to the low energy halogen doublet at the absorption edge in layer structure iodides while the lines X_3 and X_4 correspond to the two peaks of the high energy halogen doublet. The points for the two tri-iodides lie close to the lines for the di-iodides since both types of material consist of h.c.p. iodide ions. The lines Z_1 and Z_2 describe the two exciton peaks of the low energy halogen doublet in f.c.c. alkali metal iodides. Absorption data for these materials were taken from the literature (4 to 6). The Z_2 points given for KI and RbI correspond to the model proposed by Phillips (7). The X_1 peak for CaI_2 lies nearer to the line for the alkali iodides than that describing the di-iodides. This may reflect uncertainties in the value for the lattice parameter of this material or on the applicability of an excitation model (3).

The low energy halogen doublet in an isotropic iodide corresponds to the atomic excitation $5p^6 \rightarrow 5p^5\ 6s$ of the iodide ion. Two peaks are observed corresponding to the spin-orbit interaction of the p^5 hole in this ion; this situation is often compared to jj coupling in an atom, particularly when the direct electron-hole interaction is

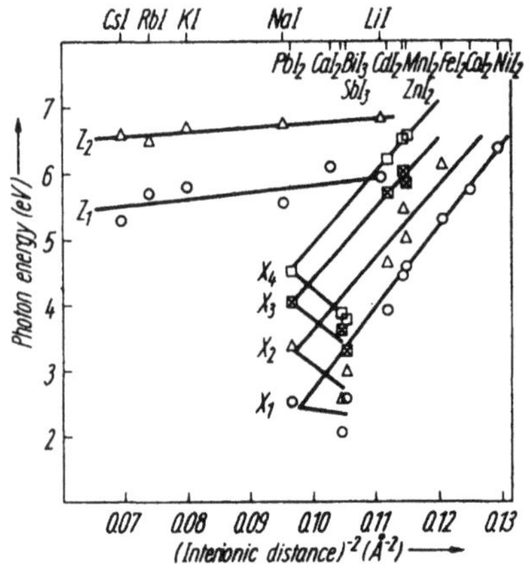

Fig. 1. The variation of photon energy with the inverse square of the interionic distance for exciton peaks in alkali iodides (Z_1 and Z_2) and layer structure di- and tri-iodides (X_1 to X_4)

smaller than the spin-orbit interaction of the hole. In an anisotropic iodide the situation is more complex since, for I^-, the 6p excited level occurs at about the same energy as the 6s. The strong electric field acting on an iodide ion in the CdI_2 structure hybridizes the 6s and 6p levels to form $5p^5$ (6s + 6p), $5p^5$ (6s - 6p) excited states (8). When the spin orbit interaction is included, allowed transitions to these excited states give four exciton peaks - X_1 to X_4 in Fig. 1 - for anisotropic metal iodides.

It is clear from Fig. 1 that the photon energies of exciton peaks in a series of isostructural materials are approximately proportional to the inverse square of the interionic distance. This type of variation is expected for a particle-in-a-box model of an exciton. In alkali halides the box dimension can reasonably be taken as the interionic distance since Dexter has pointed out that, for NaCl (9), 75% of the charge density of the Cl^- electron is contained within the equilibrium sphere of this ion. In the layer structure iodides the interionic distance is an appropriate parameter since the electron in a $5p^5$ (6s + 6p) excited state is most probably found on that side of an I^- ion adjacent to a metal ion. An excitation model is also supported by observations of an n = 2 exciton in BiI_3 (3), KI and RbI (5) which suggest that the n = 1 exciton radius is of the order of the interionic distance.

The correlation between exciton peak positions and interionic distance can thus be understood on an excitation model for Frenkel excitons. A knowledge of the form of this correlation assists in the identification of particular exciton peaks in a series of isostructural compounds and also allows exciton positions to be predicted for materials for which absorption spectra are not available.

Acknowledgement

I should like to thank the S. R. C. for their continuing support for this work.

References

(1) M. CARDONA and D.L. GREENAWAY, Phys. Rev. 133, A1685 (1964).

(2) D.L. GREENAWAY and R. NITSCHE, J. Phys. Chem. Solids 26, 1445 (1965).

(3) M.R. TUBBS, J. Phys. Chem. Solids (1968), in the press.

(4) J.E. EBY, K.J. TEEGARDEN, and D.B. DUTTON, Phys. Rev. 116, 1099 (1959).

(5) F. FISCHER, Z. Phys. 160, 194 (1960).

(6) A.A. GAVINI and M. CARDONA, Phys. Letters (Netherlands) 27A, 112 (1968).

(7) J.C. PHILLIPS, Phys. Rev. 136, A1705 (1964).

(8) M.R. TUBBS, Proc. Roy. Soc. 280A, 566 (1964).

(9) D.L. DEXTER, Phys. Rev. 108, 707 (1957).

(Received July 1, 1968)

phys. stat. sol. 28, K139 (1968)

Subject classification: 4; 5; 22.1.2

Institute of Experimental Physics, University of Warsaw

Observations of the Effects of X-Ray Polarization on Pendellösung Fringes with the Three-Crystal X-Ray Spectrometer

By

E. ZIELIŃSKA-ROHOZIŃSKA

It has been experimentally demonstrated with X-ray diffraction topographs by Hart and Lang (1) that the modulation of Pendellösung fringes may be seen if the incident beam is unpolarized. A theory of Pendellösung fringes due to the incident beam of an unpolarized spherical wave has been presented by Hattori et al. (2). It was found that the fading of Pendellösung fringes is a result of beats between two sets of fringes corresponding to the perpendicularly polarized X-rays.

In this paper Pendellösung fringes are observed for a Si single crystal free of dislocations with a three-crystal X-ray spectrometer (3). A plane-wave theory given by Max von Laue (4) for a monochromatic parallel beam in this case can be used to describe Pendellösung fringes.

When intensities of diffracted or transmitted beams are measured with a three-crystal X-ray spectrometer the intensity of X-ray radiation partially polarized after three reflections is recorded by a detector. Now we assume:

1. Rotation axes of monochromator crystals (I, II) and rotation axis (III) of the investigated crystal are exactly parallel. Reflecting crystal planes I, II, III contain the rotation axes I, III, III respectively. It is thought to be the correct adjustment of the spectrometer.

2. I and II reflections are symmetrical. In general the third reflection may

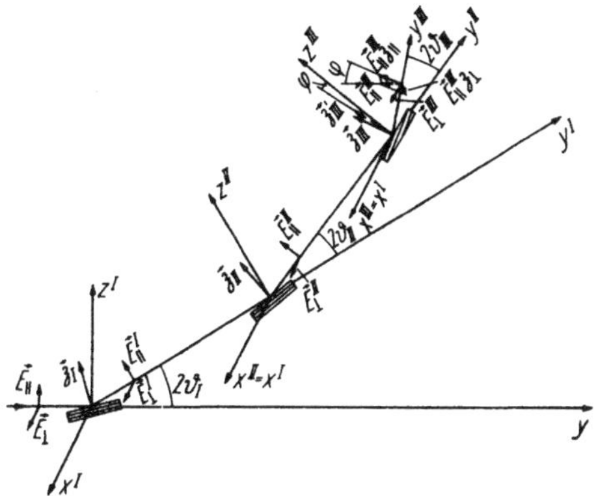

Fig. 1. Polarization of the unpolarized incident beam by a three-crystal X-ray spectrometer

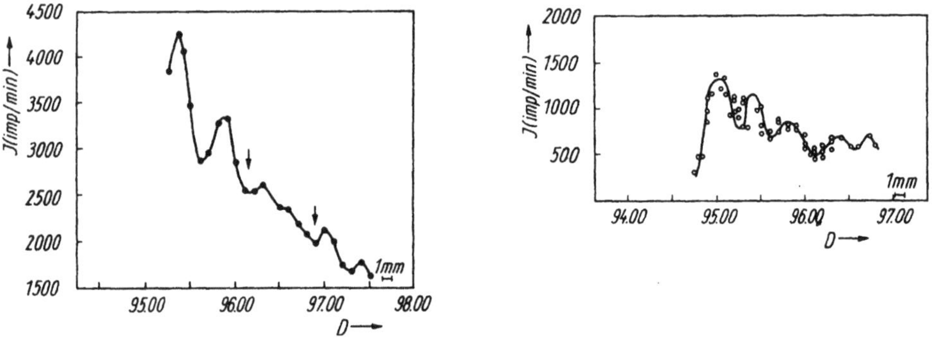

Fig. 2. Dependence of the diffracted intensity versus the crystal thickness obtained with the X-ray spectrometer. The abcissa scale indicates the shift of the crystal parallel to its surface. 1 mm of this scale corresponds to the increase in crystal thickness of 0.0038 mm.

a) Ge single crystals placed in antiparallel position (1, +1) were used as monochromators. Measurements were made for the 220 reflection of partially polarized $CuK\alpha_1$ radiation. The arrows indicate the fading region.

b) The same Ge single crystals placed in antiparallel position (3, +1) were used as monochromators. Measurements were made for 220 reflection of plane-polarized $CuK\alpha_1$ radiation

be asymmetrical, i.e. reflection at the crystal face which forms an angle $\varphi \neq 0$ with the reflecting lattice planes.

Let $\mathfrak{E}$ denotes the electric vector of the unpolarized incident beam. Moreover, $E_{\parallel}$, $E_{\perp}$ describe their independent components (Fig. 1) along two mutually perpendicular directions. According to (5) the component $E_{\perp}$ is not affected after reflections, the component $E_{\parallel}$ is attenuated by the cosine of double reflection angle 2ϑ. Hence, after the third reflection

$$E_{\perp}^{III} = AE_{\perp} \qquad E_{\parallel\vartheta\parallel}^{III} = AE_{\parallel} \cos 2\vartheta_{I} \cos 2\vartheta_{II} \cos 2\vartheta_{III} \cos\varphi \quad ,$$

$$E_{\parallel\vartheta\perp}^{III} = AE_{\parallel} \cos 2\vartheta_{I} \cos 2\vartheta_{II} \cos 2\vartheta_{III} \sin\varphi \quad .$$

(A represents all factors which are independent of the direction of polarization.) Because usually φ is quite small for the Bragg case and is close to $\pi/2$ for the

Laue case polarization factors there are respectively (4)

$$C_{\perp} = 1, \quad C_{\parallel} (-1)^{\tau_I + \tau_{II} + \tau_{III}} = \cos 2\vartheta_I \cos 2\vartheta_{II} \cos 2\vartheta_{III} \cos\varphi$$

and

$$C_{\perp} = 2, \quad C_{\parallel} (-1)^{\tau_I - \tau_{II} + \tau_{III}} = \cos 2\vartheta_I \cos 2\vartheta_{II} \cos 2\vartheta_{III} \sin\varphi$$

where $\tau_I = 1$ for $\vartheta_I > \pi/4$, $\tau_{II} = 1$ for $\vartheta_{II} > \pi/4$, $\tau_{III} = 1$ for $\vartheta_{III} > \pi/4$ and $\tau_I = 0$, $\tau_{II} = 0$, $\tau_{III} = 0$ for other cases.

In order to establish periodic or modulated periodic nature of the diffracted beam intensity versus the crystal thickness the measurements were made for wedge-shaped Si single crystals performed for 220 symmetrical reflection. CuK_{α_1} radiation was used. Measurements of the maximum of the rocking curve for the given crystal thickness have been performed for two cases:

I. Double crystal monochromator in which Ge single crystals were used with (111) reflecting planes (1, +1). Experimental data obtained in this setting are shown in Fig. 2a. As we expected, the modulation of Pendellösung fringes due to the incident unpolarized beam appears. It is a result of superposition of two sets of fringes corresponding to the partially polarized X-ray beams ($C_{\perp} = 1$, $C_{\parallel} = 0.53$).

II. Double crystal monochromator in which Ge single crystals were used with (333) and (111) reflecting planes (3, +1). Experimental data obtained in this setting are shown in Fig. 2b. In this case there is no modulation in Pendellösung fringes as we have one polarization state ($C_{\perp} = 1$, $C_{\parallel} \approx 0$) only.

These experiments show that the plane-wave theory given by Max von Laue is adequate to explain the effects of X-ray polarization of Pendellösung fringes.

I would like to express my thanks to Prof. J. Auleytner for valuable discussion and critical remarks.

References

(1) M. HART and A.R. LANG, Acta cryst. 19, 73 (1965).

(2) H. HATTORI, H. KURIYAMA, and N. KATO, J. Phys. Soc. Japan 20, 988 (1965).

(3) K. GODWOOD, M. LEFELD-SOSNOWSKA, and E. ZIELIŃSKA-ROHOZIŃSKA, phys. stat. sol. 7, 445 (1964).

(4) MAX VON LAUE, Röntgenstrahl-Interferenzen, Akademische Verlagsgesellschaft, Frankfurt/Main 1960.

(5) L. AZAROFF, Acta cryst. 8, 701 (1955).

(Received July 2, 1968)

phys. stat. sol. 28, K143 (1968)

Subject classification: 10.2; 22.5.3

Solid State Physics Branch, Physical Sciences Laboratory, Redstone Arsenal, Alabama (a), and University of Alabama (b)

Mollwo-Ivey Relation of H- and F-Bands in Alkaline Earth Fluorides

By

R. A. SHATAS, G. A. TANTON (a), and A. MUKERJI (b)

Since the early work by Görlich (1), there have been several attempts to systematize the wavelength-lattice constant relation of color centers observed in alkaline earth fluorides. A review paper by Görlich (2) discusses briefly the Mollwo-Ivey plots of additively and subtractively colored crystals. However, much of this earlier work posed interpretational difficulties caused by the appearance of bands associated with radiation-induced valence state changes of residual impurities. In particular, O'Connor and Chen (3) and Görlich et al. (4) have simultaneously recognized that the residual yttrium impurity was responsible for the characteristic four band spectrum often observed after the subtractive coloring. This observation rendered much of the earlier work obsolete. Subsequent studies by several authors (5, 6) have shown that the subtractive coloration at low temperatures induces thermally unstable centers, identified as F-centers by electron spin resonance (7) and Faraday rotation (8) experiments. Furthermore, bleaching studies (9, 10) clarified that a satellite band to the F-band is the H-band (Φ and Ω , respectively, in Görlich's notation), denoted in analogy to the F- and H-bands observed in alkali halides (11). Combining our data (10) with Görlich's recent measurements (9) we compile the wavelengths of F- and H-bands in CaF_2, SrF_2, and BaF_2 in Table 1.

Our studies of the F-band absorption and bleaching communicated earlier (10) are reproduc-

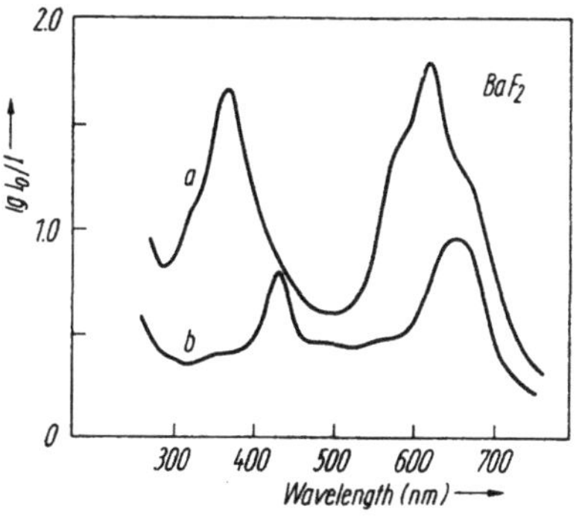

Fig. 1. Optical absorption bands of BaF_2 at liquid He temperature after a) X-raying for ≈ 25 h at liquid He temperature, and b) additively coloring at 700 to 800 °C in Ca vapor

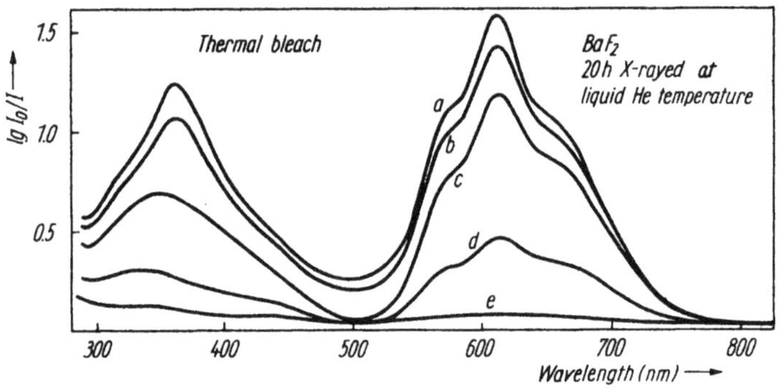

Fig. 2. Thermal bleach of BaF_2 optical absorption bands induced by X-raying at liquid He temperature. Measurements taken after ≈ 30 min at a) liquid He temperature, b) 65 $^{\circ}$K, c) 100 $^{\circ}$K, and d) 250 $^{\circ}$K

Table 1

	H-band		F-band		
	Tanton (10)	Görlich (9)	Tanton (10)	Görlich (9)	Cavenett (8)
CaF_2	314 nm (12)	285 nm	377 nm	375 nm	376 nm
SrF_2	325 nm	325 nm	442 nm	434 nm	435 nm
BaF_2	364 nm	361 nm	611 nm	545 nm	611 nm

ed in Figs. 1 and 2 since they differ from recent measurements of Görlich (9). Our specimens were typically of dimensions 20x10x1 mm^3 sawed or cleaved from single crystals grown by Stockbarger, Czochralski, and Bridgman methods. They were obtained from Harshaw, Semi-Elements, Adolph Meller, and Vinor Laboratories. Since crystals from different suppliers were grown by different methods, it was taken for granted that the residual impurity content in each batch of crystals was also different. This is justified by the fact that crystals of a given compound obtained from different sources, varied considerably in their rates of coloration when irradiated by X-rays at room temperature. At liquid He temperature, X-ray irradiation and optical absorption measurements were performed with an experimental arrangement described earlier (12). As seen in Fig. 3, the Mollwo-Ivey relation is satis-

Fig. 3. Mollwo-Ivey plot of band peak λ_m and lattice constant, a. Bands induced by X-raying at liquid He

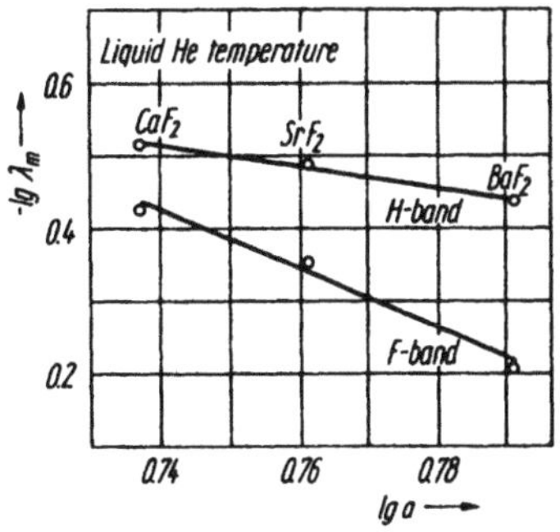

fied for H- and F-bands with respective exponents of 1.5 and 4. The latter number is somewhat larger than found by den Hartog and Arends (13).

References

(1) P. GÖRLICH and H. KARRAS, Proc. Internat. Conf. Phys. Semicond., Prague 1960 (p. 698).

(2) P. GÖRLICH, H. KARRAS, and R. LEHMANN, phys. stat. sol. 1, 389 (1961).

(3) J.R. O'CONNOR and J.H. CHEN, Phys. Rev. 130, 1790 (1963).

(4) P. GÖRLICH, H. KARRAS, W. LÜDKE, H. MOTHES, and R. REIMANN, phys. stat. sol. 3, 478 (1963).

(5) T. KAMIKAWA, Y. KAZUMATA, A. KIKUCHI, and K. OZAWA, Phys. Letters (Netherlands) 21, 126 (1966).

(6) G.A. TANTON, A. MUKERJI, and R.A. SHATAS, Bull. Amer. Phys. Soc. 11, 811 (1966).

(7) J. ARENDS, phys. stat. sol. 7, 805 (1964).

(8) B.C. CAVENETT, W. HAYES, and I.C. HUNTER, Solid State Commun. 5, 653 (1967).

(9) P. GÖRLICH, H. KARRAS, CH. SYMANOWSKI, and P. ULLMANN, phys. stat. sol. 25, 95 (1968).

(10) G.A. TANTON, A. MUKERJI, R.A. SHATAS, and J.E. WILLIAMS, Bull. Amer. Phys. Soc. 13, 254 (1968).

(11) H. DUERIG and J.J. MARKHAM, Phys. Rev. 88, 1043 (1952).

(12) A. MUKERJI, G.A. TANTON, and J.E. WILLIAMS, phys. stat. sol. 22, K19 (1967).

(13) H.W. DEN HARTOG and J. ARENDS, phys. stat. sol. 22, 131 (1967).

(Received June 18, 1968)

phys. stat. sol. 28, K147 (1968)
Subject classification: 16; 22.4.2; 22.4.3

Faculty of Mathematics and Physics, Charles University, Prague

Study of the Kinetics of X-Ray Photoconductivity in CdSe and CdTe and Determination of the Electron-Hole Generation Energy

By
M. VANĚČEK, J. TOUŠEK, and E. KLIER

The energy necessary for the generation of a pair of free carriers (E_p) is an important parameter for applications of photoconductors in the region of high energy radiation. It has been measured mostly on p-n junctions (1 to 3). On some photosensitive materials p-n junctions have not yet been prepared (CdSe) or have not the desired qualities (CdTe). In such cases it is possible to determine E_p from the photoconductivity in the high energy region, if the parameters of the kinetics are known. The present note gives the results of the evaluation of photoconductivity relaxation curves on CdSe and CdTe monocrystals excited by X-rays.

The samples in a cryostat were irradiated by the monochromatic K_α radiation of molybdenum ($h\nu$ = 17.42 keV) from a stabilized source. The absolute intensity of the incident radiation was determined by a p-i-n Si detector checked by means of an ionisation chamber and surface barrier detector. The radiation was completely absorbed by the sample. The beam was chopped by means of a rotating disc into rectangular pulses of radiation and the relaxation curves of the photocurrent recorded by a cathode ray oscillograph.

The simplest kinetics equation according to (4) is

$$\frac{dn}{dt} = \beta_f N - \frac{n}{\tau_f} \quad , \tag{1}$$

where n concentration of free carriers, N number of absorbed quanta per unit volume, β_f, τ_f phenomenological quantum efficiency and lifetime as determined from the relaxation curves.

It holds

$$\beta_f = \frac{\beta_{abs}}{1+\alpha} \qquad \tau_f = \tau_{abs}(1+\alpha) \quad ,$$

β_{abs} and τ_{abs} being absolute values of the above magnitudes and α a parameter

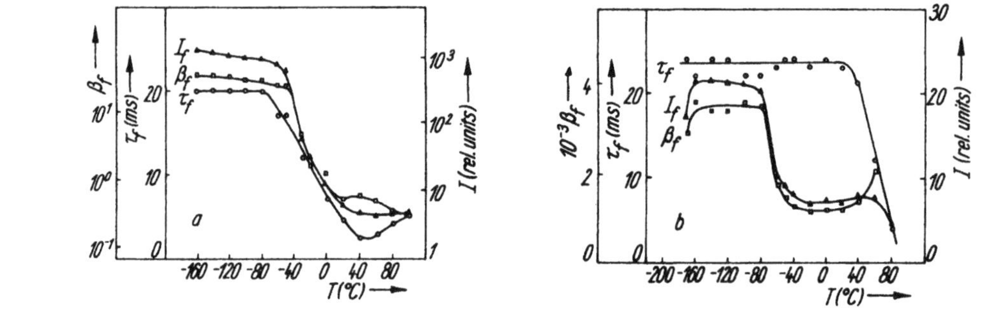

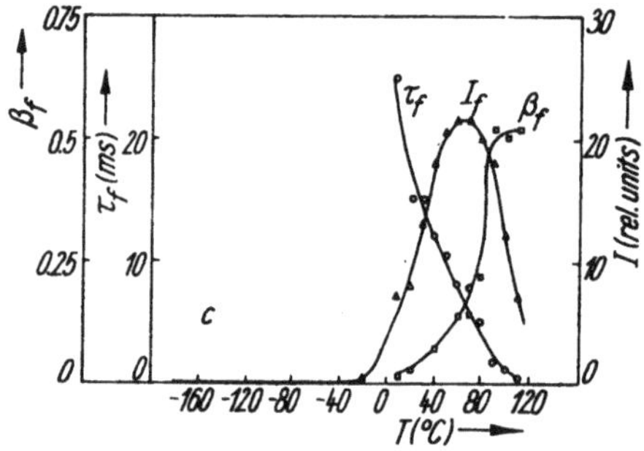

Fig. 1. Temperature dependence of the stationary photocurrent I, relaxation time τ_f, and quantum efficiency β_f.

a) Sample CdSe-1 (vapour grown). Incident flux: 1.35×10^6 quanta/s.

b) Sample CdSe 3-1 (zone molten). Incident flux: 2.9×10^6 quanta/s.

c) Sample CdTe 2. Incident flux: 5.3×10^6 quanta/s

characteristic for the trapping processes involved. In order to eliminate the effect of traps, the samples are simultaneously illuminated with visible and uv light. The intensity of the auxiliary illumination was chosen so that the X-ray quantum efficiency reached a saturation value, at which $\alpha = 0$, $\beta_f = \beta_{abs}$ can be supposed (for the analysis see (5)).

Some typical temperature dependences of β_f and τ_f in a temperature interval from -160 to +80 °C are shown in Fig. 1. Assuming that the variation of β_{abs} with temperature is negligible (within the accuracy of measurements), the following results for CdSe were obtained:

$$\beta_{abs} = 4.7 \times 10^3 \qquad E_p = 3.7 \text{ eV/pair for sample 3-1}$$
$$\beta_{abs} \doteq 4.3 \times 10^3 \qquad E_p = 4.0 \text{ eV/pair for sample 2-1 .}$$

The estimated error of measurement is ± 10 % and is caused mainly by the inaccurate determination of the intensity N. The E_p values agree well with the measurements of Svěčnikov and Čalaja (6), $E_p = 4.2$ eV, obtained by a different method.

On CdTe it was not possible to eliminate the influence of traps, probably due to their high concentration (similarly with the sample CdSe-1). The other reason was a much lower sensitivity of CdTe to X-rays as compared with CdSe.

References

(1) K.G. MC KAY, Phys. Rev. 84, 829 (1951).

(2) V.S. VAVILOV, L.S. SMIRNOV, and V.M. PACKEVIČ, Dokl. Akad. Nauk SSSR 112, 1020 (1957).

(3) J. DRAHOKOUPIL, M. MALKOVSKÁ, and J. TAUC, Czech. J. Phys. 7, 57 (1957).

(4) S.M. RYVKIN, Fotoelektricheskie yavleniya v poluprovodnikakh, Moskva 1963 (Photoelectric Effects in Semiconductors, New York 1964).

(5) M. VANĚČEK, MSc Thesis, Dept. of Solid State Physics, Math.-Phys. Faculty, Charles University, Prague 1967.

(6) S.V. SVĚČNIKOV and V.G. ČALAJA, Fiz. tverd. Tela 7, 3413 (1965).

(Received July 5, 1968)

phys. stat. sol. 28, K151 (1968)

Subject classification: 20.3; 22.4.1

Istituto di Fisica, Università di Bari (a)

and Istituto di Fisica, Università di Parma (b)

AC Modulated Green Edge Emission in CdS

By

A. CINGOLANI, A. RIZZO (a), and A. LEVIALDI (b)

In this paper we report the ac electric field effect on green edge emission of CdS at 77 $^{\circ}$K. Our high-purity CdS single crystals, $\approx$ 10 μm thick, were mounted between two conducting mica sheets in liquid nitrogen. Luminescence was excited using continuous uv light (photon energy 3.40 eV) at low intensity (excitation intensity 10^{14} photons/cm^2 s). The electric field is applied normal to the face irradiated by uv.

If I_o is the intensity of the continuous photoluminescence of CdS, the application of an ac field of adequate frequency gives rise to a modulation of the luminescence I_m. This modulated emission, I_m, has a frequency equal that of the external electric field. The modulation factor $\varrho = I_m/I_o$, depends on both amplitude and frequency of the external field. Maximum modulation (Fig. 1a) is observed for $\nu' =$ = 1500 Hz, that is, for a period $\tau' = 1/\nu'$, equal to the decay time of the green luminescence at the short wavelength of CdS at 77 $^{\circ}$K (1). Furthermore, the period τ' corresponding to each maximum of ϱ depends in turn on the amplitude of the applied field (Fig. 1b). One finds two different ranges of field values, viz. region I, $0 \leq E \leq 400$ V (ϱ is always maximum value for a period of the field $\approx 5\times10^{-4}$ s) and region II, $400 \leq E \leq 4000$ V (the maximum value of ϱ corresponds to different values of τ). There seems to be no shift of the emission band towards longer wavelengths (Fig. 2).

The polarization parameter defined by the relation

$$\alpha = \frac{I_{max}^{E \perp c}}{I_{max}^{E \parallel c}}$$

which for observations perpendicular to the (1120) crystal plane is $\approx$ 9 at 77 $^{\circ}$K,

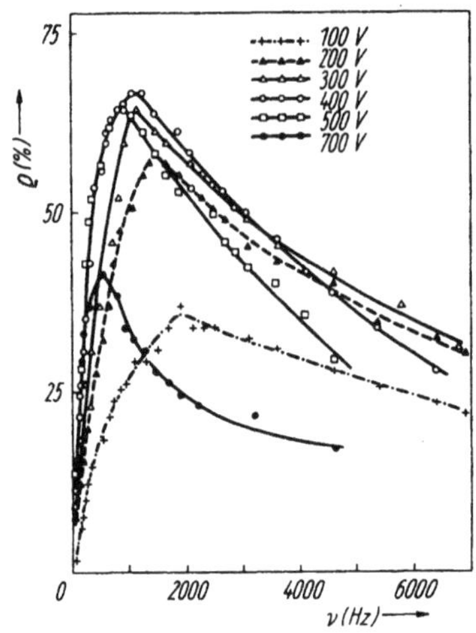

Fig. 1a. Modulation factor ρ as a function of the frequency ν of the ac electric field, for fixed values of field amplitude

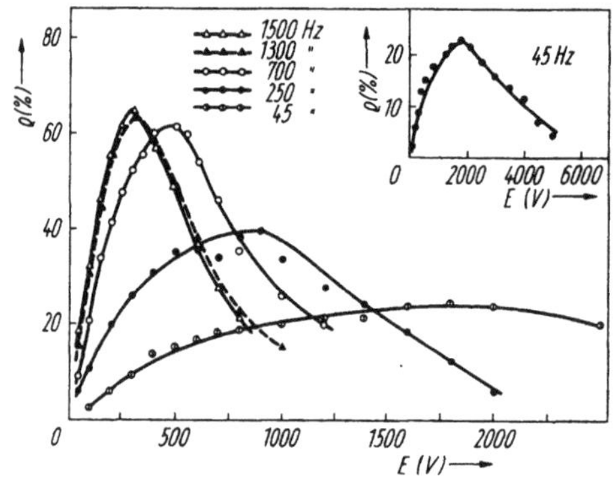

Fig. 1b. Modulation factor ρ as a function of the ac electric field amplitude, for fixed values of field frequency

remains unchanged by an external field, indeed the same α is measured analyzing the polarization state of the modulated signal I_m for $\nu = 1500$ Hz.

The variation of ρ with the field amplitude can be explained considering the model of Colbow and Nyberg (2). At low excitation intensity, the hole lifetime is very short, since they can be quickly captured by the acceptor and non-radiative recombination centers. Since the CdS surface is normally negatively charged, owing to chemiabsorbed oxygen ions, there is a strong field E^* in the region (≈ 1000 Å) where the free carriers are created. E^* will sweep free holes to the surface, where they recombine non-radiatively with trapped electrons on the surface states.

An external electric field with negative polarity clearly increase the non-radiative recombinations by sweeping holes to the surface, whereas the application of

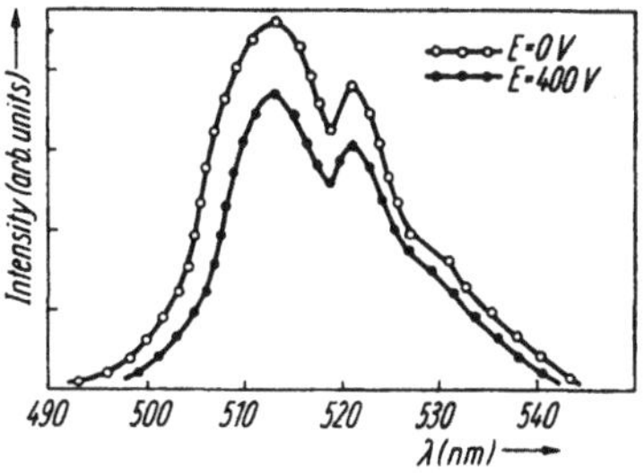

Fig. 2. CdS emission spectrum at 77 °K without electric field (upper curve) and emission spectrum of the modulated signal I_m in presence of an electric field

a positive field to the irradiated surface reduces the effect of the negative electric field E^*. Fewer holes are lost to non-radiative recombination at the surface and photoluminescence efficiency increases. Consequently a periodic variation of the electric field polarity on the irradiated surface of CdS gives rise to luminescence modulation, due to the combined effect of enhancement and quenching. This modulation will increase, by varying the ac field amplitude E, up to a maximum value of ϱ, until E does not modify the surface polarization state.

Indeed, for intensities of the field corresponding to region II, the amplitude of the ac field causes a flattening of the energy bands near the surface, reducing the number of non-radiative recombination centers and lowering the modulation factor.

From the above it is seen that ϱ depends, in addition to E, also upon the frequency of the electric field; it follows that the period τ' corresponding to maximum ϱ is a function of E. Fig. 2 shows that the electric field does not introduce into the emission spectrum any effect of the kind observed in ZnS (3) and Cu_2O (4).

Unlike the case of ZnS, the uv radiation determining the luminescence of CdS is totally absorbed on the surface in a path of ≈ 800 Å, so that the surface resistivity is much lower than in the bulk. It follows that the electric field acting on the emission centers on the surface is much weaker than the externally applied field (nominal value ≈ 10^4 V/cm). Finally the modulation, and more generally the enhancement and quenching effect of the electric field on CdS luminescence is a first order effect in comparison with that of Franz-Keldysh.

References

(1) K. COLBOW, Phys. Rev. 141, 742 (1966).

(2) D.W. NYBERG and K. COLBOW, Canad. J. Phys. 45, 2833 (1967).

(3) A. CINGOLANI, A. LEVIALDI, and P. PERILLO, Phys. Letters (Netherlands) 21, 24 (1966).

(4) S. NIKITINE, J.L. DEISS, M. CERTIER, J.C. MERLE, and A. DAUNOIS, J. Physique C2, 107 (1966).

(Received July 8, 1968)

phys. stat. sol. 28, K155 (1968)
Subject classification: 14.1; 21.7

Physico-Technical Institute of Low Temperatures,
Academy of Sciences of the Ukrainian SSR, Kharkov

The Influence of Boundary Surface on the Magnetoresistance of Bismuth

By
YU. A. BOGOD, V. V. EREMENKO, and L. K. CHUBOVA

The results of resistance measurements on Bi single crystals in transverse magnetic fields up to 20 kOe at 77 and 4.2 $^{\circ}$K are reported. The thickness of crystals was changed by electropolishing. The construction of the specimen holder allowed measurements to be made without sample remounting.

1. While studying the resistivity of the single crystal Bi-1a at $T = 4.2$ $^{\circ}$K the magnetoresistive effect was found to be dependent on the angle of the magnetic field with respect to the crystal surface even when the initial size of the cross-section was 8x3.7 mm^2. (This effect is apparently of the same nature as that found by Borovik and Lazarev on thin Bi single crystals (1).) As is seen from Fig. 1 for crystallographically equivalent directions of the magnetic field (they are realized when H turns by $\varphi = \pm 60^{\circ}$) the magnitude of the relative change of resistivity for certain values of φ appears to be different. Therefore the scattering at the sample boundaries affects the magnetoresistivity significantly. The experiments described below are devoted to the elucidation of the nature of the current carrier reflection by the sample surface.

2. It is known (2, 3) that the carrier reflection by the sample surface in Bi is specular. For specular reflection the incident and reflected electrons are strictly correlated. Due to this fact the electron motion along an open trajectory in the surface layer is possible even in metals with closed Fermi surfaces (4). In magnetic field the open orbit carriers are known to make the more significant contribution to the electric conductivity than the carriers whose orbits are closed in the momentum space (5). Therefore when an external magnetic field is present the static skin effect must occur in a conductor with finite size (4, 6), i.e. a current displacement towards the conductor surface must occur. Moreover, when the condition $r \ll l^2/d$ is satisfied (r is the electron orbit radius, l is the mean free path, d is a charac-

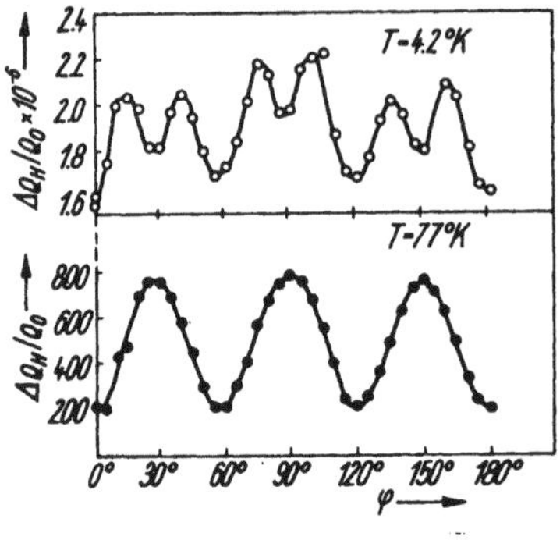

Fig. 1. The rotation diagram of the single crystal Bi-1a, H = 20 kOe, the magnetic field vector is in the binary-bisectrix plane; $\varphi = 0^o$ corresponds to H ‖ C_2 (C_2 is the binary crystallographical axis)

teristic size) the sample resistivity must increase more slowly than according to the square law (4).

Fig. 2 shows the logarithmic dependence of the resistivity of a Bi plate on magnetic field which is parallel to the trigonal axis at $T = 4.2\ ^oK$. The slope is seen to be 1.65 (at $T = 77\ ^oK$, $\Delta\varrho_H/\varrho_o \approx H^2$) in the field range investigated and is the same for different specimens and differently orientated crystals.

For Bi l is $\approx 10^{-1}$ cm at $T = 4.2\ ^oK$; therefore in magnetic fields of 10^3 to 10^4 Oe for $d \approx 1$ to 10^{-1} cm the radius of the electron orbit is two or three orders smaller than l^2/d. For $T = 77\ ^oK$ $l \approx 10^{-3}$ cm and the condition $r \ll l^2/d$ is not satisfied. Therefore the deviation of the magnetoresistance increase from the square law in Bi at $T = 4.2\ ^oK$ confirms the assumption about the specular reflection of the

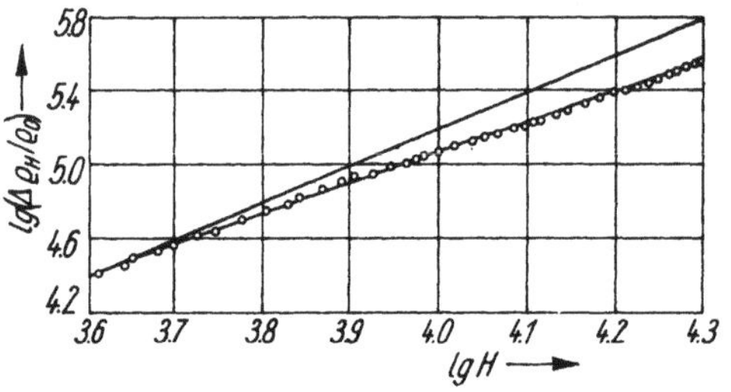

Fig. 2

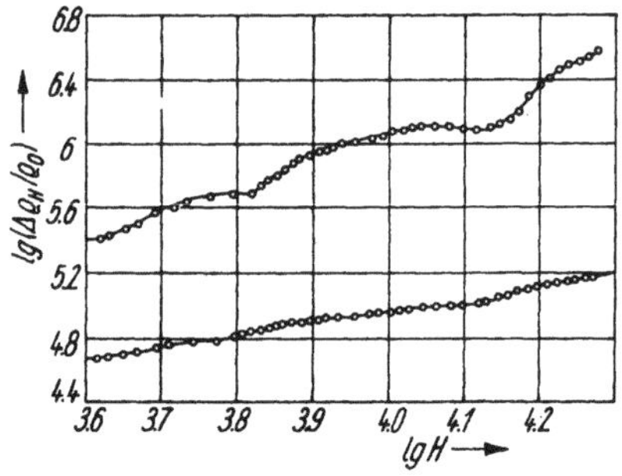

Fig. 3

Fig. 2. lg $\Delta\varrho_H/\varrho_o$ versus lg H of the sample with the cross-section (6.5 x 0.3) mm^2, the specimen axis is parallel to C_2, the magnetic field is transverse to a wide plane; $\Delta\varrho_H/\varrho_o = (\varrho_H - \varrho_o)/\varrho_o$, ϱ_H is the resistivity in the field H, ϱ_o is the same for H = 0. The solid curve is plotted for the case when $\partial(\lg \Delta\varrho_H/\varrho_o)/\partial(\lg H) = 2$

Fig. 3. Sample Bi-1b, $\emptyset$ = 4 mm, H ‖ C_2. The longitudinal axis is parallel to the trigonal one. The lower curve corresponds to the unpolished specimen

carriers by the specimen surface and indirectly proves the existence of the static skin effect.

3. As has been shown in (4), if the cylindric conductor surface has jags the high current is located near their angles, and the transverse resistivity increases linearly in magnetic field when the distance between jags is smaller or equal to the Larmor radius. (A surface of every jag is assumed to be specular.) While cutting out the sample, jags on its surface are inevitable. Therefore the magnetoresistance measurements of such a crystal before and after electropolishing when the jags disappear are undoubtedly of certain interest. Such experiments have been carried out on the sample Bi-1b.

In the initial state the magnetoresistance of the specimen at $T = 4.2\ ^{\circ}K$ increased by a linear law for all field directions. After the electropolishing the slope of the monotonous component significantly increased.

References

(1) E.S. BOROVIK and B.G. LAZAREV, Zh. eksper. teor. Fiz. 21, 857 (1951).

(2) A.N. FRIEDMAN and S.H. KOENIG, IBM J. Res. Develop. 4, 158 (1960).

(3) G.E. SMITH, Phys. Rev. 115, 1561 (1959).
M.S. KHAIKIN and B.S. EDELMAN, Zh. eksper. teor. Fiz. 47, 878 (1964).

(4) M.YA. AZBEL and V.G. PESHANSKII, Report on the 14th Symposium of Low Temperature Physics, Kharkov 1967.

(5) I.M. LIFSHITZ, M.YA. AZBEL, and M.I. KAGANOV, Zh. eksper. teor. Fiz. 31, 63 (1956).

(6) G.A. ZAITZEV, Zh. eksper. teor. Fiz. 45, 1266 (1963).

(Received July 4, 1968)

phys. stat. sol. 28, K159 (1968)
Subject classification: 20.3; 22.5

Sperry Rand Research Center, Sudbury, Massachusetts

Temperature Dependence of the Broad-Band Fluorescence of $RbMnF_3$

By

W.W. HOLLOWAY, Jr. and M. KESTIGIAN[1)]

Temperature-dependent alterations have been previously measured in the fluorescence properties of several manganese salts (1 to 4). One of the interesting features of these data was appearance of fluorescence alterations which apparently could not be described by an activation energy. In this communication we report more accurate measurements of the temperature dependence of the fluorescence lifetime of $RbMnF_3$ (3, 4) which demonstrate that the behavior in the two principal regions of fluorescence alteration, near 35 and 100 $^{\circ}$K, can in fact be described by activation energies. In addition, evidence has been found for an additional small fluorescence change at 22 $^{\circ}$K.

The fluorescence of crystals of $RbMnF_3$ was chosen for study because of the large difference in wavelength of the fluorescence maxima above and below the low temperature (35 $^{\circ}$K) fluorescence transition region. This large difference permitted the study of the lifetime of the low temperature component alone (at approximately 5800 Å) in the transition region. The experimental procedure has been described previously (4). A flow meter and a valve controlled the flow of the helium gas used in forcing the transfer of liquid helium from the dewar into the coolant reservoir. By regulating the rate of liquid helium transfer, the temperature of the sample could be effectively regulated from 15 to 77 $^{\circ}$K. This control was necessary for the more accurate measurements reported here.

The temperature dependence of the fluorescence lifetime in the two principal regions of fluorescence alteration, i.e., 30 to 40 and 80 to 120 $^{\circ}$K, is shown in Fig. 1. For the lower temperature region, the behavior of the fluorescence lifetime can be described by an activation energy $E^*(35\ ^{\circ}K) \approx 0.069$ eV, while the behavior in the higher temperature region can be described by an activation energy

---

1) Now employed at Air Force Cambridge Research Laboratory, Bedford, Mass.

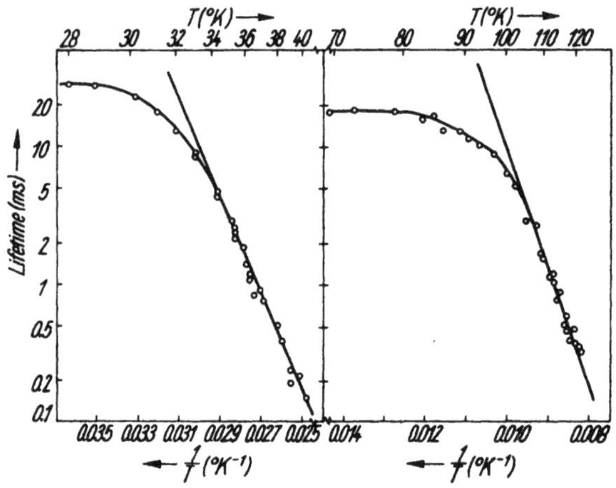

Fig. 1. The temperature dependence of the $RbMnF_3$ emission lifetimes in the two regions of principal fluorescence alteration. Activation energies of $E^*(35\ ^oK) \approx 0.069$ eV and $E^*(100\ ^oK) \approx 0.175$ eV can be determined from the slope of the asymptotic lines

$E^*(100\ ^oK) \approx 0.175$ eV.

The high temperature fluorescence change can be attributed to phonon processes which can be described by an activation energy (5). The low temperature fluorescence alteration has been attributed to a local alignment of the spin of the impurity state (i.e., the excited manganese ion) with the surrounding manganese ions. The occurrence of such an alignment below the ordering temperature of the bulk material, $T_N(RbMnF_3) \approx 82\ ^oK$, is due to the weaker interaction of the excited manganese ion, as compared to the unexcited manganese ion, with the surrounding unexcited manganese ions. A cooperative phenomenon was previously postulated to explain the apparent nonactivation-type temperature dependence (4). These new data on the broadband fluorescence of $RbMnF_3$ indicate that such a phenomenon is not necessary to explain the experimental results and that the alterations may be the result of the effect of phonons on the alignment of spins of the excited manganese ions. The relation of the changes in the spin alignment to the changes in fluorescence properties has been discussed (4). An activation-type temperature dependence has been previously reported for the sharp fluorescence transitions in fluoride salts of manganese (6).

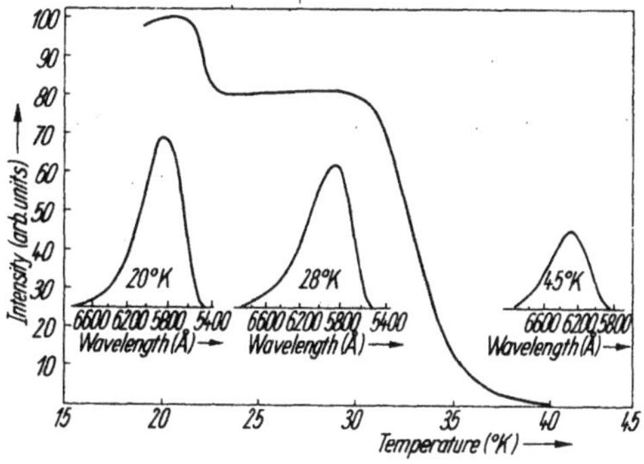

Fig. 2. The temperature dependence of the intensity of the $RbMnF_3$ fluorescence at approximately 5800 Å. The major fluorescence alteration (at $T \approx 35\ ^{o}K$) and the small change in fluorescence (at $T \approx 22\ ^{o}K$) are shown. The profiles of the observed fluorescences, typical of each temperature region, are shown. The intensity of the fluorescence profile at 45 ^{o}K has been multiplied by a factor of 10

In Fig. 2, we show the results of measurements of the fluorescence intensity at 5800 Å. In addition to the disappearance of the low temperature fluorescence component in this region, coinciding with the lifetime changes of Fig. 1, a smaller fluorescence intensity change is observed at 22 ^{o}K. The fluorescence profiles for each region are included to demonstrate the observed wavelength changes (4). Lifetime measurements could not be performed on this smaller fluorescence alteration because of lack of component separation. The shift of the fluorescence intensity maximum is found to be 0.02 eV for the change at 22 ^{o}K, while a shift of 0.14 eV has been measured for the alteration at 35 ^{o}K. The temperature dependence of the antiferromagnetic resonance of this material shows no unusual behavior in either of these regions of fluorescence change (7), suggesting that the bulk magnetism is not solely responsible for these results.

We are indebted to E.W. Prohofsky for many enlightening discussions concerning these results.

References

(1) W.W. HOLLOWAY, Jr., M. KESTIGIAN, R. NEWMAN, and E.W. PROHOFSKY, Phys. Rev. Letters 11, 82 (1963).

(2) W.W. HOLLOWAY, Jr. and M. KESTIGIAN, Phys. Rev. Letters 13, 235 (1964).

(3) E.W. PROHOFSKY, W.W. HOLLOWAY, Jr., and M. KESTIGIAN, J. appl. Phys. 36, 1041 (1965).

(4) W.W. HOLLOWAY, Jr., E.W. PROHOFSKY, and M. KESTIGIAN, Phys. Rev. 139, A954 (1965).

(5) D. CURIE, Luminescence in Crystals, John Wiley and Sons, New York 1963 (p. 202).

(6) D.D. SELL, R.L. GREENE, W.M. YEN, A.L. SCHAWLOW, and R.M. WHITE, J. appl. Phys. 37, 1229 (1966).

(7) M.J. FREISER, R.J. JOENK, P.E. SEIDEN, and D.T. TEANEY, Proc. Internat. Conf. Magnetism, Nottingham 1964 (p. 432).

(Received June 25, 1968)

phys. stat. sol. 28, K163 (1968)
Subject classification: 19; 14.4.2

Physikalisches Institut der Karl-Marx-Universität Leipzig,
Abteilung Struktur der Materie

Electric Shifts in the Paramagnetic Resonance of Ferroelectric Cu^{2+}-Doped Rochelle Salt

By

G. VÖLKEL and W. WINDSCH

The electric field induced EPR line shifts (1, 2) in Cu^{2+}-doped Rochelle salt (3, 4) can be described by an additional term of the spin Hamiltonian

$$\hat{\mathcal{H}}_{SE} = \beta_e \sum_{i,j,k} T_{ijk} E_i' H_j S_k + \sum_{i,j,k} U_{ijk} E_i' S_j I_k \; . \qquad i, j, k = x, y, z \tag{1}$$

We will discuss the first term only. E_i' are the components of the internal electric field E' acting on the paramagnetic complex and induced by the external applied electric field E. The coordinate axes of the complex system are x, y, z.

The five independent components of the third rank T tensor appropriate to the point symmetry C_{2v} of the Cu^{2+} sites have been determined from the angular dependence of the EPR line shifts shown in Fig. 1.

The plotted curves have been computed with the following values:

$$\begin{aligned} &T_{zxx} E' = 7.3\text{x}10^{-4} ; \; T_{zyy} E' = 0.3\text{x}10^{-4} ; \; T_{zzz} E' = 7.7\text{x}10^{-4} ; \\ &T_{xxz} E' = T_{xzx} E' = 9.0\text{x}10^{-4} ; \; T_{yyz} E' = T_{yzy} E' = 0.8\text{x}10^{-4} ; \end{aligned} \tag{2}$$

where we proposed the internal electric field E' to be directed along the ferroelectric a-axis of the Rochelle salt crystal. In ferroelectrics E' depends on the polarization P: $E' \approx P/\varepsilon_0$ (5). This explains the experimentally observed proportionality between EPR line shifts and polarization (2). It seems to be appropriate to determine the components T_{ijk} in terms of the macroscopically measurable polarization P. The dynamic polarization P_{max} of our samples due to an external applied electric ac field $E_{50\,Hz} = 7$ kV cm^{-1} was checked during the EPR measurements. In equation (2) E' corresponds to $P_{max} \approx 7\text{x}10^{-8}$ C cm^{-2}. From

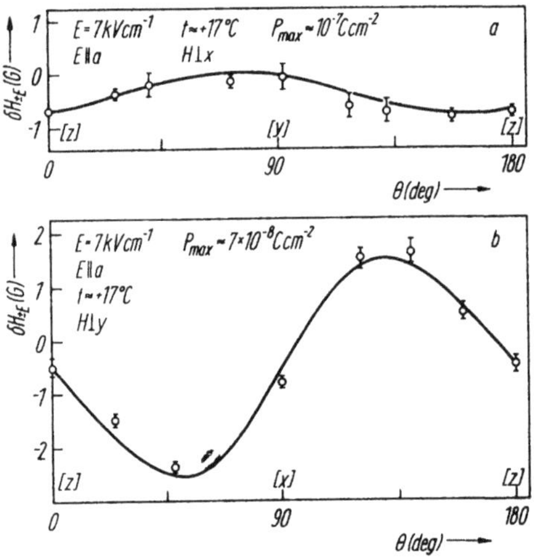

Fig. 1

theoretical considerations it follows that for $T_{zxx} \neq T_{zyy}$ the x and y axes must not lie between the ligand ions but must lie along O-Cu-O.

The Hamiltonian $\hat{\mathcal{H}}_E$ describing the electric field effect essentially consists of two parts

$$\hat{\mathcal{H}}_E = e \left[\mathfrak{E}' \mathfrak{r} + \Delta V_1\right] = e \sum_{i=x,y,z} E'_i \left[i + \sigma_i V'_i\right] = e \sum_{i=x,y,z} E'_i W_i \quad , \tag{3}$$

the electronic and the ionic term, respectively. Both terms have the same symmetry properties (6, 7). V'_i means a sum of partial differentials of the potential $V(\mathfrak{r})$ seen by the hole of the Cu^{2+} ion at $\mathfrak{r}$, derived with respect to the normal coordinates of the ligand ion deformations. $E'_i \sigma_i$ describes the ligand ion shifts.

The following equations have been obtained for the five independent T tensor components by third order perturbation theory:

$$T_{zxx(zyy)} \approx \overset{+}{(-)} 2e\lambda_o a_1 \left[\frac{2\sqrt{3}\langle 0|W_z|2\rangle}{(E_{4(3)}-E_o)(E_2-E_o)} - \frac{a_1\langle 0|W_z|0\rangle}{(E_{4(3)}-E_o)^2}\right] ,$$

$$T_{zzz} \approx -8e\lambda_o a_1^2 \left[\frac{2\beta\langle 0|W_z|2\rangle}{(E_2-E_o)(E_1-E_o)} + \frac{\langle 0|W_z|0\rangle}{(E_1-E_o)^2}\right] ,$$

$$T_{xxz(yyz)} \approx \overset{-}{(+)} 2e\lambda_o a_1 \left[\frac{\langle 0|W_{x(y)}|3(4)\rangle}{(E_{3(4)}-E_o)}\left(\frac{1}{E_{4(3)}-E_o} + \frac{2}{E_1-E_o}\right) \overset{+}{(-)} \frac{2\langle 1|W_{x(y)}|4(3)\rangle}{(E_{4(3)}-E_o)(E_1-E_o)}\right] ,$$

where $|n\rangle$ are the MO describing the magnetic Cu(II) complex with point symmetry C_{2v}. They are like those for the point symmetry D_{2h} (8), but due to the absence of a centre of inversion symmetry 4p functions of the central atom are admixed. E_n is the energy of the $|n\rangle$ state. a_1^2 and β characterize the covalency and the orthorhombic distortion, respectively, of the complex. From the hyperfine splitting we get $\beta \leq 0.025$ and $a_1^2 = 0.70$. λ_o is the spin-orbit coupling constant of the free Cu^{2+} ion.

The relative values and signs of the experimentally determined components T_{ijk} cannot be described by the electronic effect alone. Therefore the ionic effect must be important in Cu^{2+}-doped Rochelle salt. The observed EPR line shifts caused by an external applied electric field $E = 7\ kV cm^{-1}$ could be explained by a rotation of the Cu(II) tartrate complex about the y-complex axis of $\vartheta = 0.14^{o}$. An influence of the piezoelectric deformation neglected here would also be possible.

We thank Prof. Dr. A. Lösche and the collegues of the EPR group for stimulating discussions.

References

(1) R. BLINC and M. SENTJURC, Phys. Rev. Letters 19, 1231 (1967).

(2) G. VÖLKEL, phys. stat. sol. 25, K35 (1968).

(3) G. VÖLKEL and W. WINDSCH, phys. stat. sol. 17, K75 (1966).

(4) J. STANKOWSKI, Acta phys. Polon. 33, 387 (1968).

(5) F. JONA and G. SHIRANE, Ferroelectric Crystals, London 1962.

(6) C.A. BATES, private communication.

(7) C.A. BATES, Proc. Phys. Soc. 91, 359 (1967).

(8) W. WINDSCH and M. WELTER, Z. Naturf. 22a, 1 (1967).

(Received July 9, 1968)

phys. stat. sol. 28, K167 (1968)
Subject classification: 20.3; 10; 22.4.1

Bereich Halbleiteroptik der Sektion Physik der Humboldt-Universität zu Berlin
and Bereich Halbleiteroptik des Physikalisch-Technischen Instituts
der Deutschen Akademie der Wissenschaften zu Berlin

Investigation of the Emission of CdS Single Crystals in the 2.0 eV Region

By
O. GOEDE

The uv excited emission of undoped CdS single crystals in the 2.0 eV region has hardly been investigated. However, this emission should be of great interest because of the suggested connection with intrinsic defects (1). - In the present paper the existence of two broad emission bands is reported in the region near 2.0 eV, which are closely related: Crystals quenched from about 750 °C show an emission band with a maximum at 2.05 eV at 77 °K (in the following called H-band). After slow cooling of the crystals from the same temperature a band is observed at 77 °K whose maximum is shifted by about 80 meV to lower energies

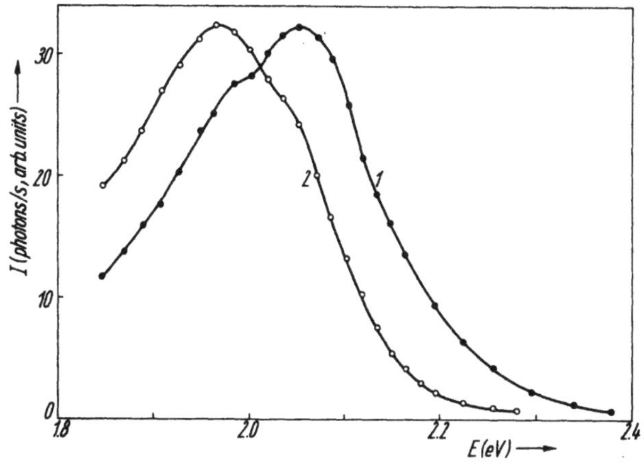

Fig. 1. Spectral distribution of the intensities of the emission bands in the 2.0 eV region. The emission is observed perpendicular to the $(11\bar{2}0)$ plane of the crystal; $\mathfrak{E} \parallel c$, T = 77 °K. Curve 1: after quenching the crystal from 750 °C (H-band), curve 2: after "slow" cooling from the same temperature (in 2 min to 400 °C) (L-band). Both curves are normalized to equal heights

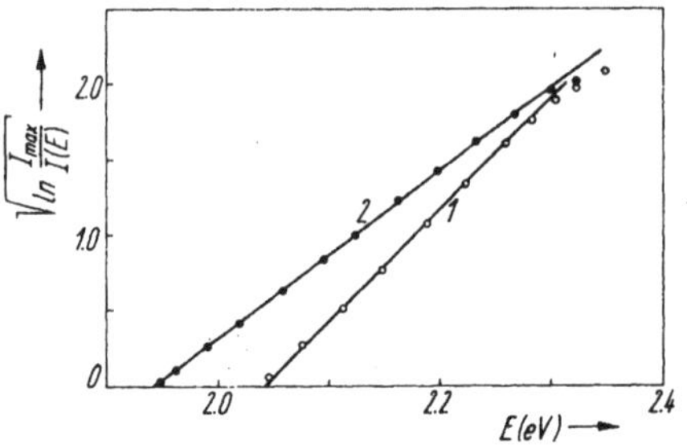

Fig. 2. $[\ln(I_{max}/I(E))]^{1/2}$ vs. E for the high-energy tail of the H-band; $\mathfrak{E} \perp c$. Curve 1: T = 78.5 °K, curve 2: T = 195 °K. I_{max} intensity in the maximum of the band

(called L-band) (see Fig. 1). The described effect is found for all investigated (undoped) crystals and is reversible, i.e. on the same crystal both the H- and the L-band could successively be produced by appropriately cooling the crystal. As can be seen in Fig. 1 under certain conditions the residual other band can be observed as distinct shoulder. Both the H- and the L-band are partially polarized $\mathfrak{E} \perp c$, in the case of the H-band the polarization ratio is $I_{max}^{\mathfrak{E} \perp c} / I_{max}^{\mathfrak{E} \parallel c} = 2.1$ at 77 °K.

In the following investigations are reported of band shape and temperature dependence of band parameters of the H-band in the temperature range 51 to 230 °K. The results are based mainly on the analysis of the high-energy tail of the H-band to avoid uncertainties due to the overlap of the residual L-band. It was found that the shape of the H-band is Gaussian for both $\mathfrak{E} \perp c$ and $\mathfrak{E} \parallel c$ in the whole temperature range investigated. Fig. 2 shows, as an example, the high-energy tail of the band for different temperatures. The experimental points lie on straight lines from I_{max} to about 0.03 I_{max}. The half-width, H, and the position of the maximum of the band, E_{max}, can be determined from the slope of the straight line and the intersection with the E-axis, respectively. The temperature dependence of H is plotted in Fig. 3 and obviously can be well represented by $H(T) = H(0)[\coth(\hbar\omega/2kT)]^{1/2}$. The best fit is obtained for $\hbar\omega$ = 20 meV and H(0) = 208 meV. The measured temperature dependence of E_{max} can be described by the expression $E_{max}(T) = A - B \coth(\hbar\omega/2kT)$ with the same $\hbar\omega$, A = 2.27 eV, and B = 176 meV. A possible difference between the positions of the maxima of the H-band for $\mathfrak{E} \perp c$ and $\mathfrak{E} \parallel c$ could not be detected and must be smaller than 10 meV. The band parameters H and E_{max} are found to be independent of excitation intensity. - The temperature dependence of the height of the H-band, I_{max}, can be approximated, for both polarization directions, by $I_{max}(T) = I_{max}(0)\,[1 + C \exp(-E_{qu}/kT)]^{-1}$, with the energy of thermal quenching $E_{qu} \approx (190 \pm 30)$ meV and $C \approx 5 \times 10^5$.

Fig. 3. Half-width H of the H-band as a function of temperature T for $\mathfrak{E} \perp c$

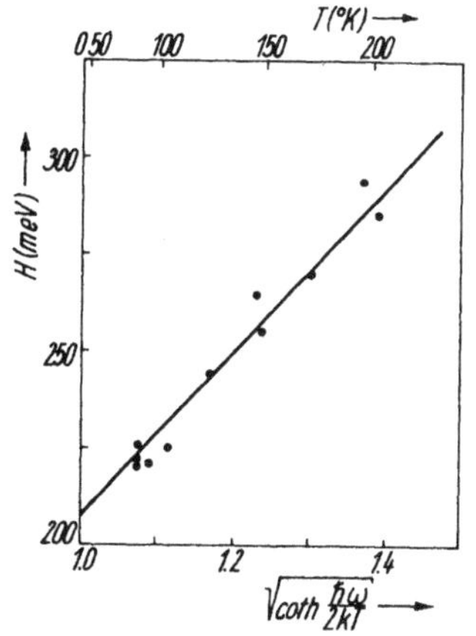

From a comparison of the observed band shape and the temperature dependence of H and E_{max} with theory it can be concluded, in an analogous manner as described in detail in (2) (see also (2) for references), that the H-band is due to an internal radiative transition between two discrete energy levels of a centre and does not involve band states. Therefore, <u>isolated</u> Cd interstitials (Cd_i) as proposed in (1) cannot be accepted to be the centres of the H-band. Since the radiative transition would have to start from an excited state with an energy about 2.0 eV above the ground state, this state in the considered case would lie in the conduction band because the distances of the ground states both of the singly and doubly ionized Cd_i from the conduction band are essentially smaller (3). - The band shape of the H-band is determined by electron-phonon interaction, $\hbar\omega$ = 20 meV being the energy of the dominant phonons. These phonons seem to be longitudinal acoustic (LA) phonons, because of the remarkable agreement with the energy of LA phonons at the boundary of the Brillouin zone (4). From the large magnitude of the phonon wave vector it may be concluded (see (5)) that the electronic wavefunctions of the centre are considerably localized (e.g., in comparison to the proposed donor-acceptor associate in the case of edge emission (2)), which is reasonable because the transition energy is significantly smaller than the band gap. From $H(0) = 2\sqrt{2}\sqrt{\ln 2}\ \hbar\omega S$ the (effective) Huang-Rhys factor can be estimated to be S = 4.4. The LA-phonon interaction in the case of the H-band therefore seems to be noticably stronger than for the high-energy series of edge emission (2). A fine structure of the H-band due to phonon interaction (in particular to LO phonons) could not be detected, even at the lowest temperature (51 °K) and with a spectral slit width of about 1 meV. However, at this temperature a distinct shoulder at the high-energy tail of the H-band at about 2.15 eV was observed indicating the occurrence of a further band which may be resolvable at lower temperatures. The Franck-Condon shift can be estimated to be $2\hbar\omega S = 176$ meV, the maximum of the absorption band corresponding to the H-band therefore should be found at about 2.23 eV at 77 ^{o}K. After (6) in the case of undoped CdS single crystals there are only two absorption bands in this spectral region with maxima at about 1.995 and 2.15 eV. It seems to be probable that the 2.15 eV absorption band corresponds to the H-band, in spite of the small deviation from the predicted position. (The 1.995 eV absorption band then might correspond to the L-emission band.)

The present results suggest that the centre responsible for the H-band is a defect associate including, at least, one intrinsic defect. This suggestion is also supported by the interpretation of the 2.15 eV absorption band given in (6). - Further investigations in particular of the relation between emission and absorption bands and measurements at lower temperatures are in progress.

Special thanks are due to Dipl.-Phys. N. Pöhner for his contribution to this work.

References

(1) B.A. KULP, Phys. Rev. 125, 1865 (1962).

(2) O. GOEDE and E. GUTSCHE, phys. stat. sol. 17, 911 (1966).

(3) R. BOYN, O. GOEDE, and S. KUSCHNERUS, phys. stat. sol. 12, 57 (1965).

(4) M. BALKANSKI, J.M. BESSON, and R. LE TOULEC, Proc. Internat. Conf. Phys. Semicond., Paris 1964 (p. 1091).

(5) M. LAX and E. BURSTEIN, Phys. Rev. 100, 592 (1955).

(6) R. BOYN, phys. stat. sol., in the press.

(Received July 12, 1968)

phys. stat. sol. 28, K171 (1968)

Subject classification: 18.2

Institut für metallische Spezialwerkstoffe

der Deutschen Akademie der Wissenschaften zu Berlin, Dresden

A Model System for Hard-Magnetic Particle Ensembles

By

S. VÖLLMAR

The interaction occurring in the systems of single-domain particles in hard-magnetic materials is taken into consideration in form of the pair model (1, 2). Moreover, consideration is given to ferromagnetic particles of equal volumes lying with their long axis side by side and one after another respectively in the direction

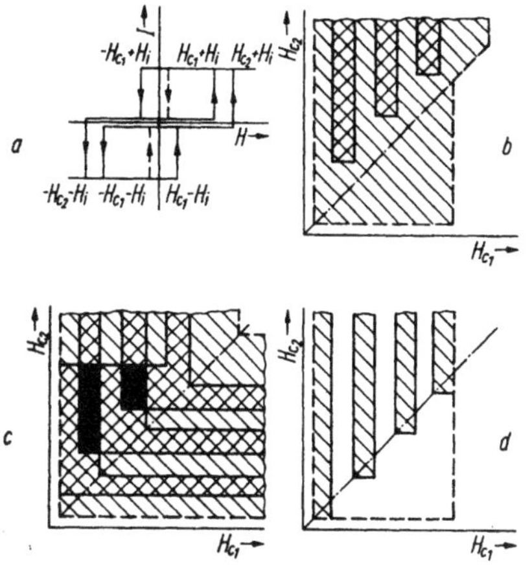

Fig. 1

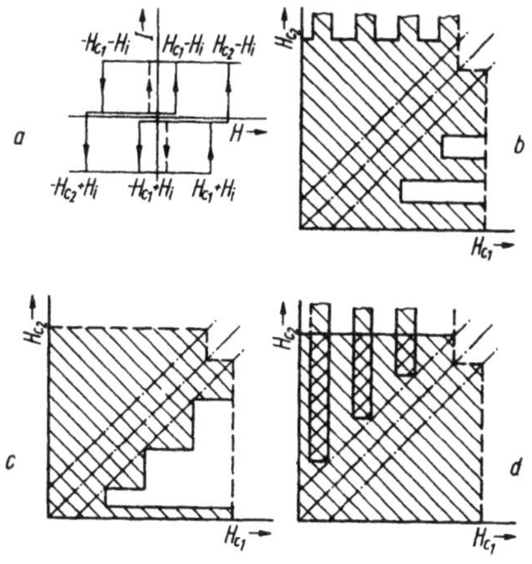

Fig. 2

Fig. 1. Negative interaction. a) Hysteresis loop of a pair with $H_{c_2} > H_{c_1}$. Representation of the states of the system in a $H_{c_1} - H_{c_2}$ plane for 4 partial steps; b) dc reverse field method j'_{DD}; c) ac field method j'_{D}; d) dc field method j'_{DR}. The broken line indicates the remanence value for one step.

Fig. 2. Positive interaction. a) Hysteresis loop of a pair with $H_{c_2} > H_{c_1} + 2H_i$. Representation of the states of the system in a $H_{c_1} - H_{c_2}$ plane for 4 partial steps; b) dc reverse field method j'_{DD}; c) ac field method j'_{D}; d) dc field method j'_{DR}. The broken line indicates the remanence value for one step.

of the applied field, which interact magnetostatically in pairs only, and the reversal of magnetization is effected according to the Stoner-Wohlfarth-mechanism (3). The hysteresis loops of the pairs in the "effective field" approximation (4) are marked in Figs. 1a and 2a where the interaction field is represented by H_i and the coercive forces of the particles by H_{c_i}.

Of special interest is the description of such experiments which yield typical deviations to non-interacting systems (5). These are the remanence curves obtained by different magnetic treatments which, according to the Wohlfarth relations,

$$j'_D = j'_{DR} = 1 - j_R = j'_{DD} = \frac{1}{2}(1 - j_D)$$

should coincide. With the partial remanence methods these curves are measured step by step. The so-called additive law means that the values of the partial remanence are equal to those of the remanence.

For a better approximation to real systems an ensemble of independent pairs is examined in which the parameters H_i and H_{c_i} are distributed.

By magnetic treatment each pair arrives at one of the 4 possible remanence states. The remanence value of the system can be calculated by averaging the distributions of the parameters. For a qualitative comparison the pairs with H_i = const which yield a positive contribution to the remanence values are to be marked in a $H_{c_1} - H_{c_2}$ plane. In the figures these are for the remanence curves the areas bordered with a broken line. Thus in the case of negative interaction

$$j'_{DR} > j'_{DD} > j'_D$$

and of positive interaction

$$j'_D > j'_{DD} = j'_{DR} .$$

In this way both experimentally observed splittings can be explained, the equality sign in the last relation being reflected by a relatively small difference of the curves (6).

The contributions to the partial remanences are indicated by hatched areas. With these methods the same pairs may give a positive contribution after various steps. These areas are marked by manifold hatching. The value of the partial remanence disagrees in every case with the appertaining remanence value obtained by one step, whereby the invalidity of the additive law is shown qualitatively. This disagreement depends on both the field strength H and the number of the partial steps. Of course, this simple model cannot describe the whole, partially very complicated, experimental results. Especially with high step number and with positive interaction it differs from the experimental behaviour.

A paper comparing quantitatively the model with the experiment is under preparation.

References

(1) L. NÉEL, C.R. Acad. Sci. (Paris) 246, 2313 (1958).

(2) E.P. WOHLFARTH, J. appl. Phys. 35, 783 (1964).

(3) E.C. STONER and E.P. WOHLFARTH, Phil. Trans. Roy. Soc. 240, 599 (1948).

(4) W.F. BROWN, Jr., in: Magnetostatic Principles in Ferromagnetism, North Holland Publ. Comp. 1966 (p. 112).

(5) O. HENKEL, IEEE Trans. Magnetics 1966 (p. 220).

(6) O. HENKEL, phys. stat. sol. 15, 211 (1966).

(Received June 20, 1968)

phys. stat. sol. 28, K175 (1968)

Subject classification: 18.1; 18.2; 22.8

Institute of Physics of Solids and Semiconductors,

Belorussian Academy of Sciences, Minsk

Magnetic Properties and Structure of $MnAs_xSb_{1-x}$ Solid Solutions

By

N.N. SIROTA and E.A. VASILEV

In the present paper we attempted to investigate the alloy $MnAs_xSb_{1-x}$ with the aim of establishing the magnetic structure of solid solutions and determining the region of anomalous change of paramagnetic susceptibility found in MnAs in which the Curie point $\Theta_1 = 318\ ^oK$ and the extrapolated Curie point $\Theta_2 = 400\ ^oK$. The assumption about sign change of the exchange interaction in MnAs above the Θ_1-point advanced in paper (1) in connection with anomalous run of paramagnetic susceptibility was not confirmed by direct diffraction investigations (2). Such a run of paramagnetic susceptibility is accounted for by the transition from B8 to B31 type structure (3). Manganese antimonide exhibits a Curie-Weiss behaviour above the Curie point.

The existence of continuous series of solid solutions between MnSb and MnAs was found in (4). It is shown that the parameters of the unit cell change practically linearly with composition, and the Curie temperature Θ_1 passes through a minimum at the composition x = 0.7. Investigation of magnetic properties of the system $MnAs_{1-y}Sb_y$ was carried out up to y = 0.2 in paper (5).

The magnetic structure of solid solutions of the system $MnAs_xSb_{1-x}$ have not been examined so far.

Investigations were carried out by us on the alloys MnSb, MnAs and solid solutions containing 25, 50, and 75 mol% MnAs. The samples were prepared in evacuated silica tubes and annealed for 24 h at 870 ^oK. X-ray, metallographic, and neutron diffraction investigations have confirmed the existence of continuous series of solid solutions.

The temperature dependence of paramagnetic susceptibility of the alloys in the temperature interval 300 to 700 ^oK has been investigated with the Faraday method. The measured values of the constants Θ_p and C_M are given in Table 1. A nonlinear

Table 1

x	1	0.75	0.5	0.25	0
C_M	1.7	0.75	0.5	1.0	1.3
Θ_p (^{o}K)	310	368	458	480	563

Curie constants C_M and paramagnetic Curie points Θ_p for the system $MnAs_xSb_{1-x}$

dependence of reciprocal susceptibility on temperature (Fig. 1) is observed below the Θ_2-points for all compositions with exception of MnSb. The Θ_2-points lie close to the straight line connecting the extrapolated Curie point Θ_2 of MnAs with the Curie point Θ_1 of MnSb (Fig. 2). This line is practically a section of the Bethe curve describing the dependence of exchange energy on the interatomic distance, since the lattice constants of the alloy change linearly with composition. The expansion of the temperature range between Θ_1 and Θ_2 is accomponied by a decrease of the effective magnetic moment defined by the Curie constant $C_M = \mu^2_{eff}/3k$.

Neutron diffraction study of alloys in the temperature range Θ_1 to Θ_2 shows the absence of magnetic moment ordering. Magnetic reflections superimpos-

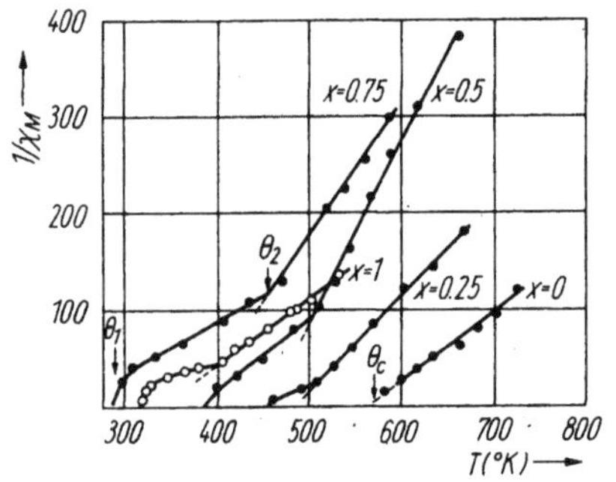

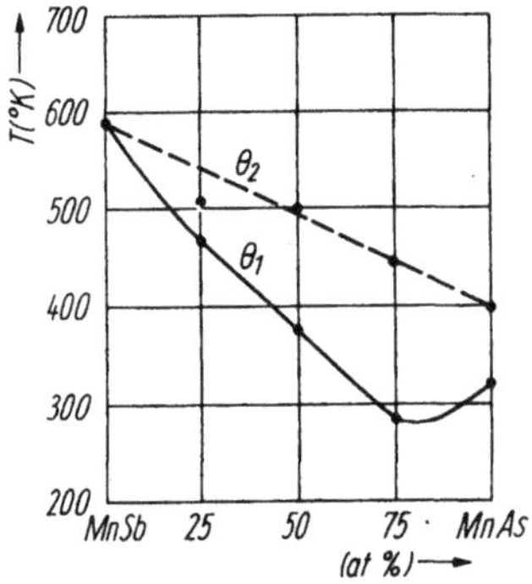

Fig. 1

Fig. 1. Reciprocal susceptibility vs. temperature for the system $MnAs_xSb_{1-x}$

Fig. 2. Curie temperatures Θ_1 and Θ_2 for the system $MnAs_xSb_{1-x}$

ing on nuclear peaks are observed on neutron patterns below the Curie point Θ_1, which shows the coincidence of the periods of magnetic and nuclear lattice. The analysis of intensities of magnetic scattering shows that ferromagnetic alignment of moments located in (000) and (001/2) positions is observed in solid solutions. The direction of magnetic moment is perpendicular to the C-axis for all alloys of the system Mn_xSb_{1-x} since only by such an arrangement coincidence between the calculated and experimental intensities is observed. Characteristic temperatures of alloys were calculated to determine the value of the Debye-Waller temperature factor. In the system $MnAs_xSb_{1-x}$ characteristic temperatures slightly depend on composition and lie in the interval 180 to 200 ^oK.

Thus, in the $MnAs_xSb_{1-x}$ solid solutions the magnetic structure does not change with composition. In this connection the appearence of the minimum in the concentration dependence of the Curie temperature Θ_1 cannot be accounted for by changing the type of magnetic ordering. Occurrence of the minimum appears to be connected with broadening of the region Θ_1 to Θ_2 which gives rise to deviation of paramagnetic susceptibility from the Curie-Weiss law in MnAs.

References

(1) C. GUILLAUD, J. Physique 12, 223 (1951).

(2) G.E. BACON and R. STREET, Nature 175, 518 (1955).

(3) J.B. GOODENOUGH, Phys. Rev. 157, 389 (1967).

(4) W. KÖSTER and E. BROWN, Ann. Phys. (Germany) 4, 66 (1959).

(5) J.B. GOODENOUGH, D.H. RIDGLEY, and W.A. NEWMAN, Proc. Internat. Conf. Magnetism, Nottingham 1964, London 1965 (p. 542).

(Received July 11, 1968)

phys. stat. sol. 28, K179 (1968)

Subject classification: 20.1; 6

Institute of Chemistry,

Academy of Sciences of the Moldavian SSR, Kishinev

Optical Bands in Electron-Vibrational Systems with Degeneracy

By

YU. B. ROSENFELD, B. G. VEKHTER, and B. S. TSUKERBLAT

The adiabatic theory of optical transitions (1 to 3) in electron-vibrational systems is confined to cases of non-degeneracy only. However, in the majority of cases of real systems (impurities in crystals, highly symmetrical molecules, etc.) the electronic levels are often degenerate which renders the division of electron and nuclear motion impossible (dynamic Jahn-Teller effect) (4). This accounts for the fact that up to now there has been no consistent theory of optical transitions between degenerate levels, while various attempts for investigating individual cases often led to contradictory results (5 to 7).

The approach suggested by Kubo (8) according to which the adiabatic approximation may in principle not be applied has recently been used by Perlin (9) who investigated the band form arising from transitions from non-degenerate into degenerate states. In the present note the results of investigation of optical bands for the general case of transitions between degenerate levels are given.

The optical absorption coefficient is given by the expression

$$K(\Omega) = \text{const} \left(1 - e^{-\frac{\hbar\Omega}{kT}}\right) \int_{-\infty}^{\infty} e^{i\Omega t} \langle\langle d^+ d(t) \rangle\rangle \, dt \,, \tag{1}$$

where $\langle\langle \ldots \rangle\rangle$ denotes the quantum-mechanical and statistical average with the Hamiltonian

$$\hat{\mathcal{H}} = \hat{\mathcal{H}}_0 + \hat{\mathcal{H}}' \,, \qquad \hat{\mathcal{H}}_0 = \hat{\mathcal{H}}_e(r) + \hat{\mathcal{H}}_p(q) \,. \tag{2}$$

$\hat{\mathcal{H}}_e(r)$ is the electronic Hamiltonian, $\hat{\mathcal{H}}_p(q)$ is the phonon Hamiltonian (in the harmonic approximation), $\hat{\mathcal{H}}' = \sum_{\varkappa} v_{\varkappa}(r) q_{\varkappa}$ is the operator of electron-phonon interaction in the approximation of linear nuclear displacements. The traces entering equation (1) are calculated by using the Feynman operator method (10) with un-significant simplification accepted in paper (9). The eigenfunctions of the Hamiltonian $\hat{\mathcal{H}}_o$ are used as a basis set. The final result is

$$\langle\langle d^+ d(t) \rangle\rangle = \frac{1}{n(\Gamma_1)} \exp\left\{ i\Omega_{21} t + \sum_\Gamma \frac{it}{2\hbar^2 \omega_\Gamma^2} \left[\frac{c^2(\Gamma_1\Gamma)}{n(\Gamma_1)} - \frac{c^2(\Gamma_2\Gamma)}{n(\Gamma_2)} \right] + \frac{1}{2} \sum_\Gamma \left[\frac{c^2(\Gamma_1\Gamma)}{n(\Gamma_1)} + \frac{c^2(\Gamma_2\Gamma)}{n(\Gamma_2)} \right] \varphi_\Gamma(t) (\hbar\omega_\Gamma)^{-2} \right\} \times$$
$$\times \sum_{\gamma'_1 \gamma'_2} \langle \Gamma_1 \gamma'_1 \Gamma_2 \gamma'_2 \mid WD \mid \Gamma_2 \gamma'_2 \Gamma_1 \gamma'_1 \rangle \; . \qquad (3)$$

Γ_1 and Γ_2 are the irreducible representations of the levels taking part in optical transitions, $n(\Gamma)$ are their dimensions, γ_i labeling the rows of representations.

$$\varphi_\Gamma(t) = \frac{ch\left(i\omega_\Gamma t + \frac{\hbar\omega_\Gamma}{2kT}\right)}{sh \frac{\hbar\omega_\Gamma}{2kT}} - \coth \frac{\hbar\omega_\Gamma}{2kT} \; , \qquad (4)$$

$$W = \exp\left\{ - \sum_{\Gamma\gamma} \frac{v_{\Gamma\gamma}^{\Gamma_1} \times v_{\Gamma\gamma}^{\Gamma_2}}{\hbar^2 \omega_\Gamma^2} \right\} \varphi_\Gamma(t) \; , \qquad (5)$$

where $v_{\Gamma\gamma}^{\Gamma_1} \times v_{\Gamma\gamma}^{\Gamma_2} \equiv V_{\Gamma_1\Gamma_2}^{\Gamma}$ denotes the direct (Kronecker) product of matrices $\| \langle \Gamma_i \gamma'_i \mid v_{\Gamma\gamma} \mid \Gamma_i \gamma'_i \rangle \|$ (Γ and γ characterize the vibrational representation). $V_{\Gamma_1 \Gamma_2}^{\Gamma} \neq 0$ only if $[\Gamma_1^2]$ and $[\Gamma_2^2]$ involve Γ simultaneously. $C(\Gamma_1\Gamma)$ is the reduced matrix element of $v_{\Gamma\gamma}$. $D = d^+ x d$ denotes the direct product of the matrices of electron operators of optical transitions (for example of dipole moment). One can arrive at some general conclusions about the character of the spectra, without concrete analysis of the spectral function using only symmetry properties, considering that according to the Wigner-Eckart theorem the matrix W is scalar on the group.

1. The spectral curve is a superposition of bands of the "Pekar" form of different halfwidth and differently shifted from the zero-phonon line which is common to all bands.

2. The number of bands is equal to the value z of different roots of the matrix W. The maximum value is equal to the number of irreducible representations in $\Gamma_1 \times \Gamma_2$ (in a particular case of extremely weak interaction with some of the active vibrational modes z may be smaller, see example for O_h symmetry in Table 1).

Table 1

Transitions	$E \rightleftharpoons E$	$E \rightleftharpoons T$	$T \rightleftharpoons T$		
Active modes	e_g	e_g	e_g	t_{2g}	$e_g + t_{2g}$
Number of bands	3	2	2	3	4

3. If the parameter of Stokes' losses of at least one of the bands is small enough (the electron-phonon interaction being however sufficiently strong) the intensity of the zero-phonon line shows anomalous temperature dependence, being sufficiently strong at high temperatures. Equation (3) implies the possibility of a new type of infrared or radiofrequency absorption in a system with degenerate ground electronic state. This case may be examined by putting in equation (3) $\Omega_{21} = 0$ and $\Gamma_1 = \Gamma_2 = \Gamma_o$. The number of bands is equal to the number of non-onedimensional representations in $\left[\Gamma_o^2\right]$.

We are grateful to Prof. Yu.E. Perlin and Prof. I.B. Bersuker for helpful discussions.

References

(1) S.I. PEKAR, Zh. eksper. teor. Fiz. 20, 510 (1950); 22, 641 (1952).
M.A. KRIVOGLAZ and S.I. PEKAR, Trudy Inst. Fiz. Akad. Nauk Ukr. SSR 4, 37 (1953).

(2) HUANG KUN and A. RHYS, Proc. Roy. Soc. A204, 406 (1950).
M. LAX, J. chem. Phys. 20, 1752 (1952).

(3) YU. E. PERLIN, Uspekhi fiz. Nauk, 80, 553 (1963).

(4) H. C. LONGUET-HIGGINS, Adv. Spectr. 2, 429 (1961).

(5) M. C. M. O' BRIEN, Proc. Phys. Soc. 86, 847 (1965).
Y. TOYOZAWA and M. INOUE, J. Phys. Soc. Japan 21, 1663 (1966).
H. C. LONGUET-HIGGINS, V. ÖPIK, M. H. L. PRYCE, and H. SACK, Proc. Roy. Soc. A2441 (1958).

(6) N. N. KRISTOFEL and G. S. ZAVT, Fiz. tverd. Tela 5, 1279 (1963).

(7) B. G. VEKHTER, Optika i Spektroskopiya 20, 258 (1966).
B. S. TSUKERBLAT, Zh. eksper. teor. Fiz. 51, 831 (1966).

(8) R. KUBO, Pictures in Theoretical Physics, Vol. I, New York 1959.

(9) YU. E. PERLIN, Fiz. tverd Tela (1968), to be published .

(10) R. FEYNMAN, Phys. Rev. 84, 108 (1951).

(Received June 8, 1968)

Pre-printed Titles

Papers to be published in "physica status solidi"

Vol. 29, No. 1

Review Article

A. SEEGER and K. P. CHIK, Stuttgart

Diffusion Mechanism and Point Defects in Silicon and Germanium

Original Papers

H. GAMARI-SEALE and P. PAPAMANTELLOS, Athen

A Neutron Diffraction Study on Gallium-Substituted Magnetite

F. DWORSCHAK, H. SCHUSTER, H. WOLLENBERGER, and J. WURM, Jülich and Aachen

Analysis of Point Defect States in Copper (I)

F. DWORSCHAK, H. SCHUSTER, H. WOLLENBERGER, and J. WURM, Jülich and Aachen

Analysis of Point Defect States in Copper (II)

R. A. BURLEY, Quebec

An Infrared Spectrophotometric Study of Vitreous Selenium Doped with Selenium Dioxide

O. V. BOGDANKEVICH, N. A. BORISOV, I. V. KRUKOVA, and B. M. LAVRUSHIN, Moscow

Temperature Dependence of Laser Threshold Current Density and Emission Spectra in Electron-Beam Pumped Gallium Arsenide Lasers

G. M. DE' MUNARI, F. GIUSIANO, and G. MAMBRIANI, Parma

A Phase Transition and the Photoemissive Quantum Yield at Low Temperatures in Ca_3Sb Films

A.A. VAIPOLIN, N.A. GORYUNOVA, K.L. KLESHCHINSKII, G.V. LOSHAKOVA, and E.O. OSMANOV, Leningrad

The Structure and Properties of the Semiconducting Compound $ZnSnP_2$

H. GENGNAGEL and U. HOFMANN, Dresden

Temperature Dependence of the Magnetocrystalline Energy Constants K_1, K_2, and K_3 of Iron

V.V. NEMOSHKALENKO, O.N. RASUMOV, and V.V. GORSKII, Kiev

Investigation of the Mössbauer Effect in Some Fe-Al Alloys

P. HUMBLE and C.T. FORWOOD, Melbourne

A Dynamical Treatment of the Stress-Induced Dissociation of Triangular Frank Dislocation Loops in F.C.C. Metals

A. SUKIENNICKI, Warsaw

On a New Type of Stripe Domains

K. SCHRÖDER and N.P. BAUM, Syracuse

Absolute Thermoelectric Power and Electrical Resistivity of Chromium-Rich Chromium-Nickel Alloys

G. LECOY and L. GOUSKOV, Montpellier

Noise in Thin Films Metal-Oxide-Metal $Al-Al_2O_3-Al$

A. CZACHOR, Swierk

On the Lattice Dynamics of Hexagonal Structure Metals

A. KUNZ, Bethlehem

Electronic Energy Bands for Rubidium Chloride and the Face-Centered Cubic Alkali Bromides

A.N. BASU and S. SENGUPTA, Calcutta

A Deformable Shell Model for the Alkali Halides

J.H. GIESKE and G.R. BARSCH, Pennsylvania

Pressure Dependence of the Elastic Constants of Single Crystalline Aluminium Oxide

W. KRESS, H. BORIK, and R.K. WEHNER, Frankfurt/Main and Zürich
Infrared Lattice Absorption of Silicon and Germanium

S.E. GUKASYAN and V.S. SHPINEL, Moscow
Mössbauer Effect in Antimony-Organic Compounds

R.C. WHELAN and D. SHAW, Hull
Evidence of a Doubly Ionized Native Donor in CdTe

G.G. KOVALEVSKAYA and S.V. SLOBODCHIKOV, Leningrad
Lifetimes of Current Carriers in InP

P.B. PICKAR and H.D. TILLER, Melbourne and Orlando
Optical Energy Gap in TISe

M.P. FONTANA, H. BLUME, and W.J. VAN SCIVER, Bethlehem
Properties of Excited States in NaI (I)

H.J. GLÄSER und D. GEIST, Clausthal
Elektrische Leitfähigkeit und Halleffekt von mangandotiertem Germanium und der Einfluß von Wärmebehandlungen auf den Einbauzustand des Mangans

K. PFEUFFER, Karlsruhe
Determination of the Electron Spin-Lattice Relaxation Time T_1 in the Direct Process

L. AHMAD, Bangor
Change in Electron Affinity of a Semiconductor due to Different Forms of Chemisorption

M. LUUKKALA, Helsinki
On Coherent Phonon Fusion in Quartz Transducers

V.M. ASNIN, A.A. ROGACHEV, and G.L. ERISTAVI, Leningrad
Effect of Weak Electric Fields on the Absorption Edge in Doped Germanium

A.M. VAN DIEPEN, H.W. DE WIJN, and K.H.J. BUSCHOW, Amsterdam and Eindhoven
Nuclear Magnetic Resonance and Susceptibility of Equiatomic Rare-Earth-Aluminum Compounds

K.H. HERRMANN and J. PUHLMANN, Berlin
Piezoresistance of Tellurium (II)

M. ALTWEIN, H. FINKENRATH, Č. KOŇÁK, J. STUKE, and G. ZIMMERER, Marburg/Lahn, Prague, and Darmstadt
The Electronic Structure of CdO (II)

P. PETRESCU, Bucharest
On the L-Bands and the First Exciton Bands in the Photoemission Spectra of Alkali Iodides

W. BRÜCKNER, R. PERTHEL, K. KLEINSTÜCK, and G.E.R. SCHULZE, Dresden and Jena
Magnetic Properties of $ZrFe_2$ and $TiFe_2$ within Their Homogeneity Range

C. SÁNCHEZ and F. AGULLÓ-LÓPEZ, Madrid
Transient Effects in the Room-Temperature F-Colouring of NaCl Irradiated with X- or γ-Rays

A. SEEGER and H. MEHRER, Stuttgart
Interpretation of Self-Diffusion and Vacancy Properties in Gold

F. WALZ, Stuttgart
Untersuchung der magnetischen Nachwirkung in Nickel nach Neutronenbestrahlung und plastischer Verformung

B. HARRIS, Brighton
Solution-Hardening and Deformation of Niobium/Tungsten Alloys

E. TAGLAUER, Jülich
Small Angle Scattering of Subthermal Neutrons from Deformed Polycrystalline Copper

W. FRANK, Stuttgart
Die kritische Schubspannung mit zweiwertigen Kationen dotierter Alkalihalogenide (I)

R.T. PASCOE and C.W.A. NEWEY, London
Deformation Modes of the Intermediate Phase NiAl

K. IBEL and R. SIZMANN, München

Energy Loss of ⟨110⟩ Channelled α-Recoil Atoms in Gold

E. NEBAUER, Berlin

Vapour-Pressure Investigations of Impurity Diffusion and Solubility in A^{II}-B^{VI} Compounds Demonstrated for the System CdS : Au : S_2

A. BOURRET and D. DAUTREPPE, Grenoble

Study of Agglomerated Defect in Irradiated Pure Nickel by Electron Microscopy

P. H. HANDEL, Garching

Instabilities and Turbulence in Semiconductors

R. BOYN, Berlin

Optical Absorption due to Intrinsic Defects in CdS Single Crystals

19. Magnetic Resonance in Solids

20. Optical Properties of Solids
    - 20.1 Spectra, Optical Constants (Colour Centres see 10.2; X-Ray Spectra see 20)
    - 20.2 Lasers
    - 20.3 Luminescence of Solids (X-Ray Spectra see 20)

21. Metals. Alloys
    - 21.1 Transition Metals (Group Sc to Zn) and their Alloys
        - 21.1.1 Fe and Fe Alloys
    - 21.2 Alkali Metals
    - 21.3 Alkali-Earth Metals
    - 21.4 Rare-Earth Metals
    - 21.5 Actinides
    - 21.6 Noble Metals
    - 21.7 Semi-Metals (Semiconducting Alloys see 22.7)

22. Semiconductors and Ionic Crystals
    - 22.1 Elements
        - 22.1.1 Germanium
        - 22.1.2 Silicon
        - 22.1.3 Group VI Elements
    - 22.2 III—V Compounds
        - 22.2.1 Arsenides
        - 22.2.2 Phosphides
        - 22.2.3 Antimonides
    - 22.3 Carbides
    - 22.4 II—VI Compounds
        - 22.4.1 Sulphides
        - 22.4.2 Selenides
        - 22.4.3 Tellurides
    - 22.5 Halides
        - 22.5.1 Silver Halides
        - 22.5.2 Alkali Halides
        - 22.5.3 Alkali-Earth Halides
    - 22.6 Oxides (CdO, ZnO see 22.4; Spinels see 18.3 or 14.4.2)
    - 22.7 Semiconducting Intermetallic Compounds
    - 22.8 Tertiary and Higher Compounds (Carbonates, Phosphates, Silicates and Mica see 22)
    - 22.9 Organic Semiconductors

23. Solidified Gases

Attention

The preceding classification scheme is closely akin to the scheme used by Physics Abstracts to classify the various fields of solid state physics. A paper will either be classified under one ore more of the sub-sections or, if none of these are relevant, it will just appear under the general section heading (i. e. sections and subsections are equivalent in this sense).

The *subject classification* given at the head of each paper begins with the main subject matter of the paper, and this is followed if necessary, by figures referring to subsiduary subject matter and the type of solids investigated.

The papers published in this issue are listed in the order of their subject classification at the end of the table of contents.

Manuscripts and letters should be sent to

physica status solidi, DDR, 102 Berlin, Neue Schönhauser Str. 20
or Internationaler Buchversand GmbH, 1 Berlin 19, Eschenalle 21.

The object of this journal is to publish rapidly important and topical results of fundamental research in the field of solid state physics. Each issue will be published on the first day of the month. Items published will fall into the following categories:

1. Review articles written by arrangement with the editors.
2. Original papers selected from those which have not been submitted for publication elsewhere, after being reviewed. (The latest date for acceptance of contributions is the 8th day of each month, 50 days before the delivery of the journal.)
3. Short notes published in the off-set part of the journal. They are, if necessary after being reviewed, selected by the editorial staff. (The latest date for acceptance of manuscripts for this part is the 12th day of each month, 16 days before the delivery of the journal.)
4. Pre-printed titles of papers to be published in the next issue.

Editorial Information

Manuscripts should be submitted preferentially in English. Besides, papers in German, French, and Russian are admitted. Authors are asked to observe the following rules:

1. Manuscripts have to be written as concisely as possible. At most, 20 and 3 pages (28 lines per page; abstracts, tables, references, captions of figures included) are admitted for original papers and short notes, respectively. All manuscripts should be typewritten on one side only. Two copies are required. Badly arranged manuscripts are returned.
2. The author's name, initials and institution should be given.
3. An abstract of about ten lines in English and a second one in German, French, or Russian are required for original papers.
4. Formulae should not be written too small and with pencil. Separate lines for formulae are desirable. Indistinct writing will lead to a return of the manuscript.
5. Drawings should be enclosed on separate sheets of paper; photographs should be submitted, on white paper, only for review articles and original papers. Duplicates are required.
6. Captions for the figures should be submitted on separate sheets of paper, and special instructions with regard to the size of illustrations should be given on these.
7. References should be numbered in order of appearance and should be given, on a separate sheet, in the form used in this journal.
8. Units, symbols, etc. should be used in accordance with the recommendations given by the IUPAP.
9. Each paper should contain a proposal for the subject classification according to the classification scheme appearing on the second and third cover of each issue.

The authors will be sent two galley proofs of review articles and original papers. Proof reading of short notes and pre-printed titles will be done by the editorial staff.

For each original paper, and short note, a total of 75 reprints is available free of charge. The number of reprints free of charge per review article is 25. In addition review articles will be remunerated.

Bei Fragen zur Produktsicherheit wenden Sie sich bitte an:
If you have any questions regarding product safety,
please contact:

Walter de Gruyter GmbH
Genthiner Straße 13
10785 Berlin
productsafety@degruyterbrill.com